THERMODYNAMICS:
FOUNDATIONS
AND APPLICATIONS

THERMODYNAMICS: FOUNDATIONS AND APPLICATIONS

Elias P. Gyftopoulos

Gian Paolo Beretta

Massachusetts Institute of Technology

Macmillan Publishing Company

New York

Collier Macmillan Canada

Toronto

Maxwell Macmillan International

New York   Oxford   Singapore   Sydney

Acquisition Editor: David Johnstone
Production Supervisor: Leo Malek
Cover Designer: Jane Edelstein
Illustrations: ECL Graphics
This book was set in Times Roman by ETP Services Co.

Macmillan Publishing Company
866 Third Avenue, New York, New York 10022

Collier Macmillan Canada, Inc.
Suite 200
1200 Eglinton Avenue, E.
Don Mills, Ontario, M3C 3N1

Library of Congress Cataloging-in-Publication Data
Gyftopoulos, E. P.
    Thermodynamics: Foundations and Applications / Elias P.
Gyftopoulos, Gian Paolo Beretta.
        p.   cm.
    ISBN 0-02-348455-1
    1.Thermodynamics.   I. Beretta, Gian Paolo.   II. Title
QC311.G994 1991
536'.7—dc20

Printing: 1 2 3 4 5 6 7 8 9 0   Year: 1 2 3 4 5 6 7 8 9 0

We dedicate this book
to our teacher of thermodynamics,
George N. Hatsopoulos

Preface

"In view of the large number of books on thermodynamics it may seem surprising that there should be any need for yet another," wrote Guggenheim in the preface of his book "Thermodynamics" in 1949. Many more books on thermodynamics have been published since then. So, why are we adding ours to the long list of entries on the subject?

During the past two decades of teaching thermodynamics to students from all over the globe, we have sensed a widespread quest for more clarity, unambiguity, and logical consistency in the exposition of the foundations than are available in traditional presentations. In response to our students' demand, we have gradually developed a set of arguments, observations, definitions, statements, and derivations that cannot be found in any of the books currently available. Continuing the effort pioneered by Keenan and Hatsopoulos, we have composed an exposition of the foundations and the applications of thermodynamics that many enthusiastic M.I.T. students have found clarifying, rewarding, and inspiring. Our class notes have grown into a coherent collection thoroughly tested in the classroom, equipped with a large number of homework problems, and mature enough for wider dissemination.

In the preface of his book "Concepts of Thermodynamics" in 1960, Obert wrote: "Most teachers will agree that the subject of engineering thermodynamics is confusing to the student, despite the simplicity of the usual undergraduate presentation." In our experience, the major source of confusion is the lack of logical consistency and completeness in the many presentations of the foundations of thermodynamics. The definition of a system as just "the subject of analysis" or "anything that is enclosed by a surface in space" is incomplete. The definition of properties in terms of the state is circular if the definition of state is in terms of the properties. The definition of heat as anything that is not work is incomplete and ambiguous. The definition of thermal equilibrium in terms of temperature is circular if the definition of temperature is in terms of thermal equilibrium. The restriction to equilibrium states is artificial if the purpose is the use of heat and cycles to define entropy. Even if entirely uninterested in the foundations, the student cannot avoid but sense this ambiguity and lack of logical consistency and develop the incorrect conviction that thermodynamics is a confusing, ambiguous, handwaving subject. Unfortunately, such a conviction is quite widespread and very difficult to dismantle.

The problem is not resolved even by recent books for engineering students such as the excellent textbooks on thermodynamics by F. F. Huang (1988 edition), K. Wark, Jr. (1988 edition), and M. J. Moran and H. N. Shapiro (1988), which tend to deemphasize the subtleties of the foundations by placing the emphasis on the applications of energy and entropy balances. The student encounters no difficulty in mastering the mechanics of energy and entropy balances. But the conviction that thermodynamics is based on ambiguous grounds remains. In our view, this hinders comprehension and discourages any desire to further explore and exploit the wide range of implications and applications of the subject.

In the novel exposition of the foundations that we present, we strive to develop each of the basic concepts in full detail and without ambiguities, at a level that assumes a sophomore background in calculus and elementary physics. Some characteristic features are as follows.

- The basic concepts and principles are introduced in a novel logical sequence that eliminates the traditional problem of incomplete definitions and vicious circles.
- The carefully worded statements of the laws of thermodynamics are presented as fundamental laws of physics that complement the laws of dynamics.
- The principles and results are developed for both macroscopic and microscopic systems, and for both equilibrium and nonequilibrium states.
- The concept of entropy is defined for any system, including a system with a single degree of freedom, and for any state, including a nonequilibrium state.
- The state principle is derived as an exact and rigorous result valid for any stable equilibrium state of any system.
- The concept of temperature is defined for the stable equilibrium states of any system, including a system consisting of one particle with a single degree of freedom.
- The concepts of work and heat are defined in terms of the concepts of energy and entropy exchanges and not vice versa.
- The third law is introduced in a novel discussion of the lowest-temperature stable equilibrium states.
- The principles and results are illustrated pictorially on a novel graph of energy versus entropy.
- A concise but precise summary of the foundations (Chapter 14) provides a suitable starting point for a first introductory undergraduate course.
- The validity of the Euler relation for relatively large values of the amounts of constituents is derived from a rigorous and novel definition of a simple system.
- The derivations of relations among properties of simple systems are made easy to follow by showing explicitly all the functional dependences.
- The chemical equilibrium equation is derived from a rigorous and novel discussion based on the definition of a simple system.
- The typical end-of-chapter problems that test each new concept are complemented by a large number of more structured problems designed to test the connections between new concepts and concepts introduced in earlier chapters.

A part of the book is intended for use as a text for an introductory undergraduate course in thermodynamics. Other parts provide material for more advanced courses, including a graduate course with emphasis on the foundations. We discuss suggestions for this partitioning at the end of Chapter 1.

In Chapter 1 we describe some salient features of the models of physical phenomena without concern as to the precise meanings of the terms we use. Our aim is to motivate the reader to pursue a systematic study of the applications and a deeper scrutiny of the foundations of thermodynamics. For this pursuit, at the end of the chapter we outline four different alternative paths to a study of thermodynamics using this book: an introductory undergraduate path, an advanced undergraduate path with emphasis on applications, a graduate path with emphasis on advanced topics, and a graduate path with emphasis on the general foundations. For each path we suggest the study of different chapters.

In Chapters 2 to 13 we discuss the key concepts and fundamental principles of thermodynamics. In doing so, we assume that concepts and principles such as space, time, reference frame, velocity, acceleration, force, mass, force field, gravitational acceleration, electrostatic field, magnetic field, momentum, kinetic energy, potential energy, Newton's laws, and Maxwell's equations are all part of the educational background of our reader in mechanics and electromagnetism at an advanced high school or introductory college level. We require this background not because we discuss any complicated issues of these facets of natural science but because it is part of the conceptual underpinning of thermodynamics.

In Chapter 2 we begin our detailed exposition of the foundations with kinematics and dynamics. In kinematics we discuss the definitions of system, property, and state. Our definitions eliminate the consideration of all the statistical arguments that lurk under all traditional expositions of thermodynamics. The concept of state that we define provides a common thread for the unification of the various branches of physics without need to modify its meaning from branch to branch. The state is the set of instantaneous values of all the amounts of constituents, all the parameters that characterize the external forces, and all the properties. Without alteration, this definition is valid for any system, macroscopic or microscopic, and any condition, changing or not changing with time.

In dynamics we discuss spontaneous and induced changes of state as functions of time, that is, we introduce the idea of the equation of motion of a system. Certain time evolutions obey Newton's equation of motion or its quantum-mechanical equivalent, the Schrödinger equation of motion. Other experimentally observed time evolutions, however, do not obey either of these equations. So the equations of motion that we have are incomplete. The discovery of the complete equation of motion that describes all physical phenomena remains a subject of research at the frontier of science—one of the most intriguing and challenging problems in physics. Among the many features of the equation of motion that have already been discovered, the most general and well established are captured by the statements of the first and the second laws of thermodynamics. From these laws we derive powerful tools for analyses of time-dependent phenomena. We discuss the two laws in Chapters 3 and 4. We conclude Chapter 2 with definitions of isolation, mode of interaction, and process.

In Chapter 3 we introduce a carefully worded and unambiguous statement of the first law in terms of the concept of state and the definition of a weight process. One of the principal implications of this statement is the existence of the property that we call energy and denote by E. Energy is defined for all systems and all states, and is an additive property. It obeys a conservation principle, that is, it remains invariant in the course of spontaneous changes of state of an isolated system. Because changes of state require a finite time to occur, the principle of energy conservation implied by the first law is an aspect of time evolution. As such, it reflects a special feature of the general equation of motion, even though the equation itself remains to be discovered.

We conclude Chapter 3 with a discussion of other implications of the first law: the impossibility of a perpetual-motion machine of the first kind, the relation of the law to special relativity and mass, and the energy balance. The energy balance is the most universal and powerful tool used in the analysis of practically every physical phenomenon.

In Chapter 4 we recognize that at each instant of time a system can be found in one of many different states. We classify each state according to its time evolution and define unsteady, steady, nonequilibrium, unstable equilibrium, metastable equilibrium, and stable equilibrium states.

We recall that equilibrium is not always stable and raise the question: Among all the states of a system that correspond to a given value of energy, are there any that are stable equilibrium? Close scrutiny of this question reveals that the answer cannot be found by means of the theory of mechanics. Yet experience shows that such stable equilibrium states exist. The answer is provided by the second law of thermodynamics. We introduce a carefully worded and unambiguous statement of the second law in terms of the concepts of energy, stable equilibrium state, and reversible process. This statement is an outgrowth of the pioneering work by Hatsopoulos and Keenan. In due course, we show that it entails all correct statements of the second law that have appeared in the literature. We emphasize that the second law implies the existence of stable equilibrium states but does not require that all states be stable equilibrium. Indeed, the vast majority of states are not stable equilibrium. The requirement that some equilibrium states must be stable reflects a most important feature of the general equation of motion, even though this equation remains to be discovered.

The two laws of thermodynamics have many important and practical implications. One of these implications, deduced in Chapter 4, is the impossibility of a perpetual-motion machine of the second kind. Other implications are explored in Chapters 5 to 12.

In Chapter 5 we show that, in general, not all the energy of a system can be transferred to a weight in a gravity field. Under the broad restrictions that define a weight process, the amount of energy that can be transferred from a system to a weight depends on the state of the system. If it is not a stable equilibrium state, at least a fraction of the energy can be transferred to the weight. But if it is a stable equilibrium state, no energy can be transferred to the weight. This limitation reflects innumerable experimental observations, but cannot be accounted for by the principles of mechanics alone. It is one of the important implications of the two laws of thermodynamics.

We conclude Chapter 5 with a proof of the existence of a novel important property that we call generalized adiabatic availability and denote by Ψ. The generalized adiabatic availability of a system in a given state is related to the optimum amount of energy that can be exchanged between the system and a weight in a weight process. Like energy, this property is well defined for all systems and all states.

In Chapter 6 we introduce a special reference system, called a reservoir, and discuss the possible weight processes that the composite of a system and a reservoir may experience. We prove the existence of another important property that we call generalized available energy with respect to a given reservoir and denote by Ω^R. The generalized available energy of a system in a given state is related to the generalized adiabatic availability of the composite of the system and the reservoir. Like energy and generalized adiabatic availability, Ω^R is a property that is well defined for all systems and all states. It is a generalization of the concept of motive power of fire first introduced by Carnot.

In Chapter 7 we disclose the existence of the property that we call entropy and denote by S. Entropy is defined in terms of energy, generalized available energy with respect to an arbitrarily selected reservoir, and a constant that depends on the reservoir. In terms of symbols the definition is

$$S = S_o + \frac{1}{c_R} \left[(E - E_o) - (\Omega^R - \Omega_o^R) \right]$$

where E_o and Ω_o^R refer to an arbitrary reference state, S_o is a constant fixed once and for all for the system, and the constant c_R is a carefully defined property of the reservoir. The definition appears to depend on the selection of the reservoir. However, we prove that the role of the reservoir is only auxiliary, that is, that the value of S is independent of any characteristics of the arbitrarily selected reservoir. Because both E and Ω^R are defined for all systems (macroscopic as well as microscopic) and all states (equilibrium as well as not equilibrium), S is also defined for all systems and all states, including a system with one degree of freedom in any of its states. The concept of entropy introduced here differs from and is more general than that of most textbooks, where, as Herbert B. Callen stresses in his "Thermodynamics," "the existence of the entropy" is postulated "only for equilibrium states" and the "postulate makes no reference whatsoever to nonequilibrium states."

A discussion of the main features of entropy completes Chapter 7. We show that entropy obeys a principle of nondecrease, that is, it either remains invariant or increases in the course of spontaneous changes of state of an isolated system. We show that the entropy increases if a spontaneous process is irreversible, and we call such increase a creation or generation of entropy within the system. Like the principle of energy conservation, the principle of nondecrease of entropy is an aspect of the time evolution and, as such, it reflects another special feature of the general equation of motion that remains to be discovered. We finally introduce the entropy balance which is another powerful tool for analyses of physical phenomena.

At this stage of our exposition, energy and entropy are fully and rigorously defined for equilibrium as well as other states. The concepts of temperature and heat have not yet been either defined or used. We emphasize this fact because it is a most distinguishing feature of the exposition of thermodynamics adopted in this book.

In Chapter 8 we focus our attention on stable equilibrium states. We show that at a stable equilibrium state the value of any property is fully and uniquely determined by the values of the energy, the amounts of constituents, and the parameters. This conclusion is known as the state principle. It is reached without any extraneous considerations, such as lack of information, difficulty associated with complicated calculations, unpredictability of initial conditions, or lack of interest in making detailed analyses of large systems.

Next, we derive the highest-entropy principle and the lowest-energy principle as useful criteria for stable equilibrium. In applications of the highest-entropy principle, a key role is played by the special form of the state principle known as the fundamental relation. It avers that the value of the entropy at a stable equilibrium state is uniquely determined by the values of the energy, the amounts of constituents, and the parameters. Again, this is a rigorous and general result, involving no approximation whatsoever.

In Chapters 9 to 11 we use the highest-entropy principle to investigate necessary conditions that must be satisfied for two systems to be in mutual stable equilibrium, that is, for the composite of the two systems to be in a stable equilibrium state. These investigations disclose the existence of properties that are defined only for stable equilibrium states, namely, temperature (Chapter 9), total potential of a constituent (Chapter 10), and pressure (Chapter 11). Each of these properties is defined in terms of a partial derivative of the fundamental relation and is readily measurable. More important, we show that necessary conditions for systems to be in mutual stable equilibrium are temperature equality, total potential equality for each constituent, and pressure equality. Each of these equalities provides the theoretical foundation for the measurement of the respective property.

In Chapter 9 we discuss a novel derivation of the absolute values of entropy. We discover that the smallest value of entropy is common to all systems, and that it corresponds to the entropy of all the states contemplated in mechanics. Because the smallest value of entropy is common to all systems, we assign to it the value zero and end up with absolute values of entropy that are all nonnegative. Moreover, we conclude that mechanics is the physics or thermodynamics of zero-entropy states.

In Chapter 9 we also investigate the question of the lowest value of temperature for given values of the amounts of constituents and parameters, and for systems with no upper bound to the value of the energy. We prove that it is nonnegative and obtains at the ground-energy stable equilibrium state. If we were using quantum-theoretical concepts, we could prove that the smallest temperature must be equal to zero. Because we do not use such concepts, we omit the proof and introduce the zero-temperature requirement as the third law of thermodynamics.

In Chapter 12 we introduce work and heat interactions. A work interaction is defined by the condition that its result be a net exchange of energy between the interacting systems involving no exchange of entropy. A heat interaction is defined by the condition that it be entirely distinguishable from work—no part of a heat interaction be mistakable as a work interaction. We prove that such an interaction exists, results in a net exchange of both energy and entropy, requires that the interacting systems be almost at the same temperature, and is such that the ratio of the amount of energy exchanged to the amount of entropy exchanged equals the almost common temperature of the interacting systems. We explain that a heat interaction is only a special case of interactions that are not work, and that we call nonwork.

We discuss the energy balance and the entropy balance for a system experiencing work and heat interactions only. The change in energy equals the sum of the work and the heat to the system. This result is just a very special consequence of the two laws of thermodynamics and hence cannot be called "the first law." The change in entropy equals the entropy supplied by the heat interaction plus the entropy generated spontaneously by irreversibility within the system. Also this result is just a very special consequence of the two laws of thermodynamics and hence cannot be called "the second law."

Work and heat are ingenious concepts. For given end states of a system, they allow the quantitative distinction between entropy generated by irreversibility and entropy exchanged via interactions with other systems. As such, these two concepts provide practical means for identifying opportunities to reduce the entropy generation by irreversibility and hence to improve the performance of the system. The identification of these opportunities would be missed if the definition of heat were, for example, just any interaction that is not work, that is, any nonwork interaction.

In Chapter 13 a novel graph of energy versus entropy is introduced. It is used repeatedly to provide pictorial illustrations for most of the ideas discussed in Chapters 2 to 12. The illustrations include the zero-entropy lines that correspond to all states encountered in mechanics; projections of the multidimensional space of all types of states; the zero-temperature states specified by the third law; the fundamental relations that account for all states encountered in classical thermodynamics; the concepts of temperature, adiabatic availability, available energy, and irreversibility; and the effects of work, heat, and nonwork interactions. In addition, energy versus entropy diagrams are used in discussions of the role of the second law on the ground-energy zero-temperature stable equilibrium states, and of systems with energy that cannot be increased indefinitely. It is shown that some of the stable equilibrium states of the latter systems have negative temperatures.

Chapter 14, the beginning chapter for the undergraduate student, is a concise but precise summary of the foundations, that is, of the concepts, principles, and results discussed in Chapters 2 to 13. The summary is especially suited for engineering students who must get on with the job of applying the theory to practical problems. It deemphasizes the intellectual subtleties, complexities, and generalities of the subject, while providing the assurance that thermodynamics is a well-founded, well-reasoned, unambiguous, and consistent science with many triumphs and yet new horizons to conquer. It provides the material for two or three introductory lectures in a first undergraduate course on thermodynamics. The practical meaning of the various definitions is gradually grasped from their applications in subsequent chapters and problem solving.

In Chapter 15 we illustrate the point that the energy balance and the entropy balance are two of the most valuable tools for analyses of practical problems by applying these balances to heat engines, heat pumps, and refrigeration units. The remaining chapters address applications to specific systems and specific classes of states.

In Chapter 16 we begin with a general study of the properties of substances in stable equilibrium states by focusing on systems with external forces that are characterized by a single parameter, volume. We introduce many stable equilibrium state properties, such as enthalpy, Helmholtz free energy, Gibbs free energy, and heat capacities, and derive many relations between them, such as the Gibbs relation and the Maxwell relations. For various processes that involve only heat and work interactions and in which the system starts in a stable equilibrium state and ends in a neighboring stable equilibrium state, we discuss the relations between heat, work, temperature of the initial state, temperature at which the heat interaction occurs, pressure, changes in energy, entropy, and volume, and entropy generated by irreversibility.

In Chapter 17 we discuss simple systems. We define a system as simple if it has volume as the only parameter and if it can be partitioned into a set of contiguous subsystems in mutual stable equilibrium so that the effects of the partitioning are negligible. The condition on partitioning is very important because it allows the extension of properties of a given amount to both smaller and larger amounts. It is satisfied only if the values of the amounts of constituents are sufficiently large, say, larger than 10 molecules or atoms per type. We derive the Euler relation and the Gibbs–Duhem relation. We introduce the concepts of extensive, intensive, partial, and specific properties, and establish the number of independent variables required for the specification of each of these properties. All these concepts provide the theoretical framework for the study of properties of a substance in stable equilibrium states.

In Chapter 18 we define homogeneous and heterogeneous states and the concept of phase. Then we examine the question: Starting from an initial stable equilibrium state consisting of several coexisting phases, how many of the intensive properties temperature, pressure, and chemical potentials can be varied independently while the system changes to another stable equilibrium state consisting of the same types of phases as the initial state? We derive the answer, the so-called Gibbs phase rule.

In Chapter 19 we discuss the stable-equilibrium-state properties of pure substances modeled as simple systems. Experimental results on energy, enthalpy, and other properties are reported in graphical and tabular forms. Much is extracted from the experimental data on a given substance by combining them with the wealth of theoretical implications that follow from the state principle, in general, and the patterns of coexisting phases, in particular.

In Chapter 20 we consider certain limited ranges of temperature and pressure in which the relation between temperature, pressure, and volume, the so-called equation of state, takes a simple and explicit mathematical form. Under such conditions, which we call ideal-behavior conditions, properties are much easier to study because they can be described by simple analytical forms rather than by charts and tabulated data. Thus we describe the ideal-gas behavior and the many analytical interrelations that apply under such ideal behavior. We also describe the ideal-incompressible fluid or ideal-solid behavior.

In Chapter 21 we discuss some of the historical equations of state that have been developed to model the behavior of substances outside the ranges of validity of the ideal-gas behavior.

In Chapter 22 we define a special class of nonequilibrium states that we call bulk-flow states. We introduce another type of nonwork interaction, the bulk-flow interaction. It involves bulk-flow states and accomplishes transfers of energy, entropy, and amounts of constituents. Bulk-flow states and bulk-flow interactions play a major role in modeling energy-processing devices, energy-conversion systems, chemical reactors, and materials-processing systems.

In Chapter 23 we consider a number of simplified models of energy-conversion devices, such as nozzles, compressors, pumps, turbines, and heat exchangers. For each of these devices, we derive interrelations between end states, bulk-flow states of inlet and outlet streams, energy and entropy transfers, entropy generated by irreversibility, and empirical characteristics of the device.

In Chapter 24 we develop the concept of optimum work in processes experienced by a composite of a system and the environmental reservoir. We discuss generalizations of the concepts of adiabatic availability and available energy that apply to various combinations of conditions. Rather than building up the vocabulary, we call each such generalization an availability or exergy. We conclude the chapter with a discussion of a thermodynamic measure of efficiency, the ratio of the availability required by a change of state of a system to the availability consumed to effect this change.

In Chapter 25 we briefly discuss energy-conversion systems, external combustion and internal combustion engines, nuclear fission reactors, and hydropower plants. Then we review the typical energy-conversion systems following the traditional classification based on the historical working-fluid cycles. Excluded from these discussions are combustion phenomena, which are covered in Chapter 31.

In Chapter 26 we discuss the stable-equilibrium-state properties of mixtures modeled as simple systems. Experimental results on energy, enthalpy, and other properties are reported in graphical and tabular forms. As for pure substances, much can be extracted from the experimental data by combining them with the wealth of theoretical implications that follow from the state principle, in general, and the patterns of coexisting phases, in particular. The main objective here is to relate properties of a mixture to properties of the pure constituents. We show that this can be done via the concepts of partial pressure and equality of chemical potentials of systems in mutual stable equilibrium.

In Chapter 27 we consider certain limited ranges of temperature and pressure in which the relations between properties of a mixture and properties of pure constituents take simple and explicit mathematical forms. Under such conditions, which we call ideal-behavior conditions, properties are much easier to study because they can be described by simple analytical forms rather than through the use of charts and tabulated data. Thus we describe the ideal Gibbs–Dalton behavior, the ideal-gas mixture behavior, and the many analytical interrelations that apply under such ideal behaviors. As a typical engineering example, we discuss moist air, the mixture of air and water vapor. We also describe the ideal-solution behavior and the rudiments of two-phase mixture behavior such as needed to design distillation columns.

In Chapter 28 we discuss some of the historical relations between temperature, pressure, chemical potentials, and mole fractions that have been developed to model the behavior of substances outside the ranges of validity of the ideal-gas mixture and the ideal-solution behaviors.

Up to this point in the book we have not considered the effects of chemical reactions. In Chapters 29 and 30 we define the concepts and terminology that are needed for a description of simple systems subject to chemical reactions, present a novel derivation of the necessary conditions for chemical equilibrium, and evaluate the effects of chemical reactions on the stable-equilibrium-state properties.

In Chapter 31 we consider the special class of chemical reactions that correspond to the combustion or oxidation of fuels, namely the reactions that represent about 90% of the energy sources we use in energy-conversion devices.

Paradoxically, our experience shows that the study of this book is easier for a reader without much background in thermodynamics than for one who has been exposed to the traditional presentation and its statistical interpretation. The inexperienced student finds here a complete, logically consistent, and self-sufficient exposition. All concepts are defined carefully and explicitly so that the learning proceeds smoothly from the foundations to the applications. No previous knowledge of the subject is needed, except for a sophomore background in calculus and introductory physics.

Experienced readers may have some difficulty. They may be appalled when they read that thermodynamics applies equally well to macroscopic and microscopic phenomena, that entropy is equally defined for equilibrium and nonequilibrium states, and that temperature is meaningful and useful also for the stable equilibrium states of a single particle. Their dismay will persist unless they realize that the new perspective requires an intellectual reorientation and a subtle and demanding reconsideration of basic premises and concepts.

For us, the need for such reconsideration of premises and concepts has not arisen only from our didactic experience. It is stimulated by the results of research on the quantum foundations of physics performed by colleagues and ourselves. The results are published in the scientific literature. However, there is no need at all to be acquainted with either such literature or with the ideas of quantum theory in order to study any part of this book. We mention these results here only to assure our doubting reader that the advanced foundations of quantum physics lurk behind our exposition, and that each of the concepts we introduce has a sound counterpart in the most advanced formulation of contemporary physics.

Even the traditional meaning of the term thermodynamics needs to be reconsidered. Physics is the science that attempts to describe all aspects of all phenomena pertaining to the perceivable universe. It can be viewed as a large tree with many branches, such as mechanics, electromagnetism, gravitation, and chemistry, each specialized in the description of a particular class of phenomena. Thermodynamics is not a branch. It pervades the entire tree. To emphasize this conception, we often use the words physics and thermodynamics as synonyms.

Acknowledgments

In preparing this text, we have been blessed with generous assistance from many sources. We are grateful to the Massachusetts Institute of Technology for the freedom, opportunity, and excitement provided by its warm and stimulating environment. We have enjoyed generous support from our endowed professorships, the Ford Professorship of Engineering and the C. R. Soderberg Professorship in Power Engineering, the Thermo Electron Corporation, George N. Hatsopoulos, and the Bernard M. Gordon Engineering Curriculum Development Fund.

Our ideas emerged and were shaped from countless probing discussions with our teacher, George N. Hatsopoulos, and the great thinker of thermodynamics, the late Professor Joseph H. Keenan. We are privileged and grateful to have been associated with them.

We are thankful to James L. Park, John P. Appleton, Enzo Zanchini, John B. Heywood, James C. Keck, Phillip Thullen, Maher El-Masri, Ernest G. Cravalho, and Michael G. O'Callahan for sharing with us the teaching of thermodynamics over the years; to Mario Silvestri, Gregory D. Botsaris, Charles Berg, Mohamad Metghalchi, Alfonso Niro, Roberto Mauri, Giacomo Elias, and Adriano Muzzio for helpful discussions; to J. Azzola, W. Chin, V. Filipenco, S. T. Free, J. Griffin, T. G. Hill, E. T. Hurlburt, R. Kach, T. P. Korakianitis, H. Najm, P. Park, J. Reilly, C. Schmidt, A. Sich, J. B. Song, A. Tangborn, C. Vlahoplus, and T. Wolf for their assistance with the problems; and to J. A. Caton, David E. Foster, Richard A. Gagglioli, David G. Goodwin, Ralph Greif, T. P. Korakianitis, M. J. Moran, E. M. Sparrow, Philip A. Thompson, and Thomas M. Weber for their reviews of selected chapters of the book.

Last but not least, we wish to express our deeply felt thanks to our families for their patience and support over the many years that we took to finish our writing.

Elias P. Gyftopoulos
Gian Paolo Beretta
Cambridge, Massachusetts
July, 1990

Contents

31 Combustion   586

A Physical Constants and Unit Conversion Tables   617

B Tables of Properties   621

1 How to Study with This Book

In thermodynamics, we model our experiences of physical phenomena in terms of concepts, mathematical symbols that represent the concepts, and relations between symbols that represent interdependences between concepts. Some aspects of this procedure are readily understandable even by people without much expertise in the subject. Other aspects are somewhat abstract and require systematic study. In any case, the results are fascinating and very practical. We use models to describe physical phenomena at one instant of time and to make predictions as a function of time. For example, we may wish to evaluate the humidity of the atmosphere on a hot summer day, or design and control the operation of a refinery that transforms crude oil into petroleum products.

In this chapter we describe some salient features of the models of physical phenomena without being concerned about the precise meanings of the terms we use. We do this especially for readers being exposed to the subject for the first time. We jump in the middle of energy and entropy balances, in much the same way that in a fluid mechanics text one jumps in the middle of mass and momentum balances, or in a chemistry text one jumps in the middle of stoichiometric balances of atomic nuclei.

After seeing how these balances play a central role in solving a few sample problems, we hope that readers will be sufficiently motivated to pursue a systematic study of the applications and a deeper scrutiny of the foundations of thermodynamics. This pursuit can be approached along four different paths: (1) an undergraduate introductory path, (2) an advanced undergraduate path with emphasis on applications, (3) a graduate path with emphasis on the general foundations, and (4) a graduate path with emphasis on advanced topics. Each path consists of different parts of the book. We describe the parts that belong to each path at the end of this chapter.

1.1 Modeling Physical Phenomena

Some concepts are essential in all models of physical phenomena. They are the basic tools with which scientists and engineers reason and report their findings. Among them are the concepts of system, property, and the law of motion or change with time. Any model of any physical phenomenon includes the definitions of a system—the object of study, its properties—the characteristics of the system that can be measured at one instant of time by means of well-defined operations, and the law of motion—the relation between values of properties at different instants of time.

The requirement that properties must be measurable provides the link between the model and the phenomena it describes. In some instances, this link is both readily understandable and fairly direct. In others, the link may be conceptually complicated and operationally indirect. For example, the position of a particle of a one-particle system is a property that we can easily visualize and readily measure. In contrast, the

1

acceleration of a particle is harder to picture, and its measurement more involved. Even more complicated is the idea of the mass of the particle and its measurement. Although some properties are more abstract than others, they are all measurable and are all linked to the phenomena they describe.

The law of motion is the necessary tool for making predictions as a function of time. It helps us not only to make forecasts about natural occurrences, but also to explore natural phenomena for the benefit of our society. For example, we use the law of motion to determine the location of a celestial body a year hence. Again, we use the law of motion to design and control the flight of an airplane from one city to another. Some models may lack a law of motion but still be very valuable because they define either specific interrelations among different properties at one instant of time, or specific facets of the law of motion, such as the time dependence of only a few of the properties, or both.

Over the centuries, a large number of different models have been developed, each for a particular class of physical phenomena. Despite the differences, however, certain properties have emerged as common features of all models, and provide an excellent degree of unity in our understanding of the workings of nature. Of course, the very ambitious and worthy undertaking of a unified interpretation of our experiences by means of an all-encompassing model must continue. In all likelihood, such an undertaking will remain forever incomplete because we will never be able to assert that we have observed all physical phenomena.

Even if the all-encompassing model were known, we would not use it in practice to solve every specific scientific problem, and every specific engineering problem because such an approach would probably be very complicated. Instead, we adapt our consideration to the problem at hand and introduce simplifications and approximations. In doing so, however, we make sure that as many of the common features of all models are retained as practicable, and that these features satisfy all the conditions that are widely accepted as valid. For example, from the various models that have been developed to date, a firm conclusion is that energy is neither created nor destroyed. So, in any modeling of a problem, we make sure that energy is a property and that energy is conserved.

Some of the well-established general features of all models and the way they influence our approach to problem solving are discussed immediately after we introduce the idea of accounting and balances in the next section.

1.2   Accounting and Balances

The procedure of performing a balance of some quantity of interest is part of our everyday experience. Outside the domain of natural phenomena, the most familiar example is what we call "accounting," whether applied to our personal bank savings or to the finances of a business. Although this example belongs to the domain of economics, we discuss it here because it brings out certain self-evident features of the properties that must be balanced and that are also features of energy and entropy.

To make these ideas specific, we consider a safe in a bank in which a family keeps its savings in the form of golden coins. We call these savings account A, denote the number of coins in the safe by N^A, and recognize that it has three useful and practical features. The first feature is that N^A can be measured at any instant of time by a simple procedure—counting. At time t_1, we denote the value of N^A by N_1^A so that if the number of coins is 2000, we write $N_1^A = 2000$ coins. It is evident that N^A must play a key role in any accounting model.

The second feature that makes the quantity N^A useful in accounting is that it is additive. This means that if N^C denotes the number of coins in a composite account C consisting of two accounts A and B with numbers of coins denoted by N^A and N^B, respectively, then at any instant of time t_1 the value N_1^C of N^C equals the sum of the values N_1^A and N_1^B, that is, $N_1^C = N_1^A + N_1^B$. The feature of additivity is useful because it allows us to keep track of a financial situation even if it consists of an intricate structure of different accounts.

The third feature is that different accounts may be made to interact with each other, that is, coins may be transferred from one account to another. If the interacting accounts are A and B, we denote the coins transferred from account B into account A by the symbol $N^{A\leftarrow}$, where we use the arrow in the superscript because an "amount transferred", like a vectorial quantity, is defined not only by a numerical value—the number of coins transferred—but also by the direction of the transfer. Thus $N^{A\leftarrow} = 1000$ coins means that 1000 coins are transferred from account B to account A, whereas $N^{A\rightarrow} = 1000$ coins means that 1000 coins are transferred from account A to account B. Because the arrow points either into A or out of A, we can use one symbol to denote both transfers, and either positive or negative values to specify whether the flow is actually into or out of A. For example, if we choose the symbol $N^{A\leftarrow}$, we assign a positive value when coins are transferred into A, and a negative value when coins are taken out of A. Clearly, if $N^{A\leftarrow} = 2000$ coins, then $N^{A\rightarrow} = -2000$ coins.

Performing a balance on the number of coins N^A of account A over a period of time from time t_1 to time t_2 means that we evaluate the change in the value of N^A during that time interval, that is, the difference $N_2^A - N_1^A$, and compare it with the net number of coins $N_{12}^{A\leftarrow}$ that were transferred into the account as a result of all the transactions that occurred between t_1 and t_2. For example, account A might have had $N_1^A = 4000$ coins at a time t_1, $N_2^A = 3000$ coins at time t_2, and have experienced two transactions, $N^{A\leftarrow} = -2500$ coins, and $N^{A\leftarrow} = 1500$ coins, so that $N_{12}^{A\leftarrow} = -2500 + 1500 = -1000$ coins over the period from t_1 to t_2.

Comparison of the values $N_2^A - N_1^A$ and $N_{12}^{A\rightarrow}$ brings out important features that have to do with the nature of the characteristic property N, and the nature of the dynamics of the system—the account. If for any time interval and any transactions, we find that the balance

$$N_2^A - N_1^A = N_{12}^{A\leftarrow} \tag{1.1}$$

is satisfied, then we say that N is conserved. Of course, we know from experience that the number of coins in a well-guarded safe is conserved. Nevertheless, this obvious conservation principle—self-evident identity—is used by accountants to make sure that all accounts are in good order.

The number of coins is not the only property of interest of each account A. Another characteristic of even greater importance is the purchasing power of the coins. We denote it by P^A. At any instant of time t_1, the purchasing power of the coins, P_1^A, represents the fraction of the gross national product (GNP) that at time t_1 can be bought with the coins in the safe. It is measured in dollars per dollar of GNP or some other currency, and has the same three basic features as the number of coins, that is, measurability, additivity, and transferability.

Over a period of time from t_1 to t_2, the purchasing power of the coins in a safe can be altered by two factors: the transfer of purchasing power $P_{12}^{A\leftarrow}$ from other accounts as a result of transfers of coins into and out of the safe; and the creation or destruction of purchasing power as a result of phenomena such as deflation or inflation of prices, and variations in the exchange rate of golden coins to currency. Denoting the latter effects by P_i, we can relate the purchasing power P_2^A at time t_2 to the purchasing power P_1^A at time t_1 by the balance

$$P_2^A - P_1^A = P_{12}^{A\leftarrow} + P_i \tag{1.2}$$

If gold appreciates as fast as prices are inflated over the period from t_1 to t_2, then P_i is zero and the purchasing power is affected only by transfer transactions. On the other hand, if gold does not appreciate as fast as prices are inflated, $P_i < 0$, and the purchasing power of a dormant account ($P_{12}^{A\leftarrow} = 0$) is being destroyed even though the number of coins in the account is conserved.

In conclusion, our simple accounting example shows that every golden-coin account A has the property N^A, representing the number of coins in the account, and the property P^A, representing the purchasing power of the coins in the account. Both properties are always defined at any instant in time and in all circumstances, and both are measurable, additive, and transferable. These features make both property N^A and property P^A very useful because we can perform balances on them over a time period of interest. Moveover, whereas property N^A is conserved, that is, its balance equation has no creation or destruction term [equation (1.1)], in general, property P^A is not conserved, that is, its balance equation contains a creation or destruction term P_i [equation (1.2)].

In the domain of natural phenomena, two properties play a similarly central role in any well-defined model: energy and entropy. Similarly to the number of coins N^A, energy is a conserved property. Similarly to the purchasing power P^A, in general, entropy is not a conserved property. We discuss these features of energy and entropy in the next section.

1.3   Energy and Entropy Balances

Most of this book is devoted to defining, understanding, and using energy and entropy in solving practical problems. Here we give a preview of how energy and entropy play such a central role in modeling the behavior of all systems under all conditions.

Both energy and entropy have the three basic features of measurability, additivity, and transferability that we discuss in the preceding section. The system that we define to model any physical phenomenon has both energy and entropy as properties. Each of these properties can be measured at any instant of time by means of well-defined procedures. We denote energy by E, entropy by S, and the values of E and S at time t_1 by E_1 and S_1, respectively. For example, upon applying the measuring procedures that define energy and entropy to a given system at time t_1, we could find $E_1 = 400$ J and $S_1 = 5$ J/K, where J stands for the units of measure of energy, and J/K for the units of measure of entropy. In due course we will see that one set of units are joules for energy and joules per kelvin for entropy. Because these measuring procedures are applicable to any well-defined system in any condition and at any instant of time, we say that energy and entropy are general properties of any model of any physical phenomenon.

Each of the properties E and S is additive. If a composite system C consists of two subsystems A and B, and we denote the energies and entropies of the three systems by E^C, E^A, E^B and S^C, S^A, S^B, respectively, then at any instant of time t_1 the value E_1^C equals the sum of the values of E_1^A and E_1^B, that is, $E_1^C = E_1^A + E_1^B$, and the value of S_1^C equals the sum of the values of S_1^A and S_1^B, that is, $S_1^C = S_1^A + S_1^B$. The additivity feature is useful because it allows us to keep track of the values of energy and entropy even in complicated systems consisting of intricate structures of subsystems, such as chemical industrial plants, power plants, or laboratory experiments.

Two systems can interact with each other and, as a result of the interaction, exchange energy and entropy. If the interacting systems are A and B, we denote the amounts of energy and entropy transferred from system B to system A by $E^{A\leftarrow}$ and $S^{A\leftarrow}$, respectively, where we use the arrow in the superscript because an amount transferred is defined not only by the numerical value—the amount of energy or entropy transferred—but also by the direction of the transfer. As we did for the accounting example, we can denote a transfer of 200 J of energy and 2 J/K of entropy into system A by writing either $E^{A\leftarrow} = 200$ J and $S^{A\leftarrow} = 2$ J/K or, equivalently, $E^{A\rightarrow} = -200$ J and $S^{A\rightarrow} = -2$ J/K.

The three features of energy and entropy just listed—measurability, additivity, and transferability—make these two properties suitable for setting up balances in much the same way that we do in the accounting example of the preceding section. Performing balances of energy and entropy for a system A over a period of time from t_1 to t_2 means that we evaluate the changes $E_2^A - E_1^A$ in energy and $S_2^A - S_1^A$ in entropy and compare them with the respective net amounts of energy $E_{12}^{A\leftarrow}$ and entropy $S_{12}^{A\leftarrow}$ transferred into the system as a result of all the interactions experienced by A during the time period between t_1 and t_2.

Comparison of the values of $E_2^A - E_1^A$ and $E_{12}^{A\leftarrow}$ brings out the important feature that energy is conserved, that is, that the energy balance is always of the form

$$E_2^A - E_1^A = E_{12}^{A\leftarrow} \tag{1.3}$$

whether $t_2 > t_1$ or $t_1 > t_2$.

Comparison of the values $S_2^A - S_1^A$ and $S_{12}^{A\leftarrow}$ brings out the important feature that entropy cannot be destroyed but can be created inside the system as time goes on. If $t_2 > t_1$, we express this feature by writing the entropy balance in the form

$$S_2^A - S_1^A = S_{12}^{A\leftarrow} + S_{\text{irr}}^A \qquad \text{for } t_2 > t_1 \tag{1.4}$$

where S_{irr}^A is such that

$$S_{\text{irr}}^A \geq 0 \tag{1.5}$$

that is, S_{irr}^A is nonnegative. Equation (1.4) may also be written in the form

$$S_2^A - S_1^A \geq S_{12}^{A\leftarrow} \qquad \text{for } t_2 > t_1 \tag{1.6}$$

In due course, we will call S_{irr}^A the entropy created or generated by irreversibility inside system A.

The energy and entropy balances, equations (1.3) and (1.4) [or (1.3) and (1.6)], are the cornerstones of the dynamical analysis of all physical phenomena. Yet they are among the most abstract concepts of the description of nature. Perusing once more the analogy with the accounting example in the preceding section, we observe that the value of the energy E^A of any system A can change only as a result of interactions that transfer energy into or out of the system, in much the same way that the value of the number of coins N^A in an account A can change only as a result of transactions that transfer coins into or out of the account. In contrast, the value of the entropy S^A of any system A can change even in the absence of any interaction, that is, spontaneously, because of the workings of nature, in much the same way that the value of the purchasing power P^A can change even in the absence of any banking transactions because of the workings of the economy. In fact, we will see that the relation between energy and entropy has several analogies to that between the number of golden coins and their purchasing power.

We never use the accounting example again, but we devote several chapters of this book to defining the meanings of energy and entropy, and explore their interrelations and far-reaching implications on the behavior of various systems. A better understanding of the interrelations and implications contributes greatly to our ability to design better equipment to control various physical phenomena.

1.4 Applications of Energy and Entropy Balances

The problem of understanding the physical meanings of energy and entropy can be tackled at various levels of depth and from different standpoints. The most basic approach would be in terms of an advanced physics course based on a general and all-encompassing model of physical phenomena. Such a course would require a good background in mathematics and a strong interest in abstract thinking. The subject is fascinating, the open questions are challenging, and the horizon for scientific contributions is exciting. But such an approach would be too basic and would hardly convey the usefulness of thermodynamics in analyses of applied physics and engineering problems.

The approach we take in this book is to start from a nonmathematical summary of concepts and statements that capture only the most general features of the all-encompassing model known to date, and to use these concepts and statements in the study of a number of broad classes of problems. The study of this approach may be pursued along four alternative paths, each providing material for a regular one-semester course, and each being outlined in Section 1.6.

In an introductory course, the meaning of energy and entropy can be grasped gradually and indirectly through use of these two properties in solving practical problems, and in describing the behavior of physical systems in what are called the thermodynamic equilibrium or stable equilibrium states. For such states, energy and entropy can be related to other properties, such as the equilibrium properties temperature and pressure, which in turn are linked to the object of study by familiar experimental instruments, such as a thermometer and a pressure gauge. For stable equilibrium states, the approach can be viewed as proceeding along two parallel but interrelated lines of analysis.

The first line of analysis focuses attention on relating energy and entropy differences to more readily measurable properties by means of models of the stable equilibrium behavior of matter in selected ranges of conditions. Examples of such models are the perfect-gas model, and the perfect-incompressible model for solids or fluids. This line of analysis addresses the left-hand sides of the energy and entropy balances [equations (1.3) and (1.4)]. For example, for stable equilibrium states of a system A that behaves according to the perfect-gas model, energy and entropy differences can be expressed as

$$E_2^A - E_1^A = C_V^A (T_2^A - T_1^A) \tag{1.7}$$

$$S_2^A - S_1^A = C_p^A \ln \frac{T_2^A}{T_1^A} - R^A \ln \frac{p_2^A}{p_1^A} \tag{1.8}$$

where C_V^A, C_p^A, and R^A are constants, and T_2^A, T_1^A and p_2^A, p_1^A are, respectively, the temperatures and the pressures of system A at times t_2 and t_1. Again, for stable equilibrium states of a system B that behaves according to the perfect-incompressible-fluid model, energy and entropy differences can be expressed as

$$E_2^B - E_1^B = C^B (T_2^B - T_1^B) \tag{1.9}$$

$$S_2^B - S_1^B = C^B \ln \frac{T_2^B}{T_1^B} \tag{1.10}$$

where C^B is a constant, and T_2^B and T_1^B are the temperatures of the system at times t_2 and t_1, respectively. In due course, we devote several chapters to defining these and other models.

The second line of analysis focuses attention on evaluating the exchanges of amounts of energy and entropy associated with selected types of interactions between systems. Examples of such types of interactions are work, heat, and bulk flow. This line of analysis addresses the right-hand sides of the energy and entropy balances. For example, for a work interaction experienced by a system A, energy and entropy transfers can be expressed as

$$E^{A\leftarrow} = W^{\leftarrow} \tag{1.11}$$

$$S^{A\leftarrow} = 0 \tag{1.12}$$

where the energy transfer $W^{\leftarrow}$ is called work. Again, for a heat interaction experienced by a system A, we have

$$E^{A\leftarrow} = Q^{\leftarrow} \tag{1.13}$$

$$S^{A\leftarrow} = \frac{Q^{\leftarrow}}{T_Q} \tag{1.14}$$

where the energy transfer $Q^{\leftarrow}$ is called heat, and T_Q is the temperature at the boundary of system A where the interaction occurs. In due course, we devote an entire chapter to defining work and heat interactions. Still another type of interaction is bulk flow, for which we have

$$E^{A\leftarrow} = m^{A\leftarrow}(h + \frac{\xi^2}{2} + gz) \tag{1.15}$$

$$S^{A\leftarrow} = m^{A\leftarrow}s \tag{1.16}$$

where $m^{A\leftarrow}$ is the amount of mass transferred into system A as a result of the interaction, and h, s, ξ, g, and z are properties of the transferred material that are defined in due course. If more than one of these interactions occurs, the energy and entropy transfers are sums of the respective amounts for the relevant interactions.

1.5 Problem Solving

The energy and entropy balances, and the two lines of analysis just cited, constitute the primary tools for the systematic approach to solving the applied physics and engineering problems discussed in this book. The approach may be subdivided into four steps.

In the first step, we define the system—the object of study. The step is important because it determines the ease of the solution. The selection of the system requires experience and a careful examination of the statement and the questions of the problem. For each question we may have to define a system that includes as many of the physical objects of the problem as possible, and that experiences at its boundary one or more of the interactions that must be evaluated.

In the second step, we model the system. Usually, we look inside the boundary of the system and partition it into a collection of subsystems for each of which we can evaluate the changes in energy and entropy. For example, we might select each subsystem so as to be able to model its behavior as that of a perfect gas, perfect incompressible fluid, or cyclic device, that is, a device that experiences zero net changes of all its properties. If successful, this step allows us to express energy and entropy differences in terms of more readily measurable properties such as temperatures and pressures. Some of the latter properties may be known; others may be unknown.

In the third step, we model the interactions occurring at the boundary of the system in terms of the transfers of energy and entropy that they contribute. For example, we may conclude that work, heat, and bulk-flow interactions are adequate to describe all the transfers of properties at the boundary of the system. The purpose of this step is to express the transfer terms $E_{12}^{A\leftarrow}$ and $S_{12}^{A\leftarrow}$ in the right-hand sides of the energy and entropy balances as sums of transfer terms, each referring to a specific interaction. Again, some of these transfer terms may be known, others may be unknown.

In the fourth step, we combine the results of the second and third steps and form the energy and entropy balances. The energy balance equation can be solved for one of the unknowns of the problem. The entropy balance can be solved for S_{irr}^A if all the other variables are known, or it can be solved for another unknown if S_{irr}^A is given as part of the problem. For example, a question might imply that $S_{\text{irr}}^A = 0$.

The foregoing four-step problem-solving scheme is succinct and captures only some of the essential ingredients of analyses based on energy and entropy balances. Proceeding along one of the four paths outlined in Section 1.6, the reader will find that the ideas so quickly summarized here are gradually clarified and, in a sense, guide the flow of the material.

We conclude this section with an example representative of the kind of practical questions that can be answered by the use of energy and entropy balances.

Example 1.1.   A block of metal, B, is immersed in a pool of liquid water, L. The block and the water interact with each other and nothing else. Temperature measurements on the block and on the liquid yield the values T_1^B and T_1^L at time t_1, and only the value T_2^B at time t_2. The engineer who performed the measurements forgot to tell us whether time t_1 was before or after t_2. Can we find the value of T_2^L, and infer the time sequence from the measured values of the temperatures?

Solution.

Step 1.   To answer the question we define as our system the composite A consisting of the block of metal B and the liquid water L.

Step 2.   We model block B as a perfect incompressible solid, and the liquid water L as a perfect incompressible liquid, each with energy and entropy differences given by equations (1.9) and (1.10). Using these equations and the fact that energy and entropy are additive properties, we can express the terms $E_2^A - E_1^A$ and $S_2^A - S_1^A$ on the left-hand sides of the energy and entropy balances [equations (1.3) and (1.4)] in terms of T_1^B, T_1^L, T_2^B, and T_2^L, so that

$$E_2^A - E_1^A = E_2^B - E_1^B + E_2^L - E_1^L = C^B (T_2^B - T_1^B) + C^L (T_2^L - T_1^L) \qquad (1.17)$$

$$S_2^A - S_1^A = S_2^B - S_1^B + S_2^L - S_1^L = C^B \ln \frac{T_2^B}{T_1^B} + C^L \ln \frac{T_2^L}{T_1^L} \qquad (1.18)$$

Step 3.   No interactions occur at the boundary of system A because the block and the liquid water do not interact with anything else than each other. Thus no energy and no entropy are transferred into or out of the system A, that is, $E_{12}^{A\leftarrow} = 0$ and $S_{12}^{A\leftarrow} = 0$.

Step 4.   Substituting equations (1.17) and (1.18) in the left-hand sides of the energy and entropy balances [equations (1.3) and (1.4)], and using $E_{12}^{A\leftarrow} = 0$ and $S_{12}^{A\leftarrow} = 0$ in the right-hand sides, we find that

$$C^B (T_2^B - T_1^B) + C^L (T_2^L - T_1^L) = 0 \qquad (1.19)$$

$$C^B \ln \frac{T_2^B}{T_1^B} + C^L \ln \frac{T_2^L}{T_1^L} = \ln \left[\left(\frac{T_2^B}{T_1^B} \right)^{C^B} \left(\frac{T_2^L}{T_1^L} \right)^{C^L} \right]$$

$$= S_{\text{irr}}^A \geqq 0 \qquad \text{for } t_2 > t_1 \qquad (1.20)$$

Upon solving equation (1.19), we determine T_2^L in terms of the measured temperatures and the two constants C^B and C^L. Substituting T_2^L in relation (1.20), we conclude that if

$$\left(\frac{T_2^B}{T_1^B} \right)^{C^B} \left(\frac{T_2^L}{T_1^L} \right)^{C^L} > 1 \qquad (1.21)$$

then $t_2 > t_1$. On the other hand, if

$$\left(\frac{T_2^B}{T_1^B}\right)^{C^B} \left(\frac{T_2^L}{T_1^L}\right)^{C^L} < 1 \qquad (1.22)$$

then $t_2 < t_1$.

1.6   Four Courses of Study

The topics included in this book are not meant to be covered either by a single course or by a sequence of interrelated courses. Different parts may be combined as one-semester offerings. Each offering serves different programmatic and student needs.

For example, an undergraduate course for engineering students may begin with the ideas that energy and entropy are two properties of a system. Rather than through the formal definitions of these properties, the student is assisted to grasp their meanings and significance by repeated applications. Chapter 14 is recommended as the starting point of such courses. It summarizes the essential concepts without emphasis on subtleties, and without compromising and misleading arguments. Applications are selected from Chapters 15 to 31 as discussed in the following suggested outlines.

1. Undergraduate Course: *Introductory Thermodynamics.* This course is for sophomores or juniors in the schools of science and engineering. From the end of the present chapter, jump directly to Chapter 14. Study Chapters 14 to 25, except for Sections 17.6 to 17.8, 20.5 and 20.6, and 21.3 to 21.6.

Heat engines, heat pumps, and refrigeration units are discussed in Chapter 15; interrelations among stable-equilibrium-state properties of systems with volume as the only parameter in Chapter 16; the simple-system model in Chapter 17; properties of pure substances, and models for ideal-gas and incompressible-fluid behaviors in Chapters 19 to 21; bulk-flow interactions in Chapter 22; devices used in energy-conversion and materials-processing systems in Chapter 23; optimum performance and thermodynamic efficiency in Chapter 24; and energy-conversion systems in Chapter 25.

2. Undergraduate Course: *Advanced Topics in Thermodynamics.* If a program foresees a second semester on thermodynamics, complete the study of Chapters 16 and 17; and study Chapters 26, 27, and 29 to 31, except for Sections 27.5 to 27.8, and 30.6 to 30.8.

Mixtures and models for ideal-gas-mixture behavior are discussed in Chapters 26 and 27, chemical reactions and chemical equilibrium in Chapters 29 and 30, and combustion in Chapter 31.

3. Graduate Course: *Thermodynamics: Foundations and Applications.* This course is for seniors and first-year graduate students in the schools of science and engineering. Its purpose is to expose the student to the full generality of thermodynamics as a subject that pervades the foundations of all of physics and much of engineering. It represents an entirely new conception. It includes Chapters 2 to 13 and selected topics from Chapters 15 to 31.

The foundations of thermodynamics are discussed in Chapters 2 to 12, and novel graphical representations on the energy versus entropy diagram in Chapter 13. Selected topics and homework problems from Chapters 15 to 31 may also serve as preparation for the doctoral qualifying exam in thermodynamics.

4. Graduate Course: *Advanced Topics in Thermodynamics.* If a program includes two semesters of thermodynamics for first-year graduate students, review Chapters 14 to 25 with emphasis on the parts not covered in the first semester, and study in depth Chapters 26 to 31.

2 Kinematics and Dynamics

The description of physical phenomena requires rigorous consideration of many basic concepts. We assume that ideas such as space, time, reference frame, velocity, acceleration, force, particle, atom, and molecule are known from introductory courses in physics and need not be reemphasized here. On the other hand, because of the breadth and depth of the scope of our exposition, the definitions of other concepts need special emphasis.

In this chapter we emphasize the definitions of system, property, and state, and discuss some of the ideas related to the causal laws governing motions that occur either spontaneously or because of interactions between systems. The branch of physics that describes the possible or allowed states of a system is called kinematics, and that which gives a causal description of the time evolution of a state is called dynamics.

2.1 Systems

In any physical study we always focus attention on a collection of constituents that are subjected to a nest of forces. When the constituents and the nest of forces are well defined, we call such a collection a *system*. Everything that is not included in the system we call the *environment* or the *surroundings* of the system. In this section we discuss the requirements for constituents and nest of forces to be well defined, namely, the requirements for the definition of a system.

A *constituent* is either a material object, such as a molecule, atom, ion, nucleus, and elementary particle, or a field, such as the electromagnetic field, that for the given description is to be considered as an elementary building block. For each system a constituent is specified by the type and the range of values of its amount. The *type* may be a specific molecule, atom, ion, nucleus, elementary particle, or a specific field. For example, in certain studies of water, the type of constituent may be just the H_2O molecule. In other studies of water, however, more than one type of constituent may be required, such as H_2, O_2, and H_2O molecules, H and O atoms, H^+ and O^- ions, and electrons, or H_3O^+, OH^-, and H_2O molecules. Again, in studies of electromagnetic radiation, the type of constituent is the electromagnetic field.

Example 2.1. In each of the following studies, is the water molecule, H_2O, the only type of constituent?

   (a) The study of a liquid solution of salt, NaCl, in water.
   (b) The study of absorption of photons by H_2O molecules in a microwave oven.
   (c) The study of steam generation in a boiler.

Solution.

(a) No. The salt molecule is also necessary.

(b) No. Here we must also consider the electromagnetic field.

(c) Yes. For a wide variety of applications and range of conditions, H_2O is the only type of molecule that need be considered.

Example 2.2. In each of the following studies, what are the types of constituents that must be specified?

(a) The study of the hydrolysis of water molecules according to the reaction mechanism $2\,H_2O = 2\,H_2 + O_2$.

(b) The study of the oxidation of hydrogen, H_2, in a mixture of nitrogen, N_2, and oxygen, O_2, according to the reaction mechanism $2\,H_2 + O_2 = 2\,H_2O$.

(c) The study of the electrolysis of water according to the cathodic reaction mechanism $2\,H_2O = 4\,H^+ + 4\,e^- + O_2$ and the anodic reaction mechanism $4\,H^+ + 4\,e^- = 2\,H_2$, where e^- stands for a negatively charged electron, and H^+ for a positively charged hydrogen ion.

(d) The study of pure water in the form of solid, liquid, or vapor in the absence of any chemical reaction.

Solution.

(a) Water molecules, H_2O, hydrogen molecules, H_2, and oxygen molecules, O_2.

(b) Same molecules as in study (a), plus nitrogen molecules, N_2.

(c) H_2O, H_2, and O_2 molecules and electrons and positively charged hydrogen atoms.

(d) H_2O molecules only.

The *amount of a constituent* of a given type can be evaluated at any instant of time by means of counting. The *value of the amount* of a constituent is the numerical result of the counting procedure and is expressed as a number of particles. It may also be expressed in terms of units of mass, based on the concept of mass that we discuss in due course. For example, we may have focused attention onto a constituent consisting of water molecules, and at a given instant of time the value of the amount may be 10^{26} molecules or about 3 kg. Again, at another instant of time the value of the amount of water may be 2 molecules or about 6×10^{-26} kg.

The unit of measure for amounts of constituents adopted by the International System of units (SI) is the *mole*, denoted by the symbol "mol" and such that 1 mol $= 6.022 \times 10^{23}$ molecules or atoms.

When modeling a specific phenomenon, during which the amount of a constituent never takes values either below a certain lower bound, or above a certain upper bound, or both, the description is sometimes simplified if as part of the system definition we specify the range of values of the amount of that constituent, that is, we specify that the model applies only for values of the amount within the selected range. If it is not specified, we say that the range of values of an amount is unbounded, that is, the model holds for all values of the amount from zero to an indefinitely large number. For example, in certain studies of water, for each of the constituents H_2, O_2, and H_2O, the range of values of each amount is bounded from below by a lower value, say, 20 molecules, and unbounded from above. In other studies, the ranges of values may be bounded from zero to 10 molecules for the amount of H_2O, from zero to 10 molecules for the amount of H_2, and from zero to five molecules for the amount of O_2. Again, in some studies the lower and upper bounds coincide and we have a fixed value of the amount, that is, the range includes just a single value of the amount, such as one molecule of H_2 or three atoms of H. Values of the amounts of constituents can be large or small, depending on the system under study.

Example 2.3. For each of the following studies, specify an appropriate range of values of the amount of H_2O.

(a) The study of the rotation of 2 kg of water in a closed, liquid-proof cylinder spinning at varying speed.

(b) The study of the filling of an initially empty tank with water from a faucet.

(c) The study of the electrolysis of 1 kg of water.

(d) The study of a water molecule in a fixed-volume enclosure.

(e) The study of water properties for all applications and ranges of conditions.

Solution.

(a) A fixed amount equal to 2 kg.

(b) A range between zero and the largest amount that the tank can contain.

(c) A range between zero and 1 kg.

(d) A fixed amount equal to one molecule.

(e) An unbounded range between zero and an indefinitely large value.

The forces required to define a system are of two kinds, internal and external. *Internal forces* describe the specified influences that hold the molecular structure of a constituent together, such as the forces between the nuclei and the electrons of H_2O, the influences between constituents, such as the forces between H_2O molecules, and the forces that promote or inhibit reaction schemes by which some constituents may combine or dissociate to give rise to other constituents. They are part of the specification of the system and may differ from study to study of the same constituents. For example, in some studies of H_2 and O_2, the forces involved in the formation of water out of hydrogen and oxygen, the chemical reaction mechanism $2H_2 + O_2 = 2H_2O$, may be neglected as unimportant, whereas in other studies, these forces may be included as important. Each internal force on a given constituent depends on the coordinates of that constituent and on the coordinates of one or more of the other constituents of the system, but not on any coordinates of constituents of bodies in the environment.

Example 2.4. What are the internal forces for the following types of constituents?

(a) A collection of positively charged oxygen ions, O^+, and negatively charged oxygen ions, O^-.

(b) A collection of H_2O molecules.

Solution.

(a) The Coulomb force on each positive and each negative charge resulting from all the other electric charges, and the forces that account for holding the parts of each of these oxygen ions together.

(b) The forces between H_2O molecules, and the forces that hold the protons, neutrons, and electrons of the H_2O molecule together.

In general, a large number of chemical and nuclear reaction mechanisms are always effective and, in principle, all should be included in the system specification. However, depending on the scope of each study, most reaction mechanisms operate on such a long time scale, that is, they are so "slow" that for all practical purposes it is as if they were not effective. Accordingly, we simplify the system specification by neglecting the extremely slow reaction mechanisms and concentrating our attention on the most effective ones. For example, in some studies of water we may specify that the reaction mechanism

$2H_2 + O_2 = 2H_2O$ is the only one to be effective. Then we deliberately neglect all the other mechanisms, such as $H_2 = 2H$, $O_2 = 2O$, and $2H_2O = H_3O^+ + OH^-$. Of course, such a modeling assumption is reflected in and deeply affects all the results of the study.

The concept of neglecting some internal reaction mechanisms, when they are much slower than others that dominate the phenomenon under study, has proved very practical and successful. Sometimes, this modeling concept is rendered more vividly by using terms such as *passive resistances, anticatalysts, reaction inhibitors, internal constraints,* or, simply, *constraints,* to suggest the idea that actual forces within the system prevent the slow reaction mechanisms from occurring at all and allow only the set of specified mechanisms. Of course, the concept of internal constraint is not just limited to the specification of which chemical or nuclear reaction mechanisms are relevant, but is used as well, for example, when we specify that two parts of a system are not allowed to exchange any amount of a given constituent.

Each *external force* describes a well-defined influence on the constituents by bodies not included in the collection under study, such as the influence of applied gravity, electric charges, magnets, and the solid walls of the container that confines the constituents within a region of space. Each external force on a given constituent depends on the coordinates of that constituent and one or more *external parameters* or, simply, *parameters* that describe the overall effect of bodies in the environment, but not on the coordinates of any other constituent, either of the system or of bodies in the environment. For example, the effects of gravity on the water molecules in a small container depend on the elevations of these molecules above sea level, and on the intensity of gravity, but not on the coordinates of the substances of the earth. The gravitational potential γ is the parameter associated with this external gravitational force, where $\gamma = gz$, g is the constant gravitational acceleration, and z the elevation above sea level. Again, the effects of the walls of an airtight container on an enclosed gas depend on the positions of the gas molecules relative to the internal surface of the container, but not on the coordinates of the molecules of the materials of the walls. For a wide variety of applications, the effects of the walls are adequately described by the volume of the container, that is, volume is a parameter. For other applications, we may need a more detailed geometrical description of the shape of the enclosure in which the constituents are confined. For example, for a cubical enclosure of side L, the length L is a parameter of the confining external forces. More formally, the forces that confine constituents within an enclosure are described by a "potential" equal to zero inside the enclosed region of space, and equal to infinity outside. Thus the gradient of the potential, that is, the force field, is zero inside the enclosure and infinite at the wall so that the particles in the enclosure cannot escape.

Each parameter, such as the gravitational potential γ, the volume V, and the length L in the three examples just cited, can be evaluated at any instant of time by means of well-defined measurement procedures. The *value of a parameter* is the numerical result of the corresponding measurement. For example, we may have focused attention onto a variable-volume enclosure, and at a given instant of time the value of the volume may be 2 cubic meters (m^3), whereas at another instant of time it may be 3 m^3. For some parameters, the description of a specific phenomenon is sometimes simplified if as part of the system definition we specify the *range of values of the parameter*, that is, if we restrict our attention on a range of values between a given lower bound and a given upper bound. In some studies, the lower and upper bounds of a parameter may coincide and then we have a fixed value of the parameter, such as a fixed value of the volume, $V = 1$ m^3. Values of parameters can be large or small, depending on the system under study.

Example 2.5. What are the parameters for the following types of external forces?

(a) The forces that confine constituents within a spherical enclosure of diameter D.

(b) The forces that confine constituents within a rectangular enclosure of sides L_1, L_2, and L_3.

(c) An applied gravity field which at elevation z has potential gz and intensity g.

(d) An applied electrostatic field which at some location in space has potential ψ and intensity E.

(e) The forces that constrain constituents onto a surface characterized by a surface area a.

Solution.

(a) The parameter is the diameter D.

(b) The parameters are the lengths L_1, L_2, and L_3.

(c) The parameter is the potential gz.

(d) The parameter is the potential ψ.

(e) The parameter is the surface area a.

For certain systems, the characterization of the external forces may require two or more parameters. For example, in a system consisting of a fixed amount of hydrogen, half of which is confined in a volume V' and the other half in a volume V'', we need the two parameters V' and V'', and their respective ranges, for the characterization of the effects of the confining walls.

Each parameter may vary either independently within its range, or be internally constrained to vary in a manner consistent with given interconnections. For example, if a system has two volumes V' and V'', these volumes may be internally constrained so as to satisfy the relation $V' + V'' =$ constant.

Each internal restriction on the changes of values of the parameters is called an *internal constraint* or, simply, a *constraint*. We use the same term as just introduced in connection with amounts of constituents and internal reaction mechanisms because also here it refers to the specification of internal restrictions that define the internal structure of the system.[1]

In summary, a system is a collection of constituents that is well defined by the following specifications: (1) the type and the range of values of the amount of each constituent, (2) the type and the range of values of each of the parameters that are needed to characterize the external forces, (3) the nature of the internal forces, including the internal reaction mechanisms, and (4) the constraints on the changes in values of amounts of constituents and parameters that the internal forces can accomplish. For a system consisting of r different types of constituents, we represent their amounts by the vector $n = \{n_1, n_2, \ldots, n_r\}$, where n_1 denotes the amount of the first type of constituent, n_2 the amount of the second type, and so on. We use the same symbols $n = \{n_1, n_2, \ldots, n_r\}$, with or without an additional subscript, to denote values of the amounts of the constituents within their respective ranges. For a system with external forces that depend on s different parameters, we represent them by the vector $\beta = \{\beta_1, \beta_2, \ldots, \beta_s\}$, where β_1 denotes the first parameter, β_2 the second, and so on. Moreover, we use the same symbols $\beta = \{\beta_1, \beta_2, \ldots, \beta_s\}$, with or without an additional subscript, to denote values of the parameters within their respective ranges. For some systems, the values of the amounts of constituents and the parameters, that is, the values of n_1, n_2, ..., n_r, β_1, β_2, ..., β_s,

---

[1]The specification of internal constraints among amounts of constituents or among parameters is sometimes done in terms of auxiliary variables called *internal variables*. Examples of such variables are the reaction coordinates that we define in Chapter 29.

are all fixed. For other systems, some or all of these values may vary over bounded or unbounded ranges. Then part of the definition of the system is the specification of the lower and upper bounds to the values of each amount $n_1, n_2, \ldots, n_r$ and each parameter $\beta_1, \beta_2, \ldots, \beta_s$.

Two systems are *identical* if they consist of the same types of constituents, experience the same internal and external forces, and have the same ranges of values of amounts of constituents and parameters, and the same constraints. If any of these identities is not valid, the two systems are *different*.

An example of a well-defined system is sketched in Figure 2.1. An amount n of water molecules is confined within a watertight container of volume V. We assume that the only internal forces are the intermolecular and those that maintain the integrity of the H_2O molecules, and that the external forces are those exerted by the fixed walls of the container on the H_2O molecules. Another example of a system is sketched in Figure 2.2. The specifications are the same as for the system in Figure 2.1 plus a movable piston. Because the piston is movable, the volume of the container has values in the range from V_1 to V_2.

A third example of a system is a thermonuclear plasma in which specified amounts of atoms and molecules experience ionizing forces and nuclear fusion reactions, release electromagnetic radiation and neutrons, and are confined by a suitably designed, time-varying electromagnetic field. Other examples of systems are the materials confined within the boundaries of the components of a pressurized-water nuclear power plant (Figure 2.3), the materials within a living cell (Figure 2.4), the neutrons within the core of a nuclear reactor, and the steam flowing through the various stages of a steam turbine. In each of the examples just cited, the amounts and types of constituents are those that exist within the boundary of the system, but the boundary may be crossed by various materials and radiation. It turns out that under certain conditions, as soon as particles enter the boundary they can be regarded as part of the system, and as soon as they exit

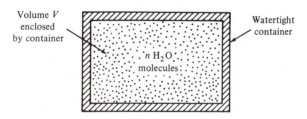

Figure 2.1 An amount n of water molecules confined within a watertight container of volume V.

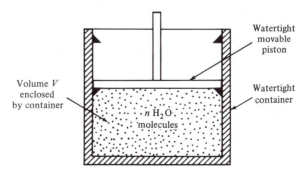

Figure 2.2 An amount n of water molecules confined within a watertight container of variable volume V ranging from V_1 to V_2.

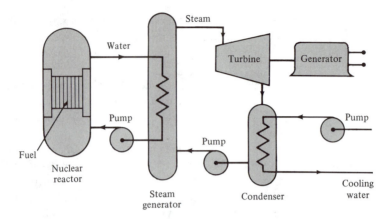

Figure 2.3   Schematic of components of a pressurized-water nuclear-reactor steam power plant.

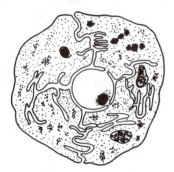

Figure 2.4   Schematic of a living cell.

from the boundary they are no longer part of the system but can be regarded as part of the environment.

Example 2.6.   Each of two rigid bottles has the same volume V and contains 1 kg of water. In bottle A all the water is in liquid form, whereas in bottle B one half is in the form of ice and the other half in the form of vapor. If the contents of each bottle are regarded as a system, are the two systems identical or different?

Solution.   The two systems are identical because each consists of the same type and amount of constituent and has the same internal and external forces. The form of the constituent—ice, liquid, and vapor—is not part of the system definition.

Example 2.7.   A piece of paramagnetic material, 5 cm × 5 cm × 25 cm, is on a bench. An identical piece of the same material is placed between the poles of a very strong magnet. Can the two pieces be regarded as identical systems?

Solution.   If the magnetic field is defined as having a specific value, the two pieces cannot be regarded as identical systems because they do not have the same external parameters: one has zero magnetic field and the other a strong magnetic field. However, if the intensity of the applied magnetic field is specified to have a range from zero to a large value at least equal to the strength of the magnet, the two pieces can be regarded as identical systems because each is composed of the same piece of matter and is subjected to forces characterized by the same parameter over the same range of values.

Because external forces are independent of coordinates of constituents in the environment, the constituents of a system are said to be *separable from the environment* or, simply, *separable*. This distinction is not trivial. If the forces exerted by a body not included in the object of study depend explicitly on the coordinates of the constituents of that body, the object of study is not separable and cannot be a system. To proceed in this case, we must redefine the collection of constituents so as to include the body in question. Thus the troublesome forces become internal, and a system may be defined. For example, this situation arises when we wish to study the oxygen atom in a molecule of water. Such an atom cannot be defined as a system because it experiences forces that depend explicitly on the coordinates of the two hydrogen atoms in the water molecule. However, by including the two hydrogen atoms in the object of study, these forces become internal, and the oxygen and hydrogen atoms bound together in a water molecule can be well defined as a system.

Example 2.8.   Two identical unrestrained magnets are near each other on a table but are unaffected by other magnets. Can we define one of the two magnets as a system?

Solution.   No. The magnetic force depends explicitly on the coordinates of both magnets. Hence the two magnets are not separable. To proceed, we must include both magnets in the system.

We emphasize that, from here on, whenever we use the term system, such as when we state the laws of thermodynamics and derive all the consequent results, we imply that each system is well defined according to all the specifications and restrictions discussed in the present section. Most of the basic concepts we develop in this book, including the concepts of energy, entropy, and temperature, would be ill defined if we attempted to apply them to a collection of constituents that, for example, does not satisfy the requirement of separability just discussed. Thus, as we proceed, we must bear in mind all the restrictions that delimit the term system.

2.2   Properties and States

At any instant of time, the amount of each type of constituent and the parameters of each external force of a system have specific values within their respective ranges. By themselves, values for all the amounts and the parameters are not sufficient to determine all the characteristic features of the system at that time. For example, they do not fix the values of the internal and external forces because these values depend on the coordinates of the constituents. To complete the characterization of a system at an instant of time, we need not only the values of the amounts of constituents and the parameters but also the values of all the properties at that instant of time. A *property* is a system attribute that can be quantitatively evaluated at any instant of time by means of a set of well-defined measurements and operations performed on the system, provided that the resulting value does not depend on the measuring devices, other systems in the environment, or other instants of time.[2] For example, the instantaneous position of a particle within a fixed container is a property. However, the distance covered by that particle over a period of 30 minutes is not a property. As we see later, other important properties are energy, adiabatic availability, and entropy.

The concept of velocity of classical mechanics provides a simple illustration of the measurements and operations involved in the definition of a property. We measure position x_1 at the instant of interest t_1, position x_2 at a later time $t_2 = t_1 + \Delta t$, divide the vector $\Delta x = x_2 - x_1$ joining the two positions by $\Delta t = t_2 - t_1$, and find the

---

[2]It is noteworthy that each amount of constituent and each parameter satisfies the definition of property. But in addition to being a property, each amount of constituent and each parameter plays a crucial role in the specification of the system. Each contributes to the characterization of a specific constituent or a specific external force. More important, amounts of constituents and parameters play a distinct role in the statement of the second law that we discuss in due course. For these reasons, we do not refer to n and β as properties.

ratio $\Delta x/\Delta t$. For any finite Δt, the distance Δx is finite, and does not correspond to a property because it depends on two instants of time. But if Δt is infinitesimal, that is, in the limit as t_2 approaches t_1, also the distance Δx becomes infinitesimal and, in general, the ratio $\Delta x/\Delta t$ assumes a finite limiting value. This limiting value depends only on the time instant of interest, t_1. It is the value of the property that we call velocity.

Although an indefinite number of properties can be attributed to a system, some of these properties are *interdependent*, that is, a change in value of one such property affects the values of some other properties. For example, the speed and the square of the speed are two interdependent properties. Nevertheless, for each well-defined system it is always possible to identify many different sets of *independent properties*, each of which satisfies the following two conditions.

(1) The values of properties in the set can be varied independently, that is, the value of each property in the set can be changed without affecting the value of any other property in the set.

(2) The values of the properties in the set are sufficient to determine the values of all the other properties of the system.

We call the properties in one such set a *basis*[3] and denote their collection by the vector $P = \{P_1, P_2, ...\}$.

For a given system, it follows that all that can be said about the results of any measurements that may be performed on the system at one instant of time is represented by: the values of the amounts of the various types of constituents, the values of the external parameters, and the values of the properties in a basis. We call the collection of all these values a possible or allowed state, or, simply, a *state*. Thus the state is a set that specifies everything about a system at one instant of time. The description of the states of a system is the subject of the branch of physics known as *kinematics*.

Two *states* of a system are *identical* only when the values of the amounts of all the constituents, the values of all the external parameters, and the values of all the properties for one state are identical to the respective values for the other state. Otherwise, the two are *different states*.

For a given system A we may use the symbol A_i to denote the ith state, $(n)_i$ to denote the set of values of all the amounts of constituents, $(\beta)_i$ to denote the set of values of all the external parameters, and $(P)_i$ to denote the set of values of all the properties in a basis. Thus A_i is determined by the sets $(n)_i$, $(\beta)_i$, $(P)_i$. Two states A_1 and A_2 are identical if each entry in $(n)_1$, $(\beta)_1$, $(P)_1$ is equal to the corresponding entry in $(n)_2$, $(\beta)_2$, $(P)_2$.

Example 2.9. A system consists of water molecules contained in an injection syringe with a volume range from 0 to 5 cubic centimeters (cm^3). Do any of the following four descriptions specify the same state? The amount of water in the syringe and its volume are, respectively: (a) 2 g and 3 cm^3, (b) 2 g and 2 cm^3, (c) 3 g and 3 cm^3, and (d) 2 g and 3 cm^3.

Solution. Descriptions (a) and (b) specify different states because the value of the parameter volume in (a) is different from that in (b). Descriptions (a) and (c) specify different states because the value of the amount of water in (a) is different from that in (c). Descriptions (b) and (c) specify different states because the values of the amount of water, and the parameter volume in (b) are different from those in (c). Descriptions (a) and (d) specify that the two

---

[3]The term *basis* is borrowed from linear algebra. There, each quantity of interest is represented by a vector in a space with many dimensions, and the basis denotes a set of linearly independent vectors which we may combine to express any vector in the space. For this reason we say that the basis spans the entire vector space.

states have identical values of amount of water, and identical values of the parameter volume, but we cannot conclude that they specify the same state because we are not told whether the values of all properties of (a) are identical to the corresponding values of (d). Finally, descriptions (b) and (d), as well as (c) and (d), differ, for obvious reasons.

Properties are sometimes called also *state variables*. The reason is that fixing the state of a system implies fixing the values of all its properties. Accordingly, it is correct to say that the value of each property depends on the state of a system. We note, however, that this is an implication of the definition of state, not the definition of a property.

We say that two sets of values of amounts of constituents and parameters $(n)_1$, $(\beta)_1$ and $(n)_2$, $(\beta)_2$ are *compatible* if the change from one set to the other can occur as a result of the internal mechanisms—reactions, interconnections, internal forces—that are allowed by the constraints of the system. For example, if a system has two volumes V' and V'' interconnected so as to satisfy the constraint $V' + V'' = $ constant, the two sets of values V_1', V_1'' and V_2', V_2'' are compatible if $V_1' + V_1'' = V_2' + V_2''$ because then the internal interconnection between the two volumes allows the change from one set of values, say, $V_1' = 3$ m^3 and $V_1'' = 5$ m^3, to the other set of values, say, $V_2' = 2$ m^3 and $V_2'' = 6$ m^3.

As we discuss later, the concept of compatibility plays a special role in the statement of the second law of thermodynamics. For this reason, here we introduce the following useful definition. We say that a state A_1 with values $(n)_1$, $(\beta)_1$, $(P)_1$ is *compatible with a given set of values n of the amounts of constituents and β of the parameters* if the two sets $(n)_1$, $(\beta)_1$ and n, β are compatible. Clearly, for systems in which the constraints disallow any internal mechanism capable of changing the values of the amounts of constituents and the parameters, two states are compatible only if they have the same values of all the amounts of constituents and all the parameters.

The number of possible states of a system is infinite. Later we show that states may be classified into different types, such as nonequilibrium, equilibrium, and stable equilibrium, and that some types are easier to describe than others. A qualitative illustration of differences between two states of a system is sketched in Figure 2.5. In Figure 2.5a, 1 kg of cold water at rest fills partially a stationary container of 1 m^3. In Figure 2.5b, 1 kg of water in the form of steam fills one-half the volume of the 1 m^3 container. The steam moves with respect to the stationary container with a bulk speed of 1 m/s along the horizontal direction. Clearly, the values of the properties of the two identical systems shown in Figures 2.5a and 2.5b differ from each other—zero momentum versus nonzero momentum—and so do the two states.

For most systems the number of properties in a basis is infinite. However, we show later that if we restrict our study to a special class of states, the number of independent properties is drastically reduced. The reason is that specifying the class of states is entirely equivalent to specifying a set of interrelations between the properties of a basis. Thus

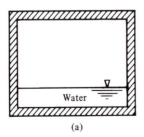
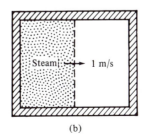

(a)   (b)

Figure 2.5 A qualitative illustration of the differences between two states of a system.

the number of independent properties sufficient to identify a state within the specified class may be much smaller than the number of independent properties in a basis. For example, we show later that the stable equilibrium states of a certain type of system have only three independent properties, even though states of that type of system require in general a basis with an infinite number of independent properties.

In many applications, we may wish to partition an overall system C in state C_1 into subsystems A and B in states A_1 and B_1, respectively, because the individual study of each of the two states A_1 and B_1 may be easier than the study of C_1. Such a partitioning is purposeful only if the study of subsystem A in state A_1 and subsystem B in state B_1 can be entirely substituted for the study of C in state C_1. In other words, the partitioning is purposeful only if the state C_1 of C is fully specified solely by the individual states A_1 of A and B_1 of B, and nothing else. If A_1 and B_1 are sufficient to specify C_1 completely, we say that they are *uncorrelated states*. Otherwise, A_1 and B_1 are *correlated states*.

For example, we show later that the entropy of the composite of two subsystems in uncorrelated states equals the sum of the entropies of the two subsystems and therefore is solely determined by their two states. On the other hand, the entropy of the composite of two subsystems in correlated states is not equal to the sum of the entropies of the two subsystems and is not determined by their states.

All statements and conclusions in this book refer to well-defined systems in uncorrelated states. For this reason, from here on whenever we use the term state we mean uncorrelated state.

2.3   Motions

A change of state as a function of time is called a *motion*. The description of the causes of a motion and the analysis of their effects are the subjects of the branch of physics known as *dynamics*. A motion is generally due to the natural tendency of a system to adapt its state to the internal and external forces. For example, a spring tends to elongate to adapt to a stretching force or, again, a mass tends to accelerate to adapt to an applied force. Such tendency is experienced either when the initial state is incompatible with all the system forces, or when the forces are altered in time or both.

2.3.1   Spontaneous Changes of State

Under certain conditions, it is a common experience that the state of a system may change as time goes on, without inducing any effects on the state of the environment. For example, a charged battery wrapped in fiberglass may discharge internally over a period of time without any effects on systems in the environment. Again, a spoonful of sugar may gradually dissolve in the hot water in a "Thermos" bottle without affecting the states of systems in the surroundings. Such changes of state are called *spontaneous*.

A system undergoing spontaneous changes of state must have parameters that are time independent, and external forces that disallow the exchange of any type of constituent between the system and its environment because then and only then the changes do not affect the states of systems in the environment. For brevity, a system that cannot induce any effect on the state of its environment will be called *isolated*.[4] Clearly, an isolated system can experience only spontaneous changes of state.

---

[4]In control theory, an isolated system is called *autonomous*. For autonomous systems, only the difference between the times of occurrence of two states is important, not each of these two times.

Example 2.10. A system consists of a nickel–cadmium battery. In both the initial and the final state of the battery, the electrodes are not connected to any electrical load. In the initial state, the battery is fully charged, whereas in the final state it is fully discharged. Each of the following descriptions specifies what happens during the time lapsed between the initial and final states.

(a) The battery is connected to and powers a radio for 3 h.

(b) The battery is immersed in a solution of water and salt for 10 h.

(c) The battery is kept under vacuum in a fiberglass bottle with perfectly reflecting internal walls for 3 months.

For each description, determine whether the change of state of the battery is spontaneous.

Solution. The changes of states described by (a) and (b) are not spontaneous because during each change the battery affects the state of systems in the environment, the radio in (a) and the water solution in (b). The change of state described by (c) is spontaneous because the vacuum and the reflecting bottle prevent the battery from affecting the states of any system in the environment, and therefore the discharge must occur internally.

A challenging question is: How do we analyze spontaneous changes of state, that is, how do we describe the dynamics of an isolated system? Formally, we need a set of equations that describe the time evolution of all the properties in a basis. We say that we need the *equation of motion* of the system.

In classical mechanics, the equation of motion is obtained from *Newton's equation of motion*, which relates the total force F on each system particle to its mass m and acceleration a so that $F = ma$. In quantum mechanics, the equation of motion is the *time-dependent Schrödinger equation*. In thermodynamics, the complete equation of motion is a subject of frontier research. Many attempts have been made to establish it, but the task is not yet accomplished. So the formal approach cannot be carried out. Nevertheless, many features of the equation of motion have already been discovered. These features provide not only a guidance for the discovery of the complete equation but also a powerful alternative procedure for analyses of many practical problems. The most general and well-established features are captured by the statements of the first and second laws of thermodynamics. We discuss these laws later.

The problems that may be completely analyzed by means of the criteria and restrictions derived from the laws of thermodynamics instead of the equation of motion are those for which both the initial and final states or the *end states* of the system under study belong to a special class. We show later that the class consists of all the stable equilibrium states of the system, plus all the states in which the system can be partitioned into two or more subsystems each in a stable equilibrium state. It turns out that this special class covers a large number of practical applications.

Of course, the dynamics of a system with end states outside the class just cited must also satisfy the criteria derived from the laws of thermodynamics. Under these circumstances, however, the analysis turns out to be incomplete and not entirely equivalent to that implied by the complete equation of motion.

2.3.2 Induced Changes of State

Under appropriate circumstances, another common experience is that systems may influence each other and thus affect their states. For example, a hot bottle may warm a person's cold feet if we establish contact between the bottle and the feet for a certain lapse of time, a battery may power a motor-driven fan if we turn the switch on (Figure 2.6a), a compressed gas may lift an elevator cage through a suitable mechanism if we let the gas expand against a piston (Figure 2.6b), and hydrogen and oxygen molecules may combine to form water molecules if we allow them to flow into a chemical reactor (Figure 2.6c). Such changes of state are said to be *induced* by the forces between the constituents of the systems that interact for a certain period of time. In general, each interaction force depends on the coordinates of the constituents of all the interacting systems. Moreover, a finite lapse of time is required because interaction forces are finite and, as such, cannot induce instantaneous changes of state.

Almost all problems of practical interest involve interactions because it is via interactions that the capability of a body to perform a useful task can be transferred to another body in order to achieve a practical objective. For example, upon contact a hot bottle warms cold feet. Again, upon closing a switch a charged battery powers a fan.

Interactions may have a great variety of effects, that is, they may change different aspects of the states of the interacting systems. In some interactions, the values of the amounts of constituents and the parameters of each interacting system remain unchanged, but the values of one or more properties change. Illustrative of such changes is the example in Figure 2.6a, where the interaction between the fan and the battery via the motor results in the angular speed of the fan being increased at the expense of a reduction of the amount of electrical charges on the electrodes of the battery. In other interactions, the values of the amounts of constituents are fixed, but the values of the parameters and some properties at the end of the interaction differ from the corresponding values at the

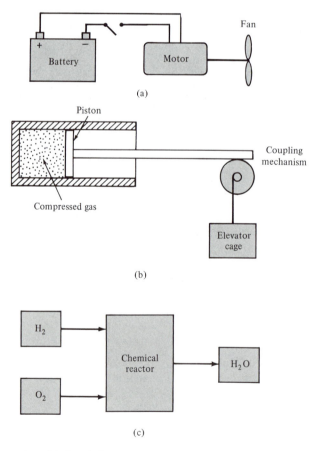

Figure 2.6   Examples of induced changes of state resulting from interactions between systems.

beginning, as in the example in Figure 2.6b, where the interaction between the gas and the elevator cage via the coupling mechanism results in the elevator being lifted at the expense of an expansion of the gas against the piston. In still other interactions, amounts of certain constituents may disappear and amounts of others may appear in their stead, as in the example in Figure 2.6c, where the interaction between the chemical reactor and the hydrogen, oxygen, and water tanks results in a flow of hydrogen and oxygen into the reactor and a flow of water out of the reactor.

Formally, we analyze induced changes of state as follows. For simplicity, we discuss only two systems A and B that are about ready to interact. At the moment just prior to the onset of the interaction forces, we define an overall isolated system C, consisting of the two systems A and B, and being in a state C_1 fully determined by the then initial states A_1 and B_1 of systems A and B, respectively. Next, we set the interaction forces in effect for a certain period of time. Clearly, these forces are now internal for isolated system C, and contribute to its spontaneous change of state. In principle, this change is described by the equation of motion. In particular, it is subject to the criteria and restrictions mentioned in the preceding section. At the end of the interaction period, that is, when the interaction forces become again ineffective, the two original systems A and B may be well defined, and their final states A_2 and B_2 may be uncorrelated. Thus we may determine the changes in values of properties of both A and B that were induced by the interaction forces.

For certain properties, it turns out that induced changes may be associated with net flows of properties from one system to another. For example, in cancer radiation therapy, photons flow from the radiation source to the malignant tumor. Each photon has momentum as a property. Thus we say that momentum is transferred from the source to the tumor. The association of induced changes with one or more flows of properties is useful in classifying interactions into different modes. Each *mode of interaction* is defined in terms of well-specified net flows of properties between the interacting systems. For example, in a later chapter we define two mutually exclusive modes of interaction, work and nonwork. We will see that work is a mode that allows the net flows of only energy and momentum between two interacting systems, whereas nonwork involves net flows other than energy and momentum. Again in subsequent chapters, we subdivide nonwork into other modes of property exchanges, such as heat, and bulk flow.

The description of a change of state in terms of the end states of a system and the modes of interaction it experiences is said to specify a *process*. As a result of the classification of interactions into different modes, processes have a corresponding classification. For example, a process for which the only type of interaction is work is called *adiabatic*. Again, a spontaneous change of state is also called a *spontaneous process*.

We will see that the classification of interactions and processes is a powerful tool for the exposition of both the principles and the applications of thermodynamics. In particular, it is a very useful technique for the definition of two properties, energy and entropy. Despite their fundamental and far-reaching importance, these two properties are not as evident and readily understandable as we would like them to be. Their definitions, methods of measurement, and time evolutions are interwoven with two fundamental physical principles, the first and second laws of thermodynamics. Beginning in Chapter 3, we introduce the two laws and study their implications.

As a final general comment, we note that a system A undergoing a process from state A_1 at time t_1 to state A_2 at time t_2 is well defined at these two times even though it may not be well defined during the lapse of time between t_1 and t_2. The reason is that the interactions which induce the change of state may involve such temporary alterations of internal and external forces that no system A can be defined during the period t_1 to t_2. Said more formally, in the course of interactions the constituents may not be separable from the constituents of external bodies with which the interactions occur or, if the constituents are separable, the states of the system may be correlated with the states of other systems. At the end of the interactions, however, the system is again well defined and its state uncorrelated.

Problems

2.1 Of the system descriptions given below, which ones specify the same system and which different systems?

(a) A rigid container of volume V, with walls impermeable to matter, is divided into two compartments by a membrane also impermeable to matter. One compartment contains 500 g of gasoline, the other 100 g of oxygen gas.

(b) Same as in part (a), except that the membrane has been ruptured and the gasoline and oxygen are mixed but not chemically combined.

(c) Same as in part (a), except that the molecules of gasoline are highly excited.

(d) Same as in part (b), except that most of the molecules of gasoline and oxygen have combined and formed water and carbon dioxide.

2.2 Are the statements that follow descriptions of the same system or of different systems? Explain your answer.

(a) An atom of uranium 235 at rest and a neutron moving at 2000 m/s toward the atom are confined in a container with walls impervious to matter.

(b) Same as in part (a), but the neutron is at rest and the uranium atom is moving toward the neutron with a speed of 130 m/s.

(c) The fission products of the nuclear reaction between an atom of uranium 235 and a neutron, all at rest and confined in the same fixed-volume container as in part (a).

(d) Same as in part (c), but the fission products move with a total kinetic energy of 200 MeV.

(e) Same as in part (c), except that the container walls are permeable to matter.

2.3 Of the descriptions given below, explain how you would define a system, that is, list the constituents, specifying whether their amounts are fixed or variable over a range of values, the parameters, specifying whether they are fixed or variable over a range of values, and give some description of the internal forces. In addition, give some description of the interactions that occur at the system boundary.

(a) Air is being compressed into a bicycle tire by means of a pump.

(b) A bottle is being filled with red wine.

(c) Water is boiling in a pot, and a bunch of spaghetti is being thrown in.

(d) A gasoline–air mixture is burning in a closed stainless-steel rigid container, giving rise to products of combustion.

2.4 A liquid-tight container, 5 cm × 5 cm × 25 cm, is filled with mercury and placed on a wooden bench. An identical container, filled with the same amount of mercury, is placed between the poles of a very strong magnet.

(a) If the container on the bench is regarded as a system, what are its parameters?

(b) What are the parameters of the other container?

(c) Can the two containers be regarded as the same system?

2.5 System A consists of a deuterium molecule, D_2, and an oxygen atom, O, confined within a fixed-volume container. System B consists of a heavy water molecule, D_2O, confined within a container identical to that of system A. System C consists of a tritium atom, T, a neutron, n, and an oxygen atom, O, within another container again identical to that of system A.

(a) Are systems A, B, and C identical? If not, can you define a system so that systems A, B, and C are identical but each is in a different state?

(b) Can you define a system that is at once identical to systems A and B, but not C?

(c) Can you define a system that is at once identical to systems A and C, but not B? If so, are the states of A and C identical?

2.6 Two electrically neutral particles are in a box in a region of space free of gravity effects from celestial bodies or any other objects, but attract each other with a force $|\boldsymbol{f}_{12}| = G_{12}/|\boldsymbol{r}_1 - \boldsymbol{r}_2|$. The force is along the direction of the vector $\boldsymbol{r}_1 - \boldsymbol{r}_2$, where $\boldsymbol{r}_1$ is the location of particle 1, $\boldsymbol{r}_2$ the location of particle 2, and G_{12} a constant for the two particles. Whenever it approaches a wall of the box, either particle is repelled by the wall almost instantly.

(a) Can you define a system consisting of only one of the two particles?

(b) Can you define a system consisting of both particles?

2.7 Which of the following statements define properties, amounts of constituents, or parameters of the system described by the words in italics?

(a) The amount of metal evaporated from an *incandescent filament*.

(b) The rate at which metal evaporates from an *incandescent filament*.

(c) The number of dollars in a *safe*.

(d) The number of dollars placed in a *safe* on the first day of the year.

(e) The number of hours a *liquid–air tank* has supplied air.

(f) The number of hours a *liquid–air tank* can supply air at a specified rate.

(g) The maximum stress a *piece of plastic material* can withstand before failure.

(h) The color of the light emitted from an *incandescent filament*.

(i) The color of a *cold tungsten filament* illuminated by solar light.

2.8 For the purposes of this problem, 10 kg of air in a 10-m^3 rigid airtight tank are assumed to have only two independent properties, A and B. At time $t_1 = 0$, measurements yield that $A = 12$ and $B = 14$, whereas at time $t_2 = 5$ h the same measurements yield $A = 10$ and $B = 14$.

(a) What is the system in this problem?

(b) What are the values of the amounts of constituents and the parameters? What are the ranges of these values?

(c) What is the state of the system at time t_1?

(d) What is the state of the system at time t_2?

2.9 Consider the descriptions given below in pairs, and indicate which pairs specify the same system and which the same state. Notice that the answers to this problem depend on how each system is defined. For each description, consider several alternative system definitions.

(a) Two kilograms of water rotating at 1 rpm in a 2-liter liquid-proof container.

(b) Two kilograms of alcohol rotating at 1 rpm in a 2-liter liquid-proof container.

(c) One kilogram of water at rest in a 1-liter liquid-proof container.

(d) Two kilograms of water in a 2-liter liquid-proof container divided into two equal compartments by a rigid partition.

(e) The products of electrolysis, H_2 and O_2, of 2 kg of water in a 2-liter liquid-proof container.

(f) Two kilograms of water in a 2-liter container with a water inflow of 1 kg/h at an inlet port and a water outflow of 1 kg/h at an outlet port.

2.10 Ten kilograms of atmospheric air is confined in an airtight balloon that has a fixed volume of 10-m^3. For the purposes of this problem, assume that air has only two independent properties

A and B under all conditions (it is a bad assumption but simplifies the discussion). Measurements of A and B have been made as a function of time over a period of 10 h, and the results are that $A(t) = A_{ave}(1 + \sin \omega t)$ and $B(t) = B_{ave}$, where $\omega = 80 \text{ s}^{-1}$, $A_{ave} = 10$ units of property A, and $B_{ave} = 10$ units of property B. The average values of A and B over the measurement period are equal to A_{ave} and B_{ave}, respectively.

(a) What is the system in this problem?

(b) What are the parameters?

(c) What is the state of the system at $t = 5$ min?

(d) Are the average values A_{ave} and B_{ave} related to any state of the system?

2.11 If a system consists of a structureless particle in a fixed-volume container and has only two independent properties P_1 and P_2, indicate the type of change of state described by the specifications that follow, that is, indicate whether there is a change of state, whether it is induced or spontaneous, and whether or not the system is experiencing interactions.

(a) The values of properties P_1 and P_2 are changing in time while two identical weights connected to the particle via two independent mechanisms are moving, one rising and the other falling at the same speed. The particle is connected to no other system in the environment.

(b) Same as in part (a), except that the values of properties P_1 and P_2 do no change in time.

(c) Same as part (a), except that each weight has zero speed.

3 Energy

Energy is a concept that underlies our understanding of all physical phenomena, yet its meaning is subtle and difficult to grasp. For some special circumstances we have a clear mental picture and a simple analytical expression for a special energy, such as when we speak of the kinetic energy $mv^2/2$ of a mass m moving with speed v. Despite the efforts of thousands and thousands of educators, scientists, and engineers, for many states of systems of both scientific and engineering interest no easily comprehensible picture and no simple analytical expression for energy are available. The meaning and definition of energy emerge from a fundamental principle known as the first law of thermodynamics, a law that prescribes the relation between changes of state of a system and external effects in the course of interactions of a special type.

In this chapter we discuss the special interactions, introduce the first law, and define energy as a property of any system in any state.

3.1 Weight Process

We consider a change of state of a system A brought about by a special interaction. The interaction is such that the only net effect caused by both the external and the interaction forces on bodies in the environment of A is mechanical. By *mechanical effect* we mean a change in velocity of a free point mass, a change in distance between two point masses at rest and interacting via a gravity force, a change in distance between two electrical charges at rest and interacting via an electrostatic force, a change in distance between two magnets at rest and interacting via a magnetostatic force, or a combination of such changes.

Pictorial representations of the end states A_1 and A_2 of processes of a system A that involve various external mechanical effects are shown in Figure 3.1. In the sketch in Figure 3.1a, the effect is a change in velocity from $\mathbf{v}_1$ to $\mathbf{v}_2$ of a mass m, in Figure 3.1b a change in distance from d_1 to d_2 between two electric charges q_1 and q_2, and in Figure 3.1c changes in both velocity from $\mathbf{v}_1$ to $\mathbf{v}_2$ of a mass m, and in elevation from z_1 to z_2 of a weight M in the gravitational field of the earth.

As is well known from the science of mechanics, any combination of mechanical effects is entirely equivalent to any one of the mechanical effects. The equivalence is in the sense that we can always find an arrangement of forces which, when applied for a certain lapse of time, annuls all the mechanical effects except one of our choice. For example, a change in elevation of a weight in a gravity field may always be used to annul a change in velocity of a free point mass.

Example 3.1. Verify the last statement.

Solution. We consider a change in velocity of a free point mass m from $\mathbf{v}_1$ to $\mathbf{v}_2$. As shown in Figure 3.2, we interconnect the mass m, in its final state, to a weight of mass M ini-

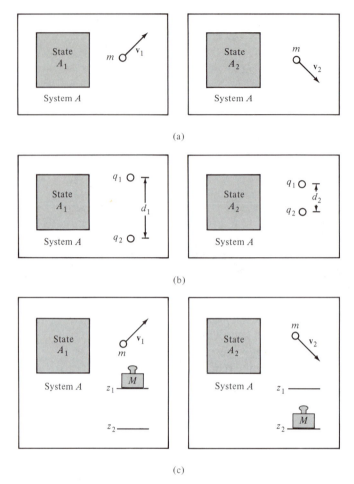

Figure 3.1 Pictorial representation of the end-states A_1 and A_2 of processes of system A that involve various mechanical effects on systems in the environment.

tially at elevation z_2 in the gravity field. The interconnection is via a massless cable and a frictionless pulley rigidly hinged on the surface of the earth. The constant force F exerted by the cable onto the point mass during the interaction time Δt is in the direction of the velocity change $\mathbf{v}_1 - \mathbf{v}_2$. By Newton's law of motion, the mass m, the magnitude of the change in velocity $|\mathbf{v}_1 - \mathbf{v}_2|$, the time of application of the force Δt, and the magnitude of the force $|F|$ satisfy the relation $m|\mathbf{v}_1 - \mathbf{v}_2|/\Delta t = |F|$. The pulley mechanism may clearly be chosen so that the magnitude $|W|$ of the vertical force W exerted by the cable onto the weight is only slightly different from the magnitude of the gravitational force Mg, that is, $||W| - Mg|/Mg \ll 1$, so that the weight changes elevation without gaining appreciable speed. At the end of the interaction time, the cable is disconnected from both the point mass and the weight. Thus the given change in velocity of the free mass has been annulled at the expense of the change in elevation of the weight. Because of the action of the pulley and the gravitational force, the momentum change of the point mass is compensated by an equal and opposite change in momentum of the earth, but the corresponding velocity change is negligible in view of the huge mass of the earth.

In view of the equivalence among mechanical effects, we select and describe in detail only one such effect as representative of all the others. To this end we give a more

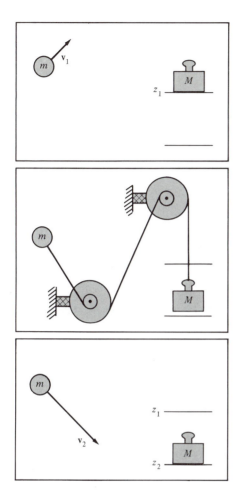

Figure 3.2 Example on the equivalence of mechanical effects.

precise and specific meaning to the term weight. We define a *weight* as an idealized system with only one independent property: the elevation z of the system in a uniform gravitational field of acceleration g. We denote by M the mass of the weight at elevation $z = 0$, that is, the force necessary to hold the weight at elevation $z = 0$ divided by the gravitational acceleration g.

The change in elevation of a weight is a mechanical effect that is not only equivalent to any other mechanical effect or combination of such effects, but is also convenient for our exposition of the foundations of thermodynamics. For these reasons and for the sake of brevity, we give a name to a change of state of a system that involves no external effects other than the change in elevation of a weight. We call it a *weight process*.

Example 3.2. The container sketched in Figure 3.3 encloses a charged battery powering an electric motor with a rope fixed on its rotor shaft. Except for a small hole, the container is perfectly insulated. Through the hole, the rope may be connected to a weight. Define a system that can experience weight processes only.

Solution. The system consisting of the perfectly insulated container, its contents, and the rope can experience changes of state that involve no external effects other than the lowering or raising of the weight, that is, weight processes only.

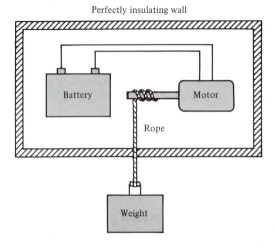
Perfectly insulating wall

Figure 3.3 Example of a system that can experience weight processes only.

3.2 First Law

In terms of concepts that we have already defined, we are now ready to state the first law as follows.

> *Any two states of a system may always be the end states of a weight process, that is, the initial and final states of a change of state that involves no net effects external to the system except the change in elevation between z_1 and z_2 of a weight. Moreover, for a given weight, the value of the quantity $Mg(z_1 - z_2)$ is fixed by the end states of the system, and independent of the details of the weight process, where M is the mass of the weight and g the gravitational acceleration.*

This is a very general statement. It applies without restriction to any well-defined system, large or small, and to all states, including states that differ drastically in their types and amounts of constituents. It is confirmed by experimental evidence at all levels of description. For example, it is observed in nuclear physics experiments that annihilation, creation, fusion, and fission reactions during a weight process can produce drastic changes of state, such as the appearance of an electron and a positron out of an initial state consisting only of radiation, the formation of a heavy nucleus and radiation out of an initial state consisting only of two lighter nuclei, and the formation of two light nuclei and radiation out of an initial state consisting only of a heavy nucleus and a neutron.

Although any two states A_1 and A_2 of a system A may be the end states of a weight process, the first law does not specify which of the two states is the initial one. Thus the weight process is possible in the direction either from state A_1 to state A_2, or from state A_2 to state A_1, but not necessarily in both directions. In general, we denote by z_1 the elevation of the weight when system A is in state A_1, and by z_2 the elevation when A is in state A_2.

A pictorial description of the statement of the first law is shown in Figure 3.4. We consider a system A, and any two of its states, A_1 and A_2. If a weight process is possible from state A_1 to state A_2, the change of state can be brought about while the only effect external to system A is the change in elevation of a given weight, and the value of $Mg(z_1 - z_2)$ is fixed by the states A_1 and A_2 only, that is, it does not depend on any other characteristics of what happens as the state changes from A_1 to A_2. If a weight process is not possible from state A_1 to A_2, such a process is possible from A_2 to A_1, and again the value of $Mg(z_1 - z_2)$ is fixed by the states A_2 and A_1 only.

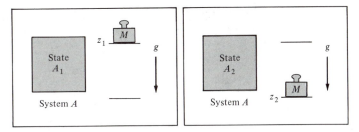

Figure 3.4   Pictorial representation of the first law. The value of $Mg(z_1 - z_2)$ is fixed by the states A_1 and A_2 only.

The pictorial description of the first law neither constitutes a proof, nor represents a precise experiment that we can carry out under all circumstances. In fact, no proof and no direct experimental verification of a physical law can ever be provided. If a physical law could be proved or verified by direct experimentation, it would cease to be a fundamental principle and would become a consequence of some other principles. We accept the first law as a fundamental principle because to date its consequences are consistent with all experience and have never been violated. As is true of every fundamental principle, the first law has many important implications. We discuss some of these implications in what follows.

3.3   Uniqueness of Values of $Mg(z_1 - z_2)$

Given two states A_1 and A_2 of a system A, many ways may exist that reproduce the transition from one state to the other, while the only external effect is the change in elevation of a weight. For example, if A is a gas in a container, the volume of the container may be either decreased initially and then increased to the value of the final state, or increased initially and then decreased to the final value. Again, gas may flow in and out of the container, but at the end of the process, the net change of the amount of gas is zero. For a given weight, the first law avers that the value of $Mg(z_1 - z_2)$ associated with a particular method of achieving a weight process with given end states A_1 and A_2 is identical to that for any other method, namely, it is uniquely determined by the end states. In particular, it is equal to the difference in values of two weight processes, one having end states A_0 and A_1, and the other end states A_0 and A_2, where A_0, A_1, and A_2 are any three states of the system.

To see the last point clearly, we consider a given weight, three states A_0, A_1 and A_2 of system A, and three weight processes as follows. The first has A_0 and A_1 as end states, and z_0 and z_1 as the corresponding elevations of the weight. The second has A_0 and A_2 as end states, and z_0 and z_2 as the corresponding elevations of the weight. The third weight process has A_1 and A_2 as end states, and z_1 and z_2' as the corresponding elevations of the weight. Using the first law, we can easily show that $z_2' = z_2$ and therefore that the value of $Mg(z_1 - z_2)$ equals the difference between the values of the first and second processes.

Example 3.3.   Prove that the statement $z_2' = z_2$ is valid.

Solution.   We assume that the change of state from A_0 to A_2 is possible in terms of two weight processes, one from A_0 to A_2 directly and the other from A_0 to A_1 and then from A_1 to A_2. For the weight process from A_0 to A_2 directly, the weight goes from z_0 to z_2, and we find the value $Mg(z_0 - z_2)$. For the two-step process from A_0 to A_2, the weight goes first from z_0 to z_1 and then from z_1 to z_2'. Thus the net change in elevation is from z_0 to z_2', and we find the value $Mg(z_0 - z_2')$. Because the first law requires that the two results be the same, we conclude that z_2' must equal z_2. Of course, the same method of proof may be used and the same conclusion be reached for each of the remaining seven combinations of possible weight processes between pairs of states selected from A_0, A_1, and A_2.

3.4   Definition of Property Energy

A very important consequence of the first law is that every system in any state must have a property that we call energy and that we define by means of the following set of operations and measurements. For the sake of convenience, we select once and for all a reference state A_0 for system A and a reference weight with mass M at elevation $z = 0$ in a uniform gravity field with acceleration g. For a given state A_1 of system A, we use a weight process with A_0 and A_1 as end states. We measure the elevation of the weight z_0 when the system is at state A_0, the elevation z_1 when the system is at state A_1, and we evaluate the expression $Mg(z_1 - z_0)$. Next, we compute the value E_1 by means of the relation

$$E_1 = E_0 - Mg(z_1 - z_0) \tag{3.1}$$

where E_0 is a constant fixed once and for all for system A. It is understood that the weight process occurs while A changes state either from A_0 to A_1, or from A_1 to A_0. Traditionally, equation (3.1) is written with a negative sign in front of $Mg(z_1 - z_0)$ to emphasize that the larger the value of elevation z_1, the smaller the value of E_1, and conversely, that the smaller the value of z_1, the larger the value of E_1.

We can easily verify that the set of operations and measurements just cited defines E as a property of system A, with value E_1 at state A_1. Indeed, from the statement of the first law we conclude the following about the value of $Mg(z_1 - z_0)$, and hence of E_1.

1. It is independent of the measuring devices, that is, of the details of the interaction forces between A and the weight.
2. It is independent of other systems in the environment, that is, of the details of the weight, which is the only external system affected during the weight process.
3. It is independent of the history of system A, that is, of the details of the changes of A during the weight process.

These conclusions are valid for any state, and therefore the set of operations and measurements fulfills the conditions defining a property (see Section 2.2). We denote this property by E and its value at state A_1 by E_1. We call E the *energy*.

Upon repeating the same set of operations and measurements for a state A_2, equation (3.1) yields the value E_2 of the energy at state A_2, that is,

$$E_2 = E_0 - Mg(z_2 - z_0) \tag{3.2}$$

Subtracting equation (3.1) from equation (3.2), we find that

$$E_2 - E_1 = -Mg(z_2 - z_1) \tag{3.3}$$

where we keep the negative sign in front of $Mg(z_2 - z_1)$ to emphasize that the larger the value of z_2, the smaller the value of E_2, and conversely.

The left-hand side of (3.3) is the difference in energy between states A_1 and A_2. The right-hand side is the unique value of the expression associated with any weight process having A_1 and A_2 as end states (see Example 3.3), as it should be.

The value E_1 of the energy at a particular state A_1 depends on the value of the constant E_0 selected for the system. We can readily verify that E_0 is the energy of the reference state A_o by substituting $z_1 = z_o$ in equation (3.1). In many studies, an arbitrary state A_o is selected, and an arbitrary value, such as zero, is assigned to E_o. Thus both positive and negative values of energy are computed for various states of the system. In other studies, the choice of A_o is prescribed so that all energy values are positive and greater than a minimum characteristic of the system. We will see later that when this choice is made, we talk about absolute values of energy.

Because the first law imposes no conditions on either the system or its states, we conclude that the property energy is defined for each system and all its states.

We see from (3.1) that energy E has the same dimensions as the expression for potential energy, Mgz, of mechanics or, equivalently, mass times length squared divided by time squared. These dimensions are denoted as $[E] = [M][L]^2[T]^{-2}$. Because of its importance, and for historical reasons, many special units of measure have been defined for this particular combination of fundamental dimensions. The unit used in the International System of units (SI) is the joule,[1] denoted by the symbol J and such that 1 J = 1 kg m²/s². Table 3.1 lists some of the other units and conversion factors between them and the joule.

The definition and existence of energy as a property of every system in any state is the most important consequence of the first law. It is not at all obvious that systems should have such a property. For example, we will see later that work is a well-defined quantity, yet it is not a property of a system. In addition, it is not possible to perform all the conceivable experiments required to prove that energy exists as a property. An infinite number of ways exist to design a weight process between two given states A_1 and A_2 of a system A. We can never hope to carry out all these ways experimentally to check that indeed we always obtain the same value of $Mg(z_1 - z_2)$. Instead, all this is guaranteed if we accept the statement of the first law.

Because energy is a property, that is, a quantity whose value depends only on the state, any change of state, spontaneous or induced, from a state A_1 to a state A_2 of a system A is always associated with a change $E_2 - E_1$ in energy. Given two end states, the change $E_2 - E_1$ is independent of the type of interaction forces that induced the change of state, that is, independent of whether or not the change of state from A_1 to A_2 is achieved by a weight process. Conversely, the change in value $E_2 - E_1$ during any process whatsoever may be evaluated by means of a weight process having the same states A_1 and A_2 as end states.

3.5 Additive Properties

For system A in state A_1, we denote the value of a property P at that state by P_1^A, for B in state B_1 by P_1^B, and for C in state C_1 by P_1^C. We assume that system C in state C_{11} is composed of system A in state A_1 and of system B in state B_1. Under these conditions, if the value P_{11}^C satisfies the relation

$$P_{11}^C = P_1^A + P_1^B \tag{3.4}$$

and this relation holds for all pairs of systems A and B, and all pairs of states A_1 and B_1 for which property P is defined, we say that P is an *additive* property. We will see that many properties, including energy and entropy, are additive, and that additivity plays an important role in analyses of physical phenomena.

Other quantities are also additive. For example, the length of a wire is additive because the length of two pieces of wire joined together equals the sum of the lengths

---

[1]This unit is named in honor of James Prescott Joule (1818–1889), English scholar, whose careful experiments during the period from 1840 to 1848 laid the foundations of the first law.

TABLE 3.1   Units for energy and conversion factors to the International System of units.[a]

Unit	Symbol	Conversion to SI units
joule	J	1 J = 1.00000 J
thermochemical calorie[b]	cal_{th}	1 cal_{th} = 4.18400 J
calorie (IT)	cal	1 cal = 4.18680 J
kilocalorie	kcal	1 kcal = 4.18680 kJ
British thermal unit	Btu	1 Btu = 1.05506 kJ
erg	erg	1 erg = 100.000 nJ
kilowatt-hour	kWh	1 kWh = 3.60000 MJ
horsepower-hour	hp h	1 hp h = 2.68452 MJ
megawatt-day	MW day	1 MW day = 86.4000 GJ
megawatt-year	MW yr	1 MW yr = 31.5360 TJ
newton-meter	N m	1 N m = 1.00000 J
kilogram-force–meter	kg_f m	1 kg_f m = 9.80665 J
gram-force–centimeter	g_f cm	1 g_f cm = 98.0665 mJ
dyne-centimeter	dyn cm	1 dyn cm = 100.000 nJ
foot-pound–force	ft lb_f	1 ft lb_f = 1.35582 J
foot-poundal	ft poundal	1 ft poundal = 42.1401 mJ
liter-atmosphere	l atm	1 l atm = 101.325 J
cubic centimeter–atmosphere	cm^3 atm	1 cm^3 atm = 101.325 mJ
cubic foot–atmosphere	ft^3 atm	1 ft^3 atm = 2.86920 kJ
electron-volt	eV	1 eV = 0.16021 aJ
ton of oil equivalent	toe	1 toe = 42 GJ
barrel of oil		= 6.33 GJ
metric ton of coal equivalent		= 29.75 GJ
kilogram of uranium		= 82.29 TJ

[a] Prefix names and symbols are defined in Table A.1 (Appendix A).

[b] The thermochemical calorie, defined exactly as 1 cal_{th} = 4.184 J, is generally preferred by physicists, whereas the International Steam Table calorie, defined exactly as 1 cal = 4.1868 J, is generally preferred by engineers and sometimes denoted by cal_{IT}. The difference between the two units is only about 0.06% and hence is negligible for most calculations. The British thermal unit is defined in terms of the International Steam Table calorie, so that 9 Btu/lb = 5 cal/g.

of the two parts. Also, values of amounts of a constituent are additive, as well as the values of certain parameters, such as volume, are additive. For example, if system A has n^A molecules of H_2O, and system B has n^B molecules of H_2O, a composite system C consisting of systems A and B has a number of molecules $n^C = n^A + n^B$.

3.6   Additivity of Energy

We consider a system C composed of two subsystems A and B, and three states of C such that in state C_{11} subsystems A and B are in states A_1 and B_1, in state C_{21} they are in states A_2 and B_1, and in state C_{22} they are in states A_2 and B_2, respectively. States A_1, A_2, B_1, and B_2 are arbitrary. The double subscript denotes the two states of the subsystems A and B that correspond to the state of C.

The change between states C_{11} and C_{21} may be accomplished by means of a weight process during which subsystem A has as end states A_1 and A_2, while subsystem B experiences no change and remains at state B_1. If $Mg(z_1 - z_2)$ denotes the effect of the process, each of $E_{21}^C - E_{11}^C$ and $E_2^A - E_1^A$ equals $Mg(z_1 - z_2)$ [see equation (3.3)] because the weight process may be associated with either system C or system A. Thus

$$E_{21}^C - E_{11}^C = E_2^A - E_1^A \tag{3.5}$$

By a similar argument, we can readily prove that the difference in energy $E_{22}^C - E_{21}^C$ between states C_{22} and C_{21} of system C equals the difference in energy $E_2^B - E_1^B$ between states B_2 and B_1, namely,

$$E_{22}^C - E_{21}^C = E_2^B - E_1^B \tag{3.6}$$

Adding equations (3.5) and (3.6), we obtain the relation

$$E_{22}^C - E_{11}^C = (E_2^A - E_1^A) + (E_2^B - E_1^B) \tag{3.7}$$

This result may easily be generalized to systems composed of more than two subsystems. It shows that the difference in energy between any two states of a composite system equals the sum of the differences in energy between the corresponding states of the component subsystems. In other words, differences in energies are additive.[2]

Moreover, energy is an additive property provided that the energy of the reference state of the composite system is selected consistent with the energies of the reference states of its subsystems.

Example 3.4. Find the condition that must be satisfied in order for energy to be an additive property.

Solution. Using a self-evident change of subscripts, we can rewrite (3.7) in the form $E_{11}^C - E_{oo}^C = (E_1^A - E_o^A) + (E_1^B - E_o^B)$ or, equivalently, $E_{11}^C = E_1^A + E_1^B + (E_{oo}^C - E_o^A - E_o^B)$, where the subscript o denotes a reference state. The last equality shows that if the energies E_{oo}^C, E_o^A, and E_o^B of the reference states C_{oo}, A_o, and B_o, respectively, are selected so that $E_{oo}^C = E_o^A + E_o^B$, then and only then

$$E_{11}^C = E_1^A + E_1^B \tag{3.8}$$

for all pairs of systems A and B and all pairs of states A_1 and B_1. In other words, then and only then, energy is additive. This conclusion may easily be extended to many subsystems.

3.7 Conservation of Energy

For a weight process of a system A from state A_1 to state A_2, the change in elevation of the weight, $z_2 - z_1$, may be positive, zero, or negative. A zero value implies that no net external effects whatsoever are left in the surroundings of the system. It does not necessarily imply, however, no change in state of the system. For example, no external effects are induced in the surroundings during a spontaneous change of state (see Section 2.3.1) during which a system evolves in time from state A_1 to a different state A_2.

Because every zero-net-external-effect process with end states A_1 and A_2 corresponds to a weight process with $z_2 = z_1$, it follows from (3.3) that

$$E_2 = E_1 \tag{3.9}$$

This result is often called the principle of *conservation of energy*. It is an implication of the definition of energy and, hence, of the first law.

The first law is often presented as equivalent to the law of conservation of energy. Energy conservation is certainly an important consequence of the first law. However,

---

[2]We recall that the requirement of separability of the constituents is part of the definition of a system. The statement of the first law, including the conclusion that energy differences are additive, applies to all systems provided, of course, that they are well defined in the sense discussed in Section 2.1.

the essence of the first law is not energy conservation but the implication that energy is a property.

Example 3.5. A system experiences a zero-effect weight process. Does this mean that during the process the system is isolated?

Solution. Not necessarily. During the process, the system may induce changes in the state of the environment, even if the net final effect is null.

Because changes of state cannot occur instantly, we conclude that equation (3.9) is a temporal relation, that is, a relation between energy values at two different instants of time for a system that at time t_1 is in state A_1, and at some later time t_2 in state A_2. For any zero-net-external-effect process, the equation requires that the energy E_2 of the system at time t_2 be equal to the energy E_1 at the earlier time t_1.

Although time dependent, equation (3.9) has been obtained without using explicitly an equation describing the time evolution of the system, that is, without using an equation of motion. Because it is an implication of the first law and we have no reason to question the validity of this law, we conclude that energy must be conserved by the equation of motion of any system in every process with no external effects. For example, during a spontaneous change of state, the equation of motion of the system must be such that $dE/dt = 0$, as required by equation (3.9).

Example 3.6. Consider an isolated composite system C consisting of two systems A and B, and a change of state for system A from state A_1 to state A_2, and for system B from state B_1 to state B_2. Show that energy changes $E_2^A - E_1^A$ of A and $E_2^B - E_1^B$ of B satisfy the relation $E_2^B - E_1^B = -(E_2^A - E_1^A)$, that is, they are equal and opposite.

Solution. Because system C is isolated, its energy is conserved and therefore $E_{22}^C - E_{11}^C = 0$. On the other hand, energy differences are additive so that $E_{22}^C - E_{11}^C = 0 = E_2^A - E_1^A + E_2^B - E_1^B$ and therefore $E_2^B - E_1^B = -(E_2^A - E_1^A)$. An identical result is valid even when C is not isolated, but its change of state has no net effects on the surroundings.

Example 3.7. Show that the relation between energy, E^W, and elevation, z, for a system consisting of a weight of mass M is such that $E_2^W - E_1^W = Mg(z_2 - z_1)$.

Solution. In a weight process of a system A, the energy change of A is $E_2^A - E_1^A = -Mg(z_2 - z_1)$ [equation (3.3)]. Because the only effect external to A is the change in elevation of the weight, the system composed of A and the weight has zero net external effects, and therefore its energy remains constant. By the additivity of energy differences, this implies that $(E_2^A - E_1^A) + (E_2^W - E_1^W) = 0$, and therefore $(E_2^W - E_1^W) = -(E_2^A - E_1^A) = Mg(z_2 - z_1)$.

Example 3.8. Show that the relation between energy, E^S, and extension, x, for a system consisting of a linear spring with a spring constant k and elongation x as the only independent property is such that $E_2^S - E_1^S = k(x_2^2 - x_1^2)/2$.

Solution. We consider a weight of mass M attached to the spring so that the gravitational force, Mg, and the elastic force, kx, are balanced, that is, $Mg = kx$, where x is the elongation of the spring from its unstretched position. Next, we consider a weight process for the spring during which the elongation changes from x to $x + dx$, where $dx \ll x$, so that the gravitational force, Mg, and the elastic force of the stretched spring, $k(x + dx)$, are still approximately balanced [i.e., $Mg \approx k(x + dx)$]. During the weight process, the change in elevation of the weight is $-dx$ and thus the change in energy of the spring [equation (3.3)] $dE^S = -Mg(-dx)$ or, equivalently, $dE^S = kx\,dx$. Because x and E^S are both properties of the spring, and the spring has only one independent property, there must exist a relation between x and E^S, that is, $E^S = E^S(x)$. Because $dE^S = kx\,dx$, we conclude that $E_2^S - E_1^S = E(x_2) - E(x_1) = \int_1^2 kx\,dx = k(x_2^2 - x_1^2)/2$.

Example 3.9. Show that the relation between energy, E^m, and velocity, $\mathbf{v}$, for a system consisting of a free point mass m in the absence of gravity and with velocity and position as the only independent properties is such that $E_2^m - E_1^m = m(|\mathbf{v}_2|^2 - |\mathbf{v}_1|^2)/2$.

Solution. We consider a weight of mass M attached to a weightless cable which, with a suitable system of frictionless and massless pulleys, is in turn attached to the point mass. The attachment occurs for a time interval dt during which the point mass changes its position and velocity from x and $\mathbf{v}$ to $x + dx$ and $\mathbf{v} + d\mathbf{v}$, respectively. By Newton's law of motion, the force $\mathbf{F}$ applied by the cable must satisfy the relation $\mathbf{F} = m \, d\mathbf{v}/dt$, because the point mass experiences no gravitational force. The magnitude of the force $|\mathbf{F}| = Mg$ and thus $Mg \, dt = m \, d|\mathbf{v}|$. The displacement of the mass $dx = \mathbf{v} \, dt$, and hence the change in elevation of the weight, is $-|\mathbf{v}| \, dt$. During this weight process, the change in energy of the free point mass [equation (3.3)] $dE^m = -Mg(-|\mathbf{v}|dt) = m|\mathbf{v}| \, d|\mathbf{v}|$. In general, because x and $\mathbf{v}$ are the only independent properties and E^m is a property, there must exist a relation $E^m = E^m(x, \mathbf{v})$. Here, however, we conclude that E^m is a function of $\mathbf{v}$ only and not of x, because $E_2^m - E_1^m = \int_1^2 m|\mathbf{v}| \, d|\mathbf{v}| = m(|\mathbf{v}_2|^2 - |\mathbf{v}_1|^2)/2 = E^m(|\mathbf{v}_2|) - E^m(|\mathbf{v}_1|)$.

Examples 3.7 to 3.9 show how equation (3.3) may be used to derive relations between energy and other properties of elementary systems of mechanics, such as weights, springs, and inertial masses. Being relations among properties of a system, the validity of these relations is independent of the special weight processes considered in the derivations. This conclusion is of paramount importance and follows from the fact that energy is a property.

Because energy is an additive property, and because any change of state of a system may always be thought of as part of a zero-effect process of a composite system consisting of all the systems with which the original system is interacting, the principle of energy conservation [equation (3.9)] is of great generality and practical importance.

For changes of state of a composite of two subsystems that occur either in isolation or without net effects on the surroundings, the result derived in Example 3.6 shows that the energy gain of one subsystem equals the energy loss of the other subsystem, and conversely. For this reason we may interpret this result as describing a transfer, an exchange, or a flow of energy between the two subsystems. The exchange of energy is a consequence of interactions between the subsystems. Later we classify interactions according to various additive properties that are transferred during each interaction.

For example, if systems A and B experience an interaction with each other, and no interaction with any other systems in their surroundings, the energy of the composite system A plus B at the end of the interaction must equal the energy at the beginning of the interaction, that is, the sum of the energies of A and B when the interaction is terminated must be the same as the sum of the energies of A and B when the interaction was initiated. This implies that the energy change of B is equal and opposite to that of A. Because the energy change of B is equal and opposite to the energy change of A, we may correctly visualize one of the effects of the interaction between A and B as a *transfer of energy* between the two systems, regardless of the type of the interaction involved, that is, regardless of whether the interaction induced exchange of constituents, physical reactions, chemical reactions, nuclear reactions, annihilation reactions, ionization reactions, separation of electric charges, or a combination of any of these reactions.

3.8 Energy Balance

The principle of energy conservation together with the idea of transfer of energy gives rise to an extremely important analytical tool used in all physics and engineering applications: the energy balance. For each system and any change of its state, we must account for the energy change of the system and the energy change of its surroundings or, equivalently, the energy transferred between the system and the surroundings. Because energy is a conserved property, any energy change of a system A as it goes from state A_1 to state A_2 must be accounted for by an energy exchange with the surroundings as expressed by the general *energy balance equation* or, simply, *energy balance*

$$(E_2 - E_1)_{\text{system}} = \begin{bmatrix} \text{net amount of energy} \\ \text{received by the system} \\ \text{from its surroundings} \end{bmatrix} = E^\leftarrow \qquad (3.10)$$

The energy balance equation can be written for all systems, all processes, and all states of a system. Later, we study different types of interactions for which the right-hand side of the equation takes special forms useful for a large variety of practical applications.

3.9   Energy Transfer in a Weight Process

In a weight process for a system A, the change in energy $E_2^A - E_1^A = -Mg(z_2 - z_1)$ [equation (3.3)]. In Example 3.7 we have shown that the change in energy of the weight $(E_2^W - E_1^W) = Mg(z_2 - z_1)$. Because the change in energy of the weight is equal and opposite to that of system A, we may visualize the effect of the process as a transfer of energy between the two systems in the amount $Mg(z_2 - z_1)$. Later we show that the interaction between A and the weight during a weight process for A falls into the class of what we call work interactions. The amount of energy transferred in a work interaction is called simply work. Thus we say that $Mg(z_2 - z_1)$ is the work done by system A on the weight during the weight process of A.

Because it represents a transfer, the work in a weight process is specified not only by the amount of energy transferred but also by the direction of the transfer, namely, by whether the energy transfer is from A to the weight or from the weight to A. To have a complete specification, we must select a reference direction and define a symbol for energy transfer such that its value is positive when the transfer is along the chosen direction, and negative when the transfer is along the opposite direction. For example, we may select the reference direction to be from the system to the weight. Then, for a system A undergoing a weight process from state A_1 to A_2, we denote the work by the symbol $W_{12}^{A\rightarrow}$, where the superscript $A\rightarrow$ implies that a positive value of $W_{12}^{A\rightarrow}$ represents an energy transfer out of system A, the subscript 12 that the state changes from A_1 to A_2, and

$$W_{12}^{A\rightarrow} = Mg(z_2 - z_1) \qquad (3.11)$$

Alternatively, we may select the opposite reference direction, from the weight to the system, and denote the work by the symbol $W_{12}^{A\leftarrow}$, where the superscript $A\leftarrow$ implies that a positive value represents an energy transfer into system A, so that

$$W_{12}^{A\leftarrow} = -Mg(z_2 - z_1) \qquad (3.12)$$

Clearly, these two specifications satisfy the relation $W_{12}^{A\rightarrow} = -W_{12}^{A\leftarrow}$.

Upon combining equations (3.3) and (3.11), for a weight process of system A from state A_1 to state A_2 we find that

$$E_2 - E_1 = -W_{12}^{A\rightarrow} \qquad (3.13)$$

Equation (3.13) is an energy balance [equation (3.10)] for a system undergoing a weight process because $-W_{12}^{A\rightarrow}$ is the net amount of energy received by the system from the surroundings. Alternatively, we may combine equations (3.3) and (3.12) to find

$$E_2 - E_1 = W_{12}^{A\leftarrow} \qquad (3.14)$$

Again this is an energy balance because $W_{12}^{A\leftarrow}$ represents the net amount of energy received by A from the weight during the weight process.

3.10 Impossibility of Perpetual-Motion Machines of the First Kind

A process in which the initial and final states of a system are identical is called a *cycle* or *cyclic process*. A *perpetual-motion machine of the first kind* (*PMM1*) is any system undergoing a cyclic process that produces no external effects except the rise or fall of a weight in a gravity field. Thus a PMM1 is a system undergoing no net change in state, yet producing a nonzero net external mechanical effect.

No matter how complicated the devices within the system are, and no matter how intricate the interconnections are between them, a PMM1 cannot exist, namely, it is impossible. To prove this assertion, we assume the contrary and show that it leads to a contradiction, that is, we assume that there exists a cyclic weight process of a system A with identical end states, $A_1 = A_2$, and external effect $Mg(z_1 - z_2) \neq 0$. Because the end states are identical and energy is a property, we conclude that $E_2 - E_1 = 0$. On the other hand, $E_2 - E_1 = 0 = -Mg(z_2 - z_1) \neq 0$ because of (3.3). This result is absurd, and therefore our assumption is false. Thus we conclude that a PMM1 is impossible.

The idea of a perpetual-motion machine of the first kind must not be confused with the idea of perpetual motion. A *perpetual motion* is a cyclic process of a system producing no external effects whatsoever. As such, it violates no known law and, in particular, is consistent with the principle of conservation of energy [equation (3.9)]. Therefore, perpetual motion is possible. For example, a powerless electric current in a superconducting coil, and a lossless pendulum oscillating in a vacuum are perpetual motions that are achieved very often in laboratories and other installations. In contrast, cyclic machines that can transfer energy to other systems without consumption of energy resources have never been built or even approached in any laboratory. Moreover, our accepting the first law as a fundamental principle entails the conclusion that such machines cannot be built.

3.11 Absolute Energy and Special Relativity

For a fully rigorous treatment consistent with special relativity, the quantity $Mg(z_1 - z_2)$ in the statements of the first law and all its consequences should be substituted by the expression $Mc^2[\exp(gz_1/c^2) - \exp(gz_2/c^2)]$, where M is the mass of the weight at $z = 0$, and c the speed of light in vacuum. In general, however, the mass M may be chosen so that for all states of interest of system A, the values of gz/c^2 are much smaller than unity. Then $\exp(gz/c^2) \approx 1 + (gz/c^2)$ and therefore, $Mc^2[\exp(gz_1/c^2) - \exp(gz_2/c^2)] \approx Mg(z_1 - z_2)$. Indeed, the speed of light in vacuum $c = 2.9979 \times 10^8$ m/s and the gravitational acceleration at the surface of the earth $g = 9.8066$ m/s^2, so that $gz/c^2 \ll 1$ whenever $z \ll 10^{16}$ m.

Regardless of the form of the statement of the first law, the concept of energy in relativistic mechanics may readily be included in the concept of energy defined here through a judicious choice of the energy value assigned to the reference state of a system. To this end, we observe the following.

According to special relativity, each state of a system A has energy E, momentum $\boldsymbol{p}$ and velocity $\boldsymbol{v}$ that satisfy the relations

$$(E^2 - |\boldsymbol{p}|^2 c^2)^{1/2} = mc^2 \tag{3.15}$$

$$\boldsymbol{p} = \boldsymbol{v}\frac{E}{c^2} \tag{3.16}$$

where m is the *mass* of the system in the given state. For $p = 0$, equation (3.15) defines an energy

$$E_r = mc^2 \tag{3.17}$$

called the *rest energy*.

The value of the mass is nonnegative and may vary from state to state. For example, for two states A_1 and A_2 with equal energies $E_1 = E_2$, and momenta $p_1 = 0$ and $p_2 \neq 0$ but such that $|p_2| \ll m_2 c$, using equations (3.15) and (3.16) we find that

$$E_1 = m_1 c^2 = E_{r1} = E_2 = m_2 c^2 \left(1 + \frac{|p_2|^2}{m_2^2 c^2} \right)^{1/2} \approx m_2 c^2 + \frac{|p_2|^2}{2m_2}$$

and therefore that $m_1 \neq m_2$. Again two states A_1 and A_2 with different energies, $E_1 \neq E_2$, but null momenta, $p_1 = p_2 = 0$, correspond to different masses, $m_1 \neq m_2$. Here, the energy of each of these two states is also a rest energy, that is, $E_1 = E_{r1}$ and $E_2 = E_{r2}$.

Among all the states of a system with given values of amounts of constituents, n, and parameters, β, at least one has the lowest rest energy. We call each such state a *ground* or *ground-energy state*, its energy a *ground-state energy*, and its mass a *ground-state mass*. We denote the ground-state energy by $E_g(n, \beta)$ or simply E_g, the ground-state mass by $m_g(n, \beta)$ or simply m_g, and recall that $E_g = m_g c^2$ because in the ground-energy state the momentum of the system is null.

For a system consisting of the smallest amount of a free constituent—a single particle, atom, molecule, or field unaffected by any external forces—the ground-state mass is a characteristic constant of the constituent that can be measured by mass-spectroscopic techniques. We call it the *ground-state mass of the free constituent* and denote it by m_g^{free}. Moreover, we call *ground-state energy of the free constituent* the rest energy $E_g^{\text{free}} = m_g^{\text{free}} c^2$. For example, the ground-state mass of a free electron $m_g^{\text{free}} = 9.1095 \times 10^{-31}$ kg, that of a free proton $m_g^{\text{free}} = 1.6726 \times 10^{-27}$ kg, and that of a free hydrogen atom $m_g^{\text{free}} = 1.6735 \times 10^{-27}$ kg. Correspondingly, the ground-state energy of a free electron $E_g^{\text{free}} = 0.511$ MeV (1 MeV $= 10^6$ eV $= 1.6022 \times 10^{-13}$ J), that of a free proton $E_g^{\text{free}} = 938.23$ MeV, and that of a free hydrogen atom $E_g^{\text{free}} = 938.74$ MeV. Again, for the free electromagnetic field—electromagnetic radiation in vacuum—the velocity $\mathbf{v}$ is always equal to the velocity of light in vacuum c, and equations (3.15) and (3.16) yield $m = 0$ for all states, regardless of the value of the energy. Therefore, the ground-state mass of the free electromagnetic field $m_g^{\text{free}} = 0$, and the ground-state energy $m_g^{\text{free}} c^2 = 0$.

To assign energy values to any state of a system in a manner consistent with special relativity and the first law, we begin with a free system F consisting of r types of constituents with values of amounts $n_1, n_2, \ldots, n_r$, and null internal forces, so that each single particle, atom, molecule, or field is free. Consistently with special relativity and the additivity of energy, we assign to the ground-state energy of this system the value $E_g^F = \sum_{i=1}^r n_i (m_g^{\text{free}})_i c^2$, where $(m_g^{\text{free}})_i$ is the ground-state mass of the ith free constituent.

Next, we consider a system A consisting of the same r types of constituents as the free system F but subject to external forces—characterized by the parameters $\beta = \{\beta_1, \beta_2, \ldots, \beta_s\}$—and to nonzero internal forces, and select a reference state A_o with the same values of the amounts of constituents as the free system, that is, $n_1, n_2, \ldots, n_r$. Without loss of generality, we can always regard the state A_o of system A and a ground-energy state F_g of the free system F as two states of a more general system with external and internal forces that may be turned on and off. Then, by virtue of the first law, these two states of the general system can be interconnected by means of a weight process, and therefore we can always evaluate the energy difference $V_o(\boldsymbol{n}, \boldsymbol{\beta}, \text{internal forces}) = E_o^A - E_g^F$. Using the value of E_g^F dictated by special relativity, we conclude that

$$E_o^A = \sum_{i=1}^{r} n_i \, (m_g^{\text{free}})_i \, c^2 + V_o(\boldsymbol{n}, \boldsymbol{\beta}, \text{internal forces}) \tag{3.18}$$

When energies are evaluated by the procedure just cited, we say that the resulting energy values are *absolute*.

3.12 Mass Balance

In general, the difference $E_g^A - E_g^F$ between the energy of a ground-energy state A_g of a system A and the ground-state energy of a free system F with the same values of the amounts of constituents as A_g is different from zero. It can be positive or negative. Indeed, whereas the sum E_g^F of the ground-state energies of the free constituents depends only on the values of the amounts of constituents, the ground-state energy E_g^A depends not only on the values of the amounts of constituents but also on the parameters and the internal forces of system A. For many applications, however, the difference between these two energies is negligible with respect to E_g^F, that is, $|E_g^A - E_g^F|/E_g^F \ll 1$. Correspondingly, the difference between the sum of the ground-state masses of the free constituents, $m_g^F = \sum_{i=1}^{r} n_i \, (m_g^{\text{free}})_i = E_g^F/c^2$, and the ground-state mass $m_g^A = E_g^A/c^2$ is negligible with respect to m_g^F, that is, $|m_g^A - m_g^F|/m_g^F \ll 1$.

For example, for the ground-energy state A_g of a system A consisting of an electron and a proton bound together to form a free hydrogen atom, accurate mass-spectroscopic data yield $E_g^A - E_g^F = -13.6$ eV, where the ground-state energy of the free system F consisting of one free electron and one free proton $E_g^F = (m_g^{\text{free}})_{\text{electron}} c^2 + (m_g^{\text{free}})_{\text{proton}} c^2 = (0.511 + 938.23)$ MeV. Although $E_g^A - E_g^F$ differs from zero, it is important to note that this difference is eight orders of magnitude smaller than either the ground-state energy E_g^A of the free hydrogen atom or the sum E_g^F of the ground-state energies of its free constituent particles—the electron and the proton.

Moreover, for many practical applications and processes that involve speeds much lower than the speed of light, the difference between the mass of any state of a system and the ground-state mass of the system corresponding to the same values of the amounts of constituents is negligible with respect to either of these two masses. So, the mass of a system A in any state with values n_1, n_2, ..., n_r of the amounts of constituents is approximately equal to the ground-state mass corresponding to the same values of the amounts, and because this ground-state mass is approximately equal to the sum of the ground-state masses of the free constituents, we have

$$m \approx \sum_{i=1}^{r} n_i \, (m_g^{\text{free}})_i \tag{3.19}$$

So, we have mass additivity.

According to our current model of physical phenomena, any molecule, atom, or elementary particle can be created or annihilated as a result of a suitable spontaneous process involving only the electromagnetic field. For example, certain spontaneous transitions of the electromagnetic field in which the absolute energy of the field decreases by at least 1.022 MeV result in the creation of an electron and a positron. Conversely, an electron and a positron may combine and disappear, giving rise to a transition of the electromagnetic field by at least 1.022 MeV. Mechanisms of this kind are called *creation and annihilation reactions*. With these reactions, massive particles can be created spontaneously from the massless electromagnetic field or annihilate each other by converting their masses into energy of the massless electromagnetic field. In addition, nuclear and chemical reactions, as well as interactions between a constituent and the electomagnetic field may also induce spontaneous changes in the value of the mass of a system.

Thus, in general, and in contrast to energy and momentum, mass is neither additive nor conserved. However, in the absence of nuclear reactions, and creation and annihilation reactions, the mass changes caused within a system by energy exchanges between constituents and the electromagnetic field or by chemical reactions are negligible with respect to the mass of the system. So mass is approximately conserved.

Because mass is approximately additive and conserved for most applications that do not involve nuclear reactions, creation and annihilation reactions, and speeds close to the speed of light, without specifying whether we mean mass or sum of ground-state masses of the free constituents, we write the *mass balance equation* or *mass balance* of system A for changes from state A_1 to state A_2 in the form

$$(m_2 - m_1)_{\text{system}} = \begin{bmatrix} \text{net amount of mass} \\ \text{received by the system} \\ \text{from its surroundings} \end{bmatrix} = m^{\leftarrow} \tag{3.20}$$

Example 3.10. A system consists initially of 2 mol or 24.0200 g of carbon, C, and 1 mol or 31.9988 g of oxygen, O_2, all at their ground states, and confined in a rigid container with volume V, and walls impermeable to matter and energy. As a result of the oxidation reaction mechanism, two molecules of carbon monoxide, CO, are produced from each pair of carbon atoms and each oxygen molecule, and the total kinetic energy of the CO molecules is 521.0287 kJ — total kinetic energy above the ground-state of CO.

(a) What is the change in energy between the initial and final states?

(b) What is the difference between the sum of the ground-state masses of the free constituents of the carbon plus oxygen and that of the carbon monoxide?

Solution. **(a)** During the oxidation process, the system is not interacting with any other system in the environment. Therefore, there is no change in energy.

(b) The difference between the sum of the ground-state masses of the free constituents of the initial state (carbon plus oxygen) and that of the final state (carbon monoxide) equals the kinetic energy divided by the square of the speed of light, $\Delta m_o = 521.0287/(2.9979 \times 10^8)^2 = 5.7973 \times 10^{-9}$ g, namely, the value of the mass of carbon monoxide is smaller than that of carbon plus oxygen by Δm_o. This difference is about nine orders of magnitude smaller than the inital mass of the system and, therefore, entirely negligible.

3.13 Comment

Partly because of the etymological origin of the term energy, which in Greek implies ability to be active, to do useful things, and partly because, as we will see, energy is transferred during work interactions, energy is often interpreted as the capacity of systems to produce work or perform useful tasks. This interpretation, however, can be easily dispelled if we consider the air around us which has lots of energy and yet no ability to heat our homes or feed our growth. More fundamentally, in due course we will see that a large amount of energy can be transferred to a system A by means of a weight process. At the end of the transfer, the system may be found in one of many different states. Each of these states has the same energy but not necessarily the same ability to perform a useful task. For example, starting from some of these states the energy could be used to effect the rise of a weight from a low to a high elevation, but for other states no such effect is possible.

The thermodynamic definition of energy differs from the popular usage of the word. The popular usage refers to something that makes electric lights work, automobiles run, and factories produce consumer goods, that is to the capacity of certain material bodies to perform useful tasks. As we will see, this usage represents what scientists and engineers call availability, which in turn is related to both energy and another inherent property of matter called entropy. We will also see that availability can be easily lost while energy is being transferred from one material to another, as in an energy transfer from hot gases to cold gases, or while matter changes from one condition to another, as in the change of a cold fuel-air mixture to hot products of combustion. So, the popular usage of the word energy does not represent a quantity that is conserved but something that is almost always degraded as we try to capture its usefulness.

Problems

3.1 Identify a suitable system undergoing a weight process, and determine whether a weight falls, stays put, or rises in bringing about the following changes.

(a) A tire pump is connected to a tire. The pump plunger is pushed down, forcing air into the tire. Assume that the walls of the tire, pump, and connecting tube are impervious to matter and perfectly insulating.

(b) A liquid fills a perfectly insulated rigid container. The liquid comes to rest from an initial state of turbulent motion.

(c) Hydrogen and oxygen are enclosed in a rigid container impermeable to matter and perfectly insulated. A spark (which may be considered infinitesimal and negligible) causes the two gases to combine.

3.2 Show that the relation between energy, E, and angular speed, ω, for a system consisting of a freewheel with moment of inertia I, having angular speed ω as the only independent property, is such that $E_2 - E_1 = \frac{1}{2}I(\omega_2^2 - \omega_1^2)$. Recall that Newton's law takes the form $\tau = I\, d\omega/dt$, where τ is the torque applied on the freewheel.

3.3 Show that the relation between energy, E, and electric potential, V, for a system consisting of a capacitor with electrical capacitance C, having electric potential V as the only independent property, is such at $E_2 - E_1 = \frac{1}{2}C(V_2^2 - V_1^2)$. Recall that the electric current $I = C\, dV/dt$, and a lossless electric motor is such that $VI = \tau\omega$, where τ is the torque and ω is the angular speed of the motor shaft.

3.4 Show that the relation between energy, E, and electric current, I, for a system consisting of an inductor with electric inductance L, having electric current I as the only independent property, is such at $E_2 - E_1 = \frac{1}{2}L(I_2^2 - I_1^2)$. Recall that the electric potential $V = L\, dI/dt$, and that $VI = \tau\omega$ for a lossless electric motor, where τ and ω are defined as in Problem 3.3.

3.5 An elevator cage moves in a vertical shaft. When fully loaded it weighs 900 kg. A 500-kg counterweight is connected to the elevator cage by means of cable and pulleys. Assume that the guide rails and pulleys are frictionless.

(a) Estimate the power required while the fully loaded elevator cage is rising at a constant speed of 2 m/s.

(b) If the empty cage weighs 200 kg, what is the power required to lower the cage at a constant speed of 2 m/s?

3.6 For the manufacturing of one day's paper products a paper mill uses the necessary coal, oil, chemicals, and wood that are already stored in the mill. In addition, some of the energy needed in the mill is obtained from a waterfall of 10^6 tons of water per day falling over a difference in elevation of 100 m. Assume that the mill experiences no interactions except that with the waterfall.

(a) What is the change in energy of the mill after 1 day of paper production?

(b) At a particular day the waterfall is interrupted and the mill uses half its wood as fuel rather than feedstock for paper. What is the change in energy of the mill during that day? The energy that is made available from the burning wood is 18,000 kJ/kg, and paper requires 40,000 kJ per kg of paper.

3.7 A well-insulated building receives only electricity from a utility. A device called a heat pump is used to transfer energy from room A to room B in the building. Initially, the two rooms are equally warm. After operating the heat pump for 1 h, room A is warmer than room B. The electric power rating of the pump is 2 kW.

(a) What is the change in energy of the building after 1 h of operation of the heat pump?

(b) At the end of 1 h, the operation of the heat pump is interrupted. What is the energy of the building 1 h after the interruption when both rooms are again equally warm?

3.8 The sun emits energy continuously into the universe. The fraction of this energy flux intercepted by the earth is 1.75×10^{11} MW. The earth–sun distance is about 1.5×10^8 km, and the earth's radius is about 6370 km. The radius of the sun is about 696,000 km and its density is about 1.4 kg/m^3.

(a) What is the overall rate at which energy is emitted from the sun in all directions?

(b) At what rate (in metric tons per second) is the sun losing mass?

(c) In how many years will the sun lose 1/10,000 of its mass due to its continuous energy emission?

3.9 A system consists of carbon atoms, C, oxygen molecules, O_2, and carbon monoxide molecules, CO, confined in a rigid container with fixed volume. The three constituents can combine according to the chemical reaction mechanism $2\,C + O_2 = 2\,CO$. The ground-state masses of free C, O_2, and CO are such that $2m_g^{\text{free}}(C) + m_g^{\text{free}}(O_2) - 2m_g^{\text{free}}(CO) = 0.4064 \times 10^{-35}$ kg. The system is initially in a ground-energy state with 4 atoms of carbon C and 3 molecules of oxygen O_2. A spark initiates the chemical reaction. After the reaction is completed, the system ends in a state with 4 molecules of CO and 1 molecule of O_2. While the system experiences this change of state, appropriate machinery transfers 2 eV of energy to the surroundings.

(a) Show that $E_g^{\text{free}}(C) + \frac{1}{2}E_g^{\text{free}}(O_2) - E_g^{\text{free}}(CO) = 1.14$ eV, where E_g^{free} denotes ground-state energy of the free constituent.

(b) What are the differences in energy and sum of the ground-state masses of the free constituents between the initial and final states?

(c) Is the final state a ground-energy state?

(d) What is the change in the value of the sum of the ground-state masses of the free constituents as a fraction of the mass of the molecules that reacted?

3.10 An isolated system A initially consists of a deuterium molecule, D_2, and an oxygen atom, O, confined within a fixed volume V. The constituents of A are in a ground-energy state. As a result of a spontaneous fusion reaction between the two D atoms, the constituents of A are transformed into a tritium atom, T, a neutron, n, and an oxygen atom, O, that have an overall kinetic energy of 3.2 MeV.

(a) What is the change of energy of system A?

(b) What is the change in value of the sum of the ground-state masses of the free constituents of system A?

3.11 Consider again the initial state of the system A defined in Problem 3.10. Here, as a result of a spontaneous chemical reaction, the deuterium molecule and the oxygen atom combine to a form a heavy water molecule, D_2O, with a kinetic energy of about 2.5 eV.

(a) How does the system in this problem differ from the system in Problem 3.10?

(b) Is the state of the D_2O molecule resulting from the chemical reaction the lowest-energy state in which a heavy water molecule can exist?

(c) What is the largest amount of energy that can be transferred out of the D_2 system?

3.12 The experimental setup for the inertial confinement approach to thermonuclear fusion has approximately the following features. A laser beam carrying about 20 kJ of energy is focused on a small pellet of a deuterium–tritium (D-T) mixture. The laser energy is delivered to the pellet in the form of a pulse of about 1 ns duration. The diameter of the pellet $d = 50\ \mu m$. As a result of the laser energy deposition, material ablates from the surface of the pellet, and the ablation, in turn, inertially compresses the pellet to a very high density. Thus the nuclei of deuterium and tritium can combine according to the nuclear fusion reaction mechanism $D + T = He + n$. The ground-state masses of free n, D, T, and He are such that $m_g^{free}(D) + m_g^{free}(T) - m_g^{free}(He) - m_g^{free}(n) = 3.137 \times 10^{-29}$ kg. As a result of the reaction, 1.8×10^{13} helium atoms and 1.8×10^{13} neutrons are formed. Consider a system consisting of deuterium, D, tritium, T, helium, He, and neutrons, n, confined inside a chamber impervious to material and energy fluxes.

(a) Show that $E_g^{free}(D) + E_g^{free}(T) - E_g^{free}(He) - E_g^{free}(n) = 17.6\,\text{MeV}$, where E_g^{free} denotes ground-state energy of the free constituent.

(b) What are the differences in energy and sum of ground-state masses of the free constituents between the initial and final states?

(c) What is the change in the sum of ground-state masses of the free constituents as a fraction of the initial mass?

(d) Next, consider the burning of a hydrocarbon fuel that makes energy available at about 45 MJ/kg. How much fuel must be burned to release the same energy as that released by the fusion reaction of 1.8×10^{13} deuterium and 1.8×10^{13} tritium nuclei?

3.13 In a pressurized water reactor (PWR) the nuclear fuel elements are kept in the reactor until 500 GW day per ton of uranium 235 is transferred from the elements to the water.

(a) If upon fission, uranium 235 makes available 1 MW day/g, what fraction of uranium 235 has fissioned when the elements are withdrawn?

(b) If the thermal power rating of the reactor is 3 GW, one-half of the power comes from uranium 235 (the other half is due to plutonium), and the reactor is opened once a year, how much uranium has fissioned when the elements are withdrawn?

(c) If coal that makes energy available at about 30 MJ/kg is used, how much coal is consumed to replace the uranium in part (b)?

3.14 Each of two identical electricity storage batteries can exchange energy with the surroundings exclusively through its terminals, that is, each battery is perfectly isolated except for connections that can be made to the electrical terminals. Each battery is initially charged with 100 MJ of electrical energy.

The terminals of battery A are connected to a resistance heating element. As a result, 75 MJ of energy is transferred out of the battery. Then battery A is left untouched for a year. Battery B is left completely untouched, also for a year. At the end of the year, both batteries are found completely discharged.

(a) What is the final energy of each battery?

(b) A perfect motor is defined as a device that can transfer to a weight 100% of the electrical energy it receives. How much energy can be transferred to a weight upon connecting a perfect motor to the terminals of either battery in the initial state? How much in the intermediate state of battery A just after it supplied energy to the heater? How much in the final state of either battery?

3.15 A European automobile is rated 120 hp (horsepower) of maximum engine power at the maximum speed of 140 km/h. The fuel consumption at maximum speed is about 0.2 $\text{kg}_{gasoline}$/km. The amount of energy received by the engine upon burning 1 kg of gasoline is about 11,000 kcal, but only a fraction of this energy is converted into motive power. Gasoline (in Europe) is priced about $1.4 per kilogram.

(a) What fraction of the fuel power is converted into useful motive power at maximum speed and maximum engine power? What is the cost per kilometer?

(b) At about one-third of maximum power and a speed of 80 km/h, the fuel consumption is about 0.07 kg/km. Compare the cost per kilometer in this regime with that of a 40-hp electrical car at its maximum speed of 80 km/h. The cost of electricity (in Europe) to recharge the batteries is $0.18 per kilowatt-hour.

(c) If the batteries in part (b) are recharged by solar photovoltaic cells in a region where the average insolation is 4 kWh/m^2 per day, and the cells convert 10% of that energy into electricity, what cell area is required to drive the 40-hp electrical car at its maximum speed for 10,000 km/yr?

(d) If the electricity to recharge the batteries comes from a hydroelectric plant that converts 90% of a 50-m waterfall into electricity, what is the daily water flow rate to operate the 40-hp electric car for 10,000 km/yr?

3.16 Hydrogen is usually produced from water by a high-temperature catalytic reaction or by electrolysis. A 100% efficient electrolytic plant would require about 34,000 kcal of electric energy to produce 1 kg of hydrogen.

An automobile engine can be modified to operate on hydrogen instead of gasoline. The amount of energy received by the engine upon burning 1 kg of hydrogen is about 34,000 kcal, but only 40% of this energy is converted by the engine into motive power.

(a) Assuming a 50% efficient electrolytic plant driven by electricity produced by the local utility with a 35% efficient nuclear power plant, what fraction of the nuclear power is finally converted into motive power by the hydrogen engine?

(b) Is it correct to view hydrogen as an energy "source" in the same sense as we view coal and oil as energy sources?

(c) If electricity for the 50% efficient electrolytic production of hydrogen comes from 10% efficient photovoltaic solar cells in a region where the insolation is 4 kWh/m^2 per day, what cell area (in square meters) is required to produce sufficient hydrogen to drive the car 500 km per week?

(d) Assume a capital investment of $1100 per kilowatt of electric power produced by the nuclear power plant, and $300 per kilowatt of electric power consumed by the electrolytic hydrogen plant; a yearly cost (except fuel cost) of 25% of the investment; a nuclear fuel cost of 0.6 cent/kWh; and the energy conversion efficiencies specified in part (a). What is the cost (in cents/kWh) of producing hydrogen if the nuclear and electrolytic plants are operated for 8000 h/yr?

(e) Repeat part (d) assuming that the electrolytic plant is operated only for 3000 h/yr corresponding to the periods of low electric power demand. Accordingly, only three-eighths of the nonfuel cost of the nuclear plant should be allocated to the hydrogen production. Compare your results in parts (d) and (e) with a cost of gasoline about 8 cents/kWh.

3.17 Cogeneration is the concurrent production of electricity and hot water or steam. A commercially available cogeneration unit known as *Tecogen* is shown schematically in Figure P3.17. It consists of a derated automobile diesel engine (mass produced) fueled by natural gas and driving an electricity generator. In addition, the waste energy in the cooling water and the exhaust gases of the engine produce hot water. The gas plus air input supplies about 50,000 kJ/kg of fuel to the engine. The engine transfers 25% of this energy to the shaft of the generator, and the generator converts 90% of the energy supplied by the shaft into electricity. Of the fuel energy not transferred to the shaft, 10% is lost to the atmosphere with the exhaust gases.

One such unit is used at an apartment complex that requires 60 kW of electricity all year long and 120 kW of hot water during the cold season (4000 h/yr). The heating load reduces to 40 kW during the mild season. The unit operates about 8000 h/yr.

The local utility sells the electricity at 15 cents/kWh and buys the excess electricity produced by the Tecogen unit at 10 cents/kWh. The capital cost of the Tecogen unit is $1100 per kilowatt

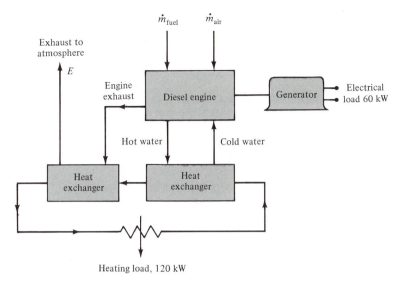

Figure P3.17

of electrical rating, and its annual charges for everything except the fuel are 25% of the capital cost. Natural gas costs about 2 cents/kWh.

(a) If the unit is designed to meet the hot water load during the cold season, what are the fuel flow rate and the electrical rating of the engine? With this design, during the mild season the excess heating energy is wasted, and if necessary, the electrical needs are supplemented by electricity purchased from the local utility.

(b) If the engine is rated to meet the electricity needs of the apartment complex, what is the required fuel flow rate? What happens then with the hot water needs?

(c) An alternative to either part (a) or (b) is to purchase all the needed electricity from the local utility and meet the hot water requirements with a 90% efficient gas burner that costs $200 per kilowatt of natural gas burned. What is the annual cost of this alternative?

(d) How does your answer in part (c) compare with the annual costs of the units in parts (a) and (b)?

3.18 A chamber contains one barrel of oil and adequate air for the complete burning of the oil, that is, the transformation of the compound molecules of carbon and hydrogen in the oil into molecules of carbon dioxide and water. The burning of 1 barrel of oil makes available about 6×10^6 kJ of energy for transfer to other systems.

(a) How would you define a system for this problem so that the energy which is usable upon burning can be transferred to other systems?

(b) If during burning the chamber is isolated, what is the change of the energy as the contents of the chamber change from oil and air to products of combustion?

(c) Make an estimate of the change in the sum of the ground-state masses of the free constituents of the chamber as oil and air change to products of combustion.

(d) If the fissionable atoms of uranium 235 are used to make available 6×10^6 kJ, what is the amount of mass that is consumed? Recall that upon fissioning, an atom of uranium 235 makes available 200 MeV of energy.

3.19 In the United States about 80 quads (1 quad $= 10^{15}$ Btu) or about 84×10^{15} kJ of energy is consumed each year. Approximately one-fourth of this energy is transformed into electricity and, on the average, about 3 units of energy are needed per unit of electricity.

(a) If all the energy were derived from coal, and each ton of coal makes available 25×10^6 kJ, what would be the amount of coal consumed per year?

(b) If all the U.S. energy consumption were to be supplied in the form of electricity, the electricity were produced from fissioning of uranium 235, and the amount of natural uranium used is about 200 times the amount of the fissioning isotope 235, what would be the annual amount of natural uranium that must be used? Recall that the fission of each atom of uranium 235 yields about 200 MeV or, equivalently, 1 MW day/g.

3.20 A system that produces both steam and electricity is shown schematically in Figure P3.20. It is called a total energy or cogeneration system, and operates as follows. Municipal solid wastes are transformed into a low-energy-content gas, a so-called "low-Btu gas," at an installation called a *gasifier*. The gas is compressed by a compressor and fed into an engine such as that used in an automobile. The gas compressor is driven by an electric motor that requires 2 kW of electricity. The engine drives a generator that produces 150 kW of electricity and is turbocharged, that is, atmospheric air for the burning of the gas is compressed in an air compressor before entering the engine. The air compressor is powered by a turbine driven by the exhaust gases from the automobile engine. Turbocharging increases the power output of the engine. The exhaust from the turbocharger turbine enters an energy exchanger where energy is transferred from the exhaust gases to water to make steam.

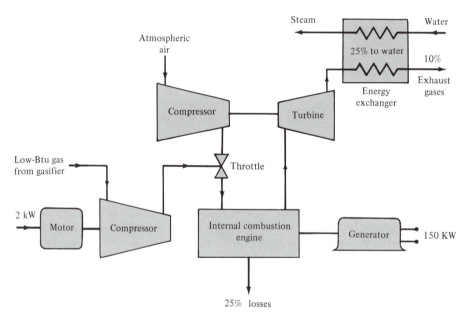

Figure P3.20

(a) What is the net electrical power produced by the total energy system?

(b) The gas supplies 5 MJ/kg. However, 25% of this energy is lost to the environment directly from the engine, 25% is transferred to the steaming water in the energy exchanger, and 10% is carried to the atmosphere by the exhaust gases exiting the energy exchanger. If no other losses occur, what is the mass flow rate of the low-Btu gas?

(c) If 2500 kJ must be transferred to the water in the energy exchanger to produce 1 kg of steam, find the flow rate of the steam.

(d) If the only operating cost to produce electricity and steam is that of the low-Btu gas, how would you allocate this cost between the two products of the plant? Briefly explain and state any additional information that you may need to make the allocation.

3.21 A remote dwelling meets its energy needs by means of the diesel-engine-generator system sketched in Figure P3.21. At full load, the diesel generator produces 100 kW of electric power.

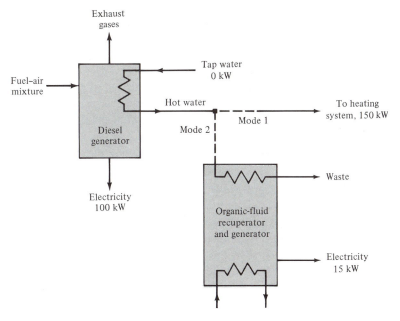

Figure P3.21

About 10% of the fuel energy is wasted in the exhaust gas stream. The fuel–air mixture supplies 40 MJ of energy per kilogram of fuel. The hot water produced by the diesel cooling system is used in either of two modes of operation. In mode 1 it is circulated in radiators that provide heating to the buildings. In mode 2 it is used to run an organic fluid recuperator and generator that generates an additional 15 kW of electric power.

(a) At what rate (in kg/s) is fuel consumed at full load by the diesel generator in either operating mode? At what rate is energy discharged in the exhaust gas stream?

(b) How would you allocate the fuel cost between the power delivered to the heating system and the electric power generated by the diesel in mode 1?

3.22 In a tanker with cargo capacity of about 800,000 barrels (bbl), the power plant consists of a turbocharged low-speed diesel engine connected to the propeller as shown in Figure P3.22. At full load, the diesel engine develops 11 MW of shaft power and consumes 1870 kg_{fuel}/h. About 15%

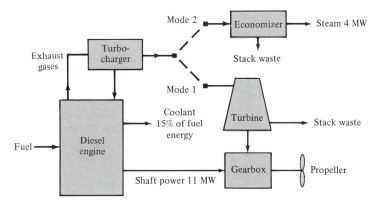

Figure P3.22

of the fuel energy goes to the cooling system. The diesel fuel costs $30/bbl (1 bbl ≈ 158 kg) and transfers into the engine an amount of energy of 11 kWh per kilogram of fuel. The power plant is operated in either of two modes.

When operating in mode 1, a turbine powered by the exhaust gases from the turbocharger provides 1.1 MW of propeller power in addition to the 11 MW provided by the engine shaft. When operating in mode 2, the exhaust gases from the turbocharger are fed into an economizer and produce 4 MW of steam that is used for heating the cargo.

(a) At what rate (in megawatts) is fuel consumed at full load by the tanker in each operating mode?

(b) At what rate is energy discharged through the stack in each mode?

(c) How would you allocate the fuel cost between the steam for heating the cargo and the propeller power when the tanker operates in mode 2?

3.23 A boiling-water nuclear power plant (Figure P3.23) is used to generate electricity and to supply process steam to a district heating system. Depending on demand, the steam from the boiling-water reactor is used either in its entirety to generate electricity (mode 1) or some for electricity and some for low-pressure process steam (mode 2). The energy rates are listed in the figure.

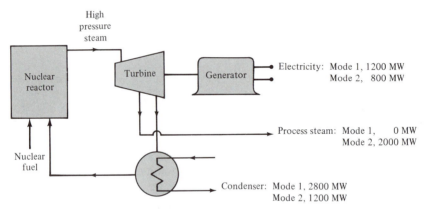

Figure P3.23

(a) At what rate (in megawatts) is nuclear fuel consumed by the nuclear power plant in each mode of operation?

(b) If nuclear fuel costs 0.6 cent/kWh, how would you allocate the fuel cost between the two products—electricity and steam—when the power plant operates in mode 2?

3.24 An existing refinery plant (Figure P3.24a) consumes 45 MW of electricity and 1150 MW of fuel to process crude oil, and produces 150,000 bbl/day of refined oil which is sold at $50/per barrel. The fuel currently used in the plant is the refined oil that the plant itself produces, which provides energy at the rate of 1760 kWh/bbl. Electricity is purchased from the local utility at 15 cents/kWh. As oil and electricity prices increase, alternative ways to meet the energy needs of the refinery are being considered.

One alternative is to produce so-called "medium-Btu gas" (MBG) from coal by means of the Texaco coal-gasification process. As shown schematically in Figure P3.24b, this process can generate 0.3 MW of MBG, 2.3 kW of electricity, and 41.2 kg/day of sulfur per 0.4 MW of coal consumed. The electricity is generated by machinery coupled to high-pressure streams in the gasifier. Coal costs 0.8 cent/kWh and sulfur can be sold at 3 cents/kg.

The gasifier is scaled to produce enough MBG to satisfy all the fuel needs of the refinery. The capital investment for the gassifier is $500 × 10^6, the annual operating cost (excluding coal) is

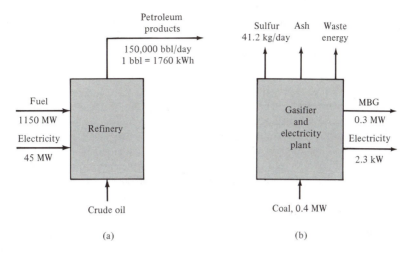

Figure P3.24

25×10^6, and the annual capital charges for everything except fuel and operating costs is 25% of the capital investment.

(a) At which rate is coal consumed by the gasifier? How much electricity must be purchased to satisfy all the electrical needs? What is the rate of energy lost through the sulfur, ash, and waste-energy streams?

(b) Compare the annual total cost when the energy needs are satisfied by the currently existing scheme to the annual total cost when the MBG unit is used.

(c) At which price for refined oil (in dollars per barrel) would the two modes of satisfying the energy needs of the refinery be equally costly?

3.25 A metal processing plant uses natural gas to heat-treat metal parts. The exhaust gases from the heat-treating unit (Figure P3.25) are fed into a waste-heat boiler that raises low-pressure steam. Depending on needs, the steam is used either directly in a metal-parts washing process or in a low-pressure turbine to generate electricity. The rates are listed in the figure. The fuel–air mixture supplies 33 MJ per kilogram of fuel to the heat-treating unit, of which 10% is lost up the stack. The heat-treated parts require 500 kW. The steam raising requires 7.4×10^6 kJ/h, and 10% of the energy of the exhaust gas stream is carried by the stream as it exits the boiler. When used for electricity, the steam generates 200 kW.

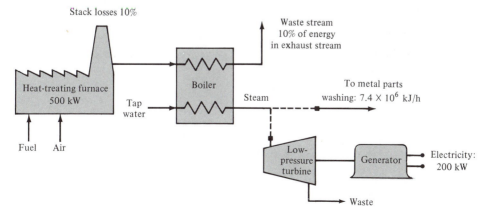

Figure P3.25

(a) At what rate is fuel consumed by the metal processing plant in either of the two modes of operation?

(b) If fuel costs 0.5 cent/kg, how should the manager of the plant allocate the fuel cost to the process steam?

3.26 An important task in everyday life is the pumping of water from underground wells. Among other methods, this can be carried out either by human effort or by a machine. We wish to compare the energy costs of these two methods of pumping 10,000 kg from a depth of 100 m. Assume that a person is fed with eggs. An egg provides 80 kcal of energy and costs about 10 cents. The human body transforms caloric food into muscle power with an efficiency of about 25%. A gasoline engine transforms fuel into a motive power with an efficiency of about 25%. Gasoline has a density of about 750 kg/m^3, can make available about 40,000 kJ/kg, and is priced at about $3.00 per gallon in most places except the United States.

(a) Find the energy cost of each of these two methods of pumping the water.

(b) By considering other foodstuffs and including in your approximate calculations both capital and labor costs, which of the two ways of pumping water would you recommend? Express your considerations and recommendations in a paragraph or so.

3.27 About 2000 years ago, the most advanced societies around the Mediterranean Sea and around the Chinese Sea used various forms of nonanimal energy for their everyday tasks, including the burning of wood for drying and heating, the burning of organic and inorganic oils for lighting, and the capture of winds for propulsion. It has been estimated that during the largest extension of the Roman Empire, the world population was about 300 million, and the energy utilization about 150 million tons of oil equivalent (toe), that is, about 0.5 toe/yr per capita. This figure did not change significantly until the industrial revolution in a few countries in the nineteenth century. For example, the energy utilization per capita in the United Kingdom changed from about 0.5 toe/yr in 1800 to about 2.8 toe/yr in 1900, to 3.3 toe/yr in 1982.

Worldwide, the population has increased from 300 million at the time of Christ, to 5 billion in 1986, and the energy utilization from 150 million toe 2000 years ago to 7.6 billion toe in 1986, that is, from 0.5 to 1.5 toe/yr per capita. Of course, worldwide averages are misleading because the energy utilization per capita varies by more than a factor of 20 between the most and the least industrialized countries. For example, the 1986 per capita values (in toe/yr) for a few countries are: United States, 7.6; Soviet Union, 5; West Germany, 4.5; Japan, 3.1; Italy, 2.7; Portugal, 1.2; China, 0.65; and Central Africa, 0.3.

(a) Comment briefly on the figures just cited and on how you think the energy utilization of a country may be influenced by demographics (increase in population), geography and climate, type of society (agricultural versus industrial, rural versus urban), technological progress, and economic activity (gross national product).

For your country, answer the following questions.

(b) What fraction of energy consumption is transformed into electricity?

(c) Of the energy consumed, what fraction is coal, oil, natural gas, and nuclear?

(d) What have been the trends in your answers in part (c) over the past few decades?

4 Stability of Equilibrium

A system can be in one of a large variety of states. Depending on the type of state, the energy of the system is more or less available to bring about changes in other systems and, therefore, specification of the type of state is of great practical importance. In this regard, a distinct role is played by equilibrium states that are stable.

In this chapter we define different types of states and investigate the existence of stable equilibrium states. The existence and importance of stable equilibrium states emerge from a fundamental principle known as the second law of thermodynamics.

4.1 Types of States

Everyday experience with physical phenomena convinces us that at any instant of time, a system can be in one of a great variety of states. By using different criteria, we may classify the states of a system into different types, such as states having the same momentum or states having the same energy. For present purposes, we use as a criterion the evolution of each state as a function of time, and classify the states of a system into four types: unsteady, steady, nonequilibrium, and equilibrium. We further distinguish three kinds of equilibrium states: unstable, metastable, and stable. In what follows we define each type of state and give some qualitative examples. Each example is more of a suggestive than a rigorous illustration of the type of state in question. We provide quantitative characteristics of the various states after we define some additional properties, such as adiabatic availability, available energy, and entropy.

4.1.1 Unsteady State

An *unsteady state* is one that changes as a function of time because of interactions of the system with other systems. An example of an unsteady state is that of a liquid in a tank being fed from a source and being depleted to a sink at inflow and outflow rates that are unbalanced (Figure 4.1). As time goes on, it is evident that the amount of liquid in the tank changes because of the unbalanced flow rates. So the state of the liquid is unsteady. Another example of an unsteady state is the state of a battery in the process of being charged by an electricity supply (Figure 4.2). As time goes on, we know from experience that the charges on the electrodes of the battery increase until the battery is fully charged.

4.1.2 Steady State

A *steady state* is one that does not change as a function of time despite interactions of the system with other systems in the environment. An example of a steady state is that of a critical nuclear reactor of a nuclear power plant that consumes fissionable material and generates electricity and fission products, all at constant and balanced rates (Figure 4.3). As time goes on, all the characteristics of the plant, such as amounts of constituents, reaction rates, and fluxes, remain invariant. The invariance is achieved by

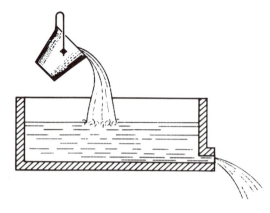

Figure 4.1 Example of an unsteady state of a liquid in a tank with unbalanced inflow and outflow rates.

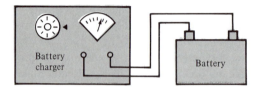

Figure 4.2 Example of an unsteady state of a battery in the process of being charged.

feeding the reactor with fissionable and nonfissionable materials, and extracting from the reactor fission products, irradiated materials, and energy, each at a rate that leaves the state of the reactor intact. Thus the nuclear reactor is in a steady state. If either the supply streams, or exit streams, or both are interrupted, the nuclear reactor is no longer critical and its state changes, and therefore it is no longer steady. So the interactions maintain the steady state.

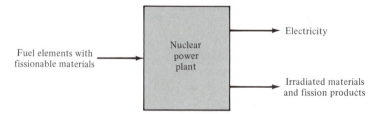

Figure 4.3 Example of a steady state of a nuclear power plant operating with balanced inflow and outflow rates.

Another example is an incompressible liquid in a tank being fed from many sources at constant flow rates and being depleted to many sinks also at constant flow rates, so that the tank remains always full (Figure 4.4). This liquid is in a steady state because all its properties remain the same at all times by virtue of the fixed flows in and out of the tank.

4.1.3 Nonequilibrium State

A *nonequilibrium state* is one that changes spontaneously as a function of time, that is, a state that evolves as time goes on without any effects on or interactions with any systems in the environment. An example of a nonequilibrium state is a gas in an isolated container such that, at some time t_0, the gas fills only part of the container (Figure 4.5). From experience we know that as time goes on after t_0, the gas spontaneously tends to fill all the container. Therefore, the gas here is in a nonequilibrium state, a state that changes with time without affecting the state of the environment. From experience we also know

Figure 4.4 Example of a steady state of a liquid in a tank with balanced inflow and outflow rates.

that by a suitable interaction mechanism—for example, by putting a mobile paddle in front of the expanding gas and connecting the paddle to a weight (Figure 4.6)—we could make the gas lift a weight without actually changing the volume of the container. In more formal language, starting from a nonequilibrium state a system can be made to lift a weight without leaving any other net changes in the state of the environment. We will see later that this is a very important general feature characteristic of nonequilibrium states.

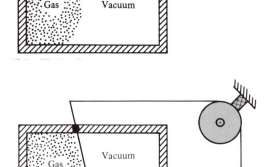

Figure 4.5 Example of a nonequilibrium state of a gas filling only part of a container.

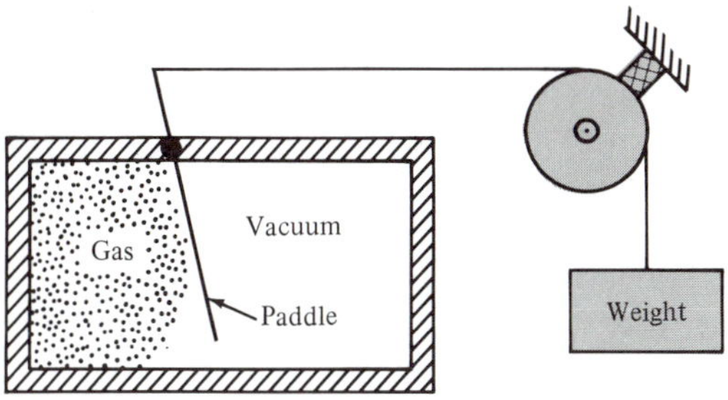

Figure 4.6 Example of a weight process involving nonequilibrium states. A mobile paddle in front of an expanding gas can raise a weight without any other net external effects.

Another example is a burning mixture of hydrocarbon and air in an isolated container. The hydrocarbon is a compound of k carbon and l hydrogen atoms, C_kH_l, and the air a mixture primarily of nitrogen, N_2, and oxygen, O_2, molecules. We know that initially we have fuel and air and after some time we end up with products of combustion. Therefore, while the hydrocarbon and air are reacting, the system is in a nonequilibrium state. Here again, we may use appropriate machinery, such as a fuel oxidation cell, and lift a weight while fuel is burning without changing the volume of the combustion chamber.

A stressed solid is in a nonequilibrium state because as time goes on the stresses relax and tend to become uniform. A composite of a battery-driven motor lifting or lowering an elevator is in a nonequilibrium state because the composite changes its condition as time goes on.

4.1.4 Equilibrium State

An *equilibrium state* is one that does not change as a function of time while the system is isolated—a state that does not change spontaneously. An example of an equilibrium state is that of an isolated room containing air and a 5 kg vessel filled with crude oil, all at ordinary room conditions. As time goes on, we know from experience that the air remains still and the crude oil stays in its vessel. So we say that the room is in an equilibrium state.

This equilibrium state, however, has an important attribute. It can be altered to an entirely different but compatible state without much net effect on any other system in the environment of the room. For example, if we interrupt temporarily the isolation of the room and introduce a spark—a negligible influence by a system in the environment of the room—we initiate a change of the contents of the room from air and crude oil to products of combustion.

In addition, the air–crude oil equilibrium state has another important attribute. Without net changes of its parameters, the system can be made to lift a weight in a gravity field. For example, if the air–crude oil room includes a fuel cell, electricity can be generated. Upon oxidizing the crude oil in the fuel cell, the electricity can be fed to a motor, and the motor can lift a weight in the earth's atmosphere, all without changing the volume of the room.

Not all equilibrium states have the two distinguishing and practical attributes just cited. For this reason, we further classify equilibrium states into three kinds: unstable, metastable, and stable. We will see that each kind of equilibrium behaves differently with respect to possible changes and ability to transfer energy to other systems.

4.1.5 Unstable Equilibrium State

An *unstable equilibrium state* is an equilibrium state that can be caused to proceed spontaneously to a sequence of entirely different states by means of a minute and short-lived interaction that has an infinitesimal temporary effect on the state of the environment. In other words, an unstable equilibrium state may be altered to an entirely different but compatible state by means of minute interactions that have practically no effect on any system in the environment.

For example, the air–crude oil equilibrium state discussed in Section 4.1.4 is unstable because a spark—a negligible influence by a system in the environment—initiates a change of air and crude oil to products of combustion.

Another example of an unstable equilibrium state is that of a supercritical nuclear reactor without any neutrons in it. Such a reactor is in an equilibrium state because no change occurs in the reactor as a function of time. If we introduce one neutron, however, we precipitate many reactions and clearly alter the state of the reactor. We know, of course, that this unstable equilibrium state has the potential of providing a lot of useful energy (or creating a lot of trouble!).

A large quantity of steam connected via a closed valve to a turbogenerator and a cold lake (Figure 4.7) is in an equilibrium state because nothing happens to any part of the system as long as the valve is closed. If we open the valve, however, the state changes and we can extract electricity from the combination steam–turbogenerator–lake. The initial state is an unstable equilibrium state because opening the valve is a minute and short-lived perturbation of the system by its environment.

A finely subdivided steel powder separated by a leakproof membrane from an atmosphere of oxygen (Figure 4.8) is in an equilibrium state. This state is unstable because a hole punched in the membrane — a minor perturbation — initiates an oxidation reaction and changes the state. In addition, energy can be transferred to other systems while the steel oxidizes.

4.1.6 Metastable Equilibrium State

A *metastable equilibrium state* is an equilibrium state that can be altered without leaving net effects in the environment of the system, and without changing the values of amounts

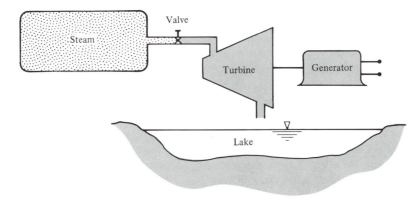

Figure 4.7 Example of an equilibrium state of a large quantity of steam connected via a closed valve to a turbogenerator and a lake.

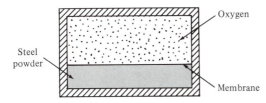

Figure 4.8 Example of an equilibrium state of a finely subdivided steel powder separated by a leak-proof membrane from an atmosphere of oxygen.

of constituents and parameters to an incompatible set of values, but this can be done only by means of interactions that have a finite temporary effect on the state of the environment. In other words, a metastable equilibrium state may be changed without net effects in the environment to an entirely different but compatible state only at the expense of a finite temporary change of state of the environment which, however, can be annulled despite the fact that the system does not return to its initial state. So at the end, the metastable equilibrium state has been changed to a different but compatible state with no net effects in the environment. It is clear that a metastable equilibrium state differs from an unstable equilibrium state because the latter may be altered by a minute perturbation, whereas the former may not.

Like unstable equilibrium states, metastable equilibrium states have the practical attribute that, from a system in such a state, energy may be supplied to a weight in a weight process without net changes in the values of the parameters of the system. An example of a metastable equilibrium state is that of an isolated room containing air, and a hermetically sealed vessel filled with natural gas, all at ordinary room conditions. As time goes on, the air remains still and the natural gas stays in the vessel. However, this state can be changed to an entirely different state as follows. We break the isolation of the room temporarily, put effort into breaking the seal of the vessel, and light a spark, that is, take actions that require a finite change in the state of the environment. Next, while the natural gas is burning, we use some of its ability to perform useful tasks to restore the environment to its initial state, and reestablish the isolation. Thus the system proceeds from its initial state to entirely different states without any net effects on the environment.

Another example is a collection of heavy hydrogen isotopes (deuterium and tritium) at low energy. At that energy the isotopes do not fuse. If, however, energy is supplied by the environment in excess of the threshold for thermonuclear reactions, the isotopes fuse and make a lot of fusion energy available. As a result, the energy supplied by the environment can be fully recovered, and the isotopes transformed into helium molecules and neutrons, that is, an entirely different state, without net effects on the environment.

Two further examples of metastable equilibrium states are those reached by liquids and vapors used in bubble and cloud chambers, respectively, for the detection of ionizing radiation. In a bubble chamber, the pressure on a hot liquid is rapidly decreased to a value at which the liquid normally begins to boil. However, the liquid does not boil until an external disturbance provides a stimulus. Such a disturbance is provided by a fast particle which, upon crossing the chamber, creates ion pairs, and these in turn become nucleation centers for bubble formation. Similarly, in a vapor chamber, the pressure is suddenly decreased to a value at which the vapor normally begins to condense. However, the vapor does not condense until a finite external disturbance triggers the condensation. Energetic radiation can provide the trigger. We discuss boiling and condensation in due course.

A final example of a metastable equilibrium state consists of a mixture of different constituents that can react with each other according to a given chemical reaction mechanism, yet coexist in equilibrium as if the chemical reaction were "frozen." For example, in the exhaust gases of an automobile engine, the amounts of carbon monoxide, CO, carbon dioxide, CO_2, and oxygen, O_2, do not correspond to the most stable proportions that could be reached by fully exploiting the recombination reaction mechanism $CO + \frac{1}{2}O_2 = CO_2$. Such a metastable state occurs because the exhaust gases are cooled too rapidly for the recombination reaction to take place effectively and, once they are cold, the reaction rate becomes practically zero. To see that the equilibrium state is metastable, we can warm the gases again, let them cool more slowly, restore the state of the environment, and observe that now the exhaust gases contain practically no CO.

4.1.7 Stable Equilibrium State

A *stable equilibrium state* is an equilibrium state that can be altered to a different state only by interactions that either leave net effects in the environment of the system, or change the values of amounts of constituents and parameters to an incompatible set of values. In other words, a stable equilibrium state cannot be changed to a different but compatible state without leaving a net change in the state of the environment of the system. Clearly, a stable equilibrium state is neither unstable nor metastable.

An example of a stable equilibrium state is that reached by the gas in Figure 4.5 after a long time. From experience we know that eventually the gas fills the entire container and from then on remains still. For zero net change of the volume of the gas, we also know from experience that we cannot change the final state of the gas without some net effect on its environment.

Another example is the final state that is reached by a burning mixture of a hydrocarbon and air in an isolated container. This state is stable equilibrium because, once it is reached, we cannot transfer energy to a weight from the combustion container without either changing its volume or leaving additional effects—additional to lifting the weight—on the environment.

The state that is reached by the system in Figure 4.7 some time after the valve is opened and electricity has been generated is a stable equilibrium state as evidenced by the experience that the electricity generation ceases (no more energy can be extracted from the system), the steam condenses into the cold lake, and no more changes occur.

The state that is reached some time after the membrane is punched in the system in Figure 4.8 is a stable equilibrium state, as indicated by the experience that by then all chemical reactions are completed and no energy can be extracted from the system while the parameters are kept fixed and no other effects (other than energy extraction) are allowed.

4.2 Reversible and Irreversible Processes

In this section we introduce the concept of a reversible process because we need it in the statement of the second law of thermodynamics. In Section 2.3.2 we define a process in terms of two end states of a system and the modes of interaction between the system and its environment during the change between the two end states. In general, during a process not only the state of the system but also that of the environment is affected.

Among other distinctions, a process may be either reversible or irreversible. A process is *reversible* if it can be performed in at least one way such that both the system and its environment can be restored to their respective initial states. A process is *irreversible* if it is impossible to perform it in such a way that both the system and its environment can be restored to their respective initial states.

As an example, we consider a system A, its environment B, and a process α_{12} as a result of which A changes from state A_1 to state A_2, and B from state B_1 to state B_2. If upon bringing the state of A from A_2 back to A_1 by any means whatsoever, we find that the state of the environment is restored from B_2 back to B_1, namely, that all changes in both system A and its environment B have been annulled, we say that process α_{12} is reversible. On the other hand, if we find that by no conceivable way can we restore A from state A_2 to state A_1 and, at the same time, the environment from state B_2 to state B_1, we conclude that process α_{12} is irreversible.

It is noteworthy that the definition of a reversible process does not impose any restrictions on the sequence of states and the modes of interaction that may be used in restoring the initial states of the system and its environment. All sequences and all interactions are admissible, including the special sequences corresponding to a reversal of all momenta of the particles of the system, provided that they lead to the restoration of the initial states.

If stable equilibrium states exist, an example of an irreversible process is a spontaneous change of state, that is, a change that causes no effects in the state of the environment, of a system starting from an initial nonequilibrium state and ending in a stable equilibrium state. Indeed, such a process cannot be reversible. For if it were, it would be possible for the system to undergo a process that starts from the stable equilibrium state and ends at the initial nonequilibrium state with no net effect on the environment, thus contradicting the definition of a stable equilibrium state.

A practical illustration of an irreversible process is the discharge of an isolated battery through its internal resistance. We will see that the battery cannot be recharged without effects on other systems in its environment. Another is the burning of a fuel inside an oxygen-containing isolated vessel. Again, we will see that the products of combustion cannot be reformed into fuel and oxygen without net effects on the environment of the vessel.

It is noteworthy that whether a process is reversible or irreversible cannot be decided solely by means of energy accounting. This point is illustrated in the following example.

Example 4.1. As a result of a process α_{12}, the state of a system A changes from A_1 to A_2 and that of the environment from B_1 to B_2. Upon restoring the state of A from A_2 back to A_1, what is the change in energy of the environment if: (**a**) the process α_{12} is irreversible; and (**b**) the process α_{12} is reversible?

Solution. (**a**) The value of the energy of the environment returns to that corresponding to state B_1 because the energy of A returns to its initial value and the overall energy of A and the environment is conserved, that is, the sum of the energy of A and the energy of the environment is a constant independent of time.

(b) The same conclusion applies when the process is reversible.

Example 4.2.   If a weight process from state A_1 to state A_2 is reversible, show that a weight process from state A_2 to state A_1 is both possible and reversible. Each of these two weight processes is said to be the reverse of the other.

Solution.   Because the process from A_1 to A_2 is reversible, both A and the environment can be restored to their initial states. The effects of any restoring process are the change of state of A from A_2 back to A_1, and the return of the weight back to its initial elevation because energy is conserved and, initially, the only effect external to A is the change in elevation of the weight. The return of the weight to its initial elevation can be regarded as a weight process from A_2 to A_1. Moreover, it is reversible because all its effects are annulled by the initial weight process without any other effects on the environment.

In applications that result in a given change of state of a system, we will see that a reversible process is the best that can be achieved, and that an irreversible process involves imperfections and penalties that either reduce the capacity of the system to perform a useful task or increase the effort required to accomplish a desired change of state. We will develop means to quantify the degree of irreversibility of a process in terms of a property called entropy. Moreover, we will see that the more irreversible the process, the larger its departure from perfection, and that each reversible process provides a limit of optimality beyond which no improvement is possible.

4.3   The Problem of Stability

Because stable equilibrium states are encountered very often in practice and because a system in a stable equilibrium state is not as capable to lift a weight in a weight process as when it is in another type of state, we examine stable equilibrium states in greater detail. For given values of the energy, the amounts of constituents, and the parameters, a system can be in any of a large variety of different states. These states differ by the values of some of the many properties other than energy. We may raise the question: Among all the states of a given system that have the same values of the energy, the amounts of constituents, and the parameters, are there any that are stable equilibrium?

Formally, both the classification of states into different types on the basis of changes with respect to time, and the answer to the question just raised require consideration of the equation of motion. For example, we solve the equation of motion of the system and find the trajectory that passes through a given state to see whether the state changes with time spontaneously or as a result of interactions. Again, we find the equilibrium states that correspond to the equation of motion of the system, and then examine their stability by means of techniques that have been specially developed for this purpose.

The problem of stability of equilibrium is not unique to thermodynamics. It arises in analyses of physical phenomena based on the principles of mechanics as well. It was first introduced by the French mathematician and astronomer Joseph Louis Lagrange (1736–1813) in the latter part of the eighteenth century, and then more carefully discussed and defined in the pioneering studies of the French mathematician Jules Henri Poincaré (1854–1912), and the Russian mathematician Aleksandr Mikhaylovich Lyapunov (1857–1918).

Within mechanics, the stability analysis yields that "among all the allowed states of a system with fixed values of amounts of constituents, and parameters, the only stable equilibrium state is that of lowest energy, that is, the ground state." An illustration of the problem of stability of equilibrium in mechanics is provided by the system shown

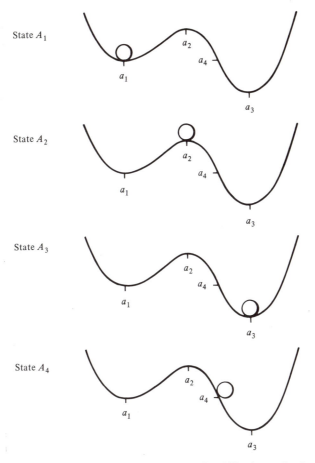

Figure 4.9 Illustration of the problem of stability in mechanics.

in Figure 4.9. The system consists of a structureless mass m—an amount of matter approximated by a point in space—that can move within a stationary bowl having one peak and two valleys, all at different elevations in a gravity field. If at zero speed the mass is positioned at point a_1, a_2, or a_3, it stays there under the influences of gravity and the equal and opposite reaction force provided by the bowl. Thus if we define states A_1, A_2, and A_3 as those that correspond to zero speed at points a_1, a_2, and a_3, respectively, each of these states is an equilibrium state. Of course, if the speed of the mass at a_1, a_2, or a_3 is different from zero, the corresponding state is nonequilibrium despite the zero force and zero acceleration. Each of the states A_1, A_2, and A_3 can be assigned an energy equal to the potential energy of the corresponding point in the gravity field. In particular, state A_3 has the least energy of all the states of the system.

Although all three states are equilibrium, A_1 is metastable, A_2 is unstable, and only A_3 is stable. Indeed, by breaking the isolation of the bowl and connecting the system to a weight, we can change the state in a weight process from A_1 to A_2, namely, lift mass m from point a_1 to point a_2 with zero change in speed. Then, by letting the weight return to its original elevation, we can bring mass m to state A_4, namely, point a_4, which is at the same elevation in the gravity field as point a_1 (Figure 4.9), and at which m has zero speed. When the system is in state A_4, we reestablish the isolation of the bowl. State A_4 is nonequilibrium and will never return to state A_1 because the force of gravity and the reaction force by the wall of the bowl are not balanced, except at point a_3, and at that point the speed of the mass is nonzero. Thus with finite energy exchanges with the environment that add up to zero—null net change in elevation of the external weight—and without net changes of the parameter of the system—shape of the bowl—we have succeeded in altering equilibrium state A_1 to nonequilibrium state A_4. It follows that A_1 is a metastable equilibrium state. We may readily verify that A_2 is an unstable equilibrium state because a negligible perturbation of that state results in the mass oscillating within the one or the other valley of the bowl, that is, in states entirely different from A_2. In contrast to both A_1 and A_2, state A_3 is a stable equilibrium state because it can be altered only at the expense of effects on other systems in the environment. For example, only by lowering a weight can we alter A_3 in a weight process that ends in some other state.

These results represent a general conclusion of the theory of mechanics to the effect that the state of lowest energy is the only stable equilibrium state. They may be formally confirmed by means of Newton's equation of motion.

In addition to providing a simple illustration of the concepts of unstable, metastable and stable equilibrium states, the example just cited points out that the principles of mechanics are not general enough to answer the question that we raised earlier. Indeed, we wish to investigate the existence of stable equilibrium states of a system for each given value of the energy, whereas in mechanics a stable equilibrium state is achieved only when the energy has its least value, that is, when the system is in a ground state.

The suggestive experiences cited in Sections 4.1.4 and 4.1.7 show that stable equilibrium states other than ground states are conceivable in practice. Therefore, we need to broaden our conceptions of states and their motions. As we stated in Chapter 2, the complete and general equation of motion that encompasses all experiences remains a subject of research at the frontiers of science. However, a fundamental principle—the second law of themodynamics—has been established which answers the question of stability of equilibrium without an explicit statement of the equation of motion. We discuss this law in the next section.

4.4   Second Law

The second law of thermodynamics is a statement of existence of stable equilibrium states. More than the first, the second law distinguishes thermodynamics from mechanics and other branches of physics. Because of its far-reaching implications concerning the properties and behavior of matter, it is frequently invoked in discussions of philosophy as well as physics.

Of the many statements of the second law, those of the German mathematical physicist Rudolf Julius Emanuel Clausius (1822–1888), the English physicist William Thomson, Lord Kelvin (1824–1907), the German physicist Max Karl Ernst Planck (1858–1947), and the Greek mathematician Constantin Carathéodory (1873–1950) are the most notable. They are given at the end of this chapter. As pointed out by George Nicholas Hatsopoulos (1926–) and Joseph Henry Keenan (1900–1977) in their pioneering textbook *Principles of General Thermodynamics* (Wiley, New York, 1965), all these statements imply the existence of a stable equilibrium state for each set of values of energy, amounts of constituents, and parameters, an implication that we take, along with Hatsopoulos and Keenan, to be the essential element of the second law. In terms of definitions and concepts already cited, we state the second law as follows.

Among all the states of a system that have a given value E of the energy and are compatible with a given set of values n of the amounts of constituents and β of the parameters, there exists one and only one stable equilibrium state. More-over, starting from any state of a system it is always possible to reach a stable equilibrium state with arbitrarily specified values of amounts of constituents and parameters by means of a reversible weight process.[1]

From here on and until Chapter 29 we restrict our discussion to systems with constraints that disallow any internal mechanism capable of changing the values of the amounts of constituents and the parameters. So, the only states that are compatible with given values n and β are the states that have the given values.[2] Accordingly, the first part of the second law can be restated in the following simpler but equivalent manner.

Among all the states of a system with a given value of the energy, and given values of the amounts of constituents and the parameters, there exists one and only one stable equilibrium state.

As already indicated, the existence of stable equilibrium states for the conditions specified and, therefore, the second law cannot be derived from the laws of mechanics. When applied to the least value of the energy for given values of the amounts of constituents and the parameters, we will see that the second law implies the existence of the stable equilibrium state of mechanics. But, in general, it asserts that a stable equilibrium state exists for each value of the energy. The many stable equilibrium states, as well as a large variety of other equilibrium and nonequilibrium states contemplated in thermodynamics, are not contemplated in mechanics.

In parallel with statements made earlier about the first law, here we note that there is no exhaustible number of experiments which would allow us to prove that a system with fixed values of energy, amounts of constituents, and parameters has a stable equilibrium state. Accordingly, the statement of existence (and uniqueness) of stable equilibrium states is the subject of a fundamental principle.

The existence of stable equilibrium states for various conditions of matter has many theoretical and practical consequences. It is a major augmentation of the principles of mechanics and is essential to understanding and explaining many natural phenonena.

In mechanics, it is relatively easy to provide a simple picture illustrating the idea of stability of equilibrium. One such picture is the schematic in Figure 4.9. When we add the second law to the other principles of physics, however, the picture of stable equilibrium is beclouded and not as readily understandable. It can no longer be one in which the particle of a one-particle system, according to classical mechanics, is at the specific position and has the specific velocity corresponding to the lowest value of the energy or, according to quantum mechanics, is described by the lowest-energy wave function. Instead, it must be regarded as something that is dispersed throughout all the space made available by the nest of forces which specify the enclosure of the system, and that is described by a special class of probability distribution functions of various measurement results. These probability distribution functions are inherent to our

---

[1]For each given set of values of amounts of constituents and parameters, in Section 9.8 we will see that the second part of the statement of the second law implies the existence of only one ground-energy state. In Section 13.17 we introduce a slight modification of the second part, and thus account for the yet-unconfirmed existence of systems that admit more than a single ground-energy state for each given set of values of amounts of constituents and parameters.

[2]All the results that we derive under this restriction can be readily generalized to less constrained systems, such as systems with chemical reaction mechanisms, by simply substituting each condition that two states have the same values of amounts of constituents and paramaters with the condition that the two states be compatible.

current understanding of the nature of matter, namely, they are of quantum theoretical origin. As such, they have nothing to do with statistics related to either ignorance, lack of information, or the inability to make accurate and precise calculations or measurements.

Because we are not using the mathematical language that is necessary to represent such probabilities, we will study the meaning of energy and stability of equilibrium indirectly by exploring a great number of implications and applications of the two laws of thermodynamics that we have stated. The next section and subsequent chapters are devoted to discussions of these implications and applications.

4.5   Impossibility of Perpetual-Motion Machines of the Second Kind

A *perpetual-motion machine of the second kind* (PMM2) is any system B undergoing a cyclic process that produces no external effects except the rise of a weight in a gravity field, and the change of another system A from an initial stable equilibrium state A_0 to a final state A_1 corresponding to the same values of amounts of constituents and parameters as state A_0. Clearly, the rise of the weight involves a decrease in the energy of system A, because the process for B is cyclic. Accordingly, an alternative definition of a PMM2 is in terms of a special weight process in which a weight rises as a system changes from an initial stable equilibrium state A_0 to a final state A_1 having less energy than but the same values of amounts of constituents and parameters as state A_0.

No matter how complicated are the devices within system B, and no matter how intricate the interconnections between them, a PMM2 cannot exist, that is, it is impossible to achieve. To prove this assertion, we assume the contrary and show that our assumption leads to a contradiction. Specifically, we assume that there exists a weight process of a system A initially in a stable equilibrium state A_0 which brings the system to a lower-energy state A_1 with the same values of amounts of constituents and parameters as state A_0. Then, starting from state A_1, we again connect the system to the weight, and let the weight fall to its initial low elevation in a weight process that causes system A to end in a nonequilibrium state A_2, for example, by setting the entire system into bodily motion. Being nonequilibrium, state A_2 cannot be identical to stable equilibrium state A_0. Thus the initial stable equilibrium state A_0 is changed into a different state A_2, yet this result obtains without net effects on the environment since the effects on the weight are annulled. Such a result is absurd because it contradicts the definition of a stable equilibrium state and therefore we conclude that a PMM2 is impossible.

From the argument just given, we see that the impossibility of a PMM2 follows from the definition of a stable equilibrium state. As such, it is intimately related to the second law, which is a statement of existence of stable equilibrium states.

In the discussion of the first law, we indicate that a weight process corresponding to two states A_0 and A_1 of system A is possible in the direction either from state A_0 to A_1 or from A_1 to A_0 but not necessarily in both directions. The impossibility of a PMM2 is one example of the restriction on the direction of the weight process when A_0 is a stable equilibrium state. Later, we find other examples of the same type of restriction.

In terms of the notation introduced in Section 3.8, the impossibility of a PMM2 can be stated as follows. Given a system A, a stable equilibrium state A_0, and any other state A_1 with the same values of the amounts of constituents and parameters as A_0, there can be no weight process for A that starts from state A_0 and ends in a state A_1 of lower energy, that is, $W_{01}^{A\rightarrow}$ and $W_{01}^{A\leftarrow}$ ($W_{01}^{A\leftarrow} = -W_{01}^{A\rightarrow}$) must satisfy the relations

$$W_{01}^{A\rightarrow} \leqq 0 \tag{4.1}$$

and

$$W_{01}^{A\leftarrow} \geqq 0 \tag{4.2}$$

Example 4.3.   Prove that if A_0 is a stable equilibrium state, the strict equality in relations (4.1) and (4.2) holds only if state A_1 coincides with state A_0.

Solution.   If $W_{01}^{A\rightarrow} = 0$, state A_1 cannot be different from A_0 because the weight process from A_0 to A_1 leaves no net effects external to A, and a stable equilibrium state, by definition, cannot be changed to a different state without net effects external to the system.

The proof that a PMM2 cannot be built is predicated on the assertion that starting from any state, the energy of a system can be increased indefinitely by means of a weight process which results in altering the initial state to some final state that is not a stable equilibrium state without altering the values of amounts of constituents and parameters. Any system with particles that can move around in space, that is, particles with translational degrees of freedom, satisfies this requirement. There exists, however, a special class of systems that do not conform to this requirement, such as nuclear-spin systems. For these special systems we establish a different impossibility result in Section 13.18. Unless we specify otherwise, in what follows we restrict our treatment to systems whose energy can be increased indefinitely. These constitute the vast majority of systems encountered in physics and engineering, and for them the statement of impossibility of a PMM2 is expressed by either (4.1) or (4.2). The impossibility of perpetual-motion machines of the second kind is the basis for the proof of existence of two important properties, adiabatic availability and available energy. These properties are discussed in the next two chapters.

4.6   Historical Statements of the Second Law

The essence of the second law of thermodynamics was already contained in the 1824 pioneering study by the French physicist Nicolas Léonard Sadi Carnot (1796–1832) entitled *Réflexions sur la puissance motrice du feu* ("Reflections on the motive power of fire"), the study that gave birth to the science of thermodynamics. The important practical questions addressed and answered by Carnot in his study are discussed in Chapter 6.

The first explicit and general statement of the second law was presented in 1850 by Clausius. Using our terminology, we can express Clausius's statement as: "No process is possible in which the sole net effect is the transfer of energy from a system in a stable equilibrium state with a lower temperature to a system in a stable equilibrium state with a higher temperature," where the term temperature is defined in Chapter 9. In Section 9.6 we show that the Clausius statement follows as a special consequence of the general statements of the laws of thermodynamics that we adopt here.

Another notable statement of the second law was presented in 1897 by Planck. It is similar to one of several statements of the second law made by Kelvin. Using our terminology, we can express the Kelvin–Planck statement as: "It is impossible to construct an engine that will work in a complete cycle and produce no effect except the raising of a weight and the transfer of energy out of a system in a stable equilibrium state." The Kelvin–Planck statement is equivalent to the statement of impossibility of a perpetual-motion machine of the second kind (PMM2). As shown in the preceding section, for systems for which the energy can be increased without limit, this impossibility is one of the consequences of the laws of thermodynamics that we adopt here.

Yet another notable statement of the second law was presented in 1909 by Carathéodory. It may be expressed as follows: "In the neighborood of any given stable equilibrium state there exist stable equilibrium states that cannot be reached by any weight process that starts from the given state." In Chapter 8 (Example 8.4) we show that the Carathéodory statement follows from the statements of the laws of thermodynamics adopted here.

4.7 Comment

Upon considering a system consisting of a particle that can move within a stationary bowl with many valleys, each at the same elevation in a gravity field (Figure 4.10), we may be tempted to conclude that the system admits many stable equilibrium states corresponding to the least value of energy and therefore violates the second law. However, such a conclusion is erroneous. Each zero-speed state corresponding to each of the locations $a_1, a_2, \ldots, a_q$ of lowest elevation is a metastable state because it can be altered to any of the others without net effects on the environment.

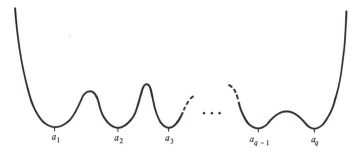

Figure 4.10 Schematic of a system consisting of one particle in a bowl with many valleys, each at the same elevation in a gravity field. None of the zero-speed states at each of the locations $a_1, a_2, \ldots, a_q$ corresponds to a stable equilibrium state. In a stable equilibrium state, a particle is dispersed throughout the available spatial extension, with a definite distribution of position probabilities.

To have a correct picture of the unique stable equilibrium state at any energy, we need the tools of quantum theory. Using such tools, we find that the stable equilibrium state of least energy is such that the particle is dispersed throughout the spatial extension of the bowl. More specifically, we find that each position measurement yields a value of the coordinate of the particle that is within the extension of the bowl, and that each measured value is associated with a definite probability of occurrence. In general, to each state there corresponds a definite distribution of such position probabilities and, in particular, to the least-energy stable equilibrium state a unique distribution of position probabilities. The latter result is fully consistent with the second law.

Problems

4.1 A system consists of a structureless particle in a fixed-volume container and has only two independent properties, P_1 and P_2. From the descriptions given below, indicate the type of state of the system.

(a) The value of properties P_1 and P_2 are not changing in time, that is, $dP_1/dt = 0$ and $dP_2/dt = 0$, and the system is experiencing interactions with its environment.

(b) The container is connected via two independent mechanisms to two identical weights, one rising at a certain speed and the other falling at the same speed. The values of properties P_1 and P_2 are changing with time.

(c) Same as in part (a). In addition, we know that it is possible to change the value of P_1 without leaving net external effects.

(d) Same as in part (a). In addition, we know that the values of properties P_1 and P_2 cannot be changed without leaving net external effects.

(e) Same as in part (b) except that the values of P_1 and P_2 are not changing with time.

(f) Same as in part (a). In addition, we know that the values of properties P_1 and P_2 can be altered without net external effects.

4.2 A system consists of a single spin experiencing a uniform external magnetic field H and no other external forces. It has only three independent properties: the magnetic moments M_x, M_y, and M_z along the x, y, and z Cartesian directions, respectively. From the descriptions given below, indicate the type of state of the spin system.

(a) The external magnetic field H is time invariant, and each of the three magnetic moments is time invariant.

(b) Same as in part (a). Moreover, we know that the magnetic moment M_x can be brought to a different value without producing any net external effect.

(c) Same as in part (a), except that M_x varies with time.

(d) The external magnetic field H is varied with time in such a way that the three magnetic moments are maintained time-invariant.

(e) Same as in part (d). Moreover, we know that the magnetic moments cannot be varied without producing some net external effect.

(f) Same as in part (d), except that M_z varies with time.

4.3 A system consists of an isolated fully charged lead storage battery. Is the system in a stable equilibrium state? If not, when is the battery in a stable equilibrium state?

4.4 An apple in good condition is placed in a room filled with ordinary air. Without much effort the apple may be sliced into two pieces.

(a) What is the type of state of the entire apple and air system at some time during the first day the apple is placed in the room and before slicing it?

(b) If the apple is sliced into two pieces, what is the type of state at some time during the next day?

(c) In view of your answer to part (b), can you improve your answer to part (a)?

4.5 A system consists of a battery in a fixed-volume container. The battery has a very large but finite internal electrical resistance and, as a result, slowly but spontaneously discharges with time. Assume that the battery has only two independent properties: energy E and charge C. Typically, the charge is measured in ampere-hours. From the descriptions below, indicate the type of state of the system. Explain your answers.

(a) The container walls are impermeable to energy. The battery is charged and its electrodes are not connected to any external device. Consider both short and long periods of time.

(b) Same as in part (a), except that the battery is now fully discharged.

(c) Same as in part (a), except that the electrodes are now connected to a battery charger that maintains the charge at a constant value.

(d) Same as in part (c), except that the battery also interacts with a system that maintains the battery energy at a constant value.

4.6 A system consists of carbon, C, oxygen, O_2, and carbon dioxide, CO_2 confined in a fixed-volume, airtight, perfectly isolated container. For each of the descriptions below, indicate the type of state of the system.

(a) Measurements are made from time to time, and it is found that the amount of free oxygen is continuously decreasing.

(b) Measurements are made from time to time, and it is found that the amounts of C, O_2, and CO_2 are invariant. Moreover, it is found that upon igniting a spark, these amounts are affected.

(c) Measurements are made from time to time, and it is found that the amounts of C, O_2, and CO_2 are invariant. Moreover, despite many efforts, including igniting a spark, these amounts are not affected.

(d) Measurements of the energy of the system made from time to time show that the energy is time invariant.

4.7 (a) If the properties of a system do not change with time, is it possible to prove that its state is stable equilibrium?

(b) A system consists of two equal weights attached to each end of a weightless cord which passes over a frictionless pulley. The weights hang from the pulley in the earth's gravitational field. Is the system in an equilibrium state? If so, what kind of equilibrium? If the cord is not weightless, what are the possible types of states of the system?

4.8 A system undergoes a cyclic process while interacting exclusively with a system in a stable equilibrium state, and produces the external effects listed below. Explain whether any of these effects violates the second law.

(a) No external effect except the rise of a weight.

(b) No external effect except the fall of a weight.

(c) No external effect except a voltage increase across an initially discharged capacitor.

(d) No external effect except a voltage drop across an initially charged capacitor.

4.9 A large amount of water (10^{11} kg) is constrained to be in a water pond at a height of 500 m above sea level. At some location, a valve connects the pond to the sea through a long duct and a water turbine.

(a) How would you define a system to study the hydropower capabilities of this water?

(b) All the water in the pond and the valve is closed. What is the type of state of the system you defined in part (a)?

(c) The valve is open and water is rushing down the duct, through the turbine, and into the sea. What is the type of state under these conditions?

(d) The pond is empty and all the water has reached the sea. What is the type of state now?

4.10 The Wilson "cloud chamber" is used to detect and visualize trajectories of charged particles or x-rays. Water vapor mixed with air and other gases is first subjected to a rapid, undisturbed expansion to a volume about 25% greater than the volume at which condensation of vapor would occur. Initiation of condensation requires a small but finite disturbance such as a nucleation center. When the rapid expansion occurs in the absence of disturbances, condensation does not occur.

After the expansion, the state of the system does not change with time until some disturbance promotes the condensation. A charged particle crossing the chamber with sufficient energy causes ionization of the water vapor along its trajectory. The disturbance promotes condensation along the trajectory, that is, the formation of liquid droplets around the ionized molecules. These liquid droplets provide an observable trace of the trajectory of the charged particle crossing the chamber. If the charged particle collides with other particles while in the Wilson chamber, the trajectory bends and so the trace of liquid droplets shows the mechanics of the collisions.

Due to the surface tension between liquid and vapor, there is a critical droplet size — of the order of a few water molecules — below which a droplet collapses and above which it grows indefinitely. Thus if all the droplets happen to have formed with size below the critical, they will all collapse and the chamber return to its initial state.

(a) What kind of equilibrium state is that of the Wilson chamber just after the rapid undisturbed expansion?

(b) What kind of equilibrium state is that of the Wilson chamber with only one droplet of exactly the critical size?

5   Adiabatic Availability

In this chapter we discuss an important and practical property, adiabatic availability. We will see that adiabatic availability is the conceptual underpinning of another practical property called entropy.

5.1   A Class of Weight Processes

Given a system A in a state A_1 with amounts of constituents $(n)_1$ and parameters $(\beta)_1$, and machinery of any desirable degree of perfection (Figure 5.1), we wish to consider the class of weight processes for system A that change its state from A_1 to any other state A_2 with values of amounts of constituents $(n)_2 = (n)_1$, and parameters $(\beta)_2 = (\beta)_1$. In particular, we wish to investigate the question: In a weight process, how much energy

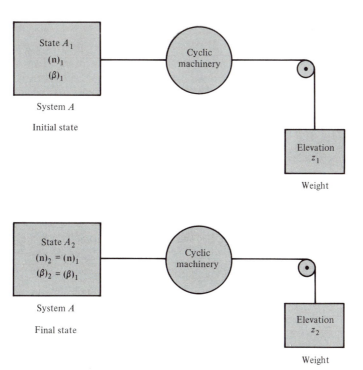

Figure 5.1   Schematic representation of the end states of a system A considered in weight processes used to investigate the question: In a weight process, how much energy can be transferred from the system to the weight without net changes in the values of amounts of constituents, and parameters of the system?

can be transferred from the system to the weight without net changes in the values of the amounts of constituents and the parameters of the system? For example, given the charged storage battery in Figure 5.2, how much electricity can be fed from the battery to a perfect motor which in turn lifts a weight from a low to a high elevation in a gravity field?

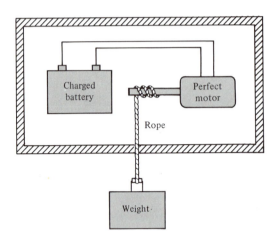

Figure 5.2 Schematic of a battery lifting a weight, as an example related to the investigation of the question: How much electricity can be fed from a charged battery to a perfect motor which in turn lifts a weight from a low to a high elevation in the gravity field?

Under these conditions it turns out that the largest amount of energy that can be transferred out of the system is: (1) limited and equal to the energy transferred in a reversible weight process in which the system starts from the initial state A_1 and ends in a stable equilibrium state A_0 with the same values of the amounts of constituents, and the parameters as state A_1; and (2) the same for all reversible weight processes that involve no net changes in the values of amounts of constituents, and parameters of A, and that begin from the given state A_1 and end in a stable equilibrium state. It also turns out that the final stable equilibrium state is the same for all these reversible processes.

The initial state A_1 that we are considering here can be of any type—unsteady, steady, nonequilibrium, or any kind of equilibrium, including stable equilibrium. Also, the system can be large or small—a large amount of matter or just one particle, confined in a small or a large box. So the assertions just made, which we are about to confirm, are of great generality.

To verify our assertions, we proceed as follows. We consider three different weight processes α, γ, and γ' of system A, all starting from the same initial state A_1, and each ending in a state of system A with the same values of both the amounts of constituents and the parameters as those of state A_1. The processes are sketched in Figure 5.3.

In process α, which may be reversible or irreversible, the system starts from initial state A_1 and ends in a final state A_2. We denote the energy exchanged between A and the weight by the symbol $W_{12}^{A\rightarrow}$, where according to the notation introduced in Section 3.9 for weight processes, the subscript 12 denotes that system A changes from state A_1 to state A_2, and the superscript $A\rightarrow$ that a positive value of $W_{12}^{A\rightarrow}$ represents an energy transfer out of system A. For this process, the energy balance yields $E_2 - E_1 = -W_{12}^{A\rightarrow}$, where E represents the energy of system A [equation (3.13)].

In process γ, which is reversible, the system starts from initial state A_1 but ends in a stable equilibrium state A_0. The energy exchanged is denoted by $(W_{10}^{A\rightarrow})_{rev}$, where the subscript rev stands for reversible and the other notations have the meanings just cited. In process γ', which is reversible, the system starts from initial state A_1 but ends

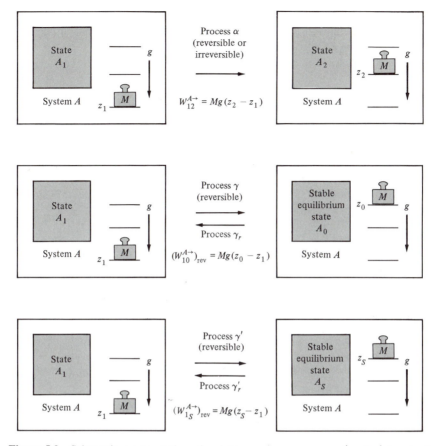

Figure 5.3 Schematic representation of weight processes α, γ, γ_r, γ' and γ'_r (see text).

in a stable equilibrium state A_S. The energy exchanged during this reversible weight process is denoted by $(W_{1S}^{A\rightarrow})_{\text{rev}}$.

We also consider two weight processes γ_r and γ'_r that are the reverse of processes γ and γ', respectively. Each of these two processes exists by virtue of the result we prove in Section 4.2 (Example 4.2). In particular, in process γ_r the system starts from A_0 and ends in A_1. So the energy exchanged, $(W_{01.}^{A\rightarrow})_{\text{rev}}$, satisfies the relation

$$(W_{01.}^{A\rightarrow})_{\text{rev}} = -(W_{10}^{A\rightarrow})_{\text{rev}} \tag{5.1}$$

because process γ_r is reversible and annuls all the effects of process γ and therefore the energy exchanged in process γ_r is equal and opposite to that in process γ. Similarly, in process γ'_r the system starts from A_S and ends in A_1. So the energy exchanged, $(W_{S1}^{A\rightarrow})_{\text{rev}}$, satisfies the relation

$$(W_{S1}^{A\rightarrow})_{\text{rev}} = -(W_{1S}^{A\rightarrow})_{\text{rev}} \tag{5.2}$$

because process γ'_r is reversible and annuls all the effects of process γ'.

Next, we consider a combined process that is a sequence of processes γ_r and α. In words, we consider a weight process in which the system goes first from state A_0 to state A_1 reversibly and then from state A_1 to state A_2 not necessarily reversibly. In this combined process, the energy exchanged between the system and the weight is $(W_{01}^{A\rightarrow})_{\text{rev}} + W_{12}^{A\rightarrow}$. Moreover, because a perpetual-motion machine of the second kind is impossible, and the process starts from a stable equilibrium state A_0, energy can be transferred only into the system and not out of the system, that is, the energy of the system can increase but not decrease. Therefore, the energy exchange involved in the combined process must satisfy the relation

$$(W_{01}^{A\rightarrow})_{rev} + W_{12}^{A\rightarrow} \leqq 0 \tag{5.3a}$$

or, equivalently,

$$W_{12}^{A\rightarrow} \leqq -(W_{01}^{A\rightarrow})_{rev} = (W_{10}^{A\rightarrow})_{rev} \tag{5.3b}$$

where in writing relation (5.3b) we use equation (5.1). Inequality (5.3b) verifies our assertion that $(W_{10}^{A\rightarrow})_{rev}$ is the largest amount of energy that can be transferred to a weight from system A in state A_1 in a weight process in which there are no net changes in the values of the amounts of constituents and the parameters of A. It is noteworthy that relations (5.3a) and (5.3b) would involve only equalities if and only if process α were reversible, and state A_2 were identical to state A_0.

Example 5.1.   Prove the last statement.

Solution.   Using the result in Example 4.1, we conclude that the strict equality in (5.3a) holds only if the end states of the combined process—sequence γ_r and α—coincide, that is, only if state A_2 coincides with state A_0. But then process α is a weight process restoring the effects of reversible weight process γ_r and, by the result in Example 4.2, it must also be reversible.

Next, we consider a combined process that is a sequence of processes γ_r and γ'. In words, we consider a weight process in which the system goes reversibly from stable equilibrium state A_0 to state A_1 and then from state A_1 to stable equilibrium state A_S. The energy exchanged in this combined process is $(W_{01}^{A\rightarrow})_{rev} + (W_{1S}^{A\rightarrow})_{rev}$ and such that

$$(W_{01}^{A\rightarrow})_{rev} + (W_{1S}^{A\rightarrow})_{rev} \leqq 0 \tag{5.4a}$$

or

$$(W_{1S}^{A\rightarrow})_{rev} \leqq -(W_{01}^{A\rightarrow})_{rev} = (W_{10}^{A\rightarrow})_{rev} \tag{5.4b}$$

Relations (5.4) are valid because a PMM2 is impossible, and the equality signs hold only if state A_0 were identical with state A_S. On the other hand, we may consider a weight process that is a sequence of processes γ'_r and γ, that is, a weight process in which the system goes reversibly first from state A_S to state A_1 and then from state A_1 to state A_0. The energy exchanged is $(W_{S1}^{A\rightarrow})_{rev} + (W_{10}^{A\rightarrow})_{rev}$ and such that

$$(W_{S1}^{A\rightarrow})_{rev} + (W_{10}^{A\rightarrow})_{rev} \leqq 0 \tag{5.5a}$$

or

$$(W_{10}^{A\rightarrow})_{rev} \leqq -(W_{S1}^{A\rightarrow})_{rev} = (W_{1S}^{A\rightarrow})_{rev} \tag{5.5b}$$

where in writing (5.5b) we use (5.2). Comparing relations (5.4b) and (5.5b), we conclude that

$$(W_{1S}^{A\rightarrow})_{rev} = (W_{10}^{A\rightarrow})_{rev} \tag{5.6}$$

Equation (5.6) confirms our assertion that the energy transferred is the same for all reversible weight processes that end in a stable equilibrium state without net changes in values of amounts of constituents and parameters.

We also conclude that states A_S and A_0 have the same energy because this energy differs from the energy E_1 of A_1 by the amount $(W_{10}^{A\rightarrow})_{rev}$. So these two stable equilibrium states correspond to the same values of energy, amounts of constituents, and parameters. As such, the two states must be identical because one and only one stable equilibrium state corresponds to each set of these values. Under the given specifications, this confirms our assertion that there is only one stable equilibrium state that can be reached reversibly in a weight process starting from a given state A_1.

5.2 Definition of Property Adiabatic Availability

The results proved in the preceding section bring out a very important consequence of the laws of thermodynamics, namely, that every system in any state must have a property that we define by means of the following set of operations and measurements.

For a given state A_1 of a system A, we use a reversible weight process that starts from A_1 and ends in a stable equilibrium state A_0 (process γ in Section 5.1). We measure the change in elevation from z_1 to z_0 experienced by the weight during the process, compute the value of the energy $(W_{10}^{A\rightarrow})_{\mathrm{rev}} = Mg\,(z_0 - z_1)$, and evaluate the quantity Ψ_1 by means of the relation

$$\Psi_1 = (W_{10}^{A\rightarrow})_{\mathrm{rev}} \tag{5.7}$$

We can easily verify that the set of operations and measurements just cited defines Ψ as a property of system A, with value Ψ_1 at any state A_1. Indeed, from the results established in Section 5.1 we conclude the following.

1. The value Ψ_1 is independent of the measuring devices, that is, of the details of the interaction forces between A and the weight during the reversible weight process.
2. The value Ψ_1 is independent of other systems in the environment, that is, of the details of the weight, which is the only external system affected during the weight process.
3. The value Ψ_1 is independent of the history of system A, that is, of the details of the changes of A during the reversible weight process, which ends in a stable equilibrium state uniquely fixed by the initial state A_1.

We call Ψ the *adiabatic availability* and denote its value at state A_1 by Ψ_1, or by Ψ_1^A when we need to refer explicitly to system A.

5.3 Features of Adiabatic Availability

5.3.1 Domain of Definition
Like energy, adiabatic availability is a property defined for all states of a system and for all systems.

5.3.2 Physical Meaning
Adiabatic availability is the largest amount of energy that can be transferred to a weight in a weight process without net changes in the values of amounts of constituents and parameters, that is, for any given state A_1

$$\Psi_1 \geqq W_{12}^{A\rightarrow} \tag{5.8}$$

where in writing (5.8) we combine (5.3b) and (5.7).

5.3.3 Nonnegativity
Adiabatic availability can take only nonnegative values, that is, for any given state A_1 of a system A

$$\Psi_1 \geqq 0 \tag{5.9}$$

and $\Psi_1 = 0$ if and only if state A_1 is stable equilibrium.

Example 5.2. Prove that Ψ_1 is nonnegative.

Solution. The value of Ψ_1 cannot be negative because if it were, that is, if $\Psi_1 = (W_{10}^{A\rightarrow})_{\text{rev}} < 0$, there would exist a weight process during which system A would start from stable equilibrium state A_0, end in state A_1, and lose energy $(W_{10}^{A\rightarrow})_{\text{rev}} = -(W_{10}^{A\rightarrow})_{\text{rev}} > 0$ to the weight. But such a process contradicts the impossibility of a PMM2 and therefore does not exist. Thus (5.9) is proved. Moreover, if state A_1 is stable equilibrium, the impossibility of a PMM2 requires that $(W_{10}^{A\rightarrow})_{\text{rev}} \leqq 0$, that is, $\Psi_1 \leqq 0$. But we just proved that $\Psi_1 \geqq 0$. Hence, here Ψ_1 can only be zero. Conversely, if $\Psi_1 = 0$, state A_1 must be stable equilibrium. If it were not, a weight process would exist that would start from stable equilibrium state A_0, end in state A_1, and have no net effects on the weight because $(W_{01}^{A\rightarrow})_{\text{rev}} = -(W_{10}^{A\rightarrow})_{\text{rev}} = -\Psi_1 = 0$. But that contradicts the definition of a stable equilibrium state, and therefore our assumption is false. Thus here, A_1 must be a stable equilibrium state.

5.3.4 Upper Bound
Adiabatic availability takes values that cannot be greater than the energy of the system in excess of the ground-state energy corresponding to the given values of amounts of constituents and parameters. In terms of symbols,

$$\Psi_1 \leqq E_1 - E_{g1} \tag{5.10}$$

where E_1 denotes the energy of state A_1, E_{g1} the ground-state energy corresponding to the values of amounts of constituents and parameters of state A_1, and Ψ_1 the adiabatic availability of state A_1.

Example 5.3. Prove (5.10).

Solution. By definition, $\Psi_1 = (W_{10}^{A\rightarrow})_{\text{rev}} = E_1 - E_0$, where in writing the second equality we use (3.13). Moreover, the ground-state energy is the lowest value of the energy compatible with the given values of the amounts of constituents and the parameters and, therefore, $E_{g1} \leqq E_0$. Upon combining these results, we obtain (5.10).

5.3.5 Changes in Reversible Weight Processes
The energy transferred to the weight in any reversible weight process that changes the state of A from A_1 to A_2 without net changes in values of amounts of constituents and parameters equals the difference $\Psi_1 - \Psi_2$ of the adiabatic availabilities of the two states, that is,

$$(W_{12}^{A\rightarrow})_{\text{rev}} = \Psi_1 - \Psi_2 \tag{5.11}$$

We prove equation (5.11) in Section 5.4.

5.3.6 Changes in Irreversible Weight Processes
The difference $\Psi_1 - \Psi_2$ represents the largest amount of energy that can be transferred to a weight out of system A in any weight process in which A changes from a state A_1 to a state A_2 with the same values of amounts of constituents and parameters as A_1. In particular, if the process is irreversible,

$$(W_{12}^{A\rightarrow})_{\text{irr}} < \Psi_1 - \Psi_2 \tag{5.12}$$

where the subscript irr denotes that the process is irreversible. We prove relation (5.12) in Section 5.4.

5.3.7 Criterion for Reversibility
Given any two states A_1 and A_2 with the same values of amounts of constituents and parameters, we can decide whether a weight process for system A from state A_1 to

state A_2 is impossible or, if possible, whether it is reversible or irreversible based on the corresponding values of the energies and adiabatic availabilities of the two states. Specifically, a weight process from A_1 to A_2 is possible only if

$$\Psi_1 - \Psi_2 \geqq E_1 - E_2 \tag{5.13}$$

and, further, the weight process is reversible only if $\Psi_1 - \Psi_2 = E_1 - E_2$. Relation (5.13) follows from (5.11) and (5.12) combined with the energy balance equation for system A during a weight process from A_1 to A_2, that is, $E_1 - E_2 = W_{12}^{A\rightarrow}$.

5.3.8 Lack of Additivity

Adiabatic availability does not satisfy the definition of an additive property given in Section 3.5. To prove this assertion, we consider two systems A and B, each in a stable equilibrium state, say, A_0 and B_0, so that the individual adiabatic availabilities are $\Psi_0^A = 0$ and $\Psi_0^B = 0$, respectively. The system C composed of the two subsystems A and B in states A_0 and B_0 is in a state C_{00} that need not be stable equilibrium, and therefore Ψ_{00}^C need not be equal to zero. It follows that Ψ_{00}^C need not be equal to $\Psi_0^A + \Psi_0^B$ and hence Ψ cannot be an additive property.

Example 5.4. A battery of the sealed type without vapor vents interacts with its surroundings only through its terminals. For the purposes of this example, we assume that each state of interest has only two independent properties, energy E and degree of charging C, and that the voltage difference between the two terminals is a constant equal to 12 V. Initially, the battery is fully discharged and has energy $E_0 = 0$ (state A_0). Upon charging, 1000 kJ of electrical energy is supplied to the battery and its degree of charging becomes $C_1 = 20$ Ah (Ah = ampere–hour). This is state A_1. Then the battery is used to power a 10-kW motor for 10 s, thus ending in a state A_2 in which the degree of charging is $C_2 = 16$ Ah. Beginning with state A_2, the battery is left perfectly isolated on a shelf for 1 year. During this time, it discharges internally completely and reaches state A_3. Find the energy and the adiabatic availability of each of the four states A_0, A_1, A_2, and A_3.

Solution. The adiabatic availability is the largest amount of energy the battery could transfer out to a weight in a weight process, as shown schematically in Figure 5.2. This largest amount of energy is given by the degree of charging C times the voltage difference between the two terminals. At state A_0, $E_0 = 0$ and $C_0 = 0$; hence $\Psi_0 = 12 \times C_0 = 0$ and therefore state A_0 is stable equilibrium. At state A_1, $E_1 = 1000$ kJ because this is the amount of energy received by the battery during the charging process, but $\Psi_1 = 12$ V $\times 20$ Ah $= 240$ Wh $= 864$ kJ. Using (5.12), we find that $\Psi_0 - \Psi_1 > E_0 - E_1$ ($-864 > -1000$) and conclude that the charging process from A_0 to A_1 is irreversible. During the weight process from A_1 to A_2 the energy transferred out to the motor $W_{12}^{A\rightarrow} = 10$ kW $\times$ 10 s $= 100$ kJ. The energy balance, $E_2 - E_1 = -W_{12}^{A\rightarrow}$, yields $E_2 = 900$ kJ. The adiabatic availability $\Psi_2 = 12$ V $\times$ 16 Ah $= 192$ Wh $= 691.2$ kJ. Again, we find that $\Psi_1 - \Psi_2 > E_1 - E_2$ ($172.8 > 100$) and conclude that the engine-powering process is irreversible. At state A_3, $E_3 = E_2 = 900$ kJ because the internal discharge is a spontaneous process. Moreover, $\Psi_3 = 0$ because $C_3 = 0$. We find that $\Psi_2 - \Psi_3 > E_2 - E_3$ ($691.2 > 0$) and conclude that the internal discharge process is also irreversible.

5.4 Proof of Relations (5.11) and (5.12)

We denote by A_{S1} the unique stable equilibrium state reached upon extraction of the adiabatic availabilty of state A_1, and by E_{S1} the energy of state A_{S1}. Similarly, we denote by A_{S2} the unique stable equilibrium state reached upon extraction of the adiabatic availability of state A_2, and by E_{S2} the energy of state A_{S2}. By the definition of adiabatic availability [equation (5.7)] and the energy balance for the reversible weight process from state A_1 to state A_{S1}, we readily verify that

$$\Psi_1 = E_1 - E_{S1} \tag{5.14}$$

and similarly, that

$$\Psi_2 = E_2 - E_{S2} \tag{5.15}$$

First, we prove that a weight process between states A_1 and A_2 is possible in the direction from A_1 to A_2 if and only if

$$E_{S1} < E_{S2} \tag{5.16}$$

and is reversible if and only if $E_{S1} = E_{S2}$. Of course, the set of values of amounts of constituents and parameters is the same for all four states A_1, A_2, A_{S1}, and A_{S2}. Then we show that (5.11) and (5.12) are valid.

To this end, we begin by assuming that $E_{S1} = E_{S2}$. Then, according to the requirements of the second law, the two stable equilibrium states A_{S1} and A_{S2} must coincide. Therefore, a sequence of two reversible weight processes, one from A_1 to A_{S1} and the other from A_{S1} to A_2, that is, from A_{S1} that coincides with A_{S2} to A_2, is possible and reversible. Hence a weight process in the direction from A_1 to A_2 is possible and reversible. Similarly, a weight process in the direction from A_2 to A_1 is possible and reversible.

Next, we assume that a weight process in the direction from A_1 to A_2 is possible and reversible. Then we consider a sequence of three reversible weight processes, one from A_{S1} to A_1, another from A_1 to A_2, and a third from A_2 to A_{S2}. In this sequence the energy of A could not decrease because A starts from a stable equilibrium state and a PMM2 is impossible. Hence the final energy, E_{S2}, must be greater than the initial energy, E_{S1}. Similarly, we consider the reverse of the preceding three reversible weight processes and conclude that E_{S1} must be greater than E_{S2}. Because E_{S1} cannot simultaneously be smaller and larger than E_{S2}, we conclude that $E_{S1} = E_{S2}$ when a reversible weight process is possible either from A_1 to A_2 or from A_2 to A_1.

Next, we assume that $E_{S1} < E_{S2}$ and consider a sequence of three weight processes, one that is reversible from A_{S1} to A_1, a second from A_1 to A_2, and a third that is reversible from A_2 to A_{S2}. This sequence is possible because it requires $E_{S1} \leqq E_{S2}$ and this relation is consistent with our assumption. Conversely, if the sequence is possible, the impossibility of a PMM2 requires that $E_{S1} \leqq E_{S2}$. Of course, equality between E_{S1} and E_{S2} would correspond to the process from A_1 to A_2 being reversible.

Finally, if $E_{S1} > E_{S2}$, a weight process in the direction from A_1 to A_2 would be impossible. Indeed, if it were possible, the impossibility of a PMM2 would have resulted in our concluding that $E_{S1} \leqq E_{S2}$, which contradicts our assumption. This completes the proof of the relation between the direction of a weight process and the difference between the energies E_{S1} and E_{S2}.

Now, we are ready to prove relations (5.11) and (5.12). If the weight process from A_1 to A_2 is reversible, we proved that (5.16) is an equality, that is, $E_{S1} = E_{S2}$. Combining this result with equations (5.14) and (5.15), that is, the relations $\Psi_1 = E_1 - E_{S1}$ and $\Psi_2 = E_2 - E_{S2}$, we find the equality $E_1 - \Psi_1 = E_2 - \Psi_2$ or $E_1 - E_2 = \Psi_1 - \Psi_2$. But $E_1 - E_2 = (W_{12}^{A\rightarrow})_{\text{rev}}$ by the energy balance for the weight process of system A, and thus equation (5.11) is proved.

On the other hand, if the weight process from A_1 to A_2 is irreversible, we proved that (5.16) holds with the strict inequality, that is, $E_{S1} < E_{S2}$. Combining this result with (5.14) and (5.15), and the energy balance, we find that $E_1 - \Psi_1 < E_2 - \Psi_2$ or $E_1 - E_2 = W_{12}^{A\rightarrow} < \Psi_1 - \Psi_2$. Thus, relation (5.12) is proved.

5.5   Generalized Adiabatic Availability

In this section we generalize the question investigated in Section 5.1 to weight processes of a system in which the values of amounts of constituents and parameters may experience net changes. The end states are listed in Figure 5.4. To proceed, we consider a system A in a state $A_{1'}$ with energy $E_{1'}$ and denote the values of the amounts of the r constituents by $n'_1, n'_2, \ldots, n'_r$, and the values of the s parameters by $\beta'_1, \beta'_2, \ldots, \beta'_s$, where the prime is used to indicate the association of a state with the primed values of amounts of constituents and parameters. We assume that machinery of any desirable degree of perfection is available, consider another set of values $n''_1, n''_2, \ldots, n''_r, \beta''_1, \beta''_2, \ldots, \beta''_s$, and investigate the question: In a weight process that starts from state $A_{1'}$ and changes the values of amounts of constituents and parameters from $n'_1, n'_2, \ldots, n'_r, \beta'_1, \beta'_2, \ldots, \beta'_s$ to $n''_1, n''_2, \ldots, n''_r, \beta''_1, \beta''_2, \ldots, \beta''_s$, how much energy can be transferred from the system to the weight? For example, given that initially a gas is in a state $A_{1'}$ that fills only part of the volume V' of the variable-volume container shown in Figure 5.5, how much energy can be exchanged with a weight in a weight process that starts from state $A_{1'}$, uses any machinery whatsoever, and ends in a state associated with volume $V'' = 2V'$?

Similarly to the results in Section 5.1, it turns out that the largest amount of energy that can be transferred out of the system under the generalized conditions just specified is: (1) limited and equal to the energy transferred in a reversible weight process in which the system starts from the initial state $A_{1'}$ and ends in a stable equilibrium state $A_{0''}$ corresponding to the values $n''_1, n''_2, \ldots, n''_r, \beta''_1, \beta''_2, \ldots, \beta''_s$; and (2) the same for all such reversible weight processes.

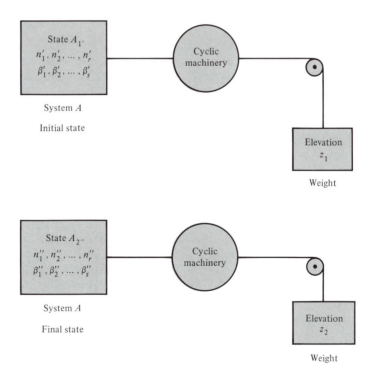

Figure 5.4   Schematic representation of the end states of a system A considered in weight processes used to investigate the question: In a weight process that starts from a state A'_1 and changes the values of amounts of constituents, and parameters from $n'_1, n'_2, \ldots, n'_r, \beta'_1, \beta'_2, \ldots, \beta'_s$ to $n''_1, n''_2, \ldots, n''_r, \beta''_1, \beta''_2, \ldots, \beta''_s$, how much energy can be trasferred from the system to the weight?

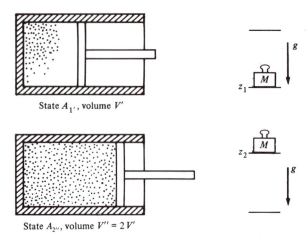

State $A_{1'}$, volume V'

State $A_{2''}$, volume $V'' = 2V'$

Figure 5.5 Schematic of the end states of a gas lifting a weight, as an example related to the investigation of the question: How much energy can be transferred from the gas to the weight in a weight process starting from a state $A_{1'}$ in which the volume is V' and ending in a state with volume $V'' = 2V'$?

The proof of these assertions is analogous to the proof given in Section 5.1. We consider two different weight processes α and γ. In process α, which may be reversible or irreversible, initially system A has values n', β', and is in a given state $A_{1'}$, and at the end the values are n'', β'' and the state is $A_{2''}$. In process γ, which is reversible, initially system A is in state $A_{1'}$, but at the end it is in a stable equilibrium state $A_{0''}$ with the values n'', β''. We denote the reverse of process γ by γ_r.

Next, we consider a process that is a sequence of processes γ_r and α. In words, we consider a weight process in which the system goes first from stable equilibrium state $A_{0''}$ to state $A_{1'}$ reversibly, and then from state $A_{1'}$ to state $A_{2''}$ not necessarily reversibly. During this sequence, the values of the amounts of constituents and parameters change first from n'', β'' to n', β'. and then from n', β' back to n'', β''. By the impossibility of a PMM2, such a sequence of processes cannot transfer energy out of the system. Therefore, the energy exchange involved in the sequence must satisfy the relation

$$(W_{0''1'}^{A\rightarrow})_{\text{rev}} + W_{1'2''}^{A\rightarrow} \leqq 0 \tag{5.17}$$

or, equivalently,

$$W_{1'2''}^{A\rightarrow} \leqq -(W_{0''1'}^{A\rightarrow})_{\text{rev}} = (W_{1'0''}^{A\rightarrow})_{\text{rev}} \tag{5.18}$$

where $(W_{0''1'}^{A\rightarrow})_{\text{rev}}$ and $W_{1'2''}^{A\rightarrow}$ denote the energies transferred to the weight in processes γ_r and α, respectively. Inequality (5.18) verifies the assertion that $(W_{1'0''}^{A\rightarrow})_{\text{rev}}$ is the largest amount of energy that can be transferred to the weight from system A in state $A_{1'}$, by means of a weight process that changes the values of the amounts of constituents and parameters from the primed to the double-primed values.

For the given state $A_{1'}$, the proof that the amount $(W_{1'0''}^{A\rightarrow})_{\text{rev}}$ would be the same for all reversible weight processes that end in a stable equilibrium state with the specified values n'', β'', and that under the given specifications there is only one stable equilibrium state that can be reached from the given state $A_{1'}$ is analogous to that given in Section 5.1, and is left as an exercise for the reader.

For a given state $A_{1'}$ of a system A, and for given values n'', β'', we conclude that the set of measurements and operations which yield quantity $(W_{1'0''}^{A\rightarrow})_{\text{rev}}$ define this quantity as the value of a property of system A in state $A_{1'}$. We call this property the *generalized adiabatic availability* with respect to values n'', β'' of a state $A_{1'}$ which is associated with values n', β'. Clearly, the adiabatic availability Ψ (Section 5.2) is a special case of the generalized adiabatic availability in the sense that the primed and double primed sets of values of amounts of constituents and parameters coincide.

5.6 Features of Generalized Adiabatic Availability

5.6.1 Physical Meaning

The generalized adiabatic availability is the largest amount of energy that can be transferred to a weight in a weight process of a system A starting from a state $A_{1'}$ with values of amounts of constituents and parameters n', β', and ending in a state $A_{2''}$ with values n'', β'', that is,

$$(W_{1'0''}^{A\rightarrow})_{\text{rev}} \geqq W_{1'2''}^{A\rightarrow} \tag{5.19}$$

This interpretation follows from (5.17) and is analogous to the meaning of adiabatic availability.

5.6.2 Positive and Negative Values

In contrast to adiabatic availability, which is always positive, generalized adiabatic availability can take both positive and negative values depending on the given state, its values n', β', and the given set of values n'', β''. We can illustrate this point by means of a simple example. We consider a gas A in a state $A_{1'}$ associated with volume V'. After transferring an amount of energy $(W_{1'0'}^{A\rightarrow})_{\text{rev}}$ to a weight, A is in a stable equilibrium state $A_{0'}$. Starting from state $A_{0'}$, we may use another reversible weight process and compress the gas to a stable equilibrium state $A_{0''}$ corresponding to a volume V''. As a result of the compression, energy $(W_{0'0''}^{A\leftarrow})_{\text{rev}}$ is transferred from the weight to the gas. This energy may be larger than $(W_{1'0'}^{A\rightarrow})_{\text{rev}}$. If it is larger, then

$$(W_{1'0'}^{A\rightarrow})_{\text{rev}} - (W_{0'0''}^{A\leftarrow})_{\text{rev}} < 0 \tag{5.20}$$

or, equivalently, using the equality $(W_{0'0''}^{A\leftarrow})_{\text{rev}} = -(W_{0'0''}^{A\rightarrow})_{\text{rev}}$,

$$(W_{1'0'}^{A\rightarrow})_{\text{rev}} + (W_{0'0''}^{A\rightarrow})_{\text{rev}} < 0 \tag{5.21}$$

The left-hand side of (5.21) is the generalized adiabatic availability of state $A_{1'}$ with respect to volume V'', and we see that it is negative here.

5.6.3 Changes in Reversible Weight Processes

Given two states of a system A, $A_{1'}$ with values n', β', and $A_{2''}$ with values n'', β'', the energy transferred to the weight in any reversible weight process that changes the state of A from $A_{1'}$ to $A_{2''}$ equals the difference of the generalized adiabatic availabilities of the two states each evaluated with respect to arbitrary but the same fixed values n, β.

Example 5.5. Verify the preceding assertion.

Solution. We consider a sequence of two reversible weight processes for system A. In the first process, the system goes from state $A_{1'}$ associated with values n', β' to stable equilibrium state A_0 corresponding to values n, β. In the second process, the system goes from state A_0 to state $A_{2''}$ associated with values n'', β''. So the net amount of energy transferred to the weight $(W_{1'2''}^{A\rightarrow})_{\text{rev}} = (W_{1'0}^{A\rightarrow})_{\text{rev}} + (W_{02''}^{A\rightarrow})_{\text{rev}}$. But $(W_{1'0}^{A\rightarrow})_{\text{rev}}$ is the generalized adiabatic availability of $A_{1'}$ with respect to values n, β, and $(W_{02''}^{A\rightarrow})_{\text{rev}}$ the negative of the generalized adiabatic availability of $A_{2''}$ with respect to the same values because $(W_{02''}^{A\rightarrow})_{\text{rev}} = -(W_{2''0}^{A\rightarrow})_{\text{rev}}$. Thus the assertion is verified.

It is clear that our conclusion holds for any arbitrary choice of the values n, β provided that they are within the ranges of values specified for the system. For this reason we may denote the generalized adiabatic availability of state $A_{1'}$ by $\Psi_{1'}$ and write

$$(W_{1'2''}^{A\rightarrow})_{\text{rev}} = \Psi_{1'} - \Psi_{2''} \tag{5.22}$$

where we understand that each of the generalized adiabatic availabilities $\Psi_{1'}$ and $\Psi_{2''}$ is evaluated with respect to the same set of values n, β, even though state $A_{1'}$ is associated with values n', β', and state $A_{2''}$ with values n'', β''. Moreover, (5.22) reduces to (5.11) when the three sets of values, primed, double-primed, and unprimed, coincide.

5.6.4 Changes in Irreversible Weight Processes

The difference $\Psi_{1'} - \Psi_{2''}$ between two generalized adiabatic availabilities represents the largest amount of energy that can be transferred to a weight out of a system A in any weight process in which A changes from a state $A_{1'}$ to a state $A_{2''}$, regardless of whether the values of amounts of constituents and parameters of $A_{1'}$ are the same or differ from the corresponding values of $A_{2''}$. In particular, if the process is irreversible,

$$(W_{1'2''}^{A\rightarrow})_{\text{irr}} < \Psi_{1'} - \Psi_{2''} \tag{5.23}$$

As shown in Section 5.6.3, (5.23) becomes an equality only if the weight process from $A_{1'}$ to $A_{2''}$ is reversible. Of course, both $\Psi_{1'}$ and $\Psi_{2''}$ are evaluated with respect to the same set of arbitrarily specified values n, β.

To prove relation (5.23), we denote by A_{S1} and A_{S2} the stable equilibrium states reached upon extracting the generalized adiabatic availability with respect to the values n and β from states $A_{1'}$ and $A_{2''}$, respectively. Denoting by $E_{1'}$, E_{S1}, $E_{2''}$, and E_{S2} the energies of states $A_{1'}$, A_{S1}, $A_{2''}$, and A_{S2}, respectively, we can readily verify that the definition of generalized adiabatic availability and the energy balance for the reversible weight process from state $A_{1'}$ to A_{S1} imply that $\Psi_{1'} = E_1' - E_{S1}$ or, equivalently, $E_{S1} = E_{1'} - \Psi_{1'}$. Similarly, we can verify that $E_{S2} = E_{2''} - \Psi_{2''}$. If a weight process from state $A_{1'}$ to state $A_{2''}$ is irreversible, we prove in Section 5.4 that $E_{S1} < E_{S2}$. Using the relations just cited, we can express the last inequality in the form $E_{1'} - \Psi_{1'} < E_{2''} - \Psi_{2''}$ or $E_{1'} - E_{2''} < \Psi_{1'} - \Psi_{2''}$. But the energy balance for the irreversible weight process of A requires that $E_{1'} - E_{2''} = (W_{1'2''}^{A\rightarrow})_{\text{irr}}$ and, therefore, relation (5.23) is proved.

5.6.5 Criterion for Reversibility

Given any two states $A_{1'}$ and $A_{2''}$, a weight process for system A from $A_{1'}$ to $A_{2''}$ is possible only if

$$\Psi_{1'} - \Psi_{2''} \geqq E_{1'} - E_{2''} \tag{5.24}$$

and, further, it is reversible only if $\Psi_{1'} - \Psi_{2''} = E_{1'} - E_{2''}$. Relation (5.24) follows from (5.22) and (5.23) combined with the energy balance equation for system A for a weight process from $A_{1'}$ to $A_{2''}$, that is, $E_{1'} - E_{2''} = W_{1'2''}^{A\rightarrow}$.

To simplify the notation, in what follows we use the same symbols for adiabatic availability and generalized adiabatic availability and make the distinction in the text as necessary.

5.7 Adiabatic Availability of a Weight

To derive the adiabatic availability of a weight we first prove that any process of a weight is reversible. We recall that a weight has only one independent property, namely, the elevation z in the gravity field. Accordingly, a weight can experience only one type of interaction—one that affects its sole independent property—and its state is fully determined by the value of z. To prove the reversibility of a process of a weight W, in which the elevation changes from z_1 to z_2, it suffices to show that the same change in elevation can be effected by at least one reversible weight process of the environment A of the weight because then both the weight and its environment can be restored to their respective initial states. The existence of such a reversible weight process of the environment A is guaranteed by the second law. It is one of the reversible weight processes of system A that connects a stable equilibrium state A_0 and a state A_1 with adiabatic availability $Mg(z_0 - z_1)$ as the weight changes from elevation z_0 to elevation z_1. So any process of a weight is reversible.

Now, for a reversible process, the energy E^W and the adiabatic availability Ψ^W of a weight must satisfy the equality (see Section 5.3.7)

$$\Psi_2^W - \Psi_1^W = E_2^W - E_1^W \tag{5.25}$$

for any two states W_1 and W_2 of the weight. Upon using this equality between any state and the ground state, we find that the adiabatic availability of the weight

$$\Psi^W = E^W - E_g^W \tag{5.26}$$

where E_g^W is the lowest or ground-state energy of the weight, and in writing (5.26) we take $\Psi_g^W = 0$.

5.8   Comment

Were we to assume that a system is subject only to the laws of mechanics, we would conclude that all the energy of the system in excess of the ground-state energy can be used to lift a weight. Said differently, we would conclude that for every state A_1 the adiabatic availability Ψ_1 is always equal to $E_1 - E_{g1}$ (see Section 5.3.4). But this conclusion is not consistent with all experimental results. Many reproducible and accurate experiences show that, in general, not all the energy can be transferred out in a weight process, that is, the adiabatic availability Ψ_1 is usually smaller than $E_1 - E_{g1}$ and, in particular, $\Psi_1 = 0$ if state A_1 is stable equilibrium. To account for these experiences, the laws of thermodynamics entail a greater variety of states than contemplated by the laws of mechanics.

The contrast between the values of the adiabatic availability in mechanics and in general is yet another way to present the fundamental differences between the domain of validity of the laws of mechanics and that of the laws of thermodynamics, which includes the domain of mechanics as a special and limiting case.

Problems

5.1 Two systems A and B interact only with each other without net changes in values of amounts of constituents and parameters. The energies and adiabatic availabilities of the end states are listed in the table.

State	E	Ψ	State	E	Ψ
A_1	100 kJ	80 kJ	A_2	50 kJ	10 kJ
B_1	50 kJ	30 kJ	B_2	100 kJ	80 kJ

(a) Are the processes for A and B reversible or irreversible?

(b) Can the changes from A_1 to A_2 and from B_1 to B_2 be reproduced by means of two weight processes, one for A and the other for B?

5.2 A system A experiences three weight processes, α, β, and γ, each without net changes in values of amounts of constituents and parameters. The characteristics of the processes are listed in the table. State A_0 is stable equilibrium.

Process	End states	Energy transfer
α	$A_1 \rightarrow A_2$	$(W_{12}^{A\rightarrow})_{\text{rev}} = 5 \text{ kJ}$
β	$A_2 \rightarrow A_0$	$(W_{20}^{A\rightarrow})_{\text{irr}} = 10 \text{ kJ}$
γ	$A_0 \rightarrow A_3$	$(W_{03}^{A\rightarrow})_{\text{rev}} = -15 \text{ kJ}$

(a) What types of states are A_1 and A_2?

(b) Is state A_3 identical to state A_1?

(c) Find a lower bound to the adiabatic availability of A_1 for the given values of amounts of constituents and parameters.

(d) What is the adiabatic availability of A_3?

5.3 Because energy E and adiabatic availability Ψ are properties defined for all states of any system A, any function of E and Ψ is a property defined for all states. For example, we can define the property

$$P = E - E_g - \Psi$$

where E_g is the ground-state energy corresponding to the values of the amounts and parameters of the given state. Property P represents the energy that cannot be transferred out of the system in any weight process without net changes in the values of the amounts of constituents and parameters. Show that, in terms of property P, the results in Section 5.3.7 can be summarized as follows.

(a) In any reversible weight process from a state A_1 to a state A_2 of a system A, the value of property P remains unchanged, that is

$$(P_2 - P_1)_{\text{rev}} = 0$$

(b) In any irreversible weight process from A_1 to A_2, the value of property P increases, that is,

$$(P_2 - P_1)_{\text{irr}} > 0$$

(c) For any state A_1, recall that $\Psi_1 \leq E_1 - E_g$, and draw on an E versus P diagram the locus of the stable equilibrium states with given values of n and β. On the same diagram and for the same values of n and β, draw the states for which $\Psi_1 = E_1 - E_g$.

(d) Is P an additive property? Prove your answer.

(e) Show that

$$P = E_S - E_g$$

where E_S denotes the energy of the unique stable equilibrium state reached upon extraction of the adiabatic availability.

5.4 A system A experiences four weight processes, α, β, γ, and δ, each without net changes in values of amounts of constituents and parameters, except volume. The characteristics of the processes are listed in the table. States $A_{0'}$ and $A_{0''}$ are stable equilibrium. The primes and double primes in the subscripts refer to the values V' and V'' of the volume, respectively.

Process	End states	Energy transfer
α	$A_{1'} \rightarrow A_{0''}$	$(W_{1'0''}^{A\rightarrow})_{\text{rev}} = 25$ kJ
β	$A_{2'} \rightarrow A_{0'}$	$(W_{2'0'}^{A\rightarrow})_{\text{rev}} = 15$ kJ
γ	$A_{0'} \rightarrow A_{3''}$	$(W_{0'3''}^{A\rightarrow})_{\text{rev}} = -15$ kJ
δ	$A_{1'} \rightarrow A_{3''}$	$(W_{1'3''}^{A\rightarrow})_{\text{irr}} = 0$ kJ

(a) Is state $A_{2'}$ identical to state $A_{3''}$?

(b) What is the adiabatic availability of state $A_{2'}$?

(c) What is the generalized adiabatic availability of $A_{1'}$ with respect to the value V'' of the volume?

(d) Is a weight process from $A_{0'}$ to $A_{0''}$ possible?

5.5 Two states $A_{1'}$ and $A_{2''}$ of a system A have values of volume, energy, and adiabatic availability as listed in the table. The two states have the same values for each amount of constituent and each parameter other than volume. Moreover, the generalized adiabatic availability of $A_{1'}$ with respect to the value $V'' = 20$ m^3 of the volume is 60 kJ.

State	V	E	Ψ
$A_{1'}$	10 m^3	100 kJ	40 kJ
$A_{2''}$	20 m^3	80 kJ	50 kJ

(a) Is a weight process from state $A_{1'}$ to state $A_{2''}$ possible?

(b) What is the generalized adiabatic availability of the state $A_{S1'}$ with respect to the value $V'' = 20$ m^3 of the volume if $A_{S1'}$ is the state reached upon extraction of the adiabatic availability starting from state $A_{1'}$?

6 Available Energy

In Chapter 5 we find that the amount of energy that can be transferred under certain conditions from a system to a rising weight is bounded and, in general, different from the energy of the system in excess of the ground-state energy. Here we wish to examine whether a similar bound exists when we relax some of the conditions. Specifically, we consider a system that may interact with a weight and other suitably selected systems in the environment.

One of many practical illustrations of this investigation is the coal-fired power plant shown schematically in Figure 6.1. In this plant, coal plus sufficient air for the burning may be regarded as a system from which energy is extracted; the motor attached to the turbogenerator as the rising weight; the boiler, circulating water, turbine, generator, condenser, and pump as machinery that undergoes cyclic changes of state; and the water in a very large lake as an additional system with which the coal–air system is interacting. The question is: How much energy can be transferred to the weight for the given initial state of the coal–air system? It is evident that the answer is of great practical importance.

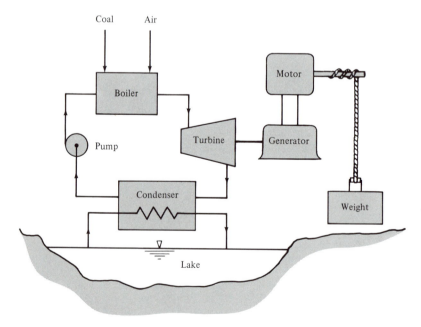

Figure 6.1 Schematic of a coal-fired power plant. It is an example of the system used to examine the question: How much energy can be transferred to a weight from a given amount of coal and sufficient air for burning the coal?

This kind of investigation is a generalization of the question raised by Sadi Carnot in his 1824 pioneering study on the motive power of fire, the study that gave birth to the science of thermodynamics. It is a generalization because Carnot assumed the initial state of the system to be restricted to a special kind — what we call a stable equilibrium state — whereas in our discussion here this assumption is not needed.

In this chapter we study the question just cited and find that it admits a definite answer in the form of a property called available energy. In addition, we introduce some other useful concepts.

6.1 Subsystems and Composite Systems

In many applications we find it convenient to combine two or more systems and regard them as subsystems of an overall composite system. For example, in Figure 6.1 the coal–air system, the lake, the fluid through the boiler, turbine, and condenser, and the weight each may be considered as a subsystem of the power plant system. The values of the amounts of constituents and the parameters of a composite system are determined by both the corresponding values of the subsystems and the constraints imposed by any specific interconnections between the subsystems. The reason is that interconnections may result in specifications of amounts of constituents, parameters and their ranges of values for the composite system that may differ from those of the subsystems.

To see the last point more clearly, we consider two oxygen containers A and B, having fixed amounts of oxygen n^A and n^B, and adjustable volumes V^A and V^B, respectively. If the two containers are unconnected (Figure 6.2a), the four quantities n^A,

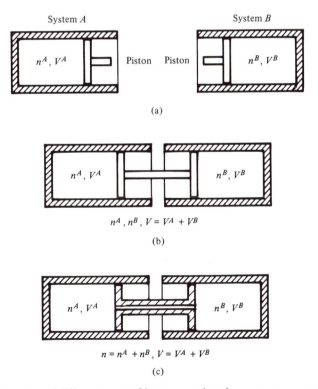

Figure 6.2 Schematics of different types of interconnections between two systems A and B. Interconnections result in specifications of the ranges of values of the amounts of constituents and the parameters of the composite of A and B that, in general, differ from those of the individual systems.

n^B, V^A, and V^B are sufficient to specify the composite of systems A and B. If the pistons of the containers are interconnected (Figure 6.2b) so that changes in the volume of A are balanced by equal and opposite changes in the volume of B, the composite of A and B is specified not only by the four quantities n^A, n^B, V^A, and V^B but also by the constraint $V^A + V^B =$ constant. And if the piston interconnection also allows an exchange of oxygen between A and B (Figure 6.2c), then in addition to four quantities n^A, n^B, V^A, and V^B, we must specify the two constraints $n^A + n^B =$ constant and $V^A + V^B =$ constant.

6.2   Mutual Stable Equilibrium

If a system is in a stable equilibrium state and is a composite of two or more subsystems, the subsystems are said to be in *mutual stable equilibrium*. For example, if a system C is in a stable equilibrium state and is the composite of two systems A and B (Figure 6.3), we say that A is in mutual stable equilibrium with B.

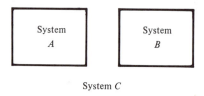

System C

Figure 6.3   Schematic of a system C consisting of subsystems A and B. If C is in a stable equilibrium state then systems A and B are in mutual stable equilibrium.

An immediate consequence of this definition is that, if two or more systems are in mutual stable equilibrium, each system must be in a stable equilibrium state. Indeed, if one system were not in a stable equilibrium state, its adiabatic availability would be nonzero. As a result, the composite of the systems would also have nonzero adiabatic availability and, therefore, could not be in a stable equilibrium state.

In contrast, if each of two systems is in a stable equilibrium state, the composite is not necessarily in a stable equilibrium state and, therefore, the systems are not necessarily in mutual stable equilibrium. Whether they are or not depends on the ranges of values over which the amounts of constituents and the parameters of the two systems can vary and on the nature of the internal forces of the composite system. For example, a bucket of a combustible but nonvolatile substance and a room full of air (Figure 6.4) can each be in a stable equilibrium state, yet the composite of the two need not be in a stable equilibrium state if the combustion reaction is not prohibited. Again, if combustion is prohibited, the same combustible substance and air may or may not be in mutual stable, equilibrium, depending on how each constituent is distributed between the two systems. In Chapters 9 to 11 we study quantitatively some important conditions that must be necessarily satisfied if two systems are in mutual stable equilibrium.

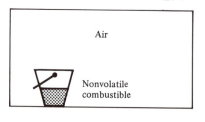

Figure 6.4   A bucket of a nonvolatile combustible and a room full of air can each be in a stable equilibrium state, and yet the combination of the two need not be in a stable equilibrium state.

In general, if a system A is in mutual stable equilibrium with a system B, and system B is in mutual stable equilibrium with a system C, system A is not necessarily in mutual stable equilibrium with system C. The reason is that the conditions that are necessary to guarantee that A and B are in mutual stable equilibrium, and that B and C are in mutual stable equilibrium may not include all the conditions that are necessary for A and C to be in mutual stable equilibrium. For example, systems A and C may be composed of the same constituent, such as oxygen, and the value of the amount of this constituent may vary over a specified range in both A and C. In contrast, system B may be composed of a fixed number of molecules of a different constituent, such as nitrogen. Then we will see in Chapter 11 that systems A and C may individually be in mutual stable equilibrium with system B, yet not in mutual stable equilibrium with each other.

6.3   Reservoirs

A *reservoir* R is an idealized kind of system that provides useful reference states both in theory and in applications, and that behaves in a manner approaching the following limiting conditions.

1. A reservoir passes through stable equilibrium states only.
2. In the course of finite changes of state at constant or varying values of its amounts of constituents and parameters, a reservoir remains in mutual stable equilibrium with a duplicate of itself that experiences no such changes.
3. At constant values of the amounts of constituents and the parameters of each of two reservoirs initially in mutual stable equilibrium, energy can be transferred reversibly from one reservoir to the other with no net effect on any other system.

We will see later that many systems in our natural environment behave almost like reservoirs. For example, in many practical applications, the earth, ocean, lake, river, or atmosphere in the immediate surroundings of a system under study may be modeled as a reservoir.

Nevertheless, we emphasize that conditions 2 and 3 that define the behavior of a reservoir are so restrictive that they must be regarded as limits that a system obeying the laws of physics can approach but cannot actually reach. To see this we consider as an example a composite system C consisting of two reservoirs in mutual stable equilibrium. System C is in a stable equilibrium state C_{00}. After a change of state during which a given amount of energy is transferred from one reservoir to the other without net changes in values of amounts of constituents and parameters of either reservoir, the composite system reaches a state different from C_{00} but with the same values of amounts of constituents and parameters as those associated with C_{00}. Because condition 3 avers that such a state can be reached with no effects external to the composite system, it constitutes a violation of the definition of a stable equilibrium state. Moreover, because the final state of the composite system has the same value of the energy as the initial state C_{00} and, by virtue of condition 2, is also a stable equilibrium state, such a change of state also constitutes a violation of the uniqueness of stable equilibrium states required by the second law.[1]

---

[1] We can restate the definition of a reservoir so as to avoid the conceptual difficulties and yet maintain, for all practical purposes, a behavior equivalent to that specified by conditions 1 to 3. Specifically, we could define a reservoir as a system that behaves as follows. 1. It passes through stable equilibrium states only. 2. In the course of finite changes of state at constant or varying values of its amounts of constituents and parameters, it departs only very slightly from the condition of mutual stable equilibrium with a duplicate of itself that experiences no such changes. 3. At constant values of the amounts of constituents and the parameters of each of two reservoirs initially in mutual stable equilibrium, energy can be transferred reversibly from one reservoir to the other at the expense of only very slight and negligible effects on any other system.

Unless we specify otherwise, in the following chapters we consider reservoirs each of which has fixed values of amounts of constituents and parameters. Other types of reservoirs are discussed later (Sections 10.3 and 11.3, and Chapter 24).

6.4   Weight Processes of a System and a Reservoir

Given a system A in a state A_1 with values of the amounts of r constituents represented by the vector $(n)_1$, and values of the s parameters represented by the vector $(\beta)_1$, a reservoir R in state R_1, and cyclic machinery of any desirable degree of perfection (Figure 6.5), we wish to consider all possible weight processes of the composite of system A and reservoir R that change the state of A from A_1 to any other state A_2 with amounts of constituents $(n)_2 = (n)_1$ and parameters $(\beta)_2 = (\beta)_1$. The question we wish to investigate is: In a weight process, how much energy can be transferred to the weight from the composite of the system and the reservoir without net changes in the values of the amounts of constituents and the parameters of either the system or the reservoir?

This question is of great practical importance especially when stated in terms of readily understandable tasks. For example, given 1 ton of coal and adequate air for combustion in a fixed-size combustion chamber, cyclic machinery, and either the North Sea or the Thames River in Great Britain, how many tons of water can we lift from the bottom of a coal mine to the ground level? The answer to the question emerges from the concept of adiabatic availability of the composite of systems A and R. In addition

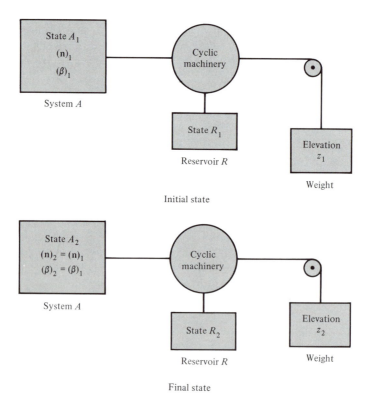

Figure 6.5   Schematic of the end states of a weight process of a composite of a system A and a reservoir R.

to its practical value, it also has the theoretical advantage that it leads to the definition of the property called entropy.

It turns out that the largest amount of energy that can be transferred to the weight out of the composite of system A and reservoir R without net changes in their individual values of amounts of constituents and parameters conforms to the following assertions.

1. It is limited and equal to the energy transferred during a reversible weight process in which system A starts from the initial state A_1, interacts with both the weight and the reservoir, and ends being in mutual stable equilibrium with the reservoir.

2. It is the same for all reversible weight processes such that system A begins from the given state A_1 and ends in a state in which A and R are in mutual stable equilibrium, and the values of the amounts of constituents and the parameters of A and R experience no net changes.

3. It is independent of the initial state of reservoir R.

4. It is the same for all reservoirs that are initially in mutual stable equilibrium with R.

To verify our assertions, we proceed as follows. We consider the composite of systems A and R, and cyclic machinery that can lift a weight in a gravity field, that is, devices that return to their respective initial states after the lifting of the weight is completed. The lifting of the weight is achieved by energy transfer either directly out of A, or via interactions with both A and R, or both. These interactions are shown schematically by the line connections in Figure 6.6, where line 1 stands for the interaction between A and machinery that cannot interact with the reservoir, and line 2 for the interaction between A and other machinery that can interact with reservoir R (line 3). The process experienced by the composite of systems A and R is a weight process, but the process experienced individually by either system A or system R need not be a weight process. In other words, the flows at the boundaries of either A or R may differ from the flows that occur in the course of a weight process.

To prove assertions 1 and 2, we consider two different weight processes α and γ for the composite of system A and reservoir R. Each of these two processes starts from the initial states A_1 of A and R_1 of R, and ends in states of A and R that correspond to the same values of the amounts of constituents and the parameters as those of states A_1 and R_1, respectively. The processes are sketched in Figure 6.7.

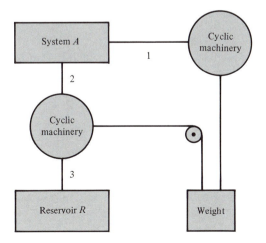

Figure 6.6 Schematic of interactions between a weight and the composite of a system A and a reservoir R. Cyclic machinery can transfer energy to the weight by interacting either directly with A, or with both A and R, or both.

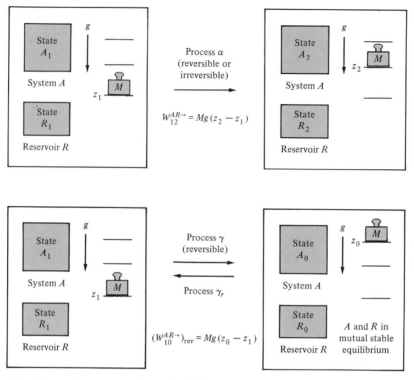

Figure 6.7   Schematic representations of weight processes α and γ for the composite of system A and reservoir R (see text).

In process α, which may be reversible or irreversible, the composite of A and R starts from initial state $(AR)_1$ in which A is in state A_1 and R in state R_1, and ends in a final state $(AR)_2$ in which A is in state A_2 and R in state R_2. We denote the energy exchanged between the composite system AR and the weight by the symbol $W_{12}^{AR\rightarrow}$, where we use the notation introduced in Section 3.9.

In process γ, which is reversible, the composite of A and R starts from initial state $(AR)_1$ but ends in a stable equilibrium state $(AR)_0$ in which A is in state A_0 and R in state R_0, and systems A and R are in mutual stable equilibrium. We denote the energy exchanged during this reversible process by $(W_{10}^{AR\rightarrow})_{\mathrm{rev}}$, and recognize that it is the adiabatic availability Ψ_1^{AR} of the composite of A and R in state $(AR)_1$. It is noteworthy that, in general, the adiabatic availability Ψ^{AR} of the composite of A and R differs from the adiabatic availability Ψ^A of A alone.

Using the results we prove in Chapter 5 for adiabatic availability, including relation (5.8), we conclude that

$$(W_{10}^{AR\rightarrow})_{\mathrm{rev}} = \Psi_1^{AR} \geq W_{12}^{AR\rightarrow} \tag{6.1}$$

Inequality (6.1) verifies the assertion that $(W_{10}^{AR\rightarrow})_{\mathrm{rev}}$ is the largest amount of energy that can be transferred to a weight from the composite of A and R in state $(AR)_1$ in a weight process in which there are no net changes in the values of the amounts of constituents and the parameters of either A or R.

Again using the results in Chapter 5, we also conclude that the energy transferred to the weight, $(W_{10}^{AR\rightarrow})_{\mathrm{rev}}$, is the same for all reversible weight processes that start from the state $(AR)_1$ and end in a stable equilibrium state $(AR)_0$. Thus we verify assertion 2, and recall that the stable equilibrium state $(AR)_0$ is the same for all such reversible weight processes.

To verify the assertion that $(W_{10}^{AR\rightarrow})_{\text{rev}}$ is independent of the initial state R_1 of reservoir R, we proceed as follows. We consider a system A initially in a state A_1, a reservoir R initially in a state $R_{1'}$ different from R_1, a reservoir Q identical to and in mutual stable equilibrium with R, initially in a state Q_1, and a reversible weight process γ' for the composite of system A and reservoir R which starts from the initial states A_1 and $R_{1'}$ and ends in states in which A and R are in mutual stable equilibrium. Process γ' consists of a sequence of three reversible weight processes. The first changes the state of R from $R_{1'}$ to R_1 and the state of Q from Q_1 to $Q_{1'}$ by transferring energy between reservoir R and reservoir Q without changing the state A_1. By the definition of a reservoir, this process involves no net effects external to the composite of R and Q. The second process is the process γ defined earlier, which changes the state of A from A_1 to A_0 and that of R from R_1 to R_0. It transfers to the weight energy $(W_{10}^{AR\rightarrow})_{\text{rev}}$. The third process in the sequence changes the state of Q from $Q_{1'}$ back to the initial state Q_1 by transferring energy between R and Q without affecting the state of A. Again, this process involves no net effects external to the composite of R and Q. Thus, the overall reversible process γ' leaves the state of reservoir Q unchanged, ends with system A and reservoir R in mutual stable equilibrium, and transfers to the weight energy $(W_{10}^{AR\rightarrow})_{\text{rev}}$ despite the fact that reservoir R was initially in an arbitrary state $R_{1'}$ different from R_1. As such, it verifies our assertion that $(W_{10}^{AR\rightarrow})_{\text{rev}}$ is independent of the initial state of R.

A similar rationale may be used to verify that $(W_{10}^{AR\rightarrow})_{\text{rev}}$ is the same for all reservoirs that are initially in mutual stable equilibrium with reservoir R.

Example 6.1. Prove the last assertion.

Solution. We consider a reversible weight process γ'' for the composite of a system A and a reservoir Q that starts from initial states A_1 of A and Q_1 of Q, and ends in states in which A and Q are in mutual stable equilibrium. To show that if reservoirs Q and R are in mutual stable equilibrium in their initial states Q_1 and R_1 then $(W_{10}^{AQ\rightarrow})_{\text{rev}} = (W_{10}^{AR\rightarrow})_{\text{rev}}$, we construe process γ'' as a sequence of the following two reversible processes. The first is the process γ defined earlier which transfers to the weight energy $(W_{10}^{AR\rightarrow})_{\text{rev}}$ and changes the state of A from A_1 to A_0 and that of R from R_1 to R_0. The second changes the state of R from R_0 back to R_1 by transferring energy between reservoirs R and Q without changing the state A_0. By the definition of a reservoir, this process involves no net effects external to R and Q, and the final state Q_0 of Q is such that system A in state A_0 is in mutual stable equilibrium with Q. Hence the energy transferred to the weight $(W_{10}^{AQ\rightarrow})_{\text{rev}} = (W_{10}^{AR\rightarrow})_{\text{rev}}$.

6.5 Definition of Property Available Energy

The results derived in the preceding section bring out another consequence of the science of thermodynamics, namely, a property that we define by means of the following set of operations and measurements. For a given system A in state A_1 and a given reservoir R in any of its states, we use cyclic machinery that can raise a weight in a gravity field while interacting with the composite of system A and reservoir R without net changes in their respective values of the amounts of constituents and the parameters. In a reversible weight process for the composite of A and R in which system A starts from state A_1 and ends in a state A_0 such that A and R are in mutual stable equilibrium, we measure the change in elevation from z_1 to z_0 experienced by the weight. Then we compute the value of the energy transferred to the weight, $(W_{10}^{AR\rightarrow})_{\text{rev}} = Mg(z_0 - z_1)$, and write the relation

$$\Omega_1^R = (W_{10}^{AR\rightarrow})_{\text{rev}} \tag{6.2}$$

In the light of the results of the preceding section, we conclude that the set of operations and measurements just cited results in the definition of Ω^R as a property of the composite of system A and reservoir R. We call Ω^R the *available energy* of system A with respect to reservoir R and denote its value at state A_1 by Ω_1^R or by $(\Omega^R)_1^A$ when the notation must refer explicitly to system A. It is noteworthy that $(\Omega^R)_1^A = \Psi_1^{AR}$, that is, the available energy of system A in state A_1 with respect to reservoir R is equal to the adiabatic availability of the composite of system A and reservoir R in state $(AR)_1$, in which A is in state A_1 and R is in any one of its states R_1.

6.6   Features of Available Energy

6.6.1   Domain of Definition
Like energy and adiabatic availability, available energy is a property defined for all states of a system — unsteady, steady, nonequilibrium, or any kind of equilibrium — and for all systems — large or small.

6.6.2   Physical Meaning
Available energy is the largest amount of energy that can be transferred to a weight out of the composite of a system A and a given reservoir R in a weight process without net changes in the values of amounts of constituents and parameters of either A or R, that is, for any given state A_1 and a given reservoir R,

$$\Omega_1^R \geqq W_{12}^{AR\rightarrow} \tag{6.3}$$

where, in writing (6.3), we combine equations (6.1) and (6.2), and $W_{12}^{AR\rightarrow}$ denotes the amount of energy transferred to a weight in any weight process in which A starts from state A_1 and ends in any other state A_2 with the same values of the amounts of constituents and the parameters as A_1.

6.6.3   Lower Bound
The available energy of a state of a system with respect to a given reservoir is usually greater than the adiabatic availability of the same state. In terms of symbols, for any given state A_1 of a system A and any given reservoir R,

$$\Omega_1^R \geqq \Psi_1 \tag{6.4}$$

Because adiabatic availability is nonnegative, it follows that the available energy is also nonnegative.

Example 6.2.   Prove (6.4) and determine when it applies with the equal sign.

Solution.   Consider a sequence of two reversible weight processes. The first is the reverse of process γ discussed in Section 6.4, that is, the composite of system A and reservoir R starts from state $(AR)_0$ and ends in state $(AR)_1$, where states $(AR)_0$ and $(AR)_1$ are as defined in Section 6.4. Thus the energy transferred to the weight as a result of this process $(W_{01}^{AR\rightarrow})_{\mathrm{rev}} = -(W_{10}^{AR\rightarrow})_{\mathrm{rev}} = -\Omega_1^R$. The second reversible weight process is such that system A by itself, that is, not in combination with R, starts from state A_1 and ends in a stable equilibrium state not necessarily the same as the state A_0 reached when A and R are in mutual stable equilibrium. Under these conditions, recall that the energy transferred to the weight is given by the adiabatic availability Ψ_1 of A_1. Hence the net amount of energy transferred to the weight as a result of the two processes is $-\Omega_1^R + \Psi_1$. Because the initial state $(AR)_0$ is stable equilibrium for the composite of A and R, and the sequence of the two weight processes may be regarded as an overall weight process for the composite of A and R, the impossibility of a PMM2 requires that the net energy transferred to the weight be nonpositive, that is, $-\Omega_1^R + \Psi_1 \leq 0$ or $\Omega_1^R \geq \Psi_1$. The equality sign in (6.4) applies when the stable equilibrium state reached during the second reversible weight process is identical to the state A_0 in which A is in mutual stable equilibrium with reservoir R, because then and only then the initial and final states of the sequence of the two weight processes are identical, and therefore the energy transferred to the weight is zero.

6.6.4 Changes in Reversible Weight Processes

For any two states A_1 and A_2 corresponding to the same values of amounts of constituents and parameters of a system A, the difference between the available energies Ω_1^R and Ω_2^R of these two states with respect to a reservoir R is equal to the energy transferred to a weight in any reversible weight process for the composite of A and R—without net changes in values of amounts of contituents and parameters—in which A changes from state A_1 to state A_2, that is,

$$(W_{12}^{AR \to})_{\text{rev}} = \Omega_1^R - \Omega_2^R \tag{6.5}$$

Example 6.3. Verify equation (6.5).

Solution. Using the notation introduced in Sections 6.4 and 6.5, we consider a sequence of two reversible weight processes for the composite of A and R. In the first process, the composite AR goes from state $(AR)_1$ to state $(AR)_0$, and in the second from state $(AR)_0$ to state $(AR)_2$. So the net amount of energy transferred to the weight $(W_{12}^{AR \to})_{\text{rev}} = (W_{10}^{AR \to})_{\text{rev}} + (W_{02}^{AR \to})_{\text{rev}}$. But $(W_{10}^{AR \to})_{\text{rev}} = \Omega_1^R$ and $(W_{02}^{AR \to})_{\text{rev}} = -(W_{20}^{AR \to})_{\text{rev}} = -\Omega_2^R$, and therefore (6.5) is verified. Because we have already proved that Ω^R is a property of system A and reservoir R, we conclude that $(W_{12}^{AR \to})_{\text{rev}}$ is the same for all reversible weight processes that satisfy the specified conditions.

6.6.5 Changes in Irreversible Weight Processes

The difference $\Omega_1^R - \Omega_2^R$ represents the largest amount of energy that can be transferred to a weight out of the composite of system A and reservoir R in any weight process in which A changes from a state A_1 to a state A_2, without any net changes in values of the amounts of constituents and the parameters of either A or R. In particular, if the process is irreversible

$$(W_{12}^{AR \to})_{\text{irr}} < \Omega_1^R - \Omega_2^R \tag{6.6}$$

As shown in Section 6.6.4, inequality (6.6) becomes an equality when the weight process from A_1 to A_2 is reversible.

To prove (6.6), we denote by $(AR)_{S1}$ and $(AR)_{S2}$ the stable equilibrium states reached by the composite of system A and reservoir R upon extraction of the available energies Ω_1^R and Ω_2^R of states A_1 and A_2 with respect to reservoir R, respectively. Denoting by E_1^{AR}, E_{S1}^{AR}, E_2^{AR}, and E_{S2}^{AR} the energies of states $(AR)_1$, $(AR)_{S1}$, $(AR)_2$, and $(AR)_{S2}$, respectively, we can readily verify that the definition of available energy [equation (6.2)], and the energy balance for the reversible weight process of the composite of A and R from state $(AR)_1$ to state $(AR)_{S1}$ imply that $\Omega_1^R = E_1^{AR} - E_{S1}^{AR}$ or $E_{S1}^{AR} = E_1^{AR} - \Omega_1^R$. Similarly, we can verify that $E_{S2}^{AR} = E_2^{AR} - \Omega_2^R$. If a weight process for the composite of A and R from state $(AR)_1$ to state $(AR)_2$ is irreversible, then $E_{S1}^{AR} < E_{S2}^{AR}$ (see Section 5.5) and therefore $E_1^{AR} - \Omega_1^R < E_2^{AR} - \Omega_2^R$ or $E_1^{AR} - E_2^{AR} < \Omega_1^R - \Omega_2^R$. Moreover, the energy balance for the irreversible weight process for the composite of A and R yields that $E_1^{AR} - E_2^{AR} = (W_{12}^{AR \to})_{\text{irr}}$ and thus (6.6) is proved.

6.6.6 Criterion for Reversibility

Given any two states A_1 and A_2 with the same values of the amounts of constituents and the parameters, we can decide whether a weight process for system A from A_1

to A_2 is impossible or, if possible, whether it is reversible or irreversible based on the corresponding values of the energies and available energies of the two states. Specifically, a weight process for system A from A_1 to A_2 is possible only if

$$\Omega_1^R - \Omega_2^R \geq E_1 - E_2 \tag{6.7}$$

and, further, the weight process is reversible only if $\Omega_1^R - \Omega_2^R = E_1 - E_2$, where R is any reservoir.

To verify (6.7) we proceed as follows. We consider a weight process for system A only. In this process system A changes from state A_1 to state A_2. The energy transferred to the weight, $W_{12}^{A\rightarrow}$, is equal and opposite to the change in energy of A, that is, [equation (3.13)],

$$W_{12}^{A\rightarrow} = E_1 - E_2 \tag{6.8}$$

regardless of whether the process is reversible or irreversible. Such a weight process, however, may also be regarded as a special weight process for the composite of system A and reservoir R—special in the sense that the participation of the reservoir is nil. Then $W_{12}^{A\rightarrow}$ may also be expressed as

$$W_{12}^{A\rightarrow} = W_{12}^{AR\rightarrow} \tag{6.9}$$

and either of (6.5) and (6.6) must hold because in the derivation of these relations we impose no restriction on the extent of participation of the reservoir in the weight process. Accordingly, upon combining either relations (6.5), (6.8) and (6.9), or relations (6.6), (6.8), and (6.9) we find that

$$\Omega_1^R - \Omega_2^R = (W_{12}^{AR\rightarrow})_{\text{rev}} = (W_{12}^{A\rightarrow})_{\text{rev}} = E_1 - E_2 \tag{6.10a}$$

$$\Omega_1^R - \Omega_2^R > (W_{12}^{AR\rightarrow})_{\text{irr}} = (W_{12}^{A\rightarrow})_{\text{irr}} = E_1 - E_2 \tag{6.10b}$$

that is, we verify (6.7).

If the change in energy from state A_1 to state A_2 is greater than the change in available energy, relation (6.10) is not satisfied and a weight process in the direction from A_1 to A_2 is impossible.

By virtue of the results in Section 5.7, the difference between the energies of any two states of a weight equals the difference between their available energies, that is, equation (6.10a) is satisfied by all pairs of states of a weight.

Example 6.4.   A system in state A_1 has energy $E_1 = 100$ kJ, and available energy with respect to a reservoir $\Omega_1^R = 200$ kJ. When in mutual stable equilibrium with reservoir R, system A is in a state A_0 corresponding to the same values of the amounts of constituents and the parameters as A_1 and having energy $E_0 = -50$ kJ. In weight processes for the composite of A and R in which A starts from state A_1 and ends in state A_0, what is the smallest value of the difference $E_2^R - E_1^R$ between the end states of the reservoir R? What is the value of $E_2^R - E_1^R$ if no energy is transferred to the weight?

Solution.   The energy balance for the composite of A and R requires that the energy transferred to the weight be equal and opposite to the sum of the energy changes of A and R, that is,

$$W_{10}^{AR\rightarrow} = E_1 - E_0 + E_1^R - E_2^R$$

or, equivalently,

$$E_2^R - E_1^R = E_1 - E_0 - W_{10}^{AR\rightarrow}.$$

The smallest value of $E_2^R - E_1^R$ obtains when the amount of energy transferred to the weight is largest, that is, when the process is reversible and $(W_{10}^{AR\rightarrow})_{\text{rev}} = \Omega_1^R$. Thus

$$(E_2^R - E_1^R)_{\text{smallest}} = E_1 - E_0 - \Omega_1^R = 100 - (-50) - 200 = -50 \text{ kJ}$$

When no energy is transferred to the weight $W_{10}^{AR\rightarrow} = 0$, and

$$E_2^R - E_1^R = E_1 - E_0 = 150 \text{ kJ}$$

In words, here all the energy change of A is transferred to the reservoir, and the weight process for the composite of A and R is irreversible, as we may readily verify using (6.10b).

6.7 Additivity of Available Energy

Available energy is an additive property. To see this, we consider two systems A and B, a system C consisting of the composite of systems A and B, and a reservoir R. For any two states A_1 and B_1 of systems A and B, respectively, we denote by C_{11} the corresponding state of C, by $(\Omega^R)_1^A$, $(\Omega^R)_1^B$, and $(\Omega^R)_{11}^C$ the available energies of A_1, B_1, and C_{11} with respect to reservoir R, and by A_0, B_0, and C_{00} the stable equilibrium states in which A, B, C, and R are in mutual stable equilibrium. To prove that available energy is additive, we must show that the relation

$$(\Omega^R)_{11}^C = (\Omega^R)_1^A + (\Omega^R)_1^B \tag{6.11}$$

is valid for any two systems A and B and any two states A_1 and B_1.

To verify (6.11), we consider a sequence of three reversible weight processes. Starting from the stable equilibrium states A_0, B_0, and C_{00} in which A, B, C, and R are in mutual stable equilibrium, we transfer energy $(\Omega^R)_1^A$ into the composite of systems A and R so that A changes from state A_0 to state A_1 while system B remains in state B_0. Then we transfer energy $(\Omega^R)_1^B$ into the composite of systems B and R so that B changes from state B_0 to state B_1 while system A remains in state A_1. We have already shown that $(\Omega^R)_1^B$ is the correct amount of energy because it is independent of the initial state of the reservoir. Finally, we transfer energy $(\Omega^R)_{11}^C$ out of the composite of A and B with the reservoir R and thus bring system A from state A_1 to state A_0, and system B from state B_1 to state B_0. Because we started from a stable equilibrium state, we conclude that the net energy exchange with the weight, $(\Omega^R)_{11}^C - (\Omega^R)_1^A - (\Omega^R)_1^B$, must be nonpositive. Carrying each of the three reversible weight processes in the reverse direction, we conclude that also $(\Omega^R)_1^A + (\Omega^R)_1^B - (\Omega^R)_{11}^C$ must be nonpositive. These two conclusions are true only if (6.11) is valid. So available energy is an additive property.[2]

6.8 Generalized Available Energy

We now generalize the question investigated in Section 6.4 to weight processes of a composite of a system and a reservoir in which the values of the amounts of constituents and the parameters of the system may experience net changes. The end states are listed in Figure 6.8. Given a system A in a state $A_{1'}$ with energy $E_{1'}$ and values $(n)'$ and $(\beta)'$, a reservoir R, machinery of any desirable degree of perfection, and the set of values $(n)''$ and $(\beta)''$ that may differ from the set $(n)'$ and $(\beta)'$ for state $A_{1'}$, we wish to investigate the question: In a weight process for the composite of system A and reservoir R, in which the state of A changes from an initial state $A_{1'}$ with values $(n)'$ and $(\beta)'$ to a final state with values $(n)''$ and $(\beta)''$, how much energy can be transferred to the weight?

---

[2]We recall that all statements and conclusions of thermodynamics, including the additivity of available energy, refer to well-defined systems in uncorrelated states, as discussed in Sections 2.1 and 2.2.

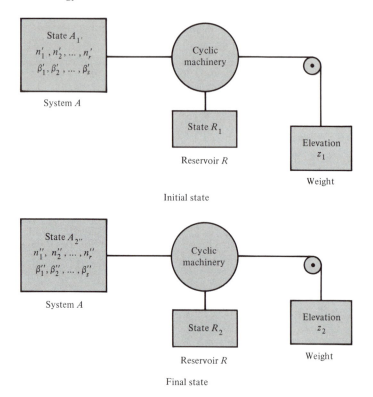

Figure 6.8 End states of weight processes involved in the discussion of generalized available energy.

Similarly to the results we find in Section 6.4, it turns out that the largest amount of energy that can be transferred out to the weight under the generalized conditions just specified is: (1) limited and equal to the energy $(W_{1'0''}^{AR\rightarrow})_{\text{rev}}$ transferred to the weight in a reversible weight process in which system A starts from the initial state $A_{1'}$ and ends in a state $A_{0''}$ corresponding to values $(n)''$ and $(\beta)''$ and such that system A and reservoir R are in mutual stable equilibrium; (2) the same for all such reversible weight processes; (3) independent of the initial state of reservoir R; and (4) the same for all reservoirs that are initially in mutual stable equilibrium with R.

With few modifications, the proof of our assertions follows that given in Section 6.4 and is left to the reader. Thus we conclude that

$$(W_{1'0''}^{AR\rightarrow})_{\text{rev}} \geqq W_{1'2''}^{AR\rightarrow} \tag{6.12}$$

where $W_{1'2''}^{AR\rightarrow}$ denotes the energy transferred to the weight in a weight process in which A starts from state $A_{1'}$ with values $(n)'$ and $(\beta)'$, and ends in state $A_{2''}$ with values $(n)''$ and $(\beta)''$.

In contrast to the result in Section 6.6.3, the value of $(W_{1'0''}^{AR\rightarrow})_{\text{rev}}$ may be positive, zero, or negative, depending on the given state $A_{1'}$ and the set of values $(n)'$, $(\beta)'$ and $(n)''$, $(\beta)''$. We can illustrate this conclusion by using an approach similar to that discussed in Section 5.6.2.

To simplify the interpretation when $(W_{1'0''}^{AR\rightarrow})_{\text{rev}}$ takes negative values, we may rewrite (6.12) in the equivalent form

$$(W_{1'0''}^{AR\leftarrow})_{\text{rev}} \leqq W_{1'2''}^{AR\leftarrow} \tag{6.13}$$

Then $(W_{1'0''}^{AR\leftarrow})_{\text{rev}}$ represents the smallest amount of energy that must be transferred from the weight to the composite of A and R in a weight process under the specified conditions.

For a given reservoir R, and for given values $(n)''$ and $(\beta)''$ of system A, we also conclude that the quantity $(W_{1'0''}^{AR\rightarrow})_{\text{rev}}$ defines the value of a property of system A in state $A_{1'}$. We call this property the *generalized available energy* of system A with respect to reservoir R and values $(n)''$ and $(\beta)''$ of A.

Clearly, the available energy Ω^R (Section 6.5) is a special case of the generalized available energy in the sense that, for any state A_1 of system A with values n and β, the available energy Ω_1^R and the generalized available energy with respect to the same values n and β coincide.

6.9   Features of Generalized Available Energy

6.9.1   Physical Meaning

The generalized available energy with respect to reservoir R and values $(n)''$ and $(\beta)''$ is the largest amount of energy that can be transferred to a weight out of the composite of a system A and a reservoir R in a weight process that starts with A in a state $A_{1'}$ with values $(n)'$ and $(\beta)'$, and ends with A in mutual stable equilibrium with R, and in a state with values $(n)''$ and $(\beta)''$.

6.9.2   Positive and Negative Values

As we discuss in Section 6.8, generalized available energy can take both positive and negative values, depending on the given state, reservoir, and changes of values of the amounts of constituents and the parameters. In contrast, available energy can take only nonnegative values.

6.9.3   Changes in Reversible Weight Processes

Given a system A and two of its states $A_{1'}$ and $A_{2''}$ with values of amounts of constituents and parameters $(n)'$, $(\beta)'$ and $(n)''$, $(\beta)''$, respectively, the energy transferred to the weight in any reversible weight process for the composite of system A and reservoir R that changes the state of A from $A_{1'}$ to $A_{2''}$ equals the difference of the generalized available energies of the two states, each evaluated with respect to reservoir R and arbitrary but the same fixed values n and β of the amounts of constituents and the parameters, respectively.

Example 6.5.   Verify the preceding assertion.

Solution.   For the composite of A and R, we consider a sequence of two reversible weight processes. The first process starts from state $(AR)_{1'}$, and ends in a state $(AR)_0$ in which the state A_0 of A has values of amounts of constituents and parameters n and β, and A and R are in mutual stable equilibrium. The second reversible weight process starts from $(AR)_0$, and ends in a state $(AR)_{2''}$ in which the state of A is $A_{2''}$. As a result of this sequence, the net amount of energy transferred to the weight $(W_{1'2''}^{AR\rightarrow})_{\text{rev}} = (W_{1'0}^{AR\rightarrow})_{\text{rev}} + (W_{02''}^{AR\rightarrow})_{\text{rev}}$. But $(W_{1'0}^{AR\rightarrow})_{\text{rev}}$ is the generalized available energy of $A_{1'}$ with respect to values n and β, and $(W_{02''}^{AR\rightarrow})_{\text{rev}}$ equals the negative of the generalized available energy of $A_{2''}$ with respect to the same values, that is, $-(W_{2''0}^{AR\rightarrow})_{\text{rev}}$. Thus the assertion is verified. Because generalized available energy is a property of system A with respect to reservoir R, we conclude that $(W_{1'2''}^{AR\rightarrow})_{\text{rev}}$ is the same for all reversible weight processes satisfying the specified conditions.

This result generalizes the conclusion we reach in Example 6.3 to states $A_{1'}$ and $A_{2''}$, which are associated with differing values of the amounts of constituents and the parameters. It also shows that the difference of the generalized available energies of two states $A_{1'}$ and $A_{2''}$ with respect to the same given values n and β depends on the states $A_{1'}$ and $A_{2''}$, and the reservoir R, but not on the arbitrarily selected values n and β. For this reason we may denote the generalized available energy of state $A_{1'}$ by $\Omega_{1'}^R$ and write

$$(W_{1'2''}^{AR\rightarrow})_{\text{rev}} = \Omega_{1'}^R - \Omega_{2''}^R \tag{6.14}$$

where we understand that each of the generalized available energies $\Omega_{1'}^R$ and $\Omega_{2''}^R$ is evaluated with respect to the same values n and β even though state $A_{1'}$ is associated with values $(n)'$ and $(\beta)'$ and state $A_{2''}$ with values $(n)''$ and $(\beta)''$.

6.9.4   Changes in Irreversible Weight Processes

The difference $\Omega_{1'}^R - \Omega_{2''}^R$ between two generalized available energies represents the largest amount of energy that can be transferred to a weight out of the composite of a system A and a reservoir R in any weight process for the composite in which A changes from a state $A_{1'}$ to a state $A_{2''}$ regardless of whether the values of amounts of constituents and parameters of $A_{1'}$ are the same or differ from the corresponding values of $A_{2''}$. In particular, if the process is irreversible,

$$(W_{1'2''}^{AR\rightarrow})_{\text{irr}} < \Omega_{1'}^R - \Omega_{2''}^R \tag{6.15}$$

As shown in Section 6.9.3, inequality 6.15 becomes an equality only if the weight process from $A_{1'}$ to $A_{2''}$ is reversible.

Example 6.6.   Prove inequality (6.15).

Solution.   As usual, we denote by $(AR)_{S1}$ and $(AR)_{S2}$ the stable equilibrium states reached by the composite of system A and reservoir R upon extraction of the generalized available energies $\Omega_{1'}^R$ and $\Omega_{2''}^R$ of states $A_{1'}$ and $A_{2''}$ with respect to reservoir R and values n and β. Denoting by $E_{1'}^{AR}$, E_{S1}^{AR}, $E_{2''}^{AR}$, and E_{S2}^{AR} the energies of states $(AR)_{1'}$, $(AR)_{S1}$, $(AR)_{2''}$, and $(AR)_{S2}$, respectively, we can readily verify that $E_{S1}^{AR} = E_{1'}^{AR} - \Omega_{1'}^R$ and $E_{S2}^{AR} = E_{2''}^{AR} - \Omega_{2''}^R$. If a weight process for the composite of A and R from state $(AR)_{1'}$ to state $(AR)_{2''}$ is irreversible, we must have $E_{S1}^{AR} < E_{S2}^{AR}$ (Section 5.4), and therefore $E_{1'}^{AR} - \Omega_{1'}^R < E_{2''}^{AR} - \Omega_{2''}^R$ or $E_{1'}^{AR} - E_{2''}^{AR} < \Omega_{1'}^R - \Omega_{2''}^R$. Moreover, the energy balance for the irreversible weight process of the composite of A and R yields that $E_{1'}^{AR} - E_{2''}^{AR} = (W_{1'2''}^{AR\rightarrow})_{\text{irr}}$, so (6.15) is proved.

6.9.5   Criterion for Reversibility

Given any two states $A_{1'}$ and $A_{2''}$ of system A, a weight process for system A alone from $A_{1'}$ to $A_{2''}$ is possible only if either of the two relations

$$\Omega_{1'}^R - \Omega_{2''}^R = (W_{1'2''}^{AR\rightarrow})_{\text{rev}} = (W_{1'2''}^{A\rightarrow})_{\text{rev}} = E_{1'} - E_{2''} \tag{6.16a}$$

$$\Omega_{1'}^R - \Omega_{2''}^R > (W_{1'2''}^{AR\rightarrow})_{\text{irr}} = (W_{1'2''}^{A\rightarrow})_{\text{irr}} = E_{1'} - E_{2''} \tag{6.16b}$$

is satisfied, where R is any reservoir and $\Omega_{1'}^R$ and $\Omega_{2''}^R$ are the generalized available energies of states $A_{1'}$ and $A_{2''}$ with respect to reservoir R and the same, arbitrarily specified values n and β. The proof of these assertions is left to the reader. It follows that the weight process is reversible only if (6.16a) is valid.

6.9.6   Additivity

Differences in generalized available energies are additive, that is, for a system C composed of two subsystems A and B, we have

$$(\Omega^R)_{2''2''}^C - (\Omega^R)_{1'1'}^C = \left[(\Omega^R)_{2''}^A - (\Omega^R)_{1'}^A\right] + \left[(\Omega^R)_{2''}^B - (\Omega^R)_{1'}^B\right] \tag{6.17}$$

for any states $A_{1'}$, $A_{2''}$, $B_{1'}$, and $B_{2''}$ of the two subsystems. The proof of additivity is based on arguments similar to those outlined in Sections 6.7 and 6.9.3.

As we show in the next chapter, relations (6.16) and (6.17) are the keys to the definition and derivation of the characteristic features of entropy, and hold for any state — unsteady, steady, nonequilibrium, equilibrium, and stable equilibrium — of any system — large or small. To simplify the notation, in what follows we use these relations without the primes and double primes, even though we refer to generalized available energy differences, that is, we rewrite them in the forms

$$\Omega_1^R - \Omega_2^R = (W_{12}^{AR\rightarrow})_{rev} = (W_{12}^{A\rightarrow})_{rev} = E_1 - E_2 \qquad (6.18a)$$

$$\Omega_1^R - \Omega_2^R > (W_{12}^{AR\rightarrow})_{irr} = (W_{12}^{A\rightarrow})_{irr} = E_1 - E_2 \qquad (6.18b)$$

$$(\Omega^R)_{22}^C - (\Omega^R)_{11}^C = \left[(\Omega^R)_2^A - (\Omega^R)_1^A\right] + \left[(\Omega^R)_2^B - (\Omega^R)_1^B\right] \qquad (6.19)$$

Problems

6.1 Three states A_1, A_2, and A_0 of a system A correspond to the same values of amounts of constituents and parameters, and have energies and available energies with respect to a reservoir R as listed in the table.

State	Energy E	Available energy Ω^R
A_1	90 kJ	100 kJ
A_2	60 kJ	20 kJ
A_0	20 kJ	—

In state A_0, system A is in mutual stable equilibrium with the reservoir R.

(a) Is a weight process for system A possible from A_1 to A_2?

(b) How much energy is transferred to a weight in a weight process for system A that has A_1 and A_2 as end states?

(c) What is the largest amount of energy that can be transferred to a weight in a weight process for the composite of A and R in which system A goes from A_1 to A_2?

(d) What is the energy change of the reservoir R if the process in part (c) occurs?

(e) How much is the available energy of state A_0?

(f) What is the least amount of energy transfer from a weight to the composite of A and R in a weight process for A and R in which the state of A changes from A_0 to A_1?

6.2 The change in available energy from iron ore to steel is 6.3×10^6 kJ/ton of steel, that is, the available energy of 1 ton of steel is higher by 6.3×10^6 kJ than the appropriate quantity of iron oxide from which it is extracated. The available energy of the fuel used in steelmaking is 48,000 kJ/kg of fuel.

(a) What is the smallest amount of fuel that is required for the production of 1 ton of steel?

(b) In some steelmaking plants, it takes about 30×10^6 kJ/ton steel. To what extent is the available energy of the fuel wasted in these plants?

(c) What is the cause of this waste?

6.3 The composite of a system A and a reservoir R experiences three weight processes, α, β, and γ, each without net changes in values of amounts of constituents and parameters of either A or R. The characteristics of the processes are listed in the table. State A_0 is such that A and R are in mutual stable equilibrium.

Process	End states of A	Energy transfer
α	$A_2 \rightarrow A_0$	$(W_{20}^{AR\rightarrow})_{rev} = 15$ kJ
β	$A_0 \rightarrow A_1$	$(W_{01}^{A\rightarrow})_{rev} = -10$ kJ
γ	$A_2 \rightarrow A_1$	$(W_{21}^{AR\rightarrow})_{irr} = 0$ kJ

(a) What types of states are A_1 and A_2?

(b) What are the adiabatic availability and the available energy of state A_1?

(c) What is the energy transfer to a weight in a reversible weight process for the composite of A and R in which A changes from A_1 to A_2?

6.4 Two states A_1 and A_2 of a system A have values of energy and available energy with respect to a given reservoir R as listed in the table. State A_0 is such that A and R are in mutual stable equilibrium.

State	E	Ω^R
A_1	120 kJ	40 kJ
A_2	180 kJ	70 kJ
A_0	100 kJ	0 kJ

(a) Is a weight process for system A from state A_1 to state A_2 possible?

(b) In a reversible weight process for the composite of A and R in which A changes from state A_2 to state A_1, what is the change in energy of the reservoir R?

6.5 With respect to a reservoir R selected as a reference once and for all, the available energy Ω^R is a property defined for all states of any system A, in the same way as energy E and adiabatic availability Ψ are properties defined for all states of A. In terms of Ω^R and Ψ, we can define the property

$$P^R = \Omega^R - \Psi$$

It represents the available energy with respect to R that remains in A upon extraction of its adiabatic availability.

(a) Show that if in state A_0, system A is in mutual stable equilibrium with R, then $P_0^R = 0$.

(b) Show that in any reversible weight process from a state A_1 to a state A_2 of a system A, the value of property P^R remains unchanged, that is,

$$(P_2^R - P_1^R)_{rev} = 0$$

(c) Denoting by E_S the energy of the unique stable equilibrium state reached upon extraction of the adiabatic availability, and by E_0 the energy of the state A_0 defined in part (a), we will see (Problem 3.11) that in an irreversible weight process from A_1 to A_2 of system A the value of P^R increases if both $E_{S1} > E_0$ and $E_{S2} > E_0$, and it decreases if both $E_{S1} < E_0$ and $E_{S2} < E_0$. If, instead, $E_{S1} < E_0$ and $E_{S2} > E_0$, the value of P^R can either decrease or increase.

6.6 Four states $A_{1'}, A_{0'}, A_{2''},$ and $A_{0''}$ of a system A have the same values for each amount of constituent and each parameter except volume. The value of the volume is V' for states $A_{1'}$ and $A_{0'}$, and V'' for states $A_{2''}$ and $A_{0''}$. In states $A_{0'}$ and $A_{0''}$, system A is in mutual stable equilibrium with a reservoir R. The available energies of states $A_{1'}$ and $A_{2''}$ with respect to R are $\Omega_{1'}^R = 100$ kJ and $\Omega_{2''}^R = 140$ kJ. The generalized available energy of state $A_{0'}$ with respect to R and the value V'' of the volume is 80 kJ.

(a) What is the generalized available energy of $A_{1'}$ with respect to R and the value V'' of the volume?

(b) What is the generalized available energy of $A_{2''}$ with respect to R and the value V' of the volume?

(c) If states $A_{1'}$ and $A_{2''}$ have the same value of the energy, is a weight process for system A from $A_{1'}$ to $A_{2''}$ possible?

(d) In a reversible weight process for the composite of A and R in which A changes from $A_{2''}$ to $A_{1'}$, what is the change in energy of the reservoir?

6.7 The composite of a system A and a reservoir R experiences five weight processes, $\alpha, \beta, \gamma, \delta,$ and ϵ, each without net changes in values of amounts of constituents and parameters of either A or R, except the volume of A. The characteristics of the processes are listed in the table. In states $A_{0'}$ and $A_{0''}$ system A and reservoir R are in mutual stable equilibrium. The primes and double primes in the subscripts refer to the values V' and V'' of the volume of A, respectively.

Process	End states of A	Energy transfer
α	$A_{2'} \rightarrow A_{0'}$	$(W_{2'0'}^{A\rightarrow})_{\text{rev}} = 0$ kJ
β	$A_{0'} \rightarrow A_{0''}$	$(W_{0'0''}^{AR\rightarrow})_{\text{rev}} = 200$ kJ
γ	$A_{2''} \rightarrow A_{0''}$	$(W_{2''0''}^{AR\rightarrow})_{\text{rev}} = 200$ kJ
δ	$A_{2''} \rightarrow A_{1''}$	$(W_{2''1''}^{A\rightarrow})_{\text{irr}} = 0$ kJ
ϵ	$A_{1''} \rightarrow A_{0''}$	$(W_{1''0''}^{A\rightarrow})_{\text{rev}} = 100$ kJ

(a) Is state $A_{2'}$ identical to state $A_{0'}$?

(b) What is the generalized adiabatic availability of state $A_{2''}$ with respect to the value V' of the volume?

(c) What is the adiabatic availability of state $A_{1''}$?

(d) What is the generalized available energy of state $A_{1''}$ with respect to R and the value V' of the volume?

7 Entropy

Observations of physical phenomena that can be represented as weight processes show that, in general, the amount of energy that can be transferred from the system to the weight differs from the energy of the system above the ground-state energy. Said differently, the capacity of a system to raise a weight is not always equal to the energy of the system in excess of the ground-state energy. For example, an initially charged battery can raise a weight via an electric motor. However, left idle and well insulated on a shelf, the battery discharges internally without transferring out any energy, and at the end, it has lost all its capacity to raise a weight via the electric motor. Formally, we find in Chapter 5 that the difference between the energy of a system and its capacity to raise a weight in a weight process is related to the difference between energy and adiabatic availability, and in Chapter 6 that the difference between the energy of a system and its capacity to raise a weight when in combination with a reservoir is related to the difference between energy and available energy. An alternative and more general way of accounting for these differences is by means of the property called entropy.

In this chapter we define the property entropy for any state of any system and discuss some of its characteristic features.

7.1 Definition of Property Entropy

We consider the following set of operations and measurements. For the sake of convenience, we select once and for all a reference state A_o for system A, an auxiliary reservoir R, and arbitrary values n_1, n_2, ..., n_r, β_1, β_2, ..., β_s of the amounts of the r constituents and the s parameters within the ranges specified for system A. For any state A_1 of system A, we evaluate the difference $E_1 - E_o$ between the energies of states A_1 and A_o, and the difference $\Omega_1^R - \Omega_o^R$ between the generalized available energies of the two states with respect to the auxiliary reservoir R and the values n and β. We note that the values of the amounts of constituents and the parameters associated with state A_1 need not be equal to either the reference values n and β, or the corresponding values of state A_o.

In terms of these differences, each of which is measurable by means of an appropriately defined weight process, we compute the value S_1 using the relation

$$S_1 = S_o + \frac{1}{c_R}[(E_1 - E_o) - (\Omega_1^R - \Omega_o^R)] \tag{7.1}$$

where S_o is a constant fixed once and for all for system A, and c_R is a positive constant property of reservoir R that we define explicitly in Section 7.4. After defining the concept of temperature in Section 9.2, we show in Section 9.7 that c_R is equal also to the constant temperature of the reservoir.

In Section 7.4, we prove that the value S_1 associated with state A_1 is independent of the reservoir R used in the set of operations and measurements just defined. In other words, no matter what reservoir is used, we conclude that we always obtain the same value S_1 for the given state A_1 and therefore that the role of the reservoir in these operations and measurements is only auxiliary. In addition, because each of the differences $E_1 - E_o$ and $\Omega_1^R - \Omega_o^R$ is independent of the values of n and β, we also conclude that $S_1 - S_o$ is independent of these values and that also their role is only auxiliary. These conclusions are valid for any state A_1, and therefore the set of operations and measurements just cited results in the definition of S_1 as the value of a property of system A in state A_1. We call S the *entropy* and denote its value at state A_1 by S_1. Because of the features discussed in Section 7.2, entropy is a property of great practical importance.

Upon repeating the same set of operations and measurements for a state A_2, equation (7.1) yields the value S_2 of the entropy at state A_2 in the form

$$S_2 = S_o + \frac{1}{c_R}\left[(E_2 - E_o) - (\Omega_2^R - \Omega_o^R)\right] \tag{7.2}$$

Moreover, upon subtracting equation (7.1) from equation (7.2), we find that

$$S_2 = S_1 + \frac{1}{c_R}\left[(E_2 - E_1) - (\Omega_2^R - \Omega_1^R)\right] \tag{7.3}$$

It is noteworthy that the constant S_o is the entropy of the reference state A_o, as can be readily verified by substituting subscript o for subscript 1 in (7.1). Moreover, because entropy is a property, its value is part of the state of the system. In general, any change of state, spontaneous or induced, from a state A_1 to a state A_2 of a system A is associated with a change $S_2 - S_1$ in entropy. The difference $S_2 - S_1$ is independent of the type of interaction forces that may effect the change between the two specified states, that is, independent of whether the change from state A_1 to state A_2 is achieved by a weight process, a weight process of the composite of system A and a reservoir, or any other kind of process. This independence has the same meaning as, for example, the independence of the energy change $E_2 - E_1$.

7.2 Features of Entropy

7.2.1 Domain of Definition

Like energy, entropy is a property defined for all states of a system — unsteady, steady, nonequilibrium, or any kind of equilibrium — and for all systems — large or small, including a single particle. This feature follows from the fact that energy and generalized available energy in (7.1) are defined for all states and all systems.

7.2.2 Additivity

Differences in entropy are additive, that is, the difference in entropy between any two states of a composite system equals the sum of the differences in entropy between the corresponding states of the component systems. Indeed, for a composite C of two systems A and B, we show in Section 3.6 that energy differences are additive, that is, that

$$E_{22}^C - E_{11}^C = (E_2^A - E_1^A) + (E_2^B - E_1^B) \tag{7.4}$$

and in Section 6.9.6 that generalized available energy differences are additive, that is, that

$$(\Omega^R)_{22}^C - (\Omega^R)_{11}^C = \left[(\Omega^R)_2^A - (\Omega^R)_1^A\right] + \left[(\Omega^R)_2^B - (\Omega^R)_1^B\right] \tag{7.5}$$

Therefore, writing the expressions for the entropies S_{11}^C, S_{22}^C, S_1^A, S_2^A, S_1^B, and S_2^B according to the definition (7.1), and using (7.4) and (7.5), we obtain the relation

$$S_{22}^C - S_{11}^C = (S_2^A - S_1^A) + (S_2^B - S_1^B) \tag{7.6}$$

which holds for any states A_1, A_2, B_1, and B_2 of the two systems A and B. This result proves that entropy differences are additive. It may be easily generalized to composite systems consisting of more than two systems.

Using equation (7.6), we can easily show that entropy is an additive property provided that the entropy of the reference state of the composite system is selected consistent with the entropies of the reference states of its parts, that is, provided that the entropies S_{oo}^C, S_o^A, and S_o^B of the reference states C_{oo}, A_o, and B_o, respectively, are selected so that $S_{oo}^C = S_o^A + S_o^B$. Then and only then,

$$S_{11}^C = S_1^A + S_1^B \tag{7.7}$$

for all pairs of systems A and B and all pairs of states A_1 and B_1. In due course (Section 9.8) we will see that, like energy, entropy can be assigned absolute values that must be nonnegative.

Example 7.1.   Prove that entropy is additive.

Solution.   Upon changing all states denoted by subscript 2 to states denoted by subscript o in equation (7.6), and using the condition $S_{oo}^C = S_o^A + S_o^B$, we find (7.7).

7.2.3   Changes in Reversible Weight Processes
In a reversible weight process for a system A that changes from a state A_1 to a state A_2, the value of the entropy remains unchanged, that is, the value S_2 of the entropy of the final state A_2 is equal to the value S_1 of the initial state A_1. Indeed, for such a process we have [equation (6.18a)]

$$\Omega_1^R - \Omega_2^R = (W_{12}^{A\rightarrow})_{rev} = E_1 - E_2 \tag{7.8}$$

and upon substituting in (7.3), we find that for a reversible weight process

$$S_2 - S_1 = 0 \tag{7.9}$$

To remind ourselves of the conditions that predicate this result, we sometimes express it in the equivalent form

$$(DS)_{rev}^{weight\ process} = 0 \tag{7.10}$$

where the subscript rev denotes that the weight process is reversible, and the symbol D a change (large or small) of the quantity that follows. Because processes do not occur instantly, (7.9) or (7.10) is a time-dependent result.

Because a system consisting of a weight can experience reversible processes only (see Section 5.7), equation (7.9) holds for any two states of a weight. Accordingly, all states of a weight have the same value of the entropy. So we conclude that a weight cannot exchange entropy with any system.

7.2.4   Changes in Irreversible Weight Processes
In an irreversible weight process in which system A changes from a state A_1 to a state A_2, the value of the entropy increases, that is, the value S_2 of the entropy in the final state A_2 is greater than the value S_1 in the initial state A_1. Indeed, for such a process we have [relation (6.18b)]

$$\Omega_1^R - \Omega_2^R > (W_{12}^{A\rightarrow})_{irr} = E_1 - E_2 \tag{7.11}$$

and upon substituting in (7.3), we find that for an irreversible weight process

$$S_2 - S_1 > 0 \tag{7.12}$$

Again, we sometimes express this result in its equivalent form

$$(DS)_{\text{irr}}^{\text{weight process}} > 0 \tag{7.13}$$

where the subscript irr denotes that the weight process is irreversible. Inequality (7.13) is a time-dependent result because a change from state A_1 to state A_2 cannot occur instantly.

We know that in a weight process for system A, between states A_1 and A_2 that differ in both energy and entropy, that is, with $E_1 \neq E_2$ and $S_1 \neq S_2$, the exchange between A and the weight involves energy flow but no entropy flow. Hence we conclude that the increase in entropy, $S_2 - S_1$, which accompanies an irreversible weight process occurs spontaneously and internally within system A. Whenever there is a spontaneous and internal increase of entropy in a system, the process is imperfect in the sense that the capacity of the system to raise a weight in a gravity field is reduced. For example, in a change of state from A_1 to A_2, the limiting amount of energy that can be transferred to a weight by combining the system with a reservoir R is $\Omega_1^R - \Omega_2^R$, yet in the presence of a spontaneous increase of entropy, the energy transferred is $(W_{12}^{A\rightarrow})_{\text{irr}}$, namely, less than $\Omega_1^R - \Omega_2^R$ [equation (7.11)].

In due course we will see that the more irreversible the process, the larger the increase in entropy and the worse the performance of the system. Conversely, the less irreversible the process, the smaller the increase in entropy and the better the performance of the system, with best or perfect performance when the process is reversible.

We may rewrite (7.13) as an equality in the form

$$(DS)_{\text{irr}}^{\text{weight process}} = S_{\text{irr}} \tag{7.14}$$

where S_{irr} is a positive quantity. We say that S_{irr} represents the *entropy created or generated by irreversibility* inside system A in the course of the irreversible weight process. Whereas energy can be neither destroyed nor created, we see that entropy cannot be destroyed but may be created.

7.2.5 Principle of Nondecrease of Entropy

We know from many experiences that changes of state can occur in isolated systems spontaneously. Each such change may be regarded as a weight process with a null effect on the weight. As such, it is a special case of the processes discussed in Sections 7.2.3 and 7.2.4. If the subscript isol denotes a spontaneous process of an isolated system, the change in entropy must satisfy the relation

$$S_2 - S_1 \geqq 0 \quad \text{or, equivalently,} \quad (DS)_{\text{isol}} \geqq 0 \tag{7.15}$$

where the equal sign applies if the spontaneous process is reversible, and the strict inequality if it is irreversible.

Because entropy is an additive property, and because any process of any system may always be regarded as a process of a composite isolated system consisting of all the systems with which the original system interacts, the conclusion represented by (7.15) is of great generality and practical importance. We call it the *principle of nondecrease of entropy*.

In some discussions the second law is presented as equivalent to the principle of nondecrease of entropy. Entropy nondecrease is certainly an important consequence of both the first and second laws. However, the quintessence of the second law is not entropy nondecrease but the more fundamental implications that adiabatic availability and entropy are properties.

Because changes of state cannot occur instantly, we conclude that (7.15) is a time dependent result, that is, a relation between entropy values at two different instants of time for a system that at time t_1 is in state A_1 and at a later time t_2 in state A_2. For any weight process, the relation requires that the entropy S_2 of the system at time t_2 be greater than or equal to the entropy S_1 at the earlier time t_1.

Although time dependent, relations (7.10), (7.13), and (7.15) are obtained without using explicitly an equation describing the time evolution of the system, that is, without using an equation of motion. Because these relations are implications of the first and second laws, and to date we have no experimental reason to question the validity of these laws, we conclude that the equation of motion of any system must not decrease the entropy in any weight process, namely, for a system undergoing a weight process the equation of motion must imply that $dS/dt \geq 0$.

An alternative way of stating the principle of nondecrease of entropy is to say that whenever any process occurs in nature, the total entropy of all systems involved in the process either increases if the process is irreversible, or remains constant if the process is reversible.

We cannot overemphasize that the principle of nondecrease of entropy does not require that a process be reversible, a little irreversible, or a lot irreversible. The principle simply states one of the consequences resulting from the process being of one or the other type.

The possibility for entropy to be created by irreversibility reflects a physical phenomenon that is sharply distinct from the great conservation principles that underlie the description of physical phenomena in mechanics. It brings forth the need to consider not only properties that are conserved, such as energy, mass, momentum, and electric charge, but also properties that may be spontaneously created, such as entropy.

In fact, even the requirement of entropy conservation in reversible processes of isolated systems introduces a radical departure from the description of physical phenomena in mechanics, a departure that would persist even if no process in nature were irreversible. The reason is that this requirement brings forth the need to describe not only states with zero entropy, such as the states encountered in mechanics, but also states with various nonzero values of entropy.

7.3 Entropy Balance

Because entropy is an additive property, the conclusions reached in Sections 7.2.4 and 7.2.5 are of great practical importance. For example, if a weight process for the composite C of two systems A and B is reversible, the change of entropy of system A must be equal and opposite to the change of entropy of system B.

Example 7.2. Show that if the composite C of systems A and B undergoes a reversible weight process that changes A from state A_1 to state A_2, and B from state B_1 to state B_2, the entropy changes $S_2^A - S_1^A$ of A and $S_2^B - S_1^B$ of B satisfy the relation $S_2^B - S_1^B = -(S_2^A - S_1^A)$, that is, they are equal and opposite.

Solution. Because the weight process for C is reversible, relation (7.10) applies, that is, $S_{22}^C - S_{11}^C = 0$. On the other hand, because entropy differences are additive, $S_{22}^C - S_{11}^C = 0 = S_2^A - S_1^A + S_2^B - S_1^B$ and, therefore, $S_2^B - S_1^B = -(S_2^A - S_1^A)$. Of course, the same result obtains if system C is isolated and undergoes a reversible process.

We may interpret the relation between the entropy changes of A and B as describing a transfer, an exchange, or a flow of entropy between the two systems. The exchange of entropy is a consequence of interactions between the systems. Later, we classify interactions according to their capability of transferring various additive properties, such as energy and entropy, between the interacting systems.

Another example of the practical importance of the results in Sections 7.2.4 and 7.2.5 is provided by an irreversible weight process for the composite C of two systems A and B. The sum of the changes of entropy of A and B must be greater than zero [relation (7.13) for nonzero net effects or (7.15) for zero net effects], that is,

$$S_2^A - S_1^A + S_2^B - S_1^B > 0 \tag{7.16}$$

We may rewrite this relation as an equality in the form

$$S_2^A - S_1^A + S_2^B - S_1^B = S_{\text{irr}} \tag{7.17}$$

by introducing the positive entropy S_{irr} generated by irreversibility. In general, the entropy S_{irr} may have been generated spontaneously in either system A or system B or both. Equation (7.17) by itself is not adequate to decide in which of the two systems the entropy is generated by irreversibility. Yet, knowing where the entropy is generated is important because, as we discuss later, steps may be taken to reduce this generation and thus improve the process — bring the process closer to its being reversible.

We can determine in which system entropy is generated by irreversibility in a large class of practical applications in which it is possible to specify the net amount of entropy that has been transferred or exchanged between systems A and B. Indeed, upon writing

$$S^{A \leftarrow B} = \begin{bmatrix} \text{net amount of entropy} \\ \text{received by system } A \\ \text{from system } B \end{bmatrix}$$

$$= - \begin{bmatrix} \text{net amount of entropy} \\ \text{received by system } B \\ \text{from system } A \end{bmatrix} = -S^{B \leftarrow A} \tag{7.18}$$

we find that the entropies S_{irr}^A and S_{irr}^B generated by irreversibility in systems A and B, respectively, are given by the relations

$$(S_{\text{irr}})^A = S_2^A - S_1^A - S^{A \leftarrow B} \tag{7.19}$$

$$(S_{\text{irr}})^B = S_2^B - S_1^B - S^{B \leftarrow A}$$

$$= S_2^B - S_1^B + S^{A \leftarrow B} \tag{7.20}$$

Of course, the sum of $(S_{\text{irr}})^A$ and $(S_{\text{irr}})^B$ equals S_{irr} [equation (7.17)], that is,

$$(S_{\text{irr}})^A + (S_{\text{irr}})^B = S_{\text{irr}} > 0 \tag{7.21}$$

It is noteworthy that (7.21) does not specify the sign of each of the quantities $(S_{\text{irr}})^A$ and $(S_{\text{irr}})^B$. However, we can prove that both $(S_{\text{irr}})^A$ and $(S_{\text{irr}})^B$ must be nonnegative. To this end, we consider one process for which (7.19) and (7.20) are valid. Then we consider a second process in which state A_1, state A_2, and the net amount of entropy received by system A from system B are the same as for the initial process, but states B_1 and B_2 of system B are such that $(S_{\text{irr}})^B = 0$. The second process may be reversible or irreversible, that is, its amount of entropy generated by irreversibility $S_{\text{irr}}' \geqq 0$, so (7.21) becomes $(S_{\text{irr}})^A = S_{\text{irr}}' \geqq 0$. Moreover, upon repeating for system B the argument just given for system A, we also conclude that $(S_{\text{irr}})^B \geqq 0$. Thus we verify that the entropy generated by irreversibility in any subsystem of a composite system is always a nonnegative quantity.

As a result of these considerations, we conclude that the entropy change of a system, $(S_2 - S_1)^{\text{system}}$, must be at least equal to the entropy received from the surroundings — because entropy cannot be destroyed — and may be greater than the entropy received if an amount $(S_{\text{irr}})^{\text{system}}$ is created inside the system by irreversibility. We use the superscript system to emphasize that the corresponding quantities pertain to the system, not to others in its surroundings. We summarize the foregoing general conclusions in the form

$$(S_2 - S_1)^{\text{system}} = S^{\leftarrow} + (S_{\text{irr}})^{\text{system}} \tag{7.22}$$

with

$$(S_{\text{irr}})^{\text{system}} = 0 \quad \text{if the process is reversible} \tag{7.23a}$$

$$(S_{\text{irr}})^{\text{system}} > 0 \quad \text{if the process is irreversible} \tag{7.23b}$$

We call (7.22) the *entropy balance equation* or, simply, the *entropy balance*. The entropy balance together with the energy balance [equation (3.10)] are the keystones of the thermodynamic analysis of physics and engineering problems in which the net amounts of energy received and entropy received from the surroundings of a system are specified. In due course we show that this is the case when the system experiences certain types of interaction, such as work, heat and bulk flow.

7.4   Definition of Constant c_R

In Section 7.1 we define the entropy S_1 of a system A in state A_1 in terms of measurements and operations that involve a reservoir R and a constant c_R (equation 7.1). In this section we define the constant c_R and show that S_1 is independent of R, that is, we prove that S is a property of system A only.[1]

To this end we consider a system A and an arbitrary reservoir R, and denote by $(DE_{12}^R)_{\text{rev}}^A$ the change in energy of R in a reversible weight process for the composite of A and R in which A changes from state A_1 to state A_2, and R from state R_1 to state R_2. We use the subscripts 1 and 2 to denote that the change implied by D is from state R_1 to state R_2, superscript R to signify that the change is in the energy of R, and the superscript A to emphasize that the reversible weight process involves the composite of systems A and R.

In addition, we consider another reservoir Q, selected once and for all and not necessarily in mutual stable equilibrium with R. We denote by $(DE_{12}^Q)_{\text{rev}}^A$ the change in energy of Q in a reversible weight process for the composite of A and Q in which A changes again from state A_1 to state A_2, and Q from state Q_1 to state Q_2. Using this notation, we prove the following.

1. The energy changes $(DE_{12}^R)_{\text{rev}}^A$ and $(DE_{12}^Q)_{\text{rev}}^A$ are given by the relations

$$(DE_{12}^R)_{\text{rev}}^A = E_1 - E_2 - (\Omega_1^R - \Omega_2^R) \tag{7.24}$$

and

$$(DE_{12}^Q)_{\text{rev}}^A = E_1 - E_2 - (\Omega_1^Q - \Omega_2^Q) \tag{7.25}$$

2. The ratio $(DE_{12}^R)_{\text{rev}}^A / (DE_{12}^Q)_{\text{rev}}^A$ is positive.

---

[1]The proof in this section was developed by Enzo Zanchini of the University of Bologna during a stay at MIT. It is a generalization to the present context of a proof found in E. Fermi, *Thermodynamics* (Prentice-Hall, Englewood Cliffs, N. J., 1936).

3. The ratio $(DE_{12}^R)_{\text{rev}}^A/(DE_{12}^Q)_{\text{rev}}^A$ is independent of the system A and its states A_1 and A_2.

4. The constant c_R, defined by the relation

$$c_R = c_Q \frac{(DE_{12}^R)_{\text{rev}}^A}{(DE_{12}^Q)_{\text{rev}}^A} \tag{7.26}$$

is a positive property of the reservoir R only, where c_Q is a positive scaling factor.

5. The use of equation (7.26) in equation (7.1) results in a value of S_1 that is independent of reservoir R.

Example 7.3.   Verify assertion 1.

Solution.   The energy balance for the reversible weight process of the composite of A and R yields $E_2 - E_1 + (DE_{12}^R)_{\text{rev}}^A = -(W_{12}^{AR\rightarrow})_{\text{rev}}$. In addition, equation (6.18a) yields $(W_{12}^{A\rightarrow})_{\text{rev}} = \Omega_1^R - \Omega_2^R$, so (7.24) is verified. A similar balance verifies (7.25).

Regarding assertion 2 we assume that the ratio $(DE_{12}^R)_{\text{rev}}^A/(DE_{12}^Q)_{\text{rev}}^A$ is negative and, therefore, that the sign of $(DE_{12}^R)_{\text{rev}}^A$ is opposite to that of $(DE_{12}^Q)_{\text{rev}}^A$. For example, if $(DE_{12}^R)_{\text{rev}}^A > 0$, then $(DE_{12}^Q)_{\text{rev}}^A < 0$. But for $(DE_{12}^R)_{\text{rev}}^A = E_1 - E_2 - (\Omega_1^R - \Omega_2^R) > 0$, the criterion for reversibility of a weight process expressed by (6.18a) yields that the weight process for A from A_1 to A_2 is impossible, whereas for $(DE_{12}^Q)_{\text{rev}}^A < 0$ the same criterion yields that the weight process for A from A_1 to A_2 is possible. Clearly, a weight process cannot be both possible and impossible, and therefore our assumption that $(DE_{12}^R)_{\text{rev}}^A$ and $(DE_{12}^Q)_{\text{rev}}^A$ have opposite signs is absurd. So, we conclude that the ratio $(DE_{12}^R)_{\text{rev}}^A/(DE_{12}^Q)_{\text{rev}}^A$ is positive.

The proof of assertion 3, that the ratio $(DE_{12}^R)_{\text{rev}}^A/(DE_{12}^Q)_{\text{rev}}^A$ is independent of the nature of system A and its states A_1 and A_2, is relatively long. To proceed, we consider two arbitrary systems A and B, two arbitrary states A_1 and A_2 of system A, two arbitrary states B_3 and B_4 of system B, two reservoirs R and Q, and cyclic machinery X_{AR}, X_{AQ}, X_{BR}, and X_{BQ} connected between systems A, B, R, and Q as shown schematically in Figure 7.1, and as specified in the following two paragraphs.

First, machinery X_{AR} is connected between A and R, and as a result of a reversible weight process for the composite of A and R, the state of A changes from A_1 to A_2, the state of R from R_1 to R_2, and the energy of R by $(DE_{12}^R)_{\text{rev}}^A$. Then machinery X_{AQ} is connected between A and Q, and as a result of another reversible weight process for the composite of A and Q, the state of A changes from A_2 back to A_1, the state of Q from Q_2 to Q_1, and the energy of Q by $(DE_{21}^Q)_{\text{rev}}^A = -(DE_{12}^Q)_{\text{rev}}^A$. At the end of the second weight process, system A is restored to its initial state A_1, that is, system A has undergone a cycle. If the sequence of the two processes is repeated N times, system A undergoes N cycles, reservoir R is in state R_4, reservoir Q in state Q_3, and the changes in energy of reservoirs R and Q are, respectively, $N(DE_{12}^R)_{\text{rev}}^A$ and $-N(DE_{12}^Q)_{\text{rev}}^A$.

Next, we consider a similar sequence of processes for system B. First, machinery X_{BQ} is connected between system B and reservoir Q in state Q_3. As a result of a reversible weight process for the composite of B and Q, the state of B changes from B_3 to B_4, the state of Q from Q_3 to Q_4, and the energy of Q by $(DE_{34}^Q)_{\text{rev}}^B$. Then machinery X_{BR} is connected between B and reservoir R in state R_4, and as a result of another reversible weight process for the composite of B and R, the state of B is restored from B_4 back to B_3, the state of R changes from R_4 to R_3, and the energy of reservoir R by $(DE_{43}^R)_{\text{rev}}^B = -(DE_{34}^R)_{\text{rev}}^B$. If this sequence is repeated M times, system B undergoes M

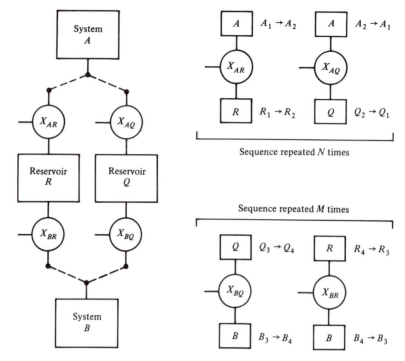

Figure 7.1 Schematic of the processes considered to prove that the ratio $(DE_{12}^R)_{\text{rev}}^A / (DE_{12}^Q)_{\text{rev}}^A$ is independent of system A and its states A_1 and A_2.

cycles, and the changes in energy of reservoirs R and Q are, respectively, $-M(DE_{34}^R)_{\text{rev}}^B$ and $M(DE_{34}^Q)_{\text{rev}}^B$.

As a result of the N cycles for A and the M cycles for B, the energy change of reservoir R is $N(DE_{12}^R)_{\text{rev}}^A - M(DE_{34}^R)_{\text{rev}}^B$, and that of reservoir Q is $M(DE_{34}^Q)_{\text{rev}}^B - N(DE_{12}^Q)_{\text{rev}}^A$. However, for given systems A and B, and states A_1, A_2, B_3, and B_4, the values of the energy changes $(DE_{12}^R)_{\text{rev}}^A$ and $(DE_{34}^R)_{\text{rev}}^B$ are fixed, and we may select the integers N and M so that

$$M \leqq N \frac{(DE_{12}^R)_{\text{rev}}^A}{(DE_{34}^R)_{\text{rev}}^B} \leqq M + 1 \qquad (7.27)$$

Then, for very large values of N, the integer M becomes very large, the ratio M/N is fairly accurately given by the relation

$$\frac{M}{N} = \frac{(DE_{12}^R)_{\text{rev}}^A}{(DE_{34}^R)_{\text{rev}}^B} \qquad (7.28)$$

and, therefore, the energy change $N(DE_{12}^R)_{\text{rev}}^A - M(DE_{34}^R)_{\text{rev}}^B$ of reservoir R is almost zero.

Because our results are independent of the initial state of either reservoir R or reservoir Q (see Section 6.8), and because to each value of the energy of a reservoir, with fixed values of the amounts of constituents and the parameters, there corresponds one and only one stable equilibrium state, we conclude that the zero energy change of R implies that R has been restored to its initial state.

In addition, it follows that the energy change of reservoir Q must equal zero because systems A, B, and R have all been restored to their initial states, and the impossibility of a PMM2 requires that no net energy can be transferred to the weight in a weight process for a reservoir Q, whereas the reversibility of the overall process in question implies that no net energy can be transferred from the weight to the reservoir Q. Thus we also conclude that for very large values of N, the net energy change of reservoir Q equals zero, that is,

$$M(DE_{34}^Q)_{\text{rev}}^B - N(DE_{12}^Q)_{\text{rev}}^A = 0 \tag{7.29}$$

or, equivalently,

$$\frac{M}{N} = \frac{(DE_{12}^Q)_{\text{rev}}^A}{(DE_{34}^Q)_{\text{rev}}^B} \tag{7.30}$$

Upon comparing (7.28) and (7.30), we find that

$$\frac{(DE_{12}^R)_{\text{rev}}^A}{(DE_{12}^Q)_{\text{rev}}^A} = \frac{(DE_{34}^R)_{\text{rev}}^B}{(DE_{34}^Q)_{\text{rev}}^B} \tag{7.31}$$

This relation proves that the ratio $(DE_{12}^R)_{\text{rev}}^A/(DE_{12}^Q)_{\text{rev}}^A$ is invariant both if states A_1 and A_2 are replaced by any other two states of A, and if system A is replaced by any other system B and any two of its states. Thus assertion 3 is verified.

Regarding assertion 4, we note that (7.31) leads to a definition of c_R that is a property of reservoir R only. To make ideas specific, we select reservoir Q once and for all as a system consisting of a large amount of H_2O in states in which solid ice, liquid water, and water vapor coexist in mutual stable equilibrium. For a given reservoir R and a system A in any two of its states A_1 and A_2, we measure the values $(DE_{12}^R)_{\text{rev}}^A$ and $(DE_{12}^Q)_{\text{rev}}^A$, and define the quantity c_R by means of the relation

$$c_R = c_Q \frac{(DE_{12}^R)_{\text{rev}}^A}{(DE_{12}^Q)_{\text{rev}}^A} \tag{7.32}$$

where c_Q is a positive constant fixed once and for all for the specially selected reservoir Q. Because (7.31) is valid, we conclude that the right-hand side of (7.32) is independent of system A and the states A_1 and A_2, and that it depends on reservoir R only.

Therefore, c_R depends on reservoir R only, and assertion 4 is proved. Later c_R is shown to be also equal to the temperature of R.

Example 7.4.   Show that the value of property c_R is the same for all the states of the given reservoir R and for all the reservoirs that are initially in mutual stable equilibrium with R.

Solution.   We already proved that the energy E and the generalized available energy Ω^R with respect to a given reservoir R are properties of a system A. Therefore, the value of $(DE_{12}^R)_{\text{rev}}^A$ determined by (7.24) depends only on the states A_1 and A_2 of system A and, in particular, is independent of the initial state of the reservoir. Hence, for all initial states of the reservoir R, equation (7.32) always yields the same value of c_R. Moreover, we also proved that Ω^R and Ω^Q are identical provided the two reservoirs R and Q are initially in mutual stable equilibrium. For two such reservoirs, it follows that $(DE_{12}^R)_{\text{rev}}^A = (DE_{12}^Q)_{\text{rev}}^A$, and therefore $c_R = c_Q$.

Now, we are ready to prove assertion 5, that the value of the property entropy is independent of the reservoir used in its definition. Indeed, upon rewriting (7.32) in the form

$$\frac{(DE_{12}^R)_{\text{rev}}^A}{c_R} = \frac{(DE_{12}^Q)_{\text{rev}}^A}{c_Q} \tag{7.33}$$

we conclude that the ratio $(DE_{12}^R)_{\text{rev}}^A/c_R$ is independent of reservoir R. Therefore, using equations (7.24) and (7.33) we can rewrite equation (7.1) in the forms

$$S_1 = S_o + \frac{1}{c_R}\left[(E_1 - E_o) - (\Omega_1^R - \Omega_o^R)\right]$$

$$= S_o + \frac{1}{c_R}(DE_{1o}^R)_{rev}^A$$

$$= S_o + \frac{1}{c_Q}(DE_{1o}^Q)_{rev}^A \tag{7.34}$$

and thus conclude that the role of the reservoir in the definition of entropy is indeed only auxiliary, and that S_1 is independent of R and depends only on the state A_1 of system A. It is noteworthy that, because it is independent of the reservoir, entropy is not affected by the approximations inherent in the definition of reservoirs (Section 6.3).

7.5   Dimensions and Units of Entropy

Equation (7.34) indicates that the dimensions of entropy are determined by the dimensions of both energy and property c_R of the auxiliary reservoir. Because the property c_R is defined by (7.26) in terms of a dimensionless ratio and an arbitrary scaling factor, the dimensions of this property are not related to the dimensions of other properties.

Traditionally, property c_R is assigned a fundamental dimension independent of all the other fundamental dimensions. We show in Chapter 8 that c_R has the same dimensions as temperature. Therefore, the traditional set of fundamental dimensions chosen in mechanics, which includes mass, length, and time, with the advent of thermodynamics is augmented by the addition of temperature.

The units in which property c_R is measured depend on the choice of the scaling factor c_Q used in 7.26 and the reservoir Q selected once and for all in the definition of c_R. In practice, in connection with a reservoir Q in mutual stable equilibrium with a system consisting of solid, liquid, and vapor water, two values are used for c_Q. One is $c_Q = 273.16$, and then we say that c_R is expressed in kelvin.[2] The other is $c_Q = 491.688$, and then we say that c_R is expressed in rankine.[3]

Having chosen a dimension for c_R, we see from (7.34) that entropy has dimensions of energy divided by temperature. Entropy values can therefore be expressed in many different units. The unit chosen by the International System of units is the joule per kelvin, denoted by the symbol J/K. Other units widely used in engineering practice are the kilocalorie per kelvin, kcal/K, and the British thermal unit per rankine, Btu/R. The conversion factor between these last two units can be obtained from the conversion factors for energy units (Table 3.1), that is, 1 kcal = 3.9683 Btu, and the relation between kelvin and rankine, that is, $c_Q = 273.16$ K $= 491.688$ R so that 1 K $= 1.8$ R. Thus 1 Btu/R $= 0.45359$ kcal/K, or recalling that 1 lb $= 0.45359$ kg, 1 Btu/lb R $= 1$ kcal/kg K.

Problems

7.1 Two blocks of metal A and B have masses $m_A = 5$ kg and $m_B = 3$ kg, respectively. Under certain conditions, the stable equilibrium states of each block may be described by the parametric relations

$$E = mc\,(x - x_o) \quad \text{and} \quad S = mc\ln\frac{x}{x_o}$$

---

[2]This unit is named in honor of William Thomson, Lord Kelvin (1824–1907), English physicist, whose pioneering work contributed to the development of the foundations of thermodynamics.

[3]This unit is named in honor of William John Macquorn Rankine (1820–1872), Scottish engineer and physicist, who made several outstanding contributions to the development of thermodynamics.

where E is the energy, S the entropy, m the mass of the block, x a parameter with dimensions of energy over entropy, and c and x_o are constants, that is, $c = 2$ kJ/kg K and $x_o = 250$ K. Moreover the same parametric relations hold also for the stable equilibrium states of the composite C of the two blocks, provided of course that $m_C = m_A + m_B$.

Assume that initially each of the two blocks is in a stable equilibrium state. In these states, the value of the parameter $x_1^A = 300$ K for block A, and $x_1^B = 700$ K for block B.

(a) What is the adiabatic availability of each block in its initial state?

(b) Is the composite C of the two blocks in a stable equilibrium state? Justify your answer numerically.

(c) If the composite C of the two blocks experiences a reversible weight process from the initial state to a final stable equilibrium state, show that the final value of the parameter x for the composite C is $x_2^C = 412.2$ K. Show that at this final state the values of the parameter x for the two blocks are $x_2^A = x_2^B = 412.2$ K.

(d) What is the adiabatic availability of the composite C of the two blocks in the initial state?

7.2 Consider three identical blocks of metal. Each block has a mass $m = 5$ kg, and its stable equilibrium states can be described by the parametric relations $E_2 - E_1 = mc\,(T_2 - T_1)$ and $S_2 - S_1 = mc\ln(T_2/T_1)$, where $c = 1$ kJ/kg K, subscripts 1 and 2 refer to any two stable equilibrium states of the block, and T is a parameter characteristic of the stable equilibrium state. Initially, each of the three blocks is in a stable equilibrium state, and the values of the parameter T are 300, 500, and 700 K, respectively.

(a) What is the adiabatic availability of each block in its initial state?

(b) What is the adiabatic availability of the three blocks considered as one system? The stable equilibrium states of this system obey the same parametric relations except for $m = 15$ kg.

(c) If interactions via cyclic machinery between blocks are allowed but interactions between the blocks and the environment are not, what is the largest value of T that can be reached by one of the blocks?

(d) Under the same conditions as in part (c), what is the smallest value of T that can be reached by one of the blocks?

(e) What are the adiabatic availabilities of the states of the three blocks found in parts (c) and (d)?

7.3 A system consists of a fixed amount of only one constituent confined in a container of fixed volume V. Volume is the only parameter of the system. Four states of the system, A_0, A_1, A_2, and A_3, have the values of energy and available energy with respect to a reservoir R listed in the following table.

State	Energy E	Available energy Ω^R
A_1	900 kJ	1000 kJ
A_2	900 kJ	700 kJ
A_3	900 kJ	300 kJ
A_0	200 kJ	0 kJ

(a) What kind of state is A_0?

(b) If the difference between the entropies of states A_3 and A_1 is $S_3 - S_1 = 2$ kJ/K, what is the value of the constant c_R of the reservoir?

(c) Assigning to the reference state for entropy a value of entropy equal to that of state A_0, what are the values of the entropies of states A_1, A_2, and A_3 with respect to that reference?

7.4 Consider the states A_0, A_1, A_2, and A_3 specified in Problem 7.3, and a reservoir $Q1$ with a constant $c_{Q1} = 300$ K.

(a) Find the difference between the available energies of states A_1 and A_3 with respect to reservoir $Q1$. How does this difference compare with that given in Problem 7.3?

(b) Repeat part (a) for a reservoir $Q2$ with a constant $c_{Q2} = 400$ K.

7.5 Consider the states A_0, A_1, A_2, and A_3, and the reservoir R specified in Problem 7.3. Assume that the state A_1 changes to state A_2, and then to state A_3 as a result of spontaneous processes.

(a) How much entropy is generated by irreversibility in the course of the process from A_1 to A_2?

(b) How much entropy is generated by irreversibility in the course of the process from A_2 to A_3? What is the smallest amount of energy that must be transferred from a weight to the composite of system A and reservoir R to return the state from A_3 to A_1? Where does this energy appear?

7.6 Consider two systems A and B, and a system C composed of the two systems A and B. The reference states for A and C have entropies $S_o^A = 5$ kJ/K and $S_{oo}^C = -2$ kJ/K, respectively.

(a) What value must be assigned to the entropy of the reference state of system B so that differences of entropies of A and B be additive?

(b) What value must be assigned to the entropy of the reference state of system B so that the entropies of A and B be additive?

(c) If $S_o^A = 10$ kJ/K, $S_o^B = -10$ kJ/K and the entropies of A and B are additive, what is the value of the entropy of the reference state of C?

7.7 Consider the property $P = E - E_g - \Psi$, where E is the energy, Ψ the adiabatic availability, and E_g the ground-state energy corresponding to the same values of the amounts of constituents and parameters as those of the state at which P is evaluated. Some interesting features of property P are derived in Problem 5.3.

(a) Show that P represents the energy that cannot be transferred out of the system in any weight process without net changes in the values of the amounts of constituents and parameters.

(b) Consider any two states A_1 and A_2 that have the same values of the amounts of constituents n and the parameters β, and that can be interconnected by means of a weight process. Prove that for such two states $P_1 > P_2$ if and only if $S_1 < S_2$, namely prove that properties P and S are in one-to-one correspondence over the entire range of states of a system.

(c) Can you think of at least one feature by which properties P and S differ? Why is entropy S more practical than property P?

7.8 Consider the entropy and property P defined in Problem 7.7. Show that the value of the entropy S of any state of any system is directly and monotonically related to the energy that cannot be transferred out of the system in any weight process. Note that this is another way to show that the entropy S is a property of the system only, independent of the auxiliary reservoir used in the definition of S.

7.9 Consider a system with fixed values of amounts of constituents and parameters. The system undergoes a weight process starting from state A_1 and ending in state A_2. The differences in energy and entropy between states A_1 and A_2 are $E_2 - E_1 = 0$ and $S_2 - S_1 = 5$ kJ/K.

(a) Is the process reversible or irreversible?

(b) Is a weight process from A_2 to A_1 possible?

(c) If the system is in combination with a reservoir R through cyclic machinery X that interacts also with a weight, prove that the least energy from the weight required to bring system A from state A_2 to state A_1 is equal to $c_R (S_2 - S_1)$, where c_R is the constant property of the reservoir used in the definition of entropy.

(d) Show that the adiabatic availabilities Ψ_1 and Ψ_2 of system A in states A_1 and A_2 are such that $\Psi_1 > \Psi_2$.

(e) For the conditions specified in part (c), find the energy transferred from the weight if the entropy generated by irreversibility is $S_{irr} = 3$ kJ/K.

7.10 Over a limited range of values of energy, the energy E, and entropy S of the stable equilibrium states of a system A with fixed values of amounts of constituents and parameters may be approximated by the parametric relations $E = mc\,(T - T_o)$ and $S = mc\ln(T/T_o)$, where the mass of the system $m = 10$ kg, $c = 1$ kJ/kg K, T is a parameter characteristic of each stable equilibrium state (we show later that T is the temperature), and T_o a reference value. The unit of T is the kelvin, K.

In a weight process starting from a stable equilibrium state with $T_1 = 300$ K, a weight transfers 1000 kJ of energy to the system. At the end of this transfer, the system is again in a stable equilibrium state.

(a) Is the weight process reversible or irreversible? If it is irreversible, how much entropy is generated by irreversibility?

(b) What is the adiabatic availability of the final state?

(c) If in the initial state the system is in mutual stable equilibrium with a reservoir R with constant $c_R = 300$ K, what is the available energy of the final state with respect to R?

(d) Why does your answer in part (c) differ from the amount of energy transferred from the weight to the system?

7.11 Explain whether the following processes are possible.

(a) A cyclic device interacts with a system in a stable equilibrium state and results in no external effect except the rise of a weight.

(b) A cyclic device interacts with a system in an equilibrium state A_1 and results in no external effect except the transfer of energy $E_1/2$ to a weight, where the energy and entropy of A_1 are E_1 and S_1, respectively, and the energy and entropy of the stable equilibrium state of the system with the same values of the amounts of constituents and parameters are $2E_1/5$ and S_1, respectively.

(c) Same as in part (b) but the energy transfer is $3E_1/4$.

7.12 As a result of an irreversible weight process, a battery A changes from state A_1 to state A_2. The energies and entropies of these two states are $E_1 = 10$ kJ, $E_2 = 1010$ kJ, $S_1 = 1$ kJ/K, and $S_2 = 3$ kJ/K. When in the initial state A_1, the battery is in mutual stable equilibrium with a reservoir R with a constant $c_R = 300$ K.

(a) How much energy is supplied by the weight?

(b) What is the largest amount of energy that the composite of the battery in state A_2 and reservoir R can transfer to a weight in a weight process?

(c) What is the largest gain in available energy of the battery A with respect to R in a weight process for the composite of A and R in which A starts from state A_1, and 1000 kJ of energy is transferred to the composite AR from a weight?

(d) Why does your answer in part (b) differ from that in part (c)?

7.13 An isolated system A consists of two compartments. Initially, one compartment is filled with very hot water and the other with very cold water. The two compartments are separated by a thin, watertight partition that allows the gradual flow of energy from one compartment to the other.

As time goes on, the hot water becomes colder, and the cold water becomes warmer. Initially, the available energy of the system with respect to a reservoir R with constant $c_R = 300$ K is $\Omega_1^R = 10$ MJ, ten minutes later it becomes $\Omega_2^R = 8$ MJ, and 20 minutes later $\Omega_3^R = 7$ MJ.

(a) Is the change within the isolated system reversible or irreversible? If it is irreversible, can you evaluate the amount of entropy generated by irreversibility during the first 10 minutes, and during the first 20 minutes?

(b) How much energy would a weight have to transfer to the composite of the system and the reservoir to restore the initial state by means of a weight process starting from the state at 20 minutes?

7.14 A system A consists of a metric ton, 1000 kg, of coal and adequate air for completely burning the coal. The energies and entropies of states A_1 and A_0 of this system are $E_1 = 24 \times 10^6$ kJ,

$S_1 = -200$ kJ/K, $E_0 = 0$, and $S_0 = 0$. In state A_0, system A is in mutual stable equilibrium with a reservoir R with constant $c_R = 300$ K.

(a) What is the available energy with respect to R of state A_1?

(b) If the change from state A_1 to state A_0 occurs as a result of a zero-net-effect weight process of the composite of A and R, how much entropy is generated by irreversibility? Assume that the changes in energy ΔE^R and entropy ΔS^R of the reservoir obey the relation $\Delta E^R = c_R \Delta S^R$.

(c) Find the ratio of your answers in parts (a) and (b), and compare it to c_R.

8 Stable-Equilibrium-State Principle

Stable equilibrium states play a prominent role in the study of physical phenomena. They occur more often and can be observed and analyzed more readily than other types of states. In this chapter we summarize some of the most important general features of stable equilibrium states.

8.1 State Principle

We consider a system specified by r different constituents, with amounts denoted by $n = \{n_1, n_2, \ldots, n_r\}$, and s different parameters denoted by $\beta = \{\beta_1, \beta_2, \ldots, \beta_s\}$. We assume that the value of each amount and each parameter can vary over a range and that constraints inhibit all the chemical and nuclear reaction mechanisms. The latter assumption indicates that the internal mechanisms cannot cause any change in the values of the amounts of constituents and the parameters, and that any such change can occur only as a result of interactions with other systems.

As we have already emphasized, the constituents and the parameters differ from system to system. Some systems consist of only one constituent, such as water, and then only one n and its range of values are needed. Also, some systems have only one parameter, such as the volume V, and then only one parameter, $\beta = V$, and its range of values are necessary. Other systems require more n's and more β's. At any given instant of time, the system is in some definite state, with corresponding values of all its properties, including the energy E, and of all the amounts n and all the parameters β. Each given set of values of E, n, β of the energy, the amounts of constituents, and the parameters is common to many different states of the system. Among these, only one is a stable equilibrium state because, according to the second law, one and only one such state corresponds to a given set of values of E, n_1, n_2, $\ldots$, n_r, β_1, β_2, $\ldots$, β_s.

Because the stable equilibrium state corresponding to the set of values E, n, β is unique, and because the state of a given system includes the values of all its properties, it follows that the values E, n, β uniquely determine the value of any property of the system in a stable equilibrium state. This is true for any stable equilibrium state and, therefore, the value of every property of the system along the stable equilibrium states is solely a function of E, n, β. Thus, for stable equilibrium states, any property P can be written as a function of the form

$$P = P(E, n_1, n_2, \ldots, n_r, \beta_1, \beta_2, \ldots, \beta_s) = P(E, n, \beta) \qquad (8.1)$$

where the explicit dependences of P on E, n, β are determined by the system, that is, the constituents, the internal forces, the external forces, and the constraints. This consequence is known as the *stable-equilibrium-state principle* or, simply, the *state principle*. Because it involves both the concept of energy and that of stable equilibrium, the state principle, that is, each expression of the form of (8.1), is a consequence of both the first and second laws of thermodynamics.

A system in general has a very large number of independent properties. When we focus on the special family of states that are stable equilibrium, however, the state principle asserts that the value of each of these properties is uniquely determined by the values of E, n, β. In contrast, for states that are not stable equilibrium, the values of E, n, β are not sufficient to specify the values of all the independent properties.

To reduce mathematical complexity, we discuss the state principle in its simplest form, that is, in connection with a system consisting of only one type of constituent, with amount n variable over a given range, and confined in a volume of size V, also variable over a given range. In general, for a given set of values of n and V and a given value of the energy E, the system may be in one of a large number of different states, some of which are equilibrium and others that are not. Each of these states requires a large number of values of independent properties, other than energy, to be specified, even though all the states in question have the same values of E, n, V. Among all these states, one and only one is a stable equilibrium state. This stable equilibrium state, the state from which no energy can be transferred to a weight without a net change in either volume or amount, or some other effect on the environment, is fully specified by the values E, n, V. This is true for all allowed combinations of values E, n, V, and therefore the value of any property P for the stable equilibrium states of the system in question must be given by an expression of the form of (8.1).

We can write expressions of the form of (8.1) for different properties $X, Y, Z, \ldots$, that is,

$$X = X(E, n, V) \qquad Y = Y(E, n, V) \qquad Z = Z(E, n, V) \tag{8.2}$$

where $X(E, n, V)$, $Y(E, n, V)$, $Z(E, n, V)$, ... are particular functions of the three independent variables E, n, and V. Among the different properties, we can select three that are independent—three properties X, Y, and Z such that the value of one may be varied without affecting the values of the other two.[1] Then the system of equations $X = X(E, n, V)$, $Y = Y(E, n, V)$, $Z = Z(E, n, V)$ can be solved for E, n, and V to yield the relations

$$E = E(X, Y, Z) \qquad n = n(X, Y, Z) \qquad V = V(X, Y, Z) \tag{8.3}$$

Using these relations, we can transform any function $P(E, n, V)$ of the independent variables E, n, V into a function of the new independent variables X, Y, Z of the form

$$P = P(X, Y, Z) \tag{8.4}$$

where $P(X, Y, Z) = P(E(X, Y, Z), n(X, Y, Z), V(X, Y, Z))$.

We cannot overemphasize that expressions of the form of either (8.2) or (8.4) are valid only for stable equilibrium states because for the systems specified, only these states are fully determined by only three independent variables. In general, for r types of constituents and s parameters, we have $r + s + 1$ independent variables, and each property is expressed by a relation of the form of (8.1). We call each equation of the form of either (8.1) or (8.4) a *stable-equilibrium-state relation*.

---

[1]Mathematically, properties X, Y, and Z are independent only if the functions $X(E, n, V)$, $Y(E, n, V)$, and $Z(E, n, V)$ are such that their Jacobian $\partial(X, Y, Z)/\partial(E, n, V)$ is nonzero. Jacobians are discussed in texts on calculus and differential equations. The usefulness of Jacobians in classical thermodynamics—thermodynamics of stable equilibrium states—was first recognized by A. Norman Shaw in "The Derivation of Thermodynamical Relations for a Simple System," *Philos. Trans. R. Soc. London*, A234, 299–328 (1934–1935). It is discussed in many textbooks on thermodynamics. See also Problem 8.7.

We will see later that all of classical thermodynamics—the physics of systems in stable equilibrium states—is based on the state principle. In contrast to other presentations of the science of thermodynamics, here the state principle is derived as a rigorous consequence of the first and second laws, not as a consequence either of difficulties related to exact calculations and lack of knowledge, or a need to describe complicated physical problems by a few gross macroscopic averages. In the present context the state principle expresses a fundamental physical feature of the stable equilibrium states of any system and implies the existence of fundamental interrelations among the properties of these states.

8.2   Criteria for Stable Equilibrium States

In general, the value of the entropy S is not the same for every state that has given values of the energy E, the amounts of constituents n, and the parameters β. For example, in the course of spontaneous changes of state in an isolated system we know that the entropy cannot decrease [relation (7.15)], and for an irreversible spontaneous change of state, such as one that starts from a nonequilibrium state and ends in a stable equilibrium state, that the entropy increases from a lower to a higher value.

In particular, among the many states of a system that have given values E, n, β, the entropy of the unique stable equilibrium state is larger than that of any other state with the same values E, n, β. We call this assertion the *highest-entropy principle*.

Example 8.1.   Prove the highest-entropy principle.

Solution.   For given values E, n and β, we denote the unique stable equilibrium state of a system A by A_0, its entropy by S_0, and any other state by A_1 and its entropy by S_1. We wish to prove that $S_0 > S_1$. Indeed, if we assume that $S_0 \leqq S_1$, a spontaneous process from state A_0 to state A_1 would be possible because it does not violate the principle of nondecrease of entropy. But such a process would alter the stable equilibrium state A_0 to a different state A_1 with no net external effects and therefore would contradict the definition of a stable equilibrium state. So our assumption is invalid and the highest-entropy principle is proved.

We conclude that a criterion for stable equilibrium is that the entropy of such a state must be larger than the entropy of any other state with the same values of the energy, the amounts of constituents, and the parameters.

We now consider all the states that have a given value of the entropy. In general, the value of the energy E is not the same for every state that has given values of the entropy S, the amounts of constituents n, and the parameters β. For example, in the course of a reversible weight process, we know that the entropy of the system remains invariant [equation (7.10)], and for a reversible weight process that starts from a state that is not stable equilibrium and ends in a stable equilibrium state, that the energy of the system decreases from a higher to a lower value by an amount equal to the adiabatic availability of the initial state [relations (5.7) and (5.9)].

In particular, among the many states of a system that have given values S, n and β, the energy of the unique stable equilibrium state is smaller than that of all the other states with the given values S, n, and β. We call this assertion the *lowest-energy principle*.

Example 8.2.   Prove that among all the states of a system that have given values of the entropy S, the amounts of constituents n, and the parameters β, there is one and only one stable equilibrium state.

Solution.   For given values of n and β, we assume that there exist two stable equilibrium states, A_1 and A_2, with the same values of the entropy, that is, such that $S_1 = S_2$. Then a reversible weight process with A_1 and A_2 as end states would be possible. However, the energies E_1 and

E_2 of these two states cannot be different. Indeed, if $E_1 < E_2$, a reversible weight process starting from stable equilibrium state A_2 and ending in A_1 would violate the impossibility of a PMM2. Similarly, if $E_1 > E_2$ a reversible weight process starting from stable equilibrium state A_1 and ending in A_2 would violate the impossibility of a PMM2. Thus we conclude that $E_2 = E_1$ and that the two stable equilibrium states A_1 and A_2 must coincide because each corresponds to the same values E, n, and β.

Example 8.3.   Prove the lowest-energy principle.

Solution.   For given values S, n, and β, we denote the unique stable equilibrium state of a system A by A_0, its energy by E_0, and any other state by A_1 and its energy by E_1. We wish to prove that $E_0 < E_1$. Indeed, if we assume that $E_0 \geq E_1$, a reversible weight process from state A_0 to state A_1 would be possible because it conforms to the requirement of conservation of entropy [equation (7.10)]. But such a process would have no other effects than to transfer energy, $E_0 - E_1$, to a weight from a system that starts in a stable equiibrium state. Such a transfer would violate the impossibility of a PMM2. So our assumption is invalid, and the lowest-energy principle is proved.

We conclude that another criterion for stable equilibrium is that the energy of such a state should be smaller than the energy of any other state with the same values of the entropy, the amounts of constituents, and the parameters.

Each of the two criteria just derived, combined with the state principle, leads to useful conditions for mutual stable equilibrium, and to the definition of practical properties that describe the stable equilibrium states of a system. These conditions and properties are discussed in Chapters 9 to 11.

8.3   The Fundamental Relation

The state principle holds as well for property entropy, that is, the entropy of the stable equilibrium states can be expressed as a function

$$S = S(E, n, \beta) \tag{8.5}$$

This relation avers that the value of the entropy of a stable equilibrium state is uniquely determined by the values of the energy, the amounts of constituents, and the parameters of that state. The actual functional dependence of S on E, n, β varies from system to system depending on the nature of the constituents, and the internal and external forces. Irrespective of the actual dependence, however, we prove that the structure of the function $S(E, n, \beta)$ must satisfy certain general restrictions.

In addition, the function $S(E, n, \beta)$ is analytic in each of its variables E, n_1, n_2, ..., n_r, β_1, β_2, ..., β_s therefore is differentiable to all orders within the range of the values of E, n, β admissible for the given system.[2]

An important implication of the existence of partial derivatives of all orders of $S(E, n, \beta)$ is that any difference between the entropies of two stable equilibrium states may be expressed in the form of a Taylor series in terms of the differences in the values of the energies, amounts of constituents, and parameters of the two states.

For example, given a system A, the function $S = S(E, n, \beta)$, and two stable equilibrium states A_0 and A_1 with the same values of the amounts of constituents, and parameters, that is, with $(n)_0 = (n)_1$ and $(\beta)_0 = (\beta)_1$, the differences in entropy, $S_1 - S_0$, and in energy, $E_1 - E_0$, are related by the Taylor series expansion

---

[2]The proof of the analyticity of $S(E, n, \beta)$ requires consideration of the so-called Hamiltonian operator of the system, within the framework of quantum theory. Although relatively simple and mathematically elegant, the proof is beyond our scope here.

$$S_1 - S_0 = \left[\left(\frac{\partial S}{\partial E}\right)_{n,\beta}\right]_0 (E_1 - E_0) + \frac{1}{2}\left[\left(\frac{\partial^2 S}{\partial E^2}\right)_{n,\beta}\right]_0 (E_1 - E_0)^2 + \cdots$$

$$= \sum_{k=1}^{\infty} \frac{1}{k!}\left[\left(\frac{\partial^k S}{\partial E^k}\right)_{n,\beta}\right]_0 (E_1 - E_0)^k \tag{8.6}$$

where $[(\partial^k S/\partial E^k)_{n,\beta}]_0$, for $k = 1, 2, \ldots, \infty$, is the number obtained by evaluating the kth partial derivative of $S(E, n, \beta)$ with respect to E at state A_0, that is, for the values E_0, $(n)_0$, $(\beta)_0$. In the notation of partial derivatives, subscripts n and β denote the variables that are kept fixed during differentiation, and subscript 0 denotes the fact that the partial derivative is evaluated at state A_0. Similar Taylor series expansions may be written to relate the difference in entropy, $S_1 - S_0$, to differences in amounts of constituents, $(n)_1 - (n)_0$, and parameters $(\beta)_1 - (\beta)_0$ between two arbitrary stable equilibrium states A_0 and A_1, as well as any combination of the differences $E_1 - E_0$, $(n)_1 - (n)_0$, and $(\beta)_1 - (\beta)_0$. We introduce these series as they are needed.

Equation (8.5) is a stable-equilibrium-state relation for entropy. As discussed in Section 8.1, upon changing the independent variables from the set E, n, β to any set of $r + s + 1$ independent properties, we may obtain another stable-equilibrium-state relation for entropy. Nevertheless, because the variables E, n, β are defined for all systems, and the state principle is derived in terms of these variables, we call (8.5) the *fundamental stable-equilibrium-state relation for entropy* or, simply, the *fundamental relation* of the system. We will see that all the results of classical thermodynamics are derived as consequences of the fundamental relation. Sometimes the term fundamental relation is used both for (8.5) and for its Taylor series expansions, such as (8.6). The fundamental relation of a system contains all the information about the stable equilibrium states of the system, that is, all attributes of the system in any such state are completely and exactly determined.

In denoting partial derivatives of a stable-equilibrium-state relation, we specify the variables that are kept constant in the differentiation by means of subscripts. Although cumbersome, this notation is not redundant. It indicates which stable-equilibrium-state relation is being differentiated, that is, which are the independent variables. For example, for a system with fundamental relation $S = S(E, n, V)$, we could also obtain another stable-equilibrium-state relation of the form $S = S(E, X, Y)$, where the properties X and Y are independent of each other and of the property energy. Clearly, the functions $S(E, n, V)$ and $S(E, X, Y)$ are different, that is, imply different functional relations between the three independent arguments even though we denote each function by S. Therefore, the symbol $\partial S/\partial E$ would be ambiguous until we specify whether $S = S(E, n, V)$ or $S = S(E, X, Y)$, and therefore whether we mean $\partial S(E, n, V)/\partial E$ or $\partial S(E, X, Y)/\partial E$. Instead of specifying explicitly the arguments, we use subscripts, that is, instead of $\partial S(E, n, V)/\partial E$ we use the notation $(\partial S/\partial E)_{n,V}$ and, similarly, instead of $\partial S(E, X, Y)/\partial E$, the notation $(\partial S/\partial E)_{X,Y}$, and recall that the variable with respect to which we differentiate plus the subscripts signify all the dependences of the function.

Example 8.4.   Prove Carathéodory's statement of the second law, namely, that in any neighborhood of a given stable equilibrium state (with energy higher than the ground-state energy) there exist stable equilibrium states that cannot be reached by any weight process that starts from the given state.

Solution.   Given a stable equilibrium state A_1 with energy E_1 greater than the ground-state energy E_{g1} corresponding to the values of the amounts of constituents and the parameters of A_1, we consider a stable-equilibrium-state relation of the form $P = P(E, n, \beta)$ for each property of the system, and evaluate each partial derivative $[(\partial P/\partial E)_{n,\beta}]_1$. Next, we fix an arbitrarily small value $\epsilon_P > 0$ for each property P and select a negative value dE such that $E_{g1} - E_1 \leqq dE < 0$ and $|[(\partial P/\partial E)_{n,\beta}]_1 dE| < \epsilon_P$ for each P. By virtue of the second law, corresponding to the values $E_2 = E_1 + dE$, n, and β there must exist a stable equilibrium state A_2, and this state is arbitrarily close to A_1 in that for each property P the absolute value of the difference $P_1 - P_2$ does not exceed the arbitrarily selected value ϵ_P. Because A_1 and A_2 are stable equilibrium states, each has an adiabatic availability equal to zero, that is, $\Psi_1 = \Psi_2 = 0$ or $\Psi_1 - \Psi_2 = 0$. Moreover, $E_1 - E_2 = -dE > 0$. Because $E_1 - E_2 > \Psi_1 - \Psi_2$, relation (5.13) is not satisfied. Therefore, no weight process starting from state A_1 can reach the neighboring state A_2, and Carathéodory's statement of the second law is seen to be a special case of the approach taken in this book.

For values of $(\partial S/\partial E)_{n,\beta} \neq 0$, the fundamental relation $S = S(E, n, \beta)$ can be solved for the energy E as a function of S, n, and β, that is,

$$E = E(S, n, \beta) \tag{8.7}$$

The energy relation $E = E(S, n, \beta)$ is analytic with respect to each of the variables S, $n_1, n_2, \ldots, n_r, \beta_1, \beta_2, \ldots, \beta_s$ for the same basic reasons that $S(E, n, \beta)$ is analytic with respect to each of its variables. As a result, given a stable equilibrium state A_0 and any other neighboring stable equilibrium state with energy $E_0 + dE$, entropy $S_0 + dS$, amounts of constituents $(n_i)_0 + dn_i$ for $i = 1, 2, \ldots, r$, and parameters $(\beta_j)_0 + d\beta_j$ for $j = 1, 2, \ldots, s$, the differences dE, dS, dn_i, and $d\beta_j$ are related by the differential of the fundamental relation, that is,

$$dS = \left[\left(\frac{\partial S}{\partial E}\right)_{n,\beta}\right]_0 dE + \sum_{i=1}^{r}\left[\left(\frac{\partial S}{\partial n_i}\right)_{E,n,\beta}\right]_0 dn_i + \sum_{j=1}^{s}\left[\left(\frac{\partial S}{\partial \beta_j}\right)_{E,n,\beta}\right]_0 d\beta_j \tag{8.8}$$

or, equivalently, by the differential of the energy relation [equation (8.7)], that is,

$$dE = \left[\left(\frac{\partial E}{\partial S}\right)_{n,\beta}\right]_0 dS + \sum_{i=1}^{r}\left[\left(\frac{\partial E}{\partial n_i}\right)_{S,n,\beta}\right]_0 dn_i + \sum_{j=1}^{s}\left[\left(\frac{\partial E}{\partial \beta_j}\right)_{S,n,\beta}\right]_0 d\beta_j \tag{8.9}$$

Equations (8.8) and (8.9) are very useful in studies of stable equilibrium states.

In Chapter 9 the partial derivative $(\partial E/\partial S)_{n,\beta}$ is defined as the temperature and denoted by the symbol T. In Chapter 10 the partial derivative $(\partial E/\partial n_i)_{S,n,\beta}$ is defined as the total potential of the ith constituent and denoted by the symbol μ_i for $i = 1, 2, \ldots, r$. In Chapter 11 the partial derivative $(\partial E/\partial \beta_j)_{S,n,\beta}$ is defined as the generalized force conjugated to the jth parameter and denoted by the symbol f_j for $j = 1, 2, \ldots, s$. In particular, if a system has volume $V = \beta_1$ as a parameter, then $(\partial E/\partial V)_{S,n,\beta}$ is defined as the negative of the pressure, which is denoted by p.

In terms of these forthcoming definitions, the differential expression of the energy relation [equation (8.9)] in the neighborhood of stable equilibrium state A_0 may be rewritten as

$$dE = T_0 dS + \sum_{i=1}^{r}(\mu_i)_0 dn_i + \sum_{j=1}^{s}(f_j)_0 d\beta_j \tag{8.10}$$

In particular, if volume V is a parameter, then

$$dE = T_0 dS + \sum_{i=1}^{r}(\mu_i)_0 dn_i - p_0 dV + \sum_{j=2}^{s}(f_j)_0 d\beta_j \tag{8.11}$$

The concepts of temperature, total potential, and generalized force may also be used in the expression of the differential form of the fundamental relation. Upon solving (8.11) for dS, we find that

$$dS = \frac{1}{T_0} dE - \sum_{i=1}^{r} \frac{(\mu_i)_0}{T_0} dn_i + \frac{p_0}{T_0} dV - \sum_{j=2}^{s} \frac{(f_j)_0}{T_0} d\beta_j \qquad (8.12)$$

and upon comparing (8.8) and (8.12) term by term, we see that

$$\left[\left(\frac{\partial S}{\partial E}\right)_{n,\beta}\right]_0 = \frac{1}{T_0}$$

$$\left[\left(\frac{\partial S}{\partial n_i}\right)_{E,n,\beta}\right]_0 = -\frac{(\mu_i)_0}{T_0}$$

$$\left[\left(\frac{\partial S}{\partial V}\right)_{E,n,\beta}\right]_0 = \frac{p_0}{T_0}$$

$$\left[\left(\frac{\partial S}{\partial \beta_j}\right)_{E,n,\beta}\right]_0 = -\frac{(f_j)_0}{T_0}$$

Moreover, recalling the definitions of T, μ_i, p, and f_j, we obtain the following relations:

$$\left(\frac{\partial S}{\partial E}\right)_{n,\beta} = \frac{1}{T} = \frac{1}{(\partial E/\partial S)_{n,\beta}} \qquad (8.13)$$

$$\left(\frac{\partial S}{\partial n_i}\right)_{E,n,\beta} = -\frac{\mu_i}{T} = -\frac{(\partial E/\partial n_i)_{S,n,\beta}}{(\partial E/\partial S)_{n,\beta}} \qquad (8.14)$$

$$\left(\frac{\partial S}{\partial V}\right)_{E,n,\beta} = \frac{p}{T} = -\frac{(\partial E/\partial V)_{S,n,\beta}}{(\partial E/\partial S)_{n,\beta}} \qquad (8.15)$$

$$\left(\frac{\partial S}{\partial \beta_j}\right)_{E,n,\beta} = -\frac{f_j}{T} = -\frac{(\partial E/\partial \beta_j)_{S,n,\beta}}{(\partial E/\partial S)_{n,\beta}} \qquad (8.16)$$

where we omit the subscript 0 because the equations are valid for any stable equilibrium state. We use these relations in subsequent chapters.

Example 8.5. Over a limited range of values of entropy S, amount of a single constituent n, and volume V, the energy E of the stable equilibrium states of a system A is given by the relation

$$E = 6.15\,n + 6150\,n \left\{ \exp\left[\frac{S - 5\,n - 8.3\,n \ln(V/5n)}{20.5\,n}\right] - 1 \right\}$$

where all variables and coefficients are expressed in a consistent set of units. Find the energy of the stable equilibrium state A_1 corresponding to $S_1 = 5$, $n_1 = 1$, and $V_1 = 5$, and write the differential form of the energy relation around state A_1.

Solution. For the given values of S, n, and V, we find that $E_1 = 6.15$. The partial derivatives of E evaluated at state A_1 are

$$\left[\left(\frac{\partial E}{\partial S}\right)_{n,V}\right]_1 = \left\{\frac{6150\,n}{20.5\,n}\exp\left[\frac{S - 5\,n - 8.3\,n\ln(V/5n)}{20.5\,n}\right]\right\}_1$$

$$= 300\exp\left[\frac{5 - 5 \times 1 - 8.3 \times 1 \times \ln(5/5 \times 1)}{20.5 \times 1}\right] = 300$$

$$\left[\left(\frac{\partial E}{\partial n}\right)_{S,V}\right]_1 = \left\{\left(6150 - \frac{6150\,n\,S}{20.5\,n^2} + 8.3\frac{6150\,n}{20.5\,n}\right)\right.$$

$$\left.\cdot \exp\left[\frac{S - 5\,n - 8.3\,n\ln(V/5n)}{20.5\,n}\right] - 6150 + 6.15\right\}_1 = 946.15$$

$$\left[\left(\frac{\partial E}{\partial V}\right)_{S,n}\right]_1 = -\left\{8.3\frac{6150\,n}{20.5\,V}\exp\left[\frac{S - 5\,n - 8.3\,n\ln(V/5n)}{20.5\,n}\right]\right\}_1 = -498$$

Thus we have

$$dE = 300\,dS + 946.15\,dn - 498\,dV$$

Problems

8.1 For certain conditions of interest, the energy e and entropy s of the stable equilibrium states of 1 mol of a substance are related by the expression

$$e = 7.54 + 8500\left\{\exp\left[\frac{s - s_0 - 8.5\ln(v/v_0)}{21.25}\right] - 1\right\} \qquad \text{(J/mol)}$$

where s_0 and v_0 are the entropy (in J/mol K) and volume (in m^3/mol), respectively, at a reference stable equilibrium state A_0. In this expression we use lower case letters to emphasize the fact that we are discussing only 1 mol of the substance.

(a) Express the differential changes of energy de as functions of the differential changes ds and dv about the reference state A_0 and about a state A_1 for which $s_1 = s_0$ and $v_1 = v_0/2$.

(b) Starting from state A_0, the substance is compressed to a volume $v_1 = v_0/2$ in a weight process. What is the least amount of energy from the weight required for this compression?

(c) If the energy from the weight in the weight process in part (b) is 3200 J/mol, what is the largest amount of entropy that could be created by irreversibility?

8.2 The relation between the energy and entropy of the stable equilibrium states of a system with fixed values of parameters is given by the expression $E = 25\mathcal{R}\exp[(S - S_0)/1.5\mathcal{R}]$ for E in cal, S in cal/K, and $\mathcal{R} = 99.3$ cal/K, where S_0 is constant. The system is in a state A_1 with energy $E_1 = 3 \times 10^4$ cal.

(a) Find the largest energy that can be transferred from the system to a weight without net changes in values of amounts of constituents and parameters if state A_1 has entropy either $S_1 = S_0$ or $S_1 - S_0 = 3\mathcal{R}$.

(b) What is the entropy of the system when its energy is E_1 and its adiabatic availability zero?

8.3 The fundamental relation of a system consisting of one constituent, n, and having volume, V, as the only parameter is given by the relation

$$S = a\,(EnV)^{1/3}$$

where a is a constant.

(a) Express small changes of entropy about the stable equilibrium state at E_0, n_0, V_0 in terms of small changes of $E, n,$ and V.

(b) Find the energy of the stable equilibrium states of the system in terms of $S, n,$ and V.

(c) Express small changes of the energy about the stable equilibrium state at S_o, n_o, V_o in terms of small changes of S, n, and V. Compare the expansion coefficients found here with those found in part (a).

8.4 The energy and entropy of the stable equilibrium states of a very special system, called a spin, are given by the relations

$$E = x_1 \epsilon_1 + x_2 \epsilon_2 \qquad \text{and} \qquad S = -k(x_1 \ln x_1 + x_2 \ln x_2)$$

where x_i is a dimensionless number between zero and unity such that

$$x_i = \frac{\exp(-\epsilon_i/kT)}{\exp(-\epsilon_1/kT) + \exp(-\epsilon_2/kT)} \qquad \text{for } i = 1, 2$$

ϵ_1 and ϵ_2 are two constants characteristic of the spin system, k is the Boltzmann constant, and T is a variable characteristic of each stable equilibrium state. The constants ϵ_1 and ϵ_2 have dimensions of energy. Assume that $\epsilon_2 - \epsilon_1 = 10^{-4}$ eV.

(a) Show that the fundamental relation is of the form $S = S(E)$.

(b) What is the range of values over which E may vary?

(c) Make a graph of S versus $(E - \epsilon_1)/(\epsilon_2 - \epsilon_1)$.

(d) For what range of values of $(E - \epsilon_1)/(\epsilon_2 - \epsilon_1)$ is the variable T positive, and for what range is it negative?

(e) How is T related to the derivative $dS(E)/dE$ of the fundamental relation?

(f) On the S versus $(E - \epsilon_1)/(\epsilon_2 - \epsilon_1)$ diagram in part (c), and with respect to the graph of the fundamental relation, where are the states of the spin system that are not stable equilibrium located?

8.5 For most systems of practical importance, the energy and the entropy of any state, equilibrium or nonequilibrium, may be expressed in the forms

$$E = \sum_i x_i \epsilon_i \qquad \text{and} \qquad S = -k \sum_i x_i \ln x_i$$

where $x_1, x_2, \ldots$ are dimensionless numbers in the range between zero and unity, $\sum_i x_i = 1$, and $\epsilon_1, \epsilon_2, \ldots$ are variables (with dimensions of energy) that are characteristic of the system and functions, in general, of the state. For the stable equilibrium states of such systems, the energy and the entropy may be expressed in the forms

$$E = \sum_i x_i^o \epsilon_i^o \qquad \text{and} \qquad S = -k \sum_i x_i^o \ln x_i^o$$

where $\epsilon_1^o, \epsilon_2^o, \ldots$ are variables characteristic of the system that depend only on the values of the amounts of constituents n and the parameters β, namely, that are constants for given values of the amounts of constituents and the parameters, $x_1^o, x_2^o, \ldots$ are given by the relations

$$x_i^o = \frac{\exp(-\epsilon_i^o/kT)}{Q} \qquad \text{for } i = 1, 2, \ldots$$

$$Q = \sum_j \exp\left(-\frac{\epsilon_j^o}{kT}\right)$$

and T is a variable characteristic of each stable equilibrium state.

(a) Verify that $T = (\partial E/\partial S)_{\epsilon^o} = (\partial E/\partial S)_{n,\beta}$, where the subscript $\epsilon^o = \{\epsilon_1^o, \epsilon_2^o, \ldots\}$.

(b) Verify that $S = E/T + k \ln Q$.

(c) Verify that $E = kT(\partial \ln Q/\partial \ln T)_{\epsilon^o}$.

(d) Verify that $S = k[\partial(T \ln Q)/\partial T]_{\epsilon^o}$.

(e) Verify that $x_i^o = -kT(\partial \ln Q/\partial \epsilon_i^o)_{T, \epsilon_{j \neq i}^o}$.

8.6 A system consists of an amount n of a single constituent and has volume V as a parameter. Its fundamental relation is

$$S = an + nR \ln\left(\frac{E^{3/2}V}{n^{5/2}}\right)$$

where E is the energy and a and R are constants.

(a) Write the difference dS in entropies of two neighboring stable equilibrium states to second order in all the differences dE, dn, and dV.

(b) Express the energy of the system in terms of the property $T = (\partial E/\partial S)_{n,V}$.

(c) For $dV = 0$, find the necessary condition for the second-order differential of dS to be sign definite for all values of dE and dn.

8.7 If S, T, and p are each a function of E, n, and V, the Jacobian is denoted by the symbol $\partial(S,T,p)/\partial(E,n,V)$, and defined as a determinant of a matrix of partial derivatives by the relation

$$\frac{\partial(S,T,p)}{\partial(E,n,V)} = \det \begin{vmatrix} (\partial S/\partial E)_{n,V} & (\partial S/\partial n)_{E,V} & (\partial S/\partial V)_{E,n} \\ (\partial T/\partial E)_{n,V} & (\partial T/\partial n)_{E,V} & (\partial T/\partial V)_{E,n} \\ (\partial p/\partial E)_{n,V} & (\partial p/\partial n)_{E,V} & (\partial p/\partial V)_{E,n} \end{vmatrix}$$

Jacobians are useful in finding the effects of changing independent variables from one set to another.

(a) Show that the Jacobian $\partial(S,n,V)/\partial(E,n,V)$ equals the partial derivative $(\partial S/\partial E)_{n,V}$.

(b) Show that the Jacobian $\partial(S,T,p)/\partial(E,n,V)$ equals the negative of the Jacobian $\partial(T,S,p)/\partial(E,n,V)$.

(c) Show that the Jacobian $\partial(S,T,p)/\partial(E,n,V)$ may be written as the product of the Jacobians $\partial(S,T,p)/\partial(x,y,z)$ and $\partial(x,y,x)/\partial(E,n,V)$.

(d) Prove that the product of the Jacobians $\partial(S,T,p)/\partial(E,n,V)$ and $\partial(E,n,V)/\partial(S,T,p)$ equals unity, that is, that the Jacobian $\partial(S,T,p)/\partial(E,n,V)$ equals the inverse of the Jacobian $\partial(E,n,V)/\partial(S,T,p)$.

(e) Use your answers in parts (a) and (d) to show that $(\partial S/\partial E)_{n,V} = 1/(\partial E/\partial S)_{n,V}$.

(f) Express $(\partial T/\partial p)_{S,n}$ as a Jacobian, and prove that

$$\left(\frac{\partial T}{\partial p}\right)_{S,n} = -\frac{(\partial S/\partial p)_{T,n}}{(\partial S/\partial T)_{p,n}}$$

9   Temperature

In this chapter we use the highest-entropy principle to derive a necessary condition for mutual stable equilibrium between systems, each of which has energy that may both increase and decrease. Other necessary conditions are derived in Chapters 10 and 11.

9.1   A Necessary Condition for Mutual Stable Equilibrium

We consider a composite system C consisting of two systems, A and B, in mutual stable equilibrium (Figure 9.1a). Thus each of the systems A, B, and C must be in a stable equilibrium state A_0, B_0, and C_{00}, respectively.

For systems A and B to be in mutual stable equilibrium, it is necessary that the state C_{00} of the composite system C be stable equilibrium and, therefore, that the entropy S_{00}^C of C in state C_{00} be larger than the entropy S_{11}^C of any other state C_{11} with the same values of E, n, β as those of state C_{00}, that is, it is necessary that $S_{11}^C - S_{00}^C < 0$.

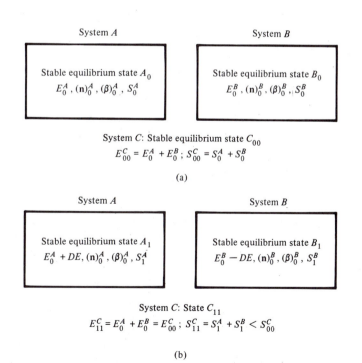

Figure 9.1   Schematics of stable equilibrium state C_{00} and state C_{11} of a composite of two systems.

The strict entropy inequality must hold for any of the specified states C_{11}, including those for which each of the systems A and B has the same values of the amounts of constituents and the parameters as state C_{00}, system A is in a stable equilibrium state A_1 with energy $E_1^A = E_0^A + DE$, and system B in a stable equilibrium state B_1 with energy $E_1^B = E_0^B - DE$, where E_0^A and E_0^B are the energies of states A_0 and B_0, respectively (Figure 9.1b). It is clear that C_{00} and C_{11} are two different states of system C with the same values of energy, $E_{11}^C = E_1^A + E_1^B = E_0^A + DE + E_0^B - DE = E_0^A + E_0^B = E_{00}^C$, amounts of constituents, and parameters.

For the choice of stable equilibrium states of A and B just cited, the entropy difference $S_{11}^C - S_{00}^C$ can be expressed in terms of the energy differences $DE = E_1^A - E_0^A = -(E_1^B - E_0^B)$. Indeed, we recall that entropy is an additive property and therefore

$$S_{11}^C - S_{00}^C = (S_1^A + S_1^B) - (S_0^A + S_0^B) = (S_1^A - S_0^A) + (S_1^B - S_0^B) \qquad (9.1)$$

On the other hand, because states A_0 and A_1, and B_0 and B_1 are all stable equilibrium, each of the two differences in the right-hand side of (9.1) can be expressed in the form of a Taylor series in terms of the corresponding energy difference [equation (8.6)]; that is,

$$S_1^A - S_0^A = \left[\left(\frac{\partial S}{\partial E}\right)_{n,\beta}\right]_0^A (E_1^A - E_0^A) + \frac{1}{2}\left[\left(\frac{\partial^2 S}{\partial E^2}\right)_{n,\beta}\right]_0^A (E_1^A - E_0^A)^2 + \cdots$$

$$= \left[\left(\frac{\partial S}{\partial E}\right)_{n,\beta}\right]_0^A (DE) + \frac{1}{2}\left[\left(\frac{\partial^2 S}{\partial E^2}\right)_{n,\beta}\right]_0^A (DE)^2 + \cdots \qquad (9.2)$$

and

$$S_1^B - S_0^B = \left[\left(\frac{\partial S}{\partial E}\right)_{n,\beta}\right]_0^B (E_1^B - E_0^B) + \frac{1}{2}\left[\left(\frac{\partial^2 S}{\partial E^2}\right)_{n,\beta}\right]_0^B (E_1^B - E_0^B)^2 + \cdots$$

$$= \left[\left(\frac{\partial S}{\partial E}\right)_{n,\beta}\right]_0^B (-DE) + \frac{1}{2}\left[\left(\frac{\partial^2 S}{\partial E^2}\right)_{n,\beta}\right]_0^B (DE)^2 + \cdots \qquad (9.3)$$

Upon combining equations (9.1) to (9.3), we find that

$$S_{11}^C - S_{00}^C = (S_1^A - S_0^A) + (S_1^B - S_0^B)$$

$$= \left\{\left[\left(\frac{\partial S}{\partial E}\right)_{n,\beta}\right]_0^A - \left[\left(\frac{\partial S}{\partial E}\right)_{n,\beta}\right]_0^B\right\}(DE)$$

$$+ \frac{1}{2}\left\{\left[\left(\frac{\partial^2 S}{\partial E^2}\right)_{n,\beta}\right]_0^A + \left[\left(\frac{\partial^2 S}{\partial E^2}\right)_{n,\beta}\right]_0^B\right\}(DE)^2 + \cdots \qquad (9.4)$$

The difference $S_{11}^C - S_{00}^C$ must be strictly negative for any value of DE. This implies that the first-order term in the expansion in the right-hand side of (9.4) must be identically zero for any value of DE. Indeed, if we assume that it is not identically zero, then for infinitesimal values of DE, that is, for $DE = dE$, the first-order term would predominate in the expansion, and the inequality $S_{11}^C - S_{00}^C < 0$ would imply the condition

$$\left\{\left[\left(\frac{\partial S}{\partial E}\right)_{n,\beta}\right]_0^A - \left[\left(\frac{\partial S}{\partial E}\right)_{n,\beta}\right]_0^B\right\}dE < 0 \qquad (9.5)$$

for both positive and negative values of the infinitesimal quantity dE. This strict inequality cannot be satisfied because for $dE > 0$ the difference $[(\partial S/\partial E)_{n,\beta}]_0^A - [(\partial S/\partial E)_{n,\beta}]_0^B$ must be negative, for $dE < 0$ the difference $[(\partial S/\partial E)_{n,\beta}]_0^A -$

$[(\partial S/\partial E)_{n,\beta}]_0^B$ must be positive and, clearly, a number cannot be simultaneously negative and positive. So whenever two systems A and B are in mutual stable equilibrium and their respective energies may both increase and decrease, a condition that must necessarily hold is

$$\left[\left(\frac{\partial S}{\partial E}\right)_{n,\beta}\right]_0^A = \left[\left(\frac{\partial S}{\partial E}\right)_{n,\beta}\right]_0^B \tag{9.6}$$

Because of the necessity of (9.6), the strict negativity of the difference $S_{11}^C - S_{00}^C$ must be guaranteed by conditions on either the second- or higher-order terms in the expansion on the right-hand side of (9.4). We study these additional conditions in Section 9.4. In general, conditions for mutual stable equilibrium are very useful in relating properties of the stable equilibrium states of one system to those of another. As we discuss in the next section, condition (9.6) is so useful that it justifies the definition of a special property of stable equilibrium states.

9.2 Definition of Absolute Temperature

For a system in a stable equilibrium state, the inverse of the partial derivative $(\partial S/\partial E)_{n,\beta}$ of the fundamental relation is denoted by the symbol T and called the *absolute temperature* or, simply, the *temperature*, that is,

$$T = \frac{1}{(\partial S/\partial E)_{n,\beta}} = \left(\frac{\partial E}{\partial S}\right)_{n,\beta} \tag{9.7}$$

where the second of equations (9.7) follows from equation (8.13).

The value of the absolute temperature T for a stable equilibrium state A_0 of a system A is denoted by T_0^A. Because its definition is in terms of the fundamental relation which holds only for the stable equilibrium states, the temperature T is a property defined only for such states.

One of the reasons for introducing the concept of temperature is that, for two systems A and B to be in mutual stable equilibrium, equation (9.6) requires their temperatures to be equal, that is, the systems must be in *temperature equality*. If system A is in stable equilibrium state A_0 with temperature $T_0^A = [(\partial E/\partial S)_{n,\beta}]_0^A$, and system B in stable equilibrium state B_0 with temperature $T_0^B = [(\partial E/\partial S)_{n,\beta}]_0^B$, then A and B are in mutual stable equilibrium only if

$$T_0^A = T_0^B \tag{9.8}$$

The necessity of temperature equality can readily be established for more than two systems which are initially in mutual stable equilibrium and each of which has energy that may both decrease and increase.

Temperature values can be assigned to a system by a method that makes use of the temperature equality condition (9.8), and capitalizes on the fact that each stable equilibrium state of certain systems called thermometers is uniquely related to some readily observable property, such as length. Upon placing a thermometer in mutual stable equilibrium with system A, and assuming the readily observable property to be a monotonic function of the temperature of the thermometer, we can use the value of that property as an indicator of the temperature of system A. For example, the length of a suitably calibrated column of mercury (Figure 9.2) is one such readily observable property of the column of mercury that can be used for temperature measurements.

Either of equations (9.7) prescribes a rigorous procedure for assigning values to the absolute temperature of a system in any of its stable equilibrium states. By making

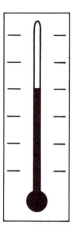

Figure 9.2   Sketch of a thermometer consisting of a suitably calibrated column of mercury.

measurements of energy and entropy of all the stable equilibrium states of the system, we can make a graph of E versus S for each set of values n and β. Then for each graph and at each point defined by a given pair of values E and S, we can evaluate the slope $(\partial E/\partial S)_{n,\beta}$ and therefore the temperature. Of course, if we have an explicit analytical form of (8.7), we readily compute the partial derivative $(\partial E/\partial S)_{n,\beta}$ and thus find the temperature of each stable equilibrium state.

It is noteworthy that absolute temperature is independent of any of the reference states used in the definitions of energy and entropy because its values depend on differences and not on the absolute values of energy and entropy.

Absolute temperature has the dimensions of energy per unit of entropy. Accordingly, the dimensions of T are the same as those of the constant reservoir property c_R that is part of the definition of entropy in Section 7.1. From the discussion in Section 7.5, it follows that the most common units of measure of absolute temperature are either the kelvin or the rankine.

Example 9.1.   Show that the absolute temperature of a reservoir is a constant, namely, show that for a reservoir R the partial derivative $[(\partial S/\partial E)_{n,\beta}]^R$ has the same value for all its states and, therefore, all the higher partial derivatives of the fundamental relation of a reservoir, $[(\partial^2 S/\partial E^2)_{n,\beta}]^R$, $[(\partial^3 S/\partial E^3)_{n,\beta}]^R$, ..., are identically equal to zero.

Solution.   Condition (9.6) must hold for any two systems in mutual stable equilibrium. In particular, if we consider two reservoirs R and R' duplicates of each other, the definition of a reservoir requires that in all of its states, reservoir R must be in mutual stable equilibrium with reservoir R' in a given fixed state (Section 6.3). Therefore, condition (9.6) implies that the partial derivative $[(\partial S/\partial E)_{n,\beta}]^R$ evaluated at any state of R must have the fixed value $[(\partial S/\partial E)_{n,\beta}]^{R'}$ corresponding to a given fixed state of R'. Hence we conclude that the partial derivative $[(\partial S/\partial E)_{n,\beta}]^R$ of reservoir R is constant and that all its derivatives, $[(\partial^2 S/\partial E^2)_{n,\beta}]^R$, $[(\partial^3 S/\partial E^3)_{n,\beta}]^R$, ..., are identically equal to zero.

9.3   Positivity of Temperature

The impossibility of a PMM2 implies that the absolute temperature of any stable equilibrium state is positive.[1] To see this implication clearly, we consider a system A and two

---

[1]Later we show that each stable equilibrium state with a ground-state energy has a zero temperature. Also, we show that for systems whose energy cannot be increased indefinitely, such as nuclear spin systems, the absolute temperature is positive for some stable equilibrium states and negative for others. For states with negative temperature, the statement of the impossibility of a perpetual-motion machine of the second kind differs from that given in Section 4.5. The modified statement is discussed later.

different stable equilibrium states A_1 and A_2 with the same values of n and β. First, we show that the two differences, $S_2 - S_1$ of the entropies and $E_2 - E_1$ of the energies of the two states A_1 and A_2, must have the same sign, namely, that

$$\left(\frac{S_2 - S_1}{E_2 - E_1}\right)_{n,\beta} > 0 \qquad (9.9)$$

Indeed, if $S_2 - S_1 > 0$, a weight process from A_1 to A_2 is possible because it conforms with the requirement on the behavior of the entropy [inequality (7.13)], and must be such that $E_2 - E_1 > 0$ so as not to violate the impossibility of a PMM2. Conversely, if $S_2 - S_1 < 0$, a weight process from A_2 to A_1 is possible and must be such that $E_2 - E_1 < 0$ so as not to violate the impossibility of a PMM2. Thus relation (9.9) is verified.

Next, we rewrite equation (8.6) in the form

$$\left(\frac{S_2 - S_1}{E_2 - E_1}\right)_{n,\beta} = \left[\left(\frac{\partial S}{\partial E}\right)_{n,\beta}\right]_1 + \frac{1}{2}\left[\left(\frac{\partial^2 S}{\partial E^2}\right)_{n,\beta}\right]_1 (E_2 - E_1) + \cdots \qquad (9.10)$$

This relation must be valid for any pair of stable equilibrium states A_1 and A_2, including pairs for which $E_2 - E_1$ is infinitesimal. Thus, for any given state A_1, we may consider a sequence of states A_2 different from A_1 and such that $E_2 - E_1$ tends to zero. Then the second term in the right-hand side tends to zero and therefore

$$\frac{1}{T_1} = \left[\left(\frac{\partial S}{\partial E}\right)_{n,\beta}\right]_1 = \lim_{E_2 \to E_1}\left(\frac{S_2 - S_1}{E_2 - E_1}\right)_{n,\beta} > 0 \qquad (9.11)$$

where in writing the inequality we use relation (9.9). This verifies our assertion that the absolute temperature T for any of the stable equilibrium states contemplated here cannot be negative. Said differently, this implies that the entropy of stable equilibrium states with given n and β is a nondecreasing function of energy. The possibility of stable equilibrium states with zero temperature is discussed in Section 9.9.

9.4 Concavity of the Fundamental Relation with Respect to Energy

We now return to the difference $S_{11}^C - S_{00}^C$ and equation (9.4) for two systems initially in mutual stable equilibrium. The first-order term in the expansion in the right-hand side of (9.4) is identically zero [condition (9.6)], and therefore the required strict negativity of $S_{11}^C - S_{00}^C$ must be satisfied by the sum of the higher-order terms for all values of DE. Unless it is identically zero, the second-order term would predominate for infinitesimal values of DE, and because $(dE)^2 > 0$, we conclude that the condition $S_{11}^C - S_{00}^C < 0$ requires that

$$\left[\left(\frac{\partial^2 S}{\partial E^2}\right)_{n,\beta}\right]_0^A + \left[\left(\frac{\partial^2 S}{\partial E^2}\right)_{n,\beta}\right]_0^B < 0 \qquad (9.12)$$

Inequality (9.12) must be satisfied for any two systems A and B in mutual stable equilibrium, including a composite of a system A and a reservoir. But for a reservoir the second-order derivative is zero (Example 9.1). So we must conclude that for any system

$$\left(\frac{\partial^2 S}{\partial E^2}\right)_{n,\beta} \leqq 0 \qquad (9.13)$$

where we have omitted the superscript A referring to the system and the subscript 0 referring to the state because relation (9.13) must be valid for any system and any of its stable equilibrium states. Mathematically, the inequality in (9.13) implies that the fundamental relation $S = S(E, n, \beta)$ is *concave*[2] in the variable E. In Section 10.3 we show that S is also concave in each of the amounts of constituents that may vary over a range. In Section 11.1 we discuss a similar result about parameters.

Of course, if both $[(\partial^2 S/\partial E^2)_{n,\beta}]_0^A$ and $[(\partial^2 S/\partial E^2)_{n,\beta}]_0^B$ happen to be equal to zero, the strict negativity of $S_{11}^C - S_{00}^C$ must be guaranteed by conditions such as

$$\left[\left(\frac{\partial^3 S}{\partial E^3} \right)_{n,\beta} \right]_0^A = \left[\left(\frac{\partial^3 S}{\partial E^3} \right)_{n,\beta} \right]_0^B \tag{9.14}$$

and

$$\left[\left(\frac{\partial^4 S}{\partial E^4} \right)_{n,\beta} \right]_0^A + \left[\left(\frac{\partial^4 S}{\partial E^4} \right)_{n,\beta} \right]_0^B < 0 \tag{9.15}$$

Because system B can be a reservoir, inequality (9.15) is equivalent to the requirement that $(\partial^4 S/\partial E^4)_{n,\beta} \leqq 0$ for any system and any stable equilibrium state. Similar conditions on higher-order partial derivatives must be considered if the value of each of the fourth-order derivatives is zero.

It is noteworthy that if A and B are two reservoirs in mutual stable equilibrium, all the higher-order partial derivatives are zero (see Example 9.1). Hence the difference $S_{11}^C - S_{00}^C$ for the two reservoirs cannot be strictly negative, in violation of the principle of highest entropy. This is yet another way to show that a reservoir cannot actually exist (see Section 6.3) even though its behavior can be approximated to any desired extent.

9.5   Convexity of the Energy Relation with Respect to Entropy

We have seen that the energy relation $E = E(S, n, \beta)$ can be obtained by solving the fundamental relation $S = S(E, n, \beta)$ in the neighborhood of each stable equilibrium state with $(\partial S/\partial E)_{n,\beta} \neq 0$, that is, with finite temperature. Hence for the states contemplated here the energy relation can always be obtained for all stable equilibrium states. In particular, we can readily show that

$$\left(\frac{\partial^2 E}{\partial S^2} \right)_{n,\beta} = -T^3 \left(\frac{\partial^2 S}{\partial E^2} \right)_{n,\beta} \tag{9.16}$$

Example 9.2.   Prove equation (9.16).

Solution.   Differentiation of $S = S(E, n, \beta)$ with respect to E yields the relations $T = T(E, n, \beta) = [(\partial S/\partial E)_{n,\beta}]^{-1}$, and differentiation of $T = T(E, n, \beta)$ the relations $(\partial T/\partial E)_{n,\beta} = -[(\partial S/\partial E)_{n,\beta}]^{-2}(\partial^2 S/\partial E^2)_{n,\beta} = -T^2(\partial^2 S/\partial E^2)_{n,\beta}$. On the other hand, differentiation of $E = E(S, n, \beta)$ with respect to S yields the relations $T = T(S, n, \beta) = (\partial E/\partial S)_{n,\beta}$. Substituting the fundamental relation into $T = T(S, n, \beta)$, we find that $T = T(S(E, n, \beta), n, \beta)$ and, by differentiation with respect to E, $(\partial T/\partial E)_{n,\beta} = (\partial T/\partial S)_{n,\beta}(\partial S/\partial E)_{n,\beta} = (\partial T/\partial S)_{n,\beta}/T = (\partial^2 E/\partial S^2)_{n,\beta}/T$. Comparing the two expressions thus obtained for $(\partial T/\partial E)_{n,\beta}$, we find (9.16). Because temperature is positive and $(\partial^2 S/\partial E^2)_{n,\beta}$ satisfies (9.13), equation (9.16) implies that

---

[2]We say that a function $f(x, \ldots)$ is *concave* in variable x if the derivative $\partial f/\partial x$ is a monotonically decreasing function of x, that is, if $\partial^2 f/\partial x^2 \leqq 0$. If instead $\partial^2 f/\partial x^2 \geqq 0$, we say that the function is *convex* in the variable x.

$$\left(\frac{\partial^2 E}{\partial S^2}\right)_{n,\beta} \geqq 0 \tag{9.17}$$

In particular, when $(\partial^2 S/\partial E^2)_{n,\beta}$ is negative, $(\partial^2 E/\partial S^2)_{n,\beta}$ must be positive and, therefore, the energy relation $E = E(S, n, \beta)$ is *convex*[3] in the variable S.

The conclusions that temperature is positive and the relation $E = E(S, n, \beta)$ is convex in the variable S imply that a graph of energy versus entropy of stable equilibrium states corresponding to given values n and β has the shape shown in Figure 9.3, that is, the slope of the curve is nonnegative at every value of S, and increases as S increases. Said differently, the conclusions imply that the energy of stable equilibrium states with given n and β is a monotonically increasing convex function of entropy.

Example 9.3.   Prove that an energy versus entropy graph with an inflection point is not possible.

Solution.   An inflection point would imply that in the neighborhood of some stable equilibrium state, the second derivative $(\partial^2 E/\partial S^2)_{n,\beta}$ changes sign, thus violating relation (9.17).

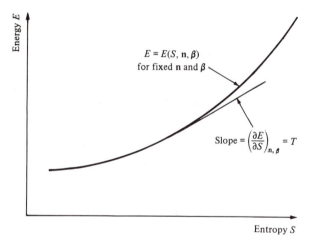

Figure 9.3   Sketch of the energy versus entropy relation for the stable equilibrium states with given values of n and β.

9.6   Temperature as an Escaping Tendency for Energy and Entropy

We consider two systems A and B that are initially in stable equilibrium states A_0 and B_0 with different temperatures T_0^A and T_0^B, respectively. We wish to study what happens if we let the two systems interact with each other so that no net effects are left in the environment of the composite of A and B, and the values of the amounts of constituents and the parameters of both A and B remain fixed (Figure 9.4). For example, system A is a hot piece of metal, and system B a cold piece of metal, and the two pieces are placed in contact with each other.

Under these conditions and irrespective of the types of interactions that are established between A and B, the process for the composite of A and B results in energy and entropy being transferred from the system that is initially at higher temperature to the system that is initially at lower temperature. For example, an amount of energy $DE > 0$ flows from system A to system B, so that the energy of A decreases from E_0^A

---

[3] See footnote 2.

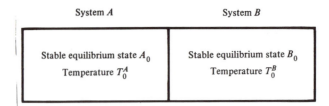

System A   System B

Stable equilibrium state A_0 Stable equilibrium state B_0

Temperature T_0^A Temperature T_0^B

Figure 9.4 When two systems A and B, initially in stable equilibrium states with different temperatures, can interact with each other and no other systems in their environment, an amount of energy $DE > 0$ can flow spontaneously from system A to system B only if $1/T_0^A < 1/T_0^B$.

to $E_0^A - DE$ and that of B increases from E_0^B to $E_0^B + DE$ only if

$$T_0^A > T_0^B \tag{9.18}$$

or, equivalently,

$$\frac{1}{T_0^A} < \frac{1}{T_0^B} \tag{9.19}$$

This result, which we are about to prove, is equivalent to the statement of the second law given by Clausius in 1850 — that no process is possible in which the sole net effect is the transfer of an amount of energy from a system in a stable equilibrium state with a lower temperature to a system in a stable equilibrium state with a higher temperature.

Relations (9.18) and (9.19) may be interpreted by thinking of either the temperature T or the inverse temperature $1/T$ as a potential that governs the tendency of energy to pass from one system to another under the specified conditions. Specifically, temperature may be regarded as an escaping tendency for energy of a system in a stable equilibrium state in the sense that energy flows from high to low temperature. Similarly, inverse temperature may be regarded as a capturing tendency in the sense that energy flows from low to high inverse temperature.

Our everyday awareness of the temperature of an object is based on the feature just cited. When we touch an object, our nervous system senses whether as a result of the contact our body is gaining or losing energy. If we gain energy, we say that the object is *hot*, and if we lose energy that the object is *cold*. Relation (9.18) tells us that if the object is hot, its temperature must be greater than the temperature of our body, whereas if it is cold, its temperature must be smaller than that of our body. Thus our assertions in this section establish a qualitative relation between the property absolute temperature of systems in stable equilibrium states and the everyday empirical notion of relative hotness of bodies.

To verify our assertions, we consider a composite C of two systems A and B. Initially, systems A and B are in stable equilibrium states A_0 and B_0, respectively, such that $T_0^A \neq T_0^B$, and therefore A and B are not in mutual stable equilibrium. We now assume that an amount of energy is exchanged between A and B without net changes in the values of the amounts of constituents and the parameters of either A or B, and without any net effects on the environment of C. In particular, we assume that the energy of the final state of A differs from the energy of the initial state by DE^A, and that the energy of the final state of B differs from the energy of the initial state by DE^B, where $DE^B = -DE^A$ by virtue of the energy balance for system C. Moreover, we denote the difference between the final and initial entropies of A by DS^A, and the corresponding difference of B by DS^B. It is noteworthy that the final state of either A, or B, or both need not be the stable equilibrium state corresponding to the new respective value of the energy.

Because entropy is additive, and the principle of nondecrease of entropy is valid for the change of state in question, we conclude that the entropy of C cannot decrease, that is,

$$DS^C = DS^A + DS^B \geq 0 \qquad (9.20)$$

On the other hand, because of the principle of highest entropy, each of the changes of entropy DS^A and DS^B must be either smaller than or, at most, equal to the entropy change that would obtain if the final state of each of the systems A and B were stable equilibrium. Therefore, DS^A and DS^B must satisfy the relations

$$DS^A \leq \left[\left(\frac{\partial S}{\partial E}\right)_{n,\beta}\right]^A_0 (DE^A) + \frac{1}{2}\left[\left(\frac{\partial^2 S}{\partial E^2}\right)_{n,\beta}\right]^A_0 (DE^A)^2 + \cdots \qquad (9.21)$$

and

$$DS^B \leq \left[\left(\frac{\partial S}{\partial E}\right)_{n,\beta}\right]^B_0 (-DE^A) + \frac{1}{2}\left[\left(\frac{\partial^2 S}{\partial E^2}\right)_{n,\beta}\right]^B_0 (DE^A)^2 + \cdots \qquad (9.22)$$

where the right-hand sides of (9.21) and (9.22) represent the Taylor series for the entropy changes corresponding to stable equilibrium final states at energies $E^A + DE^A$ and $E^B - DE^A$, respectively.

Each Taylor series is equal to a two-term expression[4] in which the second order partial derivative is evaluated at some value E_1 between the values E and $E + DE$. Accordingly, we have

$$DS^A \leq \left[\left(\frac{\partial S}{\partial E}\right)_{n,\beta}\right]^A_0 (DE^A) + \frac{1}{2}\left[\left(\frac{\partial^2 S}{\partial E^2}\right)_{n,\beta}\right]^A_1 (DE^A)^2 \qquad (9.23)$$

and

$$DS^B \leq \left[\left(\frac{\partial S}{\partial E}\right)_{n,\beta}\right]^B_0 (-DE^A) + \frac{1}{2}\left[\left(\frac{\partial^2 S}{\partial E^2}\right)_{n,\beta}\right]^B_1 (DE^A)^2 \qquad (9.24)$$

where $[(\partial^2 S/\partial E^2)_{n,\beta}]^A_1$ and $[(\partial^2 S/\partial E^2)_{n,\beta}]^A_1$ are, respectively, the values of the second partial derivatives of the fundamental relations of A and B evaluated at stable equilibrium states A_1 and B_1 that have energies E^A_1 and E^B_1, respectively, such that E^A_1 lies between E^A_0 and $E^A_0 + DE^A$, and E^B_1 between E^B_0 and $E^B_0 - DE^A$. Upon adding (9.23) and (9.24), recalling that $(\partial S/\partial E)_{n,\beta} = 1/T$, and combining the result with (9.20), we find that

$$\left(\frac{1}{T^A_0} - \frac{1}{T^B_0}\right) DE^A + \frac{1}{2}\left\{\left[\left(\frac{\partial^2 S}{\partial E^2}\right)_{n,\beta}\right]^A_1 + \left[\left(\frac{\partial^2 S}{\partial E^2}\right)_{n,\beta}\right]^B_1\right\}(DE^A)^2 \geq 0 \qquad (9.25)$$

The second term in the left-hand side of (9.25) is negative or at most zero because of the general relation (9.13). Thus we conclude that

$$\left(\frac{1}{T^A_0} - \frac{1}{T^B_0}\right) DE^A \geq 0 \qquad (9.26)$$

---

[4]We recall that the Taylor formula for the expansion of a function $f(x)$ can always be represented as a two-term expression

$$f(x) - f(x_0) = f'(x_0)(x - x_0) + \frac{1}{2} f''(x_1)(x - x_0)^2$$

where the prime and double-prime superscripts denote first- and second-order derivatives with respect to the variable x, respectively, and the second-order derivative is evaluated not at x_0 but at some other value x_1 which lies between x_0 and x.

and, therefore, if $DE^A < 0$, that is, if energy is transferred from system A to system B, then $(1/T_0^A - 1/T_0^B) \le 0$, and because temperatures are positive, $T_0^B \le T_0^A$. So if $T_0^B \ne T_0^A$, we obtain (9.18) and (9.19), and the Clausius statement of the second law emerges as a special consequence of the laws introduced in this book.

From relation (9.21) we also conclude that if $DE^A < 0$, then $DS^A < 0$ and, therefore, that both energy and entropy are transferred out of A when $T_0^A > T_0^B$. So T may be regarded as an escaping tendency for entropy as well as energy, and $1/T$ as a capturing tendency for entropy as well as energy.

9.7 Temperature of a Reservoir

We recall that a reservoir is an idealized kind of system that is assumed to pass through stable equilibrium states only, and to remain in mutual stable equilibrium with a duplicate of itself that experiences no changes of state (Section 6.3). In Example 9.1 we show that all states of a reservoir have the same value of temperature. Upon denoting the constant temperature of a reservoir R by T_R, we have that $1/T_R = [(\partial S/\partial E)_{n,\beta}]^R = $ constant, and that each of the higher-order partial derivatives of the fundamental relation of a reservoir with respect to energy is equal to zero.

For all practical purposes, a system R with fixed values of the amounts of constituents and the parameters approaches the behavior of a reservoir if the temperature of the system is approximately constant over a limited set of stable equilibrium states and, thus, the first-order term in the Taylor series expansion of the fundamental relation predominates over the higher-order terms. For any two stable equilibrium states R_1 and R_2 of such a system the relation between entropy and energy differences is to a high degree of accuracy

$$S_2^R - S_1^R = \frac{1}{T_R}(E_2^R - E_1^R) \tag{9.27}$$

Now we show that the constant property c_R defined in Section 7.4 for a reservoir R, and used in Section 7.1 in the definition of property entropy, is also equal to the temperature T_R of the reservoir. To this end we consider a reversible weight process for the composite of a system A and a reservoir R such that A goes from state A_1 to state A_2, and R from state R_1 to state R_2 with no net changes in the values of the amounts of constituents and the parameters of R. In this process, the amount of energy transferred to the weight is equal to the difference in generalized available energies, $\Omega_1^R - \Omega_2^R$ [equation (6.18a)]. Because the composite of A and R undergoes a weight process, the energy balance is

$$E_2^A - E_1^A + E_2^R - E_1^R = -(\Omega_1^R - \Omega_2^R) \tag{9.28}$$

Moreover, because the weight process is reversible, the entropy balance for the composite of A and R is

$$S_2^A - S_1^A + S_2^R - S_1^R = 0 \tag{9.29}$$

Using equation (7.3), we can express the difference in the generalized available energies as

$$\Omega_2^R - \Omega_1^R = E_2^A - E_1^A - c_R(S_2^A - S_1^A) \tag{9.30}$$

Thus, upon combining equations (9.27) to (9.30), we find that

$$c_R = T_R \qquad (9.31)$$

namely, that property c_R of the reservoir R is equal to its temperature.

This result yields yet another way by which the temperature of a stable equilibrium state of a system may be measured. Given a system in a stable equilibrium state, we find a reservoir that is in mutual stable equilibrium with the system. Then the temperature of the system must be equal to c_R, the constant property of the reservoir which can be determined using the set of operations and measurements defined in Section 7.4.

For example, if a system A in a stable equilibrium state A_0 is in mutual stable equilibrium with a reservoir Q consisting of solid, liquid, and vapor water in mutual stable equilibrium — for which $c_Q = 273.16$ K $= 491.688$ R (Section 7.5) — then by the condition of temperature equality and (9.31), we have

$$T_0^A = T_Q = c_Q = 273.16 \text{ K} = 491.688 \text{ R} \qquad (9.32)$$

Example 9.4. Can a reservoir experience a weight process? If so, can the process be reversible?

Solution. Yes, a reservoir can experience a weight process provided the energy transfer is from the weight to the reservoir. Otherwise, the process would be a violation of the impossibility of PMM2 because a reservoir is always in a stable equilibrium state. The energy change caused by the transfer must be accompanied by a simultaneous entropy change in order to bring the reservoir to another stable equilibrium state. Because the weight cannot supply any entropy, the entropy change of the reservoir is entirely due to entropy generation by irreversibility within the reservoir. Therefore, the weight process cannot be reversible. If we take the definition of a reservoir literally, this entropy generation must occur instantly because the reservoir has no other states than stable equilibrium states. Because no changes can occur instantly, this is yet another indication that no true reservoir exists in nature.

9.8 Absolute Entropy

The conclusions that the energy versus entropy relation is convex, and that temperature is positive are predicated on the consideration of stable equilibrium states having values of energy and entropy that may be both increased and decreased without net changes in the values of the amounts of constituents and the parameters. In this section we examine the value of the entropy of a ground-energy state with given values n and β for which the value of the energy may be increased but not decreased.

We pursue our examination in three steps, and show the following.

1. Among all the stable equilibrium states with given values of n and β, the ground-energy stable equilibrium state has the lowest entropy.

2. The value S_g of this lowest entropy is the same for all values of n and β.

3. No other state of the system has entropy lower than S_g.[5]

To show that each ground-energy stable equilibrium state has the lowest entropy among all the stable equilibrium states with the same values of n and β, we consider an arbitrary stable equilibrium state A_1 with energy E_1, entropy S_1, amounts $(n)_1$, and parameters $(\beta)_1$, and the ground-energy stable equilibrium state A_{g1} with energy E_{g1}, entropy S_{g1}, and the same values of $(n)_1$ and $(\beta)_1$ as state A_1. If S_{g1} were not less than S_1, a weight process from A_1 to A_{g1} would be possible and would violate the impossibility of a PMM2 because $E_1 > E_{g1}$. Thus we conclude that $S_{g1} < S_1$.

---

[5]Conclusions 1 and 2 would not hold if we adopt the modified statement of the second law discussed in Section 13.17.

To show that all ground-energy stable equilibrium states have the same entropy, we consider two ground-energy stable equilibrium states A_{g1} and A_{g2}, corresponding to different values of the amounts of constituents and the parameters $(n)_1$, $(\beta)_1$ and $(n)_2$, $(\beta)_2$, respectively. We denote by A_{S2} the stable equilibrium state with amounts $(n)_2$ and parameters $(\beta)_2$ which can be reached by means of a reversible weight process starting from state A_{g1}. Clearly, $S_{S2} = S_{g1}$ because the weight process is reversible, and $S_{S2} \geq S_{g2}$ because A_{g2} is the ground-energy state with same values $(n)_2$, $(\beta)_2$ as A_{S2}. The equality $S_{S2} = S_{g2}$ applies only if states A_{S2} and A_{g2} coincide. Thus we conclude that $S_{g1} \geq S_{g2}$.

Next, we denote by A_{S1} the stable equilibrium state with amounts $(n)_1$ and $(\beta)_1$ that can be reached by means of a reversible weight process starting from state A_{g2}. Clearly, $S_{S1} = S_{g2}$ because the weight process is reversible, and $S_{S1} > S_{g1}$ because A_{g1} is the ground-energy state with same values $(n)_1$, $(\beta)_1$ as A_{S1}. Again, the equality $S_{S1} = S_{g1}$ applies only if states A_{S1} and A_{g1} coincide. Thus we conclude that $S_{g2} \geq S_{g1}$.

Because S_{g2} cannot be simultaneously greater and smaller than S_{g1}, we conclude that $S_{g2} = S_{g1} = S_g$, namely, that the value S_g of the entropy of ground-energy stable equilibrium states is the same for all values of n and β. Moreover, we conclude that this entropy value is the same for all systems because we can apply the argument to a system consisting of all conceivable constituents and parameters, and then consider the set of values $(n)_1$ and $(\beta)_1$ such that only the amounts and parameters of a given system A are nonzero, and the set of values $(n)_2$ and $(\beta)_2$ such that only the amounts and parameters of a different system B are nonzero. In other words, each system can be considered as a special realization of such an all-inclusive single system.

Finally, to show that no state of a system has a value of entropy that is lower than the value S_g common to all ground-energy stable equilibrium states, we note that if such a state existed, namely, a state with entropy $S_1 < S_g$, the second law would require the existence of a stable equilibrium state that could be reached from the given state by means of a reversible weight process and, therefore, would have entropy equal to S_1. But such a state would contradict the fact just proved that no stable equilibrium state has entropy lower than S_g. For each given set of values of n and β, we also conclude that a system admits no ground-energy states other than the ground-energy stable equilibrium state, because the highest entropy principle would require that such states have entropy lower than S_g, which we just showed to be impossible.[6]

Because it is common to all the ground-energy states of all systems, we can assign to S_g the value zero,[7] that is,

$$S_g = 0 \quad \text{for all values of } n \text{ and } \beta \tag{9.33}$$

Moreover, if in the definition of entropy [equation (7.1)] we choose the reference state A_o so that it coincides with a ground-energy state, the resulting entropy values are positive or zero, and additive in the sense of (7.7). Entropy values thus obtained are called *absolute*.

It is noteworthy that in mechanics, each ground-energy state is stable equilibrium, and all mechanical states can be interconnected by means of reversible weight processes. It follows that all the states contemplated in mechanics have the same entropy as that of the ground-energy states. As a result, we say that mechanics is the physics of zero-entropy states.

---

[6]In Section 13.17 we discuss a slightly modified statement of the second law that captures the possibility of existence of ground-energy states that are not all stable equilibrium.

[7]If we were using quantum-theoretical concepts, we would be able to show that the minimum value of entropy is indeed equal to zero. Such concepts, however, are beyond the scope of this book.

9.9 Third Law

In this section we discuss the values of temperature of ground-energy stable equilibrium states (Figure 9.5). We recall that $(\partial E/\partial S)_{n,\beta} = T > 0$ and $(\partial^2 E/\partial S^2)_{n,\beta} \geq 0$ [relations (9.11) and (9.17)], and readily conclude that each ground-energy stable equilibrium state has the lowest temperature among all the stable equilibrium states with the same values of n and β. Indeed, at constant n and β, T is a nondecreasing function of S and therefore $T \geq T_g$, where T_g denotes the temperature of the ground-energy state corresponding to the given n and β. If we were using quantum-theoretical concepts, we would be able to show that all ground-energy states have the same temperature T_g, and that the value of this temperature is equal to zero. Such concepts, however, are beyond the scope of this book. Instead, we note that this important conclusion of the quantum-theoretical treatment of low-temperature stable equilibrium states cannot be drawn as a logical consequence of the statements of the first and second laws, but here must be presented as an additional fundamental principle. It is known as the *third law of thermodynamics* or the *Nernst principle*,[8] and can be stated as follows:

> *For each given set of values of the amounts of constituents and the parameters of a system, there exists one stable equilibrium state with zero temperature.*

The facts that the temperature can have nonnegative values only and that for given values of n and β the temperature is lowest for the ground-energy stable equilibrium state imply that the stable equilibrium state with zero temperature required by the third law is the ground-energy state. In other words, we conclude that the lowest temperature T_g is equal to zero for all values of n and β, that is , every ground-energy stable equilibrium state has zero temperature.[9]

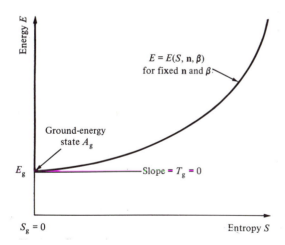

Figure 9.5 Schematic representation of the energy versus entropy relation for stable equilibrium states with given values of n and β at low temperatures.

9.10 Relative Temperatures

Historically, the notion of temperature developed quite differently from the way we have presented it here. Its origin can be traced back to the Italian astronomer and physicist Galileo Galilei (1564–1642), who invented the thermometer around the year

---

[8]In honor of Walther Hermann Nernst (1864–1941), German physical chemist, for his contributions to the study of low-temperature physics, for which he won the Nobel Prize for chemistry in 1920.

[9]Further discussion on lowest-energy and lowest-entropy implications is given in Section 13.17.

1595. Galilei's thermometer was a glass instrument containing water and air that allowed to "distinguish the variety of temperaments of places." The instrument took root with several modifications. In 1714, the German physicist Gabriel Daniel Fahrenheit (1686–1736) introduced a mercury-in-glass instrument graded with the scale that still carries his name, and in 1742, the Swedish astronomer Anders Celsius (1701–1744) devised a different scale to grade the instrument. An important feature of the Fahrenheit and Celsius thermometers is that a change in length DL of the column of liquid in the stem is directly proportional to the change in absolute temperature DT of the thermometer, that is,

$$DL = \alpha\, DT \quad \text{or} \quad L_1 - L_2 = \alpha\,(T_1 - T_2) \tag{9.34}$$

where α is a constant for a thermometer and L is the length corresponding to temperature T.

Based on this feature, the following set of measurements and operations leads to the definition of another property of stable equilibrium states. Given a system A in a stable equilibrium state A_1 with absolute temperature T_1^A, we find a thermometer that is in mutual stable equilibrium with A, measure the length L_1 of the column of liquid in the thermometer (Figure 9.6), and evaluate the quantity

$$t_1 = t_f + \frac{L_1 - L_f}{L_b - L_f}\,(t_b - t_f) \tag{9.35}$$

where L_f and L_b are the lengths of the column of liquid when the thermometer is in mutual stable equilibrium with freezing water and boiling water, respectively, and t_f and t_b are constants defining the scale of the thermometer. Upon using (9.34) and replacing length differences by temperature differences in (9.35), we find that

$$t_1 = t_f + \frac{T_1 - T_f}{T_b - T_f}\,(t_b - t_f) \tag{9.36}$$

where T_f and T_b are the absolute temperatures corresponding to freezing water and boiling water, respectively.

The condition of temperature equality, $T_1 = T_1^A$, between the thermometer and the system A in mutual stable equilibrium indicates that the set of measurements and operations just cited results in the definition of t as a property of the stable equilibrium states of system A. For each choice of t_f and t_b, we call t the *relative temperature* of system A, and denote its value at stable equilibrium state A_1 by t_1. Upon rewriting

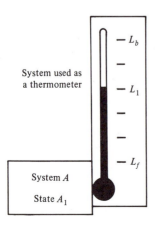

System used as
a thermometer

$-L_b$

$-L_1$

$-L_f$

System A

State A_1

Figure 9.6 When system A is in mutual stable equilibrium with a thermometer, we measure the length of the column of liquid in the thermometer, and evaluate the temperature of A.

equation (9.36) for a stable equilibrium state A_2 different from A_1,

$$t_2 = t_f + \frac{T_2 - T_f}{T_b - T_f} (t_b - t_f) \tag{9.37}$$

and subtracting (9.36) from (9.37), we find that

$$t_2 = t_1 + \frac{t_b - t_f}{T_b - T_f} (T_2 - T_1) \tag{9.38}$$

The relative temperatures that have been most used are the Fahrenheit and Celsius temperatures. Fahrenheit originally chose as fixed points on his scale the freezing temperature at ordinary atmospheric conditions of a mixture of salt, ice, and salt solution, designated as 0°F, and the temperature of the human body, designated as 96°F. Later the fixed points were changed to the freezing and boiling temperatures of water at atmospheric conditions, designated as $t_f = 32°F$ and $t_b = 212°F$, respectively. Celsius chose as fixed points on his scale the freezing and boiling temperatures of water at atmospheric conditions, designated as $t_f = 0°C$ and $t_b = 100°C$, respectively. The Celsius scale has also been called the "centigrade" scale because the stem of a thermometer graded according to this scale is divided into 100 equal intervals between the point corresponding to freezing water and that corresponding to boiling water.

The kelvin and the rankine, which are units of measure of absolute temperature, were defined after the relative temperatures. They were chosen so that $T_b - T_f = 100$ K $= 180$ R. As a result, equation (9.36) yields the following relations between absolute temperature T and either Fahrenheit or Celsius relative temperature.

The *Fahrenheit relative temperature* is defined by assuming that $t_f = 32°F$ and $t_b = 212°F$, and is related to absolute temperature by

$$t_1^{\text{Fahrenheit}} = 32°F + \frac{1°F}{1\,R} (T_1 - T_f) \tag{9.39}$$

The *Celsius relative temperature* is defined by assuming that $t_f = 0°C$ and $t_b = 100°C$ and, therefore,

$$t_1^{\text{Celsius}} = \frac{1°C}{1\,K} (T_1 - T_f) \tag{9.40}$$

Combining (9.39) and (9.40), we find a useful relation between the Celsius and Fahrenheit relative temperatures, namely,

$$t_1^{\text{Fahrenheit}} = 32°F + \frac{9°F}{5°C} t_1^{\text{Celsius}} \tag{9.41}$$

where we use the relation $1\,K/1\,R = 9/5$, which follows from the fact that $T_b - T_f = 100$ K $= 180$ R.

In order for measurements of relative temperature to be equivalent to measurements of absolute temperature, we need a single independent measurement of absolute temperature, for example, the value of T_f, the absolute temperature of freezing water at atmospheric conditions. To this end we note that the temperature T_g of any ground state is equal to zero (Section 9.8), and that among the many different measurements of relative temperature of ground states that have been made, the value on the Celsius scale is $t_g^{\text{Celsius}} = -273.15°C$. Using these observations and (9.40) for a ground state, we find that $T_f = 273.15$ K $= 491.67$ R. Substituting this value into (9.41), we then find that $t_g^{\text{Fahrenheit}} = -459.67°F$. Thus absolute temperatures can be measured with a relative temperature thermometer and, in particular, (9.39) and (9.40) yield

$$T_1 = 273.15 \text{ K} + \frac{1\,K}{1°C} t_1^{\text{Celsius}} \tag{9.42}$$

or

$$T_1 = 459.67 \text{ R} + \frac{1 \text{ R}}{1 \text{°F}} t_1^{\text{Fahrenheit}} \tag{9.43}$$

9.11 Available Energy in Terms of Energy and Entropy

For the difference in the generalized available energies of any two states A_1 and A_2, not necessarily stable equilibrium states, of a system A with respect to a reservoir R, combining equations (9.30) and (9.31) we find

$$\Omega_2^R - \Omega_1^R = E_2^A - E_1^A - T_R(S_2^A - S_1^A) \tag{9.44}$$

If Ω^R denotes the generalized available energy with respect to reservoir R and given values n, β, and A_0 the stable equilibrium state of A with the given values n, β in which A is in mutual stable equilibrium with reservoir R, then $\Omega_0^R = 0$, and therefore (9.44) yields the expression

$$\Omega_1^R = E_1^A - E_0^A - T_R(S_1^A - S_0^A) \tag{9.45}$$

where E_0^A and S_0^A are the energy and the entropy, respectively, of system A in the state A_0 just defined. State A_0 is uniquely determined by the given values of n, β, and the condition of mutual stable equilibrium with the reservoir, which requires that $T_0^A = T_R$.

9.12 Adiabatic Availability in Terms of Energy and Entropy

We consider two neighboring states of system A (neither necessarily stable equilibrium nor necessarily with the same values of the amounts of constituents and the parameters) with energy, entropy, and generalized adiabatic availability with respect to the values n, β denoted as E, S, Ψ and $E + dE$, $S + dS$, $\Psi + d\Psi$, respectively. Moreover, we denote by $E_S(S, n, \beta)$ the energy of the stable equilibrium state corresponding to the set of values S, n, β.

By the definition of generalized adiabatic availability, we have the relation

$$\Psi = E - E_S(S, n, \beta) \tag{9.46}$$

and clearly, for the neighboring state,

$$\Psi + d\Psi = E + dE - E_S(S + dS, n, \beta) \tag{9.47}$$

Using the definition of temperature, we have

$$E_S(S + dS, n, \beta) = E_S(S, n, \beta) + T_S(S, n, \beta) \, dS \tag{9.48}$$

so that combining (9.46) to (9.48), we obtain the equation

$$d\Psi = dE - T_S(S, n, \beta) \, dS \tag{9.49}$$

This equation relates the differences in energies, entropies, and adiabatic availabilities of two neighboring states to the temperature of the stable equilibrium state reached upon extraction of the adiabatic availability.

Problems

9.1 For a limited class of states and for fixed values of the amounts of constituents and the parameters, the fundamental relation of a system is $S = S_1 + 10^5 \ln(E/E_1)$, where the entropy and energy of stable equilibrium state A_1 are $S_1 = 5 \times 10^5$ kJ/K and $E_1 = 1.2 \times 10^8$ kJ, respectively.

(a) Find the temperature T_1 of state A_1, and the temperature T_3 of a state with $E_3 = 12 \times 10^8$ kJ.

(b) With respect to a reservoir at $T_R = 300$ K, the available energy of a state A_2 is $\Omega_2^R = 12 \times 10^8$ kJ and the energy $E_2 = 12 \times 10^8$ kJ. Find the entropy S_2 of state A_2.

(c) Find the adiabatic availability of state A_2.

(d) Find the largest energy that can be transferred to a weight in a weight process for the composite of the system and the reservoir in which system A goes from state A_2 to state A_1.

9.2 System A consists of 12 mol of a constituent in a rigid container of volume 24×10^{-6} m³. Its energy in stable equilibrium state A_1 is 240 J. System B consists of 4 mol of the same constituent as A, confined in a rigid container of volume 8×10^6 m³. The fundamental relation for each of these systems is $S = a(EnV)^{1/3}$, where a is a positive constant with dimensions J²ᐟ³/mol¹ᐟ³ m K.

(a) For system A in state A_1, at what value of the energy of system B are systems A and B in mutual stable equilbrium?

(b) If the energy of B in a stable equilibrium state B_1 is twice as large as your answer in part (a), and A and B can exchange energy, is the energy flow from A to B or from B to A?

(c) Is the given fundamental relation concave or convex with respect to energy?

9.3 Three identical blocks of metal are initially at temperatures $T_1^A = 300$ K, $T_1^B = 500$ K, and $T_1^C = 700$ K, respectively. Each block has a mass of 5 kg, and its stable equilibrium states can be described by the parametric relations

$$E = mc\,(T - T_o) \quad \text{and} \quad S = mc\ln\frac{T}{T_o}$$

where E is the energy, S the entropy, m the mass of the block, T the temperature, and c and T_o are constants, that is, $c = 2$ kJ/kg K and $T_o = 250$ K. Moreover, the same parametric relations also hold for the stable equilibrium states of the composite of the three blocks, provided of course that $m = 15$ kg.

(a) What is the adiabatic availability of each of the three blocks in the initial state?

(b) If the composite of the three blocks experiences a reversible weight process from the initial state to a stable equilibrium state, what is the final temperature of each of the three blocks?

(c) What is the adiabatic availability of the composite of the three blocks in the initial state?

(d) If the three blocks are allowed to interact with each other and cyclic machinery but not with any other systems in the environment, what is the largest temperature that can can be reached by one of the blocks?

(e) Under the same conditions as in part (d), what is the smallest temperature that can be reached by one of the blocks?

(f) Find the available energies with respect to a reservoir at temperature $T_R = 300$ K of each of the three blocks and of their composite when the blocks are: (1) in their initial states; (2) in the states reached in part (d); and (3) in the states reached in part (e).

9.4 Consider two large, rigid tanks A and B. Tank A is filled with 5000 kg of steam at 700°C, and tank B with 10,000 kg of a gas at 400°C. For the stable equilibrium states of interest in this problem, the energy and entropy of both the steam and the gas can be written as $E = mc\,(T - T_o)$ and $S = mc\ln(T/T_o)$, where $c_{\text{steam}} = 1.8$ kJ/kg K and $c_{\text{gas}} = 1.2$ kJ/kg K.

We can transfer energy to a weight from the composite of the rigid tanks by using a thermoelectric generator to power a motor and no interactions with any other system in the environment.

(a) What is the largest amount of energy that can be transfered to the weight in a weight process for the composite of A and B?

(b) What are the final states of A and B after the energy in part (a) has been transferred?

(c) Assume that the weight process for A and B is irreversible, that A and B end in mutual stable equilibrium, and that the weight receives half as much energy as in part (a). How much entropy is generated by irreversibility?

9.5 Consider 1 kg of products of combustion of a hydrocarbon confined in an airtight, fixed-volume combustion chamber. Assume that the energy E and entropy S of the stable equilibrium states of these products can be expressed as functions of temperature T only, so that $dE/dT = f(T)$ and $dS/dT = f(T)/T$, where $f(T) = 1003 + 0.185T$ in J/K, for T in kelvin, and in the range of values 250 K$< T <$ 850 K.

(a) What is the available energy of the products of combustion at $T = 750$ K with respect to a reservoir R at temperature $T_R = 290$ K.

(b) If the products of combustion are in state A_1 having the same energy as the stable equilibrium state at $T = 750$ K, and the same entropy as the stable equilibrium state at $T = 450$ K, what are the adiabatic availability Ψ_1 and the available energy Ω_1^R with respect to the same reservoir R as in part (a)?

(c) Starting from the state A_1 specified in part (b), can the system be brought reversibly to the stable equilibrium state at $T = 750$ K? If it can, what are the net effects on the environment, and what kind of machinery should be used?

9.6 Consider a fixed amount of a fuel–air mixture A confined in a chamber of fixed volume. If the possibility of burning of the fuel is not included in the description of the system, the initial state of the mixture is denoted by $A_{1'}$ and is stable equilibrium at temperature $T_{1'} = 300$ K. If the possibility of burning of the fuel is included, the system is different and so is the initial state. We denote this state by A_1. It is not stable equilibrium. However, states $A_{1'}$ and A_1 have the same energy E_1 and the same entropy S_1.

(a) What is the available energy of the initial state A_1' with respect to a reservoir at $T_R = 300$ K?

(b) States $A_{1'}$ and A_1 are two different descriptions of the same fuel–air mixture in the same initial condition. In what respects do the two descriptions differ from one another?

When burning of the fuel is possible, the available energy of the initial state A_1 with respect to a reservoir at $T_R = 300$ K is $\Omega_1^R = 12,948$ kJ. A reversible weight process takes the system from state A_1 to a stable equilibrium state A_{S1}. In that state the constituents are products of combustion, and the temperature $T_{S1} = 500$ K. As a result of the process, 5000 kJ of energy is transferred to a weight.

For the stable equillibrium states of the products of combustion, the relation between energy and temperature of system A is $E_2 - E_1 = 170\,(T_2 - T_1)$, where E is in kilojoule and T in kelvin.

(c) Use (8.8) and (9.7) to prove that $(\partial S/\partial T)_{n,\beta} = (\partial E/\partial T)_{n,\beta}/T$

(d) Use the relation $(\partial S/\partial T)_{n,\beta} = (\partial E/\partial T)_{n,\beta}/T$ to show that the relation between entropy and temperature of the products of combustion is $S_2 - S_1 = 170\ln(T_2/T_1)$, where S is in kJ/K and T in K.

(e) What is the adiabatic availability Ψ_1 of state A_1?

(f) Upon extraction of the available energy Ω_1^R of state A_1, what are the changes in energy and entropy of the reservoir? What is the change in entropy of system A?

(g) What is the available energy of state A_{S1}?

(h) During a weight process for the composite of system A and the reservoir in which A starts from state A_1 and ends in mutual stable equilibrium with the reservoir, an amount of entropy $S_{irr} = 10$ kJ/K is generated by irreversibility. How much energy is transfered to the weight? What are the energy and entropy changes of the reservoir? What is the change in entropy of system A?

9.7 The energy of some systems, such as spins of nuclei and laser light, may vary only over a finite range between a low and a high value, E_{min} and E_{max}. For fixed values of amounts of constituents and parameters, the states of such a system lie in the crosshatched area shown in Figure P9.7. Note that a segment of the E versus S relation for stable equilibrium states has a negative slope. This segment corresponds to negative temperatures.

(a) How would you express the impossibility of a PMM2 starting from a stable equilibrium state with negative temperature?

(b) Using your statement in part (a), prove that indeed the temperature of some states is negative.

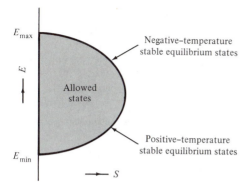

Figure P9.7

(c) Initially, each of two systems A and B is in a stable equilibrium state, the temperature T^A of A is positive and the temperature T^B of B negative. Starting from these two states, A and B are allowed to interact via cyclic machinery with a weight. Prove that in interactions that involve no net changes in the values of amounts of constituents and parameters of either A or B, and in which the weight rises, energy can flow only out of the system with the negative (initial) temperature and into the system with the positive (initial) temperature. Assume that both A and B pass through stable equilibrium states only. Consider small changes in energy.

9.8 Each of two identical containers A and B contains a spring and sulfuric acid. The spring in A is the same as that in B, and the amount of acid in A is the same as that in B. Each container is rigid and well insulated. Initially, the two systems are in states A_1 and B_1, which differ only by the fact that the spring in A is stretched whereas that in B is unstretched (Figure P9.8). The stretching is achieved by a reversible weight process that transfers 10 kJ to A. After some time, the acid has completely dissolved the two springs. A measurement on system B in its final state B_2 (acid-iron composite solution) yields a temperature $T_2^B = 300$ K. For the stable equilibrium states of the acid-iron composite solution, the following interrelations between energy E, entropy S, and temperature T hold: $E_2 - E_1 = C(T_2 - T_1)$ and $S_2 - S_1 = C \ln(T_2/T_1)$, where $C = 2$ kJ/K, and the subscripts 1 and 2 refer to any two stable equilibrium states.

(a) What types are the initial states A_1 and B_1?

(b) What is the difference between the adiabatic availability of system A in state A_1 and that of system B in state B_1?

(c) If B is viewed as an acid–iron composite solution, what is the temperature of system B in its initial state B_1?

(d) What types are the final states A_2 and B_2?

(e) Are the two systems in their final states in mutual stable equilibrium?

(f) What is the difference between the available energies of system A in state A_1 and system B in state B_1 with respect to a reservoir R at temperature $T_R = 20°C$?

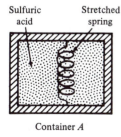

Sulfuric acid   Stretched spring

Container A

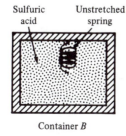

Sulfuric acid   Unstretched spring

Container B

Figure P9.8

9.9 In an intermediate stage of a steelmaking plant, steel billets of $m_0 = 10^4$ kg are initially at 1000 K and must be cooled down to 400 K. Energy and entropy of steel are functions of temperature only, such that $dE/dT = m_0 c$ and $dS/dT = m_0 c/T$, where $c = 0.4$ kJ/kg K. A nearby water pool is at environmental temperature $T_0 = 27°C$. The steelmaking company is considering a waste energy recovery investment to take advantage of the high temperature energy — high-quality energy — of the billets. The annual charges for depreciation, profit, maintenance, and taxes resulting from a capital investment in waste energy recovery can be taken as 40% of the investment. The average price of available energy from various liquid or gaseous fuels is about $7 per 10^6 kJ. The plant operates about 7500 h/yr.

(a) What is the largest amount of energy that can be transferred from the billets to a weight during the cooling process?

(b) The energy recovery plant can extract about 70% of the answer to part (a). What is the largest amount of money, in dollars per kilowatt of available energy recovered, that the company can justify to spend to purchase the waste energy recovery plant? Do you think this amount is sufficient to buy the plant?

9.10 The energy and entropy of the stable equilibrium states of each of two identical batteries are given by the relations $E_2 - E_1 = C(T_2 - T_1)$ and $S_2 - S_1 = C\ln(T_2/T_1)$, where $C = 1.4$ kJ/K. Initially, each battery is in a stable equilibrium state correponding to zero charge with $T_1 = 300$ K, $E_1 = 50$ kJ, and $S_1 = 0.2$ kJ/K, and then each battery is charged (final states A_2 and B_2) by a reversible weight process that transfers 70 kJ of energy to each battery.

(a) What types of states are the final states A_2 and B_2?

(b) What is the adiabatic availability of A_2?

(c) What is the temperature of A_2?

(d) What are the available energies of A_2 with respect to reservoirs at temperatures 270, 300, and 330 K?

(e) After being charged, battery B is left on the shelf for a long time. It discharges internally without interactions with other systems. At the end of this time, it is found in state B_3 with zero charge. What type of state is B_3?

(f) Is there any entropy generated by irreversibility in part (e)? If yes, how much?

(g) What is the adiabatic availability of B_3? What is the available energy of B_3 with respect to the reservoir at 300 K?

9.11 One kilogram of natural gas and two kilograms of atmospheric air are confined in an airtight, rigid, well-insulated combustion chamber. Upon igniting the mixture with an electrical spark, the natural gas starts burning spontaneously.

(a) Describe qualitatively how various properties of the burning mixture behave as time goes on. For example, how do the values of energy, entropy, available energy with respect to a given reservoir, adiabatic availability, and amount of constituents in the combustion chamber at time t_1 compare with the corresponding values at a later time?

(b) Whatever changes you establish in part (a), do they continue occurring forever? When do they stop?

(c) Is this spontaneous burning process efficient or inefficient?

9.12 A reservoir R at temperature $T_R = 1200$ K and a reservoir Q at temperature $T_Q = 300$ K exchange 12 kJ of energy via some cyclic machinery. The machinery interacts only with the two reservoirs.

(a) How much entropy is generated by irreversibility as a result of this energy exchange?

(b) If the machinery is perfect, how does this entropy generation occur? Is the definition of a reservoir too rigid, perhaps?

10   Total Potentials

In this chapter we again use the highest-entropy principle to derive conditions for mutual stable equilibrium between systems having common constituents with values of amounts that for each system may both increase and decrease, without changing the values of the amounts of the other constituents, the parameters, and the energy.

10.1   Additional Necessary Conditions for Mutual Stable Equilibrium

We consider two systems A and B that have in common a given constituent, with an amount that for each system may vary over a range. Without loss of generality, we assume that the common constituent is the ith in the list of constituents of system A, and the ith also in the list of constituents of system B. We denote the values of these amounts by $(n_i)^A$ and $(n_i)^B$, respectively. In addition, we consider the composite C of systems A and B. When A and B are in mutual stable equilibrium, we denote the corresponding stable equilibrium states of A, B, and C by A_0, B_0, and C_{00}, respectively.

Similarly to what we have done in Section 9.1, we compare the entropy S_{00}^C of stable equilibrium state C_{00} (Figure 10.1a) with the entropy S_{11}^C of a different but judiciously selected state C_{11} (Figure 10.1b). Specifically, we select state C_{11} so that systems A and B are, respectively, in stable equilibrium states A_1 and B_1 with energies $E_1^A = E_0^A$ and $E_1^B = E_0^B$, amounts of constituents with values that are all equal to the corresponding values of states A_0 and B_0 except for the ith constituent for which $(n_i)_1^A = (n_i)_0^A + Dn_i$ and $(n_i)_1^B = (n_i)_0^B - Dn_i$, and values of parameters that are all equal to the corresponding values of states A_0 and B_0. It is clear that C_{00} and C_{11} are two different states of system C sharing the same values of energy, $E_{11}^C = E_1^A + E_1^B = E_0^A + E_0^B = E_{00}^C$, the same values of amounts of constituents because even for the ith constituent $(n_i)_{11}^C = (n_i)_1^A + (n_i)_1^B = (n_i)_0^A + Dn_i + (n_i)_0^B - Dn_i = (n_i)_0^A + (n_i)_0^B = (n_i)_{00}^C$, and the same values of parameters. Therefore, C_{11} cannot be a stable equilibrium state and, as such, its entropy S_{11}^C must be smaller than the entropy S_{00}^C of state C_{00}, that is, the entropy difference $S_{11}^C - S_{00}^C$ must necessarily be negative.

Because entropy is additive, we can write $S_{11}^C - S_{00}^C$ in the form

$$S_{11}^C - S_{00}^C = (S_1^A - S_0^A) + (S_1^B - S_0^B) \tag{10.1}$$

Because states A_0, A_1, B_0, and B_1 are all stable equilibrium, each of the differences in the right-hand side of (10.1) can be expressed in the form of a Taylor series in terms of the respective differences in amounts $(n_i)_1^A - (n_i)_0^A = Dn_i$ and $(n_i)_1^B - (n_i)_0^B = -Dn_i$, that is,

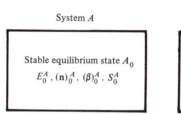
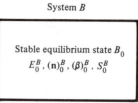

System A

Stable equilibrium state A_0
E_0^A, $(\mathbf{n})_0^A$, $(\beta)_0^A$, S_0^A

System B

Stable equilibrium state B_0
E_0^B, $(\mathbf{n})_0^B$, $(\beta)_0^B$, S_0^B

System C: Stable equilibrium state C_{00}
$(n_i)_{00}^C = (n_i)_0^A + (n_i)_0^B$, $S_{00}^C = S_0^A + S_0^B$

(a)

System A

State equilibrium state A_1
E_0^A, $(n_i)_0^A + Dn_i$, $(\beta)_0^A$
$(n_j)_0^A = (n_j)_1^A$, for $j \neq i$, S_1^A

System B

Stable equilibrium state B_1
E_0^B, $(n_i)_0^B - Dn_i$, $(\beta)_0^B$
$(n_j)_0^B = (n_j)_1^B$, for $j \neq i$, S_1^B

System C: State C_{11}
$(n_i)_{11}^C = (n_i)_{00}^C$, $(n_j)_{11}^C = (n_j)_0^A + (n_j)_0^B = (n_j)_{00}^C$ for $j \neq i$, $|S_{11}^C = S_1^A + S_1^B < S_{00}^C$

(b)

Figure 10.1 Schematics of a stable equilibrium state C_{00} and a state C_{11} of a composite system C. Both states correspond to the same values of the energy, the amounts of constituents, and the parameters.

$$S_1^A - S_0^A = \left[\left(\frac{\partial S}{\partial n_i}\right)_{E,n,\beta}\right]_0^A [(n_i)_1^A - (n_i)_0^A]$$

$$+ \frac{1}{2}\left[\left(\frac{\partial^2 S}{\partial n_i^2}\right)_{E,n,\beta}\right]_0^A [(n_i)_1^A - (n_i)_0^A]^2 + \cdots$$

$$= \left[\left(\frac{\partial S}{\partial n_i}\right)_{E,n,\beta}\right]_0^A (Dn_i) + \text{(higher-order terms)} \qquad (10.2)$$

and

$$S_1^B - S_0^B = \left[\left(\frac{\partial S}{\partial n_i}\right)_{E,n,\beta}\right]_0^B [(n_i)_1^B - (n_i)_0^B]$$

$$+ \frac{1}{2}\left[\left(\frac{\partial^2 S}{\partial n_i^2}\right)_{E,n,\beta}\right]_0^B [(n_i)_1^B - (n_i)_0^B]^2 + \cdots$$

$$= \left[\left(\frac{\partial S}{\partial n_i}\right)_{E,n,\beta}\right]_0^B (-Dn_i) + \text{(higher-order terms)} \qquad (10.3)$$

where here the subscript n denotes that all the amounts of constituents are kept fixed during differentiation except the one that appears in the partial derivative. Upon combining (10.1) to (10.3), we find that

$$S_{11}^C - S_{00}^C = \left\{ \left[\left(\frac{\partial S}{\partial n_i} \right)_{E,n,\beta} \right]_0^A - \left[\left(\frac{\partial S}{\partial n_i} \right)_{E,n,\beta} \right]_0^B \right\} Dn_i$$

$$+ \text{(higher-order terms)} \qquad (10.4)$$

Because the difference $S_{11}^C - S_{00}^C$ must be strictly negative, the coefficient of Dn_i on the right-hand side of (10.4) must be identically zero. Indeed, if we assume that it is not, then for infinitesimal values of Dn_i, that is, for $Dn_i = dn_i$, the first term predominates in the expansion, and the condition $S_{11}^C - S_{00}^C < 0$ implies that

$$\left\{ \left[\left(\frac{\partial S}{\partial n_i} \right)_{E,n,\beta} \right]_0^A - \left[\left(\frac{\partial S}{\partial n_i} \right)_{E,n,\beta} \right]_0^B \right\} dn_i < 0 \qquad (10.5)$$

for both positive and negative values of dn_i. For this to be true, the coefficient of dn_i must be negative when $dn_i > 0$ and positive when $dn_i < 0$, two results that cannot be simultaneously satisfied. So we conclude that

$$\left[\left(\frac{\partial S}{\partial n_i} \right)_{E,n,\beta} \right]_0^A = \left[\left(\frac{\partial S}{\partial n_i} \right)_{E,n,\beta} \right]_0^B \qquad (10.6)$$

In addition to temperature equality [condition (9.6) or (9.8)], equation (10.6) represents a condition—one for each common constituent—that must necessarily hold whenever two systems A and B are in mutual stable equilibrium, both contain the ith-type constituent for $i = 1, 2, \ldots, r$, and the amount of that constituent may both increase and decrease for each system. Because of the necessity of (10.6), the requirement that the entropy difference $S_{11}^C - S_{00}^C$ be negative must be assured by restrictions imposed on the higher-order partial derivatives. Such conditions are discussed in Section 10.3. Because of its utility in many practical problems, condition (10.6) justifies the introduction of yet another property of stable equilibrium states, which is discussed in the following section.

10.2   Total Potential of a Constituent

For a stable equilibrium state, the partial derivative $(\partial S / \partial n_i)_{E,n,\beta}$ is related to the partial derivative $(\partial E / \partial n_i)_{S,n,\beta}$. The latter is denoted by the symbol μ_i, is called the *total potential of the ith constituent*, and satisfies the relation

$$\mu_i = \left(\frac{\partial E}{\partial n_i} \right)_{S,n,\beta} = -T \left(\frac{\partial S}{\partial n_i} \right)_{E,n,\beta} \qquad (10.7)$$

where the second of equations (10.7) follows from (8.14). For a system that has volume V as the only parameter, that is, $\beta = V$, μ_i is called the *chemical potential of the ith constituent*.

The value of the total potential μ_i for a stable equilibrium state A_0 of a system A is denoted by $(\mu_i)_0^A$. Because it is obtained from a relation that holds only for stable equilibrium states, the total potential μ_i is a property defined only for such states.

If a constituent i is present in two systems A and B that are in mutual stable equilibrium, and if the amount of the constituent can vary in each of the systems over a range of values, equation (10.6) is applicable, and using equation (10.7), we can express it in the form

$$\frac{(\mu_i)_0^A}{T_0^A} = \frac{(\mu_i)_0^B}{T_0^B} \qquad (10.8)$$

This relation implies that the total potentials of the constituent in the two systems must be equal because for mutual stable equilibrium between systems whose energies may both increase and decrease, $T_0^A = T_0^B$, and therefore

$$(\mu_i)_0^A = (\mu_i)_0^B \tag{10.9}$$

The *total potential equality* must be true for every constituent that is present in both systems in mutual stable equilibrium, and that has an amount with a value that can both increase and decrease in each of the systems. It can readily be generalized to more than two systems in mutual stable equilibrium. For two systems that have volume as the only parameter, *chemical potential equality* must hold for every constituent that is present in both systems in mutual stable equilibrium, and that has an adjustable amount in each of the systems.

The total potential has dimensions of energy per unit amount of constituent, and assumes values that can be both positive and negative. Later we express μ in terms of more readily understandable variables. We will also see that total potentials are very useful in relating properties of mixtures to properties of pure substances, and in the study of chemical reactions.

Example 10.1. Consider two systems A and C consisting of the same single constituent, say, oxygen, and a system B consisting of a different constituent, say, nitrogen. The oxygen can increase and decrease in both systems A and C but not in system B, and nitrogen cannot vary in any system. Show that the condition that systems A and C are individually in mutual stable equilibrium with system B does not necessarily imply that A and C are in mutual stable equilibrium.

Solution. Mutual stable equilibrium between A and B requires that $T^A = T^B$ but no conditions of the form of (10.9) because the value of the amount of the constituent of B cannot vary. Similarly, mutual stable equilibrium between C and B requires that $T^C = T^B$ but no conditions of the form of (10.9). Thus we conclude that $T^A = T^B = T^C$. However, the total potentials of the common constituent of A and C are not necessarily equal, that is, μ^A and μ^C may or may not be equal, and if they are not, A and C cannot be in mutual stable equilibrium.

10.3 Total Potential as an Escaping Tendency of a Constituent

Proceeding in a way analogous to that discussed in Sections 9.4 and 9.6, we can readily verify the following assertions.

1. The fundamental relation $S = S(E, n, \beta)$ is concave in each of the variables n_i, that is, in general,

$$\sum_{i=1}^{r}\sum_{j=1}^{r}\left(\frac{\partial^2 S}{\partial n_i \partial n_j}\right)_{E,n,\beta} dn_i\, dn_j \leq 0 \qquad \text{for arbitrary } dn_i \text{ and } dn_j \tag{10.10}$$

where, of course, each dn_i must be consistent with the range specified for n_i, for $i = 1, 2, \ldots, r$. Relation (10.10) implies that the double sum on the left-hand side is a negative semidefinite quadratic form,[1] and therefore the square matrix $[(\partial^2 S/\partial n_i \partial n_j)_{E,n,\beta}]$ is negative semidefinite. A necessary condition for the matrix to be negative semidefinite is

$$\left(\frac{\partial^2 S}{\partial n_i^2}\right)_{E,n,\beta} \leq 0 \qquad \text{for } i = 1, 2, \ldots, r \tag{10.11}$$

---

[1]Discussions of quadratic forms are given in many textbooks. For example, see R. Bellman, *Introduction to Matrix Analysis*, McGraw-Hill, New York, 1960.

If a partial derivative $(\partial^2 S / \partial n_i^2)_{E,n,\beta}$ is zero, we must consider higher-order derivatives in order to guarantee the negativity of $S_{11}^C - S_{00}^C$.

2. The ratio μ_i/T may be interpreted as a potential that governs the tendency of constituent i to pass from one system to another in the course of interactions between the two systems that leave no net effects on other systems in the environment. It turns out to be a measure of the escaping tendency of the ith constituent. We can see this clearly by considering two systems A and B that are initially in stable equilibrium states A_0 and B_0 with temperatures $T_0^A = T_0^B$ but different total potentials $(\mu_i)_0^A$ and $(\mu_i)_0^B$ of a common constituent i. Interactions between A and B that leave no net effects on systems in the environment, and that affect neither the values of the parameters nor the values of the amounts of the constituents except the ith, result in an amount Dn_i of the ith constituent being transferred from system A to system B only if

$$(\mu_i)_0^A \geqq (\mu_i)_0^B \tag{10.12}$$

For example, if a system A is a bloodstream containing lots of oxygen and system B a living cell containing little oxygen, oxygen flows from the bloodstream to the cell only if the total potential of oxygen in the blood is larger than in the cell.

The verification of assertions 1 and 2 is left to the reader.

10.4 Reservoirs with Variable Amounts of Constituents

We consider a reservoir R that has fixed values of the parameters and some of the amounts of constituents, and that has r' constituents with variable amounts. Then not only the temperature T_R is a constant (Section 9.7) but also each total potential μ_{iR} is a constant, for $i = 1, 2, \ldots, r'$, where we assume that the constituents with variable amounts are the first r'. For such a reservoir, the relation between differences in values of energy, entropy, and amounts of constituents of any two stable equilibrium states R_1 and R_2 is

$$S_2^R - S_1^R = \frac{1}{T_R} \left(E_2^R - E_1^R \right) - \sum_{i=1}^{r'} \frac{\mu_{iR}}{T_R} \left[(n_i)_2^R - (n_i)_1^R \right] \tag{10.13}$$

Problems

10.1 Consider two identical systems A and B, each consisting of one constituent, and having a fixed volume as the only parameter, and a fundamental relation $S = 100 (E n V)^{1/3}$ in J/K for E in joules, n in moles and V in cubic meters. The values of the energy, amount of constituent, and volume of A are $E^A = 240$ kJ, $n^A = 12$ mol, and $V^A = 24 \times 10^{-6}$ m³. For system B, the fixed volume $V^B = 8 \times 10^{-6}$ m³.

(a) At what values of the energy E^B and number of moles n^B of system B are systems A and B in mutual stable equilibrium?

(b) What are the values of the temperature and total potential of the stable equilibrium state of A corresponding to the given values E^A, n^A, V^A?

(c) Is the given fundamental relation concave or convex with respect to the amount of constituent?

10.2 Consider the two identical systems A and B specified in Problem 10.1. Assume that initially each system is in a stable equilibrium state corresponding to the values $E^A = 240$ kJ, $n^A = 12$ mol, $V^A = 24 \times 10^{-6}$ m³, $E^B = 80$ kJ, $n^B = 2$ mol, and $V^B = 16 \times 10^{-6}$ m³.

(a) If the two systems may interact via a fixed, rigid wall permeable to the common constituent, would there be a spontaneous flow of the constituent from one system to the other?

(b) If the flow of the constituent occurs spontaneously, find the values of the amount of the constituent and the energy in each of the systems when they reach mutual stable equilibrium.

(c) If the process in part (b) occurs spontaneously, what is the amount of entropy generated by irreversibility?

10.3 Over a limited range of values of entropy S, amount of the single constituent n, and volume V, the energy E of the stable equilibrium states of each of two identical systems A and B is given by the relation

$$E = 6.15\,n + 6150\,n \left\{ \exp\left[\frac{S - 5\,n - 8.3\,n \ln(V/5n)}{20.5\,n} \right] - 1 \right\}$$

The two systems are in temperature equality at $T^A = T^B = 300$ K.

(a) Find an expression for the difference between the total potentials μ^A and μ^B as a function of n^A, n^B, V^A, and V^B for the states for which $T^A = T^B = 300$ K.

(b) If $V^A/V^B = 5$, $n^B = 2$, and A and B can exchange their common constituent, find the values of n^A for which the flow is from A to B, and the values for which the flow is from B to A.

10.4 A pure gaseous constituent is confined in a long vertical tube and is in a stable equilibrium state. The axis of the tube is along the vertical direction in a gravitational field of constant intensity g. Each thin slice of gas perpendicular to the axis of the tube at elevation z may be regarded as a system with amount of constituent $n(z)$ in a stable equilibrium state. It has the fundamental relation

$$E(z) = \frac{A\,S^3(z)}{n(z)\,V(z)} + n(z)\,M\,gz$$

where $S(z)$ and $V(z)$ are the entropy and volume of the slice at elevation z and A and M are constants.

(a) What are the conditions that must be satisfied for two slices of the column, at elevations z_1 and z_2, respectively, to be in mutual stable equilibrium?

(b) Find the difference in values of the entropy per unit amount between any two locations z_1 and z_2 as a function of A, M, g, z_1, and z_2.

10.5 A system consists of electrons, singly charged positive ions, and neutral atoms confined between two very large area, parallel electrodes, a distance Δx apart. Each neutral atom results from the combination of an electron and a positive ion. Denoting electrons, ions, and atoms by subscripts e, i, and a, respectively, and assuming that the system is in a stable equilibrium state at temperature T, the total potentials of electrons, ions, and atoms are given by the relations

$$\mu_e = kT \ln \frac{h^3 \rho_e(x)}{g_e\,(2\pi\,m_e kT)^{3/2}} + \epsilon_{oe} - q_e \psi(x)$$

$$\mu_i = kT \ln \frac{h^3 \rho_i(x)}{g_i\,(2\pi\,m_i kT)^{3/2}} + \epsilon_{oi} + q_e \psi(x)$$

$$\mu_a = kT \ln \frac{h^3 \rho_a(x)}{g_a\,(2\pi\,m_a kT)^{3/2}} + \epsilon_{oa}$$

where k is Boltzmann's constant, h is Planck's constant, q_e is the electric charge of an electron, $\rho_e(x)$, $\rho_i(x)$, and $\rho_a(x)$ are, respectively, the density of electrons, ions, and atoms at the location x between the electrodes, $\psi(x)$ is the electrostatic potential at x, and g_e, g_i, g_a, m_e, m_i, m_a, ϵ_e, ϵ_i, and ϵ_a are constants. Like temperature, each total potential is independent of x.

(a) How does each of the densities $\rho_e(x)$, $\rho_i(x)$, and $\rho_a(x)$ vary as a function of location x?

(b) If $\mu_a = \mu_e + \mu_i$ and $\epsilon_e + \epsilon_i - \epsilon_a = V_i =$ the ionization energy of the atom, find an expression for the ratio $\rho_e(x)\,\rho_i(x)/\rho_a(x)$. This expression is known as the *Saha equation*.

(c) If V_i is much larger than kT, and the electron density is about the same as the ion density, are there many ions in the system?

10.6 This problem illustrates the fact that total potentials may assume positive, zero, or negative values. An atom of atomic number Z consists of Ze units of positive charge (Z protons), N neutrons, and negative charge of electrons, where the charge of the electron $e = 1.6 \times 10^{-19}$ C. The number of electrons varies from $n_o = 0$ for a fully ionized atom, to $0 < n_i < Z$ for a partially ionized atom, to $n_Z = Z$ for a neutral atom, to $n_{Z+1} = Z+1$ for a singly, negatively charged atom.

It can be shown that the number of electrons n of an atom in a stable equilibrium state at temperature T and total potential μ is given by the relation

$$n = \frac{\sum_{i=0}^{Z+1}\sum_{j=1}^{\infty} n_i\, g_{ij}\, \exp\left[(n_i\,\mu - \epsilon_{ij})/kT\right]}{\sum_{i=0}^{Z+1}\sum_{j=1}^{\infty} g_{ij}\, \exp\left[(n_i\,\mu - \epsilon_{ij})/kT\right]}$$

where $n_i = i = 0, 1, \ldots, Z, Z+1$, $j = 1, 2, \ldots, \infty$, $g_{i1} = 1$ for all i's, $g_{ij} \geqq 1$ for $i = 0, 1, \ldots,$ Z, and $j > 1$, $g_{Z+1,j} = 0$ for $j > 1$, and the constants ϵ_{ij} (with dimensions of energy) are such that $\epsilon_{01} > \epsilon_{11} > \epsilon_{21} > \cdots > \epsilon_{Z+1,1}$, and $\epsilon_{ij} > \epsilon_{i1}$ for all i's and j's $\neq 1$.

(a) For all (positive) values of T and $n = 0$ (fully ionized atom), show that $\mu = -\infty$.

(b) For all (positive) values of T and $n = Z + 1$ (singly charged negative ion), show that $\mu = +\infty$.

(c) For small values of T, that is, in the limit as $T \to 0$, and for $n = Z$ (neutral atom), show that

$$\mu = \frac{\epsilon_{Z+1,1} - \epsilon_{Z-1,1}}{2} = -\frac{I+A}{2} < 0$$

where $I = \epsilon_{Z-1,1} - \epsilon_{Z,1} =$ the *ionization energy of the atom*, and $A = \epsilon_{Z,1} - \epsilon_{Z+1,1} =$ the *electron affinity of the atom*.

(d) For small values of T, that is, in the limit as $T \to 0$, and for $Z < n < Z + 1$, show that μ is positive.

10.7 In the region between the electrodes of a parallel-plate capacitor with negligible edge effects, the electrostatic field is uniform. If the electrodes are in vacuum and the charges on the two plates are $+q$ and $-q$, the electric field strength $\mathcal{E}$ is given by the relation

$$\mathcal{E} = \frac{\epsilon_o\, q}{a}$$

where a is the area of each plate of the capacitor, and the permittivity of vacuum $\epsilon_o = 8.854 \times 10^{-12}$ C^2/J m. If the capacitor is completely immersed in a fluid, such as a gas or a liquid,

$$\mathcal{E} = \frac{\epsilon\, q}{a}$$

where the permittivity or dielectric constant of the fluid ϵ depends, in general, on the nature of the fluid, its temperature, its density, and possibly also on the electric field strength $\mathcal{E}$, but not on the geometry of the capacitor. The ratio $\epsilon_r = \epsilon/\epsilon_o$ is called the relative permittivity or the dielectric coefficient of the fluid. Because at field strengths of ordinary laboratory experiments the dependence of the relative permittivity on $\mathcal{E}$ is negligible, ϵ can be assumed to be a function of temperature and density only.

For a system consisting of a single constituent at sufficiently high temperatures,

$$\epsilon = \epsilon_o + \frac{\rho}{M}\left(\alpha + \frac{d^2}{3\,kT}\right)$$

where ρ is the density, M the molecular weight, k the Boltzmann constant, d the electric dipole moment of each molecule, and α the molecular polarizability. Moreover, the total potential of the constituent inside a uniform electrostatic field of strength $\mathcal{E}$ is given by the relation

$$\mu = \mu_0 + RT \ln \frac{\rho\, RT}{M p_0} - \frac{1}{2}\mathcal{E}^2\left(\alpha + \frac{d^2}{3\,k\,T}\right)$$

where μ_0, R, and p_0 are constants.

(a) For the single-constituent system just described, show that the condition of mutual stable equilibrium between the region A inside a parallel-plate capacitor (with electric field strength $\mathcal{E}_A = \mathcal{E}$) and a region B outside (with $\mathcal{E}_B = 0$) implies that

$$\frac{\rho^A}{\rho^B} = \exp\left[\frac{\mathcal{E}^2}{2\,RT}\left(\alpha + \frac{d^2}{3\,k\,T}\right)\right]$$

Because α is a positive constant, it follows that the density inside the field is greater than that outside or, equivalently, that the molecules are attracted into the electrostatic field.

(b) Using the result in part (a), and for a given value ρ^B of the density outside the capacitor, show the dependence of the dielectric constant ϵ on $\mathcal{E}$, and write a condition for this dependence to be negligible.

11   Pressure

In this chapter we discuss further conditions for mutual stable equilibrium between systems with common additive parameters that for each system may both increase and decrease, without changing the values of the energy, the amounts of constituents, and the parameters that are not additive.

11.1   Further Necessary Conditions for Mutual Stable Equilibrium

We consider again the composite C of two systems A and B. When A and B are in mutual stable equilibrium, we denote the states of A, B, and C by A_0, B_0, and C_{00}, respectively. Similarly to what we have done in Sections 9.1 and 10.1, we compare the entropy S_{00}^C of stable equilibrium state C_{00} (Figure 11.1a) with the entropy S_{11}^C of

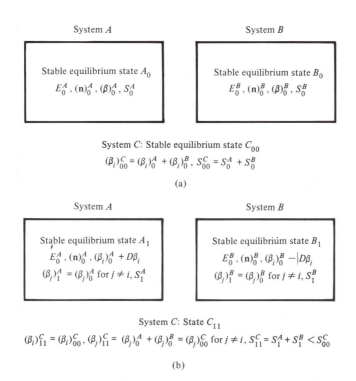

Figure 11.1   Schematics of a stable equilibrium state C_{00} and a different state C_{11} of a composite system C. Both states correspond to the same values of the energy, the amounts of constituents, and the parameters.

a different but judiciously selected state C_{11} (Figure 11.1b) for which the values of the energy E, the amounts of constituents n, and the parameters β are exactly the same as the corresponding values of state C_{00}. By the principle of highest entropy, it is necessary that the difference between the entropy S_{11}^C of C_{11} and the entropy S_{00}^C of C_{00} be negative, that is, that $S_{11}^C - S_{00}^C < 0$.

We select state C_{11} so that systems A and B (Figure 11.1b) are, respectively, in stable equilibrium states A_1 and B_1 with energies $E_1^A = E_0^A$ and $E_1^B = E_0^B$, amounts of constituents with values that are all equal to the corresponding values of states A_0 and B_0, and parameters with values that are all equal to those of states A_0 and B_0 except for the ith common additive parameter for which $(\beta_i)_1^A = (\beta_i)_0^A + D\beta_i$ and $(\beta_i)_1^B = (\beta_i)_0^B - D\beta_i$. Here we assume that parameter β_i is common to both systems, may vary over a range of values in each of the two systems, and is additive in the sense that its value $(\beta_i)^C$ for system C satisfies the relation $(\beta_i)^C = (\beta_i)^A + (\beta_i)^B$ for all states of A and B. Volume is an example of an additive parameter.

By proceeding in a manner analogous to what we have done in Section 10.1, and writing each of the differences $S_1^A - S_0^A$ and $S_1^B - S_0^B$ as a Taylor series with respect to $D\beta_i$, we conclude that the condition $S_{11}^C - S_{00}^C < 0$ implies that

$$\left[\left(\frac{\partial S}{\partial \beta_i} \right)_{E,n,\beta} \right]_0^A = \left[\left(\frac{\partial S}{\partial \beta_i} \right)_{E,n,\beta} \right]_0^B \tag{11.1}$$

where here the subscript β denotes that all parameters are kept fixed during differentiation except the one appearing in the partial derivative. In addition to temperature equality [condition (9.6) or (9.8)], equation (11.1) must necessarily hold whenever two systems A and B are in mutual stable equilibrium, provided that the ith parameter is the same for both systems, is additive, and its value for each system may both increase and decrease.

Again, because of the necessity of (11.1), the requirement that the entropy difference $S_{11}^C - S_{00}^C$ be strictly negative must be assured by restrictions on higher-order partial derivatives and, in general, we conclude that the fundamental relation must be concave with respect to each of the additive parameters β_i. A necessary condition for concavity is

$$\left(\frac{\partial^2 S}{\partial \beta_i^2} \right)_{E,n,\beta} \leq 0 \tag{11.2}$$

As we have indicated in Chapters 9 and 10, here again if any of the partial derivatives $(\partial^2 S/\partial \beta_i^2)_{E,n,\beta}$ is zero, we must consider higher-order derivatives in order to guarantee the negativity of $S_{11}^C - S_{00}^C$.

The partial derivative $(\partial S/\partial \beta_i)_{E,n,\beta}$ is related to the temperature and the partial derivative $(\partial E/\partial \beta_i)_{S,n,\beta}$. Specifically, denoting the partial derivative $(\partial E/\partial \beta_i)_{S,n,\beta}$ by the symbol f_i, we have the relation

$$f_i = \left(\frac{\partial E}{\partial \beta_i} \right)_{S,n,\beta} = -T \left(\frac{\partial S}{\partial \beta_i} \right)_{E,n,\beta} \tag{11.3}$$

where in writing the second of equations (11.3) we use (8.16). In general, property f_i is called the *generalized force conjugated to the ith parameter*. In particular, the negative of the generalized force conjugated to volume is called *pressure*. It is discussed in greater detail in Section 11.2. The dimensions and units of a generalized force depend on the conjugated parameter in question.

The value of the generalized force f_i for a stable equilibrium state A_0 of a system A is denoted by $(f_i)_0^A$. Because it is determined by a relation that holds only for stable equilibrium states, the generalized force f_i is a property defined only for such states, and only for those systems for which β_i is a parameter that may vary over a range of values.

Condition (11.1) may be expressed as an equality between generalized forces. Indeed, if each of two systems A and B has the same type of adjustable parameter β_i, if parameter β_i is additive, and if the two systems are in mutual stable equilibrium, then using (11.3), we can rewrite (11.1) in the form

$$\frac{(f_i)_0^A}{T_0^A} = \frac{(f_i)_0^B}{T_0^B} \tag{11.4}$$

and upon combining this result with the necessary condition of temperature equality, we find that

$$(f_i)_0^A = (f_i)_0^B \tag{11.5}$$

An application of the equality between generalized forces is provided by a reservoir R that has fixed values of the amounts of constituents and some of the parameters, and has s' variable parameters, each of which is additive. Then, not only the temperature T_R is a constant but also, the generalized force f_{jR} conjugated to each additive parameter is a constant, for $j = 1, 2, \ldots, s'$, where we assume that the variable parameters are the first s'. For such a reservoir, the relation between differences in values of energy, entropy, and parameters of any two stable equilibrium states R_1 and R_2 is

$$S_2^R - S_1^R = \frac{1}{T_R}(E_2^R - E_1^R) - \sum_{j=1}^{s'} \frac{f_{jR}}{T_R}\left[(\beta_j)_2^R - (\beta_j)_1^R\right] \tag{11.6}$$

If in addition to the s' additive parameters, the reservoir has r' constituents with variable amounts, we recall from Section 10.3 that each total potential μ_{iR} is a constant, for $i = 1, 2, \ldots, r'$. For such a reservoir, the relation between differences in values of energy, entropy, amounts, and parameters of any two stable equilibrium states R_1 and R_2 is

$$S_2^R - S_1^R = \frac{1}{T_R}(E_2^R - E_1^R) - \sum_{i=1}^{r'} \frac{\mu_{iR}}{T_R}\left[(n_i)_2^R - (n_i)_1^R\right] - \sum_{j=1}^{s'} \frac{f_{jR}}{T_R}\left[(\beta_j)_2^R - (\beta_j)_1^R\right] \tag{11.7}$$

11.2 Pressure

If a system has volume as a parameter with values that can be varied over a range, the volume is denoted by V instead of β, and its generalized conjugated force by $-p$ instead of f. Thus (11.3) assumes the form

$$p = -\left(\frac{\partial E}{\partial V}\right)_{S,n,\beta'} \tag{11.8}$$

Here the parameters consist of the set $\beta = \{V, \beta_2, \ldots, \beta_s\} = \{V, \beta'\}$, and p, the negative of the force conjugated to volume, is called *pressure*.

If each of two systems A and B has volume as a variable parameter (Figure 11.2), and if the two systems are in mutual stable equilibrium, (11.5) becomes a condition of pressure equality, that is,

$$p_0^A = p_0^B \tag{11.9}$$

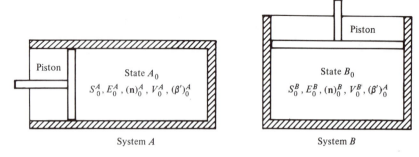

Figure 11.2   Schematics of two systems each of which has volume as a variable parameter.

Pressure, as any other generalized conjugated force, is a property of stable equilibrium states only. It has the dimensions of energy per unit volume which is equivalent to force per unit area. At a given stable equilibrium state, the value of the pressure equals the force per unit area that must be applied at every point of the surface bounding the volume of the system in order to keep the system in the specified stable equilibrium state. This force per unit area has the same value at every point on the surface.

To prove the last assertion, we consider a system A in a stable equilibrium state (Figure 11.3). At an arbitrary location on the surface of the system, there is a piston of cross-sectional area da. The piston is held in place by a small weight exerting a force $dF = f\, da$, that is, a force per unit area f times the area of the piston da. A reversible weight process for system A can be affected by displacing the piston by a small distance dx along the direction of the force, namely, by changing the volume by the amount $dV = da\, dx$. The process is selected so that each of the end states of the system is a stable equilibrium state corresponding to the same values of entropy, amounts of constituents, and parameters other than volume. Accordingly, the two end states of the process are such that $dS = 0$, $dn_i = 0$ for all i, and $d\beta_j = 0$ for $j = 2$, 3, ..., s (Figure 11.3). As a result of the energy balance for this weight process, the energy of the system changes by an amount $dE|_{S,n,\beta'}$ which is equal to the negative of the energy gain of the weight, that is, the negative of the product of the force times the displacement so that

$$dE|_{S,n,\beta'} = -dF\, dx = -f\, da\, dx = -f\, dV \qquad (11.10)$$

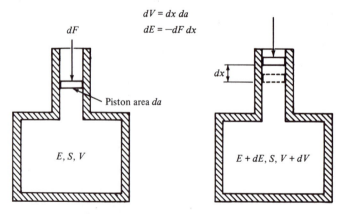

Figure 11.3   Schematics of a system with a piston of infinitesimal cross-sectional area that measures the force per unit area on the surface bounding the volume of the system.

On the other hand, because the end states of the system are stable equilibrium, the changes $dE|_{S,n,\beta'}$ and dV are related by the expression [equation (8.9)]

$$dE|_{S,n,\beta'} = \left(\frac{\partial E}{\partial V}\right)_{S,n,\beta'} dV \tag{11.11}$$

Upon comparing (11.10) and (11.11), we conclude that

$$\left(\frac{\partial E}{\partial V}\right)_{S,n,\beta'} = -f \tag{11.12}$$

and upon comparing (11.8) and (11.12), that $f = p$. Because the location of the piston was chosen arbitrarily, we further conclude that the equality $f = p$ is valid at all locations on the surface and, therefore, that the force per unit area is uniform throughout the surface.

The identification of pressure with force per unit area on the surface and the requirement of pressure equality provide a simple method for assigning values to pressure. Specifically, each stable equilibrium state of certain systems called *barometers*, *pressure gauges*, or *pressure transducers* is uniquely related to some readily observable property such as length. For example, a spring-loaded piston may be attached to the system, and the length of the spring may be used as a measure of pressure (Figure 11.4). Again, the length of a suitably calibrated vertical column of mercury (Figure 11.5) can be used for pressure measurements by placing the column in contact with a system via a movable partition.

Pressure is measured in different units, many of which and conversion factors between them are listed in Table 11.1. In SI units (the International System of units), pressure is expressed in pascal.[1] The pascal, denoted by Pa, is such that 1 Pa = 1 N/m^2.

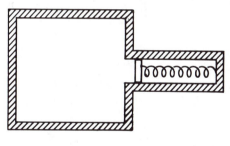

Figure 11.4 A spring-loaded piston may be attached to a system, and the length of the spring may be used as a measure of pressure.

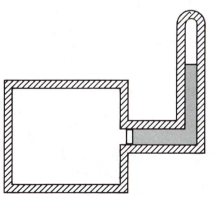

Figure 11.5 The length of a suitably calibrated column of mercury placed in contact with a system via a movable partition can be used for pressure measurements.

---

[1]This term was chosen in honor of Blaise Pascal (1623–1662), French mathematician and philosopher, whose studies contributed to the development of the concept of pressure.

A technical unit is the atmosphere, atm, which equals the pressure of atmospheric air at sea level and a temperature of 293 K.

TABLE 11.1 Units for pressure and conversion factors to the International System of units.

Unit	Symbol	Conversion to SI units
pascal	Pa	$1\ \text{Pa} = 1.00000\ \text{Pa}$
newton per square meter	N/m^2	$1\ \text{N/m}^2 = 1.00000\ \text{Pa}$
atmosphere	atm	$1\ \text{atm} = 101.325\ \text{kPa}$
bar	bar	$1\ \text{bar} = 100.000\ \text{kPa}$
torricelli	torr	$1\ \text{torr} = 133.322\ \text{Pa}$
millimeter of mercury (Hg) at $0°\text{C}$	mm Hg	$1\ \text{mm Hg} = 133.322\ \text{Pa}$
centimeter of water (H_2O) at $4°\text{C}$	cm H_2O	$1\ \text{cm}\ H_2O = 98.0665\ \text{Pa}$
pound-force per square inch	psi	$1\ \text{psi} = 6.89476\ \text{kPa}$
kilogram-force per square centimeter	kg_f/cm^2	$1\ \text{kg}_f/\text{cm}^2 = 98.0665\ \text{kPa}$
dyne per square centimeter	dyn/cm^2	$1\ \text{dyn/cm}^2 = 100.000\ \text{m Pa}$

It is noteworthy that pressure is equal to force per unit area in the sense we just discussed, but force per unit area is not pressure except for stable equilibrium states. For example, a collimated beam of particles, moving back and forth between two parallel areas of an enclosure (Figure 11.6), exerts a force on each of these two areas but nowhere else. This force is not related to a pressure because the beam is not in a stable equilibrium state. If it were, the force per unit area would have been the same at every point on the enclosure.

Collimated beam of particles
All velocities in the horizontal direction

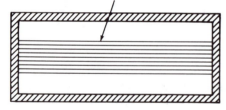

Figure 11.6 Schematic of a collimated beam of particles moving back and forth between two parallel walls.

11.3 Pressure as a Capturing Tendency for Volume

We consider two systems A and B that are initially in stable equilibrium states A_0 and B_0 with different pressures p_0^A and p_0^B, and equal temperatures T_0^A and T_0^B. We wish to study what happens if we let the two systems interact with each other in such a way that the composite of A and B remains isolated, the values of the amounts of constituents and the parameters other than volume remain fixed for both A and B, and the total volume $V^A + V^B$ remains fixed (Figure 11.7). For example, system A is pressurized air, system B is lower-pressure air, and the two are separated by a movable piston permeable to energy and entropy, but not air.

Under these conditions we show that irrespective of the types of interactions that are established between A and B, a spontaneous process for the composite of A and B results in volume being gained by the system that is initially at higher pressure. To verify this assertion, we proceed similarly to what we do in Section 9.6. We denote by C the composite of A and B, by DE^A and DE^B the changes in energy, by DS^A and DS^B the changes in entropy, and by DV^A and DV^B the changes in volume between the

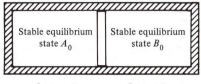

System A   System B

$T_0^A = T_0^B \qquad p_0^A \neq p_0^B$

Figure 11.7 Schematic of two system that can exchange energy, entropy and volume by means of a movable piston impermeable to matter.

initial and the final states of A and B, respectively. Because system C remains isolated and its volume remains constant, we have $DE^B = -DE^A$ and $DV^B = -DV^A$. It is noteworthy that the final state of either A, or B, or both need not be the stable equilibrium state corresponding to the new respective value of the energy. By additivity of entropy, and the principle of nondecrease of entropy, we conclude that the entropy of C cannot decrease, that is,

$$DS^C = DS^A + DS^B \geq 0 \qquad (11.13)$$

On the other hand, by the principle of highest entropy, each of the changes DS^A and DS^B is either smaller than or at most equal to the entropy change that obtains if the final state of each of the systems A and B is stable equilibrium. Therefore, DS^A and DS^B satisfy the relations [equation (8.12)]

$$DS^A \leq \frac{1}{T_0^A} DE^A + \frac{p_0^A}{T_0^A} DV^A + \text{(higher-order terms)} \qquad (11.14)$$

and

$$DS^B \leq \frac{1}{T_0^B} DE^B + \frac{p_0^B}{T_0^B} DV^B + \text{(higher-order terms)} \qquad (11.15)$$

where the higher-order terms can be shown to be negative or at most zero. The proof of this assertion is based on arguments similar to those used in Section 9.6 and on the concavity of the fundamental relation in the variables E, n and β.[2]

Upon adding (11.14) and (11.15) and neglecting the nonpositive higher-order terms, we find that

$$0 \leq DS^A + DS^B \leq \left(\frac{1}{T_0^A} - \frac{1}{T_0^B}\right) DE^A + \left(\frac{p_0^A}{T_0^A} - \frac{p_0^B}{T_0^B}\right) DV^A \qquad (11.16)$$

where in writing the right-hand side terms we use (11.13) and the relations $DE^B = -DE^A$ and $DV^B = -DV^A$. Because $T_0^A = T_0^B$, we conclude that

$$(p_0^A - p_0^B)\, DV^A \geq 0 \qquad (11.17)$$

We see that if $DV^A > 0$, that is, if system A gains volume from system B, then $p_0^A - p_0^B \geq 0$, that is, $p_0^A \geq p_0^B$. Conversely, if $DV^A < 0$, then $p_0^A \leq p_0^B$. So we interpret pressure as a capturing tendency for volume.

---

[2]By considering simultaneous changes in the values of the energies, all the constituents, and all the (additive) parameters of two systems A and B such that $dE^A = -dE^B$, $dn_i^A = -dn_i^B$ for $i = 1, 2, \ldots, r$, $d\beta_j^A = -d\beta_j^B$ for $j = 1, 2, \ldots, s$, and using the highest-entropy principle, we find that the quadratic form $\sum_{p=1}^{r+s+1} \sum_{q=1}^{r+s+1} (\partial^2 S/\partial x_p \partial x_q)_x \, dx_p \, dx_q$ must be negative semidefinite, where $S = S(x_1, x_2, \ldots, x_{r+s+1})$ is the fundamental relation for $x_1 = E$, $x_2 = n_1$, ..., $x_{r+1} = n_r$, $x_{r+2} = \beta_1$, ..., $x_{r+s+1} = \beta_s$. This requirement is satisfied if the square matrix $[(\partial^2 S/\partial x_p \partial x_q)_x]$ is negative semidefinite, i.e., if the fundamental relation is concave with respect to all the variables E, n, and β. See also footnote 1, Chapter 10.

11.4   Reservoirs with Variable Amounts of Constituents and Volume

We consider a reservoir R that has fixed values of the amounts of constituents and all the parameters except volume. Then the temperature T_R and the pressure p_R are both constants. For such a reservoir, the relation between differences in values of energy, entropy, and volume of any two stable equilibrium states R_1 and R_2 [equation (11.6)] becomes

$$S_2^R - S_1^R = \frac{1}{T_R}\left(E_2^R - E_1^R\right) + \frac{p_R}{T_R}\left(V_2^R - V_1^R\right) \tag{11.18}$$

If in addition to the volume, the reservoir has r' constituents with variable amounts, the relation between differences in values of energy, entropy, amounts of constituents, and volume of any two stable equilibrium states R_1 and R_2 [equation (11.7)] becomes

$$S_2^R - S_1^R = \frac{1}{T_R}\left(E_2^R - E_1^R\right) + \frac{p_R}{T_R}\left(V_2^R - V_1^R\right) - \sum_{i=1}^{r'}\frac{\mu_{iR}}{T_R}\left[(n_i)_2^R - (n_i)_1^R\right] \tag{11.19}$$

where μ_{iR} is the chemical potential of the ith constituent of the reservoir, for $i = 1, 2, \ldots, r'$. In addition to T_R and p_R, each of the chemical potentials μ_{iR} is a constant of the reservoir.

11.5   Partial Mutual Stable Equilibrium

Up to this point we have derived a number of necessary conditions that must be satisfied for two systems A and B to be in mutual stable equilibrium, such as temperature equality, total potential equality, and pressure equality. We have also interpreted temperature and total potential as escaping tendencies for energy and matter, respectively, and pressure as a capturing tendency for volume. If two systems are allowed to interact freely and spontaneously, no change in the state of either system occurs if the systems are initially in mutual stable equilibrium, whereas changes of the two states can occur if initially the two systems are not in mutual stable equilibrium, even if each of the systems is in a stable equilibrium state. For example, energy flows spontaneously from A to B if initially $T^A > T^B$, and this continues until temperature equality is established, provided, of course, that some interaction between A and B establishes the channel for the energy flow.

In many applications we need to consider systems composed of two or more subsystems subject to a specific set of internal constraints which prohibit certain exchanges from occurring. For example, two subsystems A and B may be isolated from one another, that is, the composite system is subject to constraints that inhibit all interactions and all flows between A and B. If A and B are not in mutual stable equilibrium but each is in a stable equilibrium state, it is clear that the state of the composite system cannot change spontaneously despite the fact that it is not a stable equilibrium state. It is also clear that the only way in which the state of the composite system can be altered without leaving net effects in its environment is by means of interactions that would violate, at least temporarily, the constraints, that is, by means of interactions that break temporarily the isolation between A and B.

We consider a system consisting of two or more subsystems subject to a specific set of constraints that prohibit certain spontaneous exchanges from occurring. We assume that the system is in an equilibrium state that can be altered without leaving net effects in the environment only by means of interactions that violate temporarily the restrictions imposed by the constraints. Under these conditions, we say that the subsystems are in *partial mutual stable equilibrium*, and that their composite system is in a *partial stable equilibrium state*.   Equivalently, when a composite system is in a partial stable equilibrium state, that is, an equilibrium state that is stable only with respect to alterations that do not violate the restrictions imposed by the internal constraints, we say that the subsystems are in partial mutual stable equilibrium

In this section we investigate the necessary conditions for partial mutual stable equilibrium between subsystems subject to constraints that prevent some flows and exchanges from occurring freely and spontaneously. We proceed in a manner similar to what we do in Sections 9.1, 10.1, and 11.1, namely, we impose the criterion that the state in which the subsystems are in partial mutual stable equilibrium has the highest entropy among all the states of the composite system that can be reached from it without violating—even temporarily—the nature of the given constraints. As a result, we conclude that this partially highest entropy state cannot be changed to any of the other states just cited without leaving net effects in the environment.

To explore the implications of the criterion, we consider two systems A and B in stable equilibrium states A_0 and B_0 with respective temperatures T_0^A and T_0^B, total potentials of the ith constituent $(\mu_i)_0^A$ and $(\mu_i)_0^B$, and pressures p_0^A and p_0^B, where the ith constituent of system A is the same as the ith constituent of system B, and the index i runs over all the constituents of each system.

It can be shown that use of the criterion just given yields the following results.

1. If they are isolated from one another (Figure 11.8), systems A and B are in partial mutual stable equilibrium even if $T_0^A \neq T_0^B$, $(\mu_i)_0^A \neq (\mu_i)_0^B$ for every i, and $p_0^A \neq p_0^B$.

2. If in contact through a fixed rigid wall permeable to energy and entropy but no constituents (Figure 11.9), systems A and B are in partial mutual stable equilibrium if $T_0^A = T_0^B$, even if $(\mu_i)_0^A \neq (\mu_i)_0^B$ for every i, and $p_0^A \neq p_0^B$.

3. If in contact through a fixed rigid wall permeable to energy, entropy, and the jth constituent, but no other types of constituents (Figure 11.10), systems A and B are in partial mutual stable equilibrium if $T_0^A = T_0^B$ and $(\mu_j)_0^A = (\mu_j)_0^B$, even if $(\mu_i)_0^A \neq (\mu_i)_0^B$ for every $i \neq j$, and $p_0^A \neq p_0^B$. This kind of rigid wall is called a *semipermeable membrane* or, *semipermeable wall*. The results can be generalized to walls semipermeable to more than one type of constituent.

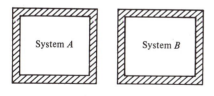

Figure 11.8   Schematics of two systems that are each in a stable equilibrium state, and that are isolated from one another. They are in partial mutual stable equilibrium even if their respective temperatures, total potentials of each constituent, and pressures are not equal.

Wall permeable to energy
and entropy but not to matter

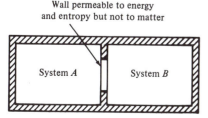

Figure 11.9   Schematics of two systems that are each in a stable equilibrium state, and that are separated from one another by a fixed rigid wall permeable to energy and entropy but no constituents. They are in partial mutual stable equilibrium if their temperatures are equal, even if their respective chemical potentials of each constituent, and pressures are not equal.

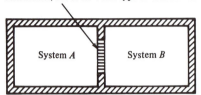

Wall permeable to energy, entropy, and the jth
constituent, but to no other type of constituent

Figure 11.10 Schematics of two systems that are
each in a stable equilibrium state, and that are separated from one another by a fixed rigid wall permeable
to energy, entropy, and one type of constituent only.
The systems are in partial mutual stable equilibrium
if their respective temperatures and chemical potentials of the exchangeable type of constituent are equal,
even if their respective chemical potentials of all the
other constituents, and their respective pressures are
not equal.

4. If in contact through a movable piston impermeable to energy, entropy, and constituents (Figure 11.11), systems A and B are in partial mutual stable equilibrium if
$p_0^A = p_0^B$, even if $T_0^A \neq T_0^B$, $(\mu_i)_0^A \neq (\mu_i)_0^B$ for every i.

5. If in contact through a movable piston permeable to energy and entropy but no
constituents (Figure 11.12), systems A and B are in partial mutual stable equilibrium
if $T_0^A = T_0^B$ and $p_0^A = p_0^B$, even if $(\mu_i)_0^A \neq (\mu_i)_0^B$ for every i.

Movable piston impermeable to
energy, entropy, and constituents

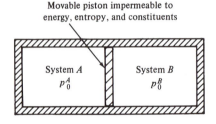

Figure 11.11 Schematics of two systems that are
each in a stable equilibrium state, and that are separated from one another by a movable piston impermeable to energy, entropy, and constituents. They are
in partial mutual stable equilibrium if their respective
pressures are equal, even if their respective temperatures, and chemical potentials of each constituent are
not equal.

Movable piston permeable to energy
and entropy but no constituents

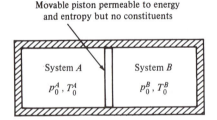

Figure 11.12 Schematics of two systems that are
each in a stable equilibrium state, are separated from
one another by a movable piston permeable to energy
and entropy but no constituents, and are in partial mutual stable equilibrium if their respective temperatures
and pressures are equal, even if their respective total
potentials of each constituent are not equal.

Problems

11.1 A particle of mass m is confined in a box of volume V. At temperature $T > 0$ but not close
to zero, the fundamental relation of the particle may be approximated by the expression

$$S = k \ln \left[\frac{V(4\pi mE/3)^{3/2}}{h^3} \right] + \frac{3}{2} k$$

where k is Boltzmann's constant, and h Planck's constant.

   (a) Find the temperature of the particle as a function of the energy.

   (b) Find the pressure of the particle as a function of temperature and volume.

11.2 The relation between entropy S and energy E for the stable equilibrium states of one electromagnetic field (a photon gas) confined in a cavity of volume V is $S = (4/3)a^{1/4}E^{3/4}V^{1/4}$, where

$a = 8\pi^5 k^4/15h^3 c^3 = 7.564 \times 10^{-16}$ J/m^3 K^4, k is Boltzmann's constant, h Planck's constant, and c the speed of light in vacuum.

(a) Prove the following relations: $S = (4/3)a\,VT^3$, $E = a\,VT^4$, and $p = (1/3)E/V$, where T is temperature, and p pressure.

(b) The pressure of the background radiation of the universe, modeled as a photon gas in a stable equilibrium state, is 13.4 fPa (f = femto, i.e., 10^{-15}). What is the temperature of the background radiation?

(c) What is the energy of the background radiation contained in a cubic light-year?

11.3 When two regions of space A and B, each filled with radiation in a stable equilibrium state but at different temperatures T_A and T_B, are placed in "contact" through an opening of cross-sectional area $\mathcal{A}$, the electromagnetic field establishes a flow of energy and entropy from the cavity at high temperature to that at low temperature at the rates

$$\dot{E}^{A \to B} = \mathcal{A}\frac{c}{4}\left[\left(\frac{E}{V}\right)_A - \left(\frac{E}{V}\right)_B\right] \qquad \text{and} \qquad \dot{S}^{A \to B} = \mathcal{A}\frac{c}{4}\left[\left(\frac{S}{V}\right)_A - \left(\frac{S}{V}\right)_B\right]$$

where E/V and S/V are the energy density and the entropy density of the radiation in each region, respectively, and c is the speed of light in vacuum.

(a) Using the results of Problem 11.2, show that the energy and entropy flow rates, in terms of the temperatures T_A and T_B of the radiation in the two regions, can be expressed as

$$\dot{E}^{E \to B} = \mathcal{A}\sigma\,(T_A^4 - T_B^4) \qquad \text{and} \qquad \dot{S}^{A \to B} = \frac{4}{3}\mathcal{A}\sigma\,(T_A^3 - T_B^3)$$

where the constant $\sigma = 5.67 \times 10^{-8}$ W/m^2 K^4.

(b) Assuming that the radiation at the surface of the sun is in contact with the interstellar space (the background radiation of the universe, assumed at a temperature 4 K) along the entire surface (the radius of the sun is $r_{sun} = 6.96 \times 10^8$ m), find the flow rates of energy and entropy emitted by the sun in order to maintain the radiation on its surface at $T_{sun} = 5760$ K.

(c) The earth (with radius $R_{earth} = 6370$ km) is placed between the sun and the interstellar space at a distance $r_{earth-sun} = 1.5 \times 10^{11}$ m from the sun. What fraction of the solar surface is exposed to that of the earth instead of the interstellar space?

(d) The radiation near the surface of the earth receives energy and entropy from the sun and emits energy and entropy to the interstellar space. Approximating the state of the radiation just outside the earth's atmosphere as a stable equilibrium state with temperature T_{earth}, what is this temperature?

(e) What are the flow rates of energy and entropy carried by the electromagnetic field from the sun to the earth? Show that the energy flux (flow rate per unit area) from the sun just outside the earth's atmosphere is about 1.35 kW/m^2. Show that the entropy flux is about 0.311 W/m^2 K.

(f) The rate at which energy comes from the hot center of the earth, 3.24×10^7 MW, is about 5000 times smaller than that coming from the sun and, thus, its contribution to determining the earth's surface temperature is negligible. What would be the earth's surface temperature without the solar contribution?

11.4 Each of two identical systems A and B consists of a fixed amount of one constituent and has volume as the only parameter. The fundamental relation for each system is $S = 100\,(EnV)^{1/3}$ in J/K for E in joules, n in moles, and V in cubic meters. The values of the energy, amount of constituent, and volume of A are $E^A = 240$ kJ, $n^A = 12$ mol, and $V^A = 24 \times 10^{-6}$ m^3. For system B, the fixed amount of the constituent is $n^B = 8$ mol.

(a) At what values of the energy E^B and volume V^B of system B are systems A and B in mutual stable equilibrium?

(b) What are the values of the temperature and pressure of A corresponding to the given values E^A, n^A, V^A?

11.5 Consider the two identical systems A and B specified in Problem 11.4. Assume that initially each system is in a stable equilibrium state corresponding to the values $E^A = 240$ kJ, $n^A = 12$ mol, $V^A = 24 \times 10^{-6}$ m^3, $E^B = 80$ kJ, $n^B = 2$ mol, and $V^B = 16 \times 10^{-6}$ m^3.

(a) If the two systems may interact via a movable piston that affects the values of the volumes of both A and B but not the values of the amounts of their contituents, is there a tendency for the volumes V^A and V^B to change?

(b) If the exchange in volume occurs spontaneously, find the final values of the volume and the entropy in each system.

12   Work and Heat

We know from experience that systems can interact with each other and, under appropriate circumstances, affect their respective states. Interactions can have a great variety of effects. Some result in exchange of energy between the interacting systems while the values of their parameters remain unchanged and neither entropy nor constituents are exchanged. Other interactions result in exchanges of energy and entropy, but neither constituents are exchanged nor the values of the parameters are affected. Still others may result in exchanges of energy, entropy, and constituents as well as in changes in values of the parameters. Such exchanges of energy, entropy, and constituents, and such changes in values of the parameters result in changing the states of the interacting systems.

In addition, if as a result of interactions a system is brought to a state that is not stable equilibrium, this state may evolve spontaneously toward a stable equilibrium state, thus causing further changes in the values of the properties of the system. For example, the spontaneous evolution of a nonequilibrium state toward a stable equilibrium state causes a spontaneous creation of entropy within the system, namely, such evolution is irreversible. Thus interactions may cause a change in entropy of a system either directly by inducing an exchange of entropy through the boundary of the system, or indirectly by inducing an irreversible spontaneous change of state, or both. Again, interactions may cause a change in the value of the amount of a constituent either directly by inducing a flow of the constituent, or indirectly by activating internal reaction mechanisms that produce the spontaneous formation or destruction of the constituent within the system, or both.

Knowing how much of the change of a property is due to exchanges with other systems and how much to spontaneous creation or destruction within the system is very important to an understanding of the performance of the system. For example, if the entropy of a system A increases only because of entropy transfers from other systems, such an increase implies no imperfections within A and no opportunities for improvements by modifications of A. On the other hand, if the same entropy increase is due solely to irreversibility within A, this increase implies imperfections within A and is subject to reduction by proper redesign of A.

We will see that a practical method for identifying the system in which entropy is generated in the course of interactions is by specifying each interaction in terms of the net exchanges that it causes at the boundary between the interacting systems. For this reason we classify interactions into different categories, depending on whether the interacting systems exchange, for example, energy and constituents but no entropy, or energy and entropy but no constituents, or energy, entropy, and constituents. We will see that when no entropy is exchanged, the interaction is called work. Otherwise, the interaction will be called nonwork, and we will see that two special types of nonwork are heat and bulk flow. Other types of interaction, such as diffusion, are beyond the scope of the book.

In this chapter we discuss work interactions and heat interactions. Bulk flow interactions are introduced later.

12.1   Work

In this section we discuss an interaction that involves a net transfer of energy with no net transfer of entropy. We consider a special process of a composite system consisting of two systems A and B (Figure 12.1). Initially, A and B are in states A_1 and B_1, with energies E_1^A and E_1^B and entropies S_1^A and S_1^B, respectively. As a result of interactions that involve no net effects on any other system in the environment, the states become A_2 and B_2, the energies $E_2^A = E_1^A + DE^A$ and $E_2^B = E_1^B + DE^B$, and the entropies remain unaffected, that is, $S_2^A = S_1^A$ and $S_2^B = S_1^B$.

For this process, the energy balances for the composite system and systems A and B yield

$$E_2^A - E_1^A + E_2^B - E_1^B = DE^A + DE^B = 0 \tag{12.1}$$

$$E_2^A - E_1^A = DE^A = E^{A\leftarrow} \tag{12.2}$$

$$E_2^B - E_1^B = DE^B = E^{B\leftarrow} \tag{12.3}$$

where $E^{A\leftarrow}$ denotes the energy exchange of A, and $E^{B\leftarrow}$ the energy exchange of B. When $E^{X\leftarrow}$ is positive, we recall that energy flows into X, and when $E^{X\leftarrow}$ is negative, that energy flows out of X. Equations (12.1) to (12.3) imply that $E^{A\leftarrow} = DE^A = -DE^B = -E^{B\leftarrow}$, namely, that an amount of energy equal to DE^A is exchanged between A and B. Specifically, if DE^A is positive, $E^{A\leftarrow}$ is positive and the energy is transferred from B to A. Conversely, if DE^A is negative, energy is transferred from A to B.

Moreover, the entropy balances for the composite system and systems A and B yield

$$S_2^A - S_1^A + S_2^B - S_1^B = (S_{irr})^A + (S_{irr})^B = 0 \tag{12.4}$$

$$S_2^A - S_1^A = S^{A\leftarrow} + (S_{irr})^A = 0 \tag{12.5}$$

$$S_2^B - S_1^B = S^{B\leftarrow} + (S_{irr})^B = 0 \tag{12.6}$$

where $(S_{irr})^A$ and $(S_{irr})^B$ denote the nonnegative amounts of entropy generated by irreversibility inside systems A and B, respectively, and $S^{X\leftarrow}$ the entropy exchange of

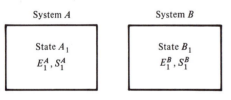

System A System B

State A_1 State B_1
E_1^A, S_1^A E_1^B, S_1^B

Initial states

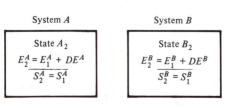

System A System B

State A_2 State B_2
$E_2^A = E_1^A + DE^A$ $E_2^B = E_1^B + DE^B$
$S_2^A = S_1^A$ $S_2^B = S_1^B$

Final states

Figure 12.1   Schematics of the end states of systems A and B before and after an interaction by which they exchange energy but no entropy.

X such that when $S^{X \leftarrow}$ is positive entropy flows into X, and when $S^{X \leftarrow}$ is negative entropy flows out of X. The relations $(S_{irr})^A \geq 0$ and $(S_{irr})^B \geq 0$ (see Section 7.3) and equations (12.4) to (12.6) imply that $(S_{irr})^A = (S_{irr})^B = 0$ and

$$S^{A \leftarrow} = S^{B \leftarrow} = 0 \qquad (12.7)$$

So the process just specified for the composite system is reversible and involves an interaction between systems A and B that results in a transfer of an amount of energy equal to DE^A but no transfer of entropy. Any interaction that results in a net exchange of energy between two systems but no exchange of entropy is called a *work interaction*, and the amount of energy thus transferred is called *work*.[1]

We denote the amount of energy transferred to or from a system A as a result of a work interaction by the symbol $W^{A \leftarrow}$ if the amount is finite, or by the symbol $\delta W^{A \leftarrow}$ if the amount is infinitesimal. We use the symbol $W^{A \leftarrow}$ instead of $W_{12}^{A \leftarrow}$ that we discussed in Section 3.9, that is, we drop the subscript 12, whenever the initial and final states of the system can be inferred without ambiguity from the context.

It is noteworthy that we use the prefix δ rather than d to denote an infinitesimal amount of a quantity that is not a property. Indeed, the amount of a property that is transferred across the boundary of a system as a result of an interaction is not a property because it depends on more than just the state of the system at one instant of time. For example, the amount of energy transferred may depend on the time history of the system, and on the states and history of the other interacting systems.

Consistently with a long-standing tradition, when the work $W^{A \leftarrow}$ or $\delta W^{A \leftarrow}$ is positive, that is, when an amount of energy is transferred into system A by means of a work interaction, we say that *work is done on the system*. Conversely, when the work $W^{A \leftarrow}$ or $\delta W^{A \leftarrow}$ is negative, that is, when an amount of energy is transferred out of system A by means of a work interaction, we say that *work is done by the system*.

Alternatively, the amount of energy transferred to or from a system A by means of a work interaction may be denoted by the symbol $W^{A \rightarrow}$ (or $\delta W^{A \rightarrow}$), and then when $W^{A \rightarrow}$ (or $\delta W^{A \rightarrow}$) is positive energy is transferred out of the system, whereas when $W^{A \rightarrow}$ (or $\delta W^{A \rightarrow}$) is negative energy is transferred into the system. Of course, $W^{A \rightarrow} = -W^{A \leftarrow}$ and $\delta W^{A \rightarrow} = -\delta W^{A \leftarrow}$.

An elementary example of a work interaction is the head-on elastic collision between two identical, rigid, and polished steel balls on a frictionless[2] horizontal plane (Figure 12.2). Initially, ball A moves with a velocity $\mathbf{v}^A$ and ball B is stationary. Upon colliding, ball A becomes stationary and ball B acquires velocity $\mathbf{v}^B = \mathbf{v}^A$. This elastic transfer of all the (kinetic) energy from A to B is accompanied by no exchange of entropy between A and B, no increase in the entropy of either ball A or ball B, and no exchange of constituents.

Work is a mode of interaction between two systems. The term work is used to denote the net amount of energy exchanged by the system as a result of such an interaction. Thus work is not a property of either of the interacting systems. In view of this remark, some expressions used in this field must be interpreted *cum grano salis*. For example, the expression "work transfer," if taken literally, is misleading and incorrect because it implies that work is something that can be transferred from one system to another, whereas work is not "contained" in any system and hence cannot be transferred between

---

[1] The exclusion of any exchange of entropy as a result of a work interaction also rules out any exchange of constituents, except for exchanges of constituents each of which is in a zero-entropy state and therefore carries no entropy.

[2] The term *frictionless plane* means that no force along the surface of the plane exists between each ball and the plane.

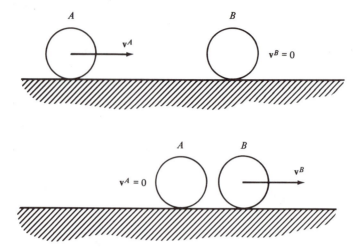

Figure 12.2   Schematic of a head-on elastic collision between two identical, rigid, polished steel balls on a frictionless horizontal plane. Such a collision is an elementary example of a work interaction.

systems. Energy is "contained" in systems and hence can be transferred. Thus we must interpret the expression "work transfer" as a shorthand statement for the concept of "energy transfer by means of a work interaction."

A process of a system that involves only work interactions with systems in its surroundings is called *adiabatic*.

The processes of the overall system and of each of the systems A and B that we consider at the beginning of this section (Figure 12.1) are examples of adiabatic processes that are also reversible. In general, however, adiabatic processes are not necessarily reversible. An example of an irreversible adiabatic process is the one discussed in Section 7.2.4. Indeed, in that process, states A_1 and A_2 differ in both energy and entropy, but the interaction between A and the weight involves no entropy exchange, and therefore the interaction is work. Nevertheless, an increase in entropy occurs spontaneously and internally within A, and therefore the process for A is adiabatic and irreversible.

Example 12.1.   Show that every weight process is adiabatic.

Solution.   We recall (Section 5.7) that a weight cannot experience any entropy change, that is, its entropy is constant, and can experience only reversible processes, that is, no entropy may be generated within a weight. Thus the entropy balance for the weight is $S_2^W - S_1^W = S^{W \leftarrow} = 0$, that is, a weight can exchange no entropy and hence can experience only work interactions. Because in any weight process a system interacts only with a weight, we conclude that the system experiences work interactions only, and therefore the process is adiabatic.

An example of an adiabatic process of interest in many applications involving rotating machinery is that of a system which interacts with its surroundings by means of a rotating shaft. The interaction through the shaft is work and equivalent to that with a weight which can be suspended on the shaft via a rope as sketched in Figure 12.3. The energy transferred by means of such an interaction is often called *shaft work*.

Another example of an adiabatic process is the displacement of a point on the boundary of a system A at which point a force is exerted in the direction of the displacement by another system (Figure 12.4). This is a work interaction because the force is entirely equivalent to and can always be provided by a suitable weight, that is, the process for A can always be simulated by a weight process. For example, the compression of a gas in

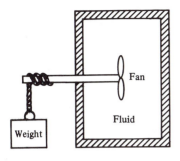

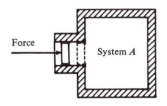

Figure 12.3   Illustration of a work interaction via a rotating shaft. The energy transferred in this way is often called shaft work.

Figure 12.4   Illustration of the work interaction involved in the displacement of a point on the boundary of a system A at which point a force is exerted in the direction of the displacement by another system.

a cylinder with a piston by a loaded spring via a lever, shown schematically in Figure 12.5a, is entirely equivalent to the force–displacement arrangements in Figure 12.5b. The latter represent work because they are equivalent to the weight process shown in Figure 12.5c.

Whereas all weight processes are adiabatic (Example 12.1), not all adiabatic processes are necessarily weight processes. For example, we can consider an isolated composite of two systems A and B such that the interaction between A and B is a work interaction only, but as a result of the interaction there is a net amount of entropy generated by irreversibility within system B. Then the process for system A is not a weight process because its effects on systems external to A are not only mechanical. These effects involve not only an energy exchange but also an entropy generation by irreversibility.

Even if an adiabatic process is not a weight process, there exists always a weight process with the same initial and final states. More generally, if two systems A and B experience a work interaction and change their respective states from A_1 and B_1 to A_2 and B_2, there must be two weight processes, one for system A and the other for system B, which can bring about the same changes of states, namely, there is a weight process that takes A from A_1 to A_2 and a weight process that takes B from B_1 to B_2.

Example 12.2.   Prove the last assertion.

Solution.   A work interaction transfers no entropy. Hence $S^{A\leftarrow} = S^{B\leftarrow} = 0$ and the entropy balances for systems A and B yield $S_2^A - S_1^A = (S_{\text{irr}})^A \geqq 0$ and $S_2^B - S_1^B = (S_{\text{irr}})^B \geqq 0$. The condition $S_2^A - S_1^A \geqq 0$ implies that a weight process for A that changes the state from A_1 to A_2 must exist [relations (7.10) and (7.13)]. Similarly, the condition $S_2^B - S_1^B \geqq 0$ implies that a weight process for B that changes the state from B_1 to B_2 must exist.

We see a posteriori that in our development of the foundations in Chapters 3 to 7 we cannot introduce the concept of adiabatic process from the outset because the definition of a work interaction requires that of entropy as well as the notion of entropy exchange. The definition of a weight process is more restrictive than that of an adiabatic process.

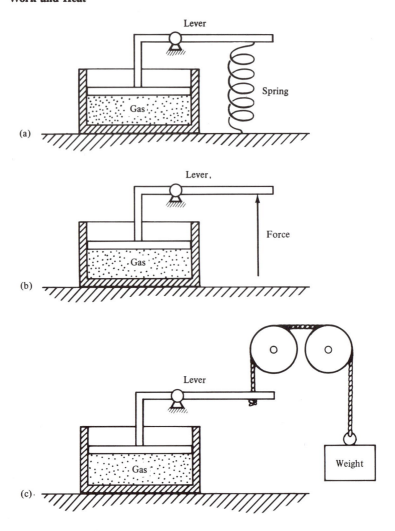

Figure 12.5   Examples of work interactions.

It is also more suitable for developing the foundations because it is expressed in terms of concepts that are well defined and exclusively mechanical.

Example 12.3.   Verify that even if the change of state of each of two interacting systems A and B can be brought about by means of a weight process, the interaction between A and B is not necessarily work.

Solution.   For example, we may consider an interaction between A and B such that the energy and entropy transfers are $E^{A\leftarrow} = -E^{B\leftarrow} = 500$ J and $S^{A\leftarrow} = -S^{B\leftarrow} = 2$ J/K, and the entropy generations due to irreversibility within A and B are $(S_{irr})^A = 1$ J/K and $(S_{irr})^B = 3$ J/K. From the entropy balances it follows that $S_2^A - S_1^A = 2 + 1 = 3$ J/K and $S_2^B - S_1^B = -2 + 3 = 1$ J/K, that is, both $S_2^A - S_1^A$ and $S_2^B - S_1^B$ are positive. Because it is positive, each entropy change satisfies (7.13) and, therefore, implies that a weight process must exist that can bring about the same change of state as that induced by the interaction. Yet the interaction between these two systems is not work because it involves both energy and entropy transfers.

Internal forces of any type do not result in any work even though they may affect the state of the system, such as setting in motion the constituents of the system. For example, for a free-falling body the gravitational force is internal to the body–earth system. This force does not result in any work because no energy is transferred into or out of the body–earth system as the body falls freely.

If a process from state A_1 to state A_2 involves work interactions only, that is, is adiabatic, the net amount of energy $E^{\leftarrow}$ exchanged by system A is denoted by $W^{\leftarrow}$ or $-W^{\rightarrow}$, the net amount of entropy exchanged by $S^{\leftarrow} = 0$, and the energy and entropy balances become

$$E_2 - E_1 = W^{\leftarrow} = -W^{\rightarrow} \tag{12.8}$$

$$S_2 - S_1 = S_{\text{irr}} \tag{12.9}$$

where we omit for simplicity the superscript A that refers to system A, and the subscript 12 that refers to states A_1 and A_2. For an adiabatic process that involves an infinitesimal transfer of energy, the net amount of energy $\delta E^{\leftarrow}$ exchanged by system A is denoted by $\delta W^{\leftarrow}$ or $-\delta W^{\rightarrow}$, and the energy and entropy balances become

$$dE = \delta W^{\leftarrow} = -\delta W^{\rightarrow} \tag{12.10}$$

$$dS = \delta S_{\text{irr}} \tag{12.11}$$

where we recall that the prefix δ denotes an infinitesimal amount of a quantity that is not a property, such as the amount of a quantity that is transferred across the boundary of a system as a result of an interaction, and the amount of a quantity that is spontaneously created within the system. Said differently, we use d to signify an infinitesimal change in a property between two neighboring states, and δ to signify an infinitesimal flow of a property across the boundary of a system as well as an infinitesimal amount of a property created spontaneously within the system. It is important to remember that equations (12.8) to (12.11) are valid for work interactions only.

In an adiabatic process, if both end states are stable equilibrium, the energy and entropy changes may be expressed in terms of changes of other properties and variables, so the work may be related to the latter changes.

Example 12.4.   Two identical blocks of metal are initially at temperatures $T_1^A = 900$ K and $T_1^B = 400$ K, respectively. For each of the two blocks energy E, entropy S, and temperature T are related so that $E_2 - E_1 = C\,(T_2 - T_1)$ and $S_2 - S_1 = C\ln(T_2/T_1)$, where the constant $C = 800$ J/K. Find the value of the adiabatic availability Ψ_1 of the composite of the two blocks in the initial state, that is, the largest amount of work that can be done using the two blocks without changes in their sizes and amounts of constituents.

Solution.   The adiabatic availability of a system in a given state is the energy that can be transferred to a weight in a reversible weight process in which the system begins from the given state and ends in a stable equilibrium state. It is work. Accordingly, if the adiabatic availability is extracted, the composite of the two blocks ends in a stable equilibrium state, that is, the two blocks end in mutual stable equilibrium and have a common temperature $T_2^A = T_2^B = T_2$. For such a reversible process of the composite of the two blocks, the energy and entropy balances yield

$$C\,(T_2 - T_1^A) + C\,(T_2 - T_1^B) = -W^{\rightarrow} \tag{12.12}$$

$$C\ln\left(\frac{T_2}{T_1^A}\right) + C\ln\left(\frac{T_2}{T_1^B}\right) = 0 \tag{12.13}$$

Equation (12.13) implies that $T_2 = (T_1^A T_1^B)^{1/2} = (900 \times 400)^{1/2} = 600$ K, and equation (12.12) yields $W^{\rightarrow} = C\,[T_1^A + T_1^B - 2(T_1^A T_1^B)^{1/2}] = 800\,[900 + 400 - 1200] = 8 \times 10^4$ J. Hence the adiabatic availability of the two blocks in the given initial state is $\Psi_1 = W^{\rightarrow} = 80$ kJ.

12.2 A Nonwork Interaction

An interaction that cannot be classified as work is called *nonwork*, and the corresponding process *nonadiabatic*. An example of a nonwork interaction is the interaction between two systems A and B that experience a process under the following specifications (Figure 12.6). Initially, each system is in a stable equilibrium state but not in mutual stable equilibrium with the other—the initial temperature T_1^A differs from the initial temperature T_1^B. At the end of the process, the energy of system B is increased, the energy of system A is decreased by an equal amount, each system experiences no net changes in values of the amounts of constituents and the parameters, and no net effects are left on any other system in the environment of the composite of systems A and B. Under these specifications, the interaction between A and B cannot be work. If it were, the change of state of each of the two interacting systems could be brought about by means of a weight process (Example 12.2), and therefore system A would decrease its energy by means of a weight process that starts from a stable equilibrium state. Such a decrease would be a violation of the impossibility of a perpetual-motion machine of the second kind (PMM2). So we conclude that under the specified conditions the interaction between the two systems is an example of a nonwork interaction. As such, it must involve an exchange of entropy in addition to the exchange of energy.

In general, the preceding interaction between the two systems A and B can be regarded as partly nonwork and partly work. To see this clearly, we reproduce the process just cited in two steps. First, we interpose between A and B a cyclic engine that produces shaft work without any entropy generated by irreversibility (Figure 12.7a). The cyclic engine X_1 can do work because the composite of systems A and B is not in a stable equilibrium state ($T_1^A \neq T_1^B$), and therefore its adiabatic availability is different from zero. The changes involved in the first step are as follows.

1. Transfers of amounts of energy and entropy out of system A and into the engine equal to the corresponding amounts of the original nonwork interaction. As a result of these transfers and perhaps internal effects, system A reaches the same final state as that reached in the original nonwork interaction.

2. Transfer of some of the extracted energy from the engine to a weight—shaft work done to lift the weight.

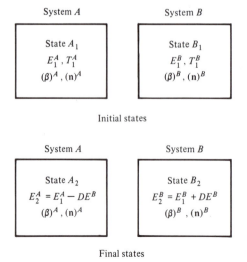

System A

State A_1
E_1^A, T_1^A
$(\beta)^A, (n)^A$

System B

State B_1
E_1^B, T_1^B
$(\beta)^B, (n)^B$

Initial states

System A

State A_2
$E_2^A = E_1^A - DE^B$
$(\beta)^A, (n)^A$

System B

State B_2
$E_2^B = E_1^B + DE^B$
$(\beta)^B, (n)^B$

Final states

Figure 12.6 Schematics of end states that correspond to a nonwork interaction.

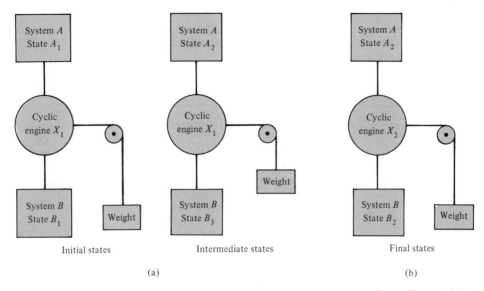

Initial states   Intermediate states   Final states

(a)   (b)

Figure 12.7 Schematics of cyclic engines that interact with two systems A and B that initially are not in mutual stable equilibrium: (a) of the energy received from A, engine X_1 transfers a fraction to the weight, and the remainder together with the entropy received from A to system B; (b) by means of engine X_2, the energy received by the weight is transferred to system B.

3. Transfer of the remaining energy extracted from A, and transfer of all the entropy exchanged between A and X_1 to system B. As a result of these transfers only, system B reaches an intermediate state B_3 different from the final state B_2 reached in the course of the original nonwork interaction because less energy is exchanged between A and B. Moreover, here the interaction for B must be viewed as nonwork because it involves both energy and entropy.

In the second step, we connect the raised weight to another cyclic engine X_2 that can do shaft work on system B without generating any entropy and while the weight is lowered to its initial elevation (Figure 12.7b). Here the weight does shaft work on B. As a result of this work and perhaps internal effects, system B is carried from state B_3 to a final state identical to the state B_2 reached in the course of the original nonwork interaction.

At the completion of the two steps, both cyclic engines X_1 and X_2, and the weight are restored to their initial states, systems A and B experience the same changes of states and exchange the same amounts of energy and entropy as in the original nonwork interaction, and we conclude that the interaction between two systems at different temperatures can be regarded as partly nonwork (step 1) and partly work (step 2).

12.3 Heat

We now discuss a special nonwork interaction that is entirely distinguishable from work in that no fraction of any such interaction can be regarded as work. To proceed, we first evaluate the work done by the cyclic engine X_1 and the systems in Figure 12.7a in a reversible weight process for the composite of systems A, B and cyclic engine X_1, in which the initial states of A and B are stable equilibrium, and in which there are no net changes in the values of the amounts of constituents and the parameters of both A and B. Under these conditions and for a given change DE^A in the energy of system A, it can be shown that the work done by cyclic engine X_1 is the largest possible when the final states of the two systems A and B are also stable equilibrium.

Example 12.5. Prove the last assertion.

Solution. We assume that the final states A_2 and B_3 are not necessarily stable equilibrium. Their energies and entropies satisfy the relations

$$E_2^A = E_1^A + DE^A \qquad \text{and} \qquad E_3^B = E_1^B + DE^B \tag{12.14}$$

and

$$S_2^A = S_1^A + DS^A \qquad \text{and} \qquad S_3^B = S_1^B + DS^B \tag{12.15}$$

By virtue of the highest entropy principle, each of the final entropies cannot be larger than the entropy of the stable equilibrium state corresponding to the same final energy. Moreover, by virtue of the state principle, the highest entropies are functions of the form $S^A(E^A)$ and $S^B(E^B)$, where we omit the dependences on the values of the amounts of constituents and the parameters because here these values experience no net changes. Thus we have

$$S_2^A \leqq S^A(E_2^A) \qquad \text{and} \qquad S_3^B \leqq S^B(E_3^B) \tag{12.16}$$

where the equal sign holds only if the final state is stable equilibrium. In addition, we have the energy and entropy balances for the reversible process of the composite of A, B, and X_1, that is,

$$DE^A + DE^B = E_2^A - E_1^A + E_3^B - E_1^B = -W^{\rightarrow} \tag{12.17}$$

and

$$DS^A + DS^B = S_2^A - S_1^A + S_3^B - S_1^B = 0 \tag{12.18}$$

Upon combining relations (12.14) to (12.18), we find that

$$S_1^A + S_1^B = S_2^A + S_3^B \leqq S_2^A + S^B(E_3^B) \leqq S^A(E_2^A) + S^B(E_3^B) \tag{12.19}$$

where the first inequality becomes an equality when state B_3 is stable equilibrium, and the second when state A_2 is stable equilibrium. From relations (12.19), we conclude further that

$$S_1^A + S_1^B - S^A(E_1^A + DE^A) \leqq S_1^A + S_1^B - S_2^A \leqq S^B(E_1^B - DE^A - W^{\rightarrow}) \tag{12.20}$$

where the first inequality becomes an equality when state A_2 is stable equilibrium, the second when state B_3 is stable equilibrium, and in writing the arguments of S^A and S^B we use relations (12.14) and (12.17). For given initial states A_1 and B_1, and a given change DE^A in the energy of A, the left-hand side of relations (12.20) has a fixed value which is equal to the lowest value that can be achieved by the difference $S_1^A + S_1^B - S_2^A$. Moreover, this value can be assumed by $S^B(E_1^B - DE^A - W^{\rightarrow})$ only if both states A_2 and B_3 are stable equilibrium. Next, we observe that the fundamental relation $S^B(E^B)$ is monotonically increasing with E^B and therefore that $S^B(E_1^B - DE^A - W^{\rightarrow})$ is monotonically decreasing with $W^{\rightarrow}$. Accordingly, the largest $W^{\rightarrow}$ is achieved when $S^B(E_1^B - DE^A - W^{\rightarrow})$ is the smallest, namely, when both states A_2 and B_3 are stable equilibrium.

For an infinitesimal change dE^A in the energy of system A, we can find an explicit expression for the largest work $(\delta W^{\rightarrow})_{\text{largest}}$. Indeed, upon using equation (8.8) with $dn_i = 0$ for $i = 1, 2, \ldots, r$ and $d\beta_j = 0$ for $j = 1, 2, \ldots, s$, and the definition of temperature, we find that

$$S^A(E_1^A + dE^A) - S_1^A = \frac{1}{T_1^A} dE^A \tag{12.21}$$

and

$$S^B(E_1^B - dE^A - \delta W^{\rightarrow}) - S_1^B = \frac{1}{T_1^B}(-dE^A - \delta W^{\rightarrow}) \tag{12.22}$$

Thus, relation (12.20) may be written in the form

$$S_1^A + S_1^B - S_1^A - \frac{1}{T_1^A} dE^A \leqq S_1^B + \frac{1}{T_1^B}(-dE^A - \delta W^{\rightarrow}) \qquad (12.23a)$$

or, equivalently,

$$\delta W^{\rightarrow} \leqq \frac{T_1^A - T_1^B}{T_1^A}(-dE^A) \qquad (12.23b)$$

Clearly, the largest work is obtained when the equal sign applies, that is, when the final states of both A and B are stable equilibrium (Example 12.5), so that

$$(\delta W^{\rightarrow})_{\text{largest}} = \frac{T_1^A - T_1^B}{T_1^A}(-dE^A) = \frac{T_1^A - T_1^B}{T_1^A} \delta E^{A\rightarrow} \qquad (12.24)$$

where in writing equations (12.24) we use the energy balance for system A, that is, $dE^A = -\delta E^{A\rightarrow}$.

Equations (12.24) imply that as long as the initial temperature T_1^A of system A differs from the initial temperature T_1^B of system B, a fraction $[(T_1^A - T_1^B)/T_1^A] \delta E^{A\rightarrow}$ of the energy $\delta E^{A\rightarrow}$ transferred out of system A can always be transferred to a weight, while the remaining fraction $(T_1^B/T_1^A) \delta E^{A\rightarrow}$ is transferred to system B. The fraction stored in the weight can always be returned to B by means of a weight process. Thus system B appears to have experienced an interaction which is partly nonwork and partly work.

However, when the two initial temperatures become almost equal, the fraction $(T_1^A - T_1^B)/T_1^A$ and therefore the ratio $(\delta W^{\rightarrow})_{\text{largest}}/\delta E^{A\rightarrow}$ approach zero, and $\delta E^{A\rightarrow} + \delta E^{B\rightarrow} = 0$. In this limit the nonwork interaction between systems A and B is entirely distinguishable from work in that no fraction of such interaction can be regarded as work, even if we interpose a cyclic engine between the two interacting systems.

Another important feature of the interaction between A and B in the limit as $T_1^A - T_1^B$ approaches zero is that it involves not only an energy transfer $\delta E^{A\rightarrow} = -\delta E^{B\rightarrow} = \delta E^{B\leftarrow}$ but also an entropy transfer which is proportional to the energy transfer, that is,

$$\delta S^{A\rightarrow} = \frac{\delta E^{A\rightarrow}}{T_Q} = \frac{\delta E^{B\leftarrow}}{T_Q} = \delta S^{B\leftarrow} \qquad (12.25)$$

where T_Q is the almost common value of the initial temperatures T_1^A and T_1^B of the interacting systems A and B. Indeed, for the process under consideration, the energy and entropy balances for systems A and B are $dE^A = -\delta E^{A\rightarrow}$, $dE^B = \delta E^{B\leftarrow}$, $dS^A = -\delta S^{A\rightarrow}$ and $dS^B = \delta S^{B\leftarrow}$, and equations (12.21) and (12.22) can be rewritten as $dS^A = dE^A/T_1^A = dE^A/T_Q$ and $dS^B = dE^B/T_1^B = dE^B/T_Q$. Combining these relations, we obtain equations (12.25).

An interaction resulting in a net exchange of energy and entropy but no constituents between two systems that are in stable equilibrium states and almost in mutual stable equilibrium, and such that the ratio of the amount of energy transferred and the amount of entropy transferred is equal to the almost common temperature of the two interacting systems is called a *heat interaction*. The amount of energy transferred to or from a system A by means of such an interaction is called *heat* and denoted by the symbol $Q^{A\leftarrow}$ or, alternatively, $Q^{A\rightarrow}$ (with $Q^{A\rightarrow} = -Q^{A\leftarrow}$) if the amount is finite, or by the symbol $\delta Q^{A\leftarrow}$ or, alternatively, $\delta Q^{A\rightarrow}$ (with $\delta Q^{A\rightarrow} = -\delta Q^{A\leftarrow}$) if the amount is infinitesimal, where the prefix δ is used for the reasons given in Section 12.1.

Heat is not a property of a system, nor is it contained in a system. It is a mode of energy transfer in the same sense that work is a mode of energy transfer. The two modes, however, are totally distinguishable from each other in that no part of heat can be confused as work, and in that heat is accompanied by a net amount of entropy transfer uniquely specified by the energy transfer and the temperature at which it occurs, whereas work is accompanied by no entropy transfer.

The term heat is used here because of the historical association of the word with thermodynamics. However, it is used in a special sense that is quite foreign to the popular concept of its meaning. The term heat is reserved here for a very special kind of interaction between systems, which only by virtue of the laws of thermodynamics is entirely distinguishable from all other kinds of interactions, including the special kind called work. If a less restrictive definition were used, the relations between heat and entropy given by equations (12.25) would not always hold.

In view of these remarks, the expression "heat transfer," which is often used in applications, should not be interpreted literally. Heat is not something that can be transferred from one system to another because heat is not "contained" in any system. Energy and entropy are "contained" in systems and hence can be transferred.

Whereas a work interaction is completely characterized by specifying the value of the work $W^{\leftarrow}$ or $\delta W^{\leftarrow}$, because the entropy transfer is null, a heat interaction is completely characterized by specifying not only the value of the heat $Q^{\leftarrow}$ or $\delta Q^{\leftarrow}$ but also the temperature T_Q at which the interaction occurs, so that the entropy transfer is unambiguously determined by $S^{\leftarrow} = Q^{\leftarrow}/T_Q$ or $\delta S^{\leftarrow} = \delta Q^{\leftarrow}/T_Q$.

It is noteworthy that the concept of a heat interaction is not a primitive one because its definition requires the concepts of energy, stable equilibrium, entropy, and temperature. Because these concepts involve both the first and second laws, it follows that the existence of a heat interaction is also a consequence of both laws.

The processes of systems A and B that we use at the beginning of this section to define heat interactions are reversible. In general, however, a process involving only a heat interaction is not necessarily reversible. For example, we may consider systems A and B that contain subsystems A' and B', respectively, as sketched in Figure 12.8. Even though systems A and B may not be in stable equilibrium states, the two subsystems A' and B' could be in stable equilibrium states, and almost in mutual stable equilibrium at an almost common temperature T_Q. If the interaction between A and B is localized to A' and B', the conditions for a heat interaction are satisfied and we can characterize the interaction between the larger systems A and B as heat. In all likelihood, this heat interaction is followed by irreversible spontaneous processes of A and B during which the energy and entropy exchanged in the course of the heat interaction are redistributed between A' and the rest of system A, and B' and the rest of system B.

A reservoir R is a system that passes only through stable equilibrium states and has a constant temperature T_R. It follows that a reservoir R can experience heat interactions at temperature T_R only. It can also experience other types of interactions, including

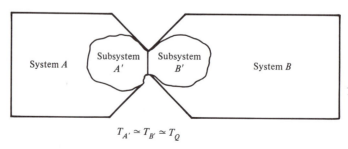

$$T_{A'} \simeq T_{B'} \simeq T_Q$$

Figure 12.8   Schematic of two systems A and B that experience a heat interaction via two subsystems A' and B' that are almost in mutual stable equilibrium at the almost common temperature T_Q.

work, but not heat at temperatures different from T_R. The interactions that are not heat at T_R are accompanied by an irreversible spontaneous creation of entropy so as to maintain the reservoir in a stable equilibrium state. So the only interactions that a reservoir can experience without a subsequent creation of entropy by irreversibility are heat interactions at the temperature of the reservoir itself.

For a process from state A_1 to state A_2 that involves only a heat interaction at temperature T_Q, that is, for a *heat-only process*, the net amount of energy $E^\leftarrow$ exchanged by system A is denoted by $Q^\leftarrow$, the net amount of entropy exchanged by $S^\leftarrow = Q^\leftarrow/T_Q$, and the energy and entropy balances become

$$E_2 - E_1 = E^\leftarrow = Q^\leftarrow \tag{12.26}$$

$$S_2 - S_1 = S^\leftarrow + S_{irr} = \frac{Q^\leftarrow}{T_Q} + S_{irr} \tag{12.27}$$

where we omit for simplicity the superscript A that refers to system A. Similarly, for an infinitesimal heat interaction, the energy and entropy balances are

$$dE = \delta Q^\leftarrow = -\delta Q^\rightarrow \tag{12.28}$$

$$dS = \frac{\delta Q^\leftarrow}{T_Q} + \delta S_{irr} = -\frac{\delta Q^\rightarrow}{T_Q} + \delta S_{irr} \tag{12.29}$$

Clearly, (12.27) and (12.29) can also be written as inequalities, that is, as

$$S_2 - S_1 \geqq \frac{Q^\leftarrow}{T_Q} \tag{12.30}$$

and

$$dS \geqq \frac{\delta Q^\leftarrow}{T_Q} \tag{12.31}$$

Here again we comment that equations (12.26) to (12.29) and relations (12.30) and (12.31) are valid for a system that experiences a heat interaction only.

Example 12.6. A metal block identical to that considered in Example 12.4 experiences a heat interaction with a reservoir at temperature $T_R = 600$ K. As a result of the interaction, the block, initially at temperature $T_1 = 300$ K ends in a state in mutual stable equilibrium with the reservoir. Find the amount of energy $Q^\leftarrow$ transferred to the block and the amount of entropy S_{irr} created inside the block by irreversibility.

Solution. The energy and entropy balances for the process of the block yield $C(T_2 - T_1) = Q^\leftarrow$ and $C \ln(T_2/T_1) = Q^\leftarrow/T_R + S_{irr}$. The temperature $T_2 = T_R$ because the block ends in a state in mutual stable equilibrium with the reservoir. Therefore, $Q^\leftarrow = 800(600 - 300) = 2.4 \times 10^5$ J, and $S_{irr} = 800 \ln(600/300) - 2.4 \times 10^5/600 = 154.4$ J/K.

Relations (12.26) to (12.31) suggest a possible way to measure entropy, provided that we can measure energy and temperature, and that we can determine whether a process is reversible. Indeed, a reversible process involving a heat interaction can be used to measure entropy differences. For example, the entropy difference dS between two neighboring states that can be interconnected by means of a reversible process involving only a heat interaction at some temperature T_Q can be measured by measuring the temperature T_Q and the energy difference between the two states, $dE = \delta Q^\leftarrow$ [equation (12.28)], so that (12.29) with $\delta S_{irr} = 0$ yields $dS = \delta Q^\leftarrow/T_Q$. The study of experimental methods to achieve energy and entropy measurements under these and similar conditions is called *calorimetry*.

As we conclude in connection with an adiabatic process, here we conclude that in a process involving heat only, if both end states are stable equilibrium, the energy and the entropy changes may be expressed in terms of changes of other properties and variables, such as temperature, pressure, and volume, so the heat may be related to the latter changes.

12.4   Work and Heat Only

As a result of work and heat interactions only, either successively or simultaneously, a system goes from state A_1 to state A_2, and its energy changes from E_1 to E_2. The energy balance requires that the energy change be equal to the net amount of energy $E^{\leftarrow}$ transferred to the system, namely, to the difference between the heat $Q^{\leftarrow}$ into the system and the work $W^{\rightarrow}$ done by the system.

$$E_2 - E_1 = E^{\leftarrow} = Q^{\leftarrow} - W^{\rightarrow} \tag{12.32}$$

or, if the changes are infinitesimal,

$$dE = \delta E^{\leftarrow} = \delta Q^{\leftarrow} - \delta W^{\rightarrow} \tag{12.33}$$

where in writing equations (12.32) and (12.33) we choose, consistently with a long-standing convention, to represent heat as positive if it is into the system and work as positive if it is done by the system. However, we can rewrite equations (12.32) and (12.33) in three other entirely equivalent ways because $E^{\leftarrow} = Q^{\leftarrow} - W^{\rightarrow} = Q^{\leftarrow} + W^{\leftarrow} = -Q^{\rightarrow} + W^{\leftarrow} = -Q^{\rightarrow} - W^{\rightarrow}$, and similarly, $dE^{\leftarrow} = \delta Q^{\leftarrow} - \delta W^{\rightarrow} = \delta Q^{\leftarrow} + \delta W^{\leftarrow} = -\delta Q^{\rightarrow} + \delta W^{\leftarrow} = -\delta Q^{\rightarrow} - \delta W^{\rightarrow}$. The choice between one form and the other is a matter of convenience and in a specific problem is typically made so that the heat and work terms assume positive values and the direction of the transfer is inferred immediately from the arrow on the symbol.

Each of equations (12.32) and (12.33) may be regarded as an energy conservation statement applied to the composite of system A and the systems with which it experiences work and heat interactions. For example, upon rewriting equation (12.33) in the form

$$dE + (-\delta Q^{\leftarrow}) + (\delta W^{\rightarrow}) = 0 \tag{12.34}$$

and recognizing that $(-\delta Q^{\leftarrow})$ equals the change in energy of the system with which the heat interaction occurred, and that $\delta W^{\rightarrow}$ equals the change of energy of the system with which the work interaction occurred, we conclude that the sum of the energy changes of all the systems participating in the interactions equals zero, as it should.

It is noteworthy that the right-hand side of either (12.32) or (12.33) is meaningful only because heat and work are totally distinguishable from one another. If they were not, each equation would simply be a statement of the arithmetic fact that the number $E^{\leftarrow}$ may be written as an algebraic sum of two other numbers $Q^{\leftarrow}$ and $W^{\rightarrow}$, where either $Q^{\leftarrow}$ or $W^{\rightarrow}$ is chosen completely arbitrarily. Such a statement would be useless because it does not relate to any physical phenomena. Hence the need for the careful and very specific definition of heat introduced in Section 12.3.

As the system goes from state A_1 to state A_2 while experiencing work and heat interactions only, its entropy changes from S_1 to S_2, and the entropy balance requires that this change be greater than or equal to the net amount of entropy $S^{\leftarrow}$ received by the system. If the heat interaction occurs at a single fixed temperature T_Q, then $S^{\leftarrow} = Q^{\leftarrow}/T_Q$. If, instead, it occurs successively at a varying temperature T_Q, then $S^{\leftarrow}$ is equal to the integral of each infinitesimal amount $\delta Q^{\leftarrow}$ divided by the corresponding value of T_Q, that is, $S^{\leftarrow} = \int \delta Q^{\leftarrow}/T_Q$, $Q^{\leftarrow} = \int \delta Q^{\leftarrow}$. If the heat interactions are at more than one different varying temperatures T_{Qk}, then $S^{\leftarrow} = \sum_k \int \delta Q_k^{\leftarrow}/T_{Qk}$, and $Q^{\leftarrow} = \sum_k \int \delta Q_k^{\leftarrow}$.

For the case of a single heat interaction at a varying temperature T_Q, the entropy balance equation is

$$S_2 - S_1 \geqq S^{\leftarrow} = \int \frac{\delta Q^{\leftarrow}}{T_Q} \tag{12.35}$$

or, equivalently,

$$S_2 - S_1 = \int \frac{\delta Q^{\leftarrow}}{T_Q} + S_{\text{irr}} \tag{12.36}$$

where S_{irr} is nonnegative. The entropy $S_{\text{irr}} = 0$ only if the process is reversible. For infinitesimal changes,

$$dS \geqq \delta S^{\leftarrow} = \frac{\delta Q^{\leftarrow}}{T_Q} \tag{12.37}$$

or, equivalently,

$$dS = \frac{\delta Q^{\leftarrow}}{T_Q} + \delta S_{\text{irr}} \tag{12.38}$$

Here we have chosen to represent heat as positive if into the system. However, we can also write the entropy balance equation in another entirely equivalent way because $S^{\leftarrow} = \int \delta Q^{\leftarrow}/T_Q = -\int \delta Q^{\rightarrow}/T_Q$ and, similarly, $\delta S^{\leftarrow} = \delta Q^{\leftarrow}/T_Q = -\delta Q^{\rightarrow}/T_Q$. If we know the change of state and the heat, we see from the entropy balance that we can identify the amount of entropy generated by irreversibility and thus determine the degree to which the process deviates from perfection.

In a process involving work and heat, if both end states are stable equilibrium, we conclude that the energy and entropy changes may be expressed in terms of changes of other properties and variables, such as temperature, pressure, and volume, so the work and the heat may be related to these other changes.

In analyses of any scientific or engineering process involving heat and work interactions only, the energy balance, equation (12.32) or (12.33), and the entropy balance, equation (12.36) or (12.38), are essential. Each balance is based on both the first and second laws, and each must be satisfied for the process to be physically possible.

12.5 Inequality of Clausius

For a system undergoing a cyclic change of state resulting from irreversible processes that involve work and heat interactions only, the sum of ratios of each heat $Q_k^{\leftarrow}$ divided by the temperature T_{Qk} at which it occurs is always less than zero, that is,

$$\sum_k \frac{Q_k^{\leftarrow}}{T_{Qk}} < (0)_{\text{irr}}^c \tag{12.39}$$

where the superscript c over the zero on the right-hand side of the inequality indicates that the inequality holds for a cyclic process, and the subscript irr indicates that the cyclic process of the system is irreversible. If each heat interaction occurs at a continuously varying temperature T_{Qk}, each term in the sum on the left-hand side of the inequality is substituted by the integral of each infinitesimal heat $\delta Q_k^{\leftarrow}$ divided by the corresponding instantaneous value of the temperature T_{Qk}, that is,

$$\sum_k \oint \frac{\delta Q_k^{\leftarrow}}{T_{Qk}} < (0)_{\text{irr}}^c \qquad (12.40)$$

where the symbol $\oint$ is used instead of $\int$ as a further reminder that the process is cyclic.

Each of the inequalities (12.39) and (12.40) is called a *Clausius inequality*, in honor of Clausius who first stated it. The Clausius inequality follows directly from relation 12.37 applied to each step of an irreversible process and summed up for all the steps of the cycle because, in a cycle, the changes of any property of the system, including entropy, add up to zero. The meaning of the Clausius inequality is that the total entropy supplied by heat to a system that undergoes a cyclic change of state is negative, that is, that the net entropy of the environment of the system has increased, a result that must be expected of an irreversible process in which the entropy change of the system is null. Of course, the inequality becomes an equality when the processes involved in the cyclic change of state of the system are reversible.

If a system passes along a cyclic sequence of stable equilibrium states at temperature T and at each infinitesimal step in the sequence experiences a heat interaction with heat $\delta Q^{\leftarrow}$ at temperature equal to T, the Clausius inequality reduces to

$$\oint \frac{\delta Q^{\leftarrow}}{T} < (0)_{\text{irr}}^c \qquad (12.41)$$

Example 12.7.   A cyclic engine experiences heat interactions with two metal blocks A and B identical to those considered in Example 12.4. Initially, the two blocks are at $T_1^A = 900$ K and $T_1^B = 400$ K. As a result of the interactions, the cyclic engine must bring block A to a stable equilibrium state at $T_2^A = 800$ K. Find the lowest final temperature T_2^B at which block B can be found under the specified conditions.

Solution.   The two heat interactions occur at the varying temperatures T^A and T^B. The energy balance for block A yields $dE^A = -\delta Q^{A\rightarrow}$ because the only interaction experienced by the block is heat. Similarly, for block B, $dE^B = -\delta Q^{B\rightarrow}$. But for each of the identical metal blocks we have $dE^A = C\,dT^A$ and $dE^B = C\,dT^B$, and, therefore, $\delta Q^{A\rightarrow} = -C\,dT^A$ and $\delta Q^{B\rightarrow} = -C\,dT^B$. Denoting the energy exchanges of the cyclic engine with blocks A and B by $\delta Q_a^{\leftarrow}$ and $\delta Q_b^{\leftarrow}$, respectively, we have that $\delta Q_a^{\leftarrow} = \delta Q^{A\rightarrow}$ and occurs at the varying temperature T^A, and that $\delta Q_b^{\leftarrow} = \delta Q^{B\rightarrow}$ and occurs at the varying temperature T^B. Hence, the left-hand side of the Clausius inequality, relation (12.39), is $\oint \delta Q_a^{\leftarrow}/T^A + \oint \delta Q_b^{\leftarrow}/T^B = -C \int dT^A/T^A - C \int dT^B/T^B = -C\ln(T_2^A/T_1^A) - C\ln(T_2^B/T_1^B) = -C\ln(T_2^A T_2^B/T_1^A T_1^B)$, and must be negative or at most zero. This requirement yields the condition $T_2^A T_2^B \geq T_1^A T_1^B$, where the equal sign holds only if the process of the engine is reversible. Numerically, $T_2^B \geq T_1^A T_1^B/T_2^A = 900 \times 400/800 = 450$ K. So the lowest final temperature T_2^B is 450 K, and the corresponding work done by the cyclic engine is $W^{\rightarrow} = -C\,(T_2^A - T_1^A) - C\,(T_2^B - T_1^B) = -800 \times (-100) - 800 \times 50 = 40 \times 10^3$ J $= 40$ kJ.

Problems

12.1 Each of two fixed-volume tanks A and B is filled with the same gas. The amounts are $m_A = 2$ kg and $m_B = 3$ kg. For the range of stable equilibrium states of interest in this problem, the energy and entropy of the gas can be written as $E = mc\,(T - T_o)$ and $S = mc\ln(T/T_o)$, where $c = 1.2$ kJ/kg K, $T_o = 300$ K, and T is temperature. At time t_1, the gas in tank A is at temperature $T_1^A = 1000$ K, and that in tank B at $T_1^B = 300$ K. The space between the two tanks is evacuated and perfectly isolated from other systems in the environment. The two tanks may interact via the electromagnetic field that fills the vacuum and carries energy and entropy between the two tanks. The rates at which energy and entropy are exchanged obey the relations

$$\dot{E}^{A \to B} = \mathcal{A}\sigma\left[(T^A)^4 - (T^B)^4\right] \quad \text{and} \quad \dot{S}^{A \to B} = \mathcal{A}\frac{4}{3}\sigma\left[(T^A)^3 - (T^B)^3\right]$$

where the dot indicates a rate, $\sigma = 5.67 \times 10^{-8}$ W/m^2 K^4 and $\mathcal{A} = 2$ m^2.

(a) At time t_1, is the interaction between the two tanks—through the electromagnetic field—work or heat?

(b) If the process for each tank between time t_1 and time $t_2 = t_1 + dt$ is reversible, what are the respective changes in energy and entropy? Are the states of the two tanks at time t_2 stable equilibrium?

(c) If the state of each tank at time $t_2 = t_1 + dt$ is stable equilibrium, how much entropy is generated by irreversibility in each tank between time t_1 and time t_2? What are the temperatures of the two tanks at time t_2?

(d) Prove that in the limit as $T^A \to T^B$ the interaction between the two tanks through the electromagnetic field is a heat interaction. Consider the fact that the interatomic space of any substance under any conditions is always filled with radiation, that is, the electromagnetic field, and comment on what is the main carrier of energy and entropy in heat interactions.

12.2 A solid block is cooled from 300 K to 100 K. The cooling is done by cyclic machinery that receives energy by means of a work interaction with a power plant and interacts with the solid and a reservoir at temperature $T_R = 500$ K. The relation between the energy E and the entropy S of the solid is $E = 3000\exp[(S - S_o)/10]$, where E is in kJ, S in kJ/K, and S_o is the entropy of the solid (in kJ/K) at 300 K.

(a) Find expressions for E and S in terms of the temperature T of the solid.

(b) Calculate the smallest work that must be done on the machinery to achieve the specified cooling of the solid.

(c) Calculate the work done on the machinery if the entropy generated by irreversibility in the machinery is 9.03 kJ/K.

12.3 A fixed amount of fuel–air mixture A is confined in a chamber having a fixed volume. If the burning of the fuel is inhibited by some anticatalyst, the state of A has a temperature $T_1 = 300$ K and is denoted by A_1. If the burning of the fuel is not inhibited, and the mixture experiences a reversible weight process starting from the initial state, which is again denoted by A_1, and ending in a stable equilibrium state A_{S1}, an amount of 5000 kJ of energy is transferred to the environment in the form of work. The temperature of A_{S1} is $T_{S1} = 500$ K. In state A_{S1}, the constituents of the system include the products of combustion. If system A (without reaction inhibitors) in state A_1 is combined with a reservoir R at $T_R = 300$ K, the available energy $\Omega_1 = 7948$ kJ. For the stable equilibrium states of the system without reaction inhibitors, the relation between energy and temperature is $E_{S2} - E_{S1} = 170(T_{S2} - T_{S1})$ for E in kilojoules and T in kelvin.

(a) How would you define each of the two states A_1 just described? Use words and values in your specifications as needed.

(b) Calculate the change in energy of the reservoir and the change in energy of A if the largest work is done by the composite system AR starting from state A_1.

(c) Calculate the changes in entropy of A, R, and the composite system at the end of the process in part (b).

(d) Write an expression for S as a function of T for the stable equilibrium states that include the products of combustion, and make a sketch of E versus S for these states.

(e) Starting from state A_1 and ending in a state in which A is in mutual stable equilibrium with R, an amount $S_{irr} = 10$ kJ/K is generated by irreversibility while work is being done by the composite of A and R. How much work is done? What is the change in entropy of the reservoir? What is the entropy change of the composite of A and R? How does the final entropy of A here differ from that of A in part (c)?

12.4 A system consists of two identical solid blocks A and B and cyclic machinery X. Each block has a mass $m = 1$ kg, and its stable equilibrium states can be described by the parametric relations $E_2 - E_1 = mc(T_2 - T_1)$ and $S_2 - S_1 = mc \ln(T_2/T_1)$, where $c = 1$ kJ/kg K and subscripts 1 and 2 denote any two stable equilibrium states. The blocks can interact with each other only via the machinery. Initially, each block is at a temperature $T_1 = 400$ K, and the machinery can only experience work interactions with other systems in the environment.

(a) If 1100 kJ of work is done on the system, what is the highest temperature that can be reached by one of the two blocks? What is then the temperature of the other block?

(b) Upon reaching the state found in part (a), what is the adiabatic availability of the system?

(c) Upon reaching the state found in part (a), the cyclic machinery is allowed an additional interaction—not necessarily work—with a reservoir of $T_0 = 300$ K. What is the largest work that can be extracted from this composite?

(d) Repeat parts (a), (b), and (c) but asssume that block A has a mass of 5 kg and block B a mass of 0.5 kg.

12.5 An elevator moves in a vertical shaft. When fully loaded it weighs 1000 kg. A 500 kg counterweight is connected to the elevator via cables and pulleys at the shaft's top. Assume that the guide rails and pulleys are frictionless.

(a) Estimate the power required while the fully loaded elevator is rising at a constant speed of 1 m/s.

(b) If the empty elevator weighs 180 kg, what power is required to lower the elevator at a constant speed of 1 m/s?

(c) What would be the power requirement in part (a) if no counterweight were used?

Now consider the descent process of the fully loaded elevator. The electric motor is disengaged by a clutch, and a friction brake is applied to give a constant speed of descent from a height of 30 m to ground level. Assume that the temperature of the elevator, including the brake, stays approximately constant and equal to the temperature of the environment at 20°C, and the kinetic energy is negligible.

(d) If instead of the brake a work extracting device is used, what is the largest work that can be extracted from the system during the descent of the elevator?

(e) Calculate the increase of entropy during the descent of the elevator due to using a brake rather than a work-extracting device. Make a graph of the path on an energy versus entropy plane.

(f) Instead of a friction brake, suppose that the motor is used as an electric generator, thus controlling the descent by producing electricity for a utility network. Show this path on the E versus S diagram of part (e) if the electric generator is perfect, that is, operates reversibly.

(g) Sketch on the E versus S diagram the paths of the system for efficiencies of the generator of part (f) of 25, 50, and 75%, respectively.

12.6 Consider two large blocks of copper, A and B, with $m^A = 2000$ kg and $m^B = 500$ kg. Initially, block A is at temperature 1100 K and block B at 450 K. For the stable equilibrium states of each block: $e_2 - e_1 = 0.42(T_2 - T_1)$ in kJ/kg, where T is in kelvin, and $e = E/m$, and $s_2 - s_1 = 0.42 \ln(T_2/T_1)$ in kJ/kg K, where $s = S/m$. For this problem, we can neglect the expansion of copper as a function of temperature.

(a) If perfect machinery is available, what is the largest work that can be done using the two copper blocks?

(b) If perfect machinery and a reservoir at 300 K are available, what is the largest work that can be done using the composite of the two copper blocks and the reservoir?

(c) How much entropy is generated by irreversibility if in part (a) the work is one-half of the largest value?

12.7 For certain conditions, the energy and entropy of stable equilbrium states of 1 mol of a substance are related by the expression

$$e = 7.54 + 6150 \left\{ \exp \left[\frac{s - s_o - 8.3 \ln(v/v_o)}{20.5} \right] - 1 \right\} \qquad \text{J/mol}$$

where s_o and v_o are a reference entropy in J/mol K and a reference volume in m^3/mol, respectively.

(a) Find expressions for the temperature T, and pressure p of these states in terms of s, s_o, v, and v_o.

(b) Find the temperature T_o of the reference stable equilibrium state A_o corresponding to s_o and v_o.

(c) Starting from state A_o, the substance is compressed from the volume v_o to a volume $v_1 = v_o/2$. The only interaction that occurs is work, and the final state is stable equilibrium. What is the least work required for the compression?

(d) If the work done on the substance in the course of the process in part (c) is 2500 J/mol, what is the largest amount of entropy that can be generated by irreversibility?

12.8 In an intermediate stage of a steelmaking process, a steel billet of mass $m_A = 15,000$ kg initially at temperature $T_1 = 1000°C$ must be cooled down to $T_2 = 200°C$. During cooling, the pressure and volume of the billet are almost constant. It is proposed that the cooling be achieved via appropriate cyclic machinery which transfers energy from the billet to warehouse B of the steel plant. The warehouse must be kept at a fixed temperture $T_B = 25°C$. In addition to the billet and the warehouse, the machinery can also interact with the environment, which is at the fixed temperature $T_0 = 10°C$. The differences in energy and entropy of the stable equilibrium states of the billet obey the relations $e_2 - e_1 = 0.4(T_2 - T_1)$ in kJ/kg, and $s_2 - s_1 = 0.4 \ln(T_2/T_1)$ in kJ/kg K, where T is in kelvin, $e = E/m$, and $s = S/m$.

(a) What is the largest amount of energy that could be transferred to the warehouse under the conditions specified?

(b) If the machinery interacts only with the billet and the warehouse, and simply transfers all the energy lost by the billet directly into the warehouse, what is the loss of available energy with respect to a reservoir at the environmental temperature?

13   Energy Versus Entropy Graphs

Because they are defined in terms of the values of the amounts of constituents, the parameters, and a complete set of independent properties, states can in principle be represented by points in a multidimensional geometrical space with one axis for each amount, parameter, and independent property. Such a representation, however, would not be enlightening because the number of independent properties of any system is indefinitely large. Nevertheless, useful information can be summarized by first cutting the multidimensional space with a hypersurface corresponding to given values of each of the amounts of constituents and each of the parameters, and then projecting the result onto a two-dimensional plane—a plane with two property axes.

In this chapter we discuss the energy versus entropy plane that illustrates many of the basic concepts of thermodynamics.

13.1   Energy Versus Entropy Plane

To simplify the discussion, we consider a system with volume, V, as the only parameter. For given values of the amounts of constituents and the volume, we project the multidimensional state space of the system onto the E versus S plane. This projection of states on the E versus S plane includes both stable equilibrium states and other states that are not stable equilibrium. It is not to be confused with the standard graphical representations of thermodynamic relations (see Chapter 19), which are restricted to stable equilibrium states only.

For any system the projection must have the shape of the crosshatched area shown in Figure 13.1, namely, all the states that share the given characteristics have property values that project on the area between the vertical line denoted as the line of the zero-entropy states and the curve of the stable equilibrium states.

A point either inside the crosshatched area or on the vertical line $S = 0$ above point E_g represents a large number of states. Each such state has the same values of amounts of constituents, volume V, energy E, and entropy S, but differing values of other properties, and is not a stable equilibrium state. It can be any type of state except a stable equilibrium state.

A point on the convex curve of the stable equilibrium states represents one and only one state. For each of these states, the value of any property is uniquely determined only by the values of the amounts of constituents, the volume, and either E or S of the point on the curve.

For the same values of the amounts of constituents but a different value of the volume, the projection of the multidimensional state space onto the E versus S plane is again an area bounded by the zero entropy line and a convex curve, but the convex curve is shifted in the vertical direction. For example, for two values of volume

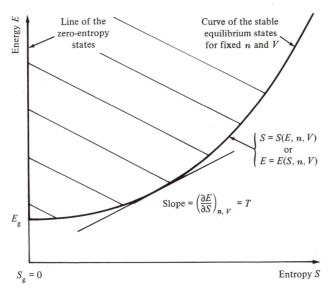

Figure 13.1 Schematic representation of the projection of the states of a system with given values of the amounts of constituents and the volume on the energy versus entropy plane.

V_1 and V_2, and such that $V_1 < V_2$, the relative position of the two convex curves is as shown in Figure 13.2a. The two projections may also be presented on different planes by using a three-dimensional representation with axes E, S, and V, as shown in Figure 13.2b.

As we discuss in the following sections, all these features are graphical illustrations of the results derived in Chapters 2 to 12.

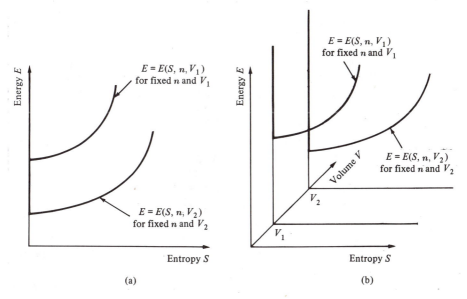

Figure 13.2 Projections of the states of a system with given values of amounts of constituents, and two different values of the volume, V_1 and V_2: (a) projections on the same E versus S plane; and (b) projections on two different planes in the E–S–V space.

13.2 Zero-Entropy Line

The line of the zero-entropy states (or the zero-entropy plane in Figure 13.2b) corresponds to all the states that have the least amount of entropy. This amount can be assigned the value zero because the laws of thermodynamics imply that no states exist with lower entropy (Section 9.8). Thus entropy has absolute values greater than or equal to zero.

It turns out that the zero-entropy line represents all the states that are defined in mechanics, classical or quantum, without concern about the implications of the laws of thermodynamics. So mechanics can be regarded as a special case of thermodynamics, namely, as zero-entropy physics.

13.3 Ground-Energy States

As shown in Figure 13.1, for given values of the amounts of constituents and the volume, the least energy of the system is E_g. It corresponds to a unique stable equilibrium state having zero entropy and zero temperature (Section 9.9). We discuss this conclusion in more detail in Section 13.17.

The energy E_g is the least energy for which the system can exist with the given values of the amounts of constituents and for the given value of the volume. For example, if the system consists of hydrogen molecules in a small fixed-volume container, the value of E_g for one hydrogen molecule is smaller than the value for two hydrogen molecules in the same container. Again, if the volume is variable, the smaller the volume, the larger becomes the value of E_g. Indeed, energy must be transferred to the system to compress the molecules from a large to a small volume. For different values of the volume the least energy varies as illustrated in Figure 13.2. Nevertheless, for all values of the amounts of constituents and the volume, the ground-energy state is always a stable equilibrium state with zero entropy and zero temperature.

13.4 The Fundamental Relation

The stable-equilibrum-state curve in Figure 13.1 represents the convex stable-equilibrium-state relation E versus S or, equivalently, the concave fundamental relation S versus E, for the given values of the amounts of constituents and the volume (Section 8.3). It is a single-valued relation because for each set of values E, n, and V there is one and only one stable equilibrium state and, therefore, a unique value of S. It has the following features (Figure 13.3).

Each stable equilibrium state is the state of least energy among all the states with the same values of S, n, and V (lowest-energy principle). For each set of values S_1, n, and V, the stable equilibrium state A_{S1} on the vertical line $S = S_1$ is the state of least energy—no states exist below A_{S1} that correspond to the same values of n and V and that lie on the line $S = S_1$ (Section 8.2). State A_{S1} can be reached starting from any state A_1 on the line $S = S_1$ by means of a reversible weight process without net changes in n and V. Indeed, in such a process the net change in the entropy of the system is zero, $S_{S1} = S_1$, and the energy $E_1 - E_{S1}$ is transferred out from the system to the weight.

Each stable equilibrium state is the state of highest entropy among all the states with the same values of E, n, and V (highest-entropy principle). For each set of values E_1, n, and V, the stable equilibrium state A_{E1} on the horizontal line $E = E_1$ is the state of highest entropy—no states exist beyond A_{E1} that correspond to the same values of n

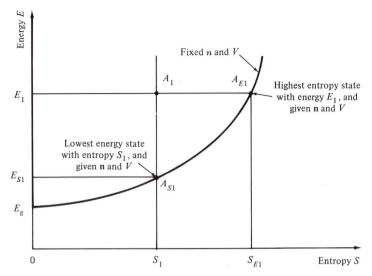

Figure 13.3   Schematic representation of the highest entropy and lowest energy principles on the energy versus entropy diagram.

and V and that lie on the line $E = E_1$ (Section 8.2). State A_{E1} can be reached starting from any state A_1 on the line $E = E_1$ by means of a zero-net-effect weight process such as a spontaneous process. Any such process is irreversible because it entails an increase in the entropy of the system without any effects on the environment.

Temperature is positive and increasing with energy. Because each stable equilibrium state is unique, the temperature $T = (\partial E/\partial S)_{n,V}$ at each point on the convex boundary is uniquely defined (Section 9.2), that is, the slope of the stable equilibrium state curve on the E versus S diagram is uniquely defined. It is positive and increasing with energy. Temperature is not defined for states that are not stable equilibrium because then E, S, n, and V are independent and, therefore, the partial derivative of E with respect to S is meaningless. More important, however, even if the partial derivative of E with respect to S were meaningful, it would be neither the expression that enters the temperature equality requirement for systems in mutual stable equilibrium, nor the quantity that is measured by the techniques of temperature measurements.

13.5   Perpetual-Motion Machine of the Second Kind

Starting from a stable equilibrium state A_{S1} on the convex boundary $E_g A_{S1} A_{E1}$ in Figure 13.3, the system cannot transfer energy to a weight without net changes in the values of the amounts of constituents and the volume because no state of lower energy exists that has an entropy equal to or larger than the entropy of state A_{S1}. Indeed, if energy is transferred to a weight, the energy of the system is reduced. But starting from state A_{S1}, all states with smaller energy also have smaller entropy. Because the weight receives only energy, and entropy cannot decrease by itself, it follows that no such transfer can occur under the conditions specified. This feature of the graph represents the impossibility of perpetual-motion machines of the second kind (Section 4.5). It is sometimes expressed as the nonexistence of a Maxwellian demon, that is, of a superbeing capable of extracting energy from a system in a stable equilibrium state without affecting the values of S, n, and V.

13.6   Equilibrium Thermodynamics

For each set of given values of the amounts of constituents and the volume, the convex boundary $E_g A_{S1} A_{E1}$ in Figure 13.3 represents the corresponding stable equilibrium states. These states are referred to in the literature as the thermodynamic equilibrium states of equilibrium thermodynamics, which is sometimes also called "classical thermodynamics" or "thermostatics." So equilibrium thermodynamics can be regarded as another special case of thermodynamics, namely, as highest-entropy physics.

13.7   Adiabatic Availability

For a given state A_1, the energy $E_1 - E_{S1}$ shown graphically in Figure 13.4 is equal to the adiabatic availability Ψ_1 of A_1 because the change of state from A_1 to A_{S1} represents the change specified in the definition of Ψ_1 (Section 5.2). We see from the figure that, in general, Ψ_1 is smaller than the energy of the system above the ground-state energy, $E_1 - E_g$. It varies from $E_1 - E_g$ to zero as the entropy S_1 of the state varies from zero to the highest value that is possible for the set of values E_1, n, and V. So entropy affects the usefulness of the energy of a system, that is, the larger the entropy for given values of E, n, and V, the smaller the adiabatic availability. This limitation on the amount of energy that can be transferred from a system to a weight in a weight process without net changes in the values of n and V is a consequence of the laws of thermodynamics of paramount theoretical importance and with many practical implications.

For given values of n and V, we see graphically from Figure 13.4 that stable equilibrium states, such as state A_{S1}, have zero adiabatic availability, and that any state with nonzero adiabatic availability cannot be stable equilibrium (Section 5.3.3).

13.8   Work in an Adiabatic Process

In an adiabatic process without net changes in the values of the amounts and the volume, the work done by the system starting from state A_1 and ending in a state different from A_{S1} (Figure 13.4) is always smaller than the adiabatic availability Ψ_1 (Section 5.3.2).

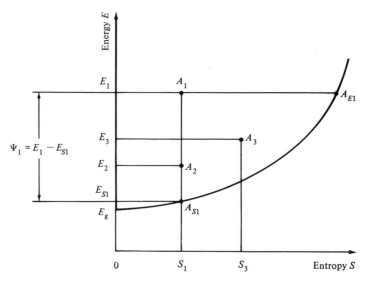

Figure 13.4 Graphical representation of adiabatic availability on the energy versus entropy diagram.

If the process is reversible, the final state $A_2 \neq A_{S1}$ has entropy $S_2 = S_1$ and energy $E_2 > E_{S1}$. Therefore,

$$(W_{12}^{A \rightarrow})_{rev} = E_1 - E_2 < E_1 - E_{S1} = \Psi_1 \tag{13.1}$$

If the process is irreversible, the final state $A_3 \neq A_{S1}$ has entropy $S_3 > S_1$. But for $S_3 > S_1$, the graph shows that A_3 has energy $E_3 > E_{S1}$, and therefore

$$(W_{13}^{A \rightarrow})_{irr} = E_1 - E_3 < E_1 - E_{S1} = \Psi_1 \tag{13.2}$$

Here the entropy increase $S_3 - S_1$ is generated by irreversibility because the process is adiabatic, and therefore no entropy is supplied by any other system.

The graph in Figure 13.4 is illustrative of an interesting point. In view of the natural tendency of a nonequilibrium state to evolve spontaneously toward a state of higher entropy, we see that in order to extract the adiabatic availability from a system in a nonequilibrium state we must act quickly. The faster we induce a change from a state A_1 toward state A_{S1}, the less time is available for spontaneous processes to generate entropy, and the larger the work that may be done on other systems. Said differently, here the faster the process from A_1 toward A_{S1}, the closer it gets to being reversible. This conclusion is contrary to a widespread conviction that a reversible process can be achieved only at infinitesimally small speed or, equivalently, over infinitely long time.

13.9   Generalized Adiabatic Availability

The generalized adiabatic availability (Section 5.5) is illustrated graphically in Figure 13.5. For a given state $A_{1'}$ with values of amounts n' and volume V', the energy $E_{1'} - E_{S1}$ is equal to the generalized adiabatic availability $\Psi_{1'}$ of $A_{1'}$ with respect to given values n of the amounts and V of the volume, which in general may differ from n' and V', respectively.

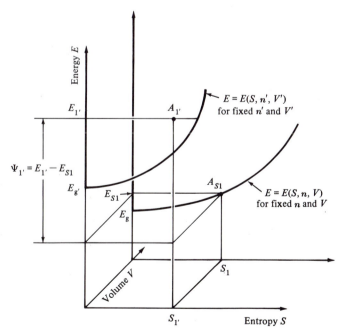

Figure 13.5   Graphical representation of the generalized adiabatic availability with respect to the values n and V on the energy versus entropy diagram.

13.10   Available Energy

The E^R versus S^R diagram of a reservoir R is just a straight line of slope T_R (Figure 13.6) because the reservoir passes through stable equilibrium states only and has constant temperature (Section 9.7). It is noteworthy that for very small values of entropy, no system can behave as a constant nonzero temperature reservoir because as entropy approaches zero, the temperature of any system must also approach zero.

Given the E versus S diagram of a system A with specified values of amounts of constituents and volume and a reservoir R at temperature T_R, we can draw a line of slope T_R tangent to the convex stable-equilibrium-state curve of system A, that is, tangent to the curve $E_g A_{S1} A_0$ in Figure 13.7. The point of tangency A_0 represents the state A_0 in which system A is in mutual stable equilibrium with the reservoir because in state A_0 the system has a temperature $T_0 = (\partial E / \partial S)_{n,V}$ and therefore equal to the temperature T_R of the reservoir. In state A_0 the system has energy E_0 and entropy S_0.

The tangent is also useful in providing a way to represent graphically the available energy of any state of A. Specifically, for a given state A_1, the vertical distance of point A_1 from the tangent, that is, the energy $E_1 - E_a$, represents the available energy Ω_1^R of A_1 with respect to reservoir R.

Indeed, with respect to reservoir R, the available energy Ω_1^R of state A_1 with energy E_1 and entropy S_1 is given by the relation [equation (9.45)]

$$\Omega_1^R = E_1 - E_0 - T_R(S_1 - S_0) \tag{13.3}$$

because the available energy Ω_0^R of state A_0 is zero, that is, the adiabatic availability of the composite of A and R in a stable equilibrium state is zero. We recall that the available energy Ω_1^R equals the work done in the course of a reversible weight process for the composite of systems A and R in which A ends in state A_0 (Figure 13.7) and R changes from state R_1 to state R_2 (Figure 13.6).

The term $E_1 - E_0$ on the right-hand side of (13.3) is the length bA_1 in Figure 13.7, that is, the negative of the change in energy of system A as it goes from state A_1 to state A_0. The term $-T_R(S_1 - S_0)$ is the length ab because $ab = (bA_0)\tan\theta = (S_0 - S_1)T_R$. Of course, ab is also equal to the negative of the change in energy $E_1^R - E_2^R$ of the reservoir as it goes from state R_1 to state R_2 (Figure 13.6). Thus the length $aA_1 = bA_1 + ab$ is indeed the negative of the energy change of the composite of A and R and therefore the available energy Ω_1^R of state A_1.

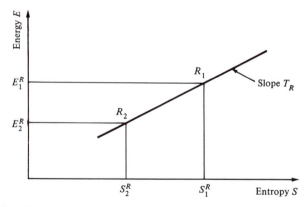

Figure 13.6 Schematic representation of the states of a reservoir with fixed values of amounts of constituents and parameters on the energy versus entropy diagram. It consists of just the line $R_1 R_2$ of slope T_R.

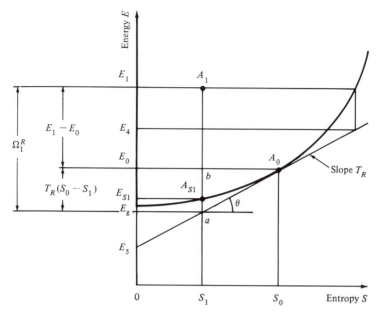

Figure 13.7 Graphical representation of the available energy with respect to a reservoir at temperature T_R.

The graphs in Figures 13.6 and 13.7 also account for the entropy changes that correspond to any reversible process that yields the available energy. They show that the change in entropy $S_0 - S_1$ of system A is equal and opposite to the change in entropy $S_2^R - S_1^R$ of the reservoir R.

By comparing the graphical representations of Ψ_1 and Ω_1^R for a given state A_1 (Figures 13.4 and 13.7), we see that in general the available energy Ω_1^R is greater than the adiabatic availability Ψ_1 [relation (6.4)]. From Figure 13.7 we also see that Ω_1^R can be greater than the energy of the system above the ground-state energy, $E_1 - E_g$. For states with energy E_1, the available energy varies from the largest value $E_1 - E_5$ to the least value $E_1 - E_4$, depending on the entropy S_1 of state A_1, that is, depending on whether the entropy is zero or the largest for the given E_1, respectively.

It is noteworthy that although the available energy can be extracted as a result of an adiabatic process for the composite of systems A and R, the processes experienced by both A and R are not necessarily adiabatic because they may involve exchanges of both energy and entropy. In fact, it is precisely the exchange of entropy between A and R that results in the available energy sometimes being greater than the energy of A, or in getting work even when A is in a stable equilibrium state, provided that A and R are not initially in mutual stable equilibrium. In this sense, the reservoir acts as a source or sink of entropy for A. Of course, this entropy exchange between A and R is always accompanied by a definite energy exchange because the reservoir must change both its entropy and its energy as it passes from stable equilibrium state to stable equilibrium state.

One of the many ways of extracting the available energy of, say, state A_1 (Figure 13.7) is as follows. We first use machinery that interacts reversibly and adiabatically with A only, and extracts the adiabatic availability Ψ_1. Thus system A is brought to stable equilibrium state A_{S1}. At this state, the system is in general at a temperature different from that of the reservoir. Next, we connect the system to the reservoir via reversible machinery that cools or heats the system to temperature T_R while doing work. Thus the total work done Ω_1^R is greater than Ψ_1.

13.11   Generalized Available Energy

For a given state $A_{1'}$, with values of amounts n' and volume V', the generalized available energy $\Omega_{1'}^R$ (Section 6.8) with respect to reservoir R and values n and V is represented graphically in Figure 13.8 and given by the relation

$$\Omega_{1'}^R = E_{1'} - E_0 - T_R(S_{1'} - S_0) \qquad (13.4)$$

where in state A_0 system A is in mutual stable equilibrium with the reservoir, and has values n and V that may differ from n' and V', respectively, of state $A_{1'}$.

13.12   Work Interactions

Graphical illustrations of processes that involve only work interactions between two systems A and B are provided by Figure 13.9. Each process for the composite of A and B is assumed to be reversible, so that the energy change of A is equal and opposite to the energy change of B, and the entropy changes of both A and B are zero because a work interaction does not transfer any entropy and the processes for both A and B are reversible (Section 12.1).

As a result of the interaction depicted in Figure 13.9a, the state of A changes from state A_1 to state A_2 and that of B from state B_1 to state B_2, none being a stable equilibrium state. Moreover, the volume of either system A or system B or both may or may not change.

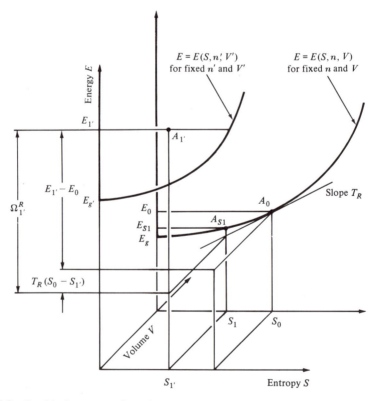

Figure 13.8 Graphical representation of the generalized available energy with respect to a reservoir at T_R and the values n and V on the energy versus entropy diagram.

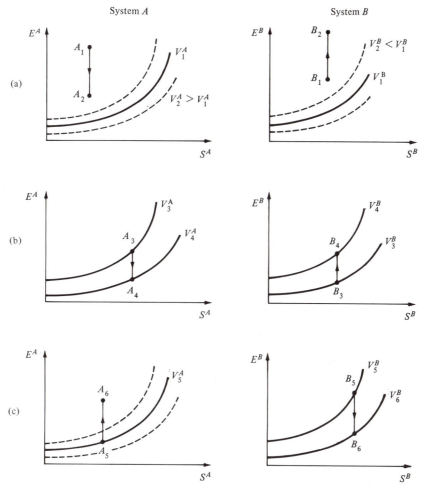

Figure 13.9   Graphical illustration of work interactions on the energy versus entropy diagram. Each process for the composite of systems A and B is assumed to be reversible.

As a result of the interaction shown in Figure 13.9b, the state of A changes from A_3 to A_4 and that of B from B_3 to B_4, all being stable equilibrium states. Here the volume of system A changes from V_3^A to V_4^A, and the volume of system B from V_3^B to V_4^B.

As a result of the interaction shown in Figure 13.9c, the state of A changes from stable equilibrium state A_5 to state A_6, which is not stable equilibrium and may or may not have a different volume than A_5, whereas the state of B changes from state B_5 to state B_6 both being stable equilibrium states, but with different volumes V_5^B and V_6^B.

In the example of Figure 13.9a, irreversibility can occur in either A, or B, or both because either state A_2, or state B_2, or both can evolve spontaneously toward the corresponding stable equilibrium states. In the example of Figure 13.9c, irreversibility can occur in A but not in B because only state A_6 can evolve spontaneously, whereas stable equilibrium state B_6 cannot. The processes in Figure 13.9b cannot become irreversible because the final states of both A and B are stable equilibrium states, and therefore each has the highest entropy compatible with the corresponding energy.

These simple examples illustrate the well-known fact that spontaneous generation of entropy by irreversibility can occur if and only if the system is in a state that is not stable equilibrium.

13.13   Heat Interactions

Graphical illustrations of processes that involve only heat interactions between two systems A and B are provided in Figures 13.10 and 13.11. Each process for the composite of A and B is assumed to be reversible, so that the changes in energy and entropy of system A are equal and opposite to the changes in energy and entropy of system B, respectively (Section 12.3).

As a result of the heat interaction shown in Figure 13.10, system A changes from state A_1 to state A_2, system B from state B_1 to state B_2, all being stable equilibrium states, without net changes in values of amounts of constituents and volumes. The temperatures of A and B are almost equal to T_Q. The two systems exchange energy and entropy. The ratio of the energy exchanged and the entropy exchanged is equal to the common temperature. Because the final states are stable equilibrium, no spontaneous changes of state can occur, and therefore no entropy can be generated by irreversibility.

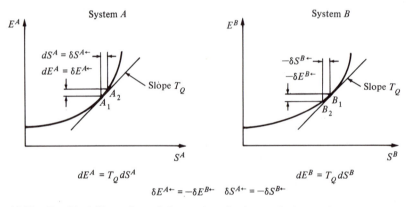

Figure 13.10 Graphical illustration of the results of a heat-only interaction at temperature T_Q between two systems A and B initially at temperatures almost equal to T_Q.

As a result of the interaction shown in Figure 13.11, system A changes from state A_1 to state A_2, system B from state B_1 to state B_2, none of which is a stable equilibrium state. However, systems A and B contain subsystems A' and B', respectively, which change from stable equilibrium states A'_1 and B'_1 to stable equilibrium states A'_2 and B'_2, all with temperatures almost equal to T_Q. Thus the interaction between subsystems A' and B' is of the same kind as that sketched in Figure 13.10. When viewed as an interaction between systems A and B, however, it is clear that the interaction may be followed by irreversible spontaneous rearrangements of energy and entropy between either A' and other subsystems of A, or B' and other subsystems of B, or both.

13.14   Nonwork Interactions

Processes that involve nonwork interactions (Section 12.2) that are not heat between two systems A and B are illustrated in Figure 13.12. Each process for the composite of A and B is assumed to be reversible, so that the changes in energy and entropy of system A are equal and opposite to the changes in energy and entropy of system B, respectively.

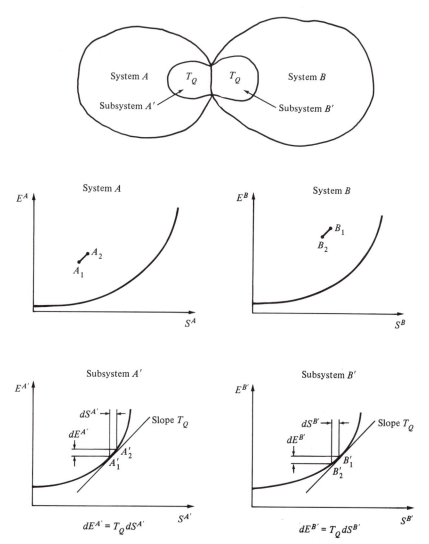

Figure 13.11 Graphical illustration of a heat interaction at temperature T_Q between two systems A and B neither of which is in a stable equilibrium state but such that each contains a subsystem initially at a temperature almost equal to T_Q. The process for the composite of A and B is assumed to be reversible.

As a result of the interaction shown in Figure 13.12a, the energy of system A decreases but its entropy increases as A changes from state A_1 to state A_2 and, correspondingly, the energy of system B increases but its entropy decreases as B changes from state B_1 to state B_2. It is clear that this interaction is not heat in the strict sense of the example in Figure 13.10 because neither system A nor system B passes through stable equilibrium states.

As a result of the interaction shown in Figure 13.12b, the energy and the entropy of system A are both decreased as A changes from nonequilibrium state A_3 to stable equilibrium state A_4, and correspondingly, the energy and entropy of system B are both increased as B changes from stable equilibrium state B_3 to nonequilibrium state B_4. Assuming that the temperatures T_4 and T_3 of stable equilibrium states A_4 and B_3 are not equal, the interaction cannot be heat in the strict sense illustrated in Figure 13.10.

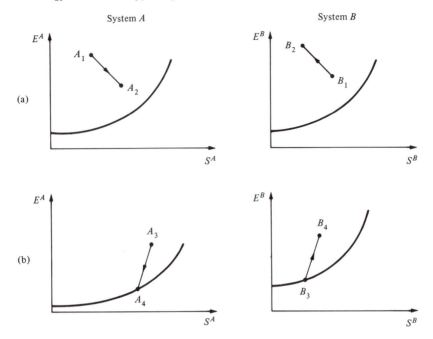

Figure 13.12 Graphical illustration of nonwork interactions that are not heat interactions. The process for the composite of systems A and B is assumed to be reversible.

It is not heat even in the generalized sense represented in Figure 13.11 because even if the exchanges occurred between two subsystems A' and B' passing through stable equilibrium states, the temperatures of these two subsystems are not almost equal to each other. The reason is that the temperature of A' must be equal to T_A, and that of B' to T_B because in state A_4 subsystem A' is in mutual stable equilibrium with the other subsystems of A, and in state B_3 subsystem B' is in mutual stable equilibrium with the other subsystems of B.

All the processes represented in Figure 13.12 can evolve into irreversible processes. For example, in Figure 13.12a, irreversibility can occur in either system A, or system B, or both because the final states of both A and B are not stable equilibrium.

The need for the distinction between heat and other types of nonwork interactions is illustrated by the changes of state shown in Figure 13.13. System A is initially in a stable equilibrium state A_1 at temperature T_Q. As a result of interactions involving no net changes in values of amounts of constituents and volume, A decreases its energy by an amount $\delta E^{A\rightarrow}$. As the graph illustrates, this change in energy is consistent with each of the final states on the line A_2A_3. Except for state A_2, every state on this line corresponds to a transfer of entropy $\delta S^{A\rightarrow}$ different from $\delta E^{A\rightarrow}/T_Q$. Therefore, either we call heat all the interactions that involve an exchange of both energy and entropy, but then we cannot use the relation $\delta E^{A\rightarrow} = \delta Q^{\rightarrow} = T_Q\,\delta S^{A\rightarrow}$ for all these interactions, or we reserve the term heat for interactions for which $\delta Q^{\rightarrow} = T_Q\,\delta S^{\rightarrow}$, and then we need the term nonwork for interactions that involve exchanges of both energy and entropy, and we must realize that heat is only one special kind of nonwork interaction. It is the latter choice that has been made in the present exposition of thermodynamics. In fact, this choice is the only one that allows the identification of the entropy generated by irreversibility as expressed in the entropy balance equations (12.27), (12.29), (12.36), and (12.38).

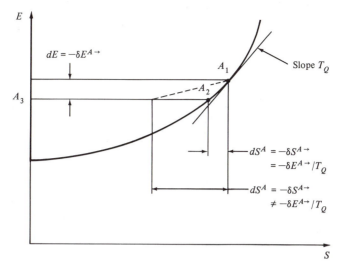

Figure 13.13   Graphical illustration of the need for the distinction between heat and other types of nonwork interactions (see text).

13.15   Optimum Changes in Available Energy

Figure 13.14 is a graphical illustration of the result that the optimum amount of energy that can be exchanged between a weight and a composite of a system A and a reservoir R as A changes from state A_1 to state A_2 equals the change in available energy between these two states (Section 6.6.4).   Available energies are evaluated with respect to the values n and V of the amounts of constituents and the volume, for which system A is in mutual stable equilibrium with reservoir R in state A_0.   We have already shown

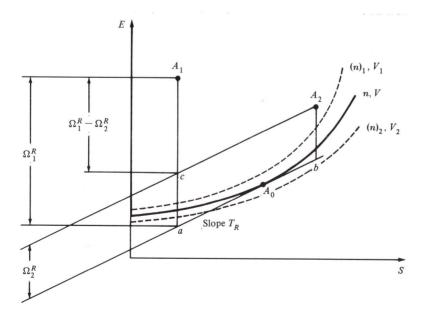

Figure 13.14   Graphical illustration of the optimum amount of energy that can be exchanged between a weight and a system A in combination with a reservoir R as A changes from state A_1 to state A_2.

that the length $A_1a = \Omega_1^R$ and is the largest amount of energy that can be transferred to the weight as system A changes from state A_1 to state A_0 in a weight process for the composite of A and the reservoir R. Similarly, the negative of the length A_2b equals the negative of Ω_2^R and is the smallest amount of energy that must be transferred from the weight to the composite of A and R in order to change system A from state A_0 to state A_2. Accordingly, the difference $\Omega_1^R - \Omega_2^R$ depicted by the length A_1c is the optimum amount of energy exchanged with the weight as A changes from A_1 to A_2, where point c is determined as the intersection of the vertical line A_1ca and the line A_2c that passes through A_2 and has slope equal to T_R, the temperature of the reservoir.

It is noteworthy that states A_1 and A_2 need not have the same values of amounts of constituents and volume. Indeed, Figure 13.14 must be viewed as the superposition of three E versus S diagrams for system A, corresponding, respectively, to the values $(n)_1$, V_1 of state A_1, the values $(n)_2$, V_2 of state A_2, and the values n, V chosen as reference values to evaluate generalized available energies.

The optimum amount of energy exchanged with the weight, $\Omega_1^R - \Omega_2^R$, can be positive, negative, or zero. If it is positive, $\Omega_1^R - \Omega_2^R$ corresponds to the largest work that the composite of A and R can do as a result of an adiabatic process for the composite in which A changes from state A_1 to state A_2. If it is negative, $\Omega_1^R - \Omega_2^R$ corresponds to least work that must be done on the composite of A and R in an adiabatic process for the composite in order to change system A from A_1 to A_2.

13.16 Effects of Irreversibility on the Capacity to Do Work

In Section 13.8 we illustrate the adverse effects of irreversibility in adiabatic processes for system A. Figure 13.15 provides a graphical illustration of the adverse effects of irreversibility on the capacity to do work of a composite of a system A and a reservoir R (Section 6.6.5).

To make ideas specific, we consider a reversible adiabatic process for the composite of A and R in which the state of A changes from state A_1 to state A_3 having energy $E_3 = E_2$. The work done by the composite AR is given by the length A_1d on the vertical line A_1ca. Because state A_3 is not stable equilibrium, it can change spontaneously to

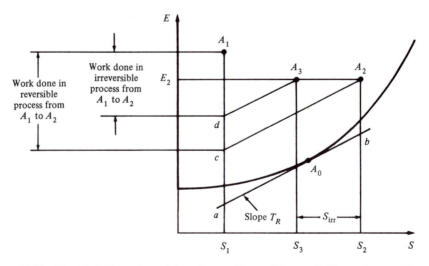

Figure 13.15 Graphical illustration of the adverse effects of irreversibility on the capacity to do work of a composite of a system A and a reservoir R.

state A_2, thus generating an amount of entropy $S_{\text{irr}} = S_2 - S_3$ due to irreversibility. During this spontaneous change of state, no work is done by either A or the composite of A and R, and the energies of both A and R remain fixed. However, if state A_2 is reached as a result of a reversible adiabatic process for the composite of A and R, the work done is given by the length A_1c, which is greater than A_1d. We can readily verify that the lost opportunity for doing work is represented by the difference $cd = A_1c - A_1d$, and that it is equal to $T_R S_{\text{irr}}$. So we confirm again the adverse effects of irreversibility.

13.17   Third Law

For a system with constraints that prevent the internal reaction mechanisms from being capable of altering the values of the amounts of constituents and the parameters, we recall that the second part of the statement of the second law avers that:

> *Starting from any state of a system it is always possible to reach a stable equilibrium state with arbitrarily specified values of amounts of constituents and parameters by means of a reversible weight process*

With this statement of the second law, we conclude the following (Section 9.8).

1. Among all the stable equilibrium states with given values of the amounts of constituents n and the parameters β, the ground-energy (least-energy) stable equilibrium state has the least entropy and the lowest temperature.

2. The value S_g of this least entropy is the same for all values of n and β of a system, and hence it is the same for all ground-energy states of all systems.

3. No other state of the system has entropy lower than S_g.

Because it is common to all the ground-energy states of all systems, we can assign to S_g the value zero, that is,

$$S_g = 0 \qquad \text{for all values of } n \text{ and } \beta \qquad (13.5)$$

It is noteworthy that, in mechanics, each ground-energy state is stable equilibrium, and all mechanical states can be interconnected by means of reversible weight processes. It follows that all the states contemplated in mechanics have the same entropy as that of the ground-energy states, that is, it follows that mechanics is the physics of zero-entropy states.

The second law also implies that the temperature T_g of a ground-energy stable equilibrium state is the least for the given values of n and β, but the value of T_g remains unspecified. To resolve this question without resorting to the formalism of quantum theory, we introduce the third law. It asserts that for each given set of values of the amounts of constituents and the parameters, the ground-energy stable equilibrium state has zero temperature (Section 9.9).

Within the mathematical framework of quantum theory, the third law assertion just stated follows as a theorem of the explicit expression of the stable equilibrium states obtained by applying the highest entropy principle to the explicit expressions of the energy, the amounts of constituents, and the entropy.[1]

---

[1]For example, the stable equilibrium state energy is given by the well-known canonical formula

$$E(n, \beta) = \frac{\text{Tr } H(n, \beta) \exp[-H(n, \beta)/kT]}{\text{Tr } \exp[-H(n, \beta)/kT]}$$

where $H(n, \beta)$ is the Hamiltonian operator of the system. For any set of values n of the amounts and β of the parameters, it can be shown that the ground-energy stable equilibrium state has temperature equal to zero, i.e., $T_g(n, \beta) = 0$. The same follows from the well-known grand-canonical formula.

In addition, the quantum theoretical formalism implies that, in general, the value of the entropy of the ground-energy stable equilibrium state is given by $S_g(n, \beta) = k \ln D_g(n, \beta)$, where $D_g(n, \beta)$ is sometimes called the *degeneracy* of the ground-energy value for the given values of n and β. Alternatively, we say that the ground-energy value for the given n and β is either $D_g(n, \beta)$–*fold degenerate* or simply *degenerate* if $D_g(n, \beta) \neq 1$, or *nondegenerate* if $D_g(n, \beta) = 1$.

With the statement of the second part recalled at the beginning of this section, the second law implies equation (13.5) and therefore requires that for any system every ground-energy value be nondegenerate, that is, that

$$D_g(n, \beta) = 1 \qquad \text{for all values of } n \text{ and } \beta \qquad (13.6)$$

Many authors[2] have argued that (13.6) is too restrictive because we lack conclusive experimental evidence that excludes the existence of systems in nature for which the entropy of the ground-energy stable equilibrium state is nonzero, and a function of n and β. As a result, these authors suggest that, in general, $D_g(n, \beta) \geq 1$.

To account for the existence of such systems, the second part of the statement of the second law should be modified to read:

Starting from any state of a system it is always possible to reach either a stable equilibrium state or a ground-energy state with arbitrarily specified values of amounts of constituents and parameters by means of a reversible weight process.

With this modification, the second law is consistent with the possibility that for a given set of values of amounts of constituents and parameters a system admits more than a single ground-energy state, namely, it is consistent with the possibility that a ground-energy value be degenerate. The third law remains the same and still implies that each ground-energy stable equilibrium state has zero temperature.

Indeed, with the modified statement of the second law, we would conclude that the curved boundary of the projection onto the E versus S plane could take the shape shown in Figure 13.16 rather than the shape shown in Figure 13.1. Specifically, the horizontal line $E_g A_g$ represents the E versus S relation for all the states that are not stable equilibrium but have the ground-state energy E_g, and the curve $A_g A_{E1}$ the E versus S relation for the stable equilibrium states. Each point on the line $E_g A_g$, except A_g, is the projection of many states none of which is stable equilibrium, whereas each point on the curve $A_g A_{E1}$ is a unique stable equilibrium state.

To verify the last assertion, we note that given a set of values of amounts of constituents and parameters for which A_g is the ground-energy stable equilibrium state, a reversible weight process starting from a state A_1 of system A reaches a stable equilibrium state if the entropy of A_1 is greater than or equal to that of A_g; otherwise, it reaches a ground-energy state that is not stable equilibrium. Not being stable equilibrium, none of the states with $E = E_g$ and $0 \leq S < S_g$ can be assigned a temperature.

Whether equation (13.6) is satisfied by all systems or not, that is, whether the modified statement of the second law must be adopted or not, can only be decided by experiments on systems in stable equilibrium states at very low temperatures and with very small values of the amounts of constituents. Indeed, for most systems it can be shown with the tools of quantum theory that the modification of the statement of the

---

[2]For example, H. B. G. Casimir, *Z. Phys.*, *171*, 246 (1963); M. J. Klein, *Rendiconti S.I.F.*, 10*th Course*, 1 (1960); and R. B. Griffiths, in *A Critical Review of Thermodynamics*, E. B. Stuart, B. Gal-Or, and A. J. Brainard, eds., Mono Book Co., Baltimore, 1970.

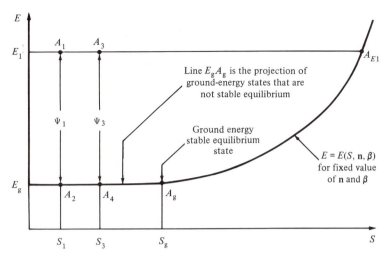

Figure 13.16   Shape of the energy versus entropy diagram consistent with the statement of the second law modified to account for the possibility that for a given set of values of amounts of constituents and parameters a system admits more that a single ground-energy state.

second law has nonnegligible measurable implications only for very-low-temperature and very-few-particle states.

The modified statement of the second law results in an interesting exception to our general understanding of irreversibility. Starting from any state A_1 with entropy $S_1 < S_g$ and using a reversible weight process, we can reach a ground-energy state A_2 that is not stable equilibrium (Figure 13.16). The adiabatic availability Ψ_1 is represented by the length A_1A_2. Similarly, starting from any state A_3 with energy $E_3 = E_1$ and entropy $S_1 < S_3 < S_g$, and using a reversible weight process, we can reach a state A_4 that is not stable equilibrium. Now the adiabatic availability $\Psi_3 = \Psi_1$ and is represented by the length A_3A_4. But state A_1 can evolve spontaneously into A_3, and then the increase in entropy $S_3 - S_1$ is generated by irreversibility. We conclude that, for states with entropy between zero and S_g, irreversibility does not affect the values of the adiabatic availability, a conclusion that is an exception to our general understanding of the adverse effects of irreversibility.

13.18   Negative Temperatures

Many of our conclusions, including the positivity of temperature (Section 9.3), are based on the impossibility of a perpetual-motion machine of the second kind (PMM2). The proof that a PMM2 cannot be built is predicated on the assertion that, starting from any state, the energy of a system can be increased indefinitely by means of a weight process, which results in altering the initial state to some final state that is not a stable equilibrium state without altering the values of the amounts of constituents and the parameters (Section 4.5).

However, a special class of systems exists, such as nuclear spin systems and atomic laser systems, in which energy cannot be increased indefinitely by means of weight processes that do not change the values of the amounts of constituents and the parameters. For a system in this class, the energy of some stable equilibrium states can be decreased but cannot be increased at all by means of a weight process at fixed amounts of constituents and parameters. For such systems the statement of the impossibility of a PMM2 must be modified as follows. Given a stable equilibrium state A_0 and a weight process that starts from A_0 and ends in any state A_1 different from A_0 but with the same values of amounts of constituents and parameters, the work done in the process, $W_{01}^{A\rightarrow}$, must satisfy either the relation

$$W_{01}^{A\rightarrow} \leqq 0 \tag{13.7}$$

or the relation

$$W_{01}^{A\rightarrow} \geqq 0 \tag{13.8}$$

If relation (13.7) is satisfied, we say that state A_0 is a *normal stable equilibrium state*. If, instead, relation (13.8) is satisfied, we say that state A_0 is a *special stable equilibrium state*.

Systems with particles that are free to move from one place to another—systems with translational degrees of freedom—have no upper bound to the energy and therefore have normal stable equilibrium states only. They constitute the vast majority of systems considered in engineering applications.

Systems with an upper bound to the energy, instead, have both normal and special stable equilibrium states. Although we do not discuss such systems in the preceding chapters, the following conclusions can be verified. Among all the states that have given values of the entropy, the amounts of constituents, and the parameters, systems with an upper bound to the energy admit two stable equilibrium states, one with lowest energy and one with highest energy. The state with lowest energy is a normal stable equilibrium state and has positive temperature. The state with highest energy is a special stable equilibrium state and has negative temperature. For given values of the amounts of constituents and the parameters, the energy versus entropy diagram has the form shown in Figure 13.17. It is clear that both types of states are consistent with the second law of thermodynamics because to each set of values of the energy, the amounts of constituents, and the parameters there corresponds one and only one stable equilibrium state.

It is noteworthy that energy can flow spontaneously from a system at negative temperature to a system at positive temperature, but the converse is impossible. To see this clearly, we consider two systems A and B that are initially in stable equilibrium states,

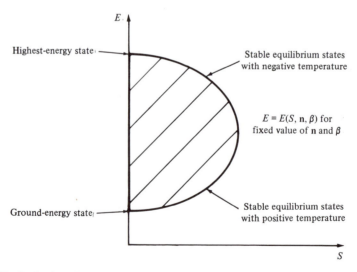

Figure 13.17 Projection of the states of a system with an upper bound to the energy on an energy versus entropy plane.

one special at negative temperature T_0^A and the other normal at positive temperature T_0^B. We let the two systems interact with each other so that no effects are left in the environment of the composite of A and B, and the values of the amounts of constituents and the parameters of both A and B remain fixed (Figure 13.18). Following exactly the same procedure as in Section 9.6, we conclude that

$$\left(\frac{1}{T_0^A} - \frac{1}{T_0^B}\right) DE^A \gtreqless 0 \tag{13.9}$$

Because T_0^A is negative and T_0^B is positive, relation (13.9) is satisfied only if $DE^A < 0$, that is, only if energy flows from A to B.

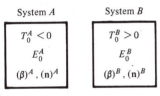

Initial states

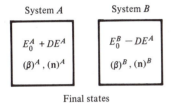

Final states

Figure 13.18 Schematics of end states of an interaction between a system A initially in a special stable equilibrium state with $T_0^A < 0$ and a system B initially in a normal stable equilibrium state with $T_0^B > 0$. Energy can flow spontaneously only from A to B.

Problems

13.1 Consider a weight process that changes the state of a system A from A_1 to A_2 without net changes in values of amounts of constituents and parameters, and the energy versus entropy diagram for A.

(a) Show graphically that the adiabatic availabilities are such that

$$(W_{12}^{A\rightarrow})_{\text{rev}} = \Psi_1 - \Psi_2$$

if the process is reversible, and

$$(W_{12}^{A\rightarrow})_{\text{irr}} < \Psi_2 - \Psi_1$$

if the process is irreversible.

(b) Show graphically the analogous results that hold for the generalized adiabatic availabilities.

13.2 Consider a weight process that changes the state of a system A from A_1 to A_2 without net changes in values of amounts of constituents and parameters, and the energy versus entropy diagram for A.

(a) Show graphically that the process is possible only if the adiabatic availabilities and energies satisfy the relation

$$\Psi_1 - \Psi_2 \gtreqless E_1 - E_2$$

(b) Show graphically the analogous result that holds for the generalized adiabatic availabilities.

13.3 Consider the energy versus entropy diagrams of a system.

(a) Show graphically that adiabatic availability is not an additive property.

(b) Show graphically that available energy is additive.

13.4 Consider a weight process that changes the state of a system A from A_1 to A_2 without net changes in values of amounts of constituents and parameters, a reservoir R and the energy versus entropy diagram of A.

(a) Show graphically that the available energies are such that

$$(W_{12}^{A \rightarrow})_{\text{rev}} = \Omega_1^R - \Omega_2^R$$

if the process is reversible, and

$$(W_{12}^{A \rightarrow})_{\text{irr}} < \Omega_1^R - \Omega_2^R$$

if the process is irreversible.

(b) Show graphically the analogous results that hold for the generalized available energies.

13.5 Consider a weight process that changes the state of a system A from A_1 to A_2 without net changes in values of amounts of constituents and parameters, a reservoir R, and the energy versus entropy diagram of A.

(a) Show graphically that the process is possible only if the available energies and energies satisfy the relation

$$\Omega_1^R - \Omega_2^R \geq E_1 - E_2$$

(b) Show graphically the analogous result that holds for the generalized available energies.

13.6 Consider the ratio $(DE_{12}^R)_{\text{rev}}^A / (DE_{12}^Q)_{\text{rev}}^A$ defined in Section 7.4. On an energy versus entropy diagram for a system A show graphically that the ratio remains invariant if states A_1 and A_2 are replaced by any other two states of A.

13.7 Consider a three-dimensional space with energy, entropy, and volume as axes.

(a) On suitably chosen planes in this space, show graphically the temperature and pressure of a stable equilibrium state. Comment briefly on the fact that these two properties are defined only at the stable equilibrium states and have no meaning whatsoever at any other state.

(b) What space would you use to represent graphically the total potential of a constituent?

13.8 Give a graphical representation of the discussion in Section 9.12, in general, and of the differential relation $d\Psi = dE - T_S dS$, in particular.

13.9 A system A consists of a fixed amount of oxygen molecules confined in a cylinder with variable volume V. The volume of the cylinder cannot exceed $1.4 V_0$, where V_0 is a known reference value. For stable equilibrium states with energy larger than a known reference value E_0, the relation between entropy S, energy E, and volume V is

$$S = S_0 + C \ln \left(1 + \frac{E - E_0}{CT_0} \right) + R \ln \frac{V}{V_0}$$

where the constant $C = 3$ J/K, $T_0 = 250$ K, $R = 8$ J/K, and S_0 is a known reference constant.

(a) Make a graph of energy E versus entropy S of the stable equilibrium states of this system in the given range of values of energy, and for three values of the volume: V_0, $1.2 V_0$, and $1.4 V_0$. Go up to $E - E_0 \approx 10 CT_0$.

(b) For $V = V_0$ and the given range of values of energy, show the allowed states of the system on your graph.

(c) Find an expression for the temperature as a function of energy and volume.

(d) Find an expression for the temperature as a function of entropy and volume.

(e) Are there any allowed states of the system that on your graph would be represented by points below the stable-equilibrium-state curve corresponding to $V = 1.2\,V_o$? Are there any below the curve corresponding to $V = 1.4\,V_o$?

(f) Indicate on a separate E versus S diagram all the allowed states of a reservoir at $T_R = 300$ K.

(g) For a state A_1 of system A with energy $E_1 = E_o + 2.5$ kJ, entropy $S_1 = S_o + 3.67$ J/K and volume $V_1 = V_o$, find the adiabatic availability Ψ_1 and the available energy Ω_1^R with respect to a reservoir at $T_R = 300$ K. What would be the values of Ψ_1 and Ω_1^R if the volume $V_1 = 1.4\,V_o$?

(h) Consider states A_2 and A_3 such that $T_2 = 300$ K and $V_2 = 1.4\,V_o$, and $T_3 = 400$ K and $V_3 = V_o$. Use entropy considerations to determine whether a weight process from state A_2 to state A_3 is possible. If your answer is yes, is the process reversible or irreversible?

(i) Consider stable equilibrium states A_4 and A_5 with entropies $S_4 = S_5 = S_o + 2$ J/K and volumes $V_4 = V_o$ and $V_5 = 1.4\,V_o$. Is a weight process from A_4 to A_5 possible? If yes, evaluate the work done by the gas. Is this a violation of the impossibility of perpetual-motion machine of the second kind?

13.10 A given amount of a fuel–air mixture is confined in a fixed volume V. For the products of combustion of this mixture, the energy E and entropy S of the stable equilibrium states are given by the relations $E = 5x$ and $S - S_o = 5\ln(x/x_o)$, respectively, where E is in kJ, S in kJ/K, x is a parameter measured in kelvin, $x_o = 280$ K, and S_o is a constant.

Prior to combustion the mixture is in a state A_1 with energy E_1 and entropy S_1 such that $E_1 = 10,000$ kJ and $S_1 - S_o = 1$ kJ/K. Starting from state A_1, 1180 kJ of work and 1320 kJ of heat from a source at $57°$ C are supplied to the fuel–air mixture. Thus the system is found in state A_2. A reservoir is available at 300 K.

(a) Express the temperature of the products of combustion as a function of the parameter x.

(b) Make a rough sketch of the E versus S relation for the stable equilibrium states of the products of combustion between $S = S_o$ and $S = S_o + 15$ kJ/K. Three or four pairs E, S are adequate to define the curve.

(c) Show the locus of the possible states A_2 on your graph in part (b), and calculate the largest adiabatic availability that state A_2 could have.

(d) If upon spontaneous combustion starting from state A_2, the products of combustion are found in a stable equilibrium state, what is their available energy with respect to the given reservoir?

(e) What is the change of energy of the products of combustion per unit temperature change at constant volume?

13.11 Consider the property $P^R = \Omega^R - \Psi$ defined in Problem 6.5. Show graphically the validity of the statement of part (c) of Problem 6.5.

14   Summary of Basic Concepts

Because of its breadth and depth, the exposition of thermodynamics requires unambiguous consideration of many basic concepts. Some of these concepts, such as space, time, reference frame, velocity, acceleration, mass, force, kinetic energy, and potential energy are known from introductory courses in physics and need not be reexamined here. On the other hand, other concepts, such as system, property, state, process, energy, and entropy are sometimes not clearly defined and need special emphasis.

In this chapter we provide a brief summary of these concepts, the laws of thermodynamics, and their implications. Definitions are summarized and results are stated without proofs. More elaborate discussions of the definitions and proofs of the results are presented in Chapters 2 to 13. Because it skips the discussion of many subtleties of definitions, concepts, and results, this chapter is particularly suited for a first exposure to thermodynamics. It provides an introduction to the science of thermodynamics for readers who are concerned primarily with applications and only to a lesser degree with the foundations of the subject. It does so, however, without compromises, circular arguments, and artificial and unnecessary restrictions to special phenomena.

14.1   Systems, Properties, and States

A well-defined *system* is a collection of constituents determined by the following specifications.

1. The type and the range of values of the *amount* of each *constituent*, for example, one water molecule or between 5 and 10 kg of atmospheric air.

2. The type and the range of values of the *parameters* that fully characterize the *external forces* exerted on the constituents by bodies other than the constituents, such as the parameters that describe the size and geometrical shape of an airtight container, and an applied electrostatic field.

3. The *internal forces* between constituents, such as the forces between two water molecules and the forces that separate constituents in one region of space from constituents in another region, such as a partition of the spatial extension of the system into two parts.

4. The *internal constraints* that characterize the interconnections between separated parts, such as the condition that the overall volume of two variable-volume parts be fixed, and that define the modeling assumptions, such as the condition that some or all chemical reactions are inactive.

Everything that is not included in the system is called the *environment* or the *surroundings* of the system.

For a system consisting of r different types of constituents, we denote their amounts by the vector $n = \{n_1, n_2, \ldots, n_r\}$, where n_1 stands for the amount of the first type of constituent, n_2 for the amount of the second, and so on. For example, the different types of constituents could be three specific molecules, such as the H_2, O_2, and H_2O molecules, with amounts denoted, respectively, by n_1, n_2, and n_3; two specific atoms, such as the H and O atoms, with amounts denoted, respectively, by n_1 and n_2; four specific ions, such as the H^+, O^-, H_3O^+, and OH^- ions, with amounts denoted, respectively, by n_1, n_2, n_3, and n_4; three specific elementary particles, such as the electron, proton, and neutron particles, with amounts denoted by n_1, n_2, and n_3; or a single specific field, such as the electromagnetic radiation field, with amount denoted by n and equal to unity, $n = 1$. The value of each amount of constituent may be variable over a range. With or without an additional subscript, we use the same symbols $n = \{n_1, n_2, \ldots, n_r\}$ to denote the values of the amounts of constituents within their respective ranges.

It is clear that for each set of different types of constituents there may be different arrangements of internal forces between constituents. For example, if only H_2O molecules are considered, the only internal force is that between the H_2O molecules. Again, if H_2, O_2, and H_2O molecules are considered, and the chemical reaction $H_2 + (1/2) O_2 = H_2O$ occurs, the intermolecular forces between all types of molecules must be specified as well as the forces that control the chemical reaction.

For a system with external forces described by s parameters, we denote the parameters by the vector $\beta = \{\beta_1, \beta_2, \ldots, \beta_s\}$, where β_1 stands for the first parameter, β_2 for the second, and so on. For example, one of the parameters could be the side L or the volume V of a three-dimensional cubic container which encloses the constituents that belong to the system and separates them from all the others that do not and are outside the enclosure. Another parameter could be the potential γ of a uniform gravitational field in which the constituents are immersed, the potential of an electrostatic field in which the constituents are floating, or the area of a two-dimensional surface in space on which the constituents are constrained. The value of each parameter may be variable over a range. With or without an additional subscript, we use the same symbols $\beta = \{\beta_1, \beta_2, \ldots, \beta_s\}$ to denote the values of the parameters within their respective ranges.

Two *systems are identical* if they consist of the same types of constituents, experience the same internal and external forces, and have the same ranges of values of amounts of constituents and parameters, and the same constraints. If any of these identities is not valid, the two *systems are different*.

At any instant in time, the amount of each type of constituent and the parameters of each external force have specific values within the corresponding ranges of the system. By themselves, these values do not suffice to characterize completely the condition of the system at that time. We also need the values of all the properties at the same instant in time. Each *property* is an attribute that can be evaluated at any given instant of time by means of a set of measurements and operations that are performed on the system and result in a numerical value — the *value of the property*. This value is independent of the measuring devices, other systems in the environment, and other instants in time. For example, the instantaneous position of each molecule of a constituent is a property of a system. However, the distance covered by a molecule over a finite period of time is not a property because the measurement of such a distance depends necessarily on two different instants of time.

Properties are denoted by symbols, such as E for energy, S for entropy, Ψ for adiabatic availability, and Ω^R for available energy, and their values are denoted by the same symbols with or without subscripts. The meanings of the terms energy, entropy, adiabatic availability, and available energy are given later.

Two properties are *independent* if the value of one can be varied without affecting the value of the other. Otherwise, the two properties are *interdependent*. For example, position and velocity of a molecule are independent properties, whereas speed and kinetic energy of a molecule are interdependent properties.

For a given system, the values of the amounts of all the constituents, the values of all the parameters, and the values of a complete set of independent properties encompass all that can be said about the system at an instant in time and about the results of any measurements or observations that may be performed on the system at that same instant in time. As such, the collection of all these values constitutes a complete characterization of the system at that instant in time. We call this characterization at an instant in time the *state* of the system. Because fixing the state of a system implies fixing the values of all its properties, it is correct to say that the value of each property depends exclusively on the state of the system. But this is an implication of the definition of state, not the definition of a property.

For a given system, two *states are identical* only when the values of the amounts of all the constituents, the values of all the parameters, and the values of all the properties for one state are identical to the respective values for the other state. Otherwise, the two *states are different*.

Two states of a system are said to be *compatible* if the corresponding sets of values of amounts of constituents and parameters are such that a change from one set to the other can occur as a result of the allowed internal mechanisms of the system, such as chemical reactions, interconnections, and internal forces.

Until Chapter 29, we restrict our discussion to systems with internal mechanisms that are incapable of changing the values of amounts of constituents and parameters, so that two states are compatible only if they have the same values of all the amounts of constituents and all the parameters.

14.2 Changes of State with Time

The state of a system may change with time either spontaneously due to the internal dynamics of the system or as a result of interactions with other systems, or both. For example, a charged battery wrapped in fiberglass may discharge internally over a period of time without any effects on systems in the environment. Again, the same battery discharges over a period of time if it powers an electric train — a system in the environment.

A system that can experience only spontaneous changes of state, that is, a system that cannot induce any effects on the state of the environment, is called *isolated*. Systems that are not isolated can interact with each other in a number of different ways, some of which may result in net flows of properties from one system to another. For example, an interaction by means of elastic collisions results in the flow or transfer of momentum from one system to the other.

The relation that describes the evolution of the state of a system as a function of time is the *equation of motion*. The discovery of the complete equation of motion — the equation of motion that describes all physical phenomena contemplated by thermodynamics — remains a subject of research at the frontier of science — one of the most intriguing and challenging problems in physics. Among the many features of the complete equation of motion that have already been discovered, the most general and well established are captured by the statements of the laws of thermodynamics — the subject of our study.

Rather than through the explicit time dependence, which requires the complete equation of motion, we describe a change of state in terms of the *end states*, that is, the initial and final states of the system, the *modes of interaction* that are active during the change of state, and conditions on the values of properties of the end states that are consequences of the laws of thermodynamics, that is, conditions that express not all, but the most general and well-established features of the complete equation of motion. Each mode of interaction is characterized by means of well-specified net flows of properties across the boundaries of the interacting systems. For example, after defining the properties energy and entropy, we will see that some modes of interaction involve the flow of energy across the boundaries of the interacting systems without any flow of entropy, whereas other modes of interaction involve the flow of both energy and entropy. Among the conditions on the values of properties of the end states that are consequences of the laws of thermodynamics — conditions that express well-established features of time-dependent behavior of systems — we will see that the energy change of a system must equal the energy transferred into the system, and that its entropy change must be greater than or at most equal to the entropy transferred into the system.

The end states and the modes of interactions associated with a change of state of a system specify a *process*. The modes of interactions may be used to classify processes into different types. For example, a process that involves no interactions and, therefore, no flows across the boundary of the system is called a *spontaneous process*. Again, a process that involves interactions that result in no external effects other than the change in elevation of a weight (or an equivalent mechanical effect) is called a *weight process*.

Another important classification of processes is in terms of the possibility of annulling all their effects. A process may be either reversible or irreversible. A process is *reversible* if it can be performed in at least one way such that both the system and its environment can be restored to their respective initial states. A process is *irreversible* if it is impossible to perform it in such a way that both the system and its environment can be restored to their respective initial states. For example, any electron of the current in a superconducting ring returns to an arbitrarily specified initial position without any effects whatsoever on systems in the environment of the superconductor. So, the process of the electron going from one position to another is reversible. On the other hand, the products of burning methane in the air of a perfectly insulated burner cannot be reformed into methane and air without a definite and permanent effect on systems in the environment of the burner. So, the process of burning is irreversible.

We will see that any irreversible process involves the irrecoverable degradation of a valuable resource, whereas no reversible process involves such a degradation. For this reason, putting aside all economic, social, and environmental considerations, we sometimes say that a reversible process is the "best possible." For example, for a given change of energy of a system, an irreversible weight process results in either a smaller raise or a larger drop of the weight than the corresponding results of a reversible process.

In general, a system A that undergoes a process from state A_1 at time t_1 to state A_2 at time t_2 is well defined at these two times but is not necessarily well defined during the lapse of time between t_1 and t_2. The reason is that the interactions which induce the change of state may involve such temporary alterations of internal and external forces that no system A can be defined during the period t_1 to t_2. Said more formally, in the course of interactions the constituents of a system may not be separable from the environment or, if they are, the states of the system may be correlated with the states of other systems. Nevertheless, at the end of the process the system becomes again well-defined, and its state is again uncorrelated.[1]

14.3   Energy and Energy Balance

Energy is a concept that underlies our understanding of all physical phenomena, yet its meaning is subtle and difficult to grasp. It emerges from a fundamental principle known as the first law of thermodynamics.

The *first law* asserts that *any two states of a system may always be interconnected by means of a weight process and, for a given weight subject to a constant gravitational acceleration, that the change in elevation during such a process is fixed uniquely by the two states of the system.*

The main consequence of this law is that every system A in any state A_1 has a property called *energy*, and denoted by the symbol E_1. The energy E_1 of any state A_1 can be evaluated by means of an auxiliary weight process that interconnects state A_1 and a reference state A_0 to which is assigned a fixed reference value E_0, and the expression

$$E_1 - E_0 = -Mg\,(z_1 - z_0) \tag{14.1}$$

where M is the mass of the weight, g the gravitational constant, and z the elevation of the weight. The energy E_2 of another state A_2 can be evaluated by a similar procedure so that

$$E_2 - E_0 = -Mg\,(z_2 - z_0) \tag{14.2}$$

Moreover, subtracting (14.1) from (14.2) we find

$$E_2 - E_1 = -Mg\,(z_2 - z_1) \tag{14.3}$$

where we keep the negative sign in front of $Mg\,(z_2 - z_1)$ in order to emphasize that the larger the value of z_2 the smaller the value of E_2, and conversely.

The unit of energy in the International System of units (SI) is the joule, denoted by J. A list of various other units and conversion factors between them is given in Appendix A.

Energy is an *additive* property, namely, the energy of a system consisting of two or more subsystems — a composite system — equals the sum of the energies of the subsystems and this holds for all combinations of states of the subsystems. Moreover, energy has the same value at the final time as at the initial time whenever the system experiences a zero-net-effect weight process, or remains invariant in time whenever the process is spontaneous. In either process, $z_2 = z_1$ and $E(t_2) = E(t_1)$ for time t_2 greater than t_1, that is, energy is *conserved*.

Because of additivity, and because any process of a system can always be thought of as part of a zero-net-effect weight process of a composite system consisting of all the interacting systems, the conclusion that as time proceeds energy is invariant — it can neither be created nor destroyed — is of great generality and practical importance. It is known as the *principle of energy conservation*.

Energy can be transferred between systems by means of interactions. Denoting by $E^{A\leftarrow}$ the net amount of energy transferred from the environment to system A as a result

---

[1]We say that a system is well defined and its constituents are *separable from the environment* if the forces exerted on the constituents by a body not included in the system do not depend explicitly on the coordinates of constituents of that body. We say that a state is *uncorrelated* from the state of the environment if none of the values of the properties of the system depends on the values of properties of systems in the environment. All statements and conclusions in this book refer to well-defined systems in uncorrelated states.

of all the interactions involved in a process that changes the state of A from A_1 to A_2, we derive an extremely important analytical tool, the *energy balance equation* or, simply, the *energy balance*. This equation is based on the additivity of energy and on the principle of energy conservation. It requires that, as a result of a process, the change in the energy of the system from E_1 to E_2 must be equal to the net amount of energy $E^{A\leftarrow}$ transferred into the system, namely,

$$E_2 - E_1 = E^{A\leftarrow} \tag{14.4}$$

Said differently, the energy gained by the system must be equal to the energy transferred into the system from the environment.

Among all the states of a system with given values of amounts of constituents and parameters, the states of least energy are called the *ground-energy states* and their energy the *ground-state energy*. We denote it by E_g.

According to special relativity, the ground-state energy E_g is related to the *ground-state mass* m_g of the system by the relation $E_g = m_g c^2$, where c is the speed of light in vacuum.

For a system consisting of the smallest amount of a free constituent — a single particle, atom, molecule, or field unaffected by any external forces, such as a free hydrogen atom — the ground-state mass is a characteristic constant of the constituent that can be measured by mass-spectroscopic techniques. We call it the *ground-state mass of the free constituent* and denote it by m_g^{free}. The corresponding energy, $E_g^{\text{free}} = m_g^{\text{free}} c^2$, is called the *ground-state energy of the free constituent*. For example for a free hydrogen atom $m_g^{\text{free}} = 1.6735 \times 10^{-27}$ kg and $E_g^{\text{free}} = m_g^{\text{free}} c^2 = 938.74$ MeV.

For a given system, using as a reference state that which corresponds to the ground-state of all the constituents regarded as free of all external and internal forces and constraints, with energy equal to the sum of the ground-state energies of the free constituents, we find that energy values defined by the first law conform to the requirements of special relativity, and are always nonnegative. They are called *absolute energy values*.

Another implication of special relativity is that associated with each state A_1 of a system A is a *mass* m_1. In general, this mass differs from the sum of the ground-state masses of the constituents of the system in that state considered as free. However, for most practical applications that do not involve nuclear reactions, creation and annihilation reactions, and speeds close to that of light, the difference is so small that we can consider the mass of the system as fixed and equal to the sum of the ground-state masses of the constituents considered as free. In these applications, the mass is approximately additive and conserved. So, if there is exchange of constituents with other systems, we must use a *mass balance equation* or, simply, *mass balance* of the form

$$m_2 - m_1 = m^{A\leftarrow} \tag{14.5}$$

where m_1 and m_2 are the masses of states A_1 and A_2, respectively, and $m^{A\leftarrow}$ is the mass flow into system A from other systems in the environment. In the mass balance we do not specify whether m refers to the mass or the sum of the ground-state masses of the constituents considered as free because the two values are approximately equal when the mass balance equation holds.

14.4   Types of States

Because the number of independent properties of a system is infinite even for a system consisting of a single particle with a single translational degree of freedom — a single variable that fixes the configuration of the system in space — and because most properties can vary over a range of values, the number of possible states of a system is infinite.

To facilitate the discussion of these states, we classify them into different categories according to their time evolutions, that is, according to the way each state changes as a function of time. This classification brings forth many important aspects of physics, and provides a readily understandable motivation for the introduction of the second law of thermodynamics. We consider four types of states: unsteady, steady, nonequilibrium, and equilibrium. Moreover, we further classify equilibrium states into three types: unstable, metastable, and stable.

An *unsteady state* is one that changes as a function of time because of interactions of the system with other systems. For example, a battery connected to a load is in an unsteady state. A *steady state* is one that does not change as a function of time, despite interactions of the system with other systems in the environment. A *nonequilibrium state* is one that changes spontaneously as a function of time, that is, a state that evolves as time goes on without any effects on or interactions with any other systems. For example, an ignited amount of gasoline in the air of an airtight and isolated combustion chamber is in a nonequilibrium state. An *equilibrium state* is one that does not change as a function of time while the system is isolated — a state that does not change spontaneously. For example, a disconnected battery without internal discharging is in an equilibrium state. An *unstable equilibrium state* is an equilibrium state that may be caused to proceed spontaneously to a sequence of entirely different states by means of a minute and short-lived interaction that has only an infinitesimal temporary effect on the state of the environment. A *metastable equilibrium state* is an equilibrium state that may be changed to an entirely different but compatible state without leaving net effects in the environment of the system, but this can be done only by means of interactions that have a finite temporary effect on the state of the environment. A *stable equilibrium state* is an equilibrium state that can be altered to a different but compatible state only by interactions that leave net effects in the environment of the system.

Starting either from a nonequilibrium state or from an equilibrium state that is not stable, a system can be made to raise a weight without leaving any other net changes in the state of the environment. We find examples of these conclusions in our everyday experiences with gasoline-powered engines and battery-driven equipment. In contrast, experience shows that from some other types of states — they turn out to be stable equilibrium states — such a raise of a weight is impossible. This impossibility is one of the most striking consequences of the first and second laws of thermodynamics.

14.5 Adiabatic Availability

The existence of stable equilibrium states is not self-evident. It is the essence of the second law. In the absence of internal mechanisms, such as chemical reactions or internal interconnections, capable of causing spontaneous changes in the values of the amounts of constituents and the parameters, the *second law* asserts that *among all the states of a system with given values of the energy, the amounts of constituents and the parameters, there exists one and only one stable equilibrium state. Moreover, starting from any state of a system it is always possible to reach a stable equilibrium state with arbitrarily specified values of amounts of constituents and parameters by means of a reversible weight process.*

The existence of stable equilibrium states for various conditions of matter has many theoretical and practical consequences. One consequence is that, starting from any stable equilibrium state of any system, no energy can be transferred to a weight in a weight process in which the values of amounts of constituents and parameters of the system experience no net changes. This consequence is often referred to as the *impossibility of a perpetual-motion machine of the second kind* (PMM2). In some expositions of thermodynamics, it is taken as the statement of the second law. In this book, it is only one aspect of the second law.

Another consequence of the existence of stable equilibrium states is that not all states of a system can be changed to a ground-energy state by means of a weight process and so, in general, not all the energy above the ground-state energy can be transferred to a weight in a weight process. This consequence is a generalization of the impossibility of a PMM2.

Upon investigating carefully the fundamental question "how much of the energy of a system can be transferred to a weight in a weight process?" we find that every system A in any state has a property called *adiabatic availability* and is denoted by the symbol Ψ. The value Ψ_1 at state A_1 equals the energy transferred to a weight in the course of a reversible weight process that interconnects state A_1 and a stable equilibrium state A_{S1} with the same values of amounts of constituents and parameters as A_1. Because it is defined in terms of a reversible process, Ψ_1 is the largest energy that can be transferred to the weight in a weight process starting from state A_1.

The value of Ψ_1 ranges from zero corresponding to A_1 being a stable equilibrium state (impossibility of a PMM2) to a value equal to the difference between the energy E_1 of A_1 and the ground-state energy E_{g1} corresponding to the values of amounts of constituents and parameters of A_1.

The investigation of the energy transfer just cited also reveals the existence of another property called *generalized adiabatic availability* which is determined in the same manner as the adiabatic availability except that the values of the amounts of constituents and the parameters of the final stable equilibrium state are assigned arbitrarily and differ in general from those of state A_1.

14.6   Available Energy

Because neither the adiabatic availability nor the generalized adiabatic availability are additive properties, their practical usefulness is limited. We gain additivity by considering the adiabatic availability of a composite system consisting of a system A and reservoir R. A *reservoir* is an idealized kind of system with a behavior that approaches the following three limiting conditions.

1. It passes through stable equilibrium states only.
2. In the course of finite changes of state, it remains in mutual stable equilibrium with a duplicate of itself that experiences no such changes.
3. At constant values of amounts of constituents and parameters of each of two reservoirs initially in mutual stable equilibrium, energy can be transferred reversibly from one reservoir to the other with no net effects on any other system.

Two systems are in *mutual stable equilibrium* if their composite system is in a stable equilibrium state.

Given a reservoir R with fixed values of amounts of constituents and parameters, the adiabatic availability of the composite of a system A and reservoir R is an additive property of system A called *available energy with respect to reservoir R*, and denoted by the symbol Ω^R.

The available energy Ω_1^R — with respect to reservoir R — of a system A in any state A_1 is the largest amount of energy that can be transferred to a weight in a weight process for the composite of system A and the reservoir R without net changes in the values of the amounts of constituents and the parameters of the system and the reservoir. When

Ω_1^R is transferred out, systems A and R are left in mutual stable equilibrium, that is, their composite system is in a stable equilibrium state.

The first scientist who raised the question about the largest amount of energy that can be transferred to a weight in a weight process for the composite of a system A and a reservoir R was Carnot. He restricted his investigation, however, to A also being a reservoir. His results constitute the seminal ideas — the conception event — of the science of thermodynamics. The disclosure of the available energy Ω_1^R as a property is a generalization of the results of Carnot in that system A need not be a reservoir, and state A_1 need not be a stable equilibrium state. Available energy can be assigned to any system in any state.

Another property, the *generalized available energy*, may also be defined as a property of a system A in any state A_1. Its definition is identical to that of available energy except that the final state of system A corresponds to arbitrarily assigned values of the amounts of constituents and parameters, that differ in general from those of state A_1. The generalized available energy of a state A_1 is defined with respect to a reservoir R and the arbitrarily assigned values of the amounts of constituents and parameters. For simplicity, we denote it by the same symbol Ω_1^R as the available energy. We distinguish it from the available energy of state A_1 with respect to reservoir R by name and context.

The difference between the generalized available energies, $\Omega_1^R - \Omega_2^R$, of two states A_1 and A_2 is equal to the energy that can be exchanged with a weight in a reversible weight process of the composite AR of system A and reservoir R, as system A goes from state A_1 to state A_2. Upon denoting the energy exchanged with the weight by $(W_{12}^{AR\rightarrow})_{\text{rev}}$, we have

$$(W_{12}^{AR\rightarrow})_{\text{rev}} = \Omega_1^R - \Omega_2^R \tag{14.6}$$

The value of $(W_{12}^{AR\rightarrow})_{\text{rev}}$ is positive when energy is transferred from the composite AR to the weight, and then it is the largest energy transfer to the weight that can be achieved as system A goes from state A_1 to state A_2. It is negative when energy is transferred from the weight to the composite AR, and then it is the least energy transfer that is required to achieve the change of A from state A_1 to state A_2.

Two important relations exist between the energies E_1 and E_2, and the generalized available energies Ω_1^R and Ω_2^R of any two given states A_1 and A_2 of a system A. By virtue of the first law, the two states can always be interconnected by means of a weight process for system A alone. But the first law determines neither the direction of the weight process, nor its reversibility. By contrast, a comparison between the difference in energies and the difference in generalized available energies of the two states determines both the direction and the reversibility of the process. Specifically, if

$$\Omega_1^R - \Omega_2^R = E_1 - E_2 \tag{14.7}$$

then a weight process for A alone is possible both from A_1 to A_2 and from A_2 to A_1, and is reversible. However, if

$$\Omega_1^R - \Omega_2^R > E_1 - E_2 \tag{14.8}$$

then a weight process for A alone is possible only from A_1 to A_2, and is irreversible.

For spontaneous or zero-net-effect weight processes, energy conservation implies $E_2 = E_1$ or, emphasizing the time dependence, $E(t_2) = E(t_1)$ for $t_2 > t_1$. When applied to these processes, equation (14.7) and relation (14.8) reveal the following results. If the process is reversible then $\Omega_2^R = \Omega_1^R$ or, emphasizing the time dependence, $\Omega^R(t_2) = \Omega^R(t_1)$ for $t_2 > t_1$, namely, the generalized available energy is conserved. If the spontaneous or zero-net-effect weight process is irreversible then $\Omega_2^R < \Omega_1^R$ or $\Omega^R(t_2) < \Omega^R(t_1)$ for $t_2 > t_1$, namely, the generalized available energy is not conserved.

Said differently, in the course of an irreversible, zero-net-effect weight process a system loses some of its potential ability to transfer energy to a weight. Whereas energy is conserved, the amount of energy that can be transferred to a weight in a weight process — the potential of a system to perform useful tasks — is not conserved. This potential cannot be created but may be dissipated to a lesser or larger degree depending on whether the process is a little or a lot irreversible. A quantitative measure of irreversibility can be expressed in terms of the property entropy discussed in the next section.

A noteworthy feature of energy, adiabatic availability, generalized adiabatic availability, available energy, and generalized available energy is that each of these properties is defined for any state of any system, regardless of whether the state is steady, unsteady, equilibrium, nonequilibrium, or stable equilibrium, and regardless of whether the system has many degrees of freedom or one degree of freedom, or whether its size is large or small.

Being properties, energy, adiabatic availability, generalized adiabatic availability, available energy, and generalized available energy are measurable. Except for differences in detail, the measurement of each of these properties is achieved by means of a weight process.

As compared with adiabatic availability and generalized adiabatic availability, available energy and generalized available energy with respect to a given reservoir have the important advantage of being additive, but the disadvantage of depending not only on the state of the system but also on the given reservoir. As discussed in the next section, we gain independence of the reservoir, without losing additivity, by considering the difference between energy and generalized available energy divided by a constant property of the reservoir.

14.7 Entropy and Entropy Balance

An important consequence of the two laws of thermodynamics is that every system A in any state A_1, with energy E_1 and generalized available energy Ω_1^R with respect to an auxiliary reservoir R, has a property called *entropy*, denoted by the symbol S_1. Entropy is a property in the same sense that energy is a property, or momentum is a property. It can be evaluated by means of the auxiliary reservoir R, a reference state A_o, with energy E_o and generalized available energy Ω_o^R, to which is assigned a fixed reference value S_o, and the expression

$$S_1 = S_o + \frac{1}{c_R} \left[(E_1 - E_o) - (\Omega_1^R - \Omega_o^R) \right] \tag{14.9}$$

where c_R is a well-defined positive constant. For the given auxiliary reservoir R, c_R is selected in such a way that the values of entropy found by means of (14.9) are independent of the reservoir.[2] In other words, despite the dependence of the value of the difference of generalized available energies, $\Omega_1^R - \Omega_o^R$, in equation (14.9) on the selection of the reservoir R, we can show that there is a constant property c_R of reservoir R that makes the right-hand side of equation (14.9) independent of R. Thus, S is a property of system A only, in the same sense as energy E and adiabatic availabiliy Ψ are properties of system A only. In due course, the concept of temperature is defined as a property of stable equilibrium states. Then we show that the temperature of a reservoir is constant, and that c_R is also equal to the constant temperature of the reservoir R.

---

[2]The precise definition of c_R and the proof that S_1 is independent of the reservoir are not summarized here for brevity. They are given in Section 7.4. However, for the purposes of the present summary, it is sufficient to know that such definition and proof exist, and therefore that we are proceeding on safe ground.

The entropy S_2 of a state A_2 is given by an expression similar to that of A_1, namely,

$$S_2 = S_o + \frac{1}{c_R} \left[(E_2 - E_o) - (\Omega_2^R - \Omega_o^R) \right] \tag{14.10}$$

Moreover, subtracting equation (14.9) from (14.10), we find

$$S_2 - S_1 = \frac{1}{c_R} \left[(E_2 - E_1) - (\Omega_2^R - \Omega_1^R) \right] \tag{14.11}$$

or, equivalently,

$$\Omega_2^R - \Omega_1^R = E_2 - E_1 - c_R (S_2 - S_1) \tag{14.12}$$

Like energy, entropy is an additive property, namely, the entropy of a system consisting of two or more subsystems equals the sum of the entropies of the subsystems and this holds for all combinations of states of the subsystems. Whereas energy remains constant in time whenever the system experiences either a spontaneous process or a zero-net-effect weight process, equations (14.7) and (14.11) show that the entropy remains constant in time when the process is reversible. In the course of an irreversible either spontaneous or zero-net-effect weight process, equations (14.8) and (14.11) show that the entropy increases with time, and part of the potential ability of the system to transfer energy to a weight is destroyed. Because of additivity, and because any process of a system can always be thought of as part of a spontaneous process of a composite system consisting of all the interacting systems, the conclusion that as time proceeds entropy can either be created, if the process is irreversible, or remain constant, if the process is reversible, but can never be destroyed is of great generality and practical importance. It is known as the *principle of entropy nondecrease*. The entropy created as time proceeds during an irreversible process is called *entropy generated by irreversibility* or *entropy production due to irreversibility*. It is positive.

Like energy, entropy can be transferred between systems by means of interactions. Denoting by $S^{A \leftarrow}$ the net amount of entropy transferred from systems in the environment to system A as a result of all the interactions involved in a process in which the state of A changes from A_1 to A_2, we derive another extremely important analytical tool, the *entropy balance* equation. This equation is based on the additivity of entropy and on the principle of entropy nondecrease. It requires that the change in the entropy of the system from S_1 to S_2 must be equal to the net amount of entropy $S^{A \leftarrow}$ transferred into the system, plus the positive amount of entropy S_{irr} generated by irreversibility inside A in the course of the process, that is,

$$S_2 - S_1 = S^{A \leftarrow} + S_{\text{irr}} \tag{14.13}$$

The value of $S^{A \leftarrow}$ is positive when entropy is transferred into A and negative when entropy is transferred out of A.

It is worth repeating that S is defined for any state of any system because energy E and generalized available energy Ω^R are defined for any state of any system. Thus, like energy, entropy is defined for all states, steady, unsteady, equilibrium, nonequilibrium, and stable equilibrium states, and for all systems, that is, systems with many degrees of freedom and systems with few degrees of freedom, including a single particle with a single translational degree of freedom, because both energy and generalized available energy are defined for all these states and all these systems.

The dimensions of entropy are determined by the dimensions of both energy and the property c_R of the auxiliary reservoir. We can show that the dimension of c_R is independent of the dimensions of other properties, and the same as the dimension of temperature (defined later). The unit of c_R chosen in the International System of units is the kelvin, denoted by K. Another unit is the rankine, denoted by R, where 1 R = 1.8 K. Thus entropy values are expressed in many different units such as the joule per kelvin (J/K), the kilocalorie per kelvin (kcal/K), and the British thermal unit per rankine (Btu/R). In particular, it turns out that 1 Btu/lb R = 1 kcal/kg K.

14.8   Stable Equilibrium States

In the absence of internal mechanisms capable of altering the values of the amounts of constituents and the parameters, that is, in the absence of chemical reactions, nuclear reactions, and other types of internal interconnections, a system admits an indefinite number of states that have given values of the energy E, the amounts of constituents n_1, n_2, ..., n_r, and the parameters β_1, β_2, ..., β_s. Most of these states are nonequilibrium, some are equilibrium, and according to the second law, only one is a stable equilibrium state. It follows that the value of any property P of the system in a stable equilibrium state is uniquely determined by the values of E, n_1, n_2, ..., n_r, and β_1, β_2, ..., β_s, that is, it can be written as a function of the form

$$P = P(E, n_1, n_2, \ldots, n_r, \beta_1, \beta_2, \ldots, \beta_s) \qquad (14.14)$$

This result, known as the *stable-equilibrium-state principle* or, simply, the *state principle*, expresses a fundamental physical feature of the stable equilibrium states of the system, and implies the existence of interrelations among the properties at these states.

A system in general has a very large number of independent properties. When we focus on the special family of states that are stable equilibrium, however, the state principle asserts that the value of each of these properties is uniquely determined by the values of E, $\boldsymbol{n}$, $\boldsymbol{\beta}$. In contrast, for states that are not stable equilibrium the values of E, $\boldsymbol{n}$, $\boldsymbol{\beta}$ are not sufficient to specify the values of all the independent properties.

When written for the entropy S of stable equilibrium states, equation (14.14) becomes

$$S = S(E, n_1, n_2, \ldots, n_r, \beta_1, \beta_2, \ldots, \beta_s) \qquad (14.15)$$

It is known as the *fundamental stable-equilibrium-state relation for entropy* or, simply, the *fundamental relation*. We can show that the function $S(E, \boldsymbol{n}, \boldsymbol{\beta})$ admits partial derivatives of all orders and, therefore, that any difference between the entropies of two stable equilibrium states may be expressed in the form of a Taylor series in terms of the partial derivatives of $S(E, \boldsymbol{n}, \boldsymbol{\beta})$ at one stable equilibrium state, and differences in the values of the energy, amounts of constituents, and parameters of the two stable equilibrium states.

The function $S(E, \boldsymbol{n}, \boldsymbol{\beta})$ is concave in each of the variables E, n_1, n_2, ..., n_r. It is concave in each of the parameters β_1, β_2, ..., β_s, which are additive, like volume, and it is also concave collectively with respect to all the variables E, n_1, n_2, ..., n_r, and the parameters β_1, β_2, ..., β_s, which are additive. Concavity implies that $(\partial^2 S/\partial E^2)_{n,\beta} \leqq 0$, $(\partial^2 S/\partial n_i^2)_{E,n,\beta} \leqq 0$ for each i, $(\partial^2 S/\partial \beta_j^2)_{E,n,\beta} \leqq 0$ for each additive β_j, and some other necessary conditions on all the second-order derivatives of the fundamental relation.

Using the second law, we assert that the entropy of each unique stable equilibrium state is larger than that of any other state with the same values of E, n_1, n_2, ..., n_r, and β_1, β_2, ..., β_s. This assertion is known as the *highest-entropy principle*. This principle is extremely useful in establishing conditions that must be satisfied by properties of systems in stable equilibrium states.

Equation (14.15) may be solved for E as a function of S, n_1, n_2, ..., n_r, and β_1, β_2, ..., β_s so that

$$E = E(S, n_1, n_2, \ldots, n_r, \beta_1, \beta_2, \ldots, \beta_s) \qquad (14.16)$$

The function $E(S, n, \beta)$ admits partial derivatives of all orders and, therefore, any difference between the energies of two stable equilibrium states may be expressed in the form of a Taylor series in terms of the partial derivatives of $E(S, n, \beta)$ at one of the stable equilibrium states, and differences in the values of the entropy, amounts of constituents, and parameters of the two stable equilibrium states.

Among all the partial derivatives, each first order partial derivative of either the function $S(E, n, \beta)$ or the function $E(S, n, \beta)$ represents an important and practical property of the family of stable equilibrium states of a system. It is important because each such property enters a condition for mutual stable equilibrium with other systems, and practical because it can be relatively easily related to simple measurements. We emphasize that each such property is defined only for the stable equilibrium states of the system.

14.9   Temperature

The partial derivative of $E(S, n, \beta)$ with respect to entropy, or the inverse of the partial derivative of $S(E, n, \beta)$ with respect to energy, that is,

$$T = \left(\frac{\partial E}{\partial S}\right)_{n,\beta} = \frac{1}{(\partial S/\partial E)_{n,\beta}} \tag{14.17}$$

is defined as the *absolute temperature* or, simply, the *temperature*. The first of equations (14.17) defines T as a function of E, n, β, and the second as a function of S, n, β. Two units of temperature are the kelvin and the rankine.

If two systems A and B in states A_0 and B_0 are in mutual stable equilibrium, then the temperature T_0^A of system A must be equal to the temperature T_0^B of system B. Said differently, equality of temperatures of the two systems is a necessary condition for the two systems to be in mutual stable equilibrium.

If the temperature T_0^A differs from the temperature T_0^B, and A and B are allowed to interact with each other so that no net effects are left in the environment of their composite system, and the values of the amounts of constituents and the parameters of both A and B remain unaltered, then energy can flow from A to B only if $T_0^A > T_0^B$, and from B to A only if $T_0^A < T_0^B$.

Our everyday awareness of the temperature of an object is based on the feature just cited. When we touch an object, our nervous system senses whether as a result of the contact our body is gaining or losing energy. If we gain energy, we say that the object is hot, and if we lose energy we say that the object is cold. In the former case, the temperature of the object is higher than that of our body, and in the latter case it is lower than the temperature of our body.

14.9.1   Temperature of a Reservoir

A reservoir R is an idealized kind of system that passes through stable equilibrium states only, and remains in mutual stable equilibrium with a duplicate of itself that experiences no changes of state. These specifications imply that all states of a reservoir have the same temperature T_R, and that this temperature is equal to the constant property c_R introduced in equation (14.9).

14.9.2   Relative Temperatures

Historically, the notion of temperature developed quite differently from the way we have presented it here. Its origin can be traced to the Italian astronomer and physicist Galileo Galilei (1564–1642), who invented the thermometer around the year 1595. Galilei's thermometer was a glass instrument containing water and air that allowed to "distinguish the variety of temperaments of places." The instrument took root with several modifications. In 1714, the German physicist Gabriel Daniel Fahrenheit (1686–1736) introduced a mercury-in-glass instrument graded with the scale that still carries his name, and in 1742, the Swedish astronomer Anders Celsius (1701–1744) devised a different scale to grade the instrument. An important feature of the Fahrenheit and Celsius thermometers is that a change in length DL of the column of liquid in the stem is directly proportional to the change in absolute temperature DT of the thermometer; that is,

$$DL = \alpha \, DT \quad \text{or} \quad L_1 - L_2 = \alpha \, (T_1 - T_2) \tag{14.18}$$

where α is a constant for a thermometer and L is the length corresponding to temperature T.

Based on this feature, the following set of measurements and operations leads to the definition of another property of stable equilibrium states. Given a system A in a stable equilibrium state A_1 with absolute temperature T_1^A, we find a thermometer that is in mutual stable equilibrium with A, measure the length L_1 of the column of liquid in the thermometer, and evaluate the quantity

$$t_1 = t_f + \frac{L_1 - L_f}{L_b - L_f} \, (t_b - t_f) \tag{14.19}$$

where L_f and L_b are the lengths of the column of liquid when the thermometer is in mutual stable equilibrium with freezing water and boiling water, respectively, and t_f and t_b are constants defining the scale of the thermometer. Upon using (14.18), and replacing length differences by temperature differences in (14.19), we find

$$t_1 = t_f + \frac{T_1 - T_f}{T_b - T_f} \, (t_b - t_f) \tag{14.20}$$

where T_f and T_b are the absolute temperatures corresponding to freezing water and boiling water, respectively.

The condition of temperature equality, $T_1 = T_1^A$, between the thermometer and the system A in mutual stable equilibrium indicates that the set of measurements and operations just cited results in the definition of t as a property of the stable equilibrium states of system A. For each choice of t_f and t_b, we call t the *relative temperature* of system A, and denote its value at stable equilibrium state A_1 by t_1. Upon rewriting (14.20) for a stable equilibrium state A_2 different from A_1,

$$t_2 = t_f + \frac{T_2 - T_f}{T_b - T_f} \, (t_b - t_f) \tag{14.21}$$

and subtracting (14.20) from (14.21) we find

$$t_2 = t_1 + \frac{t_b - t_f}{T_b - T_f} \, (T_2 - T_1) \tag{14.22}$$

The relative temperatures that have been most used are the Fahrenheit and Celsius temperatures. Fahrenheit originally chose as fixed points on his scale the freezing temperature at ordinary atmospheric conditions of a mixture of salt, ice, and salt solution, designated as 0°F, and the temperature of the human body, designated as 96°F. Later the fixed points were changed to the freezing and boiling temperatures of water at atmospheric conditions, designated as $t_f = 32°\text{F}$ and $t_b = 212°\text{F}$, respectively. Celsius chose as fixed points on his scale the freezing and boiling temperatures of water at atmospheric conditions, designated as $t_f = 0°\text{C}$ and $t_b = 100°\text{C}$, respectively. The Celsius scale has also been called the "centigrade" scale because the stem of a thermometer graded according to this scale is divided into 100 equal intervals between the point corresponding to freezing water and that corresponding to boiling water.

The kelvin and the rankine, that is, the units of measure of absolute temperature, were defined after the relative temperatures. They were chosen so that $T_b - T_f = 100$ K $= 180$ R. As a result, equation (14.20) yields the following relations between absolute temperature T and either Fahrenheit or Celsius relative temperature.

The *Fahrenheit relative temperature* is defined by assuming $t_f = 32°$F and $t_b = 212°$F, and is related to absolute temperature by

$$t_1^{\text{Fahrenheit}} = 32°\text{F} + \frac{1°\text{F}}{1\text{ R}}(T_1 - T_f) \tag{14.23}$$

The *Celsius relative temperature* is defined by assuming $t_f = 0°$C and $t_b = 100°$C and, therefore,

$$t_1^{\text{Celsius}} = \frac{1°\text{C}}{1\text{ K}}(T_1 - T_f) \tag{14.24}$$

Combining (14.23) and (14.24), we find a useful relation between the Celsius and Fahrenheit relative temperatures, namely,

$$t_1^{\text{Fahrenheit}} = 32°\text{F} + \frac{9°\text{F}}{5°\text{C}} t_1^{\text{Celsius}} \tag{14.25}$$

where we use the relation $1\text{ K}/1\text{ R} = 9/5$, which follows from the fact that $T_b - T_f = 100$ K $= 180$ R.

In order for measurements of relative temperature to be equivalent to measurements of absolute temperature, we need a single independent measurement of absolute temperature, for example, the value of T_f, the absolute temperature of freezing water at atmospheric conditions. Such measurements yield $T_f = 273.15$ K $= 491.67$ R. Thus, absolute temperatures can be measured with a relative temperature thermometer and, in particular, (14.23) and (14.24) yield

$$T_1 = 273.15\text{ K} + \frac{1\text{ K}}{1°\text{C}} t_1^{\text{Celsius}} \tag{14.26}$$

or

$$T_1 = 459.67\text{ R} + \frac{1\text{ R}}{1°\text{F}} t_1^{\text{Fahrenheit}} \tag{14.27}$$

14.10   Total Potentials

The *total potential of the ith constituent*, μ_i, is defined by either of the two relations

$$\mu_i = \left(\frac{\partial E}{\partial n_i}\right)_{S,n,\beta} = -T\left(\frac{\partial S}{\partial n_i}\right)_{E,n,\beta} \tag{14.28}$$

The dimensions of total potential are energy per unit of amount. The first of equations (14.28) defines μ_i as a function of S, n, β, and the second as a function of E, n, β because $S = S(E, n, \beta)$.

If volume is the only parameter, each total potential is called a *chemical potential*.

If two systems A and B in states A_0 and B_0 are in mutual stable equilibrium, both contain the ith type of constituent, for $i = 1, 2, \ldots, r$, and the amount of that constituent may both increase and decrease in each system, then the total potential $(\mu_i)_0^A$ of the ith constituent of A must be equal to the total potential $(\mu_i)_0^B$ of the ith constituent of B. Said differently, in addition to temperature equality, also equality of total potentials for every constituent is a necessary condition for two systems to be in mutual stable equilibrium.

If the temperature $T_0^A = T_0^B$, but $(\mu_i)_0^A \neq (\mu_i)_0^B$, then interactions between A and B that leave no net effects in systems in the environment, and that affect neither the values of the parameters nor the values of the amounts of constituents except the ith, can result in a transfer of an amount of the ith constituent from A to B only if $(\mu_i)_0^A > (\mu_i)_0^B$, and from B to A only if $(\mu_i)_0^A < (\mu_i)_0^B$.

14.11   Pressure

The *generalized force conjugated to the jth parameter*, f_j, is defined by either of the two relations

$$f_j = \left(\frac{\partial E}{\partial \beta_j} \right)_{S,n,\beta} = -T\left(\frac{\partial S}{\partial \beta_j} \right)_{E,n,\beta} \tag{14.29}$$

If volume V is a parameter, the negative of the generalized force conjugated to V is called *pressure*, denoted by p, and given by either of the two relations

$$p = -\left(\frac{\partial E}{\partial V} \right)_{S,n,\beta} = T\left(\frac{\partial S}{\partial V} \right)_{E,n,\beta} \tag{14.30}$$

where here $\beta = \{V, \beta_2, \beta_3, \ldots, \beta_s\}$. The first of equations (14.30) defines p as a function of S, n, β, and the second as a function of E, n, β because $S = S(E, n, \beta)$. The dimensions of pressure are energy per unit volume, or force per unit area. One of the units of pressure is the pascal, denoted by Pa, such that $1 \text{ Pa} = 1 \text{ N/m}^2$. Another unit is the atmosphere, denoted by atm, which equals the pressure of atmospheric air at sea level and a temperature of 293.15 K. Other units are listed in Appendix A.

If two systems A and B in states A_0 and B_0 are in mutual stable equilibrium, and each has volume as a variable parameter, then they must also be in pressure equality, that is, $p_0^A = p_0^B$.

If the temperature $T_0^A = T_0^B$, then interactions between A and B that leave no net effects on systems in the environment, and that affect neither the values of the amounts of constituents nor the values of the parameters except the volumes of both A and B, can result in an increase of volume V_0^A at the expense of volume V_0^B only if $p_0^A > p_0^B$, or in an increase of volume V_0^B at the expense of volume V_0^A if $p_0^A < p_0^B$.

Pressure, as any other generalized conjugated force, is a property of stable equilibrium states only. At a given stable equilibrium state, the value of the pressure equals the force per unit area that must be applied at every point of the surface bounding the volume of the system in order to keep the system in the specified stable equilibrium state. This force per unit area has the same value at every point on the surface.

To prove the last assertion, we consider a system A in a stable equilibrium state. At an arbitrary location on the surface of the system, there is a piston of cross-sectional area da. The piston is held in place by a small weight exerting a force $dF = f\,da$, that is, a force per unit area f times the area of the piston da. A reversible weight process for system A can be affected by displacing the piston by a small distance dx along the direction of the force, namely, by changing the volume by the amount $dV = da\,dx$. The process is selected so that each of the end states of the system is a stable equilibrium state corresponding to the same values of entropy, amounts of constituents, and parameters other than volume. Accordingly, the two end states of the process are such that $dS = 0$, $dn_i = 0$ for all i, and $d\beta_j = 0$ for $j = 2, 3, \ldots, s$. As a result of the energy balance for this weight process, the energy of the system changes by an amount $dE\,\big|_{S,n,\beta}$ which is equal to the negative of the energy gain of the weight, that is, the negative of the product of the force times the displacement so that

$$dE \big|_{S,n,\beta} = -dF \, dx = -f \, da \, dx = -f \, dV \tag{14.31}$$

On the other hand, because the end states of the system are stable equilibrium, the changes $dE \big|_{S,n,\beta}$ and dV are related by the expression

$$dE \big|_{S,n,\beta} = \left(\frac{\partial E}{\partial V}\right)_{S,n,\beta} dV \tag{14.32}$$

Upon comparing equations (14.31) and (14.32) we conclude that

$$\left(\frac{\partial E}{\partial V}\right)_{S,n,\beta} = -f \tag{14.33}$$

and upon comparing equations (14.30) and (14.32), that $f = p$. Because the location of the piston was chosen arbitrarily, we further conclude that the equality $f = p$ is valid at all locations on the surface and, therefore, that the force per unit area is uniform throughout the surface.

The identification of pressure with force per unit area on the surface, and the requirement of pressure equality provide a simple method for assigning values to pressure. Specifically, each stable equilibrium state of certain systems called *barometers*, *pressure gauges*, or *pressure transducers* is uniquely related to some readily observable property such as length. For example, a spring-loaded piston may be attached to the system, and the length of the spring may be used as a measure of pressure. Again, the length of a suitably calibrated vertical column of mercury can be used for pressure measurements by placing the column in contact with a system via a movable partition.

14.11.1 First-Order Taylor Series Expansions

In terms of T_0, p_0, $(\mu_i)_0$, and $(f_j)_0$ of an arbitrary stable equilibrium state A_0 of a system A, small differences in energy, $dE = E_1 - E_0$, entropy, $dS = S_1 - S_0$, volume, $dV = V_1 - V_0$, other parameters, $d\beta_2 = (\beta_2)_1 - (\beta_2)_0$, $d\beta_3 = (\beta_3)_1 - (\beta_3)_0$, ..., $d\beta_s = (\beta_s)_1 - (\beta_s)_0$, and amounts of constituents, $dn_1 = (n_1)_1 - (n_1)_0$, $dn_2 = (n_2)_1 - (n_2)_0$, ..., $dn_r = (n_r)_1 - (n_r)_0$ between two neighboring stable equilibrium states are related by a first-order Taylor series expansion or *differential energy relation*

$$dE = T_0 \, dS - p_0 \, dV + \sum_{i=1}^{r} (\mu_i)_0 \, dn_i + \sum_{j=2}^{s} (f_j)_0 \, d\beta_j \tag{14.34}$$

On solving equation (14.34) for dS, and writing dS as a first-order Taylor series expansion in terms of dE, dV, dn_i, and $d\beta_j$, we find

$$dS = \frac{1}{T_0} dE + \frac{p_0}{T_0} dV - \sum_{i=1}^{r} \frac{(\mu_i)_0}{T_0} dn_i - \sum_{j=2}^{s} \frac{(f_j)_0}{T_0} d\beta_j \tag{14.35a}$$

$$= \left[\left(\frac{\partial S}{\partial E}\right)_{V,n,\beta}\right]_0 dE + \left[\left(\frac{\partial S}{\partial V}\right)_{E,n,\beta}\right]_0 dV + \sum_{i=1}^{r} \left[\left(\frac{\partial S}{\partial n_i}\right)_{E,V,n,\beta}\right]_0 dn_i$$

$$+ \sum_{j=2}^{s} \left[\left(\frac{\partial S}{\partial \beta_j}\right)_{E,V,n,\beta}\right]_0 d\beta_j \tag{14.35b}$$

Upon comparing the coefficients of dE, dV, dn_i, and $d\beta_j$ in (14.35a) and (14.35b) we find

$$\left(\frac{\partial S}{\partial E}\right)_{V,n,\beta} = \frac{1}{T} \tag{14.36}$$

$$\left(\frac{\partial S}{\partial V}\right)_{E,n,\beta} = \frac{p}{T} \tag{14.37}$$

$$\left(\frac{\partial S}{\partial n_i}\right)_{E,V,n,\beta} = -\frac{\mu_i}{T} \qquad \text{for } i = 1, 2, \ldots, r \tag{14.38}$$

$$\left(\frac{\partial S}{\partial \beta_j}\right)_{E,V,n,\beta} = -\frac{f_j}{T} \qquad \text{for } j = 2, 3, \ldots, s \tag{14.39}$$

where in writing these equalities we simplify them by dropping the subscript 0 which specifies the particular stable equilibrium state about which we make the Taylor series expansion, and at which we evaluate the partial derivatives. Each of equations (14.36), (14.37), (14.38), and (14.39) proves the second of equations (14.17), (14.30), (14.28), and (14.29), respectively.

14.11.2 Energy Relation of a Reservoir

We recall that a reservoir is an idealized kind of system that passes through stable equilibrium states only, and remains in mutual stable equilibrium with a duplicate of itself that experiences no changes of state. If reservoir R has only volume as a parameter, the specifications just cited imply that all states of R have the same value of the temperature, T_R, the same value of the pressure, p_R, and the same values of the chemical potentials of the r constituents, μ_{1R}, μ_{2R}, $\ldots$, μ_{rR}, so that the necessary conditions of temperature equality, pressure equality, and chemical potential equality for all constituents are satisfied. It follows that, for a reservoir, equation (14.34) may be stated in terms of differences — large or small — between properties of any two states R_1 and R_2 so that

$$E_2^R - E_1^R = T_R (S_2^R - S_1^R) - p_R (V_2^R - V_1^R) + \sum_{i=1}^{r} \mu_{iR} \left[(n_i)_2^R - (n_i)_1^R \right] \tag{14.40}$$

14.12 Work and Heat Interactions

Interactions result in the exchange of properties across the boundaries of the interacting systems. Various combinations of exchanges are used to classify interactions into different categories. An interaction between two systems that results in a transfer of energy between the two systems without any transfer of entropy is classified as a *work interaction*. The amount of energy exchanged as a result of such an interaction is called *work*. All interactions that result in the exchange of entropy between the interacting systems are called *nonwork interactions*.

A process of a system experiencing only work interactions is called an *adiabatic process*. Any process that involves nonwork interactions is called a *nonadiabatic process*.

In the course of an adiabatic process, system A changes from state A_1 to state A_2, the energy exchange $E_{12}^{A \leftarrow}$ is work, that is, $E_{12}^{A \leftarrow} = -W_{12}^{A \rightarrow}$, where $W_{12}^{A \rightarrow}$ denotes the *work done by* system A on systems in its surroundings with which it interacts. In the course of an adiabatic process, the entropy exchange $S_{12}^{A \leftarrow} = 0$. Therefore, the energy and entropy balances are

$$E_2 - E_1 = -W_{12}^{A\rightarrow} \tag{14.41}$$

$$S_2 - S_1 = S_{\text{irr}} \tag{14.42}$$

where S_{irr} denotes the entropy generated by irreversibility inside A during the process.

A special example of a nonwork interaction is entirely distinguishable from work. It is an interaction between two systems, initially differing infinitesimally in temperature. It results in no other effects except a transfer of energy and a transfer of entropy between the two systems such that the ratio of the amount of energy transferred and the amount of entropy transferred equals the almost common temperature of the interacting systems. It is called a *heat interaction*. The amount of energy transferred as a result of such an interaction is called *heat*.

Often, in applications, a system A consists of many subsystems, one of which A' is in a stable equilibrium state at a temperature T_Q. Similarly, a system B consists of many subsystems, one of which B' is in a stable equilibrium state at temperature almost equal to T_Q. If the two subsystems A' and B' experience a heat interaction, we say that systems A and B experience a heat interaction at temperature T_Q, even though A and B are not necessarily in stable equilibrium states.

In the course of a process that involves only a heat interaction at temperature T_Q, system A changes from state A_1 to state A_2, the energy exchange $E_{12}^{A\leftarrow}$ is heat, and is denoted by $Q_{12}^{A\leftarrow}$, that is, $E_{12}^{A\leftarrow} = Q_{12}^{A\leftarrow}$, and the entropy exchange $S_{12}^{A\leftarrow} = Q_{12}^{A\leftarrow}/T_Q$. Therefore, the energy and entropy balances are

$$E_2 - E_1 = Q_{12}^{A\leftarrow} \tag{14.43}$$

$$S_2 - S_1 = \frac{Q_{12}^{A\leftarrow}}{T_Q} + S_{\text{irr}} \tag{14.44}$$

where S_{irr} is the entropy generated by irreversibility inside A during the process.

If a process of a system A involves both work and heat but no other interactions, the energy exchange is $E_{12}^{A\leftarrow} = Q_{12}^{A\leftarrow} - W_{12}^{A\rightarrow}$, and the entropy exchange $S_{12}^{A\leftarrow} = Q_{12}^{A\leftarrow}/T_Q$. Therefore, the energy and entropy balances for A are

$$E_2 - E_1 = Q_{12}^{A\leftarrow} - W_{12}^{A\rightarrow} \tag{14.45}$$

$$S_2 - S_1 = \frac{Q_{12}^{A\leftarrow}}{T_Q} + S_{\text{irr}} \tag{14.46}$$

where S_{irr} denotes again the entropy generated by irreversibility inside A during the process. Upon dropping some self-evident subscripts and superscripts, we may rewrite (14.45) and (14.46) in the forms

$$E_2 - E_1 = Q^{\leftarrow} - W^{\rightarrow} \tag{14.47}$$

$$S_2 - S_1 = \frac{Q^{\leftarrow}}{T_Q} + S_{\text{irr}} \tag{14.48}$$

or, for differential changes,

$$dE = \delta Q^{\leftarrow} - \delta W^{\rightarrow} \tag{14.49}$$

$$dS = \frac{\delta Q^{\leftarrow}}{T_Q} + \delta S_{\text{irr}} \tag{14.50}$$

It is noteworthy that the prefix d denotes infinitesimal differences between the values of a property at two different states of the system, whereas the prefix δ denotes infinitesimal amounts of quantities that are not properties, such as work, heat, and entropy generation by irreversibility.

For processes in which the end states of the system are stable equilibrium states, energy and entropy changes, and therefore work, heat, and entropy generation by irreversibility may be related to changes of other properties and variables, such as temperature, pressure, and volume. For example, if the changes are infinitesimal, (14.49) and (14.50) can be combined with (14.34) to find relations between the transfers of energy and entropy by means of the interactions, that is, $\delta Q^{\leftarrow}$ and $\delta W^{\rightarrow}$, the changes in energy, entropy, volume, other parameters, and amounts of constituents, that is, dE, dS, dV, $d\beta_2$, $d\beta_3$, ..., $d\beta_s$, and dn_1, dn_2, ..., dn_r, the entropy generated by irreversibility δS_{irr}, and the values of temperature, pressure, other generalized forces, and total potentials, that is, T, p, f_2, f_3, ..., f_s, and μ_1, μ_2, ..., μ_r of the initial stable equilibrium state.

Work and heat interactions are most frequently encountered in engineering applications. Other interactions, involving transfers of energy, entropy, and amounts of constituents, are discussed later. By knowing the end states and the interactions of a process, we can identify the entropy generated by irreversibility and, thus, begin to address the question of improvement of the process.

14.13 Energy Versus Entropy Graphs

Because they are defined in terms of the values of the amounts of constituents, the parameters, and a complete set of independent properties, states can in principle be represented by points in a multidimensional geometrical space with one axis for each amount, parameter and independent property. Such a representation, however, would not be enlightening because the number of independent properties of any system is indefinitely large. Nevertheless, useful information can be summarized by first cutting the multidimensional space with a hypersurface corresponding to given values of each of the amounts of constituents and each of the parameters, and then projecting the result onto a two-dimensional plane — a plane with two property axes. In this section, we discuss the energy versus entropy plane that illustrates many of the basic concepts of thermodynamics.

We consider a system with volume, V, as the only parameter. For given values of the amounts of constituents and the volume, we project the multidimensional state space of the system onto the E versus S plane. This projection of states on the E versus S plane includes both stable equilibrium states and other states that are not stable equilibrium. The projection must have the shape of the cross-hatched area shown in Figure 14.1, namely, all the states that share the given characteristics have property values that project on the area between the vertical line denoted as the line of the zero-entropy states, and the curve of the stable equilibrium states.

A point either inside the cross-hatched area or on the vertical line $S = 0$ represents a large number of states. Each such state has the same values of amounts of constituents, volume V, energy E, and entropy S, but differing values of other properties, and is not a stable equilibrium state. It can be any type of state except a stable equilibrium state.

A point on the convex curve of the stable equilibrium states represents one and only one state. For each of these states, the value of any property is uniquely determined only by the values of the amounts of constituents, the volume, and either E or S of the point on the curve.

14.13.1 Zero-Entropy Line

The line of the zero-entropy states (Figure 14.1) corresponds to all the states that have the least amount of entropy. This amount can be assigned the value zero because the two laws of thermodynamics imply that no states exist with lower entropy. Thus, entropy has

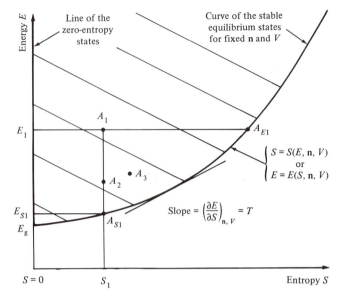

Figure 14.1 Schematic representation of the projection of the states of a system with given values of the amounts of constituents and the volume on the energy versus entropy plane.

absolute values greater than or equal to zero. For each set of values n, V, it turns out that the zero-entropy line represents all the states that are defined in mechanics. So mechanics can be regarded as a special case of thermodynamics, namely, as zero-entropy physics.

14.13.2 Lowest-Energy States

For given values of the amounts of constituents and the volume, the lowest energy of the system is E_g (Figure 14.1). It corresponds to a unique stable equilibrium state having zero entropy and zero temperature. The energy E_g is the lowest energy for which the system can exist with the given types and amounts of constituents, and for the given value of the volume. For different values of the amounts of constituents and the volume, the lowest energy state is different but still a stable equilibrium state with zero entropy and zero temperature.

14.13.3 The Fundamental Relation

The stable-equilibrum-state curve represents the convex stable equilibrium state relation E versus S for given values of the amounts of constituents and the volume (Figure 14.1). It is a single-valued relation because for each set of values E, n, and V there is one and only one stable equilibrium state and, therefore, a unique value of S. It has the following features.

   Each stable equilibrium state is the state of lowest energy among all the states with the same values of S, n, and V (lowest-energy principle). For each set of values S_1, n, and V, the stable equilibrium state A_{S1} on the vertical line $S = S_1$ is the state of lowest energy — no states exist below A_{S1} that correspond to the same values of n and V, and that lie on the line $S = S_1$. State A_{S1} can be reached starting from any state A_1 on the line $S = S_1$ by means of a reversible weight process without net changes in n and V. Indeed, in such a process the net change in the entropy of the system is zero, $S_{S1} = S_1$, and the energy $E_1 - E_{S1}$ is transferred out from the system to the weight.

   Each stable equilibrium state is the state of highest entropy among all the states with the same values of E, n, and V (highest-entropy principle). For each set of values E_1, n, and V, the stable equilibrium state A_{E1} on the horizontal line $E = E_1$ is the state of highest entropy—no states exist beyond A_{E1} that correspond to the same values of n and V and that lie on the line $E = E_1$. State A_{E1} can be reached starting from any state A_1 on the line $E = E_1$ by means of a zero-net-effect weight process such as a spontaneous process. Any such process is irreversible because it entails an increase in the entropy of the system without any effects on the environment.

Temperature is positive and increasing with energy. Because each stable equilibrium state is unique, the temperature $T = (\partial E/\partial S)_{n,V}$ at each point on the convex boundary is uniquely defined. Temperature is not defined for states that are not stable equilibrium because then E, S, n, and V are independent and, therefore, the partial derivative of E with respect to S is meaningless.

14.13.4   Perpetual-Motion Machine of the Second Kind

Starting from a stable equilibrium state A_{S1} on the convex boundary $E_g A_{S1} A_{E1}$ (Figure 14.1), the system cannot transfer energy to a weight without net changes in the values of the amounts of constituents and the volume because no state of lower energy exists that has an entropy equal to or larger than the entropy of state A_{S1}. Indeed, if energy were transferred to a weight, the energy of the system would be reduced. But starting from state A_{S1} all states with smaller energy have also smaller entropy. Because the weight receives only energy, and entropy cannot decrease by itself, it follows that no such transfer can occur under the conditions specified. This feature of the graph represents the impossibility of perpetual-motion machines of the second kind. It is sometimes expressed as the nonexistence of a *Maxwellian demon*.

For each set of given values of the amounts of constituents and the volume, the convex boundary $E_g A_{S1} A_{E1}$ represents the corresponding stable equilibrium states. These states are referred to in the literature as the thermodynamic equilibrium states of equilibrium thermodynamics, which is sometimes also called "classical thermodynamics" or "thermostatics." So, equilibrium thermodynamics can be regarded as another special case of thermodynamics, namely, as highest-entropy physics.

14.13.5   Adiabatic Availability

For a given state A_1, the energy $E_1 - E_{S1}$ is equal to the adiabatic availability Ψ_1 of A_1 because the change of state from A_1 to A_{S1} represents the change specified in the definition of Ψ_1 (Figure 14.1). We see from the figure that, in general, Ψ_1 is smaller than the energy of the system above the ground-state energy, $E_1 - E_g$. It varies from $E_1 - E_g$ to zero as the entropy S_1 of the state varies from zero to the highest value that is possible for the set of values E_1, n, and V. So entropy affects the usefulness of the energy of a system, that is, the larger the entropy for given values of E, n, and V, the smaller the adiabatic availability.

14.13.6   Work in an Adiabatic Process

In an adiabatic process without net changes in amounts and volume, the work done by the system starting from state A_1 and ending in a state different from A_{S1} is always smaller than the adiabatic availability Ψ_1 (Figure 14.1). If the process is reversible, the final state $A_2 \neq A_{S1}$ must have entropy $S_2 = S_1$, energy $E_2 > E_{S1}$, and therefore

$$(W_{12}^{A\rightarrow})_{\text{rev}} = E_1 - E_2 < E_1 - E_{S1} = \Psi_1 \qquad (14.51)$$

If the process is irreversible, the final state $A_3 \neq A_{S1}$ must have entropy $S_3 > S_1$. But for $S_3 > S_1$, the graph shows that A_3 must have energy $E_3 > E_{S1}$, and therefore

$$(W_{13}^{A\rightarrow})_{\text{irr}} = E_1 - E_3 < E_1 - E_{S1} = \Psi_1 \qquad (14.52)$$

14.13.7   Available Energy

The E^R versus S^R diagram of a reservoir R is just a straight line of slope T_R because the reservoir passes through stable equilibrium states only, and can be shown to have constant temperature.

Given the E versus S diagram of a system A with specified values of amounts of constituents and volume, and a reservoir R at temperature T_R, we can draw a line of slope T_R tangent to the convex stable-equilibrium-state curve of system A, that is, tangent to the curve $A_{S1}A_0A_{E1}$ in Figure 14.2. The point of tangency A_0 represents the state A_0 in which system A is in mutual stable equilibrium with the reservoir because in state A_0 the system has a temperature $T_0 = (\partial E/\partial S)_{n,V}$ and, therefore, equal to the temperature T_R of the reservoir. In state A_0 the system has energy E_0, entropy S_0, and available energy $\Omega_0^R = 0$. For any other state A_1, using equation (14.12) with $c_R = T_R$ we have

$$\Omega_1^R = E_1 - E_0 - T_R(S_1 - S_0) \tag{14.53}$$

The tangent is also useful in providing a way to represent graphically the available energy of any state of A. For a given state A_1, we can easily show that the vertical distance of point A_1 from the tangent, that is, the energy $E_1 - E_a$ is equal to the available energy Ω_1^R of A_1 with respect to reservoir R.

14.13.8   Work Interactions

Graphical illustrations of reversible processes involving work-only interactions between two systems A and B are provided by Figure 14.3. In each process, the energy change of A is equal and opposite to the energy change of B, and the entropy changes of both A and B are zero because a work interaction does not transfer any entropy and the processes for both A and B are reversible.

As a result of the interaction depicted in Figure 14.3a, the state of A changes from state A_1 to state A_2 and that of B from state B_1 to state B_2, none being a stable equilibrium state. Moreover, the volume of either system A, or system B or both may or may not change.

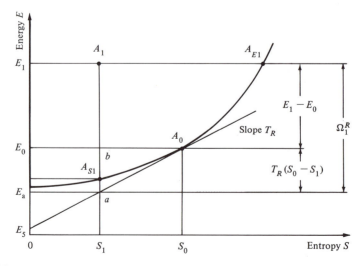

Figure 14.2 Graphical representation of the available energy with respect to a reservoir at temperature T_R.

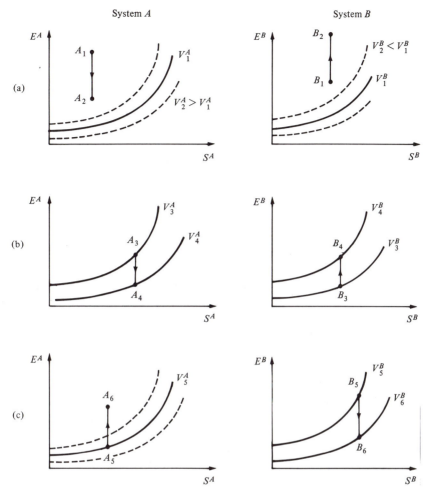

Figure 14.3   Graphical illustrations of work interactions on the energy versus entropy diagram. Each process for the composite of systems A and B is assumed to be reversible.

As a result of the interaction shown in Figure 14.3b, the state of A changes from A_3 to A_4 and that of B from B_3 to B_4, all being stable equilibrium states. Here the volume of system A changes from V_3^A to V_4^A, and the volume of system B from V_3^B to V_4^B.

As a result of the interaction shown in Figure 14.3c, the state of A changes from stable equilibrium state A_5 to state A_6 that is not stable equilibrium and may or may not have a different volume than A_5, whereas the state of B changes from state B_5 to state B_6 both being stable equilibrium states, but with different volumes V_5^B and V_6^B.

In the example of Figure 14.3a, irreversibility could occur in either A, or B, or both because either state A_2, or state B_2, or both could evolve spontaneously towards the corresponding stable equilibrium states. In the example of Figure 14.3c, irreversibility could occur in A but not in B because only state A_6 could evolve spontaneously, whereas stable equilibrium state B_6 could not. The processes in Figure 14.3b cannot become irreversible because the final states of both A and B are stable equilibrium states and, therefore, each has the highest entropy compatible with the corresponding energy.

14.13.9   Heat Interactions

A graphic illustration of a reversible process involving heat-only interactions between two systems A and B is provided in Figure 14.4. System A changes from state A_1 to

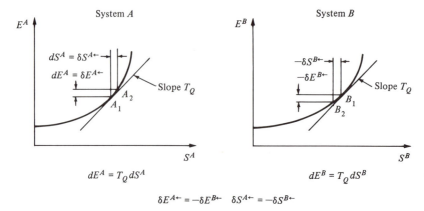

Figure 14.4 Graphical illustration of a heat-only interaction at temperature T_Q between two systems A and B initially at temperatures almost equal to T_Q.

state A_2, system B from state B_1 to state B_2, all being stable equilibrium states, without net changes in values of amounts of constituents and volumes. The temperatures of A and B are almost equal to T_Q. The two systems exchange energy and entropy. The ratio of the energy exchanged and the entropy exchanged is equal to the common temperature. Because the final states are stable equilibrium, no spontaneous changes of state can occur and, therefore, no entropy can be generated by irreversibility.

The need for the distinction between heat and other types of nonwork interactions is illustrated by the changes of state shown in Figure 14.5. System A is initially in a stable equilibrium state A_1 at temperature T_Q. As a result of interactions involving no net changes in values of amounts of constituents and volume, A decreases its energy by an amount $\delta E^{A\rightarrow}$. As the graph illustrates, this change in energy is consistent with each of the final states on the line $A_2 A_3$. Except for state A_2, every state on this line corresponds to a transfer of entropy $\delta S^{A\rightarrow}$ different from $\delta E^{A\rightarrow}/T_Q$. Therefore, either we call heat all the interactions that involve an exchange of both energy and entropy, but

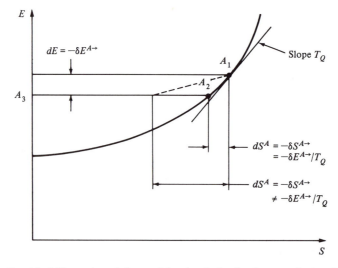

Figure 14.5 Graphical illustration of the need for the distinction between heat and other types of nonwork interactions.

then we cannot use the relation $\delta E^{A\rightarrow} = \delta Q^{\rightarrow} = T_Q\,\delta S^{A\rightarrow}$ for all these interactions, or we reserve the term heat for interactions for which $\delta Q^{\rightarrow} = T_Q\,\delta S^{\rightarrow}$, and then we need the term nonwork for interactions that involve exchanges of both energy and entropy, and we must realize that heat is only one special kind of nonwork interaction.

Problems

14.1 Do Problem 2.1.

14.2 Do Problem 2.3.

14.3 Do Problem 2.4.

14.4 Do Problem 2.7.

14.5 Do Problem 2.8.

14.6 Do Problem 2.10.

14.7 Do Problem 3.1.

14.8 Do Problem 3.5.

14.9 Do Problem 3.6.

14.10 Do Problem 3.7.

14.11 Do Problem 3.20.

14.12 Do Problem 3.21.

14.13 Do Problem 3.29.

14.14 Do Problem 4.1.

14.15 Do Problem 4.3.

14.16 Do Problem 4.5.

14.17 Do Problem 4.6.

14.18 Do Problem 5.1.

14.19 Do Problem 5.2.

14.20 Do Problem 6.1.

14.21 Do Problem 6.2.

14.22 Do Problem 6.3.

14.23 Do Problem 6.4.

14.24 Do Problem 7.1.

14.25 Do Problem 7.3.

14.26 Do Problem 7.5.

14.27 Do Problem 7.9.

14.28 Do Problem 7.11.

14.29 Do Problem 7.13.

14.30 Do Problem 8.1.

14.31 Do Problem 8.3.

14.32 Do Problem 8.6.

14.33 Do Problem 9.1.

14.34 Do Problem 9.2.

14.35 Do Problem 9.4.

14.36 Do Problem 9.8.

14.37 Do Problem 9.10.

14.38 Do Problem 9.11.

14.39 Do Problem 10.1.

14.40 Do Problem 10.3.

14.41 Do Problem 11.1.

14.42 Do Problem 11.2.

14.43 Do Problem 11.4.

14.44 Do Problem 12.1.

14.45 Do Problem 12.4.

14.46 Do Problem 12.5.

14.47 Do Problem 12.7.

14.48 Do Problem 12.8.

14.49 Do Problem 13.1.

14.50 Do Problem 13.7.

14.51 Do Problem 13.9.

14.52 Do Problem 13.10.

15 Heat Engines

In this chapter we provide a simple illustration of work and heat interactions and of the usefulness of energy and entropy balances by analyzing a special class of engines.

15.1 Definition of a Heat Engine

A *heat engine* is a device that experiences cyclic changes of its state, and heat interactions with two reservoirs A and B, each at a different constant temperature, and does work $W^{\rightarrow}$ on systems in the environment (Figure 15.1). Because of the cyclic changes of its state, the device is called *cyclic*.

The energy and entropy balances for the cyclic device are

$$0 = -W^{\rightarrow} + Q^{A\rightarrow} - Q^{B\leftarrow} \tag{15.1}$$

and

$$0 = S^{A\rightarrow} - S^{B\leftarrow} + S_{\text{irr}} \tag{15.2}$$

where $Q^{A\rightarrow}$ and $S^{A\rightarrow}$ are the energy and entropy transferred out of reservoir A, $Q^{B\leftarrow}$ and $S^{B\leftarrow}$ the energy and entropy received by reservoir B, and S_{irr} is the entropy generated by irreversibility in the cyclic device. The energy and entropy exchanged by each reservoir due to the respective heat interaction are related by

$$Q^{A\rightarrow} = T_A S^{A\rightarrow} \qquad \text{and} \qquad Q^{B\leftarrow} = T_B S^{B\leftarrow} \tag{15.3}$$

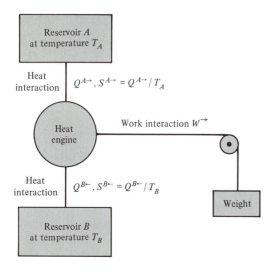

Figure 15.1 Schematic of a heat engine, that is, a device that experiences cyclic changes of its state, and heat interactions with two reservoirs A and B at different constant temperatures, and does work on systems in the environment, for example, by lifting a weight.

where T_A and T_B are the constant temperatures of reservoirs A and B, respectively. Eliminating $Q^{B \leftarrow}$, $S^{A \rightarrow}$, and $S^{B \leftarrow}$ from (15.1) to (15.3), we find that the work $W^{\rightarrow}$ done by the engine is given by the relation

$$W^{\rightarrow} = \frac{T_A - T_B}{T_A} Q^{A \rightarrow} - T_B S_{\text{irr}} \tag{15.4}$$

We see from equation (15.4) that work done ($W^{\rightarrow} > 0$) is the largest when the process is reversible, namely, when $S_{\text{irr}} = 0$ because the term $T_B S_{\text{irr}}$ is nonnegative. Moreover, energy and entropy must flow from the hotter reservoir into the engine so that $Q^{A \rightarrow} > 0$ if $T_A > T_B$ (or $Q^{B \leftarrow} < 0$ if $T_B > T_A$). This energy is used partly for the work done on the environment, and partly to dispose of the entropy received from the hotter reservoir and the entropy generated inside the engine into the colder reservoir so that $Q^{B \leftarrow} > 0$ if $T_A > T_B$ (or $Q^{A \rightarrow} < 0$ if $T_B > T_A$).

For $T_A > T_B$, the largest work $(W^{\rightarrow})_{\text{largest}}$ that can be done is

$$(W^{\rightarrow})_{\text{largest}} = \frac{T_A - T_B}{T_A} Q^{A \rightarrow} \tag{15.5}$$

In honor of Sadi Carnot, who first recognized its significance, we define the *Carnot coefficient*, η_{Carnot}, by the relation

$$\eta_{\text{Carnot}} = \frac{T_A - T_B}{T_A} \tag{15.6}$$

The value of the Carnot coefficient represents the largest fraction of the energy out of the hot reservoir at T_A that can be tranferred by the heat engine to other systems as work when the colder reservoir is at temperature T_B, that is, the fraction of the energy $Q^{A \rightarrow}$ that is available for conversion into work with respect to reservoir B. An engine that yields this fraction as work would be the best conceivable for the conditions specified, namely, a reversible heat engine, $S_{\text{irr}} = 0$. Because it yields in the form of work the entire fraction of the energy $Q^{A \rightarrow}$ that is available with respect to reservoir B, we say that this engine is 100% efficient.

The term $Q_A^{\rightarrow} (1 - T_B/T_A)$ can be interpreted as the *work equivalent with respect to a reservoir at temperature* T_B of the heat $Q_A^{\rightarrow}$ from a heat source at temperature T_A.

The engine is imperfect if it is irreversible, that is, if $S_{\text{irr}} > 0$. Then for a given amount $Q^{A \rightarrow}$ of energy taken out of reservoir A, the work done is less than the energy available for conversion to work by the amount $T_B S_{\text{irr}}$ [equation (15.4)], and the more the entropy generation by irreversibility S_{irr}, the smaller the work. The irreversible engine has a thermodynamic efficiency of less than 100%. The *thermodynamic efficiency of a heat engine* is defined as the ratio of the work done by the actual engine [equation (15.4)] and the work done if there is no entropy generated by irreversibility [equation (15.5)]. Accordingly, denoting the efficiency by ϵ we find that

$$\epsilon = \frac{W^{\rightarrow}}{(W^{\rightarrow})_{\text{largest}}} = \frac{(W^{\rightarrow})_{\text{largest}} - T_B S_{\text{irr}}}{(W^{\rightarrow})_{\text{largest}}} = 1 - \frac{T_B S_{\text{irr}}}{(W^{\rightarrow})_{\text{largest}}} \tag{15.7}$$

We see from (15.7) that if the amount of entropy generated by irreversibility in the engine

$$S_{\text{irr}} = \frac{(W^{\rightarrow})_{\text{largest}}}{T_B} = \frac{T_A - T_B}{T_A T_B} Q^{A \rightarrow} = \frac{\eta_{\text{Carnot}} Q^{A \rightarrow}}{T_B} \tag{15.8}$$

then the engine efficiency is equal to zero, the work done by the engine is zero, and all the capacity of $Q^{A \rightarrow}$ to yield work is wasted. If S_{irr} is even greater than given by equation (15.8), the engine efficiency is negative, the work is actually done on the engine instead of by the engine, and all the capacity of $Q^{A \rightarrow}$ to yield work as well as the work done on the engine are wasted.

The energy $Q^{A\rightarrow}$ supplied by reservoir A to the engine is accompanied by an amount of entropy $S^{A\rightarrow} = Q^{A\rightarrow}/T_A$. The engine does work and, therefore, transfers energy but no entropy to systems in the environment. Because the engine is cyclic, at least the entropy $S^{A\rightarrow}$ received from reservoir A must be transferred to reservoir B. But a transfer of entropy to reservoir B must be accompanied by an amount of energy because $Q^{B\leftarrow} = T_B S^{B\leftarrow}$, and therefore this energy is not available for transfer to the weight by the engine. Thus the larger the irreversibility the larger the values of $S^{B\leftarrow}$ and $Q^{B\leftarrow}$, and the smaller the fraction of $Q^{A\rightarrow}$ that remains for work done by the engine.

Example 15.1. A heat engine operates between a waste-energy source that we can model as a reservoir at temperature $T_A = 60°C = 333$ K, and the ocean that we can model as a reservoir at temperature $T_B = 10°C = 283$ K. The engine operates by flowing a pressurized organic fluid (Figure 15.2) through a heat exchanger (evaporator) where the fluid receives energy from the waste-energy source at the rate of 5000 kJ/s and is fully evaporated. The vapor, flowing through a turbine connected to an electric generator by means of a shaft, expands and exits the turbine at a pressure lower than the inlet pressure. It is then condensed by flowing into another heat exchanger (condenser), where it transfers energy to the ocean. The fluid is then recompressed by a pump that is also connected on the turbine shaft. If the thermodynamic efficiency of the heat engine is $\epsilon = 80\%$, find the rate at which work is done by the engine on the electric generator.

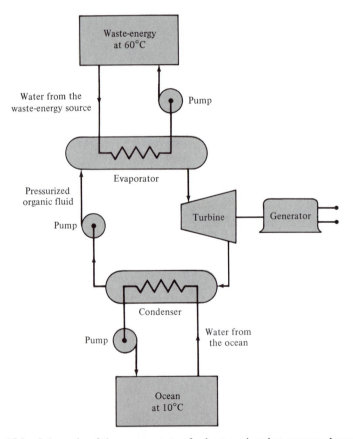

Figure 15.2 Schematic of the components of a heat engine that operates between a waste-energy source at 60°C and the ocean at 10°C.

Solution.   Every second, the engine receives an amount of energy $Q^{A\rightarrow} = 5000$ kJ from reservoir
A. The largest work that a heat engine can do under the specified conditions is given by
(15.5), that is, $(W^{\rightarrow})_{\text{largest}} = 5000(333 - 283)/333 = 750.7$ kJ, and therefore the work done
by the actual engine is $W^{\rightarrow} = 0.8 \times 750.7$ kJ $= 600.6$ kJ, and the rate of work done is
600.6 kW.

15.2   Heat Pumps and Refrigeration Units

A *heat pump* is a cyclic device that operates in a mode reverse to that of a heat engine. It
transfers energy out of a cold reservoir B at constant temperature T_B into a hot reservoir
A at constant temperature T_A by receiving work $W^{\leftarrow}$ from systems in the environment
(Figure 15.3). It is called a heat pump when its purpose is to transfer energy into the
hot reservoir. It is called a *refrigeration unit* when its purpose is to transfer energy out
of the cold reservoir.

Using the energy and entropy balances for the cyclic device [equations (15.1) and
(15.2)] and proceeding as in Section 15.1, we find that the work done on the device is
given by the relation

$$W^{\leftarrow} = \frac{T_A - T_B}{T_A}Q^{A\leftarrow} + T_B S_{\text{irr}} \tag{15.9}$$

Moreover, upon solving this relation for $Q^{A\leftarrow}$, we find that

$$Q^{A\leftarrow} = \frac{T_A}{T_A - T_B}W^{\leftarrow} - \frac{T_A T_B}{T_A - T_B}S_{\text{irr}} \tag{15.10}$$

We see from equation (15.10) that for a given work $W^{\leftarrow}$ done on the device ($W^{\leftarrow} > 0$),
the largest amount of energy $(Q^{A\leftarrow})_{\text{largest}}$ is transferred to the hot reservoir when the
process is reversible, $S_{\text{irr}} = 0$. This largest heat $(Q^{A\leftarrow})_{\text{largest}}$ is given by the relation

$$(Q^{A\leftarrow})_{\text{largest}} = \frac{T_A}{T_A - T_B}W^{\leftarrow} \tag{15.11}$$

The heat pump is imperfect if it is irreversible, that is, if $S_{\text{irr}} > 0$. Then for a given
work $W^{\leftarrow}$ done on the heat pump, the energy $Q^{A\leftarrow}$ transferred to the hot reservoir is

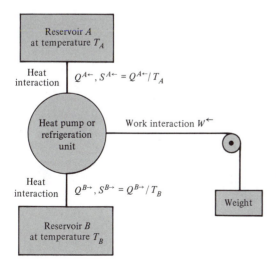

Figure 15.3   Schematic of a cyclic device
that operates in a mode reverse to that of a
heat engine. It is called a heat pump when
its purpose is to transfer energy into the
hot reservoir, or a refrigeration unit when
its purpose is to transfer energy out of the
cold reservoir.

less than the largest amount $(Q^{A\leftarrow})_{\text{largest}}$. We define the *coefficient of performance of a heat pump*, $\text{COP}_{\text{h.p.}}$, by the relation

$$\text{COP}_{\text{h.p.}} = \frac{Q^{A\leftarrow}}{W^{\leftarrow}} = \frac{T_A}{T_A - T_B}\left(1 - \frac{T_B S_{\text{irr}}}{W^{\leftarrow}}\right) \qquad (15.12)$$

The largest value of $\text{COP}_{\text{h.p.}}$, obtained when the heat pump is reversible, is equal to $T_A/(T_A - T_B)$. When the process is irreversible, then the $\text{COP}_{\text{h.p.}}$ is smaller than $T_A/(T_A - T_B)$ and can even be zero.

The thermodynamic efficiency of a heat pump is usually less than 100%. For a given $W^{\leftarrow}$, the *thermodynamic efficiency of a heat pump* is defined as the ratio of the heat $Q^{A\leftarrow}$ of the actual heat pump and the largest heat $(Q^{A\leftarrow})_{\text{largest}}$ achieved by the reversible heat pump, both operating between the same reservoirs, and denoted by $\epsilon_{\text{h.p.}}$ so that

$$\epsilon_{\text{h.p.}} = \frac{Q^{A\leftarrow}}{(Q^{A\leftarrow})_{\text{largest}}} = 1 - \frac{T_B S_{\text{irr}}}{W^{\leftarrow}} \qquad (15.13)$$

Example 15.2.   A heat pump operates between the air in a building that we model as a reservoir at ambient temperature $T_A = 20°C = 293$ K, and a cold water supply that we model as a reservoir at temperature $T_B = 7°C = 280$ K. The heat pump operates by flowing compressed vapor of an organic fluid (Figure 15.4) through a heat exchanger (condenser) where the vapor

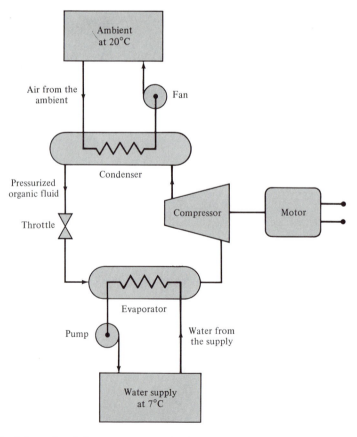

Figure 15.4 Schematic of the components of a heat pump that operates between the air in a building at 20°C and a cold water supply at 7°C.

condenses by transferring energy to the ambient air at the rate of 300 kJ/s. The condensed fluid is expanded through a throttle to a lower pressure and then flows through a heat exchanger (evaporator), where it evaporates by receiving energy from the cold water supply. The vapor is finally recompressed by a compressor on which work is done by the motor. If the efficiency of the heat pump is $\epsilon_{\text{h.p.}} = 60\%$, find the rate at which work is done on the compressor and the rate at which entropy is generated by irreversibility.

Solution. Every second, the energy transferred to the ambient A is $Q^{A\leftarrow} = 300$ kJ. The largest heat is $(Q^{A\leftarrow})_{\text{largest}} = Q^{A\leftarrow}/\epsilon_{\text{h.p.}} = 300/0.6 = 500$ kJ; hence, by equation (15.11) we have $W^{\rightarrow} = (T_A - T_B)(Q^{A\leftarrow})_{\text{largest}}/T_A = 13 \times 500/293 = 22.2$ kJ, and the rate of work on the compressor is 22.2 kW. By equation (15.13), we have $S_{\text{irr}} = (1 - \epsilon_{\text{h.p.}})W^{\rightarrow}/T_B = 0.4 \times 22.2/280 = 31.7$ J/K, and the rate of entropy generated by irreversibility is 31.7 W/K.

Alternatively, eliminating $Q^{A\leftarrow}$, $S^{A\leftarrow}$, and $S^{B\rightarrow}$ from the energy and entropy balances, we find that

$$W^{\leftarrow} = \frac{T_A - T_B}{T_B}Q^{B\rightarrow} + T_A S_{\text{irr}} \tag{15.14}$$

or, equivalently,

$$Q^{B\rightarrow} = \frac{T_B}{T_A - T_B}W^{\leftarrow} - \frac{T_A T_B}{T_A - T_B}S_{\text{irr}} \tag{15.15}$$

We see from (15.15) that for a given work $W^{\leftarrow}$ done on the device, the largest amount of energy $(Q^{B\rightarrow})_{\text{largest}}$ is transferred out of the cold reservoir when the process is reversible, $S_{\text{irr}} = 0$. This largest heat $(Q^{B\rightarrow})_{\text{largest}}$ is given by the relation

$$(Q^{B\rightarrow})_{\text{largest}} = \frac{T_B}{T_A - T_B}W^{\leftarrow} \tag{15.16}$$

We define the *coefficient of performance of a refrigeration unit*, $\text{COP}_{\text{r.u.}}$, by the relation

$$\text{COP}_{\text{r.u.}} = \frac{Q^{B\rightarrow}}{W^{\leftarrow}} = \frac{T_B}{T_A - T_B}\left(1 - \frac{T_A S_{\text{irr}}}{W^{\leftarrow}}\right) \tag{15.17}$$

When the process is irreversible, the $\text{COP}_{\text{r.u.}}$ is smaller than the largest value $T_B/(T_A - T_B)$, and can even be zero.

The *thermodynamic efficiency of a refrigeration unit* is defined as the ratio of the heat $Q^{B\rightarrow}$ of the actual unit and the largest heat $(Q^{B\rightarrow})_{\text{largest}}$ of a reversible refrigeration unit operating with equal $W^{\leftarrow}$ between the same reservoirs, and denoted by $\epsilon_{\text{r.u.}}$ so that

$$\epsilon_{\text{r.u.}} = \frac{Q^{B\rightarrow}}{(Q^{B\rightarrow})_{\text{largest}}} = 1 - \frac{T_A S_{\text{irr}}}{W^{\leftarrow}} \tag{15.18}$$

Example 15.3. A commercial freezer maintains the temperature of a refrigerated cell at $T_B = -18°C = 255$ K while the ambient temperature is $T_A = 27°C = 300$ K. The refrigeration unit operates only 20% of the time, and consumes electricity at the rate of 400 W. Find the average rate of energy extracted from the refrigerated cell if the efficiency $\epsilon_{\text{r.u.}} = 50\%$.

Solution. Every second of operation, the unit consumes 400 J of energy from electricity. On the average, the consumption is $W^{\leftarrow} = 0.2 \times 400 = 80$ J and, by equation (15.16), $(Q^{B\rightarrow})_{\text{largest}} = 255 \times 80/(300 - 255) = 453$ J. Therefore, $Q^{B\rightarrow} = 453 \times 0.5 = 226.5$ J, and the average rate at which energy is extracted from the refrigerated cell in order to maintain it at $-18°C$ is 226.5 W. By equation (15.18), $S_{\text{irr}} = (1 - \epsilon_{\text{r.u.}})W^{\leftarrow}/T_A = 0.5 \times 80/300 = 0.133$ J/K, and the average rate of entropy generation by irreversibility is 0.133 W/K.

Problems

15.1 A cyclic heat engine operating between an energy source at constant temperature $T_1 = 900$ K, and a reservoir at $T_0 = 300$ K produces half as much work as the energy extracted from the source.

(a) How much entropy is generated by the engine per unit of energy extracted from the source?

(b) If the engine operates between the same source as in part (a) and a reservoir at 150 K, and generates the same amount of entropy per unit of energy extracted from the source as in part (a), what is the work output of the engine per unit of energy extracted from the source?

15.2 An inventor claims to have solved the electricity problems of the United States by using the reservoirs and cyclic machinery shown in Figure P15.2. Energy Q_2 is pumped out of a reservoir R_2 at temperature T_2 to a reservoir R_3 at temperature $T_3 = 400$ K. The work input of the heat pump is generated by supplying an amount of energy also equal to Q_2 to a heat engine operating between the reservoirs R_2 and R_1. Reservoir R_1 is at the environmental temperature $T_1 = 300$ K.

(a) If the heat pump and the heat engine are perfect, what is the value of the temperature T_2?

(b) For your answer in part (a), compare the largest work we could produce using the energy supplied to reservoir R_3 with the largest work we could produce using the energy consumed from reservoir R_2, all with respect to the environmental reservoir R_1. What have we gained by using the invention with perfect machinery?

(c) If the temperatures are as those in part (a), but for the given energy Q_2 the heat engine produces only half as much work as a perfect heat engine, and the heat pump pumps only half as much energy from R_2 as a perfect heat pump, how much entropy is generated by irreversibility?

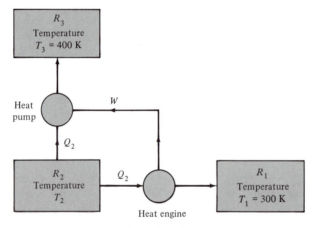

Figure P15.2

15.3 Solar energy maintains seawater at different depths at different temperatures. An ocean thermal energy conversion (OTEC) plant exploits this temperature difference to generate electricity. The plant (Figure P15.3) operates between the surface-level seawater at 20°C and seawater at a depth where the temperature is 0°C. The rate of energy transfer to the plant from the top-level seawater is 2400 MW, and from the plant to the bottom-level seawater 2300 MW. Because of the small temperature difference between the top and bottom levels, large amounts of seawater must be pumped through the plant's heat exchangers, and about 25% of the electricity output of the plant's turbo-generator is consumed in the plant by the circulation pumps of the sea water. The work into the pump of the turbine circuit is negligible.

(a) What is the net electric power output of the plant?

(b) What is the largest electric power that can be produced in the specified ocean location and for the specified energy intake in the upper level?

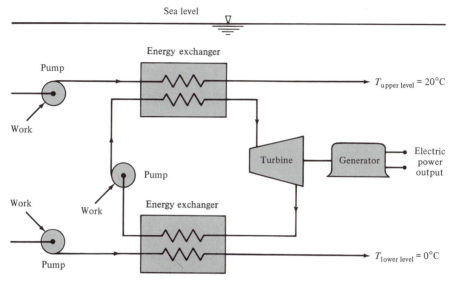

Figure P15.3

(c) What is the rate of entropy generation by irreversibility in the OTEC plant?

(d) What is the efficiency of the plant? What reduction of entropy generation would be necessary to increase the efficiency by one percentage point?

(e) Considering that the annual cost of capital, including minimal profits and maintenance and operating costs, is 25% of the investment, that electricity can be sold to the local utility at 5 cents/kWh, and that the OTEC plant operates 4000 h/yr, what is the largest investment (in dollars per kilowatt) that you could justify for the construction of the OTEC plant? How much would you be willing to invest to increase the efficiency of the OTEC plant by one percentage point?

15.4 A perfect heat engine operates between two reservoirs, but the transfer of energy and entropy between each reservoir and the engine is through a thermal resistance (Figure P15.4). Some authors call such an engine endoreversible. The two reservoirs are at temperatures T_H and T_L, respectively. The heat from the reservoir at T_H passes through one of the thermal resistances, so that the heat rate $\dot{Q}_H^{\rightarrow}$ is proportional to the temperature difference across the resistance, that is, $\dot{Q}_H^{\rightarrow} = K(T_H - T_1)$, where T_1 is the temperature of the heat interaction between the re-

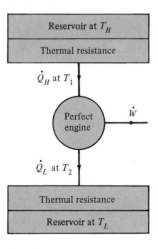

Figure P15.4

sistance and the perfect heat engine. Similarly, the heat to the reservoir at T_L passes through the other thermal resistance, so that the heat rate $\dot{Q}_L^{\leftarrow}$ is proportional to the temperature difference across the resistance, that is, $\dot{Q}_L^{\leftarrow} = K(T_2 - T_L)$, where T_2 is the temperature of the heat interaction between the perfect heat engine and the resistance. Notice that the two thermal resistances have identical thermal conductivity K. Answer all questions in terms of T_H, T_L, and K.

(a) What is the value of $\dot{Q}_H^{\rightarrow}$ for which there is no irreversibility?

(b) What is the value of $\dot{Q}_H^{\rightarrow}$ for which the entropy generation by irreversibility is largest?

(c) Find an expression for the temperature T_2 in terms of T_1, T_H, and T_L when the engine operates in neither of the extremes in parts (a) and (b).

(d) Express the engine output power $\dot{W}^{\rightarrow}$ in terms of T_1, T_H, T_L, and K.

(e) Verify that $\dot{W}^{\rightarrow}$ is largest when $T_2/T_1 = (T_L/T_H)^{1/2}$.

15.5 A food-processing plant requires that two large process areas A and B be kept at constant temperatures $T_A = -8°C$ and $T_B = 12°C$. The refrigeration load of area A is 175 kW or 49.7 tons of refrigeration (1 ton of refrigeration = 12,000 Btu/h = 12,660 kJ/h). The heating load of area B is 210 kW. Conditioning of the two spaces is achieved by an electric air-conditioning and heating unit, as shown in Figure P15.5. In addition to A and B, the electric unit is also interacting with the environment at $T_0 = 5°C$.

(a) What is the rate of energy exchanged between the unit and the environment under the best conditions of operation?

(b) What is the least amount of electric power required to operate the air-conditioning and heating unit in part (a)?

(c) If the unit is perfect but requires a temperature difference of $15°C$ in order to transfer energy between itself and each of the three constant temperature systems, A, B, and environment, what is the electric power required by the unit?

(d) How much entropy is generated by irreversibility in part (c)?

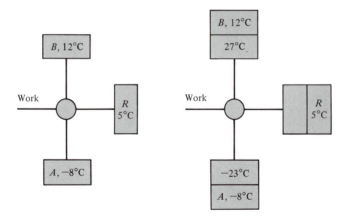

Figure P15.5

15.6 A heat pump transfers 36 MJ/h of energy to a heated space at 450 K from an environment at 300 K. The thermodynamic efficiency of the pump is 75%.

The pump is powered by a heat engine that operates between a reservoir at 1000 K and the heated space at 450 K, and has a thermodynamic efficiency of 60%. Determine the following.

(a) The power input to the heat pump.

(b) The rate at which energy is rejected to the heated space by the heat engine.

(c) The rate of entropy generation by irreversibilitry in both the heat engine and the heat pump.

(d) The heat rate from the high-temperature reservoir if both the heat engine and the heat pump are reversible. How does this answer differ from the answer when the two devices are imperfect as specified?

15.7 A building is kept at 20°C by electric resistance heaters. It requires an average of 50 kW during a cold month. If the building were heated by a heat pump that interacts with the environment at −5°C and has a coefficient of performance of 5.5, find the following.

(a) The power input to the heat pump.

(b) The entropy generated by irreversibility.

(c) The energy difference between the two methods of heating, assuming that power plants consume 3 units of energy to generate 1 unit of electrical energy.

16 Systems with Volume as the Only Parameter

An important application of thermodynamics is the study of properties of substances in stable equilibrium states. Under ordinary atmospheric conditions, substances exist naturally in such states. Under conditions different from atmospheric, substances can be placed in stable equilibrium states with relative ease.

For stable equilibrium states, experimental results can be obtained more easily than for other types of states because the values of all properties do not change with time. Moreover, analytical interrelations between properties assume their simplest mathematical forms because, by virtue of the stable equilibrium state principle, the number of independent variables is minimal.

We begin our study of properties of substances by considering situations for which the substance of interest can be modeled fairly accurately as a system having volume as the only parameter, that is, volume is the only quantity necessary to describe the effects of the external forces on the constituents of the system.

In this chapter we provide a general discussion of such systems and many interrelations between their stable equilibrium state properties.

16.1 General Remarks

We consider a system that (1) is confined in a region of space of variable volume and (2) has volume as the only parameter and is not influenced by external forces due to gravity, electricity, magnetism, shear deformation, capillarity, and other surface effects.

A system is not influenced by applied or external forces or force fields due to gravity, electricity, and magnetism when the intensity of each of these fields is zero. If an applied field has a fixed intensity different from zero, its effects can be superimposed to those found in its absence provided that the system has a small spatial extension in the direction of the field so that the resulting effects are uniform throughout the extension of the system, and that the position of the system is fixed with respect to the systems that generate the force field.

A system is not subject to shear deformation when the stresses generated by the external forces on its boundaries are isotropic—the same in all directions. If the stresses are anisotropic, their effects can be superimposed to those found in the absence of stresses provided that the system has a small spatial extension so that the resulting deformation field is uniform throughout the extension of the system.

A system is not influenced by capillarity when surface tension phenomena—phenomena that are particularly pronounced at interfaces between liquids and solids—are negligible with respect to phenomena occurring in the bulk of the system.

Surface effects are negligible when the amount of matter within a few monolayers in the vicinity of the surface enclosing the system volume is small compared with the total amount of the matter within the volume.

The systems defined by conditions (1) and (2) are very useful for modeling stable equilibrium states of substances. In this chapter we discuss some of the relations that result from this modeling.

16.2 Independent Properties

In studying experimentally the properties of various substances in stable equilibrium states, we are faced with two questions: (1) what quantities should we measure so as to fully describe each state, that is, what is a proper set of independent variables or independent properties which fully describe these states? and (2) how can we reduce the number of measurements, and yet obtain values of properties for all stable equilibrium states?

We explore the answers to these questions by studying the stable equilibrium states of a system with volume as the only parameter as a model of a real substance. According to the state principle, one answer is that each property of such a system in a stable equilibrium state can be expressed as a single-valued function of the $r + 2$ independent variables: energy E, volume V, and amounts of the r constituents $n_1, n_2, \ldots, n_r$ of the system. In particular, if P_1 and P_2 are two independent properties, we may write

$$P_1 = P_1(E, V, n) \tag{16.1}$$

$$P_2 = P_2(E, V, n) \tag{16.2}$$

where the form of each of the functions $P_1(E, V, n)$ and $P_2(E, V, n)$ depends on the system. We recall that the definition of a system requires not only the types and amounts of its constituents, but also the internal forces and constraints. Here we assume that the constraints inhibit all internal reaction mechanisms capable of altering the values of the amounts of constituents.

Because properties P_1 and P_2 are independent of one another, we can solve (16.1) and (16.2) for E and V, and find

$$E = E(P_1, P_2, n) \tag{16.3}$$

$$V = V(P_1, P_2, n) \tag{16.4}$$

Thus any expression for any other property P_3 of the form

$$P_3 = P_3(E, V, n) \tag{16.5}$$

may also be written as an expression of the form

$$P_3 = P_3(P_1, P_2, n) \tag{16.6}$$

by using equations (16.3) and (16.4) to express E and V in terms of P_1, P_2, and n. Equation (16.6) is a simple illustration of the fact that we can use $r + 2$ independent variables other than E, V, and n.

Traditionally, the energy of a system with volume as the only parameter is called *internal energy* and is denoted by the symbol U instead of E. Thus the fundamental relation is expressed in the form

$$S = S(U, V, n) \tag{16.7}$$

where the explicit form of the function $S(U, V, n)$ is determined by the system. Moreover, equation (16.7) can be solved for U so that

$$U = U(S, V, n) \tag{16.8}$$

where again the explicit form of the function $U(S, V, n)$ depends on the system.

Knowing the form of $U(S, V, n)$, we can readily obtain many other properties of stable equilibrium states as functions of S, V, and n. For example, we can readily evaluate the temperature, pressure, and chemical potentials at each stable equilibrium state by the relations

$$T(S, V, n) = \left(\frac{\partial U}{\partial S}\right)_{V,n} \tag{16.9}$$

$$p(S, V, n) = -\left(\frac{\partial U}{\partial V}\right)_{S,n} \tag{16.10}$$

$$\mu_i(S, V, n) = \left(\frac{\partial U}{\partial n_i}\right)_{S,V,n} \qquad \text{for } i = 1, 2, \ldots, r \tag{16.11}$$

and therefore we can also evaluate the properties

$$H = U + pV \qquad A = U - TS \qquad G = U - TS + pV$$

As we discuss later, properties H, A, and G appear in many practical studies.

We can make similar comments about the results we obtain from the form $S(U, V, n)$, that is, by using U, V, and n as independent variables.

We see that the questions posed at the beginning of this section would be answered if the explicit form of either function $S(U, V, n)$ or function $U(S, V, n)$ in (16.7) or (16.8) were known. With very few exceptions, the form of either of these functions is not available for any system with volume as the only parameter because of mathematical complexities arising from the transcendental nature of the quantum theoretical interrelations between the properties of a system in a stable equilibrium state. Only for some very special limiting conditions the forms of equations (16.7) and (16.8) can be written explicitly.

For example, we show later that the forms are explicit when a substance behaves as an incompressible fluid or solid, or as an ideal gas. Again, for the relatively high temperature stable equilibrium states of the electromagnetic field in a cavity of volume V that confines all frequency modes, the form of the fundamental relation is known. In this system there is only one constituent, the electromagnetic field, and its amount is fixed and equal to unity. The fundamental relation is given by the expression

$$S = (4/3)(a V U^3)^{1/4} \tag{16.12}$$

where $a = 8\pi^5 k^4/15h^3 c^3 = 7.565 \times 10^{-16}$ J/m^{3}K^4, k is the Boltzmann constant ($k = 1.38066 \times 10^{-23}$ J/K), h the Planck constant ($h = 6.6260 \times 10^{-34}$ J s), and c the speed of light in vacuum ($c = 2.9979 \times 10^8$ m/s). Note that S is not written as a function of n because here n is fixed and equal to unity.

Example 16.1.   Find expressions for the internal energy $U(S, V)$, temperature $T(S, V)$, and pressure $p(S, V)$ of the electromagnetic field just defined.

Solution.   Solving (16.12) for U, and using the definitions of T and p, we find that

$$U = \left(\frac{3}{4}\right)^{4/3} \left(\frac{S^4}{aV}\right)^{1/3}$$

$$T = \left(\frac{\partial U}{\partial S}\right)_V = \left(\frac{3S}{4aV}\right)^{1/3} = \left(\frac{U}{aV}\right)^{1/4}$$

$$p = -\left(\frac{\partial U}{\partial V}\right)_S = \frac{1}{3}\left(\frac{3}{4}\right)^{4/3}\frac{1}{V}\left(\frac{S^4}{aV}\right)^{1/3} = \frac{U}{3V} = \frac{1}{3}aT^4$$

where in writing the expressions $T(U, V)$, $p(U, V)$, and $p(T)$ we use (16.12).

Another example of a system for which the fundamental relation is known is the electromagnetic field at relatively high temperature in a cavity of volume V that confines only the modes with frequencies between ν and $\nu + \Delta\nu$. In the limit as $\Delta\nu$ tends to zero, the fundamental relation for this system is shown to be expressible as

$$ S = \frac{kU}{h\nu}\left[\left(1 + \frac{bV\nu^3\Delta\nu}{U}\right)\ln\left(1 + \frac{bV\nu^3\Delta\nu}{U}\right) - \frac{bV\nu^3\Delta\nu}{U}\ln\frac{bV\nu^3\Delta\nu}{U}\right] \qquad (16.13) $$

where $b = 8\pi h/c^3 = 5.553 \times 10^{-57}$ J s^4/m^3.

Example 16.2. Find expressions for the temperature T and the pressure p of the electromagnetic field under the conditions just defined. For a given value of the temperature T, show that U, S, and p are proportional to $\Delta\nu$ and therefore infinitesimal. Thus find expressions for the energy per unit frequency $U_\nu = U/\Delta\nu$, the entropy per unit frequency $S_\nu = S/\Delta\nu$, and the pressure per unit frequency $p_\nu = p/\Delta\nu$ in terms of V, T, and ν.

Solution. Using (16.13) and the definition of T, we find that

$$ \frac{1}{T} = \left(\frac{\partial S}{\partial U}\right)_V = \frac{k}{h\nu}\ln\left(1 + \frac{bV\nu^3\Delta\nu}{U}\right) $$

or, equivalently,

$$ U_\nu = \frac{U}{\Delta\nu} = \frac{bV\nu^3}{\exp(h\nu/kT) - 1} $$

Again, using (16.13) and the definition of p, we find

$$ p = T\left(\frac{\partial S}{\partial V}\right)_U = \frac{kT\,b\nu^2\Delta\nu}{h}\ln\left(1 + \frac{U}{bV\nu^3\Delta\nu}\right) $$

$$ p_\nu = \frac{p}{\Delta\nu} = \frac{kT\,b\nu^2}{h}\ln\frac{1}{1 - \exp(-h\nu/kT)} $$

$$ S_\nu = \frac{S}{\Delta\nu} = \frac{k\,b\nu^2 V}{h}\left[\frac{h\nu/kT}{\exp(h\nu/kT) - 1} + \ln\frac{1}{1 - \exp(-h\nu/kT)}\right] = \frac{1}{T}\left(U_\nu + p_\nu V\right) $$

where in writing the expressions for p_ν and S_ν we use the relation between U_ν and T. It is noteworthy that the energy, entropy, and pressure defined by $U = \int_0^\infty U_\nu\,d\nu$, $S = \int_0^\infty S_\nu\,d\nu$, and $p = \int_0^\infty p_\nu\,d\nu$ satisfy all the relations obtained in Example 16.1.

For most conditions for which no explicit analytical expressions are available, the state principle and its implications provide guidance about the number of properties or variables that need be considered, about interpolations between and extrapolations of experimental results that are consistent with the laws of physics, and about procedures for carrying out measurements of properties that are of practical importance. Experimental data are presented in the forms of tables, charts, or approximate correlations. These forms are obtained by an interplay between theory and experiments.

For example, sparse measurements of values of some properties of a given amount of water are made as functions of two independent variables, temperature and volume. The values of these properties are then completed, and other properties are evaluated by using the general interrelations between properties of neighboring states implied by the state principle. The combined experimental and analytical results are then presented in either tabular form, or graphical form, or both. The properties of water are discussed later.

Some of the techniques used in the interplay between theory and experiment are discussed in the following sections.

16.3   Characteristic Functions

In the preceding section we note that if the function $U(S, V, n_1, n_2, \ldots, n_r)$ is known, the properties $T(S, V, n)$, $p(S, V, n)$, $\mu_1(S, V, n)$, $\mu_2(S, V, n)$, $\ldots$, $\mu_r(S, V, n)$ can be derived by differentiation of the function with respect to each of its independent variables. In addition, a large number of other properties can be derived as combinations of the function $U(S, V, n)$, its independent variables S, V, n, and its derivatives. A function such as U from which all the stable equilibrium state properties can be derived by differentiation and algebraic manipulations is called a *characteristic function*.[1] Clearly, in view of the state principle, any function from which we can reconstruct the function $U(S, V, n)$ by using only differentiation and algebra is a characteristic function, because the values of U, V, and n determine the values of all the properties of a stable equilibrium state.[2] Accordingly, the function $S(U, V, n)$ is characteristic because we can solve it for $U(S, V, n)$.

There are many characteristic functions other than the functions $S(U, V, n)$ and $U(S, V, n)$. One is the *enthalpy* $H(S, p, n)$ defined by the relation

$$H = U + pV \tag{16.14}$$

To prove this assertion, we begin by considering the total differential of H, that is,

$$dH = dU + p\,dV + V\,dp \tag{16.15}$$

The term dU may be expressed as a total differential of equation (16.8), that is,

$$dU = T\,dS - p\,dV + \mu_1\,dn_1 + \mu_2\,dn_2 + \cdots + \mu_r\,dn_r \tag{16.16}$$

where in writing this differential we use the definitions of T, p, μ_1, μ_2, $\ldots$, μ_r. Relation 16.16 is called the *Gibbs relation*. Combining equations (16.15) and (16.16), we obtain

$$dH = T\,dS + V\,dp + \mu_1\,dn_1 + \mu_2\,dn_2 + \cdots + \mu_r\,dn_r \tag{16.17}$$

Because H is a function of at most $r+2$ variables, equation (16.17) for the total differential dH implies that H can be regarded as a function of S, p, n_1, n_2, $\ldots$, n_r. As such, the coefficients of the differentials in the right-hand side must be related to the partial derivatives of H, that is,

$$T = \left(\frac{\partial H}{\partial S}\right)_{p,n} \tag{16.18}$$

$$V = \left(\frac{\partial H}{\partial p}\right)_{S,n} \tag{16.19}$$

$$\mu_i = \left(\frac{\partial H}{\partial n_i}\right)_{S,p.n} \qquad \text{for } i = 1,\ 2,\ \ldots,\ r \tag{16.20}$$

Equation (16.19) and the relation $U = H - pV$ show that we can derive V and U from $H(S, p, n)$ by differentiation and algebraic manipulations only, that is, we derive the relations $V = V(S, p, n)$ and $U = U(S, p, n)$. Eliminating p from these relations, we

---

[1]The term characteristic function was introduced by H. M. Massieu, *C. R. Acad. Sci. Paris*, 69, 858, 1057 (1869).

[2]The distinction between functions that are characteristic and others that are not may be illustrated by comparing $U(S, V, n)$ and $U(T, V, n)$. The function $U(T, V, n)$ may be derived from $U(S, V, n)$ because $T = (\partial U/\partial S)_{V,n} = T(S, V, n)$, and upon solving the last relation for $S = S(T, V, n)$ we can substitute the result in $U(S, V, n)$ to find $U(T, V, n)$. On the other hand, the function $U(S, V, n)$ cannot be found from $U(T, V, n)$ by differentiation, because S can be obtained only by integration of $(\partial S/\partial T)_{V,n} = (1/T)(\partial U/\partial T)_{V,n}$.

obtain $U = U(S, V, \boldsymbol{n})$, and therefore we can deduce expressions for all the properties of stable equilibrium states. It follows that $H(S, p, \boldsymbol{n})$ is a characteristic function. We will see that enthalpy appears in many practical problems.

Another characteristic function is the *Helmholtz free energy*[3] $A(T, V, \boldsymbol{n})$, defined by the relation

$$A = U - TS \tag{16.21}$$

Indeed, proceeding in a manner analogous to what we have just done for the enthalpy, we find that

$$dA = dU - T \, dS - S \, dT$$
$$= -S \, dT - p \, dV + \mu_1 \, dn_1 + \mu_2 \, dn_2 + \cdots + \mu_r \, dn_r \tag{16.22}$$

where in writing the second of equations (16.22) we use equation (16.16). From the last equation, we conclude that A can be regarded as a function of T, V, n_1, n_2, ..., n_r, and that

$$S = -\left(\frac{\partial A}{\partial T}\right)_{V,n} \tag{16.23}$$

$$p = -\left(\frac{\partial A}{\partial V}\right)_{T,n} \tag{16.24}$$

$$\mu_i = \left(\frac{\partial A}{\partial n_i}\right)_{T,V,n} \qquad \text{for } i = 1, 2, \ldots, r \tag{16.25}$$

Again, equation (16.23) and the relation $U = A + TS$ show that we can derive S and U from $A(T, V, \boldsymbol{n})$ by differentiation and algebraic manipulations only, that is, we derive the relations $S = S(T, V, \boldsymbol{n})$ and $U = U(T, V, \boldsymbol{n})$. Eliminating T from these relations, we obtain $U = U(S, V, \boldsymbol{n})$, and therefore we can deduce all the properties of stable equilibrium states. It follows that $A(T, V, \boldsymbol{n})$ is a characteristic function. In Section 16.5.3 we show that the Helmholtz free energy is related to the optimum work associated with isothermal changes between stable equilibrium states.

Another important characteristic function is the *Gibbs free energy*[4] $G(T, p, \boldsymbol{n})$, defined by the relation

$$G = U - TS + pV = H - TS \tag{16.26}$$

Here, too, we can proceed as in the case of enthalpy, and find

$$dG - dII - T \, dS - S \, dT$$
$$= -S \, dT + V \, dp + \mu_1 \, dn_1 + \mu_2 \, dn_2 + \cdots + \mu_r \, dn_r \tag{16.27}$$

where in writing the second of equations (16.27) we use (16.17). The last equation indicates that, as long as T and p can be varied independently,[5] G can be regarded as a

---

[3]This property is named in honor of Hermann Ludwig Ferdinand von Helmholtz (1821–1894), German physicist, anatomist, and physiologist, who first introduced the term free energy and used it in evaluating the work involved in isothermal reversible processes. He introduced the term free energy in a paper published in *Sitzungsber. K. Preuss. Akad. Wiss., I*, 22 (1882).

[4]This property is named in honor of Josiah Willard Gibbs (1839–1903), American mathematician and physicist, who first demonstrated the great utility of both functions A and G in the interpretation of the most diverse physicochemical phenomena. His comprehensive work was published in *Trans. Acad. Arts Sci., II*, 309, 382 (1873); *III*, 108, 343 (1875).

[5]For example, we show later that for systems composed of a single type of constituent, T and p cannot always be varied independently.

function of T, p, and n, and that

$$S = -\left(\frac{\partial G}{\partial T}\right)_{p,n} \tag{16.28}$$

$$V = \left(\frac{\partial G}{\partial p}\right)_{T,n} \tag{16.29}$$

$$\mu_i = \left(\frac{\partial G}{\partial n_i}\right)_{T,p,n} \qquad \text{for } i = 1, 2, \ldots, r \tag{16.30}$$

Clearly, equations (16.28) and (16.29), and the relation $U = G + TS - pV$ show that we can derive S, V, and U by differentiation and algebraic manipulations only, that is, we derive the relations $S = S(T,p,n)$, $V = V(T,p,n)$, and $U = U(T,p,n)$. Eliminating T and p from these relations we obtain $U = U(S,V,n)$ and, therefore, we conclude that $G(T,p,n)$ is a characteristic function. In Section 16.5.3 we show that the Gibbs free energy is related to the optimum work associated with changes between stable equilibrium states under conditions of constant temperature and pressure.

More generally, characteristic functions can be derived by means of Legendre transformations of the fundamental relation $S = S(U,V,n)$.[6]

Equations (16.9) to (16.11), (16.18) to (16.20), (16.23) to (16.25), and (16.28) to (16.30) are a small sample of the numerous interrelations that exist between properties, volume, and amounts of constituents for stable equilibrium states. These interrelations can be used for a variety of purposes, such as to evaluate unmeasured properties in terms of others that have been measured, to interpolate and extrapolate experimental data, to disclose new effects, and to check the consistency of observations. Therefore, they are both delimiting and practical. Another sample of interrelations with similar practical uses can be obtained by considering second-order partial derivatives of characteristic functions. They are discussed in the next section.

16.4   Maxwell Relations

Given a function of several independent variables, a second partial derivative with respect to any two of these variables is independent of the order of differentiation. For example, given $f(x,y,z)$ we have

$$\frac{\partial}{\partial x}\left[\frac{\partial f(x,y,z)}{\partial y}\right] = \frac{\partial}{\partial y}\left[\frac{\partial f(x,y,z)}{\partial x}\right] \tag{16.31}$$

Applying this mathematical rule to the characteristic functions $U(S,V,n)$, $H(S,p,n)$, $A(T,V,n)$, and $G(T,p,n)$, and using equations (16.9), (16.10), (16.18), (16.19), (16.23), (16.24), (16.28), and (16.29) for the first partial derivatives, we find that

$$\left(\frac{\partial T}{\partial V}\right)_{S,n} = -\left(\frac{\partial p}{\partial S}\right)_{V,n} \tag{16.32}$$

$$\left(\frac{\partial T}{\partial p}\right)_{S,n} = \left(\frac{\partial V}{\partial S}\right)_{p,n} \tag{16.33}$$

$$\left(\frac{\partial S}{\partial V}\right)_{T,n} = \left(\frac{\partial p}{\partial T}\right)_{V,n} \tag{16.34}$$

---

[6]Legendre transformations are discussed in many texts. For example, see M. Modell and R. C. Reid, *Thermodynamics and Its Applications*, Prentice-Hall, Englewood Cliffs, N.J., 1974, pp. 119–137, and H. B. Callen, *Thermodynamics and an Introduction to Thermostatistics*, Wiley, New York, 1985, pp. 137–148.

$$\left(\frac{\partial S}{\partial p}\right)_{T,n} = -\left(\frac{\partial V}{\partial T}\right)_{p,n} \tag{16.35}$$

Each of equations (16.32) to (16.35) is called a *Maxwell relation*.[7]

Maxwell relations have many practical applications. For example, equation (16.34) relates changes in entropy to relatively easily measurable changes in temperature, pressure, and volume and is therefore useful in measuring entropy changes between stable equilibrium states.

Example 16.3. Temperature, pressure, and volume measurements performed on 1 kg of H_2O in three different stable equilibrium states A_1, A_2, and A_3 yield the following results: $T_1 = T_2 = 673$ K, $T_3 = 774$ K, $V_1 = V_3 = 0.09935$ m^3, $V_2 = 0.08453$ m^3, $p_1 = 3$ MPa, and $p_2 = 3.5$ MPa. Estimate the difference in entropy $S_2 - S_1$, and compare it to the measured value of -81 J/K.

Solution. Modeling the water as a system with volume as the only parameter, and considering the measured changes as small, we can express differences in terms of partial derivatives and, conversely, partial derivatives as ratios of differences. Specifically, because states A_1 and A_2 have the same temperature, we can write the approximate relation $S_2 - S_1 \approx (\partial S/\partial V)_{T,n}(V_2 - V_1) = (\partial p/\partial T)_{V,n}(V_2 - V_1)$, where in writing the second equation we use the Maxwell relation given by (16.34). Using the data for states A_1 and A_3 which have the same volume, we can estimate $(\partial p/\partial T)_{V,n}$ as a ratio of differences, that is, $(\partial p/\partial T)_{V,n} \approx (p_3 - p_1)/(T_3 - T_1)$. Thus $S_2 - S_1 \approx (p_3 - p_1)(V_2 - V_1)/(T_3 - T_1) = 0.5 \times 10^6$ N/m$^2(-0.01482$ $m^3)/101$ K $= -73.4$ J/K. The estimate differs from the measured value of -81 J/K by 9%. The error is due to our assumption that the measured changes are small.

Many more Maxwell relations can be derived by considering second-order derivatives that involve the amounts of constituents. For example,

$$\left[\frac{\partial}{\partial T}\left(\frac{\partial G}{\partial n_i}\right)_{T,p,n}\right]_{p,n} = \left[\frac{\partial}{\partial n_i}\left(\frac{\partial G}{\partial T}\right)_{p,n}\right]_{T,p,n} \tag{16.36a}$$

or, equivalently,

$$\left(\frac{\partial \mu_i}{\partial T}\right)_{p,n} = -\left(\frac{\partial S}{\partial n_i}\right)_{T,p,n} \tag{16.36b}$$

where in writing equation (16.36b) we use equations (16.28) and (16.30). Again,

$$\left[\frac{\partial}{\partial p}\left(\frac{\partial G}{\partial n_i}\right)_{T,p,n}\right]_{T,n} = \left[\frac{\partial}{\partial n_i}\left(\frac{\partial G}{\partial p}\right)_{T,n}\right]_{T,p,n} \tag{16.37a}$$

or, equivalently,

$$\left(\frac{\partial \mu_i}{\partial p}\right)_{T,n} = \left(\frac{\partial V}{\partial n_i}\right)_{T,p,n} \tag{16.37b}$$

where in writing (16.37b) we use equations (16.29) and (16.30). Later, we find that equations (16.36b) and (16.37b) are useful in the description of the stable-equilibrium-state properties of multiconstituent systems which are traditionally called mixtures.

In general, by equating second partial derivatives of a characteristic function, such as $U(S, V, n)$, $H(S, p, n)$, $A(T, V, n)$, and $G(T, p, n)$, with respect to any pair of its $r + 2$ independent variables, we obtain a Maxwell relation. Thus each characteristic function generates $(r + 2)(r + 1)/2$ Maxwell relations.

---

[7]These relations are named in honor of James Clerk Maxwell (1831–1879), Scottish physicist, who first stated them and recognized their importance in his book *Theory of Heat*, Longmans, London, 1885, p. 169.

16.5   Heat and Work Interactions

Many changes of states of a system with volume as the only parameter can be brought about by means of heat and work interactions that do not change the values of the amounts of constituents. For processes that involve heat and work interactions only, and that start from a stable equilibrium state and end in a neighboring stable equilibrium state, the change in internal energy dU, the change in entropy dS, the heat $\delta Q^{\leftarrow}$ at temperature T_Q, and the work $\delta W^{\rightarrow}$ satisfy the energy and entropy balances

$$dU = \delta Q^{\leftarrow} - \delta W^{\rightarrow} \tag{16.38}$$

and

$$dS = \frac{\delta Q^{\leftarrow}}{T_Q} + \delta S_{\text{irr}} \tag{16.39}$$

where δS_{irr} denotes the nonnegative entropy generated by irreversibility within the system. Moreover, the energy and entropy changes are also interrelated by (16.16). For neighboring states that do not differ in their values of the amounts of constituents, namely, for $dn_i = 0$ for $i = 1, 2, \ldots, r$, we have

$$dU = T \, dS - p \, dV \tag{16.40}$$

where dV is the change in volume of the system, and T and p are the temperature and pressure of the initial stable equilibrium state.

For some special processes, changes of properties can be related directly to the heat and the work. For example, in a reversible process involving both heat and work, either simultaneously or consecutively, with heat occurring at a temperature T_Q equal to the temperature T of the initial stable equilibrium state, equations (16.38) to (16.40) yield the relations $\delta Q^{\leftarrow} = T \, dS$ and $\delta W^{\rightarrow} = p \, dV$ because $\delta S_{\text{irr}} = 0$ and $T_Q = T$. However, if either the process is irreversible, or $T_Q \neq T$, or both, the heat and the work cannot be related directly to $T \, dS$ and $p \, dV$, namely, $\delta Q^{\leftarrow} \neq T \, dS$ and $\delta W^{\rightarrow} \neq p \, dV$. Indeed, then equations (16.38) to (16.40) imply that

$$\delta W^{\rightarrow} = p \, dV + \left(1 - \frac{T}{T_Q} \right) \delta Q^{\leftarrow} - T \, \delta S_{\text{irr}} \tag{16.41}$$

and

$$\delta Q^{\leftarrow} = T_Q \, dS - T_Q \, \delta S_{\text{irr}} \tag{16.42}$$

More examples of relations between changes in properties and heat and work are discussed in Sections 16.5.1 to 16.5.4.

16.5.1   Constant-Volume Processes

By means of an interaction, a system initially in a stable equilibrium state at temperature T and confined in a chamber with fixed volume V can be changed to a neighboring stable equilibrium state having the same values of amounts of constituents. The interaction may be either work, or heat with another system at temperature T_Q. For example, the work can be done by a falling weight connected to a rotating paddle wheel (Figure 16.1a) or by a dynamo connected to an electrical resistor (Figure 16.1b). A constant-volume process of a system is called *isochoric*.

Modeling the system as one with volume as the only parameter, we find that for $dV = 0$, and $dn_i = 0$ for $i = 1, 2, \ldots, r$, equation (16.16) reduces to

$$dU = T \, dS \tag{16.43}$$

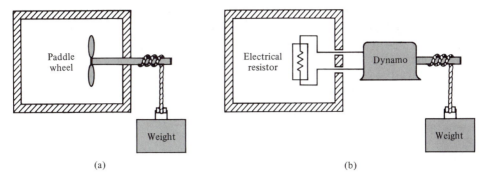

Figure 16.1 Schematics of a constant volume system connected to a falling weight via (a) a rotating paddle wheel and (b) an electric resistor and a dynamo.

If the interaction is only heat, the energy and entropy balances yield

$$dU = \delta Q^{\leftarrow} \tag{16.44}$$

and

$$\delta S_{\text{irr}} = dS - \frac{\delta Q^{\leftarrow}}{T_Q} = \left(\frac{1}{T} - \frac{1}{T_Q} \right) \delta Q^{\leftarrow} > 0 \tag{16.45}$$

where in writing the second equation (16.45) we use (16.43) and (16.44). From relations (16.45), we see that the process is reversible if $T_Q = T$, and irreversible if $T_Q \neq T$. Moreover, the latter can occur with $\delta Q^{\leftarrow} > 0$ only if $T_Q > T$ or with $\delta Q^{\leftarrow} < 0$ only if $T_Q < T$. In either case we have that

$$dS = \frac{\delta Q^{\leftarrow}}{T} \geqq \frac{\delta Q^{\leftarrow}}{T_Q} \tag{16.46}$$

If the interaction is work only, the energy and entropy balances yield

$$dU = -\delta W^{\rightarrow} \tag{16.47}$$

and

$$\delta S_{\text{irr}} = dS = -\frac{\delta W^{\rightarrow}}{T} > 0 \tag{16.48}$$

where in writing the second of equations (16.48) we use (16.43) and (16.47). We conclude that the process is always irreversible and can occur only if $\delta W^{\rightarrow} < 0$, that is, only if work is done on the system.

Processes involving only work interactions at constant volume are used to measure another useful property of stable equilibrium states. Specifically, the work in such a process satisfies (16.47). On the other hand, by differentiation of the fundamental relation $S = S(U, V, n)$ we find the temperature T as a function of U, V, and n, namely, $T = T(U, V, n)$, and then the internal energy U as a function of T, V, and n, namely, $U = U(T, V, n)$. Thus for two neighboring stable equilibrium states with the same values of amounts of constituents and volume, we may express the difference in internal energy dU as a Taylor series of $U(T, V, n)$ truncated at the first term, that is,

$$dU = \left(\frac{\partial U}{\partial T} \right)_{V,n} dT \tag{16.49}$$

The partial derivative $(\partial U / \partial T)_{V,n}$ is called the *heat capacity under constant volume* and denoted by C_V, that is,

$$C_V = \left(\frac{\partial U}{\partial T}\right)_{V,n} = \left(\frac{\partial U}{\partial S}\right)_{V,n}\left(\frac{\partial S}{\partial T}\right)_{V,n} = T\left(\frac{\partial S}{\partial T}\right)_{V,n} \tag{16.50}$$

where in writing the second and third of equations (16.50) we regard U as a function of S, V, and n, and S as a function of T, V, and n. Thus, upon combining equations (16.47), (16.49) and (16.50), we conclude that

$$C_V\, dT = (\delta W^{\leftarrow})_V \tag{16.51}$$

or, equivalently,

$$C_V = \frac{(\delta W^{\leftarrow})_V}{dT} \tag{16.52}$$

where the subscript V of $(\delta W^{\leftarrow})_V$ is a reminder that the process is at constant-volume. Equation (16.52) indicates that the heat capacity C_V may be measured by finding the work required in a constant-volume process per unit of temperature raise of the system. It is noteworthy that the right-hand side of (16.52) is just a ratio of $(\delta W^{\leftarrow})_V$ and dT, not a derivative. The reason is that $(W^{\leftarrow})_V$ is not a function of T.

Example 16.4.   Measurements of temperature, volume, and pressure on 1 kg of H_2O in states A_1 and A_2 yield $T_1 = 693$ K, $T_2 = 915$ K, $V_1 = V_2 = 0.10279$ m^3, and $U_2 - U_1 = 369$ kJ. Estimate the value of C_V.

Solution.   We assume that the system has volume as the only parameter. Because states 1 and 2 have the same volume, we can estimate C_V by $C_V = (\partial U/\partial T)_V \approx (U_2 - U_1)/(T_2 - T_1) = 369$ kJ/222 K = 1.66 kJ/K.

It is noteworthy that the heat capacity C_V is not defined at states for which the function $T = T(U, V, n)$ cannot be inverted, that is, when $(\partial T/\partial U)_{V,n} = -T^2(\partial^2 S/\partial U^2)_{V,n} = 0$. We discuss such states later.

16.5.2   Constant-Pressure Processes

We consider a given amount of matter confined in a chamber closed at the top by a weighted piston which can move up and down without friction and without clearance (Figure 16.2). We model this substance as a system with volume as the only parameter, initially in a stable equilibrium state at temperature T and pressure p. In addition to its interaction with the weight, the system experiences a heat interaction $\delta Q^{\leftarrow}$ at temperature T_Q, and ends in a stable equilibrium state with pressure p equal to the initial pressure because the weight is fixed. A constant-pressure process of a system is called *isobaric*.

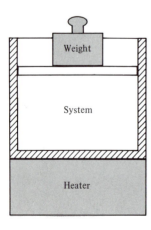

Figure 16.2   An amount of matter confined in a chamber closed at the top by a weighted piston that can move up and down without friction and without clearance.

During an infinitesimal isobaric process, the volume changes by dV, and the piston and the weight each changes its elevation by dz. The interaction with the weight through the piston is a work interaction, and the work done on the weight is $\delta W^{\rightarrow} = p\,dV$.

Example 16.5. Prove that the work done on the weight is $\delta W^{\rightarrow} = p\,dV$.

Solution. The force exerted by the weighted piston on the system in stable equilibrium is equal to $p\,a$, where p is the pressure of the system and a the area of the piston surface. For a given weight, the force is a constant. If the piston is displaced by a distance dz, the work done on the weight is $p\,a\,dz$ or $p\,dV$ because $a\,dz = dV$.

For $dp = 0$ and $dn_i = 0$ for $i = 1, 2, \ldots, r$, equation (16.16) can be written as

$$T\,dS = dU + p\,dV = d(U + pV)_p = dH \qquad (16.53)$$

where the subscript p is a reminder of the condition of constant pressure. If the system experiences no interactions other than heat and work through the piston, the energy and entropy balances yield

$$dU = \delta Q^{\leftarrow} - \delta W^{\rightarrow} = \delta Q^{\leftarrow} - p\,dV \qquad (16.54)$$

$$\delta S_{\text{irr}} = dS - \frac{\delta Q^{\leftarrow}}{T_Q} = \frac{dH}{T} - \frac{\delta Q^{\leftarrow}}{T_Q} \geqq 0 \qquad (16.55)$$

where in writing the second equation (16.55) we use (16.53). Combining equations (16.53) and (16.54), we obtain

$$\delta Q^{\leftarrow} = dU + p\,dV = d(U + pV)_p = dH \qquad (16.56)$$

So we conclude that in a constant-pressure or isobaric process with no other work than $\delta W^{\rightarrow} = p\,dV$, the heat equals the change in enthalpy of the system. Moreover, the process is reversible if $T_Q = T$, and irreversible if $T_Q \neq T$. Again, the latter can occur with $\delta Q^{\leftarrow} > 0$ only if $T_Q > T$ or with $\delta Q^{\leftarrow} < 0$ only if $T_Q < T$.

If in addition to the work interaction through the piston, and a heat interaction at a temperature T_Q equal to the temperature T of the initial stable equilibrium state, the system experiences also other work interactions, such as those induced by a weight connected to a paddle wheel (Figure 16.3), the energy and entropy balances may be written in the forms

$$dU = \delta Q^{\leftarrow} - \delta W^{\rightarrow} = dH - p\,dV = T\,dS - p\,dV \qquad (16.57)$$

$$\delta Q^{\leftarrow} = T(dS - \delta S_{\text{irr}}) = dH - T\,\delta S_{\text{irr}} \qquad (16.58)$$

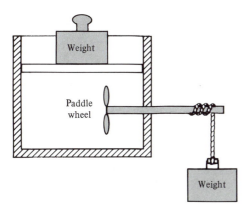

Figure 16.3 Schematic of an apparatus to measure the heat capacity at constant pressure (see text).

where $\delta W^{\rightarrow}$ is the total work done by the system. So, if the process is reversible, we conclude that $\delta Q^{\leftarrow} = dH$ and $\delta W^{\rightarrow} = p\,dV$, that is, that the work done via interactions other than through the piston is zero or, equivalently, $\delta W_w^{\leftarrow} = -(\delta W^{\rightarrow} - p\,dV) = 0$. On the other hand, if the process is irreversible, equations (16.57) and (16.58) yield

$$\delta S_{\text{irr}} = \frac{1}{T}(dH - \delta Q^{\leftarrow}) = -\frac{1}{T}(\delta W^{\rightarrow} - p\,dV) = \frac{1}{T}\delta W_w^{\leftarrow} \geq 0 \qquad (16.59)$$

and therefore we conclude that $\delta W_w^{\leftarrow}$ can be only nonnegative, that is, work interactions other than the work $p\,dV$ through the piston can transfer energy only into the system, not out of it.

Experiments involving processes at constant pressure are used to measure another property of stable equilibrium states. Specifically, for the system in Figure 16.3, if $\delta Q^{\leftarrow} = 0$, then $\delta W_w^{\leftarrow}$ equals the change in enthalpy [equation (16.59)]. On the other hand, if we can write the enthalpy as a function of T, p, and n, namely, $H = H(T, p, n)$,[8] then for two neighboring stable equilibrium states with the same values of amounts of constituents and pressure, we can express the difference in enthalpy dH as a Taylor series of $H(T, p, n)$ truncated at the first term; that is,

$$dH = \left(\frac{\partial H}{\partial T}\right)_{p,n} dT \qquad (16.60)$$

The partial derivative $(\partial H/\partial T)_{p,n}$ is called the *heat capacity under constant pressure*, and denoted by C_p, that is,

$$C_p = \left(\frac{\partial H}{\partial T}\right)_{p,n} = \left(\frac{\partial H}{\partial S}\right)_{p,n}\left(\frac{\partial S}{\partial T}\right)_{p,n} = T\left(\frac{\partial S}{\partial T}\right)_{p,n} \qquad (16.61)$$

where in writing the second and third of equations (16.61) we regard H as a function of S, p, and n, and S as a function of T, p, and n. Thus, upon substituting $dH = (\delta W_w^{\leftarrow})_p$ in (16.61), we find that

$$C_p\,dT = (\delta W_w^{\leftarrow})_p \qquad (16.62)$$

or, equivalently,

$$C_p = \frac{(\delta W_w^{\leftarrow})_p}{dT} \qquad (16.63)$$

where the subscript p of $(\delta W_w^{\leftarrow})_p$ is a reminder that the process is at constant pressure. The last relation indicates that the heat capacity under constant pressure may be measured by finding the work $(\delta W_w^{\leftarrow})_p$ required in a constant-pressure process to obtain a unit raise of the temperature of the system. Again, we note that $(\delta W_w^{\leftarrow})_p/dT$ is a ratio of two small quantities and not a derivative because $(W_w^{\leftarrow})_p$ is not a function of T.

Example 16.6.   Measurements on a system A consisting of 1 kg of water in four different stable equilibrium states A_1, A_2, A_3, and A_4 yield: $T_1 = 693$ K, $T_2 = 913$ K, $T_3 = 793$ K, $T_4 = 713$ K, $p_1 = p_2 = p_3 = p_4$, $V_1 = 0.10279$ m^3, $H_2 - H_1 = 497$ kJ, $H_3 - H_1 = 225$ kJ, and $H_4 - H_1 = 45.2$ kJ. Estimate the value of C_p.

Solution.   Because states A_1, A_2, A_3, and A_4 have the same pressure, we can estimate C_p by each of the following three relations: $C_p = (\partial H/\partial T)_{p,n} \approx (H_2 - H_1)/(T_2 - T_1) = 497$ kJ/220 K = 2.26 kJ/K; $C_p \approx (H_3 - H_1)/(T_3 - T_1) = 225$ kJ/100 K = 2.25 kJ/K; and $C_p \approx (H_4 - H_1)/(T_4 - T_1) = 45.2$ kJ/20 K = 2.26 kJ/K.

In general, both the heat capacity under constant volume and the heat capacity under constant pressure are functions of all the independent properties or independent variables

---

[8]We will see later that this can be done for a large variety of stable equilibrium states, but not all.

of the stable equilibrium states, such as T, p, and n. We show later, however, that in some special ranges of T and p, the heat capacities become independent of p, and in others independent of both T and p.

16.5.3 Constant-Temperature Processes

We consider again the system defined in the preceding section and sketched in Figure 16.2. Here, however, we assume that the heat interaction $\delta Q^{\leftarrow}$ occurs at temperature $T_Q = T$, and that the final state is also at temperature T, where T is the temperature of the initial state. The weight is not necessarily fixed, so that the initial and final pressures of the system are not necessarily the same. Under these conditions, the process experienced by the system is called constant-temperature or *isothermal*. For $dT = 0$ and $dn_i = 0$ for $i = 1, 2, \ldots, r$, equation (16.16) can be written as

$$p\,dV = -dU + T\,dS = -d(U - TS)_T = -dA \tag{16.64}$$

where subscript T is a reminder of the condition of constant temperature. If the system experiences no work interactions other than that through the weighted piston, then $\delta W^{\rightarrow} = p\,dV$, and therefore

$$\delta W^{\rightarrow} = -dA \tag{16.65}$$

that is, in an isothermal process involving no other work than that done through the piston, the work $\delta W^{\rightarrow}$ equals the negative of the change in the Helmholtz free energy of the system. In particular, if the process is reversible, $\delta Q^{\leftarrow} = T\,dS$ because $T_Q = T$, and $\delta W^{\rightarrow}$ of course is equal to $p\,dV$. For isothermal processes which are also isobaric, equation (16.16) can be written as

$$0 = dU - T\,dS + p\,dV = d(U - TS + pV)_{T.p} = dG \tag{16.66}$$

where the subscripts T and p remind us of the conditions of constant temperature and constant pressure. We see that, in constant-temperature and constant-pressure processes that do not change the values of the amounts of constituents, the Gibbs free energy is conserved. Such processes are also called *isothermobaric*.

16.5.4 Constant-Entropy Processes

We consider a constant-entropy or *isentropic* process of a system. For $dS = 0$ and $dn_i = 0$ for $i = 1, 2, \ldots, r$, equations (16.16), (16.38), and (16.39) become

$$dU = -p\,dV \tag{16.67}$$

$$dU = \delta Q^{\leftarrow} - \delta W^{\rightarrow} \tag{16.68}$$

$$\delta S_{\text{irr}} = -\frac{\delta Q^{\leftarrow}}{T_Q} \tag{16.69}$$

We see that the isentropic process cannot be reversible if $\delta Q^{\leftarrow} \neq 0$ or, equivalently, $\delta W^{\rightarrow} \neq p\,dV$, and it cannot be adiabatic if it is not reversible. In other words, whereas a reversible and adiabatic process is isentropic by virtue of (16.39), an isentropic process is not necessarily reversible and adiabatic.

16.6 Comment

All the results in this chapter are valid for any values of the amounts of constituents, including a single particle confined in a box of volume V. In Chapter 17 we impose the restriction that the values of the amounts of constituents are large and derive a number of explicit interrelations between properties such as the Euler and Gibbs–Duhem relations.

Problems

16.1 A particle of mass M is confined in a parallelepiped of sides L_1, L_2, L_3, and volume $V = L_1L_2L_3$. The energy U and entropy S of a stable equilibrium state of the particle can be shown to satisfy the relations

$$U = \sum_{l=1}^{\infty} \sum_{m=1}^{\infty} \sum_{n=1}^{\infty} x_{lmn}\, \epsilon_{lmn}$$

$$S = -k \sum_{l=1}^{\infty} \sum_{m=1}^{\infty} \sum_{n=1}^{\infty} x_{lmn} \ln x_{lmn}$$

where

$$x_{lmn} = \frac{\exp(-\epsilon_{lmn}/kT)}{Q}$$

$$Q = \sum_{l=1}^{\infty} \sum_{m=1}^{\infty} \sum_{n=1}^{\infty} \exp\left(-\frac{\epsilon_{lmn}}{kT}\right)$$

$$\epsilon_{lmn} = \frac{h^2}{8\,M}\left(\frac{l^2}{L_1^2} + \frac{m^2}{L_2^2} + \frac{n^2}{L_3^2}\right)$$

k is the Boltzmann constant, h the Planck constant, T the temperature of the stable equilibrium state determined by the value of U, and l, m, and n are integers ranging from 1 to infinity.

(a) Show that S can be expressed in the form $S = k \ln Q + U/T$.

(b) Show that Q can be expressed in terms of the Helmholtz free energy A, that is, $Q = \exp(-A/kT)$ or, equivalently, $A = -kT \ln Q$.

(c) Show that $U = -T(\partial(A/T)/\partial \ln T)_{L_1, L_2, L_3}$.

16.2 Consider the system specified in Problem 16.1, for $L_1 = L_2 = L_3 = L$.

(a) Show that

$$Q = \left[\sum_{n=1}^{\infty} \exp\left(-\frac{h^2 n^2}{8MV^{2/3}kT}\right)\right]^3$$

(b) Consider values of M, V, and T such that $h^2/8MV^{2/3}kT \ll 1$, and justify the proposition that the sum in part (a) can be replaced by an integral, that is,

$$\sum_{n=1}^{\infty} \exp\left(-\frac{h^2 n^2}{8MV^{2/3}kT}\right) = \int_0^{\infty} \exp\left(-\frac{h^2 x^2}{8MV^{2/3}kT}\right) dx$$

(c) Using the result in part (b), show that

$$Q = \left(\frac{2\pi\, M\, kT}{h^2}\right)^{3/2} V$$

16.3 Consider the system in Problem 16.1 and the expression for Q found in Problem 16.2, part (c).

(a) Find the Helmholtz free energy A as a function of V and T.

(b) Use the Helmholtz free energy to find the pressure of the particle. Show that it satisfies the simple relation

$$pV = kT$$

Is this relation valid at very low temperatures? Is it valid at very small values of the volume?

(c) Use the Helmholtz free energy to show that

$$U = \frac{3}{2} kT$$

$$S = \frac{3}{2} k + k \ln \frac{(2\pi MkT)^{3/2} V}{h^3}$$

Are these relations valid at either very low temperatures or very small values of the volume?

(d) Find the heat capacities of the particle under constant volume and under constant pressure.

16.4 Consider the system in Problem 16.1 and the stable equilibrium states at very low temperatures, namely, in the limit as $T \to 0$. At such low temperatures, only the values of ϵ_{lmn} for $(l = 1, m = 1, n = 1)$, $(l = 1, m = 1, n = 2)$, $(l = 1, m = 2, n = 1)$, and $(l = 2, m = 1, n = 1)$ are important. Assume that $L_1 = L_2 = L_3$, and find the energy and the entropy as functions of T. How do your answers here compare with the expressions found in Problem 16.3, part (c)?

16.5 The one-dimensional *harmonic oscillator* is a system for which the *frequency of the oscillator* ν is the only parameter, and the energy E and the entropy S of the stable equilibrium states are given by the relations

$$E = \sum_{i=0}^{\infty} x_i \, \epsilon_i$$

$$S = -k \sum_{i=0}^{\infty} x_i \ln x_i$$

where

$$x_i = \frac{\exp(-\epsilon_i/kT)}{Q}$$

$$Q = \sum_{i=0}^{\infty} \exp\left(-\frac{\epsilon_i}{kT}\right)$$

$$\epsilon_i = \left(i + \frac{1}{2}\right) h\nu$$

i is an integer ranging from 0 to infinity, k the Boltzmann constant, h the Planck constant, and T the temperature.

(a) Show that

$$Q = \frac{\exp(-h\nu/2\,kT)}{1 - \exp(-h\nu/kT)} = \frac{2}{\sinh(h\nu/2\,kT)}$$

(b) Show that

$$E = \frac{1}{2}h\nu + \frac{h\nu}{\exp(h\nu/kT) - 1}$$

$$S = k\left\{\frac{h\nu/kT}{\exp(h\nu/kT) - 1} - \ln\left[1 - \exp\left(-\frac{h\nu}{kT}\right)\right]\right\}$$

(c) Write the differentials dE and dS to first order with respect to $d(h\nu)$ and $d(1/kT)$.

(d) Find the derivative $(\partial E/\partial S)_{h\nu}$, and show that it is equal to T.

16.6 For a system consisting of a single structureless particle of mass M confined in a box of volume V, the stable equilibrium states are relatively simple to describe provided that the value of $h^2/8MV^{2/3}kT$ is much smaller than unity, where h is the Planck constant, k the Boltzmann constant, and T the temperature. Under such conditions, the Gibbs free energy is given by the relation

$$G = kT\left[1 - \ln \frac{(2\pi M)^{3/2}(kT)^{5/2}}{h^3 p}\right]$$

where p is the pressure.

(a) Find the expressions for the entropy S, volume V, energy U, heat capacity at constant volume C_V, and heat capacity at constant pressure C_p.

(b) Verify the Maxwell relation $(\partial S/\partial p)_T = -(\partial V/\partial T)_p$.

(c) If the temperature changes from T_1 to T_2 in a constant-volume weight process, how much entropy is generated by irreversibility?

(d) If the pressure changes from p_1 to p_2 in a reversible constant-temperature process using an energy source at the same temperature as that of the single particle, what are the change in volume of the particle and the heat?

(e) If the volume changes from V_1 to V_2 in a constant-entropy process, what are the changes in pressure and temperature?

16.7 Consider a system consisting of various constituents confined in a volume V. The system contains a region of volume V_c in which there is a uniform electrostatic field of strength $\mathcal{E}$, whereas in the remaining volume the electrostatic field is zero. For this system, a characteristic function $G(T, p, \mathcal{E}, V_c, n)$ is defined by the relation

$$G = U - TS + pV - V_c \epsilon \mathcal{E}^2$$

where the *dielectric constant* ϵ is a function of temperature T and pressure p only, that is, $\epsilon = \epsilon(T, p)$.

(a) Show that

$$\left(\frac{\partial S}{\partial \mathcal{E}}\right)_{T, p, V_c, n} = 2V_c \mathcal{E}\left(\frac{\partial \epsilon}{\partial T}\right)_p$$

(b) Show that

$$\left(\frac{\partial V}{\partial \mathcal{E}}\right)_{T, p, V_c, n} = -2V_c \mathcal{E}\left(\frac{\partial \epsilon}{\partial p}\right)_T$$

This relation shows that an increase in the electrostatic field strength $\mathcal{E}$ in the region of volume V_c at constant temperature and pressure is accompanied by a decrease in volume. The phenomenon is called *electrostriction*.

16.8 The function $J(T, V, n)$ defined by the relation

$$J = S - \frac{U}{T}$$

is called the *Massieu function* and is a characteristic function.

(a) Prove the assertion that $J(T, V, n)$ is a characteristic function.

(b) Obtain the Maxwell relations that follow from this characteristic function.

16.9 The function $K(T, p, n)$ defined by the relation

$$K = S - \frac{U}{T} - \frac{pV}{T}$$

is called the *Planck function* and is a characteristic function.

(a) Prove the assertion that $K(T, p, n)$ is a characteristic function.

(b) Obtain the Maxwell relations that follow from this characteristic function.

16.10 The inequalities

$$\left[\frac{\partial(1/T)}{\partial U}\right]_{V, n} \lesseqgtr \left[\frac{\partial(1/T)}{\partial U}\right]_{p/T, n} < 0$$

$$\left[\frac{\partial(p/T)}{\partial V}\right]_{U, n} \lesseqgtr \left[\frac{\partial(p/T)}{\partial V}\right]_{1/T, n} < 0$$

known as *Le Châtelier–Braun inequalities*, follow from the concavity of the fundamental relation $S = S(U, V, n)$ with respect to U and V. These inequalities have the following interesting interpretation. For stable equilibrium states, a change to a higher-energy state, $dU > 0$, causes an increase in the escaping tendency for energy, $dT > 0$, that is, the system tends to moderate the increase in energy by enhancing its tendency to give up energy. Similarly, the system responds to an increase in volume, $dV > 0$, by enhancing its tendency to give up volume, $d(p/T) < 0$, thus contrasting the increase. This natural tendency to react to a perturbation by favoring the reestablishment of the initial state, that is, by moderating the perturbation, is a typical manifestation of stability of equilibrium.

To prove the Le Châtelier–Braun inequalities, we express the concavity of the fundamental relation $S = S(U, V, n)$ with respect to U and V by stating that the quadratic form

$$(d^2 S)_n = \frac{\partial^2 S}{\partial U^2} (dU)^2 + 2 \frac{\partial^2 S}{\partial U \partial V} dU \, dV + \frac{\partial^2 S}{\partial V^2} (dV)^2$$

is negative definite, that is, negative for all possible pairs of values of dU and dV. For this to happen, the matrix

$$\begin{bmatrix} \dfrac{\partial^2 S}{\partial U^2} & \dfrac{\partial^2 S}{\partial U \partial V} \\ \dfrac{\partial^2 S}{\partial U \partial V} & \dfrac{\partial^2 S}{\partial V^2} \end{bmatrix}$$

must be negative definite. The determinant D of this matrix is called the *Hess determinant* of the function $S = S(U, V, n)$ with respect to U and V.

(a) Using the results on Jacobian determinants found in Problem 8.7, show that the Hess determinant D is equal to the Jacobian $\partial(1/T, p/T)/\partial(U, V)$, and can be expressed in either of the three forms

$$D = \left[\frac{\partial(1/T)}{\partial U} \right]_{V,n} \left[\frac{\partial(p/T)}{\partial V} \right]_{U,n} - \left[\frac{\partial(1/T)}{\partial V} \right]_{U,n} \left[\frac{\partial(p/T)}{\partial U} \right]_{V,n}$$

$$D = \left[\frac{\partial(1/T)}{\partial U} \right]_{p/T,n} \left[\frac{\partial(p/T)}{\partial V} \right]_{U,n}$$

$$D = \left[\frac{\partial(p/T)}{\partial V} \right]_{1/T,n} \left[\frac{\partial(1/T)}{\partial U} \right]_{V,n}$$

(b) Using the results in part (a), show that

$$\left[\frac{\partial(1/T)}{\partial U} \right]_{V,n} = \left[\frac{\partial(1/T)}{\partial U} \right]_{p/T,n} + \left[\frac{\partial(1/T)}{\partial V} \right]_{U,n}^2 \left[\frac{\partial(p/T)}{\partial V} \right]_{U,n}^{-1}$$

$$\left[\frac{\partial(p/T)}{\partial V} \right]_{U,n} = \left[\frac{\partial(p/T)}{\partial V} \right]_{1/T,n} + \left[\frac{\partial(p/T)}{\partial U} \right]_{V,n}^2 \left[\frac{\partial(1/T)}{\partial U} \right]_{V,n}^{-1}$$

(c) Verify that the quadratic form $(d^2 S)_n$ may be written in either of the forms

$$(d^2 S)_n = \left[\frac{\partial(1/T)}{\partial U} \right]_{V,n} (dU + c \, dV)^2 + \left[\frac{\partial(p/T)}{\partial V} \right]_{1/T,n} (dV)^2$$

$$(d^2 S)_n = \left[\frac{\partial(1/T)}{\partial U} \right]_{p/T,n} (dU)^2 + \left[\frac{\partial(p/T)}{\partial V} \right]_{U,n} (dV + c' \, dU)^2$$

that are called the canonical forms of the quadratic form. What are c and c'?

(d) Use the result in part (c) to show that because $(d^2 S)_n$ is negative for all possible values of dU and dV, the derivatives

$$\left[\frac{\partial(1/T)}{\partial U}\right]_{V,n}, \quad \left[\frac{\partial(p/T)}{\partial V}\right]_{1/T,n}, \quad \left[\frac{\partial(1/T)}{\partial U}\right]_{p/T,n}, \quad \left[\frac{\partial(p/T)}{\partial V}\right]_{U,n}$$

are all negative.

(e) Use the results in parts (b) and (d) to prove the Le Châtelier–Braun inequalities.

(f) By a procedure analogous to that just outlined, prove that the convexity of the energy relation $U = U(S, V, n)$ implies the Le Châtelier–Braun inequalities

$$\left(\frac{\partial T}{\partial S}\right)_{V,n} \geqq \left(\frac{\partial T}{\partial S}\right)_{p,n} > 0$$

$$-\left(\frac{\partial p}{\partial V}\right)_{S,n} \geqq -\left(\frac{\partial p}{\partial V}\right)_{T,n} > 0$$

17   Simple Systems

In studying stable equilibrium states of a substance that may be modeled as a system having volume as the only parameter, we are faced with a number of practical questions in addition to those discussed in Chapter 16. Some of these questions are as follows. What are the appropriate values of the amounts of constituents and the volume that we must investigate? Can the results obtained for a particular set of values be readily extended to other values of the amounts of constituents and the volume? Is it possible to establish some explicit analytical expressions between various properties and variables of stable equilibrium states? For example, if we have studied the properties of 1 kg of water, can we use the results to evaluate properties of 1 g or 1 ton of water, and can we write some explicit relation between internal energy, entropy, temperature, pressure, and volume?

In this chapter we define a special class of systems which we call simple and which provide good and fairly accurate answers to the questions just raised.

17.1   Definition of a Simple System

We define a system as *simple* if it has volume as the only parameter (Section 16.1) and if it satisfies the following two additional requirements.

1. If in any of its stable equilibrium states it is partitioned into a set of contiguous subsystems in mutual stable equilibrium, the system is such that the effects of the partitions are negligible.

2. In any of its stable equilibrium states, the instantaneous "switching on or off" of one or more internal reaction mechanisms causes negligible instantaneous changes in the values of the energy, the entropy, the volume, and the amounts of constituents.

When conditions 1 and 2 are satisfied, we find that results obtained on stable equilibrium states of a simple system with one set of values of amounts of constituents and volume can be readily extended to other sets of values, and that certain properties and variables are explicitly interrelated.

In the next section we explore the implications of the validity of condition 1. The discussion of condition 2 is postponed to Chapter 30, where we study the stable equilibrium states of simple systems with chemical reactions. Until then, we consider simple systems with constraints that inhibit all internal mechanisms capable of altering the amounts of constituents.

17.2   Implications of Partitioning

We consider systems A and B, which are shown schematically in Figures 17.1a and 17.1b. System A is in a stable equilibrium state with energy U^A, volume V^A, and amounts of constituents $n_1^A, n_2^A, \ldots, n_r^A$. System B consists of two subsystems, each identical to system A but in a stable equilibrium state with energy $U^A/2$, volume $V^A/2$, and amounts of constituents $n_1^A/2, n_2^A/2, \ldots, n_r^A/2$. Being in identical stable equilibrium states, the two

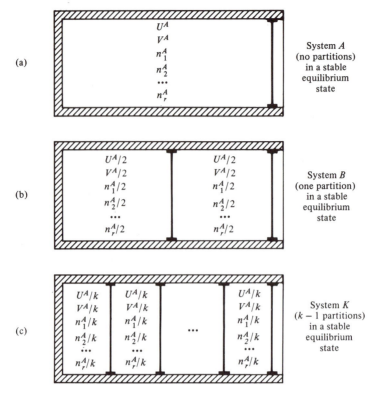

Figure 17.1 Schematics of a simple system A and its partitioning into two or k contiguous subsystems.

subsystems have identical values of temperature T, pressure p, and chemical potentials μ_1, μ_2, ..., μ_r, namely, they are in mutual stable equilibrium, and therefore system B is in a stable equilibrium state. It is noteworthy that system B is not identical to system A because it requires two parameters, that is, a volume for each of the two subsystems rather than just one volume.

By virtue of the fundamental relation for systems with volume as the only parameter [equation (16.7)], and the additivity of entropy, systems A and B have entropies $S^A = S^A(U^A, V^A, n^A)$ and $S^B = 2\ S^A(U^A/2, V^A/2, n^A/2)$.

In general, S^A is not equal to S^B because of the presence of the partition that separates the two subsystems of system B. For example, if each of the two subsystems of B contains only one particle, it can be shown with the tools of quantum theory that the particle is not uniformly distributed in the available space but is more rarefied near the confining walls, and therefore the wall partitioning system B into the two subsystems imposes a significant difference between the properties of systems A and B. However, it can also be shown with the tools of quantum theory that such differences become less and less important, and negligible for all practical purposes, as the values of the amounts of constituents in each subsystem increase beyond the order of 10. Hence, if the amounts of constituents are large, we can neglect the effects of the partition, write $S^B = S^A$ without appreciable error, neglect the differences between systems A and B, and conclude that the stable equilibrium state of A is the same as the state of B, and that the temperature, pressure, and chemical potentials of A have the same values as the respective properties of each subsystem of B.

We can repeat the preceding reasoning for the systems A and K shown in Figures 17.1a and 17.1c. System K consists of a large number k of subsystems, each in a stable equilibrium state with energy U^A/k, volume V^A/k, and amounts of constituents $n_1^A/k, n_2^A/k, \ldots, n_r^A/k$. For large amounts of constituents, we conclude again that $S^K = k\, S^A(U^A/k, V^A/k, n_1^A/k, n_2^A/k, \ldots, n_r^A/k) \approx S^A$, and that the effects of the partitions can be neglected.

Of course, we reiterate that the influence of partitions is not negligible if the number of particles in any of the subsystems is very small. Hence the results just cited cannot hold for arbitrarily large values of k. However, because we are usually concerned with amounts of constituents that correspond to very large numbers of particles, the effect of partitions is negligible up to a very large number of subdivisions, so that each subsystem resulting from such a subdivision can be considered for all practical purposes infinitesimal as compared to the overall system.

Subject to the restriction just cited, we can write the relation between the entropies of A and each subsystem of K in Figure 17.1 in the form

$$S(U, V, \boldsymbol{n}) = k\, S\!\left(\frac{U}{k}, \frac{V}{k}, \frac{\boldsymbol{n}}{k}\right) \tag{17.1}$$

where we drop the superscript A for simplicity.

Using equation (16.8), we write $U^A = U^A(S^A, V^A, \boldsymbol{n}^A)$. Because each of the subsystems of K is in a stable equilibrium state with entropy S^A/k, volume V^A/k, and amounts of constituents $\boldsymbol{n}^A/k$, its energy must be $U^A(S^A/k, V^A/k, \boldsymbol{n}^A/k)$. Because energy is additive, $U^A = U^K$ and, therefore,

$$U(S, V, \boldsymbol{n}) = k\, U\!\left(\frac{S}{k}, \frac{V}{k}, \frac{\boldsymbol{n}}{k}\right) \tag{17.2}$$

where again we have dropped the superscript A for simplicity.[1] In words, if each of the values of the additive variables S, V, $\boldsymbol{n}$ is altered by a factor k, the value of the internal energy of a simple system is altered by the same factor.

From equation (17.2) we see that the stable equilibrium properties of a simple system with entropy S, volume V, and amounts of constituents $\boldsymbol{n}$ are identical to those of a composite of k subsystems each of which is identical to the simple system but in a stable equilibrium state with entropy, volume, and amounts of constituents k times smaller. Being identical and in identical states, all such subsystems have the same temperature, pressure, and chemical potentials, and these are also the temperature, pressure, and chemical potentials of the overall system. Because the number k can be chosen very large, a simple system in a stable equilibrium state can always be viewed as a composite of a contiguous collection of infinitesimal simple subsystems, all in mutual stable equilibrium.

Either equation (17.1) or equation (17.2) indicates that results on stable equilibrium states of a simple system corresponding to values U, V, $\boldsymbol{n}$ can be readily extended to any other larger or smaller values. We discuss this result in more detail in the following sections.

17.3   Gibbs, Euler, and Gibbs–Duhem Relations

We consider a simple system in a stable equilibrium state having entropy S, volume V, amounts of constituents $\boldsymbol{n}$, energy $U(S, V, \boldsymbol{n})$, temperature T, pressure p, and chemical potentials $\mu_1, \mu_2, \ldots, \mu_r$. We subdivide this system into a contiguous collection of

---

[1]Mathematically, a function of many variables that changes by a factor k if each of some variables changes by the same factor is called *homogeneous of degree one* with respect to these variables. It is noteworthy that if the homogeneity holds for any integer number k, it also holds for any real number k because then k can be approximated by a sequence of ratios of integers such that k_1/k_2 tends to k.

infinitesimal subsystems, each having the same temperature T, pressure p, and chemical potentials $\mu_1, \mu_2, \ldots, \mu_r$ as the system itself, and entropy dS, volume dV, amounts of constituents $dn_1, dn_2, \ldots, dn_r$, and energy dU.

Alternatively, the differential dU may be regarded as the difference in energy between two neighboring stable equilibrium states differing in entropy by dS, in volume by dV, and in amounts of constituents by $dn_1, dn_2, \ldots, dn_r$. As such, it is given by (16.16), the Gibbs relation, which we repeat here for convenience:

$$dU = T\,dS - p\,dV + \mu_1 dn_1 + \mu_2 dn_2 + \cdots + \mu_r dn_r \tag{17.3}$$

This relation imposes a restriction on the differential changes dU, dS, dV, and dn_i for $i = 1, 2, \ldots, r$ between two neighboring stable equilibrium states and the values of T, p, and μ_i for $i = 1, 2, \ldots, r$ of one of these states.

The additive properties energy U and entropy S, the additive volume V, and the additive amounts $n_1, n_2, \ldots, n_r$ of the simple system can each be viewed as the sum or integral of the energies, entropies, volumes, and amounts, respectively, of all the infinitesimal subsystems in a given partition of the spatial extension of the simple system. Thus, for given values T, p, $\mu_1, \mu_2, \ldots, \mu_r$, we can think of the simple system as resulting from successive additions of infinitesimal parts that build up the value of each of U, S, V, $n_1, n_2, \ldots, n_r$ from zero to the value of the simple system, while maintaining each of the properties T, p, $\mu_1, \mu_2, \ldots, \mu_r$ unchanged throughout all the additions. At each step the increments dU, dS, dV, $dn_1, dn_2, \ldots, dn_r$ are related by the Gibbs relation. Hence, upon integration of this relation at constant T, p, $\mu_1, \mu_2, \ldots, \mu_r$, we find that

$$U = TS - pV + \mu_1 n_1 + \mu_2 n_2 + \cdots + \mu_r n_r \tag{17.4}$$

Clearly, equation (17.4) is valid for any set of values T, p, $\mu_1, \mu_2, \ldots, \mu_r$ consistent with the values of S, V, $n_1, n_2, \ldots, n_r$, and therefore it is valid for all such values. It is called the *Euler relation*.[2] It is one of the explicit forms we are seeking.[3]

The Euler relation may be used to derive analogous expressions for characteristic functions other than $U(S, V, \boldsymbol{n})$. Specifically, upon combining equations (16.14), (16.21), and (16.26) with (17.4), we find that

$$H = U + pV = TS + \mu_1 n_1 + \mu_2 n_2 + \cdots + \mu_r n_r \tag{17.5}$$

$$A = U - TS = -pV + \mu_1 n_1 + \mu_2 n_2 + \cdots + \mu_r n_r \tag{17.6}$$

$$G = H - TS = \mu_1 n_1 + \mu_2 n_2 + \cdots + \mu_r n_r \tag{17.7}$$

Equations (17.5) to (17.7) represent more of the explicit forms we are seeking.

Upon writing the differential of (17.4) in the form

$$dU = T\,dS + S\,dT - p\,dV - V\,dp$$
$$+ \mu_1 dn_1 + n_1 d\mu_1 + \mu_2 dn_2 + n_2 d\mu_2 + \cdots + \mu_r dn_r + n_r d\mu_r \tag{17.8}$$

---

[2] In honor of Leonhard Euler (1707–1783), Swiss mathematician who made major contributions to mathematical analysis.

[3] We can derive the Euler relation also by a different procedure. We define the variables $S' = S/k$, $V' = V/k$, and $\boldsymbol{n}' = \boldsymbol{n}/k$, so that (17.2) becomes $U(S, V, \boldsymbol{n}) = k\,U(S', V', \boldsymbol{n}')$. Differentiating this equation with respect to k at constant S', V', $\boldsymbol{n}'$, we find that

$$\left(\frac{\partial U}{\partial S}\right)_{V, \boldsymbol{n}} S' + \left(\frac{\partial U}{\partial V}\right)_{S, \boldsymbol{n}} V' + \sum_i \left(\frac{\partial U}{\partial n_i}\right)_{S, V, \boldsymbol{n}} n_i' = U(S', V', \boldsymbol{n}')$$

This equation holds for all values of k. If we set $k = 1$, then $S' = S$, $V' = V$, $\boldsymbol{n}' = \boldsymbol{n}$, and recalling the definitions of T, p, and μ_i for $i = 1, 2, \ldots, r$, we obtain (17.4).

and substituting dU from (17.3), we find another important and useful result, that is,

$$SdT - Vdp + n_1 d\mu_1 + n_2 d\mu_2 + \cdots + n_r d\mu_r = 0 \qquad (17.9)$$

Equation (17.9) is known as the *Gibbs–Duhem*[4] *relation*. It imposes another restriction between the differential changes dT, dp, and $d\mu_i$ for $i = 1, 2, \ldots, r$ of two neighboring stable equilibrium states, and the values of S, V, and n_i for $i = 1, 2, \ldots, r$ of one of these states.

17.4   Extensive and Intensive Properties

For a simple system in a stable equilibrium state, and for any number k, the temperature $T(S, V, n) = (\partial U/\partial S)_{V,n}$, pressure $p(S, V, n) = -(\partial U/\partial V)_{S,n}$, and chemical potentials $\mu_i(S, V, n) = (\partial U/\partial n_i)_{S,V,n}$, for $i = 1, 2, \ldots, r$, satisfy the relations

$$T(S, V, n_1, n_2, \ldots, n_r) = T\left(\frac{S}{k}, \frac{V}{k}, \frac{n_1}{k}, \frac{n_2}{k}, \ldots, \frac{n_r}{k}\right) \qquad (17.10)$$

$$p(S, V, n_1, n_2, \ldots, n_r) = p\left(\frac{S}{k}, \frac{V}{k}, \frac{n_1}{k}, \frac{n_2}{k}, \ldots, \frac{n_r}{k}\right) \qquad (17.11)$$

$$\mu_i(S, V, n_1, n_2, \ldots, n_r) = \mu_i\left(\frac{S}{k}, \frac{V}{k}, \frac{n_1}{k}, \frac{n_2}{k}, \ldots, \frac{n_r}{k}\right) \quad \text{for } i = 1, 2, \ldots, r \qquad (17.12)$$

namely, if each of the values of the additive variables S, V, n_1, n_2, $\ldots$, n_r is altered by the factor k, the values of the temperature T, the pressure p, and the chemical potentials μ_1, μ_2, $\ldots$, μ_r remain unaltered.[5] In contrast to T, p, and μ_i, other properties of a stable equilibrium state change by a factor of k if each of the additive variables S, V, n changes by the same factor. For example, energy behaves in this manner [equation (17.2)]. Again, the enthalpy $H(S, p, n)$, the Helmholtz free energy $A(T, V, n)$, and the Gibbs free energy $G(T, p, n)$ exhibit the same behavior because they satisfy the relations

$$H(S, p, n_1, n_2, \ldots, n_r) = k H\left(\frac{S}{k}, p, \frac{n_1}{k}, \frac{n_2}{k}, \ldots, \frac{n_r}{k}\right) \qquad (17.13)$$

$$A(T, V, n_1, n_2, \ldots, n_r) = k A\left(T, \frac{V}{k}, \frac{n_1}{k}, \frac{n_2}{k}, \ldots, \frac{n_r}{k}\right) \qquad (17.14)$$

$$G(T, p, n_1, n_2, \ldots, n_r) = k G\left(T, p, \frac{n_1}{k}, \frac{n_2}{k}, \ldots, \frac{n_r}{k}\right) \qquad (17.15)$$

Example 17.1.   Prove equation (17.13).

Solution.   We consider the simple system as a composite of k identical simple subsystems all in mutual stable equilibrium with temperature T, pressure p, chemical potentials μ_i, for $i = 1, 2, \ldots, r$, entropy S/k, volume V/k, and amounts n_i/k for $i = 1, 2, \ldots, r$. By equations (17.1) and (17.10) to (17.12), the composite simple system with entropy S, volume V, and amounts n_i, for $i = 1, 2, \ldots, r$, has energy $U(S, V, n)$, temperature T, pressure p, and chemical

---

[4]Pierre Duhem (1861–1916), French physicist and historian of French science.

[5]We discuss these equalities in Section 17.2. To provide another proof, we define the variables $S' = S/k$, $V' = V/k$ and $n' = n/k$ so that (17.2) is $U(S, V, n) = k U(S', V', n')$. Differentiating this equation with respect to S at constant V, n and k, we find that

$$T(S, V, n) = \left(\frac{\partial U}{\partial S}\right)_{V,n} = k\left(\frac{\partial U}{\partial S'}\right)_{V',n'} \frac{1}{k} = T(S', V', n')$$

i.e., equation (17.10). Similarly, we obtain (17.11) by differentiating $U(S, V, n)$ with respect to V at constant S, n, and k, and (17.12) by differentiating with respect to n_i at constant S, V, k and the remaining n's.

potentials μ_i, for $i = 1, 2, \ldots, r$. Therefore, the enthalpy of the composite system is given by $H(S, p, n) = U(S, V, n) + pV = k\left[U(S/k, V/k, n/k) + pV/k\right] = k\,H(S/k, p, n/k)$. Similar proofs yield (17.14) and (17.15) for the Helmholtz free energy and the Gibbs free energy, respectively.

It is noteworthy that the enthalpy, the Helmholtz free energy, and the Gibbs free energy are defined only for stable equilibrium states. For example, if a system is composed of two subsystems each in a stable equilibrium state but not in mutual stable equilibrium, the enthalpy, the Helmholtz free energy, and the Gibbs free energy are defined for each subsystem but not for the composite system.

Any property of a stable equilibrium state that changes by a factor of k if each of the additive variables S, V, and n changes by a factor k is called *extensive*. For example, energy and enthalpy are extensive properties because of (17.2) and (17.13), respectively. In this book, the concept of extensiveness is used exclusively for stable equilibrium states of simple systems and not synonymously with the concept of additivity. Indeed, our definitions imply that every additive property is also extensive, but an extensive property is additive only if restricted to systems in mutual stable equilibrium and not for all systems in all states. For example, energy is both extensive and additive, but enthalpy is only extensive and not additive because, in general, the sum of the enthalpies of two systems in stable equilibrium states does not represent an enthalpy of the composite of the two systems unless they are in mutual stable equilibrium.

Any property of a stable equilibrium state of a simple system that remains unchanged if each of the additive variables S, V, and n changes by a factor k is called *intensive*.[6] For example, temperature, pressure, and chemical potentials are all intensive properties of stable equilibrium states of any simple system because of (17.10) to (17.12). Again, any partial derivative of an extensive property with respect to another extensive property is intensive because both the numerator and the denominator of the partial derivative change by the same factor as the common factor that denotes the change in S, V, and n.

Other properties of stable equilibrium states of simple systems that are intensive are all the ratios of any two extensive quantities such as U, S, V, n_1, n_2, $\ldots$, n_r, H, A, G, or ratios of any linear combination of these quantities such as U and either the total amount of constituents $n = n_1 + n_2 + \cdots + n_r$ or the total mass m. Indeed, if each of S, V, n changes by a factor k, each extensive quantity changes by the same factor, and therefore the ratio of any two extensive quantities remains unchanged.

A ratio of two extensive quantities or properties is also called a *specific property*. Each specific property is denoted by a lowercase letter. It is a *number-specific property* if it is found by dividing an extensive quantity by the total amount n of constituents. It is a *mass-specific property* if it is found by dividing an extensive quantity by the total mass m of the system.

Examples of such properties are the *specific energy* $u = U/n$ or U/m, the *specific entropy* $s = S/n$ or S/m, the *specific volume* $v = V/n$ or V/m, the *specific enthalpy* $h = H/n$ or H/m, and the *specific Gibbs free energy* $g = G/n$ or G/m. In this book we use the same lowercase letter for both a number-specific and the corresponding mass-specific property, and rely on the context to indicate which is the case.

Extensive and intensive properties are not all inclusive. Some properties fail to conform to either definition. For example, the square of the energy is a property that is neither extensive nor intensive. No special name is given to such properties because they are not encountered very often in practice.

---

[6]Mathematically, a function of many variables that remains unchanged if some of the variables change by a factor k is called *homogeneous of degree zero* with respect to these variables.

17.5   Dependences of Intensive Properties

In general, a property of a stable equilibrium state of a simple system with r constituents depends on $r + 2$ independent variables such as S, V, n_1, n_2, ..., n_r. In contrast, each intensive property depends at most on $r + 1$ independent variables because each such property is independent of the total amount of constituents.

We can verify the last assertion by considering any intensive property. For example, if in equation (17.10) for temperature we assume that $k = n$, where n is the total amount of constituents, $n = n_1 + n_2 + \cdots + n_r$, we find that

$$T(S, V, n_1, n_2, \ldots, n_r) = T(s, v, y_1, y_2, \ldots, y_r) \tag{17.16}$$

where $s = S/n$, $v = V/n$, $y_1 = n_1/n$, $y_2 = n_2/n$, ..., $y_r = n_r/n$. The function $T(s, v, y_1, y_2, \ldots, y_r)$ depends at most on $r + 1$ independent intensive variables because we can use the relation $y_1 + y_2 + \cdots + y_r = 1$ to eliminate one of the fractions y_i. By examining equations (17.11) and (17.12), we reach the same conclusion for the pressure and the chemical potentials.

A consequence of these results is that the $r + 2$ intensive properties T, p, μ_1, μ_2, ..., μ_r cannot all be varied independently because they depend at most on $r + 1$ independent variables. We reach the same conclusion by recognizing that the Gibbs–Duhem relation [equation (17.9)] imposes a general condition on the possible changes in these properties. In Chapter 18, we will see the number of the $r + 2$ intensive properties that can be varied independently may be reduced by other conditions in addition to the Gibbs–Duhem relation, and that, depending on the type of stable equilibrium state, this number may range from $r + 1$ to zero.

Another illustration of the reduced number of independent variables necessary to describe intensive properties is provided by the dependences of ratios of extensive properties. Using the Gibbs–Duhem relation, and evaluating it for different combinations of differentials, we find that

$$\frac{S}{V} = \left(\frac{\partial p}{\partial T} \right)_\mu \qquad \frac{S}{n_i} = -\left(\frac{\partial \mu_i}{\partial T} \right)_{p,\mu} \qquad \frac{V}{n_i} = \left(\frac{\partial \mu_i}{\partial p} \right)_{T,\mu} \tag{17.17}$$

where subscript μ stands for all the μ_j's being kept fixed except for the one that appears in the partial derivative. We see that each of the ratios of the extensive quantities S, V, n_1, n_2, ..., n_r is determined by a partial derivative, and that each function that is being differentiated depends at most on $r + 1$ independent variables. For example, S/V depends on T, μ_1, μ_2, ..., μ_r.

For simple systems, another important aspect of the reduced dependences of intensive and specific properties is that they are independent of the total amount n. To verify this assertion, we consider the specific internal energy u defined by the relation

$$u = \frac{1}{n} U(S, V, n_1, n_2, \ldots, n_r) = \frac{1}{n} U(ns, nv, ny_1, ny_2, \ldots, ny_r) \tag{17.18}$$

Upon differentiating this relation, we find that

$$\begin{aligned}
du &= \frac{1}{n} dU - \frac{U}{n^2} dn \\
&= \frac{1}{n} \left[\left(\frac{\partial U}{\partial S} \right)_{V,n} (n\,ds + s\,dn) + \left(\frac{\partial U}{\partial V} \right)_{S,n} (n\,dv + v\,dn) \right. \\
&\quad \left. + \sum_{i=1}^r \left(\frac{\partial U}{\partial n_i} \right)_{S,V,n} (n\,dy_i + y_i\,dn) \right] - \frac{U}{n^2} dn
\end{aligned}$$

$$= T\,ds - p\,dv + \sum_{i=1}^{r} \mu_i\,dy_i - \frac{1}{n^2}\left(U - TS + pV - \sum_{i=1}^{r} \mu_i n_i\right) dn \quad (17.19)$$

Each of the coefficients of the $r+3$ differentials in the right-hand side of the third of equations (17.19) represents a partial derivative with respect to the variable in the corresponding differential. In particular, the coefficient of dn equals $(\partial u/\partial n)_{s,v,y}$, where the subscript y denotes that all mole fractions are kept fixed. But $U - TS + pV - \sum_{i=1}^{r} \mu_i n_i = 0$ because of the Euler relation [equation (17.4)], and therefore $(\partial u/\partial n)_{s,v,y} = 0$,

$$du = T\,ds - p\,dv + \sum_{i=1}^{r} \mu_i\,dy_i \qquad (17.20)$$

and

$$u = u(s, v, y_1, y_2, \ldots, y_r) \qquad (17.21)$$

subject to the relation $y_1 + y_2 + \cdots + y_r = 1$.

A similar procedure can be used to show that the specific Gibbs free energy $g = G/n$ is a function of T, p, and y only, that is, $g = g(T, p, y_1, y_2, \ldots, y_r)$. Indeed, upon differentiating the relation $g = G(T, p, \boldsymbol{n})/n$, we find that

$$dg = -s\,dT + v\,dp + \sum_{i=1}^{r} \mu_i\,dy_i - \frac{1}{n^2}\left(G - \sum_{i=1}^{r} \mu_i n_i\right) dn$$

$$= -s\,dT + v\,dp + \sum_{i=1}^{r} \mu_i\,dy_i \qquad (17.22)$$

where in writing the second of equations (17.22) we use equation (17.7). Thus we verify that g is a function of T, p, and y only, independent of n.

In particular, for a single-constituent simple system, it also follows that $G = \mu n$ or, equivalently,

$$\mu = \frac{G}{n} = g(T, p) = u + p\,v - Ts = h - Ts \qquad (17.23)$$

Here the chemical potential is equal to the specific Gibbs free energy, and can be expressed in terms of the specific enthalpy h and the specific entropy s. Moreover, because for such a system $y = 1$ and $dy = 0$, equations (17.20) and (17.22) become

$$du = T\,ds - p\,dv \qquad (17.24)$$

$$d\mu = -s\,dT + v\,dp \qquad (17.25)$$

where in writing (17.25) we use the result $\mu = g$ [equation (17.23)]. The last equation is the Gibbs–Duhem relation of the single-constituent simple system in terms of its specific entropy and specific volume.

17.6 Convexity of Specific Energy

In subsequent chapters we need the relative magnitudes and sign definiteness of various practical properties of simple systems, such as the relative magnitudes of the heat capacities C_p and C_V. Such results are based on the convexity of the relation $u = u(s, v, y)$ with respect to s and v.

To prove that $u = u(s, v, y)$ is convex, we consider two states of a simple system. One is a stable equilibrium state in which the system is subdivided into two subsystems in mutual stable equilibrium, each having amounts of constituents ny, specific volume v, specific entropy s, and specific internal energy u. The total internal energy is $2nu(s, v, y)$. The other is also a state in which the system is subdivided into two subsystems each in a stable equilibrium state. Here, however, the two subsystems are not in mutual stable equilibrium. Specifically, one subsystem has amounts of constituents ny, specific volume $v + dv$, specific entropy $s + ds$, and internal energy $n(u + du_1)$, whereas the respective values of the other subsystem are ny, $v - dv$, $s - ds$, and $n(u + du_2)$.

According to the lowest-energy principle, the difference in energy between these two states must be positive, that is,

$$du_1 + du_2 > 0 \qquad (17.26)$$

because they have the same values of the amounts of constituents, the volume, and the entropy. Moreover, upon expanding the relation $u = u(s, v, y)$ for each of the subsystems, and truncating each expansion at the second-order terms, we find that

$$du_1 = T\,ds - p\,dv + \frac{1}{2}\left[\left(\frac{\partial^2 u}{\partial s^2}\right)_{v,y} ds^2 + 2\left(\frac{\partial^2 u}{\partial s \partial v}\right)_y ds\,dv + \left(\frac{\partial^2 u}{\partial v^2}\right)_{s,y} dv^2\right] \qquad (17.27)$$

and

$$du_2 = -T\,ds + p\,dv + \frac{1}{2}\left[\left(\frac{\partial^2 u}{\partial s^2}\right)_{v,y} ds^2 + 2\left(\frac{\partial^2 u}{\partial s \partial v}\right)_y ds\,dv + \left(\frac{\partial^2 u}{\partial v^2}\right)_{s,y} dv^2\right] \qquad (17.28)$$

where T and p are the temperature and the pressure of the initial stable equilibrium state. Thus, upon substituting equations (17.27) and (17.28) into inequality (17.26), we find that

$$\left(\frac{\partial^2 u}{\partial s^2}\right)_{v,y} ds^2 + 2\left(\frac{\partial^2 u}{\partial s \partial v}\right)_y ds\,dv + \left(\frac{\partial^2 u}{\partial v^2}\right)_{s,y} dv^2 \geqq 0 \qquad (17.29)$$

where the equality sign may hold because we truncated the expansion at the second-order term and, if it holds, the strict inequality in (17.26) is satisfied by the higher-order terms that we neglect in (17.27) and (17.28).

Because relation (17.29) must hold for any combination of the values of the differentials ds and dv, the left-hand side is a positive semidefinite quadratic form,[7] and we conclude that

$$\left(\frac{\partial^2 u}{\partial s^2}\right)_{v,y} = \left(\frac{\partial T}{\partial s}\right)_{v,y} \geqq 0 \qquad (17.30)$$

$$\left(\frac{\partial^2 u}{\partial s^2}\right)_{v,y}\left(\frac{\partial^2 u}{\partial v^2}\right)_{s,y} - \left(\frac{\partial^2 u}{\partial s \partial v}\right)_y^2 \geqq 0 \qquad (17.31)$$

$$\left(\frac{\partial^2 u}{\partial v^2}\right)_{s,y} = -\left(\frac{\partial p}{\partial v}\right)_{s,y} \geqq 0 \qquad (17.32)$$

In view of relations (17.30) to (17.32) we say that $u(s, v, y)$ is convex with respect to s and v and use these relations in subsequent discussions.

17.7 Gibbs Free Energy

In this section we express properties of simple systems in terms of chemical potentials because the results are useful in relating properties of multiconstituent systems to properties of one-constituent systems.

---

[7]See also footnote 1 in Chapter 10 and footnote 2 in Chapter 11.

We use the Gibbs free energy $G(T, p, \boldsymbol{n})$ because it is a characteristic function (Section 16.3) and we can express it in the form [equation (17.7)]

$$G = \sum_{i=1}^{r} n_i \, \mu_i(T, p, \boldsymbol{y}) \tag{17.33}$$

subject to the condition that $y_1 + y_2 + \cdots + y_r = 1$. We recall that each chemical potential μ_i is an intensive property, function of at most $r + 1$ independent variables. Equation (17.33) can be used to evaluate the entropy S, volume V, enthalpy H, and internal energy U by means of the relations

$$S = -\left(\frac{\partial G}{\partial T}\right)_{p,n} = -\sum_{i=1}^{r} n_i \left(\frac{\partial \mu_i}{\partial T}\right)_{p,y} \tag{17.34}$$

$$V = \left(\frac{\partial G}{\partial p}\right)_{T,n} = \sum_{i=1}^{r} n_i \left(\frac{\partial \mu_i}{\partial p}\right)_{T,y} \tag{17.35}$$

$$H = G + TS = G - T\left(\frac{\partial G}{\partial T}\right)_{p,n} = \left[\frac{\partial(G/T)}{\partial(1/T)}\right]_{p,n}$$

$$= \sum_{i=1}^{r} n_i \left[\frac{\partial(\mu_i/T)}{\partial(1/T)}\right]_{p,y} \tag{17.36}$$

$$U = H - pV = \sum_{i=1}^{r} n_i \left\{ \left[\frac{\partial(\mu_i/T)}{\partial(1/T)}\right]_{p,y} - p\left(\frac{\partial \mu_i}{\partial p}\right)_{T,y} \right\} \tag{17.37}$$

Equations (17.34) to (17.37) show that we can evaluate all the stable equilibrium properties of the multiconstituent system if we know the μ_i's as functions of temperature, pressure, and mole fractions. It turns out that such functional dependences can be found by performing measurements or calculations on r judiciously chosen one-constituent systems. We discuss these measurements and calculations when we treat multiconstituent systems in more detail.

17.8 Partial Properties

For a simple system with r constituents, each extensive variable or property, such as S, V, H, and U in (17.34) to (17.37), can be expressed in terms of yet another set of partial derivatives which, in turn, suggest further simple interrelations between properties of a multiconstituent system and properties of one-constituent systems. To introduce these derivatives, we express any extensive variable or property Z as a function of T, p, $\boldsymbol{n}$, that is,

$$Z = Z(T, p, \boldsymbol{n}) \tag{17.38}$$

We denote by the symbol z_i the partial derivative

$$z_i = \left(\frac{\partial Z}{\partial n_i}\right)_{T,p,n} \qquad \text{for } i = 1, 2, \ldots, r \tag{17.39}$$

and call it the *mole partial Z* or simply the *partial Z* of the ith constituent. For example, if Z represents the entropy S, the volume V, the enthalpy H, the internal energy U, or the Gibbs free energy G, then z_i represents the *partial entropy* s_i, the *partial volume* v_i, the *partial enthalpy* h_i, the *partial internal energy* u_i, or the *partial Gibbs free energy* g_i, respectively. Specifically, for $i = 1, 2, \ldots, r$, we have

$$s_i = \left(\frac{\partial S}{\partial n_i}\right)_{T,p,n} \tag{17.40}$$

$$v_i = \left(\frac{\partial V}{\partial n_i}\right)_{T,p,n} \tag{17.41}$$

$$h_i = \left(\frac{\partial H}{\partial n_i}\right)_{T,p,n} \tag{17.42}$$

$$u_i = \left(\frac{\partial U}{\partial n_i}\right)_{T,p,n} \tag{17.43}$$

$$g_i = \left(\frac{\partial G}{\partial n_i}\right)_{T,p,n} = \mu_i \tag{17.44}$$

where in (17.44) we use (16.30) and, therefore, conclude that the chemical potential μ_i of the ith constituent is equal to its partial Gibbs free energy g_i, for each $i = 1, 2, \ldots, r$. It is also noteworthy that relations $H = U + pV$ and $G = H - TS$, and equations (17.40) to (17.44) imply that for each $i = 1, 2, \ldots, r$,

$$h_i = u_i + p v_i = g_i + T s_i = \mu_i + T s_i \tag{17.45}$$

$$g_i = \mu_i = h_i - T s_i = u_i + p v_i - T s_i \tag{17.46}$$

Using equations (17.40) to (17.45) and the Maxwell relations $(\partial \mu_i / \partial T)_{p,n} = -(\partial S / \partial n_i)_{T,p,n}$ and $(\partial \mu_i / \partial p)_{T,n} = (\partial V / \partial n_i)_{T,p,n}$ [equations (16.36) and (16.37)], we find that

$$s_i = \left(\frac{\partial S}{\partial n_i}\right)_{T,p,n} = -\left(\frac{\partial \mu_i}{\partial T}\right)_{p,n} = -\left(\frac{\partial \mu_i}{\partial T}\right)_{p,y} \tag{17.47}$$

$$v_i = \left(\frac{\partial V}{\partial n_i}\right)_{T,p,n} = \left(\frac{\partial \mu_i}{\partial p}\right)_{T,n} = \left(\frac{\partial \mu_i}{\partial p}\right)_{T,y} \tag{17.48}$$

$$h_i = \mu_i + T s_i = \mu_i - T\left(\frac{\partial \mu_i}{\partial T}\right)_{p,y} = \left[\frac{\partial(\mu_i/T)}{\partial(1/T)}\right]_{p,y} \tag{17.49}$$

$$u_i = h_i - p v_i = \left[\frac{\partial(\mu_i/T)}{\partial(1/T)}\right]_{p,y} - p\left(\frac{\partial \mu_i}{\partial p}\right)_{T,y} \tag{17.50}$$

where in writing the second of equations (17.47) and (17.48) we use the fact that μ_i depends on T, p, and y only, and for the second of (17.50) we use (17.49) and (17.48).

Combining equations (17.47) to (17.50) with (17.34) to (17.37), we conclude that the extensive variables and properties S, V, H, and U can also be expressed in terms of partial properties, that is, $S = \sum_{i=1}^{r} n_i s_i$, $V = \sum_{i=1}^{r} n_i v_i$, $H = \sum_{i=1}^{r} n_i h_i$, and $U = \sum_{i=1}^{r} n_i u_i$. Writing explicitly the dependences, subject to the relation $y_1 + y_2 + \cdots + y_r = 1$, we have

$$S(T, p, n_1, n_2, \ldots, n_r) = \sum_{i=1}^{r} n_i s_i(T, p, y_1, y_2, \ldots, y_r) \tag{17.51}$$

$$V(T, p, n_1, n_2, \ldots, n_r) = \sum_{i=1}^{r} n_i v_i(T, p, y_1, y_2, \ldots, y_r) \tag{17.52}$$

$$H(T, p, n_1, n_2, \ldots, n_r) = \sum_{i=1}^{r} n_i h_i(T, p, y_1, y_2, \ldots, y_r) \tag{17.53}$$

$$U(T, p, n_1, n_2, \ldots, n_r) = \sum_{i=1}^{r} n_i \, u_i(T, p, y_1, y_2, \ldots, y_r) \qquad (17.54)$$

$$G(T, p, n_1, n_2, \ldots, n_r) = \sum_{i=1}^{r} n_i \, \mu_i(T, p, y_1, y_2, \ldots, y_r) \qquad (17.55)$$

These relations show that partial properties represent the contribution per unit amount of each constituent to the corresponding extensive property of the multiconstituent simple system. Later, we discuss measurement and calculation procedures to determine the values of partial properties. Under special ranges of conditions, we find that each partial property is directly related to a specific property of a one-constituent simple system.

Problems

17.1 For a simple system consisting of a single constituent, the entropy S, the volume V, and the chemical potential μ are given by the relations

$$S = nS_o + \frac{3}{2} nR \ln \frac{3 \, (RT)^{5/3}}{2 \, U_o \, (pV_o)^{2/3}}$$

$$V = \frac{nRT}{p}$$

$$\mu = \frac{5}{2} RT - TS_o - \frac{3}{2} RT \ln \frac{3 \, (RT)^{5/3}}{2 \, U_o \, (pV_o)^{2/3}}$$

where R, U_o, V_o, and S_o are constants, T is the temperature, p the pressure, and n the amount of the constituent.

(a) Find an expression for the internal energy U in terms of the given constants and T, p, and n only.

(b) Find the expression for the Gibbs free energy G as a function of T, p, and n.

(c) Verify that $(\partial T/\partial S)_{V,n} > 0$ and $(\partial p/\partial V)_{S,n} < 0$.

17.2 For a simple system consisting of two constituents, the Gibbs free energy may be expressed in the form

$$G = \frac{5}{2} nRT - \frac{3}{2} nRT \ln \frac{3 \, (RT)^{5/3}}{2 \, U_o \, (pV_o)^{2/3}} - nTS_o + nRT \, (y_1 \ln y_1 + y_2 \ln y_2 + A \, y_1 y_2)$$

where R, U_o, V_o, S_o, and A are constants, T is the temperature, p the pressure, $n = n_1 + n_2$, n_1 the amount of constituent 1, n_2 the amount of constituent 2, $y_1 = n_1/n$ and $y_2 = n_2/n$, that is, the mole fraction of constituent 1 and the mole fraction of constituent 2.

(a) Find expressions for the entropy S, the volume V, the enthalpy H, and the internal energy U.

(b) Find expressions for the partial entropy s_i, the partial volume v_i, the partial enthalpy h_i, and the partial internal energy u_i of constituent i for $i = 1, 2$, and verify (17.51) to (17.54).

(c) Find expressions for the chemical potentials μ_1 and μ_2, and verify (17.55).

(d) Using the results in parts (a) and (c), verify the Euler relation.

17.3 A system consists of two constituents experiencing negligible interparticle forces and confined in a volume V. It can be shown that at relatively high temperatures, T, low pressures, p, and large amounts of constituents, n_1 and n_2, the Gibbs free energy of the system satisfies the relation

$$G(T, p, n_1, n_2) = -\sum_{i=1}^{2} n_i kT \ln \left[d_i \left(\frac{2\pi M_i kT}{h^2} \right)^{3/2} \frac{kT}{p} \frac{n}{n_i} \right]$$

where $n = n_1 + n_2$, d_i is a constant characteristic of constituent i, M_i the molecular weight of constituent i, k the Boltzmann constant, and h the Planck constant. For a one-constituent system consisting of either constituent 1 only or constituent 2 only, the Gibbs free energy satisfies the relation

$$G_{ii}(T, p, n_i) = -n_i kT \ln \left[d_i \left(\frac{2\pi M_i kT}{h^2} \right)^{3/2} \frac{kT}{p} \right] \qquad \text{for } i = 1, 2$$

(a) Find the entropy, volume, chemical potentials, and internal energy of the two-constituent system as functions of T, p, n_1, and n_2.

(b) Find the differences in entropies, volumes, and internal energies of the two-constituent system at T, p, n_1, n_2 and the composite of two one-constituent systems, one at T, p, n_1, and the other at T, p, n_2.

(c) Verify the Euler relation and (17.15), and thus prove that the given Gibbs free energy describes a simple system.

(d) Show that the difference in entropies found in part (b) reduces abruptly to zero as the two dissimilar constituents become identical. This discontinuous change in entropy as the constituents change from different to alike is known as the *Gibbs paradox*. The paradox is explained and rationalized using concepts and tools of quantum thermodynamics that are beyond our scope here.

17.4 A simple system that instead of volume has surface area a as the only parameter can be used to model constituents that have only two translational degrees of freedom such as constituents that are constrained onto a given surface. For such a simple system, the fundamental relation has the form $S = S(U, a, \boldsymbol{n})$ or, equivalently, the energy relation is $U = U(S, a, \boldsymbol{n})$, and the generalized force conjugated to the parameter a is called *surface tension* and denoted by σ, that is,

$$\sigma = \left(\frac{\partial U}{\partial a} \right)_{S, \boldsymbol{n}} = -T \left(\frac{\partial S}{\partial a} \right)_{U, \boldsymbol{n}}$$

(a) Show that the function $\mathcal{G} = \mathcal{G}(T, \sigma, \boldsymbol{n})$ defined by the relation $\mathcal{G} = U - TS + \sigma a$ is a characteristic function.

(b) Find the Maxwell relations that follow from the characteristic function in part (a).

(c) Find the Maxwell relations that follow from the characteristic function $\mathcal{H} = \mathcal{H}(S, \sigma, \boldsymbol{n})$ defined by the relation $\mathcal{H} = U + \sigma a$.

17.5 Consider a surface C that separates two regions A and B of a fixed-volume container. If the principal radii of curvature of the surface are r_1 and r_2 and constant throughout the surface, and region A contains the centers of curvature, changes in volumes of the two regions that accompany a change of area da of the surface can be shown to satisfy the relation

$$dV^A = -dV^B = \frac{r_1 r_2}{r_1 + r_2} da$$

Assume that A, B, and C are simple systems. If they consist of the same constituent, are in mutual stable equilibrium, and the surface C is described as in Problem 17.4, use the relation between dV^A, dV^B, a and the principle of highest entropy to show that

$$p^A - p^B = \sigma^C \left(\frac{1}{r_1} + \frac{1}{r_2} \right)$$

where σ^C is the surface tension of C (see Problem 17.4).

18 Phase Rule

Matter may exist in different forms of aggregation, such as solid, liquid, or vapor. For example, at atmospheric pressure ordinary water is solid at temperatures below 273.15 K, liquid at room temperature, and vapor at temperatures above 373.15 K. Moreover, these forms may be juxtaposed as when ice is floating in a lake, or water and steam are present in a boiler.

In this chapter we consider substances that can be modeled as simple systems and establish conditions that must be satisfied for two or more forms of aggregation to be in mutual stable equilibrium.

18.1 Homogeneous and Heterogeneous States

In Chapter 17 we define as simple a system with r constituents that has volume as the only parameter and that can be regarded as a composite of a contiguous collection of infinitesimal subsystems, all in mutual stable equilibrium. Mutual stable equilibrium requires that all the subsystems have the same temperature T, pressure p, and chemical potentials $\mu_1, \mu_2, \ldots, \mu_r$. In fact, because each subsystem can be very small—almost infinitesimal—we can think of temperature, pressure, and the chemical potentials as being defined locally and having uniform values throughout the spatial extension of the system.

In contrast to the equalities satisfied by T, p, and the μ_i's, the condition of mutual stable equilibrium between subsystems imposes no restrictions on the values of other intensive properties, such as the amount of matter per unit volume and the energy per unit amount of matter. For this reason the values of such intensive properties may differ from subsystem to subsystem within the simple system and hence need not be uniform throughout the extension of the system. For example, in a state consisting of ice floating on water, the density of the ice (the mass of H_2O per unit volume of ice) differs from that of the liquid water, even though ice and water are in mutual stable equilibrium.

A stable equilibrium state of a simple system is called a *homogeneous state* if the value of every intensive property is common to all the subsystems into which the system can be subdivided. Otherwise, the stable equilibrium state is called a *heterogeneous state*. Thinking in terms of a continuum in space, we see that a stable equilibrium state is homogeneous when all the intensive properties have uniform values throughout the extension of the system, and heterogeneous when not all the intensive properties have uniform values.

In subsequent chapters we show that properties of systems in heterogeneous states can be investigated in terms of properties of systems in homogeneous states. Here we develop a tool for such investigations.

18.2 Phases

To establish the dependence of a heterogeneous state on its homogeneous parts, we consider a subdivision of a simple system into subsystems each of which is in a homogeneous state, and call the collection of all the subsystems that have the same values of all the intensive properties a *phase*. The subsystems that make up a phase need not be contiguous to one another. Their collection, however, can be modeled as a simple system with values of energy, volume, and amounts of constituents equal to the values of the respective sums over the subsystems of the phase, and values of the intensive properties identical to the respective values of any of the subsystems that make up the phase. Thus a phase can be modeled as a simple system in a homogeneous state. Clearly, a simple system in a homogeneous state consists, by definition, of one phase only, whereas a simple system in a heterogeneous state consists of at least two phases.

Example 18.1. Prove the last assertion.

Solution. By definition, if a simple system is in a homogeneous state, all its subsystems are in homogeneous states and share the same values of all the intensive properties. Hence we can identify only one set of values of the intensive properties and thus only one phase coinciding with the overall system. If a simple system is in a heterogeneous state, not all its subsystems share the same values of all the intensive properties. Hence we can identify at least two different sets of values of the intensive properties and thus at least two phases.

As an example of a heterogeneous state, we consider a system consisting of the single constituent H_2O in a stable equilibrium state in which liquid water droplets are dispersed throughout water vapor. The collection of all the liquid droplets constitutes one phase, and the collection of all the vapor another phase. The boundary between each liquid droplet and the vapor is not a geometric surface but rather a thin film across which some intensive properties exhibit a steep gradient. We assume, however, that the conditions are such for the simple system model to be a good approximation. Thus we regard all the intensive properties as being uniform within each phase, and some intensive properties as exhibiting an absolute discontinuity across the interfaces between phases. The degree of accuracy of the simple system model is higher the larger the radii of curvature of the separating surfaces between liquid and vapor.

As another example we consider a metal that may solidify in two different interdispersed crystalline structures such as a cubic crystal and a center-cubic crystal. Because a cubic crystal is less dense than a center-cubic crystal, the intensive property mass per unit volume is not uniform, and hence the state of the metal is heterogeneous. It consists of two phases, one being the collection of all cubic microcrystals, and the other the collection of all the center-cubic microcrystals.

When two or more phases are present, we say that they *coexist*. Thus a heterogeneous state is a stable equilibrium state of *coexisting phases*. When q phases coexist, we also say that the simple system is in a *q-phase heterogeneous state*, or simply, a *q-phase state*. Because each phase can be modeled as a simple system in a homogeneous state, a simple system in a q-phase heterogeneous state can be modeled as a composite of simple systems each corresponding to a phase and all in mutual stable equilibrium with each other.

The terms phase and form of aggregation are sometimes used synonymously, although the latter is not as restrictive as the former. The term form of aggregation refers to some broad characteristics of a substance, and is not limited to stable equilibrium states. For example, the terms *solid* and *fluid* refer to the behavior of the substance under a shearing stress (a solid can sustain it, a fluid cannot). The terms *liquid* and *gas* (or vapor) refer to the behavior in a container under the action of the earth's gravitational field (a liquid collects at the bottom, a gas fills the entire volume available). The terms *vapor* and *gas* are sometimes used interchangeably, but in general refer to the behavior in different ranges of temperature and pressure. In certain ranges, a vapor is a substance that can be condensed to a liquid by either isobaric cooling or isothermal compression. In other ranges, a gas is a substance that cannot be condensed by either of the two processes just cited. The terms *crystal* and *amorphous solid* (or *glass*) refer to the long-range order and symmetry of the geometrical arrangement of the nuclei of a substance (in a crystal, nuclei are distributed in large ordered and symmetrical patterns, in an amorphous solid the nuclei are distributed in irregular patterns). The term *liquid crystal* refers to the concomitance of features that are typical of liquids and crystals.

In general, a phase consists of only one form of aggregation. However, a form of aggregation may consist either of many phases (such as a solid mixture of two different crystals), or not be a phase (such as a solid in a nonequilibrium state).

An interesting issue pertaining to phases is the number of independent intensive properties of a *q*-phase state. We discuss this issue in the next section.

18.3 Gibbs Phase Rule

We consider a simple system consisting of r types of constituents in a heterogeneous state with q coexisting phases. The values of the temperature T, pressure p, and chemical potentials μ_1, μ_2, ..., μ_r are common to all phases because they are all in mutual stable equilibrium. We wish to examine the following question. Starting from an initial q-phase state, how many of the $r + 2$ intensive properties T, p, μ_1, μ_2, ..., μ_r can be varied independently while the system changes to a q-phase state of the same types of phases as those of the initial state? For example, if a two-phase state of H_2O consists of liquid water and water vapor how many of the three intensive properties T, p, and μ can be varied independently while the system changes to another two-phase state consisting also of liquid water and water vapor?

The answer to the general question just cited is $r + 2 - q$, and is known as the *Gibbs phase rule*. The number $r + 2 - q$ is sometimes called *variance* and denoted by the symbol F so that[1]

$$F = r + 2 - q \qquad (18.1)$$

To prove the Gibbs phase rule, we consider two neighboring q-phase states and denote the differences in temperature, pressure, and chemical potentials between these two states by dT, dp, $d\mu_1$, $d\mu_2$, ..., $d\mu_r$, respectively. Each of these differences is common to all phases because the q-phases remain in mutual stable equilibrium. Moreover, each phase must satisfy a Gibbs–Duhem relation [equation (17.9)], so that

$$S^{(1)}dT - V^{(1)}dp + n_1^{(1)}d\mu_1 + n_2^{(1)}d\mu_2 + \cdots + n_r^{(1)}d\mu_r = 0$$
$$S^{(2)}dT - V^{(2)}dp + n_1^{(2)}d\mu_1 + n_2^{(2)}d\mu_2 + \cdots + n_r^{(2)}d\mu_r = 0$$
$$\vdots \qquad (18.2)$$
$$S^{(q)}dT - V^{(q)}dp + n_1^{(q)}d\mu_1 + n_2^{(q)}d\mu_2 + \cdots + n_r^{(q)}d\mu_r = 0$$

where $S^{(j)}$, $V^{(j)}$, $n_1^{(j)}$, $n_2^{(j)}$, ..., $n_r^{(j)}$, for $j = 1, 2, ..., q$, are the entropy, volume, and amounts of constituents of the jth phase. Because there are q Gibbs–Duhem relations that must be satisfied, only $r + 2 - q$ of the $r + 2$ differences dT, dp, $d\mu_1$, $d\mu_2$, ..., $d\mu_r$ are independent, and therefore only $r + 2 - q$ of the intensive properties T, p, μ_1, μ_2, ..., μ_r can be varied independently.

---

[1]Equation (18.1) is valid in the absence of chemical reactions between the r constituents. In the presence of chemical reactions, the Gibbs phase rule has a different expression. We discuss it in Chapter 30.

This conclusion holds only if none of the phases disappears as a result of the variations in T, p, μ_1, μ_2, ..., μ_r, because if any phase disappeared the corresponding Gibbs–Duhem relation would not be needed. It also holds only if no phase is substituted by another of a different type, because if there is a substitution the two q-phase states could not be regarded as neighboring. Indeed, if a phase is substituted for another, some intensive property varies by a finite amount, and therefore the differential Gibbs–Duhem relation is no longer valid.

The Gibbs phase rule has many practical implications. For $q = 1$, that is, for homogeneous states, the phase rule implies that only $r+1$ of the $r+2$ intensive properties T, p, μ_1, μ_2, ..., μ_r can be varied independently, a result that we already derived in Section 17.5. Specifically, we can express these properties as

$$T = T(s, v, y_1, y_2, \ldots, y_r)$$
$$p = p(s, v, y_1, y_2, \ldots, y_r)$$
$$\mu_1 = \mu_1(s, v, y_1, y_2, \ldots, y_r)$$
$$\mu_2 = \mu_2(s, v, y_1, y_2, \ldots, y_r)$$
$$\vdots$$
$$\mu_r = \mu_r(s, v, y_1, y_2, \ldots, y_r)$$

(18.3)

where $s = S/n$, $v = V/n$, $y_i = n_i/n$ for $i = 1, 2, \ldots, r$, $n = n_1 + n_2 + \cdots + n_r$, and we recognize that the sum of the y_i's equals unity.

For $r = 1$, that is, systems with only one constituent, such as H_2O, the phase rule implies that if the number of coexisting phases is 1, 2, or 3, then the variance is 2, 1, or 0, respectively. Thus, for ice, water, or steam, each by itself in a one-phase state ($q = 1$), temperature and pressure may be regarded as two independent intensive properties, but the chemical potential is a function of T and p. For ice, water, and steam present in pairs in a two-phase state ($q = 2$), temperature T may be regarded as an independent property, but then the pressure and chemical potential are functions of T; alternatively, pressure p may be regarded as an independent property, but then the temperature and chemical potential are functions of p. In general, for a given pair of coexisting phases, the relation between temperature and pressure is such that to a given value of temperature there corresponds a unique value of pressure and, conversely, to a given value of pressure there corresponds a unique value of temperature. For ice, water, and steam all present in a three-phase state ($q = 3$), none of the intensive properties T, p and μ can be varied independently without reducing the number of coexisting phases. In other words, for a given constituent, the solid, liquid, and vapor phases can coexist only for a specific triplet of values of T, p, and μ. For this reason, such a three-phase state is called a *triple-point state*. Details about the properties of simple systems consisting of one constituent are discussed in subsequent chapters.

For $r = 2$, that is, systems consisting of two constituents, such as O_2 and N_2, and in the absence of chemical reactions, the phase rule implies that if the number of coexisting phases is 1, 2, 3, or 4, the number of independent intensive properties T, p, μ_1, and μ_2 is 3, 2, 1, or 0, respectively. For a one-phase, solid, liquid, or gaseous mixture of O_2 and N_2, temperature, pressure, and chemical potential of one constituent may be regarded as the independent intensive properties, in contrast to the result for single-constituent, single-phase states discussed earlier. For two coexisting phases, such as liquid and vapor, temperature and pressure can be varied independently, and they in turn fix the chemical potentials of the constituents in each phase. Analogous conclusions can be reached for three and four coexisting phases. The properties of simple systems composed of more than one constituent with and without chemical reactions are studied in detail in later chapters.

It is noteworthy that a stable equilibrium state of a system consisting of two or more constituents mixed in gaseous or liquid form must be considered as homogeneous. Otherwise, to distinguish between types of molecules, we would have to subdivide the system down to the molecular level where the simple system model and its implications, including the Gibbs phase rule, are not valid.

Problems

18.1 Two substances are said to be *miscible in the liquid phase* if they form a single liquid phase when their pure liquid phases are mixed in any proportion. Many pairs of organic substances are miscible in the liquid phase. In general, in a two-phase liquid–vapor state of a simple system consisting of two constituents that are miscible in the liquid phase, the composition of the liquid and the vapor phases are different. An important exception is the liquid–vapor state of a mixture, called an *azeotropic mixture*, which at one particular composition, called *azeotropic composition*, behaves like a single-constituent simple system, in the sense that for a given value of pressure the liquid phase and the vapor phase coexist at a unique value of temperature and, conversely, for a given value of temperature they coexist at a unique value of the pressure.

(a) What is the variance F for a liquid–vapor state of a two-constituent simple system with composition different from azeotropic?

(b) What is the variance F for a liquid–vapor state with the azeotropic composition?

18.2 Two substances are said to be *immiscible in the liquid phase* if they remain in two phases when their pure liquid phases are mixed in any proportion. For example, mercury, Hg, and water, H_2O, are practically immiscible in the liquid phase.

(a) What is the variance F of a two-phase liquid–vapor state of a mercury–water simple system in which only the mercury is present in the liquid phase?

(b) What is the variance F of a liquid–vapor state in which both mercury and water are present as liquid phases?

18.3 Two substances are said to be *partially miscible in the liquid phase* if they form a single liquid phase when their pure liquid phases are mixed in certain proportions, but remain in two liquid phases when mixed in other proportions. What is the variance F for a two-constituent simple system in a liquid–vapor state in which the vapor phase coexists with the liquid phase of one constituent and with a minute liquid droplet of the other constituent?

19 Thermophysical Properties of Pure Substances

In this chapter we apply the results of Chapters 17 and 18 to a system consisting of a single constituent in one-phase, two-phase, and three-phase states. The experimental and theoretical results on such a system are known as the thermophysical properties of a *pure substance* or, simply, a *substance*.

19.1 Specific Properties

Prior to discussing experimental results, we review briefly some of the relevant terminology. We consider an amount n of molecules or atoms of a single constituent confined in a volume V under conditions that can be modeled by the stable equilibrium states of a one-constituent simple system. Each specific property of such a system depends at most on two independent variables (see Section 17.5). For example, the specific internal energy u depends on the specific entropy s and the specific volume v, so that

$$u = u(s, v) \tag{19.1}$$

We use the same lowercase letters to denote specific properties either with respect to the amount of constituent n, such as $u = U/n$, $s = S/n$, and $v = V/n$, or with respect to the mass m, such as $u = U/m$, $s = S/m$, and $v = V/m$, and infer which is the case from the context. The ratio of a number-specific property and the corresponding mass-specific property is m/n, that is, the mass per unit amount of the constituent.

19.2 Molecular Weight

Amounts of constituents are expressed in terms of the unit of amount of matter adopted by the International System of units (SI), that is, the *mole* or *gram mole*, abbreviated as mol, and defined as "the amount of substance of a system which contains as many elementary entities (atoms or molecules) as there are atoms in 12 grams of carbon 12." Carbon 12 is the isotope of C that has six protons and six neutrons in its nucleus. This amount is

$$1 \text{ mol} = 1 \text{ gram mole} = 6.022 \times 10^{23} \text{ molecules or atoms} \tag{19.2}$$

The conversion factor between mole and number of molecules, $N_A = 6.022 \times 10^{23}$ molecules/mol, is called *Avogadro's number*. By analogy, kilomole (kmol), pound mole (lbmol), and slug mole (slugmol) are related to carbon 12 amounts of 12 kg, 12 lbm, and 12 slugs, respectively.

The mass $m = nM$, where M is the mass per unit amount.[1] Historically, when n is expressed in moles, M is called the *molecular weight*, although a more appropriate term

---

[1]According to the discussion in Section 3.12, the difference $m - m_g$ between the mass m and the ground-state mass m_g of a given molecule is practically negligible when compared to the value of either m or m_g and, therefore, we make no distinction here between mass and ground-state mass. In terms of the molecular weight M, the ground-state mass $m_g = M/N_A$, and the ground-state energy $E_g = m_g c^2 = Mc^2/N_A$.

would be the *molar mass*. The units of molecular weight are grams per mole (g/mol), kilograms per kilomole (kg/kmol), and so on. For example, for H_2O,

$$M = 18.012 \text{ g/mol} = 18.012 \text{ kg/kmol}$$

Molecular weights for various substances are listed in Table 19.1. Each molecular weight may also be expressed as a mass per molecule by dividing M by Avogadro's number N_A. For water,

$$M = \frac{18.012 \text{ kg/kmol}}{6.022 \times 10^{26} \text{ molecules/kmol}} = 2.991 \times 10^{-26} \text{ kg/molecule}$$

19.3 Experimental Results

A system consisting of a single constituent and such that its stable equilibrium state properties are accurately described by a simple system model is called, for brevity, a *pure substance*. Pure substances, such as hydrogen, H_2, water, H_2O, carbon dioxide, CO_2, methane, CH_4, and refrigerant 12, CCl_2F_2, exhibit patterns of coexisting phases and forms of aggregation that are common to all constituents. In what follows, we describe such patterns for water because they are representative of those of other pure substances, except for differences in detail.

To make ideas specific, we consider n molecules of H_2O confined in a volume V of a cylinder with a movable piston (Figure 19.1). The piston is loaded with a weight. For each value of its mass, the weight exerts a definite force per unit area. Thus, for a fixed weight the pressure of the system remains constant.

We measure the temperature T, pressure p, and specific volume $v = V/m$, and for each value of p we make a graph of temperature T versus the specific volume v. For $p = 1$ atm, the T versus v graph has the form shown in Figure 19.2. At a temperature T_1 less than $0°C$, the H_2O is solid—ice—and has specific volume v_1. This is represented as state 1 in the figure. As energy is added to the system at the constant pressure $p = 1$ atm, both the temperature and the specific volume increase. When the temperature reaches $0°C$ (state 2), further addition of energy results in no change in temperature but a decrease in specific volume. The ice starts melting and both ice and liquid water coexist in a two-phase state. Water is representative of substances that contract upon melting. As more energy is added, more ice melts at the constant temperature of $0°C$, and the constant pressure of 1 atm until only the liquid phase is present (state 3). Further addition of energy beyond state 3 increases both the temperature and the specific volume

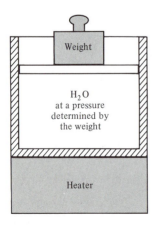

Figure 19.1 A heater adds energy to a simple system consisting of n water molecules confined in a volume V. The system passes through stable equilibrium states and is maintained at a constant pressure by a movable piston loaded with a fixed weight.

TABLE 19.1. Molecular weights and critical-state properties for various substances.

Substance	Formula	M kg/kmol	T_c K	p_c MPa	v_c m^3/kmol
Acetic acid	$C_2H_4O_2$	60.052	594.4	5.79	0.171
Acetone	CH_3COCH_3	58.08	508.1	4.70	0.209
Acetylene	C_2H_2	26.038	308.3	6.14	0.113
Ammonia	NH_3	17.031	405.6	11.28	0.0725
Argon	Ar	39.948	150.8	4.87	0.0749
Benzene	C_6H_6	78.114	562.1	4.89	0.259
Carbon dioxide	CO_2	44.01	304.2	7.38	0.094
Carbon monoxide	CO	28.01	132.9	3.50	0.0931
Chlorine	Cl_2	70.906	417.0	7.70	0.124
Chloroform	$CHCl_3$	119.378	536.4	5.47	0.239
Ethane	C_2H_6	30.07	305.4	4.88	0.148
Ethanol	C_2H_5OH	46.069	516.2	6.38	0.167
Ethylene	C_2H_4	28.054	282.4	5.04	0.129
Fluorine	F_2	37.997	144.3	5.22	0.0662
Freon 12	CCl_2F_2	120.914	385.0	4.12	0.217
Freon 13	$CClF_3$	104.459	302.0	3.92	0.18
Freon 21	$CHCl_2F$	102.923	451.6	5.17	0.197
Freon 22	$CHClF_2$	86.469	369.2	4.98	0.165
Freon 114	$C_2Cl_2F_4$	170.922	418.9	3.26	0.293
Genetron 100	CH_3CHF_2	66.051	386.6	4.50	0.181
Genetron 101	CH_3CClF_2	100.496	410.2	4.12	0.231
Hydrogen	H_2	2.016	33.2	1.30	0.065
Hydrogen chloride	HCl	36.461	324.6	8.31	0.081
Isooctane	C_8H_{18}	114.232	543.9	2.56	0.468
Methane	CH_4	16.043	190.6	4.60	0.099
Methanol	CH_3OH	32.042	512.6	8.10	0.118
Methylene chloride	CH_2Cl_2	84.933	510.0	6.08	0.193
Naphthalene	$C_{10}H_8$	128.174	748.4	4.05	0.41
Neon	Ne	20.183	44.4	2.76	0.0417
Nitric oxide	NO	30.006	180.0	6.48	0.058
Nitrogen	N_2	28.013	126.2	3.39	0.0895
Nitrogen dioxide	NO_2	46.006	431.4	10.13	0.17
Nitrous oxide	N_2O	44.013	309.6	7.24	0.0974
n-Octane	C_8H_{18}	114.232	568.8	2.48	0.492
Oxygen	O_2	31.999	154.6	5.05	0.0734
Ozone	O_3	47.998	261.0	5.57	0.0889
Propane	C_3H_8	44.097	369.8	4.25	0.203
Propylene	CH_2CHCH_3	42.081	365.0	4.62	0.181
Water	H_2O	18.015	647.3	22.05	0.056

Source: Data from R. C. Reid, J. M. Prausnitz, and T. K. Sherwood, *The Properties of Gases and Liquids*, McGraw-Hill, New York, 1977.

of the liquid until the temperature reaches 100°C (state 4). From then on, as energy is added, the temperature remains fixed at 100°C, and only the specific volume increases. Now part of the liquid vaporizes. The liquid and vapor phases coexist in mutual stable equilibrium up to state 5, at which only vapor is present. Addition of energy beyond state 5 results in raising both the temperature and the specific volume of the vapor. The vapor temperature may be raised to relatively high values until chemical dissociation destroys

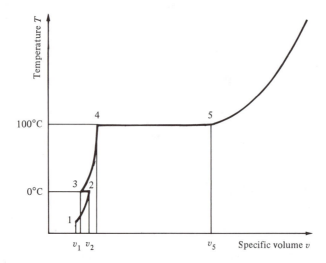

Figure 19.2 Graph of temperature versus specific volume obtained by heating water at atmospheric pressure.

the assumed purity of the water molecules. Similar results are obtained at pressures different from 1 atm.

For a substance that expands upon melting, a typical T versus v trace at constant pressure is shown in Figure 19.3, and many such traces, each at a different pressure, in Figure 19.4. The p–v–T results can be represented as a surface in a three-dimensional space with axes p, v, and T. Such surfaces for two substances, one that contracts and the other that expands upon melting, are shown in Figure 19.5. Each of the T versus v traces discussed earlier, such as in Figure 19.3, is essentially an intersection of the p–v–T surface with a plane at a fixed p value. Several of these traces superimposed, such as in Figure 19.4, represent a projection of the p–v–T surface on the T–v plane.

As pressure is raised to a value p_c called *critical pressure*, the liquid and vapor phases become identical, and the vaporization process contracts to an infinitesimal process occurring at a value of temperature T_c, called *critical temperature*, and a value of specific

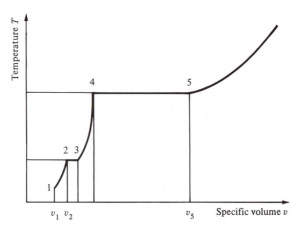

Figure 19.3 Graph of T versus v for constant-pressure heating of a substance that expands upon melting.

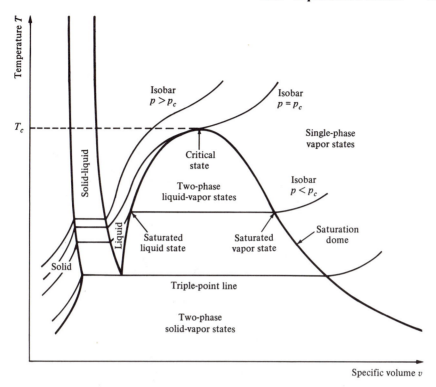

Figure 19.4 Typical T versus v diagram for a substance that expands upon melting.

volume v_c, called *critical volume*. Values of critical temperatures, pressures, and volumes for various substances are listed in Table 19.1. The three critical values of temperature, pressure, and specific volume determine the *critical point* or *critical state*. At higher pressures no vaporization occurs. It becomes clear, then, that the liquid and vapor forms of aggregation differ in degree rather than in kind, because a change from liquid to vapor may occur without passing through any two-phase state. The same cannot be said of any change from solid to liquid or from solid to vapor. At relatively low pressures, a solid phase may coexist in mutual stable equilibrium with a vapor phase, and change from solid to vapor may occur without passing through the liquid form of aggregation.

Figures 19.4 and 19.5 show curves, called *saturation curves*, which partition the projections on the T–v plane, and the p–v–T surface into regions corresponding to single-phase states (solid, liquid, vapor), and into two-phase states (solid–liquid, liquid–vapor, solid–vapor). A state on such a curve is called a *saturated state*. Each saturated state represents the onset of a transition from a single-phase to a two-phase state. The dome-shaped region bounded by the locus of saturated-liquid states and the locus of saturated-vapor states (Figure 19.4) is sometimes called the *saturation dome*. The states to the right of the saturation dome are called the *superheated vapor states* and those to the left the *subcooled liquid states*. The saturated state with the highest pressure and highest temperature is the critical state. States at pressures higher than the critical pressure and temperatures higher than the critical temperature are called *supercritical states*.

Projections of the p–v–T surface on planes other than the T–v plane are also illuminating. Two such projections, one on the p–v plane and another on the p–T plane, are shown in Figures 19.6 and 19.7.

A three-phase state at which solid, liquid, and vapor coexist in mutual stable equilibrium is called a *triple-point state*. Such states correspond to a single point on the p–T

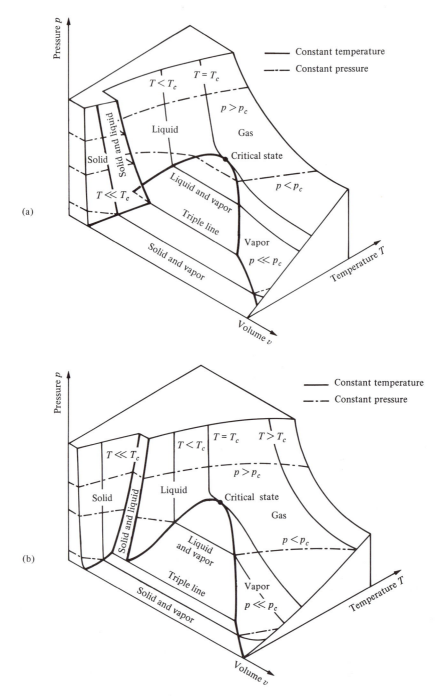

Figure 19.5 (a) p–v–T surface for a substance, like water, that contracts upon melting; (b) p–v–T surface for a substance, like carbon dioxide, that expands upon melting.

diagram, consistently with the Gibbs phase rule discussed in Chapter 18, to a line — the *triple-point line* — on the p–v diagram, and to a triangle — the *triple-point triangle* — on the u–v diagram discussed in Section 19.6. Values of triple-point temperature and pressure for various substances are listed in Table 19.2.

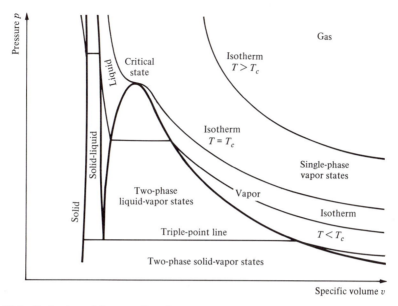

Figure 19.6 Projection of the p–v–T surface on the p–v plane for a substance that expands upon melting.

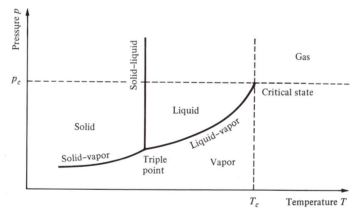

Figure 19.7 Projection of the p–v–T surface on the p–T plane for a substance that expands upon melting.

A locus of states having the same temperature is called an *isotherm*, a locus of states with the same pressure an *isobar*, and a locus of states with the same volume an *isochor*. A locus of states having the same temperature and pressure is called an *isothermobar*.

From Figure 19.6 we note that on the p–v diagram the isotherm at the critical temperature — the *critical isotherm* — has an inflection point at the critical point, that is, at the critical state we have $(\partial p/\partial v)_T = 0$, $(\partial^2 p/\partial v^2)_T = 0$, and $(\partial^3 p/\partial v^3)_T < 0$.

A process in which a change at constant pressure occurs from solid to liquid, liquid to vapor, and solid to vapor is called *fusion* or *melting*, *vaporization*, and *sublimation*, respectively. A process in which a change at constant pressure occurs from liquid to solid, vapor to liquid, and vapor to solid is called *solidification* or *freezing*, *condensation*, and *desublimation*, respectively.

TABLE 19.2.   **Triple-point temperatures and pressures of various substances.**

Substance	Formula	T_{tp} K	p_{tp} kPa
Acetylene	C_2H_2	192.4	120
Ammonia	NH_3	195.40	6.076
Argon	A	83.81	68.9
Carbon (graphite)	C	3900	10,100
Carbon dioxide	CO_2	216.55	517
Carbon monoxide	CO	68.10	15.37
Deuterium	D_2	18.63	17.1
Ethane	C_2H_6	89.89	8×10^{-4}
Ethylene	C_2H_4	104.0	0.12
Helium 4 (λ point)	He	2.19	5.1
Hydrogen	H_2	13.84	7.04
Hydrogen chloride	HCl	158.96	13.9
Mercury	Hg	234.2	1.65×10^{-7}
Methane	CH_4	90.68	11.7
Neon	Ne	24.57	43.2
Nitric oxide	NO	109.50	21.92
Nitrogen	N_2	63.18	12.6
Nitrous oxide	N_2O	182.34	87.85
Oxygen	O_2	54.36	0.152
Palladium	Pd	1825	3.5×10^{-3}
Platinum	Pt	2045	2.0×10^{-4}
Sulfur dioxide	SO_2	197.69	1.67
Titatium	Ti	1941	5.3×10^{-3}
Uranium hexafluoride	UF_6	337.17	151.7
Water	H_2O	273.16	0.61
Xenon	Xe	161.3	81.5
Zinc	Zn	692.65	0.065

Source: Data from *Natl. Bur. Stand. (U.S.) Circ., 500* (1952).

Measurements of energy, enthalpy, heat capacities, and other properties are performed by the procedures discussed earlier and the results are presented in either tabular or graphic form. For example, pressure versus temperature graphs for H_2O and CO_2 are shown in Figure 19.8.

Pure substances may admit several kinds of solid forms of aggregation each being a different type of crystal. For example, Figure 19.9 shows the p–v–T surface and the p–T diagram for H_2O at relatively high pressures and low temperatures where several different phases of ice obtain.

To a very high degree of accuracy, the experimental results obtained from systems with relatively large amounts of a substance conform to the assumption that the stable equilibrium states of such systems can be modeled as stable equilibrium states of a simple system.

For a few substances of wide practical use, data are presented in extensive tables. For example, there are tables for substances such as water, H_2O, ammonia, NH_3, and carbon dioxide, CO_2. For most other substances, data are available only over limited regions of states that are relevant to specific applications. In either case, however, much can be extracted from the experimental data on a given substance by combining them with the wealth of theoretical implications that follow from the stable-equilibrium-state

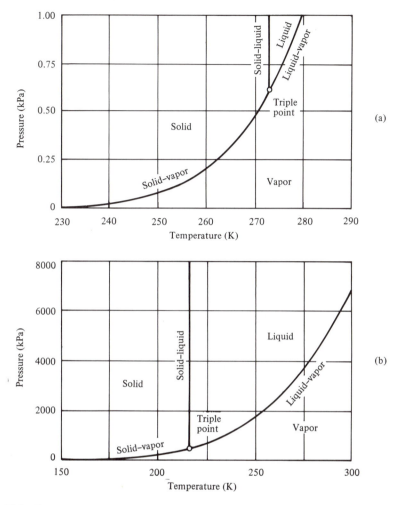

Figure 19.8 Pressure versus temperature graphs near the triple point for: (a) water; and (b) carbon dioxide (adapted from J. B. Jones, and G. A. Hawkins, *Engineering Thermodynamics*, Wiley, New York, 1986).

principle for simple systems, in general, and the patterns of coexisting phases exhibited by all pure substances, in particular.

19.4 Specific Latent Heats

A change of a substance from one form of aggregation to another usually proceeds through a path of heterogeneous states, and then is called a *phase transition* or *phase change*. For example, we have seen that a change of a pure substance from solid to liquid always involves a path of two-phase states. Again, a change from liquid to vapor may be effected by a phase transition, through a path of two-phase states. It can also be effected, however, without a phase transition such as when the path on the p–v or T–v diagram goes around the saturation dome, passing through supercritical states. Clearly, such a path involves only homogeneous states.

When a phase change occurs during a constant-pressure heating or cooling, the temperature remains unchanged as long as both phases are present. If such a process is reversible, the heat required to effect the phase change of a pure substance is equal, ac-

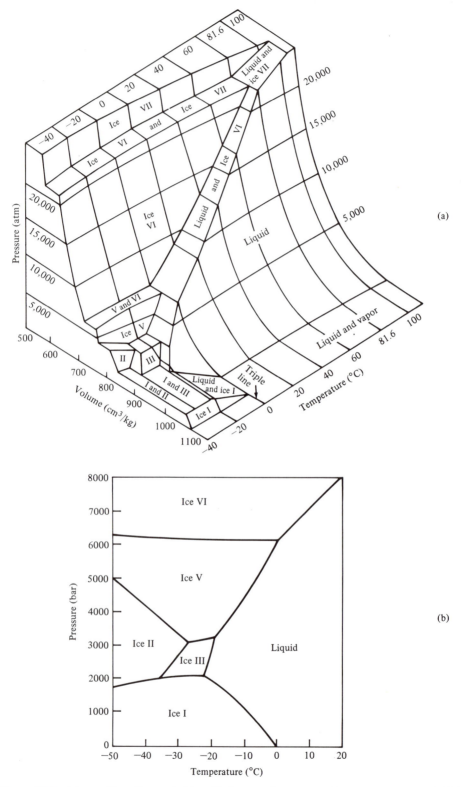

Figure 19.9 (a) p–v–T surface and (b) p–T diagram for water at low temperatures and high pressures showing the coexistence regions of several different solid forms of aggregation denoted as ice I to ice VII (from M. W. Zemansky, *Heat and Thermodynamics*, 5th ed., McGraw-Hill, New York, 1968).

cording to (16.58), to the change in enthalpy of the substance, and is called the *enthalpy* or the *latent heat of the phase change*. Three different specific enthalpies or specific latent heats are distinguished. Denoting the solid, liquid, and vapor phases by subscripts j, f, and g, respectively, we have the *specific enthalpy of fusion*, h_{jf}, for a phase change from solid to liquid, the *specific enthalpy of sublimation*, h_{jg}, for a phase change from solid to vapor, and the *specific enthalpy of vaporization*, h_{fg}, for a phase change from liquid to vapor. These are also called *specific latent heats of fusion*, *sublimation*, and *vaporization*, respectively. Associated with each of these phase changes we also have: a specific entropy, such as the *specific entropy of fusion*, s_{jf}, the *specific entropy of sublimation*, s_{jg}, and the *specific entropy of vaporization*, s_{fg}; a specific volume, such as the *specific volume of fusion*, v_{jf}, the *specific volume of sublimation*, v_{jg}, and the *specific volume of vaporization*, v_{fg}; and a specific internal energy, such as the *specific internal energy of fusion*, u_{jf}, the *specific internal energy of sublimation*, u_{jg}, and the *specific internal energy of vaporization*, u_{fg}. Each of these quantities is defined as a difference between the corresponding specific properties of the two coexisting phases, that is, denoting the specific enthalpies, specific entropies, specific volumes, and specific internal energies of the saturated solid, saturated liquid, and saturated vapor phases, respectively, by h_j, s_j, v_j, u_j; h_f, s_f, v_f, u_f; and h_g, s_g, v_g, u_g, we have

$$h_{jf} = h_f - h_j \qquad h_{jg} = h_g - h_j \qquad h_{fg} = h_g - h_f$$
$$s_{jf} = s_f - s_j \qquad s_{jg} = s_g - s_j \qquad s_{fg} = s_g - s_f$$
$$v_{jf} = v_f - v_j \qquad v_{jg} = v_g - v_j \qquad v_{fg} = v_g - v_f$$
$$u_{jf} = u_f - u_j \qquad u_{jg} = u_g - u_j \qquad u_{fg} = u_g - u_f$$

Relations exist between specific properties of saturated single phases and specific properties associated with phase changes. Indeed, two coexisting phases must satisfy the condition of equality of chemical potentials or, equivalently, equality of specific Gibbs free energies since the chemical potential equals the specific Gibbs free energy [equation (17.23)]. Accordingly, for coexisting solid and liquid phases at temperature T, the condition $\mu_j = \mu_f$ implies that $g_j = g_f$ or $h_j - Ts_j = h_f - Ts_f$ and, therefore,

$$h_{jf} = h_f - h_j = T(s_f - s_j) = Ts_{jf} \qquad (19.3)$$

In words, the specific enthalpy of fusion h_{jf} is related to the specific entropy of fusion s_{jf} so that the ratio h_{jf}/s_{jf} equals the temperature at which the two phases coexist.

Similarly, for coexisting solid and vapor phases at temperature T,

$$h_{jg} = h_g - h_j = T(s_g - s_j) = Ts_{jg} \qquad (19.4)$$

and for coexisting liquid and vapor phases at temperature T,

$$h_{fg} = h_g - h_f = T(s_g - s_f) = Ts_{fg} \qquad (19.5)$$

By measuring the heat in a reversible constant-pressure process that changes a given amount of saturated liquid at temperature T into saturated vapor, we measure the specific enthalpy of vaporization h_{fg} and, by virtue of (19.5), also the specific entropy of vaporization s_{fg}. Moreover, by measuring the change in volume during the process, we measure the specific volume of vaporization, $v_{fg} = v_g - v_f$, and, by virtue of the relation $h = u + pv$, also the specific internal energy of vaporization,

$$u_{fg} = u_g - u_f = (h_g - pv_g) - (h_f - pv_f) = h_{fg} - pv_{fg} \qquad (19.6)$$

Analogous relations hold between the specific enthalpies, specific volumes, and specific internal energies of fusion and sublimation.

Values of the specific enthalpy of fusion, h_{jf}, and the specific enthalpy of vaporization, h_{fg}, at atmospheric pressure for various substances are given in Table 19.3. It is noteworthy that for carbon dioxide and acetylene the triple-point pressure is higher than atmospheric, and therefore only solid and vapor phases can coexist at atmospheric pressure. The specific enthaplies of sublimation, h_{jg}, at 1 atm for CO_2 and C_2H_2 are 574 kJ/kg and 820 kJ/kg, respectively.

19.5 Two-Phase Mixtures

A two-phase state is also called a *two-phase mixture* or, sometimes, simply a *mixture* because often in practice the two phases are interdispersed, such as when steam is dispersed in boiling water. In this section we derive expressions for some properties of the mixture in terms of properties of the two coexisting phases. We analyze a liquid–vapor-mixture, but the results can be readily adapted to any other pair of coexisting phases such as solid–liquid, solid–vapor, and solid–solid for substances that admit more than one solid form of aggregation.

We consider two neighboring states of a two–phase liquid–vapor mixture, one at temperature T, pressure p, and chemical potential μ, and the other at temperature $T+dT$, pressure $p+dp$, and chemical potential $\mu+d\mu$. Consistently with the Gibbs phase rule, we recall that only one of the three differences dT, dp, and $d\mu$ can be chosen independently. Indeed, the Gibbs–Duhem relations for the two neighboring liquid phases and the two neighboring vapor phases require that

$$s_f\, dT - v_f\, dp + d\mu = 0 \tag{19.7}$$

and

$$s_g\, dT - v_g\, dp + d\mu = 0 \tag{19.8}$$

Eliminating $d\mu$ from (19.7) and (19.8), we find that

$$\frac{dp}{dT} = \frac{s_g - s_f}{v_g - v_f} = \frac{s_{fg}}{v_{fg}} \tag{19.9}$$

or, using (19.5),

$$\frac{dp}{dT} = \frac{h_g - h_f}{T(v_g - v_f)} = \frac{h_{fg}}{Tv_{fg}} \tag{19.10}$$

Each of equations (19.9) and (19.10) is called a *Clausius–Clapeyron*[2] *relation*.

Equation (19.10) links three commonly measured quantities: the slope dp/dT of the pressure–temperature relation for the two-phase mixture — here the liquid–vapor mixture — the specific latent heat h_{fg}, and the specific volume v_{fg} involved in the transition from one phase to the other. For two-phase liquid–vapor mixtures we denote the pressure–temperature relation by writing $T = T_{sat}(p)$ or $p = p_{sat}(T)$ where, clearly, the slope of the function $p_{sat}(T)$ is equal to dp/dT. Writing the chemical potentials of the liquid phase and the vapor phase as $\mu_f = \mu_f(T,p)$ and $\mu_g = \mu_g(T,p)$, respectively, when the two phases are in mutual stable equilibrium the functions $p_{sat}(T)$ and $T_{sat}(p)$ are such that $\mu = \mu_g(T,p_{sat}(T)) = \mu_f(T,p_{sat}(T))$ or, equivalently, $\mu = \mu_g(T_{sat}(p),p) = \mu_f(T_{sat}(p),p)$. Values of $T_{sat}(1\text{ atm})$ for various substances are listed in Table 19.3. The table also lists values of $T_{j,sat}(1\text{ atm})$, that is, the saturated-solid temperature at atmospheric pressure. As shown in the table, $T_{j,sat}(1\text{ atm}) = T_{sat}(1\text{ atm})$ for C_2H_2 and for CO_2. This is because for C_2H_2 and CO_2 the triple-point pressure is higher than the atmospheric pressure.

---

[2]Emile Clapeyron (1799–1864), French engineer and physicist, one of the pioneers of thermodynamics.

TABLE 19.3. Values of saturated-solid temperature, $T_{j,sat}$, specific enthalpy of fusion, h_{jf}, saturated-vapor temperature, T_{sat}, and specific enthalpy of vaporization, h_{fg}, at pressure $p_o = 1$ atm for various substances.

Substance	Formula	$T_{j,sat}(p_o)$ K	$h_{jf}(p_o)$ kJ/kg	$T_{sat}(p_o)$ K	$h_{fg}(p_o)$ kJ/kg
Acetic acid	$C_2H_4O_2$	289.8	195.4	391.1	394.6
Acetone	CH_3COCH_3	178.2	98.0	329.4	501.7
Acetylene	C_2H_2	189.2	—	189.2	—
Ammonia	NH_3	195.4	332.4	239.7	1371.8
Argon	Ar	83.8	30.4	87.3	163.5
Benzene	C_6H_6	278.7	126.0	353.3	394.1
Carbon dioxide	CO_2	194.7	—	194.7	—
Carbon monoxide	CO	68.1	29.9	81.7	215.8
Chlorine	Cl_2	172.2	90.4	238.7	288.2
Chloroform	$CHCl_3$	209.6		334.3	249.0
Ethane	C_2H_6	89.9	95.1	184.5	489.4
Ethanol	C_2H_5OH	159.1	107.9	351.5	841.6
Ethylene	C_2H_4	104	119.5	169.4	483.1
Fluorine	F_2	53.5		85	171.9
Freon 12	CCl_2F_2	115.4	34.3	243.4	165.2
Freon 13	$CClF_3$	92		191.7	148.5
Freon 21	$CHCl_2F$	138		282	242.4
Freon 22	$CHClF_2$	113		232.4	233.7
Freon 114	$C_2Cl_2F_4$	179.3		276.9	136.2
Genetron 100	CH_3CHF_2	156.2		248.4	323.3
Genetron 101	CH_3CClF_2	142		263.4	285.4
Hydrogen	H_2	14	58.2	20.4	448.6
Hydrogen chloride	HCl	159	54.7	188.1	443.2
Isooctane	C_8H_{18}	165.8	79.2	372.4	271.6
Methane	CH_4	90.7	58.7	111.7	510.2
Methanol	CH_3OH	175.5	99.2	337.8	1101.0
Methylene chloride	CH_2Cl_2	178.1		313	329.8
Naphthalene	$C_{10}H_8$	353.5	44.7	491.1	337.8
Neon	Ne	24.5	16.0	27	91.3
Nitric oxide	NO	109.5	76.7	121.4	460.5
Nitrogen	N_2	63.3	25.7	77.4	199.2
Nitrogen dioxide	NO_2	261.9		294.3	414.5
Nitrous oxide	N_2O	182.3	148.7	184.7	376.2
n-Octane	C_8H_{18}	216.4	181.6	398.8	301.5
Oxygen	O_2	54.4	13.9	90.2	213.3
Ozone	O_3	80.5		161.3	232.9
Propane	C_3H_8	85.5	79.9	231.1	426.0
Propylene	CH_2CHCH_3	87.9	71.4	225.4	437.8
Water	H_2O	273.15	333.7	373.15	2258.3

Source: Data mainly from R. C. Reid, J. M. Prausnitz, and T. K. Sherwood, *The Properties of Gases and Liquids*, McGraw-Hill, New York, 1977.

The Clausius–Clapeyron relation is used frequently to evaluate one of the quantities dp/dT, h_{fg}, and v_{fg} when the other two have been measured, or to check the consistency of experimental data when all three quantities have been measured. At each temperature, the slope dp/dT of the pressure–temperature relation for two-phase states corresponds in

Figure 19.7 to the slope of the coexistence curves that separate the one-phase regions of the graph.

Denoting generically by Z any of the extensive properties of the two-phase mixture, such as U, S, V, and H, and by $z = Z/n$ the corresponding number-specific property, for a liquid–vapor mixture we have

$$Z = Z_g + Z_f \tag{19.11}$$

where Z_g and Z_f are the extensive properties of the vapor phase and liquid phase, respectively. In view of the extensive character of these properties (Sections 17.4 and 17.5), we have $Z_g = n_g z_g$ and $Z_f = n_f z_f$, where n_g denotes the number of moles of vapor in the mixture and n_f that of liquid, and

$$z = \frac{Z}{n} = \frac{n_g}{n} z_g + \frac{n_f}{n} z_f \tag{19.12}$$

where $n = n_g + n_f$. Denoting by x the ratio n_g/n, namely, the mole fraction that is vapor, we rewrite (19.12) in the form

$$z = x z_g + (1 - x) z_f \tag{19.13}$$

or

$$x = \frac{z - z_f}{z_g - z_f} \tag{19.14}$$

Equations (19.13) and (19.14) hold for z being replaced by u, s, v, and h. They also hold for mass-specific properties, that is, $z = Z/m$, $Z_g = m_g z_g$, $Z_f = m_f z_f$, $m = m_g + m_f$, because $x = n_g/n = m_g/m$. For liquid–vapor mixtures, the fraction x of vapor in the mixture is sometimes called the *quality* of the mixture.

It is noteworthy that here z_f and z_g are functions of T only or, alternatively, of p only, that is, we can write either $z_f = z_f(T)$ and $z_g = z_g(T)$ or $z_f = z_f(p)$ and $z_g = z_g(p)$. In general, z_f and z_g represent specific properties of single-phase states and, as such, are not fully specified by giving only the value of T or only the value of p. However, here each of these specific properties is evaluated at a saturated state. At such a state, we must have equality of the chemical potentials of the liquid and vapor phases, namely, $\mu_f(T, p) = \mu_g(T, p)$ and this implies a definite relation between T and p. Thus, fixing T also fixes p (and vice versa), as well as the specific properties of the single-phase saturated liquid and saturated vapor states with the same value of T (or p). From (19.13) it follows that either $z = z(T, x)$ or $z = z(p, x)$, namely, the intensive properties of a two-phase mixture are fully specified by giving T and x, p and x, T (or p) and any of the properties u, s, v, and h, or any pair of the properties u, s, v, and h.

19.6   Tables and Charts of Properties

Data on the values of properties of pure substances are presented in the form of tables, charts, or semiempirical correlations. The properties of interest in applications are T, p, v, u, h, and s. Tables are structured in two sections. One section refers to single-phase states and gives values of v, u, h, s for various pairs of values of T and p. The value of T or p in each pair is varied in even increments so that linear interpolation between adjacent entries yields reliable results (Table 19.4). The other section refers to saturated states. For saturated liquid and saturated vapor states, this section gives values of v_f, v_g, u_f, u_g, u_{fg}, h_f, h_g, h_{fg}, s_f, s_g, s_{fg}, for various values of T (or p), where again T (or p) is varied in even increments (Table 19.5). More tables are included in Appendix B.

The same data may be presented in a compact chart form. Several types of charts are used in practice. For example, the *Mollier chart* is a diagram of h versus s on which are reported isotherms, isobars, and, within the liquid–vapor two-phase region, the loci of states with the same vapor fraction x (Figure 19.10). Several features of the

TABLE 19.4.   Sample of data on water vapor.[a]

	p (T$_{sat}$)											
	0.90 (96.71)				**0.95** (98.20)				**1.0** (99.63)			
T	v	u	h	s	v	u	h	s	v	u	h	s
Sat.	1869.5	2502.6	2670.9	7.3949	1777.3	2504.4	2673.2	7.3767	1694.0	2506.1	2675.5	7.3594
100	1887.3	2507.7	2677.5	7.4128	1786.5	2507.2	2676.9	7.3865	1695.8	2506.7	2676.2	7.3614
110	1941.1	2522.9	2697.6	7.4659	1837.6	2522.5	2697.0	7.4398	1744.5	2522.0	2696.5	7.4149
120	1994.5	2538.1	2717.6	7.5174	1888.4	2537.7	2717.1	7.4915	1792.9	2537.3	2716.6	7.4668
130	2047.7	2553.2	2737.5	7.5674	1938.9	2552.9	2737.1	7.5416	1840.9	2552.5	2736.6	7.5170
140	2100.6	2568.3	2757.4	7.6160	1989.1	2568.0	2756.9	7.5903	1888.7	2567.7	2756.5	7.5659
150	2153.	2583.3	2777.1	7.6634	2039.	2583.0	2776.8	7.6377	1936.4	2582.8	2776.4	7.6134
160	2206.	2598.4	2796.9	7.7095	2089.	2598.1	2796.6	7.6839	1983.8	2597.8	2796.2	7.6597
170	2258.	2613.4	2816.6	7.7545	2139.	2613.1	2816.3	7.7290	2031.1	2612.9	2816.0	7.7048
180	2311.	2628.4	2836.3	7.7985	2188.	2628.2	2836.1	7.7731	2078.2	2627.9	2835.8	7.7489
190	2363.	2643.4	2856.0	7.8415	2238.	2643.2	2855.8	7.8162	2125.3	2643.0	2855.5	7.7921
200	2415.	2658.4	2875.8	7.8837	2287.	2658.2	2875.5	7.8583	2172.	2658.1	2875.3	7.8343
210	2467.	2673.5	2895.5	7.9249	2337.	2673.3	2895.3	7.8996	2219.	2673.1	2895.1	7.8756
220	2519.	2688.6	2915.2	7.9654	2386.	2688.4	2915.0	7.9401	2266.	2688.2	2914.8	7.9162
230	2571.	2703.7	2935.0	8.0051	2435.	2703.5	2934.8	7.9799	2313.	2703.4	2934.6	7.9559
240	2623.	2718.8	2954.8	8.0441	2484.	2718.7	2954.7	8.0189	2359.	2718.5	2954.5	7.9949
250	2674.	2734.0	2974.7	8.0824	2533.	2733.9	2974.5	8.0572	2406.	2733.7	2974.3	8.0333
260	2726.	2749.2	2994.6	8.1200	2582.	2749.1	2994.4	8.0949	2453.	2749.0	2994.3	8.0710
270	2778.	2764.5	3014.5	8.1571	2631.	2764.4	3014.3	8.1319	2499.	2764.3	3014.2	8.1080
280	2829.	2779.8	3034.5	8.1935	2680.	2779.7	3034.3	8.1684	2546.	2779.5	3034.2	8.1445
290	2881.	2795.2	3054.5	8.2294	2729.	2795.1	3054.4	8.2043	2592.	2795.0	3054.2	8.1804
300	2933.	2810.6	3074.6	8.2647	2778.	2810.5	3074.4	8.2396	2639.	2810.4	3074.3	8.2158
320	3036.	2841.6	3114.8	8.3338	2876.	2841.5	3114.7	8.3087	2732.	2841.5	3114.6	8.2849
340	3139.	2872.9	3155.4	8.4010	2973.	2872.8	3155.3	8.3759	2824.	2872.7	3155.2	8.3521
360	3242.	2904.3	3196.1	8.4664	3071.	2904.3	3196.0	8.4413	2917.	2904.2	3195.9	8.4175
380	3345.	2936.0	3237.1	8.5301	3169.	2936.0	3237.9	8.5051	3010.	2935.9	3236.9	8.4813
400	3448.	2968.0	3278.3	8.5923	3266.	2967.9	3278.2	8.5672	3103.	2967.9	3278.9	8.5435
420	3551.	3000.2	3319.8	8.6530	3364.	3000.2	3319.7	8.6279	3195.	3000.1	3319.6	8.6042
440	3654.	3032.7	3361.5	8.7123	3461.	3032.6	3361.4	8.6873	3288.	3032.6	3361.4	8.6636
460	3756.	3065.4	3403.5	8.7704	3558.	3065.4	3404.4	8.7454	3380.	3065.3	3403.4	8.7216
480	3859.	3098.4	3445.7	8.8272	3656.	3098.4	3445.7	8.8022	3473.	3098.3	3445.6	8.7785
500	3962.	3131.7	3488.2	8.8829	3753.	3131.6	3488.2	8.8579	3565.	3131.6	3488.1	8.8342
520	4065.	3165.2	3531.0	8.9375	3851.	3165.1	3531.0	8.9125	3658.	3165.1	3530.9	8.8888
540	4167.	3199.0	3574.0	8.9911	3948.	3198.9	3574.0	8.9661	3750.	3198.9	3574.0	8.9424
560	4270.	3233.0	3617.4	9.0437	4045.	3233.0	3617.3	9.0188	3843.	3233.0	3617.3	8.9950
580	4373.	3267.4	3660.9	9.0954	4143.	3267.4	3660.9	9.0704	3935.	3267.3	3660.9	9.0467
600	4476.	3302.0	3704.8	9.1462	4240.	3302.0	3704.8	9.1213	4028.	3301.9	3704.7	9.0976
620	4578.	3336.9	3748.9	9.1962	4337.	3336.9	3748.9	9.1712	4120.	3336.8	3748.9	9.1475
640	4681.	3372.1	3793.3	9.2454	4434.	3372.0	3793.3	9.2204	4213.	3372.0	3793.3	9.1967
660	4784.	3407.5	3838.0	9.2938	4532.	3407.5	3838.0	9.2688	4305.	3407.4	3838.0	9.2451
680	4886.	3443.2	3883.0	9.3415	4629.	3443.2	3883.0	9.3165	4397.	3443.2	3882.9	9.2928
700	4989.	3479.2	3928.2	9.3884	4726.	3479.2	3928.2	9.3635	4490.	3479.2	3928.2	9.3398
750	5245.	3570.5	4042.6	9.5030	4969.	3570.5	4042.5	9.4780	4721.	3570.4	4042.5	9.4544
800	5502.	3663.5	4158.7	9.6138	5212.	3663.5	4158.7	9.5888	4952.	3663.5	4158.6	9.5652
850	5759.	3758.3	4276.6	9.7212	5455.	3758.3	4276.6	9.6962	5183.	3758.3	4276.5	9.6725
900	6015.	3854.8	4396.2	9.8253	5698.	3854.8	4396.1	9.8004	5414.	3854.8	4396.1	9.7767
950	6272.	3953.0	4517.4	9.9266	5941.	3953.0	4517.4	9.9016	5644.	3953.0	4517.4	9.8779
1000	6528.	4052.8	4640.3	10.0251	6184.	4052.8	4640.3	10.0001	5875.	4052.8	4640.3	9.9764
1100	7041.	4257.3	4891.0	10.2195	6670.	4257.3	4891.0	10.1896	6337.	4257.3	4891.0	10.1659
1200	7554.	4467.7	5147.6	10.3939	7156.	4467.7	5147.6	10.3699	6799.	4467.7	5147.6	10.3463
1300	8067.	4683.5	5409.5	10.5669	7642.	4683.5	5409.5	10.5419	7260.	4683.5	5409.5	10.5183

Source: J. H. Keenan, F. G. Keyes, P. G. Hill, and J. G. Moore, *Steam Tables*, Wiley, New York, 1969.

[a]T is temperature in °C, p pressure in bar, v specific volume in liter/kg (1 liter = 10^{-3} m^3), u specific internal energy in kJ/kg, h specific enthalpy in kJ/kg, s specific entropy in kJ/kg K.

TABLE 19.5.   Sample of data on saturated liquid water and saturated water vapor.[a]

Temp. °C T	Press. bars p	Specific Volume		Internal Energy			Enthalpy			Entropy		
		Sat. Liquid v_f	Sat. Vapor v_g	Sat. Liquid u_f	Vapori- zation u_{fg}	Sat. Vapor u_g	Sat. Liquid h_f	Vapori- zation h_{fg}	Sat. Vapor h_g	Sat. Liquid s_f	Vapori- zation s_{fg}	Sat. Vapor s_g
55	.15758	1.0146	9568.	230.21	2219.9	2450.1	230.23	2370.7	2600.9	.7679	7.2234	7.9913
56	.16529	1.0151	9149.	234.39	2217.0	2451.4	234.41	2368.2	2602.6	.7807	7.1940	7.9747
57	.17331	1.0156	8751.	238.57	2214.2	2452.7	238.59	2365.8	2604.4	.7933	7.1649	7.9582
58	.18166	1.0161	8372.	242.75	2211.3	2454.0	242.77	2363.4	2606.1	.8060	7.1357	7.9419
59	.19036	1.0166	8013.	246.93	2208.4	2455.3	246.95	2360.9	2607.9	.8186	7.1071	7.9157
60	.19940	1.0172	7671.	251.11	2205.5	2456.6	251.13	2358.5	2609.6	.8312	7.0784	7.9096
61	.20881	1.0177	7346.	255.29	2202.7	2458.0	255.31	2356.0	2611.3	.8437	7.0499	7.8936
62	.21860	1.0182	7037.	259.47	2199.8	2459.3	259.49	2353.6	2613.1	.8562	7.0216	7.8778
63	.22877	1.0188	6743.	263.66	2196.9	2460.6	263.68	2351.1	2614.8	.8686	6.9934	7.8621
64	.23934	1.0194	6463.	267.84	2194.0	2461.8	267.86	2348.7	2616.5	.8811	6.9654	7.8465
65	.2503	1.0199	6197.	272.02	2191.1	2463.1	272.06	2346.2	2618.3	.8935	6.9375	7.8610
66	.2617	1.0205	5943.	276.21	2188.2	2464.4	276.23	2343.7	2620.0	.9058	6.9098	7.8156
67	.2736	1.0211	5701.	280.39	2185.3	2465.7	280.42	2341.3	2621.7	.9181	6.8822	7.8004
68	.2859	1.0217	5471.	284.58	2182.4	2467.0	284.61	2338.8	2623.4	.9304	6.8548	7.7852
69	.2986	1.0222	5252.	288.76	2179.5	2468.3	288.80	2336.3	2625.1	.9427	6.8275	7.7702
70	.3319	1.0228	5042.	292.95	2176.6	2469.6	292.98	2333.8	2626.8	.9549	6.8004	7.7553
71	.3256	1.0234	4843.	297.14	2173.7	2470.8	297.17	2331.4	2628.5	.9671	6.7734	7.7405
72	.3399	1.0240	4652.	301.33	2170.8	2472.1	301.36	2328.9	2630.2	.9792	6.7466	7.7258
73	.3546	1.0247	4470.	305.51	2167.9	2473.4	305.56	2326.4	2631.9	.9914	6.7199	7.7113
74	.3699	1.0253	4297.	309.70	2165.0	2472.7	309.74	2323.9	2633.6	1.0034	6.6934	7.6968
75	.3858	1.0259	4131.	313.90	2162.0	2475.9	313.93	2321.4	2635.3	1.0155	6.6669	7.6824
76	.4022	1.0265	3973.	318.09	2159.1	2477.2	318.13	2318.9	2637.0	1.0275	6.6404	7.6682
77	.4192	1.0272	3822.	322.28	2156.2	2478.4	322.32	2316.3	2638.7	1.0395	6.6145	7.6540
78	.4368	1.0278	3677.	326.47	2153.2	2479.7	326.51	2313.8	2640.3	1.0515	6.5885	7.6400
79	.4550	1.0285	3539.	330.66	2150.3	2481.0	330.72	2311.3	2642.0	1.0634	6.5627	7.6260
80	.4739	1.0291	3407.	334.86	2147.4	2482.2	334.91	2308.8	2643.7	1.0753	6.5369	7.6122
81	.4937	1.0298	3281.	339.05	2144.4	2483.5	339.10	2306.2	2645.3	1.0871	6.5113	7.5985
82	.5136	1.0305	3160.	343.25	2141.5	2484.7	343.30	2303.7	1647.0	1.0990	6.4858	7.5848
83	.5345	1.0311	3044.	347.45	2138.5	2485.9	347.50	2301.1	2648.6	1.1108	6.4605	7.5713
84	.5560	1.0318	2934.	351.64	2135.5	2487.2	351.70	2298.6	2650.3	1.1225	6.4353	7.5578
85	.5783	1.0325	2828.	355.84	2132.6	2488.4	355.90	2296.0	2651.9	1.1343	6.4102	7.5445
86	.6014	1.0332	2726.	360.04	2129.6	2489.6	360.10	2293.5	2653.6	1.1460	6.3852	7.5312
87	.6252	1.0339	2629.	364.24	2126.6	2490.9	364.30	2290.9	2655.2	1.1577	6.3604	7.5180
88	.6498	1.0346	2536.	368.44	2123.7	2492.1	368.51	2288.3	2656.9	1.1693	6.3356	7.5050
89	.6752	1.0353	2446.	372.64	2120.7	2493.3	372.71	2285.8	2658.5	1.1809	6.3110	7.4920
90	.7014	1.0360	2361.	376.85	2117.7	2494.5	376.92	2283.2	2660.1	1.1925	6.2866	7.4791
91	.7284	1.0367	2278.	381.05	2114.7	2495.8	381.12	2280.6	2661.7	1.2041	6.2622	7.4663
92	.7564	1.0375	2200.	385.26	2111.7	2497.0	385.33	2278.0	2663.3	1.2156	6.2379	7.4539
93	.7852	1.0382	2124.	389.46	2108.7	2498.2	389.54	2275.4	2664.9	1.2271	6.2138	7.4409
94	.8149	1.0389	2052.	393.67	2105.7	2499.4	393.75	2272.8	2666.5	1.2386	6.1898	7.4284
95	.8455	1.0397	1981.9	397.88	2102.7	2500.6	397.96	2270.2	2668.1	1.2500	6.1659	7.4159
96	.8771	1.0404	1915.0	402.09	2099.7	2501.8	402.17	2267.6	2669.7	1.2615	6.1421	7.4036
97	.9097	1.0412	1850.8	406.30	2096.7	2503.0	406.39	2264.9	2671.3	1.2728	6.1184	7.3913
98	.9433	1.0420	1789.1	410.51	2093.6	2504.1	410.61	2262.3	2672.9	1.2842	6.0948	7.3791
99	.9778	1.0427	1729.9	414.72	2090.6	2505.3	414.83	2259.7	2674.5	1.2956	6.0714	7.3669
100	1.0135	1.0435	1672.9	418.94	2087.6	2506.5	419.04	2257.0	2676.1	1.3069	6.0480	7.3549
101	1.0502	1.0443	1618.2	423.15	2084.5	2507.7	423.26	2254.4	2677.6	1.3181	6.0248	7.3429
102	1.0880	1.0451	1565.5	427.37	2081.5	2508.9	427.48	2251.7	2679.2	1.3294	6.0016	7.3310
103	1.1269	1.0459	1514.9	431.58	2078.5	2510.0	431.71	2249.0	2680.7	1.3406	5.9786	7.3192
104	1.1669	1.0467	1466.2	435.80	2075.4	2511.2	435.92	2246.4	2682.3	1.3518	5.9557	7.3075
105	1.2082	1.0475	1419.4	440.02	2072.3	2512.4	440.15	2243.7	2683.8	1.3630	5.9328	7.2958
106	1.2506	1.0483	1374.3	444.24	2069.3	2513.5	444.37	2241.0	2685.4	1.3741	5.9101	7.2843
107	1.2942	1.0491	1330.9	448.47	2066.2	2514.7	448.60	2238.3	2686.9	1.3853	5.8875	7.2728
108	1.3391	1.0499	1289.1	452.69	2063.1	2515.8	452.83	2235.6	2688.4	1.3964	5.8650	7.2613
109	1.3853	1.0508	1248.9	456.92	2060.0	2517.0	457.06	2232.9	2690.0	1.4047	5.8425	7.2500
110	1.4327	1.0516	1210.2	461.14	2057.0	2518.1	461.30	2230.2	2691.5	1.4185	5.8202	7.2387
111	1.4815	1.0525	1172.8	465.37	2053.9	2519.2	465.53	2227.5	2693.0	1.4295	5.7980	7.2275
112	1.5317	1.0533	1136.9	469.60	2050.8	2520.4	469.76	2224.7	2694.5	1.4405	5.7758	7.2163
113	1.5832	1.0542	1102.2	473.84	2047.7	2521.5	474.01	2222.0	2696.0	1.4515	5.7538	7.2052
114	1.6362	1.0550	1068.8	478.07	2044.5	2522.6	478.24	2219.2	2697.5	1.4624	5.7318	7.1942

Source: J. H. Keenan, F. G. Keyes, P. G. Hill, and J. G. Moore, *Steam Tables*, Wiley, New York 1969.

[a] T is temperature in °C, p pressure in bar, v specific volume in liter/kg (1 liter = 10^{-3} m³), u specific internal energy in kJ/kg, h specific enthalpy in kJ/kg, s specific entropy in kJ/kg K.

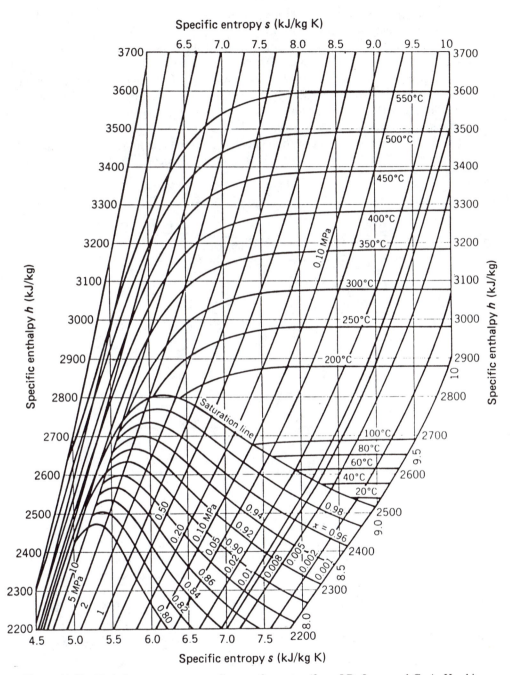

Figure 19.10 Enthalpy versus entropy diagram for water (from J.B. Jones and G. A. Hawkins, *Engineering Thermodynamics*, 2nd ed., Wiley, New York, 1986).

Mollier chart can be proved in general for all substances. For example, equation (19.14) applied to $z = h$ and $z = s$ yields the relation $(h - h_f)/(h_g - h_f) = (s - s_f)/(s_g - s_f)$. For liquid–vapor mixtures, this relation implies that $h = h(s, T)$ or $h = h(s, p)$ because h_f, h_g, s_f, and s_g are functions of T (or p) only. It also implies that the slope of the isothermobars in the two-phase region, $(\partial h/\partial s)_T = (\partial h/\partial s)_p = (h_g - h_f)/(s_g - s_f) = T$, where in writing the last equation we use (19.5). Hence these isothermobars are straight lines.

Another general feature of the Mollier chart is that at any point in the single- phase vapor region the slope of the isochor, $(\partial h/\partial s)_v$, is greater than the slope of the isobar, $(\partial h/\partial s)_p$, which in turn is greater than the slope of the isotherm, $(\partial h/\partial s)_T$. Proofs of these assertions are given in Example 19.3 (Section 19.10).

A diagram of u versus v is shown in Figure 19.11. Here again we see that in the two-phase region the isothermobars are straight lines. Moreover, by virtue of (19.14) we can readily verify that for a mixture represented by point P on such a line, the ratio of the amounts of the two phases is given by the ratio of the lengths of the segments of the line interconnecting the point with the points corresponding to the saturated states at the same temperature. Thus, in Figure 19.11 point P represents a mixture in which the proportion of liquid to vapor is $PQ : RP$.

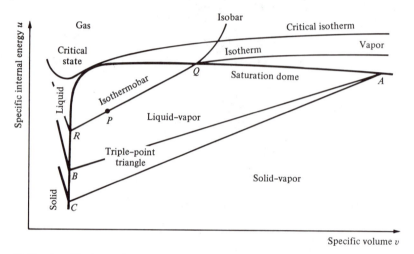

Figure 19.11 Specific internal energy versus specific volume diagram showing regions of solid–vapor and liquid–vapor two-phase states, and the triangular region of solid–liquid–vapor three-phase states.

The diagram includes the triple-point triangle ABC, which delimits the region of u versus v within which are all the triple-point states of a unit amount of the substance. Point A represents the state of a unit amount of the vapor phase, point B of the liquid phase, and point C of the solid phase, which coexist in mutual stable equilibrium at any triple-point state.

Typical charts of T versus s and p versus h are shown in Figures 19.12 and 19.13. A three-dimensional surface representing the relation $u = u(s, v)$ and exhibiting the convexity of such a relation is shown in Figure 19.14.

19.7 Specific Heats

Similarly to the definitions of heat capacities given in Section 16.5, the *specific heat at constant volume* and the *specific heat at constant pressure* are defined as

$$c_v = \left(\frac{\partial u}{\partial T} \right)_v = T \left(\frac{\partial s}{\partial T} \right)_v \tag{19.15}$$

$$c_p = \left(\frac{\partial h}{\partial T} \right)_p = T \left(\frac{\partial s}{\partial T} \right)_p \tag{19.16}$$

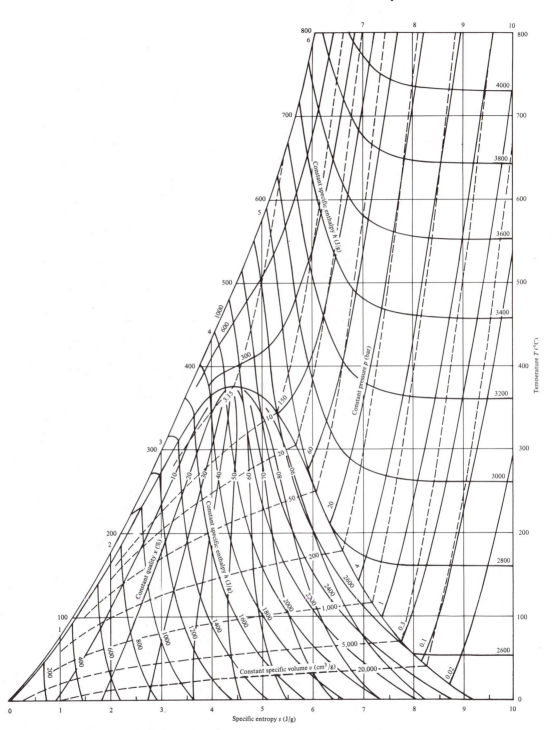

Figure 19.12   Temperature versus entropy chart for water (from J. H. Keenan, F. G. Keyes, P. G. Hill, and J. G. Moore, *Steam Tables*, International Edition, Metric Units, Wiley, New York, 1969).

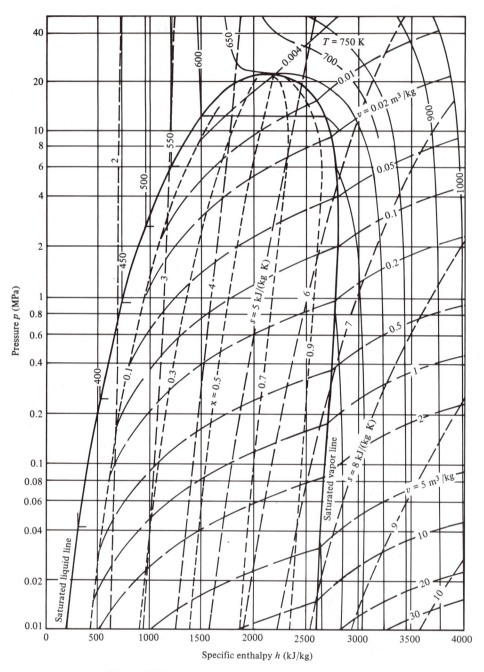

Figure 19.13   Pressure versus enthalpy chart for water.

These definitions are not valid for all states. Indeed, c_v is a partial derivative of $u = u(T, v)$ which can be obtained by eliminating s from the relations $u = u(s, v)$ and $T = T(s, v)$. This elimination is possible, however, only for states for which $T \neq 0$ and $(\partial T/\partial s)_v = (\partial^2 u/\partial s^2)_v \neq 0$. Because $(\partial^2 u/\partial s^2)_v = 0$ for states within the triple-point triangle (Figure 19.14) as well as for all the states of a reservoir (Example 9.1), we conclude that the specific heat at constant volume is undefined for such states. For all other states, equation (19.15) implies that

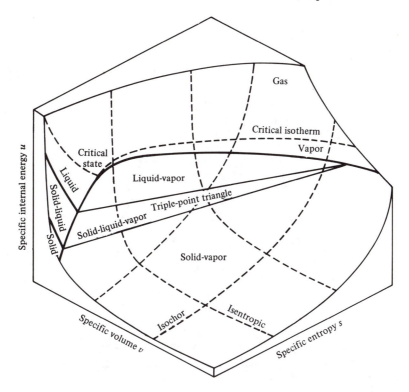

Figure 19.14 Three-dimensional sketch of the convex relation $u = u(s, v)$.

$$c_v = \frac{T}{(\partial T/\partial s)_v} = \frac{T}{(\partial^2 u/\partial s^2)_v} > 0 \tag{19.17}$$

where the inequality follows from the convexity of relation $u = u(s, v)$, that is, from relations (17.29) and (17.30). Again, c_p is a partial derivative of $h = h(T, p)$, a relation that holds only for single-phase states. For two-phase states, T and p are not independent of each other, and h depends not only on T (or p) but also on the fraction x of one of the phases. Similarly, three-phase states do not depend on T and p. We conclude that the specific heat at constant pressure is defined only for single-phase states.

In view of the discussion on heat capacities in Section 16.5, we conclude that the specific heats are readily measurable because they involve measurements of temperature, pressure, volume, and energy. A chart of values of c_p for water vapor as a function of temperature and pressure is given in Figure 19.15.

Later we show that specific heats are important because they allow us to evaluate entropies from measurements of temperature, pressure, volume, and energy.

19.8 Equation of State

Temperature, pressure, and specific volume of a pure substance are interrelated because each depends at most on two independent intensive properties. We may express the relation between T, p, and v in the form

$$f(T, p, v) = 0 \tag{19.18}$$

This, or an equivalent relation, is called an *equation of state* of the pure substance and is especially useful because measurements of temperature, pressure, and specific volume are relatively easy to perform. The qualitative features of the equation of state are reflected

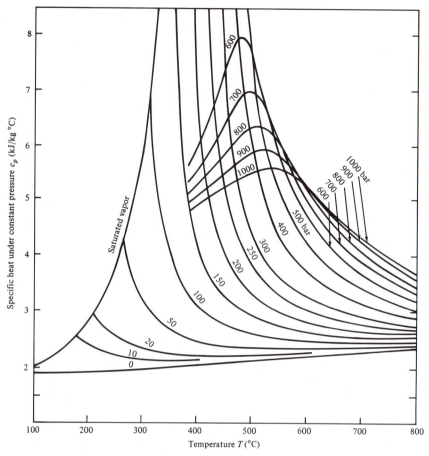

Figure 19.15 Specific heat at constant pressure as a function of temperature and pressure for water (from J. H. Keenan, F. G. Keyes, P. G. Hill, and J. G. Moore, *Steam Tables*, Wiley, New York, 1969).

in the graphs in Figures 19.3 to 19.6. We discuss these features in more detail in Chapters 20 and 21.

For single-phase states, T and p are independent of each other and sufficient to determine v, so that the equation of state can be expressed in the form

$$v = v(T, p) \tag{19.19}$$

For two-phase states, T and p are not independent of each other, and v is a function not only of T (or p) but also of the relative amount of each phase. For three-phase states, v depends only on the relative amounts of the phases.

It is noteworthy that $v(T, p)$ is not a characteristic function because it does not determine all the intensive properties of the substance by means of differentiation and algebraic manipulations. For example, from equations (17.23) and (17.25), we conclude that $\partial \mu(T, p)/\partial p = \partial g(T, p)/\partial p = v(T, p)$. Hence $\mu(T, p)$ can be determined from $v(T, p)$ only by integration which requires that we also know the other partial derivative of $\mu(T, p)$, namely, $\partial \mu(T, p)/\partial T = -s(T, p)$. Thus to fully determine the intensive properties of the single-phase states of a pure substance as functions of T and p only, we need to know the equation of state $v = v(T, p)$ and the relation

$$s = s(T, p) \tag{19.20}$$

or, alternatively, the relation

$$c_p = c_p(T, p) \tag{19.21}$$

Example 19.1.   Prove that if $v = v(T, p)$ and $c_p = c_p(T, p)$ are known, $s = s(T, p)$ can be determined.

Solution.   From equation (19.16) we conclude that $\partial s(T, p)/\partial T = c_p(T, p)/T$, and from the Maxwell relation (16.35) that $\partial s(T, p)/\partial p = -\partial v(T, p)/\partial T$ because here $S = ns$ and $V = nv$, and n is independent of T and p. We see that the two partial derivatives of $s(T, p)$ are determined by $v(T, p)$ and $c_p(T, p)$ and therefore that upon integration we can find $s(T, p)$.

Example 19.2.   Show how the equation of state $v = v(T, p)$ and the function $c_p = c_p(T, p)$ can be integrated to yield the functions $s = s(T, p)$ and $h = h(T, p)$.

Solution.   From the results in Example 19.1, we have[3]

$$ds = \frac{c_p(T, p)}{T} \, dT - \frac{\partial v(T, p)}{\partial T} \, dp$$

and

$$s(T, p) = s(T_0, p_0) + \int_{T_0}^{T} \frac{c_p(T', p)}{T'} \, dT' - \int_{p_0}^{p} \left. \frac{\partial v(T, p')}{\partial T} \right|_{T = T_0} dp'$$

or, equivalently,

$$s(T, p) = s(T_0, p_0) - \int_{p_0}^{p} \frac{\partial v(T, p')}{\partial T} \, dp' + \int_{T_0}^{T} \frac{c_p(T', p_0)}{T'} \, dT'$$

where T_0 and p_0 can be selected arbitrarily as long as, in the range from p_0 to p and from T_0 to T, the pure substance admits only single-phase states. This result is an example of the use of a specific heat to evaluate the specific entropy. To find $h = h(T, p)$, we combine the relation $dh = T \, ds + v \, dp$ with that for ds just given. The result is

$$dh = c_p(T, p) \, dT + \left[v(T, p) - T \frac{\partial v(T, p)}{\partial T} \right] dp$$

and

$$h(T, p) = h(T_0, p_0) + \int_{T_0}^{T} c_p(T', p) \, dT' + \int_{p_0}^{p} \left[v(T_0, p') - T_0 \left. \frac{\partial v(T, p')}{\partial T} \right|_{T = T_0} \right] dp'$$

or, equivalently,

$$h(T, p) = h(T_0, p_0) + \int_{p_0}^{p} \left[v(T, p') - T \frac{\partial v(T, p')}{\partial T} \right] dp' + \int_{T_0}^{T} c_p(T', p_0) \, dT'$$

---

[3]We recall that if the total differential of a function $f(x, y)$ is

$$df = X(x, y) \, dx + Y(x, y) \, dy$$

we have $X(x, y) = \partial f(x, y)/\partial x$, $Y(x, y) = \partial f(x, y)/\partial y$, $\partial X(x, y)/\partial y = \partial^2 f(x, y)/\partial x \partial y = \partial Y(x, y)/\partial x$, and

$$f(x, y) = f(x_0, y_0) + \int_{x_0}^{x} X(x', y_0) \, dx' + \int_{y_0}^{y} Y(x, y') \, dy'$$

or, equivalently,

$$f(x, y) = f(x_0, y_0) + \int_{x_0}^{x} X(x', y) \, dx' + \int_{y_0}^{y} Y(x_0, y') \, dy'$$

19.9   Coefficients of Isothermal Compressibility and Isobaric Expansion

Rather than giving the equation of state in the form of (19.19), it is often convenient to provide information related to the two partial derivatives of $v(T, p)$. Two stable-equilibrium-state properties related to these derivatives are the *coefficient of isothermal compressibility* κ_T, and the *coefficient of isobaric expansion* α_p defined by

$$\kappa_T = -\frac{1}{v}\left(\frac{\partial v}{\partial p}\right)_T \tag{19.22}$$

$$\alpha_p = \frac{1}{v}\left(\frac{\partial v}{\partial T}\right)_p \tag{19.23}$$

As the equations indicate, the coefficient of isothermal compressibility represents the fractional, isothermal change in volume per unit pressure change, and the coefficient of isobaric expansion the fractional isobaric change in volume per unit temperature change. Similarly to c_p, the coefficients κ_T and α_p are not defined for two-phase or three-phase states because then v is a function not only of T (or p) but also of the relative amount of each phase. Values of κ_T and α_p at 25°C for a few constituents in the liquid form of aggregation are listed in Table 19.6.

TABLE 19.6.   **Values of specific volume, v, specific heat, c_p, coefficient of isothermal compressibility, κ_T, coefficient of isobaric expansion, α_p, ratio of specific heats, $\gamma = c_p/c_v$, and speed of sound, c_s, for various liquids at 25°C.**

Liquid	Formula	v m^3/kmol	c_p kJ/kmol K	κ_T $10^{-10}Pa^{-1}$	α_p $10^{-20}K^{-1}$	γ	c_s m/s
Acetone	CH_3COCH_3	0.07352	126.4	12.4	4.84	1.35	1174
Benzene	C_6H_6	0.08979	135.6	9.7	4.05	1.41	1295
Chloroform	$CHCl_3$	0.08012	115.5	9.7	3.04	1.41	987
Ethanol	C_2H_5OH	0.05832	113.0	11.4	4.74	1.31	1207
Methanol	CH_3OH	0.04051	81.6	12.6	4.29	1.21	1103
Water	H_2O	0.01805	75.4	4.6	1.18	1.02	1497

Source: Data for v, c_p, κ_T, and c_s from R. C. Weast, ed., *CRC Handbook of Chemistry and Physics*, 66th ed., CRC Press, Boca Raton, Fla., 1985; values of α_p and γ from equations (19.26) and (19.36).

Clearly, knowledge of the two expressions $\kappa_T = \kappa_T(T, p)$ and $\alpha_p = \alpha_p(T, p)$ is equivalent to that of the equation of state. Accordingly, the results in the preceding section imply that knowledge of the expressions $\kappa_T = \kappa_T(T, p)$, $\alpha_p = \alpha_p(T, p)$, and $c_p = c_p(T, p)$ is sufficient to deduce all the intensive properties of the single-phase states of a pure substance, including u, s, h, and v, as functions of T and p.

This conclusion may also be reached by expressing the differentials of (19.19) and (19.20) in terms of κ_T, α_p, and c_p. Indeed,

$$dv = \left(\frac{\partial v}{\partial T}\right)_p dT + \left(\frac{\partial v}{\partial p}\right)_T dp = \alpha_p v\, dT - \kappa_T v\, dp \tag{19.24}$$

$$ds = \left(\frac{\partial s}{\partial T}\right)_p dT + \left(\frac{\partial s}{\partial p}\right)_T dp = \frac{c_p}{T}\, dT - \alpha_p v\, dp \tag{19.25}$$

where in writing the second of equations (19.25) we use (19.16) and the Maxwell relation (16.35). We see that knowledge of the expressions for α_p, κ_T, and c_p specifies v and s and, therefore, all the other intensive properties.

The specific heat c_p under constant pressure is related to the specific heat c_v under constant volume by the expression

$$c_p = c_v + \frac{T\alpha_p^2 v}{\kappa_T} \tag{19.26}$$

which is known as the *Mayer relation*.[4] Indeed, upon eliminating dp from (19.24) and (19.25), we find that

$$
\begin{aligned}
ds &= \left(\frac{c_p}{T} - \frac{\alpha_p^2 v}{\kappa_T}\right) dT + \frac{\alpha_p}{\kappa_T}\, dv \\
&= \left(\frac{\partial s}{\partial T}\right)_v dT + \left(\frac{\partial s}{\partial v}\right)_T dv \\
&= \frac{c_v}{T}\, dT + \frac{\alpha_p}{\kappa_T}\, dv
\end{aligned}
\tag{19.27}
$$

where in writing the second of equations (19.27) we express s as $s(T, v)$, and in writing the third of equations (19.27) we use the second of equations (19.15). Comparison of the coefficients of dT in the first and the third of equations (19.27) yields (19.26). The Mayer relation is particularly useful for solid and liquid phases, for which it is more difficult to measure c_p than c_v, α_p, and κ_T. From measurements of T, v, κ_T, α_p, and c_p or c_v, the relations $s = s(T, p)$ and $s = s(T, v)$ can be obtained by integration of (19.25) and (19.27), respectively. Similarly, the relation $s = s(p, v)$ can be obtained by integration of the relation that follows upon eliminating dT from (19.24) and (19.25), that is,

$$
\begin{aligned}
ds &= \frac{c_p}{T\alpha_p v}\, dv + \left(\frac{\kappa_T c_p}{T\alpha_p} - \alpha_p v\right) dp \\
&= \frac{c_p}{T\alpha_p v}\, dv + \frac{\kappa_T c_v}{T\alpha_p}\, dp
\end{aligned}
\tag{19.28}
$$

where in writing the second of equations (19.28) we use (19.26).

For most single-phase states, c_p, c_v, κ_T, and α_p are well defined, and relations (17.30) to (17.32) apply with the strict inequality, that is, for a pure substance,

$$\left(\frac{\partial^2 u}{\partial s^2}\right)_v > 0 \tag{19.29}$$

$$\left(\frac{\partial^2 u}{\partial s^2}\right)_v \left(\frac{\partial^2 u}{\partial v^2}\right)_s - \left(\frac{\partial^2 u}{\partial s \partial v}\right)^2 > 0 \tag{19.30}$$

$$\left(\frac{\partial^2 u}{\partial v^2}\right)_s > 0 \tag{19.31}$$

We have already used (19.29) to show that $c_v > 0$ [relation (19.17)]. Regarding κ_T, we note that $(\partial^2 u/\partial s^2)_v = (\partial T/\partial s)_v = T/c_v$, $(\partial^2 u/\partial v^2)_s = -(\partial p/\partial v)_s = c_p/\kappa_T v c_v$, and $(\partial^2 u/\partial s \partial v) = (\partial T/\partial v)_s = -T\alpha_p/\kappa_T c_v$, where in writing the last two equalities we use (19.28) and (19.27) with $ds = 0$, respectively. Upon substituting these results in relation (19.30), we find that

---

[4] In honor of Julius Robert von Mayer (1814–1878), German physicist and physician, one of the pioneers of thermodynamics.

$$\frac{T}{c_v}\frac{c_p}{\kappa_T v c_v} - \left(\frac{T\alpha_p}{\kappa_T c_v}\right)^2 = \frac{T}{\kappa_T v c_v} > 0 \tag{19.32}$$

where in writing the equality in (19.32) we use equation (19.26). Because T, v, and c_v are positive, it follows that also κ_T is positive. In contrast, the sign of the coefficient α_p is not definite. It is positive for most states, including the vapor stable equilibrium states, but is negative for other states, such as the liquid states of H_2O at atmospheric pressure and temperatures between 0 and 4°C.

From the positivity of T, v, and κ_T, and (19.26) we find that, in general,

$$c_p > c_v \tag{19.33}$$

or, equivalently, the *ratio of specific heats* $\gamma = c_p/c_v > 1$.

19.10 Speed of Sound

Another property related to specific volume and relatively easy to measure is the speed of sound, namely, the speed at which an infinitesimal disturbance propagates across a pure substance initially in a stable equilibrium state. For a single-phase state, the *speed of sound* c_s is defined as

$$c_s^2 = \left(\frac{\partial p}{\partial \rho}\right)_s = -\frac{v^2}{M}\left(\frac{\partial p}{\partial v}\right)_s \tag{19.34}$$

where ρ is the density, namely, $\rho = M/v$, M the molecular weight, and v the number-specific volume. It can be related to other properties as follows. For $ds = 0$, equation (19.28) yields

$$\left(\frac{\partial p}{\partial v}\right)_s = -\frac{c_p}{c_v}\frac{1}{v\kappa_T} \tag{19.35}$$

and, therefore,

$$c_s^2 = -\frac{v^2}{M}\left(\frac{\partial p}{\partial v}\right)_s = \frac{c_p}{c_v}\frac{1}{\rho\kappa_T} = \frac{\gamma}{\rho\kappa_T} \tag{19.36}$$

that is, the speed of sound is related to the ratio of specific heats γ, the density ρ, and the isothermal compressibility κ_T. Because T and κ_T are positive definite, c_s^2 is indeed positive definite. Values of c_s at 25°C for a few constituents in the liquid form of aggregation are listed in Table 19.6.

Example 19.3.   Prove that at every point where $\alpha_p \geq 0$ in a single-phase region of a Mollier chart (h versus s diagram), such as the vapor region, the slope of the isochor is greater than the slope of the isobar, which in turn is greater than the slope of the isotherm.

Solution.   From the relations $h = u + pv$ and $du = Tds - pdv$, we find $dh = Tds + v\,dp$, so that $(\partial h/\partial s)_v = T + v(\partial p/\partial s)_v$, $(\partial h/\partial s)_p = T$, and $(\partial h/\partial s)_T = T + v(\partial p/\partial s)_T$. From equation (19.28) for $dv = 0$, we find $(\partial p/\partial s)_v = T\alpha_p/\kappa_T c_v \geq 0$, where the inequality follows from the nonnegativity of T, κ_T, and c_v, and the assumption that $\alpha_p \geq 0$. From equation (19.25) for $dT = 0$, we find $(\partial p/\partial s)_T = -1/\alpha_p v \leq 0$. Thus we conclude that $(\partial h/\partial s)_v = T + T\alpha_p v/\kappa_T c_v \geq (\partial h/\partial s)_p = T \geq (\partial h/\partial s)_T = T - 1/\alpha_p$, that is, that the slope $(\partial h/\partial s)_v$ of the isochor is greater than the slope $(\partial h/\partial s)_p$ of the isobar, and this in turn is greater than the slope $(\partial h/\partial s)_T$ of the isotherm.

Problems

19.1 A tank is divided into two compartments each containing 1 kg of liquid water at atmospheric pressure. Initially, the water in compartment A is at temperature $T_i^A = 20°C$, and that in compartment B at temperature $T_1^B = 80°C$. While the tank experiences no interactions with the environment, the two compartments eventually reach mutual stable equilibrium.

(a) What is the final temperature of the water in the two compartments?

(b) How much entropy is generated by irreversibility during the process?

(c) What is the least work required to reestablish the initial state of the water in the two compartments by means of machinery that interacts only with a weight and the water in the tank?

(d) What is the answer to part (c) if machinery can also interact with the environment at 30°C?

19.2 Consider 5 kg of water having energy values in the range 10,000 to 25,000 kJ, and confined in a volume of either $V_1 = 0.5$ m³ or $V_2 = 0.125$ m³.

(a) Make a graph of the energy U versus entropy S of the stable equilibrium states of this system in the given energy range and for each of the given volumes. Show on each graph the locus of all possible states that have $U = 22,000$ kJ.

(b) If the initial state of the water has energy $U_1 = 22,000$ kJ and entropy $S_1 = 36$ kJ/K, find its adiabatic availability when $V = V_1$, and $V = V_2$.

(c) A reservoir is available at 260°C. If the initial state is that specified in part (b) with volume V_1, what is the largest work that can be done by the composite of the water and the reservoir without net changes in volume?

(d) Repeat part (c) if the initial volume is V_2. Show your answers in parts (c) and (d) on your graphs.

(e) If imperfect equipment is used, and only one-half of the available energy in part (c) for $V = V_1$ is obtained, how much entropy is generated by irreversibility?

(f) What is the temperature and pressure of the state specified in part (b)?

19.3 A system consists of 10 kg of H_2O in a fixed-volume container. Initially, the water is in a stable equilibrium state at a temperature $T_1 = 1100°C$ and a pressure $p_1 = 6$ MPa. The system is surrounded by thermal insulation in an environment at $T_0 = 25°C$. The insulation is not perfect. After some time it is observed that the system is again in a stable equilibrium state but its pressure $p_2 = 2$ MPa.

(a) What is the temperature of the water in state 2?

(b) How much entropy is generated by irreversibility as the system changes from state 1 to state 2? Where does this entropy generated by irreversibility appear?

(c) What is the largest work that can be done as the water changes its state from 1 to 2?

19.4 One ton of H_2O is in a fixed-volume container. Initially, the temperature $T_1 = 190°C$ and the pressure $p_1 = 10$ bar. It is desired to raise the pressure to $p_2 = 30$ bar. This is achieved by interactions with another system that transfer energy and 1000 kJ/K of entropy.

(a) How much energy is transferred?

(b) How much entropy is generated by irreversibility?

(c) What are the possible types of interactions that could result in the given transfers of energy and entropy?

19.5 Each of two identical rigid containers A and B has a fixed volume of 1 m³ and is filled with 1.4 kmol of Freon 12, CCl_2F_2. The pressure in container A is 30 kPa, whereas that of container B is 1 MPa. Consider the composite system C consisting of subsystems A and B in the states just cited.

(a) What are the values of the energy E^C and entropy S^C of the composite system C?

(b) What is the value of the available energy Ω^{RC} with respect to a reservoir at $T_R = 20°C$?

(c) What is the value of the adiabatic availability Ψ^C?

19.6 An amount of 1000 kg of H_2O is in a fixed-volume container at a pressure of $p_1 = 5$ bar and an internal energy 800 kJ/kg (property values taken from steam tables in Appendix B). The water pressure is raised to $p_2 = 20$ bar by means of a process involving work interactions only.

(a) How much energy is transferred to the water during the work interactions?

(b) Is the process reversible or irreversible? Specify the amount of entropy change involved.

(c) What are the types of possible interactions, in addition to work, that could result in the given change of state of H_2O?

19.7 A system consists of 10 kg of H_2O initially in a stable equilibrium state at 35 bar and 300°C. The system is compressed isothermally to another stable equilibrium state at 50 bar. The system is surrounded by thermal insulation which permits enough exchange of energy with the environment so as to keep the H_2O temperature constant. The environment is at constant temperature $T_0 = 20$°C and constant pressure $p_0 = 1$ bar.

(a) Find the useful work done by the system, namely, the work done by the system minus the work done on the atmosphere.

(b) Find the optimum useful work that could be done under the given change of state of the system.

(c) Calculate the amount of entropy generated by irreversibility in part (a). Where is this entropy generated?

19.8 One kilogram of water is heated at constant pressure p from $T_1 = 20$°C to $T_2 = 180$°C. For the specific heats at constant pressure of liquid water and steam assume, respectively, $(c_p)_f = 1$ kcal/kg K and $(c_p)_g = 0.48$ kcal/kg K.

(a) Write the expressions for the changes in enthalpy, $h_2 - h_1$, and in entropy, $s_2 - s_1$.

(b) If $p = 1$ atm, how much energy and how much entropy are required to heat the water reversibly from $T_1 = 20$°C to $T_2 = 180$°C?

(c) If $p = 1.05$ atm, make an estimate of the saturation temperature T_s of water in terms of values of properties at $p = 1$ atm.

(d) For given T_1 and T_2, make an estimate of $d(h_2 - h_1)/dp$ assuming that h_f and h_g do not vary appreciably with the saturation temperature.

19.9 An electrical resistor is immersed in a mixture of liquid water and ice contained in a cylinder and maintained at a constant temperature of 0°C and pressure of 1 atm by a piston that seals the cylinder. At these conditions the specific enthalpy and specific volume of saturated-solid and saturated-liquid water are, respectively, $h_j = -333.7$ kJ/kg, $h_f = 0$ kJ/kg, $v_j = 1.09 \times 10^{-3}$ m^3/kg and $v_f = 1.00 \times 10^{-3}$ m^3/kg. The piston–cylinder combination is well insulated from the environment. A voltage of 24 V is applied to the resistor and a current of 2 A flows through it for 1 h.

(a) If there is still ice in the cylinder after this 1-h period, what is the amount of ice that has melted?

(b) What are the work and the heat for the resistor system, and the ice–water system?

(c) Determine the amount of entropy generated by irreversibility, if any, by this process.

19.10 The reactor core of a pressurized-water nuclear reactor is inside a core vessel, which in turn is inside a containment building. The pressurized water transfers energy and entropy from the reactor core to the secondary coolant via a heat exchanger. A large break in a pipe interrupts the circulation of the primary coolant and puts the core vessel in communication with the containment building. Make some reasonable assumptions and compute the final pressure in the containment building. Prior to the rupture, the primary coolant is saturated liquid at 160 atm and occupies 350 m^3, and the containment building may be regarded as evacuated. The volume of the containment building is 50,000 m^3.

19.11 An insulated cylinder is fitted with a frictionless, insulated piston with an area of 13 cm^2 and a mass of 15 kg (Figure P19.11). The environmental pressure is 1 atm. The volume trapped between the piston face and the end of the cylinder is partitioned into two unequal parts by a rigid membranelike insulator. The volume V^A between the piston face and the membrane contains 4.5 kg of saturated liquid, while the volume V^B between the membrane and the end of the cylinder contains 0.45 kg of H_2O at pressure of 7 bar and a temperature of 205°C. At some instant of time, the membrane is ruptured and the system comes to a stable equilibrium state.

(a) Determine the final state of the H_2O, including its quality and volume.

(b) Is there any work experienced by the H_2O as it reaches its final stable equilibrium state? If so, how much?

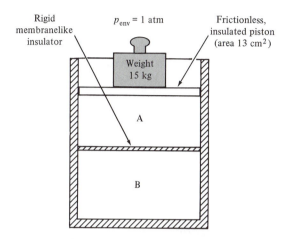

Figure P19.11

19.12 It is desired to cool a given quantity of material quickly to 5°C. The energy transfer is 2000 kJ and is accompanied by an entropy transfer of 5.7 kJ/K. One possibility is to immerse the material in a mixture of ice and water, thus resulting in melting ice. Another possibility is to cool the material by evaporating Freon 12 at −18°C. Then the energy transfer results in changing Freon 12 from saturated liquid to saturated vapor. A third possibility is to cool the material with liquid nitrogen at a pressure of 1 atm.

(a) Calculate the entropy generated by irreversibility in each of the three cooling processes.

(b) If the environment is at 25°C, calculate the loss of available energy in each of the three cooling processes.

19.13 Find the proportions by volume of liquid and vapor H_2O in a sealed tube at atmospheric pressure that will pass through the critical state when heated. Find the necessary energy transfer to the contents of the tube between atmospheric pressure and the critical state if the enclosed volume is 10 cm³.

19.14 A stream of Freon 12 from a storage tank at 3 atm and 25°C is compressed to 10 atm and 80°C by a compressor, condensed to saturated liquid at 10 atm in a condenser, and then injected into an initially empty bottle. At the end of the filling process, the bottle is cooled by the surroundings at 25°C so that the temperature of the Freon 12 at completion of the filling is 25°C, and the bottle contains 50 kg of Freon 12, 85% liquid and 15% vapor by volume.

(a) What is the volume of the bottle?

(b) What is the compressor work during the filling process?

(c) What is the heat from the condenser?

(d) What is the heat to the bottle from the surroundings?

(e) What is the overall entropy flow to the surroundings during the filling process?

19.15 To study the behavior of liquid and vapor phases of carbon dioxide near the critical point, we place CO_2 in a quartz vial, seal it, and bring it to the critical state by heating. Assuming that the vial is filled at room temperature $T = 25°C$, and using property data for CO_2, find the following.

(a) The pressure in the vial at room temperature.

(b) The percent liquid mass at room temperature to ensure passage through the critical point upon heating.

(c) The percent liquid volume associated with part (b).

(d) For this heating process, make a graph of the ratio of specific volume of the vapor and specific volume of the liquid as a function of pressure divided by the critical pressure.

19.16 Using the triple-point data for water, and assuming that changes in enthalpy and specific volume upon melting are independent of pressure, evaluate the melting temperature of ice at a pressure of 1000 atm.

19.17 System A has a volume of 6290 cm^3 and contains 1 kg of superheated steam at 289°C. System B has the same volume, mass, and internal energy as A but consists of two identifiable parts. One part is in a stable equilibrium state and has a volume of 2930 cm^3, contains 0.5 kg of water vapor, and is at 93°C. The other part is also in a stable equilibrium state.

(a) What are the energy and entropy of A? What is the entropy of B?

(b) If $Q_A = 94.2$ kJ of energy is extracted from A without change in volume and mass, what is the largest work that can be done with respect to a reservoir at 20°C?

(c) Express your answer in part (b) as a Carnot coefficient times Q_A.

(d) What is the change of entropy of the reservoir for your answer in part (b)?

(e) If $Q_B = 94.2$ kJ of energy is extracted from B without change in volume and mass, what is the largest work that can be done by using the same reservoir as in part (b)?

(f) What is the change of entropy of the reservoir for your answer in part (e)?

(g) Compare and discuss your answers in parts (b) and (e). In other words, explain whether the Carnot coefficient is an upper fraction that defines the largest work that can be obtained from a given amount of energy.

19.18 One kilogram of H_2O is contained in a cylinder with a piston exerting a constant pressure $p = 7$ bar. Initially, the volume of the container is $v_1 = 250 \times 10^3$ cm^3. The H_2O is converted into solid ice at $T_2 = -20$°C and $p_2 = p = 7$ bar. The conversion is achieved by interactions between the H_2O, the piston, and a connection to the cyclic engine as shown in Figure P19.18. The cyclic engine does shaft work $W_s^{\rightarrow}$ and interacts with a solid block of mass $m_s = 150$ kg, specific heat $c_s = 0.13$ kJ/kg K, and initial temperature $T_1^s = 25$°C.

(a) Write the initial and final energies, enthalpies, and entropies of H_2O, as well as the final volume v_2 of H_2O.

(b) If all processes are optimal, find the final temperature of the solid block. Is your answer a maximum or a minimum?

(c) What is the optimum work done by the cyclic engine?

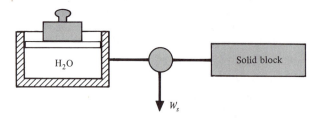

Figure P19.18

19.19 A waste-energy source in the form of saturated liquid water provides 1000 MW at 100°C. The approximate temperature of the environment at the location of the source is 20°C. The price of electricity from a utility is 10 cents/kWh. The annual cost of capital equipment for depreciation, profit, maintenance, and taxes is about 30% of the initial investment.

(a) What is the available power (available energy per unit time) from the waste-energy source?

(b) If a waste-energy-recovery electricity turbogenerator can capture 75% of the available power in part (a) for 7500 h/yr and costs about $1000 per kilowatt, what is the cost of the electricity (cents/kWh) generated from this energy conservation scheme?

(c) Is the investment in part (b) a good energy productivity investment?

19.20 For single-phase states of a pure substance for which c_p, c_v, κ_T, and α_p are well defined, prove the following relations.

(a) $\left(\dfrac{\partial u}{\partial v}\right)_T = \dfrac{T\alpha_p}{\kappa_T} - p$ (b) $\left(\dfrac{\partial u}{\partial p}\right)_T = p\kappa_T v - T\alpha_p v$

(c) $\left(\dfrac{\partial u}{\partial p}\right)_v = \dfrac{c_v \kappa_T}{\alpha_p}$ (d) $\left(\dfrac{\partial u}{\partial v}\right)_p = \dfrac{c_p}{\alpha_p v} - p$

(e) $\left(\dfrac{\partial u}{\partial T}\right)_p = c_p - p\alpha_p v$ (f) $\left(\dfrac{\partial u}{\partial p}\right)_s = \dfrac{pc_v \kappa_T v}{c_p}$

(g) $\left(\dfrac{\partial u}{\partial T}\right)_s = \dfrac{pc_v \kappa_T}{T\alpha_p}$ (h) $\left(\dfrac{\partial T}{\partial p}\right)_u = \dfrac{p\kappa_T v - T\alpha_p v}{p\alpha_p v - c_p}$

(i) $\left(\dfrac{\partial p}{\partial v}\right)_u = \dfrac{p\alpha_p v - c_p}{c_v \kappa_T v}$ (j) $\left(\dfrac{\partial v}{\partial T}\right)_u = \dfrac{c_v \kappa_T}{p\kappa_T - T\alpha_p}$

19.21 For single-phase states of a pure substance for which c_p, c_v, κ_{T_s} and α_p are well defined, prove the following relations.

(a) $\left(\dfrac{\partial s}{\partial p}\right)_u = \dfrac{pc_v \kappa_T v}{pT\alpha_p v - Tc_p}$ (b) $\left(\dfrac{\partial s}{\partial T}\right)_u = \dfrac{pc_v \kappa_T}{pT\kappa_T - T^2\alpha_p}$

(c) $\left(\dfrac{\partial s}{\partial v}\right)_T = \dfrac{\alpha_p}{\kappa_T}$ (d) $\left(\dfrac{\partial s}{\partial p}\right)_v = \dfrac{c_v \kappa_T}{T\alpha_p}$

(e) $\left(\dfrac{\partial s}{\partial v}\right)_p = \dfrac{c_p}{T\alpha_p v}$

19.22 Show how the functions $\alpha_p = \alpha_p(T, p)$ and $\kappa_T = \kappa_T(T, p)$ can be integrated to yield the equation of state $v = v(T, p)$.

19.23 Show that

$$\left(\frac{\partial \alpha_p}{\partial p}\right)_T = -\left(\frac{\partial \kappa_T}{\partial T}\right)_p \quad \text{and} \quad \left(\frac{\partial p}{\partial T}\right)_v = \frac{\alpha_p}{\kappa_T}$$

19.24 Show that

(a) $du = (c_p - p\alpha_p v)\,dT + (p\kappa_T - T\alpha_p)v\,dp$ (b) $dh = c_p\,dT + (1 - T\alpha_p)v\,dp$

(c) $\left(\dfrac{\partial T}{\partial p}\right)_h = \dfrac{v}{c_p}(T\alpha_p - 1)$

The partial derivative $(\partial T/\partial p)_h$ is known as the Joule–Thomson coefficient.

19.25 Show that

$$\left(\frac{\partial c_p}{\partial p}\right)_T = -T\left(\frac{\partial^2 v}{\partial T^2}\right)_p$$

and therefore that the pressure dependence of the specific heat c_p can be determined from the equation of state.

19.26 The specific enthalpy of vaporization of H_2O in the range between 0.01 and 300°C is approximated to within ±1.5% by the relation

$$h_{fg} \approx 2162.3 + 3.29\,T - 80\left(\frac{T}{100}\right)^2$$

for h_{fg} in kJ/kg and T in kelvin. In the range between 20 and 200°C, h_{fg} is approximated to within ±1% by the linear relation

$$h_{fg} \approx 3328.4 - 2.9\,T$$

The specific volume of vaporization of H_2O in the range between 0.01 and 100°C is approximated to within ±1.5% by the relation

$$v_{fg} = \frac{0.46151\,T}{p}$$

for v_{fg} in m³/kg, p in kPa, and T in kelvin.

(a) Using these approximations, and the fact that $p_{sat}(100°C) = 1$ atm, show that the pressure–temperature saturation relation $p = p_{sat}(T)$ is approximated by either of the relations (p in kPa)

$$p = 101.3\exp\left(-23.19 - \frac{4685}{T} + 7.129\ln T - 0.01733T\right)$$

and

$$p = 101.3\exp\left(56.54 - \frac{7212}{T} - 6.284\ln T\right)$$

(b) Estimate the accuracy of the relations in part (a) by comparing them with tabulated values from the steam tables.

19.27 The worldwide average atmospheric temperature and pressure as functions of altitude z are approximated by the relations

$$T = 288.15 - 6.5\,z \qquad \text{and} \qquad p = 101.325\,(1 - 0.02256\,z)^{5.256}$$

for T in kelvin, p in kilopascal, z in kilometers, and $z = 0$ at sea level. Based on these relations and those in Problem 19.26, estimate the average temperature, pressure, and boiling temperature of water at the top of Mount Everest at $z = 8.848$ km. What is the boiling temperature of water at $z = 1.5$ km?

19.28 At atmospheric pressure, the enthalpy of fusion of ice is $h_{jf} = 333.7$kJ/kg, the density of ice at 0°C is 917 kg/m³, and that of liquid water is 999.8 kg/m³.

(a) Show that the pressure-temperature relation for two-phase solid–liquid states of water is well approximated by the relation

$$\frac{dT}{dp} = -\frac{1}{13,527}\frac{K}{kPa}$$

(b) Using the relations given in Problem 19.27, estimate the freezing temperature of water at atmospheric pressure on top of Mount Everest.

19.29 To illustrate the behavior of solid, liquid, and vapor phases at the triple point, one can place H_2O in a quartz vial at room temperature (25°C) and appropriate conditions, seal the device, and cool the H_2O to states at the triple-point temperature $T_{tp} = 273.16$ K and pressure $p_{tp} = 611.3$ Pa.

(a) Using the triple-point data for H_2O summarized in the following table, determine the state of the H_2O in the quartz vial at room temperature that ensures passage, upon cooling the fixed volume device, through the state at triple-point temperature and pressure, and mass fractions of solid, liquid, and vapor all equal to 1/3.

	Specific volume	Specific enthalpy	Specific entropy
Solid	1.0908×10^{-3} m³/kg	−333.40 kJ/kg	−1.221 kJ/kg K
Liquid	1.0002×10^{-3} m³/kg	0.01 kJ/kg	0.000 kJ/kg K
Vapor	206.1 m³/kg	2501.4 kJ/kg	9.156 kJ/kg K

(b) During the cooling process just described, what are the mass fractions of solid, liquid, and vapor when the H_2O enters and exits the triple-point triangle?

19.30 Using data from Table 19.1 and recalling the discussion in Section 3.11, evaluate the ground-state mass m_g (in kg) and the ground-state energy E_g (in MeV) of the free constituents CO_2, CO, H_2, O_2, H_2O. Show the validity of the relations

$$\frac{m_g}{n} \left[\text{in } \frac{\text{kg}}{\text{molecule}} \right] = 1.66058 \times 10^{-27} \times M \left[\text{in } \frac{\text{kg}}{\text{kmol}} \right]$$

$$\frac{E_g}{n} \left[\text{in } \frac{\text{MeV}}{\text{molecule}} \right] = 931.49 \times M \left[\text{in } \frac{\text{kg}}{\text{kmol}} \right]$$

where M is the molecular weight.

19.31 A perfectly insulated cylinder is subdivided into two equal-volume parts A and B by a frictionless piston permeable to energy and entropy but not matter (Figure P19.31). Initially, each part contains 1 kg of a two-phase mixture of water at $p = 2$ bar, with quality $x = 0.7$. In a reversible process, the piston is moved to the left until the contents of A consist of compressed liquid water.

(a) Find the pressure of A when its contents consist of saturated liquid water.

(b) Find the work done on A and B, and the heat between A and B in the course of the process in part (a).

(c) Find the work done on A and B if the final pressure of A is five times as large as your answer in part (a).

(d) For the process in part (c), make a sketch of the states of A and B on the temperature versus specific entropy diagram.

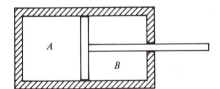

Figure P19.31

19.32 A phase transition of a pure substance is called a *first-order phase transition* if it involves two phases which, when coexisting, have different values of the first-order derivatives of the chemical potential with respect to temperature and pressure. We recall that the first-order derivatives of the chemical potential are related to the specific entropy and the specific volume, i.e., $s = -(\partial\mu/\partial T)_p$ and $v = (\partial\mu/\partial p)_T$. Because in general a liquid phase and a coexisting vapor phase have different specific volumes, as well as different specific entropies, it follows that all the liquid–vapor phase transitions are of the first order. For analogous reasons, all the solid–liquid phase transitions are of the first order. However, not all transitions are first-order phase transitions.

Many pure substances admit pairs of phases which, when coexisting, do not differ in the values of the specific volume and the specific entropy but differ in symmetry, crystallographic configuration, magnetic-moment configuration (ferromagnetism), or other features, including the values of the second-order derivatives of the chemical potential with respect to temperature and pressure. A phase transition involving such a pair of phases is called a *second-order phase transition*.

(a) Show the relation between the specific heat at constant pressure, c_p, the coefficient of isothermal compressibility, κ_T, the coefficient of isobaric expansion, α_p, and the second-order derivatives of the chemical potential μ with respect to temperature and pressure.

(b) Using a reasoning analogous to that leading to the Clausius–Clapeyron relation for a first-order phase transition, for a second-order phase transition between a phase a and a phase b prove that the changes in temperature and pressure between two neighboring two-phase states satisfy the relation

$$\frac{dp}{dT} = \frac{c_{p,ab}}{Tv\alpha_{p,ab}}$$

known as the *Ehrenfest relation*, where $c_{p,ab} = (c_p)_b - (c_p)_a$ and $\alpha_{p,ab} = (\alpha_p)_b - (\alpha_p)_a$.

19.33 The pressure on a 1-kg block of copper at a temperature of 0°C is increased isothermally and reversibly from 1 atm to 1000 atm. Assume a coefficient of isobaric expansion $\alpha_p = 5 \times 10^{-5}$ K^{-1}, a coefficient of isothermal compressibility $\kappa_T = 9 \times 10^{-9}$ kPa^{-1}, a density $\rho = 8000$ kg/m^3, a specific heat at constant pressure $c_p = 0.3$ kJ/kg K, and that all four are constant.

(a) Calculate the work and the heat experienced by the copper.

(b) What are the rise in temperature and the work if the process is isentropic instead of isothermal?

20   Ideal Gases, Liquids, and Solids

In certain limited ranges of temperature and pressure, the equation of state of any substance is well approximated by a simple and explicit mathematical form that in turn results in explicit and readily understandable interrelations among properties. When such a simple equation of state is valid, we say that it describes an *ideal behavior* of the substance or that the substance is *ideal*.

In this chapter we define ideal gases, ideal liquids, and ideal solids, and derive a number of practical relations between properties for each of these ideal substances.

20.1   Ideal-Gas Behavior

At relatively high temperatures T, and sufficiently low pressures p, every ordinary substance behaves as a single-phase fluid in which each molecule is so weakly coupled to the other molecules that it hardly experiences their presence. Under such conditions, it can be shown with the tools of quantum thermodynamics, and it is verified experimentally to a very high degree of accuracy, that the equation of state is given by the relation

$$pv = RT \qquad \text{or} \qquad pV = nRT \tag{20.1a}$$

where n is the amount of the constituent, v the mole-specific volume, namely, $v = V/n$, and R a constant called the *universal gas constant*. When equation (20.1a) is valid, we say that the substance *behaves as an ideal gas* or, simply, that it is an *ideal gas*.[1]

When the amount n of the constituent is expressed in terms of the number of molecules, the universal gas constant R is called the Boltzmann constant and denoted by the symbol k. Its value is

$$R = k = 1.38066 \times 10^{-23} \text{ J/K molecule}$$

When the amount n is expressed in terms of the number of moles, the symbol R is used and

$$R = N_A k = 8.3145 \text{ J/mol K}$$

where $N_A = 6.022 \times 10^{23}$ molecules/mol. In terms of other practical units,

$$R = 1.986 \text{ kcal/kmol K}$$
$$= 1.986 \text{ Btu/lbmol R}$$
$$= 82.057 \text{ cm}^3 \text{atm/mol K}$$

---

[1] We emphasize that the term "ideal" refers strictly to the simplicity of the equation of state, not to any thermodynamic optimality such as a reduction of the detrimental effects of irreversibility, or improved performance in particular processes.

It is noteworthy that at given values of temperature and pressure, equation (20.1a) implies that the mole-specific volume is fixed and equal for all substances that behave as ideal gases. For example, at $T = 0°C$ and $p = 1$ atm, the mole-specific volume of any ideal gas is $v = RT/p = 0.022414$ m^3/mol.

In terms of mass-specific properties, the equation of state of an ideal gas is

$$pv = R_M T \qquad \text{or} \qquad pV = mR_M T \tag{20.1b}$$

where m is the mass, v the mass-specific volume, namely, $v = V/m$, $R_M = R/M$, and M is the molecular weight or molar mass of the substance. The constant R_M is called the *gas constant of the substance* in question. Its value differs from substance to substance because so does the molecular weight. For example, for H_2O, $M = 18.015$ g/mol (Table 19.1), so that $R_M = 8.314/18.015 = 0.461$ kJ/kg K. Again, for H_2, $M = 2.016$ g/mol so that $R_M = 8.314/2.016 = 4.124$ kJ/kg K.

From either of equations (20.1), it follows that

$$\kappa_T = -\frac{1}{v}\left(\frac{\partial v}{\partial p}\right)_T = \frac{1}{p} \tag{20.2}$$

and

$$\alpha_p = \frac{1}{v}\left(\frac{\partial v}{\partial T}\right)_p = \frac{1}{T} \tag{20.3}$$

Example 20.1. Prove equations (20.2) and (20.3).

Solution. Upon rewriting (20.1a) as $v = RT/p$, we find that $(\partial v/\partial p)_T = -RT/p^2$ and $(\partial v/\partial T)_p = R/p$, so that $\kappa_T = RT/p^2 v = 1/p$ and $\alpha_p = R/pv = 1/T$.

By virtue of (20.2) and (20.3), the Mayer relation [equation (19.26)] becomes either

$$c_p - c_v = R \tag{20.4a}$$

or

$$c_p - c_v = R_M \tag{20.4b}$$

depending on whether the specific heats c_p and c_v are expressed on a mole basis or on a mass basis, respectively.

Example 20.2. Verify (20.4).

Solution. Substituting (20.2) and (20.3) into (19.26), we find that $c_p = c_v + T\alpha_p^2 v/\kappa_T = c_v + T(1/T)^2 v/(1/p) = c_v + pv/T = c_v + R$ or $c_v + R_M$, where in the last step we use equation (20.1a) or (20.1b).

Although the difference $c_p - c_v$ is a constant, in general each of the specific heats c_p and c_v is a function of temperature. Indeed, by virtue of the equation of state, the specific internal energy and the specific enthalpy of an ideal gas are functions of temperature only, that is,

$$u = u(T) \tag{20.5}$$

$$h = h(T) \tag{20.6}$$

and therefore the specific heats are also functions of temperature only, that is,

$$c_v = \frac{du}{dT} = c_v(T) \tag{20.7}$$

$$c_p = \frac{dh}{dT} = c_p(T) \tag{20.8}$$

Example 20.3. For an ideal gas, prove that the relation $u = u(T, v)$ reduces to $u = u(T)$.

Solution. Upon combining relations $du = T\,ds - p\,dv$ and $ds = c_v\,dT/T + \alpha_p\,dv/\kappa_T$ [equation (19.27)], we obtain $du = c_v\,dT + (T\alpha_p/\kappa_T - p)\,dv = c_v\,dT$ because, by virtue of (20.2) and (20.3), the coefficient $T\alpha_p/\kappa_T - p = 0$. But this coefficient is also equal to $(\partial u/\partial v)_T$, and therefore it is null only if u is independent of v, that is, if $u = u(T)$.

Example 20.4. For an ideal gas, prove that the relation $h = h(T, p)$ reduces to $h = h(T)$.

Solution. Upon combining relations $dh = T\,ds + v\,dp$ and $ds = c_p\,dT/T - \alpha_p v\,dp$ [equation (19.25)], we find $dh = c_p\,dT + (1 - T\alpha_p)v\,dp = c_p\,dT$ because, by virtue of (20.3), $(\partial h/\partial p)_T = 1 - T\alpha_p = 0$. Accordingly, h is independent of p, that is, $h = h(T)$. Of course, we can also use the result $u = u(T)$ of Example 20.3 and show that $h = u + pv = u(T) + RT = h(T)$.

Because the difference $c_p - c_v$ is a constant independent of temperature and pressure, it follows that the functional dependence of $c_p(T)$ on T must be the same as the functional dependence of $c_v(T)$ on T.

Upon integrating (20.7) from temperature T_o to temperature T, we find that the specific internal energy is given by the relation

$$u(T) = u(T_o) + \int_{T_o}^{T} c_v(T')\,dT' \tag{20.9}$$

Hence experimental data on the specific heat under constant volume, $c_v(T)$, can be used with (20.9) to calculate differences of specific internal energy between any pair of stable equilibrium states of the ideal gas.

Similarly, upon integrating (20.8) from T_o to T, we find that the specific enthalpy is given by the relation

$$h(T) = h(T_o) + \int_{T_o}^{T} c_p(T')\,dT' \tag{20.10}$$

Again, experimental values of the specific heat under constant pressure $c_p(T)$ can be used with (20.10) to evaluate differences in specific enthalpy between any pair of stable equilibrium states of the ideal gas.

In addition, and perhaps more important, specific heats can be used to evaluate the specific entropy. Indeed, upon combining (19.25), (19.27), and (19.28) with (20.1) to (20.3), we find that

$$ds = \frac{c_p(T)}{T}\,dT - \frac{R}{p}\,dp \tag{20.11a}$$

$$ds = \frac{c_v(T)}{T}\,dT + \frac{R}{v}\,dv \tag{20.11b}$$

$$ds = \frac{c_p(T)}{v}\,dv + \frac{c_v(T)}{p}\,dp \tag{20.11c}$$

Upon integrating (20.11a) from T_o, p_o to T, p, we find that

$$s(T, p) = s(T_o, p_o) + \int_{T_o}^{T} \frac{c_p(T')}{T'}\,dT' - R\ln\frac{p}{p_o} \tag{20.12a}$$

Upon integrating (20.11b) from T_o, v_o to T, v, we find that

$$s(T, v) = s(T_o, v_o) + \int_{T_o}^{T} \frac{c_v(T')}{T'}\,dT' + R\ln\frac{v}{v_o} \tag{20.12b}$$

Equation (20.11c) can also be readily integrated from v_0, p_0 to v, p provided that we substitute T by pv/R in the expressions of $c_v(T)$ and $c_p(T)$. Then we find that

$$s(v,p) = s(v_0,p_0) + \int_{v_0}^{v} \frac{c_p(pv'/R)}{v'}\, dv' + \int_{p_0}^{p} \frac{c_v(p'v_0/R)}{p'}\, dp' \tag{20.12c}$$

We see that experimental values of $c_v(T)$ and $c_p(T)$ can be used with any of equations (20.12) to evaluate differences in specific entropy between any pair of stable equilibrium states of the ideal gas.

In applications for which the absolute values of properties such as u, h, and s are not relevant, namely, applications that require only the evaluation of differences in these properties between different states, the constants of integration in (20.9), (20.10), and (20.12) can be specified by selecting a reference temperature T_0 and a reference pressure p_0, and arbitrarily assigning the values $u(T_0) = 0$ and $s(T_0,p_0) = 0$. Then the relation $h = u + pv = u + RT$ implies that $h(T_0) = RT_0$, and the relation $pv = RT$ that $s(T_0,v_0) = s(v_0,p_0) = 0$ provided that $v_0 = RT_0/p_0$. However, later we show that in applications that involve changes in the amounts of substances, such as in chemical reactors, the simplification just cited cannot be adopted.

Data for use in (20.9), (20.10), and (20.12) are usually found either as an algebraic relation expressing one of the specific heats as a function of temperature, or as a tabulation of values of properties obtained by a combination of experimental data and theoretical interrelations.

For example, an expression for $c_p(T)$ that correlates data with good accuracy over a range of temperatures between 300 and 1000 K is given by

$$c_p = a + bT + cT^2 + dT^3 \tag{20.13a}$$

where a, b, c, and d are constants. Values of these constants for various substances are listed in Table 20.1. Again, another expression that correlates data with good accuracy not only in the range of temperatures between 300 and 1000 K but also between 1000 and 4000 K is given by

$$c_p = a + bT^{1/4} + cT^{1/2} + dT^{3/4} \tag{20.13b}$$

where the values of the constants a, b, c, and d differ from those in (20.13a), and are listed in Table 20.2 for various substances.

We can use each of the correlations (20.13) in equations (20.8) and (20.11a) to resolve the indefinite integrals $h(T) = \int c_p(T)\, dT$ and $s(T,p) = \int c_p(T)\, dT/T - R \ln p$. The resolutions yield the expressions for $h(T)$ and $s(T,p)$ listed in Tables 20.1 and 20.2, which in turn can be used to evaluate specific enthalpy differences, $h(T_2) - h(T_1)$, and specific entropy differences, $s(T_2,p_2) - s(T_1,p_1)$.

For a few gases, a typical tabulation of experimental values of c_p, c_v, and $\gamma = c_p/c_v$ as functions of temperature is shown in Table 20.3.

The tabulation of values of u, h, and s for ideal-gas behavior is less involved than that for simple systems in general. The simplification arises from the fact that u and h are functions of T only, and the equation of state is a simple analytic expression. Moreover, although specific entropy is a function of both temperature and pressure, this dependence can be separated in two additive terms, each of which involves only one independent variable. Specifically, we may rewrite (20.12a) in the form

$$s = s_0 + \phi(T) - \phi(T_0) - R \ln \frac{p}{p_0} \tag{20.14}$$

where $\phi(T)$ is a function of temperature only and its derivative is $c_p(T)/T$. Values of $\phi(T)$ and $h(T)$ versus T are tabulated. The change in specific entropy between two

TABLE 20.1. Values of the constants a, b, c, and d for use in the approximate expressions

$$c_p(T) = a + bT + cT^2 + dT^3$$

$$h(T) = aT + \tfrac{1}{2} bT^2 + \tfrac{1}{3} cT^3 + \tfrac{1}{4} dT^4$$

$$s(T, p) = a \ln T + bT + \tfrac{1}{2} cT^2 + \tfrac{1}{3} dT^3 - R \ln p$$

of various substances.[a] See also Table 20.2.

Substance	Formula	a	$10^3 b$	$10^6 c$	$10^9 d$
Acetic acid	$C_2H_4O_2$	1.74	319	−235	69.8
Acetone	CH_3COCH_3	6.3	261	−125	20.4
Acetylene	C_2H_2	26.8	75.8	−50.1	14.1
Ammonia	NH_3	27.3	23.8	17.1	−11.9
Argon	Ar	20.8	0	51.7	0
Benzene	C_6H_6	−33.9	474	−302	71.3
Carbon dioxide	CO_2	19.8	73.4	−56.0	17.2
Carbon monoxide	CO	30.9	−12.9	27.9	−12.7
Chlorine	Cl_2	26.9	33.8	−38.7	15.5
Chloroform	$CHCl_3$	24	189	−184	66.6
Ethane	C_2H_6	5.41	178	−69.4	8.71
Ethanol	C_2H_5OH	9.01	214	−83.9	1.37
Ethylene	C_2H_4	3.81	157	−83.5	17.6
Fluorine	F_2	23.2	36.6	−34.6	12
Freon 12	CCl_2F_2	31.6	178	−151	43.4
Freon 13	$CClF_3$	22.8	191	−158	44.6
Freon 21	$CHCl_2F$	23.7	158	−120	32.6
Freon 22	$CHClF_2$	17.3	162	−117	30.6
Freon 114	$C_2Cl_2F_4$	38.8	344	−295	85.1
Genetron 100	CH_3CHF_2	8.68	240	−146	33.9
Genetron 101	CH_3CClF_2	16.8	276	−199	53.1
Hydrogen	H_2	27.1	9.3	−13.8	7.65
Hydrogen chloride	HCl	30.3	−7.2	12.5	−3.9
Isooctane	C_8H_{18}	−7.46	778	−429	91.7
Methane	CH_4	19.3	52.1	12.0	−11.3
Methanol	CH_3OH	21.2	70.9	25.9	−28.6
Methylene chloride	CH_2Cl_2	13	162	−130	42.1
Naphthalene	$C_{10}H_8$	−68.8	850	−651	198
Nitric oxide	NO	29.04	−0.9	9.7	−4.19
Nitrogen	N_2	31.2	−13.6	26.8	−11.7
Nitrogen dioxide	NO_2	24.2	48.4	−20.8	0.29
Nitrous oxide	N_2O	21.6	72.8	−57.8	18.3
n-Octane	C_8H_{18}	−6.10	771	−420	88.6
Oxygen	O_2	28.1	0.0	17.5	−10.7
Ozone	O_3	20.5	80.1	−62.4	17
Propane	C_3H_8	−4.22	306	−159	32.2
Propylene	CH_2CHCH_3	3.71	235	−116	22.1
Water	H_2O	32.2	1.9	10.6	−3.6

Source: Data from R. C. Reid, J. M. Prausnitz, and T. K. Sherwood, *The Properties of Gases and Liquids*, McGraw-Hill, New York, 1977.

[a] c_p in kJ/kmol K, h in kJ/kmol, s in kJ/kmol K, and T in kelvin between 300 and 1000 K.

TABLE 20.2.   Values of the constants a, b, c, and d for use in the approximate expressions

$$c_p(T) = a + bT^{1/4} + cT^{1/2} + dT^{3/4}$$

$$h(T) = aT + \tfrac{4}{5}bT^{5/4} + \tfrac{2}{3}cT^{3/2} + \tfrac{4}{7}dT^{7/4}$$

$$s(T, p) = a\ln T + 4bT^{1/4} + 2cT^{1/2} + \tfrac{4}{3}dT^{3/4} - R\ln p$$

of various substances.[a] See also Table 20.1.

Substance	Formula	a	b	c	d
Acetylene	C_2H_2	−72.4	36.2	−1.98	0
Ammonia	NH_3	82.8	−46	11.1	−0.665
Carbon dioxide	CO_2	−55.6	30.5	−1.96	0
Carbon monoxide	CO	62.8	−22.6	4.6	−0.272
Chloroform	$CHCl_3$	−252	135	−17.1	0.732
Ethylene	C_2H_4	−239	90.8	−5.56	0
Freon 12	CCl_2F_2	−357	195	−27.4	1.28
Freon 21	$CHCl_2F$	−274	141	−17.5	0.732
Hydrogen	H_2	79.5	−26.3	4.23	−0.197
Hydrogen (atomic)	H	20.8	0	0	0
Hydronium ion	H_3O^+	131	−71.1	15.3	−0.904
Hydroxyl	OH	119	−47.3	7.91	−0.409
Hydroxyl ion	OH^-	104	−40.5	6.9	−0.36
Methane	CH_4	104	−77.8	20.1	−1.3
Nitric oxide	NO	49.4	−15.6	3.51	−0.217
Nitrogen	N_2	72	−26.9	5.19	−0.298
Nitrogen (atomic)	N	25.4	−1.7	0.158	0
Nitrogen dioxide	NO_2	−92.5	51	−5.65	0.202
Nitrous oxide	N_2O	−95	52.3	−5.73	0.202
Oxygen	O_2	10.3	5.4	−0.18	0
Oxygen (atomic)	O	29.4	−2.7	0.21	0
Oxygen ion	O^-	33	−5.4	0.787	−0.0382
Ozone	O_3	−163	90	−12.3	0.569
Proton	H^+	20.8	0	0	0
Water	H_2O	180	−85.4	15.6	−0.858

Source: Regression of data from the *JANAF Thermochemical Tables*, 2nd ed., D. R. Stull and H. Prophet, project directors, NSRDS–NBS37, 1971.

[a] c_p in kJ/kmol K, h in kJ/kmol, s in kJ/kmol K, and T in kelvin between 300 and 4000 K.

stable equilibrium states 1 and 2 at temperatures and pressures T_1, p_1 and T_2, p_2 is found by means of the relation

$$s_2 - s_1 = \phi(T_2) - \phi(T_1) - R\ln\frac{p_2}{p_1} = \phi_2 - \phi_1 - R\ln\frac{p_2}{p_1} \qquad (20.15)$$

Tables structured in this fashion are called *gas tables*. A sample of data from such tables is given in Table 20.4. The tabulated values are calculated from the values of specific heats measured at many levels of temperature.

It is noteworthy that if two states 1 and 2 have the same entropy, equation (20.15) yields

$$\left(\ln\frac{p_2}{p_1}\right)_{s_2 = s_1} = \frac{\phi(T_2) - \phi(T_1)}{R} \qquad (20.16)$$

TABLE 20.3. Values of specific heats c_p, c_v, and the ratio $\gamma = c_p/c_v$ for a few gases as functions of temperature.[a]

Temp. K	c_p	c_v	c_p/c_v	c_p	c_v	c_p/c_v	c_p	c_v	c_p/c_v	Temp. K
	Air			Carbon dioxide, CO_2			Carbon monoxide, CO			
250	1.003	0.716	1.401	0.791	0.602	1.314	1.039	0.743	1.400	250
300	1.005	0.718	1.400	0.846	0.657	1.288	1.040	0.744	1.399	300
350	1.008	0.721	1.398	0.895	0.706	1.268	1.043	0.746	1.398	350
400	1.013	0.726	1.395	0.939	0.750	1.252	1.047	0.751	1.395	400
450	1.020	0.733	1.391	0.978	0.790	1.239	1.054	0.757	1.392	450
500	1.029	0.742	1.387	1.014	0.825	1.229	1.063	0.767	1.387	500
550	1.040	0.753	1.381	1.046	0.857	1.220	1.075	0.778	1.382	550
600	1.051	0.764	1.376	1.075	0.886	1.213	1.087	0.790	1.376	600
650	1.063	0.776	1.370	1.102	0.913	1.207	1.100	0.803	1.370	650
700	1.075	0.788	1.364	1.126	0.937	1.202	1.113	0.816	1.364	700
750	1.087	0.800	1.359	1.148	0.959	1.197	1.126	0.829	1.358	750
800	1.099	0.812	1.354	1.169	0.980	1.193	1.139	0.842	1.353	800
900	1.121	0.834	1.344	1.204	1.015	1.186	1.163	0.866	1.343	900
1000	1.142	0.855	1.336	1.234	1.045	1.181	1.185	0.888	1.335	1000
	Hydrogen, H_2			Nitrogen, N_2			Oxygen, O_2			
250	14.051	9.927	1.416	1.039	0.742	1.400	0.913	0.653	1.398	250
300	14.307	10.183	1.405	1.039	0.743	1.400	0.918	0.658	1.395	300
350	14.427	10.302	1.400	1.041	0.744	1.399	0.928	0.668	1.389	350
400	14.476	10.352	1.398	1.044	0.747	1.397	0.941	0.681	1.382	400
450	14.501	10.377	1.398	1.049	0.752	1.395	0.956	0.696	1.373	450
500	14.513	10.389	1.397	1.056	0.759	1.391	0.972	0.712	1.365	500
550	14.530	10.405	1.396	1.065	0.768	1.387	0.988	0.728	1.358	550
600	14.546	10.422	1.396	1.075	0.778	1.382	1.003	0.743	1.350	600
650	14.571	10.447	1.395	1.086	0.789	1.376	1.017	0.758	1.343	650
700	14.604	10.480	1.394	1.098	0.801	1.371	1.031	0.771	1.337	700
750	14.645	10.521	1.392	1.110	0.813	1.365	1.043	0.783	1.332	750
800	14.695	10.570	1.390	1.121	0.825	1.360	1.054	0.794	1.327	800
900	14.822	10.698	1.385	1.145	0.849	1.349	1.074	0.814	1.319	900
1000	14.983	10.859	1.380	1.167	0.870	1.341	1.090	0.830	1.313	1000

Source: K. Wark, *Thermodynamics*, McGraw-Hill, New York, 1983.
[a] Both c_p and c_v are in kJ/kg K.

For an ideal gas, equation (20.16) indicates that the ratio of the pressures of two states having different temperatures but the same entropy, namely, two states along an isentropic, is fixed by the values of the two temperatures and is the same for all isentropics. Hence, if the values of pressure for one isentropic are tabulated for all values of the temperature, the values of the pressure for any other isentropic are proportional to those tabulated. Accordingly, denoting the pressures along a reference isentropic by p_r, we

TABLE 20.4. Sample of data from the gas tables for gaseous nitrogen at low pressures.[a]

T	t	h	p_r	u	v_r	φ	T	t	h	p_r	u	v_r	φ	T	t	h	p_r	u	v_r	φ
100	-173.15	2902.2	.0219	2070.8	37.998	159.667	850	576.85	25296.2	43.366	18229.0	.163	222.789	1600	1326.85	50571.4	555.53	37268.4	.0239	243.993
110	-163.15	3193.3	.0305	2278.7	29.941	162.441	860	586.85	25614.2	45.350	18463.8	.158	223.161	1610	1336.85	50922.8	570.35	37536.6	.0235	244.212
120	-153.15	3484.3	.0414	2486.6	24.087	164.974	870	596.85	25932.8	47.404	18699.3	.153	223.529	1620	1346.85	51274.4	585.48	37805.1	.0230	244.429
130	-143.15	3775.4	.0548	2694.5	19.717	167.303	880	606.85	26252.1	49.531	18935.5	.148	223.894	1630	1356.85	51626.2	600.93	38073.8	.0226	244.646
140	-133.15	4066.4	.0711	2902.4	16.382	169.460	890	616.85	26572.1	51.732	19172.3	.143	224.255	1640	1366.85	51978.4	616.70	38342.7	.0221	244.861
150	-123.15	4357.5	.0905	3110.3	13.786	171.468	900	626.85	26892.6	54.009	19409.7	.139	224.614	1650	1376.85	52330.7	632.79	38612.0	.0217	245.075
160	-113.15	4648.5	.1134	3318.2	11.732	173.347	910	636.85	27213.9	56.365	19647.8	.134	224.969	1660	1386.85	52683.4	649.22	38881.4	.0213	245.288
170	-103.15	4939.6	.1402	3526.1	10.081	175.111	920	646.85	27535.7	58.801	19886.5	.130	225.320	1670	1396.85	53036.2	665.98	39151.2	.0208	245.500
180	-93.15	5230.6	.1713	3734.1	8.739	176.775	930	656.85	27858.2	61.319	20125.8	.126	225.669	1680	1406.85	53389.3	683.08	39421.1	.0204	245.711
190	-83.15	5521.7	.2070	3942.0	7.633	178.349	940	666.85	28181.3	63.921	20365.7	.122	226.015	1690	1416.85	53742.7	700.53	39691.3	.0201	245.921
200	-73.15	5812.8	.2477	4149.9	6.714	179.842	950	676.85	28504.9	66.609	20606.3	.119	226.357	1700	1426.85	54096.3	718.33	39961.8	.0197	246.129
210	-63.15	6103.9	.2938	4357.8	5.943	181.262	960	686.85	28829.2	69.386	20847.4	.115	226.697	1710	1436.85	54450.1	736.48	40232.5	.0193	246.337
220	-53.15	6394.9	.3458	4565.8	5.290	182.616	970	696.85	29154.2	72.253	21089.2	.112	227.033	1720	1446.85	54804.1	755.00	40503.4	.0189	246.543
230	-43.15	6686.0	.4040	4773.7	4.734	183.910	980	706.85	29479.6	75.214	21331.5	.108	227.367	1730	1456.85	55158.4	773.88	40774.5	.0186	246.749
240	-33.15	6977.1	.4689	4981.7	4.256	185.149	990	716.85	29805.7	78.269	21574.5	.105	227.698	1740	1466.85	55512.9	793.14	41045.9	.0182	246.953
250	-23.15	7268.2	.5409	5189.6	3.843	186.337	1000	726.85	30132.4	81.421	21818.0	.102	228.027	1750	1476.85	55867.7	812.77	41317.4	.0179	247.156
260	-13.15	7559.3	.6205	5397.6	3.484	187.479	1010	736.85	30459.6	84.673	22062.1	.099	228.352	1760	1486.85	56222.6	832.78	41589.3	.0176	247.359
270	-3.15	7850.5	.7082	5605.6	3.170	188.578	1020	746.85	30787.5	88.027	22306.8	.096	228.675	1770	1496.85	56577.8	853.18	41861.3	.0172	247.560
280	6.85	8141.6	.8044	5813.6	2.894	189.637	1030	756.85	31115.8	91.485	22552.0	.094	228.995	1780	1506.85	56933.1	873.97	42133.5	.0169	247.760
290	16.85	8432.8	.9096	6021.7	2.651	190.658	1040	766.85	31444.8	95.050	22797.8	.091	229.313	1790	1516.85	57288.7	895.17	42406.0	.0166	247.959
300	26.85	8724.1	1.0243	6229.7	2.435	191.646	1050	776.85	31774.2	98.723	23044.1	.088	229.629	1800	1526.85	57644.5	916.77	42678.6	.0163	248.158
310	36.85	9015.3	1.1490	6437.9	2.243	192.601	1060	786.85	32104.3	102.508	23291.0	.086	229.941	1810	1536.85	58000.5	938.77	42951.5	.0160	248.355
320	46.85	9306.7	1.2842	6646.1	2.072	193.526	1070	796.85	32434.8	106.408	23538.4	.084	230.252	1820	1546.85	58356.8	961.20	43224.5	.0157	248.551
330	56.85	9598.1	1.4304	6854.3	1.918	194.423	1080	806.85	32765.9	110.423	23786.4	.081	230.560	1830	1556.85	58713.2	984.04	43497.8	.0155	248.746
340	66.85	9889.6	1.5882	7062.7	1.780	195.293	1090	816.85	33097.6	114.558	24034.9	.079	230.865	1840	1566.85	59069.8	1007.31	43771.3	.0152	248.941
350	76.85	10181.2	1.7582	7271.1	1.655	196.138	1100	826.85	33429.7	118.81	24283.9	.0770	231.169	1850	1576.85	59426.6	1031.02	44044.9	.0149	249.134
360	86.85	10472.9	1.9409	7479.7	1.542	196.960	1110	836.85	33762.4	123.20	24533.4	.0749	231.470	1860	1586.85	59783.6	1055.16	44318.8	.0147	249.326
370	96.85	10764.8	2.1368	7688.4	1.440	197.759	1120	846.85	34095.5	127.70	24783.4	.0729	231.769	1870	1596.85	60140.8	1079.75	44592.9	.0144	249.518
380	106.85	11056.8	2.3466	7897.3	1.346	198.538	1130	856.85	34429.2	132.34	25033.9	.0710	232.065	1880	1606.85	60498.2	1104.79	44867.1	.0141	249.709
390	116.85	11349.0	2.5709	8106.3	1.261	199.297	1140	866.85	34763.3	137.11	25284.9	.0691	232.360	1890	1616.85	60855.8	1130.28	45141.5	.0139	249.898
400	126.85	11641.3	2.8103	8315.6	1.183	200.037	1150	876.85	35098.0	142.02	25536.4	.0673	232.652	1900	1626.85	61213.5	1156.24	45416.1	.0137	250.087
410	136.85	11933.9	3.0654	8525.0	1.112	200.760	1160	886.85	35433.1	147.06	25788.4	.0656	232.942	1910	1636.85	61571.5	1182.67	45690.9	.0134	250.275
420	146.85	12226.8	3.3370	8734.7	1.046	201.466	1170	896.85	35768.7	152.24	26040.8	.0639	233.230	1920	1646.85	61929.6	1209.57	45965.9	.0132	250.462
430	156.85	12519.9	3.6256	8944.7	.986	202.155	1180	906.85	36104.8	157.57	26293.8	.0623	233.516	1930	1656.85	62287.9	1236.96	46241.1	.0130	250.648
440	166.85	12813.3	3.9320	9155.0	.930	202.830	1190	916.85	36441.3	163.05	26547.1	.0607	233.800	1940	1666.85	62646.4	1264.83	46516.4	.0128	250.833

322

T	t	h	p_r	u	v_r	φ
450	176.85	13107.0	4.2568	9365.5	.879	203.490
460	186.85	13401.0	4.6009	9576.4	.831	204.136
470	196.85	13695.4	4.9649	9787.7	.787	204.769
480	206.85	13990.2	5.3497	9999.3	.746	205.390
490	216.85	14285.4	5.7560	10211.3	.708	205.998
500	226.85	14581.0	6.1846	10423.8	.672	206.596
510	236.85	14877.0	6.6364	10636.6	.639	207.182
520	246.85	15173.5	7.1122	10850.0	.608	207.758
530	256.85	15470.5	7.6130	11063.8	.579	208.323
540	266.85	15767.9	8.1395	11278.1	.552	208.879
550	276.85	16065.9	8.6928	11493.0	.526	209.426
560	286.85	16364.4	9.2737	11708.4	.502	209.964
570	296.85	16663.5	9.8833	11924.3	.480	210.493
580	306.85	16963.1	10.5226	12140.8	.458	211.014
590	316.85	17263.3	11.1926	12357.8	.438	211.528
600	326.85	17564.1	11.894	12575.5	.419	212.033
610	336.85	17865.5	12.629	12793.8	.402	212.531
620	346.85	18167.6	13.397	13012.6	.385	213.022
630	356.85	18470.2	14.201	13232.1	.369	213.507
640	366.85	18773.5	15.040	13452.3	.354	213.984
650	376.85	19077.4	15.917	13673.0	.340	214.455
660	386.85	19381.9	16.833	13894.4	.326	214.920
670	396.85	19687.2	17.788	14116.5	.313	215.379
680	406.85	19993.0	18.784	14339.2	.301	215.833
690	416.85	20299.6	19.823	14562.6	.289	216.280
700	426.85	20606.8	20.906	14786.7	.278	216.722
710	436.85	20914.7	22.033	15011.5	.268	217.159
720	446.85	21223.2	23.207	15236.9	.258	217.590
730	456.85	21532.5	24.428	15463.0	.248	218.017
740	466.85	21842.4	25.699	15689.7	.239	218.439
750	476.85	22153.0	27.021	15917.2	.231	218.856
760	486.85	22464.3	28.394	16145.3	.223	219.268
770	496.85	22776.2	29.822	16374.1	.215	219.676
780	506.85	23088.9	31.304	16603.6	.207	220.079
790	516.85	23402.2	32.844	16833.8	.200	220.478
800	526.85	23716.2	34.442	17064.7	.193	220.873
810	536.85	24030.8	36.100	17296.2	.187	221.264
820	546.85	24346.2	37.819	17528.4	.180	221.651
830	556.85	24662.2	39.602	17761.2	.174	222.034
840	566.85	24978.9	41.451	17994.8	.168	222.413

T	t	h	p_r	u	v_r	φ
1200	926.85	36778.3	168.67	26801.0	.0592	234.082
1210	936.85	37115.7	174.45	27055.3	.0577	234.362
1220	946.85	37453.6	180.38	27310.0	.0562	234.640
1230	956.85	37791.9	186.48	27562.2	.0548	234.916
1240	966.85	38130.7	192.73	27820.8	.0535	235.191
1250	976.85	38469.9	199.15	28076.9	.0522	235.463
1260	986.85	38809.5	205.74	28333.3	.0509	235.734
1270	996.85	39149.5	212.50	28590.2	.0497	236.003
1280	1006.85	39489.9	219.43	28847.5	.0485	236.270
1290	1016.85	39830.8	226.55	29105.2	.0473	236.535
1300	1026.85	40172.0	233.84	29363.3	.0462	236.798
1310	1036.85	40513.6	241.32	29621.8	.0451	237.060
1320	1046.85	40855.7	248.99	29880.6	.0441	237.320
1330	1056.85	41198.1	256.85	30139.9	.0431	237.579
1340	1066.85	41540.9	264.91	30399.6	.0421	237.835
1350	1076.85	41884.0	273.16	30659.6	.0411	238.091
1360	1086.85	42227.6	281.62	30920.0	.0402	238.344
1370	1096.85	42571.5	290.28	31180.7	.0392	238.596
1380	1106.85	42915.7	299.16	31441.8	.0384	238.846
1390	1116.85	43260.3	308.25	31703.3	.0375	239.095
1400	1126.85	43605.3	317.55	31965.1	.0367	239.343
1410	1136.85	43950.6	327.08	32227.3	.0358	239.588
1420	1146.85	44296.3	336.83	32489.8	.0351	239.833
1430	1156.85	44642.2	346.81	32752.6	.0343	240.075
1440	1166.85	44988.6	357.03	33015.8	.0335	240.317
1450	1176.85	45335.2	367.48	33279.3	.0328	240.557
1460	1186.85	45682.2	378.17	33543.1	.0321	240.795
1470	1196.85	46029.5	389.11	33807.3	.0314	241.032
1480	1206.85	46377.1	400.29	34071.8	.0307	241.268
1490	1216.85	46725.0	411.73	34336.5	.0301	241.502
1500	1226.85	47073.2	423.43	34601.6	.0295	241.735
1510	1236.85	47421.7	435.39	34867.0	.0288	241.967
1520	1246.85	47770.6	447.62	35132.7	.0282	242.197
1530	1256.85	48119.7	460.11	35398.6	.0276	242.426
1540	1266.85	48469.1	472.88	35664.9	.0271	242.653
1550	1276.85	48818.8	485.93	35931.5	.0265	242.880
1560	1286.85	49168.8	499.27	36198.3	.0260	243.105
1570	1296.85	49519.0	512.89	36465.4	.0255	243.329
1580	1306.85	49869.6	526.80	36732.8	.0249	243.551
1590	1316.85	50220.4	541.02	37000.5	.0244	243.772

T	t	h	p_r	u	v_r	φ
1950	1676.85	63005.0	1293.19	46791.9	.0125	251.018
1960	1686.85	63363.9	1322.06	47067.6	.0123	251.201
1970	1696.85	63722.9	1351.43	47343.5	.0121	251.384
1980	1706.85	64082.0	1381.32	47619.5	.0119	251.566
1990	1716.85	64441.4	1411.72	47895.7	.0117	251.747
2000	1726.85	64800.9	1442.65	48172.1	.0115	251.927
2010	1736.85	65160.5	1474.12	48448.6	.0113	252.107
2020	1746.85	65520.4	1506.12	48725.3	.0112	252.285
2030	1756.85	65880.4	1538.67	49002.1	.0110	252.463
2040	1766.85	66240.5	1571.77	49279.1	.0108	252.640
2050	1776.85	66600.8	1605.43	49556.3	.0106	252.816
2060	1786.85	66961.3	1639.66	49833.6	.0104	252.991
2070	1796.85	67321.9	1674.47	50111.1	.0103	253.166
2080	1806.85	67682.7	1709.85	50388.7	.0101	253.340
2090	1816.85	68043.6	1745.82	50666.5	.0100	253.513
2100	1826.85	68404.6	1782.4	50944.4	.00980	253.685
2110	1836.85	68765.8	1819.6	51222.5	.00964	253.857
2120	1846.85	69127.2	1857.3	51500.7	.00949	254.028
2130	1856.85	69488.7	1895.7	51779.0	.00934	254.198
2140	1866.85	69850.3	1934.7	52057.5	.00920	254.367
2150	1876.85	70212.1	1974.4	52336.2	.00905	254.536
2160	1886.85	70574.0	2014.7	52614.9	.00891	254.704
2170	1896.85	70936.1	2055.6	52893.8	.00878	254.871
2180	1906.85	71298.3	2097.2	53172.9	.00864	255.038
2190	1916.85	71660.6	2139.4	53452.1	.00851	255.204
2200	1926.85	72023.1	2182.4	53731.4	.00838	255.369
2210	1936.85	72385.7	2226.0	54010.8	.00825	255.533
2220	1946.85	72748.4	2270.2	54290.4	.00813	255.697
2230	1956.85	73111.2	2315.2	54570.1	.00801	255.860
2240	1966.85	73474.2	2360.9	54850.0	.00789	256.022
2250	1976.85	73837.3	2407.2	55129.9	.00777	256.184
2260	1986.85	74200.6	2454.3	55410.0	.00766	256.345
2270	1996.85	74563.9	2502.1	55690.2	.00754	256.506
2280	2006.85	74927.4	2550.7	55970.6	.00743	256.665
2290	2016.85	75291.0	2600.0	56251.0	.00732	256.825
2300	2026.85	75654.7	2650.0	56531.6	.00722	256.983
2310	2036.85	76018.6	2700.8	56812.3	.00711	257.141
2320	2046.85	76382.5	2752.4	57093.1	.00701	257.298
2330	2056.85	76746.6	2804.7	57374.0	.00691	257.455
2340	2066.85	77110.8	2857.8	57655.1	.00681	257.611

Source: J. H. Keenan and J. Kaye, *Gas Tables*, Wiley, New York, 1945.

^{a}T in kelvin, t in °C, h in J/mol, p_r dimensionless, u in J/mol, v_r dimensionless, ϕ in J/mol K.

have that the pressures p along any other isentropic satisfy the relation

$$\left(\frac{p_2}{p_1}\right)_{s_2=s_1} = \frac{p_{r2}}{p_{r1}} \tag{20.17}$$

where subscripts 1 and 2 refer to states with temperatures T_1 and T_2, respectively. Given p_1, T_1, and T_2, then p_2 may be found from (20.17) by substituting tabulated values. Given p_1, p_2, and T_2, then p_{r2} may be found from (20.17) and the corresponding value of T_2 from the tables. In Chapter 23 we show that these computations are useful in engineering applications involving compressors and gas turbines.

Property p_r is called the *relative pressure*. An entirely analogous property, the *relative specific volume* v_r, is also tabulated. Starting from equation (20.12b) and proceeding in a manner analogous to that used for pressures, we conclude that the specific volumes v along any isentropic must conform to the relation

$$\left(\frac{v_2}{v_1}\right)_{s_2=s_1} = \frac{v_{r2}}{v_{r1}} \tag{20.18}$$

where again subscripts 1 and 2 refer to states with temperatures T_1 and T_2, respectively.

Example 20.5. Nitrogen initially at $T_1 = 1900$ K and $p_1 = 10$ atm is expanded to $p_2 = 1$ atm. If the expansion is isenthalpic, that is, $h_2 = h_1$, find T_2 and $s_2 - s_1$. If the expansion is isentropic, find T_2 and $h_1 - h_2$. Repeat the problem for $T_1 = 600$ K.

Solution. Assume that the behavior of nitrogen is ideal. For the isenthalpic expansion, $T_2 = T_1$ because $h_2 = h_1$ and h is a function of T only. Using equation (20.15), we find that $s_2 - s_1 = -R\ln(p_2/p_1) = 19.144$ J/mol K. For the isentropic expansion, equation (20.17) with $p_1/p_2 = 10$ and the value $p_{r1} = 1156.24$ taken from Table 20.4 yields $p_{r2} = 115.624$. Thus, by interpolating data in the table between 1090 and 1100 K, we find $T_2 = 1092.5$ K and $h_2 = 33,180.9$ J/mol, so that $h_1 - h_2 = 61,213.5 - 33,180.9 = 28,032.6$ J/mol. If, instead, $T_1 = 600$ K, then $p_{r1} = 11.894$ and $p_{r2} = 1.1894$. Now, by interpolating data in the table between 310 and 320 K, we find that $T_2 = 313$ K and $h_2 = 8808$ J/mol, so that $h_1 - h_2 = 17,564.1 - 8808 = 8756.1$ J/mol.

It is noteworthy that some of the substances in Tables 20.1 and 20.2, such as carbon dioxide, behave as ideal gases in the specified temperature ranges even at ordinary pressures of the order of 1 atm. Other substances, however, do not behave as ideal gases at 300 K unless the pressure is relatively small. Water is an example of such a substance.

For some applications it is expedient to assign fictitious values to properties of a pure substance by using the ideal gas relations beyond the ranges of temperature and pressure in which the substance behaves as an ideal gas (see Section 29.10). In doing so, we define properties of fictitious stable equilibrium states, each of which is called an *ideal-gas state*. For example, the specific enthalpy of an ideal-gas state of a substance at T_o and p_o may be found from the relation $h(T_o, p_o) = h(T, p_o) - \int_{T_o}^{T} c_p(T')\, dT'$, where $h(T, p_o)$ is the specific enthalpy of the substance at conditions at which it behaves as an ideal gas, and $c_p(T)$ is the specific heat obtained from either Table 20.1 or 20.2.

20.2 Perfect-Gas Model

The *perfect-gas*[2] model is based on the observation that, over various limited ranges of temperature, each specific heat of an ideal gas is a weak function of T and, to a good

---

[2]Again we emphasize that the term "perfect" refers strictly to the simplicity of the stable-equilibrium-state interrelations, not to any thermodynamic optimality such as a reduction of the detrimental effects of irreversibility, or improved performance in particular processes.

degree of accuracy, can be considered as a constant. The value of the constant depends on the range of temperature over which the weak dependence on T is observed. Thus the perfect-gas model satisfies the relations $pv = RT$, $c_v = $ constant, or equivalently $c_p = $ constant, $c_p - c_v = R$, and $c_p/c_v = \gamma = $ constant. It is noteworthy that the value of the ratio γ depends on the range of T over which c_p and c_v are approximated by constants.

Because here c_v and c_p are constants, all the integrals in the equations in Section 20.1 can readily be resolved. For example, the expressions for the specific internal energy, specific enthalpy, and specific entropy of a perfect gas are

$$u(T) - u_0 = c_v (T - T_0) = \frac{1}{\gamma - 1}(pv - p_0 v_0) \tag{20.19}$$

$$h(T) - h_0 = c_p (T - T_0) = \frac{\gamma}{\gamma - 1}(pv - p_0 v_0) \tag{20.20}$$

$$s(T, p) - s_0 = c_p \ln \frac{T}{T_0} - R \ln \frac{p}{p_0} \tag{20.21}$$

$$s(T, v) - s_0 = c_v \ln \frac{T}{T_0} + R \ln \frac{v}{v_0} \tag{20.22}$$

$$s(v, p) - s_0 = c_p \ln \frac{v}{v_0} + c_v \ln \frac{p}{p_0} \tag{20.23}$$

where u_0, h_0, and s_0 refer to the state at T_0, p_0, and $v_0 = RT_0/p_0$, and we use the relations

$$c_v = \frac{R}{\gamma - 1} \quad \text{and} \quad c_p = \frac{\gamma R}{\gamma - 1} \tag{20.24}$$

It is noteworthy that none of the ideal-gas relations is valid at low temperatures $(T \to 0)$ because no substance behaves as an ideal gas at such temperatures.

The simplicity of the equation of state and the constancy of the specific heats also result in explicit expressions for changes in values of properties between various stable equilibrium states of a perfect gas. For example, for two states 1 and 2 having the same entropy, (20.21) to (20.23) yield

$$c_p \ln \frac{T_2}{T_1} = R \ln \frac{p_2}{p_1} \quad \text{or} \quad T_2\, p_2^{-(\gamma-1)/\gamma} = T_1\, p_1^{-(\gamma-1)/\gamma} \tag{20.25}$$

$$c_v \ln \frac{T_2}{T_1} = -R \ln \frac{v_2}{v_1} \quad \text{or} \quad T_2\, v_2^{\gamma-1} = T_1\, v_1^{\gamma-1} \tag{20.26}$$

$$c_p \ln \frac{v_2}{v_1} = -c_v \ln \frac{p_2}{p_1} \quad \text{or} \quad p_2\, v_2^{\gamma} = p_1\, v_1^{\gamma} \tag{20.27}$$

More generally, along any isentropic of a perfect gas we have

$$T\, p^{-(\gamma-1)/\gamma} = \text{constant} \qquad T\, v^{\gamma-1} = \text{constant} \qquad p\, v^{\gamma} = \text{constant} \tag{20.28}$$

Various types of interactions can result in isentropic changes of state, that is, in isentropic processes. For example, we show in Section 16.5 that a reversible adiabatic process is isentropic. Again, a process involving two different heat interactions is isentropic if one carries into the system a certain amount of entropy, and the other out of the system the same amount of entropy plus whatever amount of entropy may have been generated within the system by irreversibility.

For isentropic processes, the energy and entropy balances can be combined with (20.19) to (20.27) to yield explicit expressions for the work, heat, and other types of interactions in terms of properties of the perfect gas. For example, for a reversible adiabatic process from a state 1 to a state 2, the only interaction involved is work and the energy balance yields

$$w^{\to} = u_1 - u_2 = c_v \, (T_1 - T_2) = \frac{1}{\gamma - 1}(p_1 v_1 - p_2 v_2)$$

$$= \frac{1}{\gamma - 1}p_1 v_1 \left[1 - \left(\frac{p_2}{p_1}\right)^{(\gamma - 1)/\gamma} \right] \tag{20.29}$$

where $w^{\to}$ denotes the work done per mole of the gas, and v_1 is the mole-specific volume of state 1.

Example 20.6.   Repeat Example 20.5 by modeling nitrogen as a perfect gas with $\gamma = 1.4$.

Solution.   For the isenthalpic expansion the answer is the same as given in Example 20.5. For the isentropic expansion, equation (20.25) yields $T_2 = T_1 \, (p_2/p_1)^{(\gamma - 1)/\gamma} = 1900(1/10)^{0.4/1.4} = 984$ K so that $h_1 - h_2 = c_p \, (T_1 - T_2) = 29.1(1900 - 984) = 26{,}654.7$ J/mol, where we use $c_p = \gamma R/(\gamma - 1) = 29.1$ J/mol K. For $T_1 = 600$ K we find $T_2 = 600(1/10)^{0.4/1.4} = 310.8$ K and $h_1 - h_2 = 29.1(600 - 310.8) = 8415.7$ J/mol. Comparing these results with those in Example 20.5, we see that the perfect-gas model implies a large error, $(984 - 1092)/1092 = -0.10$ or -10% at high temperatures, but a smaller error, $(310.8 - 313)/313 = -0.007$ or -0.7%, at lower temperatures. As discussed in the next section, the reason for the large error is that the constant value $\gamma = 1.4$ is inappropriate for nitrogen in this range of temperatures.

In view of the ease of use of equations (20.28) for isentropic processes, it is sometimes practical to consider changes of stable equilibrium states of a perfect gas that follow special paths called *polytropic*, and that obey the relations

$$T p^{-(j-1)/j} = \text{constant} \qquad T v^{j-1} = \text{constant} \qquad p v^j = \text{constant} \tag{20.30}$$

where, in general, j is a number different from the ratio of specific heats γ. Graphs of polytropic paths on a p–v diagram are shown in Figure 20.1, where we see that for $j = 1$ the path is along an isotherm, for $j = 0$ along an isobar, for $j = \pm\infty$ along an isochor, and for $j = \gamma$ along an isentropic.

Example 20.7.   A mole of a perfect gas undergoes a reversible process from state 1 to state 2 along a polytropic path, while experiencing a work and a heat interaction. Evaluate the work $w^{\to}$, the heat $q^{\gets}$, and the specific entropy change $s_2 - s_1$, as functions of the temperatures, pressures, and specific volumes of the end states and the polytropic exponent j.

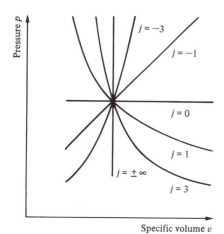

Figure 20.1   Graphs of polytropic paths on a p–v diagram.

Solution.   Because the process is reversible and passes through stable equilibrium states (Section 16.5), the work $\delta w^{\rightarrow} = p\,dv$ and, therefore, using (20.30) for $j \neq 1$ and (20.1a), we find that

$$w^{\rightarrow} = \int_1^2 p\,dv = p_1 v_1^j \int_{v_1}^{v_2} \frac{dv}{v^j} = \frac{p_1 v_1}{j-1}\left[1 - \left(\frac{v_1}{v_2}\right)^{j-1}\right]$$

$$= \frac{1}{j-1}(p_1 v_1 - p_2 v_2) = \frac{R}{j-1}(T_1 - T_2)$$

Next, using the energy balance and (20.19), we find that

$$q^{\leftarrow} = u_2 - u_1 + w^{\rightarrow} = c_v(T_2 - T_1) + \frac{R}{j-1}(T_1 - T_2)$$

$$= \frac{jc_v - c_p}{j-1}(T_2 - T_1)$$

Finally, using (20.22) and (20.28), we find that

$$s_2 - s_1 = c_v \ln\frac{T_2}{T_1} + R\ln\frac{v_2}{v_1} = c_v \ln\frac{T_2}{T_1} - \frac{R}{j-1}\ln\frac{T_2}{T_1}$$

$$= \frac{jc_v - c_p}{j-1}\ln\frac{T_2}{T_1}$$

For $j = 1$, namely, for an isothermal path, we find that $u_2 - u_1 = 0$, $w^{\rightarrow} = q^{\leftarrow} = p_1 v_1 \ln(v_2/v_1)$, and $s_2 - s_1 = R\ln(v_2/v_1)$.

20.3   Specific Heat and Molecular Structure

The temperature dependence of the specific internal energy and correspondingly the specific heat under constant volume can be understood in terms of the methods of quantum thermodynamics. Because we do not cover these methods here, we only discuss some of the main features of the results.

Under conditions of ideal-gas behavior, the intermolecular forces of a pure substance are so weak that practically all its properties are determined by the individual degrees of freedom of each individual molecule of the substance and not by the degrees of freedom of the collection of the molecules. In particular, the internal energy can be thought of as residing within each individual molecule, that is, each molecule has its own private energy, which is accommodated among the various individual degrees of freedom.

The degrees of freedom of a molecule are related to its kinematics. They refer to the independent variations that can be made in the coordinates which specify the position and configuration of the molecule. They are classified as translational, rotational, vibrational, and electronic as follows.

A *translational degree of freedom* relates to the ability of the center of mass of a molecule to move about in space. Each dimension of space provides one translational degree of freedom, and therefore the center of mass of each molecule has $f_t = 3$ such degrees of freedom.

A *rotational degree of freedom* relates to the ability of a molecule to rotate as a rigid structure about one of its principal axes of rotation and, as a result, increase or stretch the relative distance between nuclei within the molecule. The increased distance contributes to increasing the moment of inertia of the molecule with respect to the axis under consideration. In general, a polyatomic molecule can be stretched by rotation about any of its three principal axes passing through its center of mass. Hence it has three rotational degrees of freedom. However, in a molecule with all its nuclei aligned along one axis, such as O_2, N_2, CO, and CO_2, the distances between nuclei are not affected by rotation around this axis. Accordingly, such a molecule has only two rotational degrees of freedom because no stretching is possible about the third axis. Finally, a monatomic molecule has no rotational degrees of freedom because none of its rotations can contribute to an increase of its moments of inertia. In summary, a molecule has f_r rotational degrees of freedom, where f_r equals 3, 2, or 0.

A *vibrational degree of freedom* relates to the ability of a nucleus in a molecule to vibrate about a mean position with respect to the other nuclei of the molecule. Clearly, a monatomic molecule has no vibrational degree of freedom. A diatomic molecule such as H_2 has one vibrational degree of freedom. In general, a polyatomic molecule with l nuclei has $f_v = 3l - 6$ vibrational degrees of freedom if the nuclei are not arranged along a line, and $f_v = 3l - 5$ if they are arranged along a line. For example, water, H_2O, has three nuclei not along a line, and therefore $f_v = 3 \times 3 - 6 = 3$. Again, carbon dioxide, CO_2, has three nuclei along a line, and therefore $f_v = 3 \times 3 - 5 = 4$. Associated with each vibrational degree of freedom is a characteristic frequency of vibration.

The number of vibrational degrees of freedom is obtained as follows. For a molecule with l nuclei, we require $3l$ position coordinates or, more generally, generalized coordinates to describe its configuration in space. Of these, three are necessary to describe the position of the center of mass of the molecule, and therefore $3l - 3$ remain to account for the relative configuration of the nuclei inside the molecule. Specifically, if the nuclei are not arranged along a line, three angular coordinates are necessary to describe the rotations of the molecule about its principal axes, and therefore $3l - 6$ coordinates remain to be associated with the vibrational degrees of freedom. If, however, the nuclei are arranged along a straight line, only two angular coordinates are necessary to describe the rotations, and therefore $3l - 5$ coordinates remain to be associated with the vibrational degrees of freedom.

Up to this point, our discussion has not considered the configuration of the electrons of the molecule that hold the nuclei together. These are the valence electrons—the outermost and most loosely bound electrons of the molecule. An *electronic degree of freedom* relates to the ability of each valence electron of a molecule to assume different configurations around the molecule, configurations that differ from that of the ground-energy state of the molecule. Each valence electron provides one electronic degree of freedom.

In principle, all types of degrees of freedom are present under all conditions, and accommodate some of the energy of the molecule. Numerically, however, the energy residing in each nontranslational degree of freedom is negligible at low temperatures, and becomes comparable to that in a translational degree of freedom only at relatively high temperatures.

At low temperatures, the internal energy of an ideal gas is uniformly distributed principally among the translational degrees of freedom and increases linearly with temperature. Each such degree accounts for $k/2$ units of energy per degree of temperature, where k is the Boltzmann constant. So the specific heat per mole under constant volume is

$$c_v = 3N_A \frac{k}{2} = 3\frac{R}{2} \tag{20.31}$$

because there are $3N_A$ translational degrees of freedom per mole, where N_A is Avogadro's number.

At higher temperatures, the energy accommodated in the rotational degrees of freedom becomes comparable to that in the translational degrees of freedom. In fact, at temperatures higher than a threshold value Θ_r characteristic of each rotation of the molecule, the energy accommodated in the corresponding rotational degree of freedom increases linearly with temperature, and accounts for $k/2$ units of energy per degree of temperature. So at temperatures higher than the largest threshold value Θ_r of all the f_r rotations of the molecule, the specific heat per mole under constant volume is

$$c_v = (3 + f_r)\frac{R}{2} \qquad (20.32)$$

Measured values of c_v for H_2 as a function of T are shown in Figure 20.2. We see that up to 80 K the value of c_v is equal to $3R/2$. It corresponds to the three translational degrees of freedom. Between 300 and 800 K, the value of c_v is $5R/2$ because the two rotational degrees of freedom—$f_r = 2$ for H_2—are fully effective.

Threshold values Θ_r corresponding to the rotational degrees of freedom of various molecules are listed in Table 20.5. With the exception of H_2, we see that the values of Θ_r are much smaller than the corresponding saturation temperatures at atmospheric pressure, $T_{sat}(p = 1 \text{ atm})$. At conditions of pressure and temperature that are close to saturation, the intermolecular forces become strong and the pure substance ceases to behave as an ideal gas. Therefore, for most substances, the ideal gas behavior is evidenced at temperatures comparable with the typical values of Θ_r only at very low pressures for which $T_{sat}(p)$ is lower than Θ_r. We conclude that, for ordinary substances and ordinary pressures, the ideal-gas behavior is exhibited at temperatures for which the rotational contributions are fully effective.

At even higher temperatures, the energy accommodated in each vibrational degree of freedom becomes comparable to that in a translational degree of freedom. Specifically, at temperatures higher than a threshold value Θ_v characteristic of each vibration of the molecule, the energy accommodated in the corresponding vibrational degree of freedom increases linearly with temperature, and accounts for k units of energy per degree of temperature. For example, hydrogen, H_2, has only one vibrational degree of freedom, and at temperatures larger than about 6000 K (Figure 20.2), this degree of freedom stores R units of energy per mole per degree of temperature. For molecules with many vibrational degrees of freedom, we see from Table 20.5 that many substances have threshold values Θ_v that fall within the range of temperatures that are typical of engineering applications. In other words, at moderately high temperatures the vibrational degrees of freedom are active but not fully effective, and therefore the energy they accommodate does not increase linearly with temperature. As a result, the specific heat at constant volume is

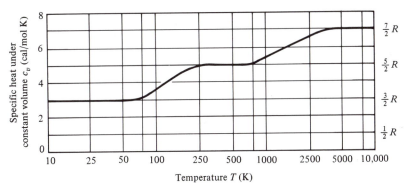

Figure 20.2 Graph of measured values of c_v for hydrogen as a function of temperature (from F. K. Richtmyer, E. H. Kennard and T. Lauritsen, *Introduction to Modern Physics*, McGraw-Hill, New York, 1955).

TABLE 20.5. Values of the rotational and vibrational threshold temperatures, and the saturation temperature at atmospheric pressure for various molecules.

Molecule	Θ_r K	Θ_v K	T_{sat}(1 atm) K	Molecule	Θ_r K	Θ_v K	T_{sat}(1 atm) K
H_2	85.0	5980	20.4	NO_2	0.59	1900	294.3
D_2	42.5	4300	23.5		0.624	1980	
HCl	15.0	4150	188.1		11.5	2330	
N_2	2.84	3350	77.4	NH_3	8.92	1370	239.7
CO	2.74	3080	81.7		13.6	2340	
NO	2.42	2700	121.4		13.6	2340	
O_2	2.06	2240	90.2			4800	
Cl_2	0.346	796	238.7			4910	
CO_2	0.56	960	194.7			4910	
		960	(subl.)	CH_4	7.54	1870	111.7
		2000			7.54	1870	
		3380			7.54	1870	
N_2O	0.60	847	184.7			2180	
		847				2180	
		1850				4170	
		3200				4320	
H_2O	13.4	2290	373.15			4320	
	20.9	5250				4320	
	40.1	5400					

Source: Data mainly from R. H. Fowler and E. A. Guggenheim, *Statistical Thermodynamics*, Cambridge University Press, London, 1965.

not constant. Theoretical results show that under such conditions the specific heat at constant volume per mole of an ideal gas is given by the relation

$$c_v(T) = (3 + f_r)\frac{R}{2} + \sum_{j=1}^{f_v} \frac{(\Theta_{vj}/2T)^2}{\sinh^2(\Theta_{vj}/2T)} R \tag{20.33}$$

where Θ_{vj} is the threshold temperature characteristic of the jth vibrational degree of freedom of the molecule, for $j = 1, 2, \ldots, f_v$, and $\sinh(x) = [\exp(x) - \exp(-x)]/2$.

For temperatures larger than the largest threshold value Θ_{vj}, equation (20.33) reduces to

$$c_v = (3 + f_r)\frac{R}{2} + f_v R \tag{20.34}$$

because then each of the ratios $(\Theta_{vj}/2T)^2/\sinh^2(\Theta_{vj}/2T)$ approaches unity. However, for temperatures comparable with any of the threshold values Θ_{vj}, the specific heat c_v is not independent of temperature T, and the perfect-gas model does not yield accurate results.

Similar remarks can be made about the electronic degrees of freedom. Each contributes to the specific heat under constant volume after some very high temperature of the order of 10,000 K.

Example 20.8. Estimate the value of c_v for H_2O at 1000 K using (20.33) and data from Table 20.5. Compare the result with data from Figure 19.15.

Solution. Being polyatomic with nuclei that are not aligned, the H_2O molecule has three rotational degrees of freedom, that is, $f_r = 3$, and three vibrational degrees of freedom, that is, $f_v = 3$. From Table 20.5 we see that the threshold temperatures for H_2O are $\Theta_{v1} = 2290$ K, $\Theta_{v2} =$

5250 K, and $\Theta_{v3} = 5400$ K. Using these values in (20.33), we find that $c_v(1000 \text{ K}) = (3 + 3)R/2 + (0.656 + 0.146 + 0.133)R = 32.71$ J/mol K or, on a mass-specific basis, $c_v = 3.935R/M = 1.82$ kJ/kg K. Correspondingly, $c_p = c_v + R = 4.935R/M = 2.28$ kJ/kg K, a value in very good agreement with the low-pressure value at 1000 K $= 727°$C that we read in Figure 19.15.

20.4 Ideal Incompressible Behavior

Under conditions for which either the liquid or the solid form of aggregation prevails, the effect of intermolecular forces on the behavior of a substance cannot be neglected. Each molecule cannot move freely because it is encircled by its nearest neighbors and interacts strongly with them. These interactions involve a balance of strong cohesive and repelling forces and, as a result, both a liquid and a solid can sustain very large changes in pressure without appreciable changes in volume. Because of the strength of the intermolecular forces, the specific volume of either a liquid or a solid is relatively insensitive to temperature and especially insensitive to pressure. Of course, this is true for a liquid only for temperatures well below the critical temperature.

Quantitatively, these observations are equivalent to saying that both the coefficient of isobaric expansion, α_p, and the coefficient of isothermal compressibility, κ_T, are so small that, for all practical purposes, the equation of state of various substances either in the liquid or in the solid form of aggregation may be approximated by the relation

$$v \approx \text{constant, independent of } T \text{ and } p \tag{20.35}$$

Moreover, experimental data show that measured values of T, v, α_p, κ_T, and the specific heat c_p are such that the ratio $T\alpha_p^2 v / \kappa_T c_p$ is much smaller than unity and, for all practical purposes, can be assumed equal to zero, that is,

$$\frac{T\alpha_p^2 v}{\kappa_T c_p} \approx 0 \tag{20.36}$$

For example, the data for water at atmospheric pressure shown in Table 20.6 confirm the validity of (20.36).

In the ranges of temperature and pressure in which (20.35) and (20.36) are valid, we say that the substance behaves as an *ideal incompressible liquid* or an *ideal incompressible solid*.

Using approximation (20.36) in equation (19.26), we find that

$$c_p = c_v + \frac{T\alpha_p^2 v}{\kappa_T} \approx c_v \tag{20.37}$$

Moreover, using (20.35) and (20.37) in the general relations for du (Example 20.3), ds [the third equation (19.27)], and $dh = T ds + v dp$, we find that

$$du = c_v \, dT + \left(\frac{T\alpha_p}{\kappa_T} - p \right) dv \approx c_v \, dT \tag{20.38}$$

$$ds = \frac{c_v}{T} \, dT + \frac{\alpha_p}{\kappa_T} \, dv \approx \frac{c_v}{T} \, dT \tag{20.39}$$

$$dh = T \, ds + v \, dp \approx c_v \, dT + v \, dp \tag{20.40}$$

where in approximation (20.40) we use approximation (20.39).

From the approximations $du \approx c_v \, dT$ and $ds \approx c_v \, dT/T$ we conclude that, to a good degree of accuracy, u, s, and c_v are functions of T only, namely, $u = u(T)$, $s = s(T)$,

TABLE 20.6.   Data for liquid H_2O at atmospheric pressure.

T	v	α_p	κ_T	c_p	$T\alpha_p^2 v / \kappa_T c_p$
°C	10^{-3} m³/kg	10^{-6} K⁻¹	10^{-11}/Pa⁻¹	kJ/kg K	10^{-2}
0	1.00016	−68.05	50.88	4.2177	0.06
1	1.00010	−50.09	50.51	4.2141	0.03
2	1.000060	−32.74	50.15	4.2107	0.01
3	1.000036	−15.97	49.81	4.2077	0.003
4	1.000028	0.27	49.48	4.2048	0.000
5	1.000036	16.00	49.17	4.2022	0.003
10	1.00030	87.97	47.81	4.1922	0.10
20	1.00180	206.8	45.89	4.1819	0.65
30	1.0044	303.2	44.77	4.1785	1.5
40	1.0078	385.3	44.24	4.1786	2.5
50	1.0118	457.6	44.17	4.1807	3.7
60	1.0171	523.1	44.50	4.1844	5.0
70	1.0227	583.7	45.16	4.1896	6.3
80	1.0290	641.1	46.14	4.1964	7.7
90	1.0359	696.2	47.43	4.2051	9.1
100	1.0434	750.1	49.02	4.2160	10.6

Source: Data from R. C. Weast, editor, *CRC Handbook of Chemistry and Physics*, 66th ed., CRC Press, Boca Raton, Fla., 1985.

and $c_v = du/dT = T\,ds/dT$. Thus for ideal incompressible behavior we can write the approximate relations

$$du = c_v(T)\,dT \tag{20.41}$$

$$ds = \frac{c_v(T)}{T}\,dT \tag{20.42}$$

$$dh = c_v(T)\,dT + v\,dp \tag{20.43}$$

Perfect incompressible behavior is a special case of the ideal incompressible behavior referring to conditions under which the specific heat can be approximated by a constant c independent of T. For example, over a limited range of temperatures, c can be taken as the average of $c_v(T)$. Then, we can integrate equations (20.41) to (20.43) and find

$$u(T) - u_0 = c\,(T - T_0) \tag{20.44}$$

$$s(T) - s_0 = c\,\ln\frac{T}{T_0} \tag{20.45}$$

$$h(T, p) - h_0 = c\,(T - T_0) + v\,(p - p_0) \tag{20.46}$$

where u_0, s_0, and h_0 refer to the state at T_0 and p_0.

For solids, approximations 20.38 to 20.40 yield reasonable results even though approximation 20.37 does not hold for all temperatures. Theoretical results show that the temperature dependence of the specific heat at constant volume is given by the relation

$$c_v(T) = 3R\left[12\left(\frac{T}{\Theta_D}\right)^3 \int_0^{\Theta_D/T} \frac{x^3}{e^x - 1}\,dx - \frac{3\,\Theta_D/T}{e^{\Theta_D/T} - 1}\right] \tag{20.47}$$

where Θ_D is a characteristic constant of the solid called the *Debye temperature*[3] and R the universal gas constant.[4]   Data for a variety of solids versus the dimensionless temperature T/Θ_D are shown in Figure 20.3. The values of Θ_D that best correlate these data according to (20.47) are listed in Table 20.7.

For $T \gg \Theta_D$, (20.47) is well approximated by the relation

$$c_v(T) \approx 3R\left[1 - \frac{1}{20}\left(\frac{\Theta_D}{T}\right)^2\right] \qquad (T \gg \Theta_D) \qquad (20.48)$$

The fact that $c_v(T) \to 3R$ for temperatures much larger that the Debye temperature is known as the *Dulong and Petit law*.[5]   For $T \ll \Theta_D$ equation (20.47) is well approximated by the relation

$$c_v(T) \approx 3R\,\frac{4\pi^4}{5}\left(\frac{T}{\Theta_D}\right)^3 \qquad (T \ll \Theta_D) \qquad (20.49)$$

For values of Θ_D/T between 0.5 and 10 equation (20.47) cannot be approximated by any simpler relation. Values of $c_v(T)/3R$ for Θ_D/T between 0.5 and 10 obtained from equation (20.47) are listed in Table 20.8.

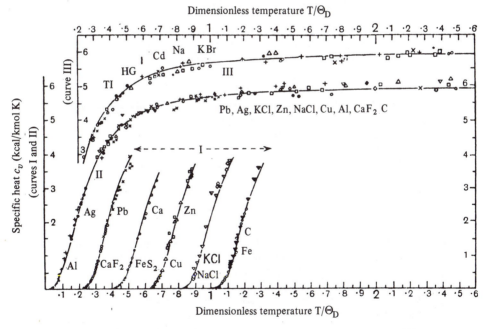

Figure 20.3   Graph of $c_v(T)$ in kcal/kmol K for various solids as a function of T/Θ_D where the values of Θ_D are given in Table 20.6. If they were not shifted to avoid overlapping of experimental data, curves I and III would be congruent with II (from R. H. Fowler and E. A. Guggenheim, *Statistical Thermodynamics*, Cambridge University Press, London, 1965).

---

[3]In honor of Peter Joseph Wilhelm Debye (1884–1966), Dutch physical chemist, who developed the theory of heat capacity of solids, and received the Nobel Prize in physics in 1936.

[4]We note in Section 20.1 that R is a universal physical constant, also called the Boltzmann constant when expressed on a per atom or per molecule basis. Of course, its role is not restricted to the gaseous form of aggregation just because it is called the universal "gas" constant.

[5]In honor of Pierre Louis Dulong (1785–1838) and Alexis Thérèse Petit (1791–1820), French physicists, who established this law in 1819.

TABLE 20.7.   **Values of the Debye temperature Θ_D for various solids.**

Solid	Formula	Θ_D [K]
Lead	Pb	88
Thallium	Tl	96
Mercury	Hg	97
Iodine	I	106
Cadmium	Cd	168
Sodium	Na	172
Potassium bromide	KBr	177
Silver	Ag	215
Calcium	Ca	226
Sylvine	KCl	230
Zinc	Zn	235
Rocksalt	NaCl	281
Copper	Cu	315
Aluminum	Al	398
Iron	Fe	453
Fluorspar	CaF_2	474
Iron pyrites	FeS_2	645
Diamond	C	1860

Source: R. H. Fowler and E. A. Guggenheim, *Statistical Thermodynamics*, Cambridge University Press, London, 1965.

TABLE 20.8.   **Values of $c_v(T)/3R$ for Θ_D/T between 0.5 and 10 obtained from equation (20.47).**

Θ_D/T	0.5	1	1.5	2	3	4
$c_v(T)/3R$	0.9876	0.9517	0.8960	0.8254	0.6628	0.5031
Θ_D/T	5	6	7	8	9	10
$c_v(T)/3R$	0.3686	0.2656	0.1909	0.1382	0.1015	0.07582

The specific heat at constant pressure of a solid is approximately equal to $c_v(T)$ for temperatures up to the Debye temperature Θ_D. For higher temperatures, whereas $c_v(T)$ remains constant at the value $3R$, $c_p(T)$ increases with T.

Example 20.9.   Estimate the values of the specific heat at constant volume c_v at $T = 310$ K for diamond, C, and iron, Fe. Compare the results with data from Figure 20.3.

Solution. From Table 20.7 we find that $\Theta_D = 1860$ K for diamond. For $\Theta_D/T = 1860/310 = 6$, equation (20.47) yields $c_v(T)/3R = 0.2656$, and therefore, $c_v = 0.2656 \times 3 \times 1.986 = 1.58$ kcal/kmol K. For iron, $\Theta_D = 453$ K, $\Theta_D/T = 453/310 = 1.46 \approx 1.5$, $c_v(T)/3R = 0.8960$, and $c_v = 0.8960 \times 3 \times 1.986 = 5.34$ kcal/kmol K. The values found for c_v are in very good agreement with the data in Figure 20.3 corresponding to $T/\Theta_D = 310/1860 = 0.166$ for diamond and $T/\Theta_D = 310/453 = 0.684$ for iron.

20.5   Fugacity and Activity

In preparation for the study of multiconstituent simple systems in later chapters, we introduce two auxiliary properties, each of which is directly related to the chemical potential, and is based on the fact that at sufficiently low pressures and relatively high temperatures every substance behaves as an ideal gas.

For an ideal gas, the chemical potential and pressure satisfy the relation

$$\left(\frac{\partial \mu}{\partial p}\right)_T = v = \frac{RT}{p} \tag{20.50a}$$

or, equivalently,

$$\left[\frac{\partial(\mu/RT)}{\partial \ln p}\right]_T = 1 \tag{20.50b}$$

Integrating this relation from arbitrary pressure p_0 to pressure p—within the range of ideal-gas behavior—we find that

$$\mu(T,p) = \mu(T,p_0) + RT \ln \frac{p}{p_0} \tag{20.51a}$$

or, equivalently,

$$p = p_0 \exp\left[\frac{\mu(T,p) - \mu(T,p_0)}{RT}\right] \tag{20.51b}$$

Equation (20.51) can be extended outside the range of ideal-gas behavior by replacing the pressure with the function $\pi(T,p)$ called the *fugacity* and defined by the relation

$$\pi(T,p) = \pi(T,p_0)\exp\left[\frac{\mu(T,p) - \mu(T,p_0)}{RT}\right] \tag{20.52}$$

together with the condition

$$\pi(T,p) = p \qquad \text{in the limit of ideal-gas behavior} \tag{20.53}$$

The fugacity is just a transformation of the chemical potential. It has dimensions of pressure and reduces to the pressure under ideal-gas behavior. In particular, an indicator of the departure from ideal-gas behavior is the ratio $\phi = \pi/p$. It is called the *fugacity coefficient*.

Another useful function is the *activity*, $a(T,p,p_0)$, defined by the relation

$$a(T,p,p_0) = \exp\left[\frac{\mu(T,p) - \mu(T,p_0)}{RT}\right] = \frac{\pi(T,p)}{\pi(T,p_0)} \tag{20.54}$$

The activity is dimensionless and, at a given state with temperature T and pressure p, measures the excess chemical potential relative to a state at the same temperature but reference pressure p_0. The terminology "fugacity" and "activity" arises from the interpretation of the chemical potential as an escaping tendency (Section 10.3).

Example 20.10.   Evaluate the fugacity π of saturated water vapor at $T = 200°C$ using data from the steam tables.

Solution.   First we select p_0 in the range of ideal-gas behavior, so that $\pi(T,p_0) = p_0$ for $T = 200°C$. By inspection of the Mollier chart (Figure 19.10) we see that the isotherm at 200°C becomes flat at pressures below 20 kPa. This indicates that the enthalpy is a function of T only, i.e., that the vapor behaves as an ideal gas. Thus we select $p_0 = 10$ kPa and set $\pi(200°C, 10 \text{ kPa}) = 10$ kPa. Now, from the steam tables we read that saturated vapor at 200°C is at pressure $p = p_{sat}(200°C) = 1554$ kPa and has $h_g = 2790.9$ kJ/kg and $s_g = 6.4276$ kJ/kg K. Thus $\mu(T,p) = h_g - Ts_g = 2790.9 - 473 \times 6.4276 = -249.35$ kJ/kg. On the other hand, vapor at 200°C and 10 kPa has $h = 2879.5$ kJ/kg, $s = 8.9040$ kJ/kg K, and therefore, $\mu(T, p_o) = h - Ts = 2879.5 - 473 \times 8.9040 = -1332.1$ kJ/kg. We conclude that the activity $a(T, p, p_o) = \exp\{[\mu(T, p) - \mu(T, p_o)]/RT\} = \exp[(-249.35 + 1332.1)/0.461 \times 473] = 143.4$, the fugacity $\pi(T, p) = \pi(T, p_o)a(T, p, p_o) = 10 \times 143.4 = 1434$ kPa, and the fugacity coefficient $\phi = \pi(T, p)/p = 1434/1554 = 0.92$.

20.6   Effect of Pressure on the Fugacity of a Liquid

At a temperature T lower than the critical value, we recall from Section 19.5 that the form of aggregation of a one-constituent simple system is liquid or solid if the pressure $p > p_{sat}(T)$, and is vapor if $p < p_{sat}(T)$, where $p_{sat}(T)$ is the saturation pressure at temperature T. As an application of the ideas introduced in this chapter, we estimate the fugacity of a liquid at temperature T and pressure $p > p_{sat}(T)$. In the liquid form of aggregation we assume that the simple system behaves as an ideal incompressible fluid for any pressure $p' > p_{sat}(T)$, and in the vapor form as an ideal gas for any pressure $p' < p_{sat}(T)$. We use some of the results of this section in the discussion of liquid–vapor two-phase states of multiconstituent systems.

Equation (20.52) for $p > p_{sat}(T)$ and $p_o < p_{sat}(T)$ may be written as

$$\pi_f(T, p) = \pi_g(T, p_o) \exp\left[\frac{\mu_f(T, p) - \mu_g(T, p_o)}{RT}\right] \tag{20.55}$$

where we use the subscripts f and g as reminders of whether the form of aggregation is liquid or vapor, respectively. For ideal-gas behavior, the relation $\mu = h - Ts$ [equation (17.23)] combined with equations (20.10) and (20.12a) yields

$$\mu_g(T, p_o) = \mu_g(T, p') + RT \ln \frac{p_o}{p'} \tag{20.56}$$

For ideal incompressible fluid behavior, the relation $\mu = h - Ts$ combined with (20.45) and (20.46) yields

$$\mu_f(T, p) = \mu_f(T, p') + v_f(p - p') \tag{20.57}$$

Subtracting equation (20.56) from equation (20.57), and selecting $p' = p_{sat}(T)$ so that $\mu_f(T, p') = \mu_g(T, p')$, we find that

$$\mu_f(T, p) - \mu_g(T, p_o) = v_f\left[p - p_{sat}(T)\right] + RT \ln \frac{p_{sat}(T)}{p_o} \tag{20.58}$$

Moreover, using equation (20.58) and $\pi_g(T, p_o) = p_o$ (ideal-gas behavior) in equation (20.55), we have

$$\pi_f(T, p) = p_{sat}(T) \exp\left\{\frac{v_f}{RT}\left[p_f - p_{sat}(T)\right]\right\} \tag{20.59}$$

The term

$$\exp\left\{\frac{v_f}{RT}\left[p_f - p_{sat}(T)\right]\right\} \approx 1 + \frac{v_f}{RT}\left[p_f - p_{sat}(T)\right] \tag{20.60}$$

is called the *Poynting correction*.

For ordinary substances the Poynting correction is close to unity, and hence the fugacity of the liquid is close to $p_{sat}(T)$ and independent of pressure. For example, for water at $T = 303$ K $= 30°$C, $p_{sat}(T) = 4.246$ kPa, $v_f = 10^{-3}$ m^3/kg, $R = 0.461$ kJ/kg K, $v_f p_{sat}(T)/RT = 3 \times 10^{-5}$, and we can verify that the value of π_f differs very little from that of $p_{sat}(T)$ even if p is 1000 times higher than $p_{sat}(T)$.

At temperatures close to the critical temperature, the liquid ceases to behave as incompressible and the vapor ceases to behave as an ideal gas. Then the difference between π_f and $p_{sat}(T)$ can be appreciable. To see this, we integrate the general relation $(\partial\mu/\partial p)_T = v$ for both the vapor phase and the liquid phase, and find that

$$\mu_g(T,p_o) = \mu_g(T,p_{sat}(T)) + \int_{p_{sat}(T)}^{p_o} v_g(T,p')\,dp' \tag{20.61}$$

$$\mu_f(T,p) = \mu_f(T,p_{sat}(T)) + \int_{p_{sat}(T)}^{p} v_f(T,p')\,dp' \tag{20.62}$$

Substituting these equations into (20.55), we find the following equivalent expressions for the fugacity of the liquid:

$$\pi_f(T,p) = \pi_g(T,p_o)\exp\left[\frac{1}{RT}\int_{p_o}^{p_{sat}(T)} v_g(T,p')\,dp'\right]\exp\left[\frac{1}{RT}\int_{p_{sat}(T)}^{p} v_f(T,p')\,dp'\right]$$

$$= \pi_g(T,p_{sat}(T))\exp\left[\frac{1}{RT}\int_{p_{sat}(T)}^{p} v_f(T,p')\,dp'\right]$$

$$= p_{sat}(T)\,\phi_g(T,p_{sat}(T))\exp\left[\frac{1}{RT}\int_{p_{sat}(T)}^{p} v_f(T,p')\,dp'\right] \tag{20.63}$$

where in writing the second of equations (20.63) we use the definition of fugacity for saturated vapor, and in the third equation we use the fugacity coefficient of the saturated vapor. We see that the fugacity of the liquid is given, in general, by $p_{sat}(T)$ times two correction terms, one being the fugacity coefficient of saturated vapor, the second being the exponential term that we still call the Poynting correction. For relatively small pressure deviations from $p_{sat}(T)$, equation (20.63) can be approximated by

$$\pi_f(T,p) \approx p_{sat}(T)\,\phi_g(T,p_{sat}(T))\left\{1 + \frac{v_f(T)\,[p - p_{sat}(T)]}{RT}\right\} \tag{20.64}$$

where $v_f(T)$ is the specific volume of saturated liquid at temperature T. When the Poynting correction can be neglected, (20.63) becomes

$$\pi_f(T,p) \approx p_{sat}(T)\,\phi_g(T,p_{sat}(T)) = \pi_g(T,p_{sat}(T)) \tag{20.65}$$

and we conclude that the fugacity of the liquid is independent of pressure and is close to the fugacity of saturated vapor at $p_{sat}(T)$.

For water at $T = 473$ K $= 200°C$, $p_{sat}(T) = 1.554$ MPa, $v_f(T) = 1.1565 \times 10^{-3}$ m^3/kg, $\phi_g = 0.92$ (Example 20.10), $v_f/RT = 0.0053$ /MPa, the Poynting correction is negligible, and $\pi_f \approx \pi_g = 1.554 \times 0.92 = 1.43$ MPa, that is, the deviation from $p_{sat}(T)$ is entirely due to the nonideality of saturated vapor. For relatively small pressure deviations from the saturation pressure, the Poynting correction can in general be neglected even in the vicinity of the critical temperature. For example, for water at $T = 643$ K $= 370°C$, $p_{sat}(T) = 21.03$ MPa, $v_f(T) = 2.213 \times 10^{-3}$ m^3/kg, $v_f/RT = 0.0075$ /MPa, and the Poynting correction is negligible for $p - p_{sat}(T)$ not much larger than a few megapascal. Following the procedure outlined in Example 20.10, we find that $\phi_g = 0.663$ and, therefore, $\pi_f \approx \pi_g = 21.03 \times 0.663 = 13.94$.

As an example of application of the results of this section, we consider the conditions of mutual stable equilibrium between a liquid and a vapor across a curved surface. We consider the system in Figure 20.4 consisting of H_2O confined in two variable-volume cylinders interconnected by means of a capillary tube, that is, a tube of very small diameter, and we study a two-phase state of the system with the following features: (1) the H_2O is in the vapor phase in one cylinder and in the liquid phase in the other cylinder; and (2) the surface of separation between the liquid and the vapor is inside

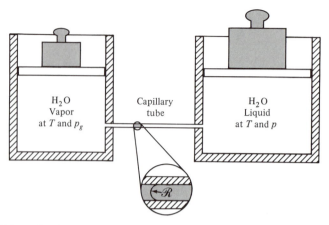

Figure 20.4   Schematic of a system consisting of H_2O confined in two variable-volume cylinders interconnected by means of a capillary tube, in a two-phase state in which the vapor phase is all in one cylinder, the liquid phase is all in the other cylinder, and the surface of separation between the liquid and the vapor phase is inside the capillary tube.

the capillary tube. Because of surface effects, within the capillary tube the surface of separation between the liquid and the vapor is curved and is capable of sustaining a difference in pressure between the liquid side and the vapor side. For the purposes of the present example, we assume that, when the liquid and the vapor are in mutual stable equilibrium, the relation between the liquid pressure p in one cylinder, the vapor pressure p_g in the other cylinder, and the radius of curvature $\mathcal{R}$ of the separating surface in the capillary tube is

$$p - p_g = \frac{2\sigma}{\mathcal{R}} \tag{20.66}$$

where σ is a known constant.

Because fugacity is a single-valued function of the chemical potential, the chemical potential equality, $\mu_g(T, p_g) = \mu_f(T, p)$, which is necessary for the liquid and the vapor phases to be in mutual stable equilibrium, implies equality of fugacities of the two phases, $\pi_g(T, p_g) = \pi_f(T, p)$. Hence, using (20.63), we find that

$$\pi_g(T, p_g) = \pi_f(T, p) = \pi_g(T, p_{sat}(T)) \exp\left[\frac{1}{RT} \int_{p_{sat}(T)}^{p} v_f(T, p')\, dp'\right] \tag{20.67}$$

or, neglecting the Poynting correction,

$$\pi_g(T, p_g) \approx \pi_g(T, p_{sat}(T)) \tag{20.68}$$

which, in turn, implies that

$$p_g \approx p_{sat}(T) \tag{20.69}$$

For the system in Figure 20.4, we conclude that the pressure of the vapor phase is approximately independent of the pressure of the liquid phase and equal to the saturation pressure corresponding to the given temperature T. Moreover, combining (20.66) and (20.69), we find that $p = p_{sat}(T) + 2\sigma/\mathcal{R}$. For a given T, we conclude that the surface of separation between the liquid and the vapor phase is inside the capillary tube only if the pressure p of the liquid phase is adjusted so as to match the characteristics σ and $\mathcal{R}$ of the capillary tube and the surface of separation.

Problems

20.1 For certain conditions of interest, the energy relation of 1 mol of a substance satisfies the expression $u = 6.15 + 6150(\exp\{[s - s_0 - 8.3\ln(v/v_0)]/20.5\} - 1)$, where u is in J/mol, s_0 is a reference specific entropy in J/mol K, and v_0 is a reference volume in m^3.

(a) Does this substance behave as a perfect gas?

(b) Is the given expression of the energy relation valid for all values of u and v?

20.2 (a) Starting from (20.19) and (20.22), derive the fundamental relation for perfect-gas behavior, that is, a relation of the form $s = s(u, v; c_v, R, T_0, p_0, u_0, s_0)$.

(b) Sketch $(u - u_0)/c_v T_0$ versus $(s - s_0)/c_v$ for different values of v/v_0.

(c) For a fixed v, select a stable equilibrium state with temperature T^*, and sketch qualitatively on your graph the shape of the fundamental relation for temperatures larger than T^*, and behavior that is ideal but not perfect.

20.3 (a) Starting from (20.44) and (20.45), derive the fundamental relation for perfect incompressible behavior, that is, derive a relation of the form $s = s(u; c, T_0, u_0, s_0)$.

(b) Sketch $(u - u_0)/cT_0$ versus $(s - s_0)/c$.

(c) Select a stable equilibrium state with temperature T^* and sketch qualitatively on your graph the shape of the fundamental relation for temperatures higher than T^*, and behavior that is ideal but not perfect.

20.4 An inventor has proposed a new device for powering automobiles that uses compressed air as a source of motive power. The compressed air is stored in a tank on the vehicle at environmental temperature $T_0 = 25°C$ and pressure of $p_0 = 100$ atm, and behaves as a perfect gas.

The gasoline tank of a typical automobile contains 60 kg of gasoline, which has an available energy relative to the atmosphere of about 40 MJ/kg. Only about 25% of the available energy is converted into motive energy in a conventional gasoline-fueled automobile engine. Gasoline has a density of 0.8 g/cm^3.

(a) If the compressed air is 100% effective, write an expression for the mass of air required to replace a conventional tank filled with gasoline.

(b) Find the ratio of the volume of the compressed air in part (a) to the volume of the corresponding gasoline tank.

20.5 Each of three identical containers A, B, and C is filled with 1 kg of water. Initially, the water in each container is at temperature T_1^A, T_1^B, and T_1^C, respectively, and can be treated as an incompressible fluid. Work producing machinery is connected to the three containers and no other systems or reservoirs.

(a) For the given initial temperatures, find an expression for the largest work that the machinery could produce.

(b) Find an expression for the final temperature of the water in the three containers when no more work can be done, and the work actually produced—starting from the given initial conditions—is only one-half of the answer to part (a).

20.6 One kilogram of H_2O is contained in a cylinder with a piston exerting a constant pressure $p = 700$ kPa. Initially, the volume of the container is $V_1 = 0.25$ m^3. A solid block of mass $m_s = 150$ kg and specific heat $c_s = 0.13$ kJ/kg K is initially at temperature $T_1^s = 25°C$. A cyclic engine is connected between the cylinder and the block and does shaft work on a weight while the H_2O is converted into solid ice at $T_2 = -20°C$ and $p_2 = p = 700$ kPa.

(a) Find the values of the specific energy, specific enthalpy, specific entropy and specific volume of the H_2O at its initial and final states.

(b) If the process is optimal, what is the final temperature of the solid block? Is the answer a maximum or a minimum?

(c) What is the optimum work done by the cyclic engine for the specified change of state of the H_2O?

20.7 System A consists of 1 mol of H_2 that is confined in a container with variable volume, and that behaves as a perfect gas with $c_p = 28.8$ J/mol K and $c_v = 20.6$ J/mol K. For specific volume $v = v_0$, the system is in state A_0 with temperature $T_0 = 300$ K, specific internal energy $u_0 = 6.15$ kJ/mol, and specific entropy $s_0 = 10$ J/mol K.

(a) For specific volume $v_1 = v_0/2$, find the values of the specific internal energy u_1, and specific entropy s_1 corresponding to a state A_1 at temperature $T_0 = 300$ K.

(b) Starting from state A_0, the system receives energy $\Delta u = 24$ kJ/mol, entropy $\Delta s = 6$ J/mol K, and decreases its volume from v_0 to $v_1 = v_0/2$. If the process is reversible, what kind of state is the final state of the system?

(c) What is the available energy of the final state in part (b) with respect to a reservoir at $T_0 = 300$ K, and for a system volume v_1?

(d) If three-fourths of the energy exchange in part (b) is heat, what is the temperature of the energy source?

(e) If the process in part (b) is irreversible, what is the largest amount of entropy that can be generated in the H_2 system?

20.8 An ice-making factory next to a steelmaking plant is considering to satisfy its energy needs by using steel billets that in an intermediate stage of the steelmaking process have to be cooled from $T_1^b = 1000°C$ to $T_2^b = 200°C$. Each steel billet has mass $m_b = 15,000$ kg and specific heat $c_b = 0.4$ J/kg K. The ice-making factory processes a stream of 1 kg/s of water at the environmental temperature $T_0^w = 25°C$ and converts it into ice cubes at $T_i^w = 0°C$. The enthalpy of melting of ice at 0°C is 333.7 kJ/kg, and the specific heat of liquid water is 4.2 kJ/kg K.

(a) What is the largest work that could be produced by machinery that cools one steel billet while interacting with the environment?

(b) What is the smallest work rate required to operate machinery that converts the stream of water into ice cubes while interacting with the environment?

(c) What is the largest amount of water that could be converted into ice by machinery that cools one billet and interacts with the environment?

(d) What type of machinery would you use for the recovery of energy from the billets and the making of ice?

20.9 One ton of metal parts (1000 kg, $c = 0.13$ kJ/kg K) must be cooled from a high temperature $T_1 = 1500$ K to the low temperature $T_2 = 280$ K relatively quickly. Three possible methods are considered: (1) cooling the parts by melting ice in a mixture of ice and water at atmospheric pressure; (2) cooling the parts by evaporating saturated liquid Freon 12 at $-18°C$; and (3) cooling the parts by interaction with liquid nitrogen at atmospheric pressure.

(a) For each of the three cooling methods, calculate the entropy generated by irreversibility.

(b) With respect to an environment at 280 K, what is the largest work that could have been done while reducing the temperature of the metal parts from T_1 to T_2?

(c) How would you compare your answers in part (a) with your result in part (b), and how do you justify the differences you observe?

20.10 System A consists of 4.5 kg of air contained in a cylinder with a frictionless piston of cross-sectional area 0.01 m^2. The air behaves as a perfect gas with $R = 287$ J/kg K and $c_p = 1$ kJ/kg K, and is initially at temperature $T_1^A = 4°C$. The piston is loaded with a weight of 700 kg. System B consists of a metal block. Its heat capacity $C_B = 10$ kJ/K and its initial temperature $T_1^B = 120°C$.

(a) Upon interacting with each other and the weight, the air and the metal block reach mutual stable equilibrium. How much energy is transferred from the air to the block? And how much entropy is generated by irreversibility?

(b) Is the interaction between A and B in part (a) work, heat, or other type of nonwork?

(c) If the air with the weighted piston and the block are connected with cyclic machinery and no other systems in the environment, what is the largest work that can be done by the machinery?

20.11 A 0.3-m^3 rigid and well-insulated vessel contains air at a pressure of 1 atm and a temperature of 25°C. The air is stirred by a paddle wheel until the pressure is 4 atm. The paddle wheel is actuated by an electric motor. Assume air to behave as a perfect gas with constant specific heats $c_p = 1$ kJ/kg K and $c_v = 0.73$ kJ/kg K.

   (a) Find the work done by the motor.

   (b) Find the available energy of the air with respect to a reservoir at 25°C.

   (c) Why are your answers in parts (a) and (b) different?

20.12 A cylinder, closed at both ends, is initially separated into two equal compartments of volume 0.03 m^3 by means of a frictionless, perfectly conducting piston. One compartment contains air at 15°C, and the other contains ammonia at 15°C and a quality of 0.80. The two compartments are in partial mutual stable equilibrium. The cylinder experiences a heat interaction just sufficient to evaporate all of the ammonia. Assume that the air can be modeled as a perfect gas with $c_v = 0.714$ kJ/kg K and $R = 0.286$ kJ/kg K.

   (a) Find the volume occupied by the ammonia vapor in the initial state.

   (b) Find the final pressure and temperature of the contents of the cylinder.

   (c) Find the total heat to the contents of the cylinder.

20.13 A cylindrical enclosure is subdivided into two parts A and B by a perfectly conducting piston. Side A contains 1 m^3 of a perfect gas ($\gamma = 1.4$) at a pressure $p_A = 10$ atm, and temperature $T_A = 300$ K. Side B contains 1 m^3 of the same perfect gas at $p_B = 1$ atm, $T_B = 300$ K. Initially, the piston is clamped in the center of the enclosure. Then the piston is released and the system comes to a stable equilibrium state.

   (a) What is the final volume on each side of the piston?

   (b) What are the temperature and the pressure of the gas?

   (c) What is the entropy change for the entire system?

   (d) Is the process reversible?

20.14 A rigid metal tank of volume 1 m^3 and very large heat capacity connects to the atmosphere through a valve. Initially, the valve is closed and the tank is empty and at the temperature of the atmosphere. The valve is then opened and air flows into the tank. Two phases of this airflow are observed. In the first phase, which lasts for a few seconds, the air flows rapidly into the tank at a decreasing rate as the pressure in the tank increases from zero to a constant value. In the second phase, which lasts about 1 h, air continues to flow into the tank at a very low rate while the pressure remains constant. Eventually, this second phase ends when the airflow goes to zero. Atmospheric conditions are $p = 10^5$ N/m^2 and $T = 300$ K. Devise a suitable model to answer the following questions.

   (a) What are the pressure and the temperature of the air in the tank at the end of the first phase?

   (b) What are the pressure and the temperature of the air in the tank at the end of the second phase?

   (c) How much entropy is generated by irreversibility during the first phase of the process, and how much during the second phase?

20.15 A compression machine used to produce high-temperature gases consists of a perfectly insulated, leakless vertical container closed by a frictionless heavy piston of mass M and surface area $A = 0.001$ m^2. Initially, the piston is supported by a pin at an elevation $z_1 = 3$ m, where z_1 is measured from the bottom of the cylinder. The gas is perfect with $R = 0.28$ kJ/kg K and $c_v = 0.71$ kJ/kg K, and at its initial state the temperature $T_1 = 300$ K, the pressure is atmospheric at $p_1 = p_0 = 0.1$ MPa, and the mass is m. After removal of the support pin, the gas reaches a stable equilibrium state at the pressure exerted by the piston plus the atmosphere and the piston reaches elevation z_2.

(a) Find an expression for the mass M as a function of z_1, z_2, T_1, T_2, p_0, A, c_v, R, and the gravity constant g.

(b) Find the smallest piston mass M_{min} required to produce a final gas temperature $T_2 = 1000$ K.

(c) What is the final piston height z_2 corresponding to your answer in part (b)?

20.16 The system shown in Figure P20.16 consists of a heat source A at temperature $T_A = 727°C$, a cyclic device X, and two rigid containers B and C, each filled with 10 kg of a perfect gas ($c_p = 1.4$ kJ/kg K, $c_v = 1$ kJ/kg K). Initially, the gas in each container is at a temperature $T_0 = 40°C$.

(a) In one process, the cyclic device transfers 500 kJ to the gas in container B by means of a work interaction, and 500 kJ to the gas in container C by means of a nonwork interaction. If the process is reversible, what kind of states are the final states of B and C?

(b) In another process, the energy transfers are as in part (a) but irreversibility occurs in B and C. What is the largest amount of entropy that can be generated by irreversibility?

(c) For 1000 kJ taken out of A, what is the largest work that can be done on B?

(d) For a 1000 kJ taken out of A, what is the largest work that can be produced with respect to an environment at 40°C?

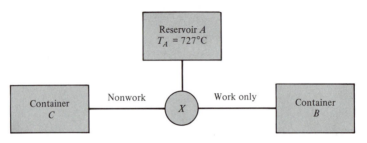

Figure P20.16

20.17 A steel billet A (mass $m_A = 15 \times 10^3$ kg, specific heat $c_A = 0.4$ kJ/kg K) initially at 1000°C is to be cooled down to 200°C in an intermediate phase of a steelmaking process. It is proposed that the cooling be done using appropriate cyclic machinery X operating between the billet A and a nearby water pool B. The water pool (mass $m_B = 10^5$ kg, specific heat $c_B = 4.2$ kJ/kg K) is initially at 40°C. Changes in volumes of both A and B are negligible.

(a) What is the largest work that the cyclic machinery X operating only between A and B can produce?

(b) What is the final temperature of the water?

(c) Is the temperature calculated in part (b) the largest or the smallest that could be attained by the water in the proposed configuration?

(d) If a reservoir at 20°C were available and machinery X could interact only with the billet (from 1000 to 200°C), the water and the reservoir, write the equation that defines the largest temperature that could be attained by the water.

20.18 For methane gas, use data from Table 20.1 to answer the following questions.

(a) If the enthalpy at 298.15 K is $-17,890$ kcal/kmol, what is the enthalpy at 600 K?

(b) What is the specific heat under constant volume at 600 K?

(c) What is the internal energy at 600 K with respect to the same reference state as that used for enthalpy in part (a)?

20.19 A system A consists of 100 kg of a metal at temperature 400 K. Its specific heat is 2 kJ/kg K. A system B consists of 100 kg of a perfect gas at a temperature of 500 K and pressure 0.1 atm.

Its specific heat at constant volume is 1 kJ/kg K, and its gas constant 0.4 kJ/kg K. Both A and B are in an environment at 300 K and 1 atm.

(a) Which of the two systems has the greater energy relative to the state in which it is in mutual stable equilibrium with the environment?

(b) Which of the two systems can do the larger work?

(c) What devices would you use to achieve the result in part (b) as closely as practically possible?

20.20 Consider a spherical surface separating a liquid from its vapor and assume that the liquid behavior is incompressible and the vapor behavior is ideal-gas. The surface consists of the same constituent as the liquid and the vapor and sustains a difference between the pressure p_i in the region inside the sphere and the pressure p_e outside given by the relation (Problem 17.4)

$$p_i - p_e = \frac{2\sigma}{r}$$

where r is the radius of the sphere, and σ the surface tension.

(a) Show that for a drop of liquid in mutual stable equilibrium with the surrounding vapor, the vapor pressure p_g is related to temperature by means of the *Kelvin–Helmholtz relation*

$$p_g = p_{sat}(T) \exp\left(\frac{2\sigma v_f}{rRT}\right)$$

(b) What is the form of the Kelvin–Helmholtz relation for a bubble of vapor in mutual stable equilibrium with the surrounding liquid?

21   Equations of State

Outside the ranges of relatively low pressures and relatively high temperatures in which any substance behaves as an ideal gas, that is, the equation of state is $pv = RT$, the interrelations between stable equilibrium properties depend not only on the private degrees of freedom of each molecule but also on the intermolecular forces. As a result of these additional dependences, the equation of state is much more complicated.

In this chapter we discuss equations of state that apply when a substance does not behave as an ideal gas.

21.1   Compressibility Factor

To illustrate the departure of a pure substance from the ideal-gas behavior, we consider the *compressibility factor* $\mathcal{Z}$ defined by the relation

$$\mathcal{Z} = \frac{pv}{RT} \tag{21.1}$$

and examine its deviations from unity. In general, the compressibility factor is a function of only two of the variables T, p, v because the third is determined from these two by the equation of state. Experimental results on the compressibility factor of nitrogen versus pressure at different temperatures are shown in Figure 21.1. We see that, at relatively low pressures and relatively high temperatures, $\mathcal{Z}$ approaches unity, that is, nitrogen behaves as an ideal gas. For other values of p and T, however, the behavior of nitrogen deviates appreciably from ideal.

Instead of the variables T, p, v, it is customary to use the *reduced temperature* $T_R = T/T_c$, the *reduced pressure* $p_R = p/p_c$, and the *reduced volume* $v_R = v/v_c$, and graph traces of $\mathcal{Z}$ either at constant reduced temperature or at constant reduced volume versus reduced pressure, where T_c, p_c, and v_c are the critical values of temperature, pressure, and specific volume of the substance, respectively. For a number of substances critical values are listed in Table 19.1.

Experimental p–v–T data on several substances are shown in Figure 21.2, where the compressibility factor is graphed as a function of T_R and p_R. It is noteworthy that the data for all substances fall on a single set of curves. Such experimental evidence is the basis of the development of the generalized compressibility charts discussed in Section 21.6. In addition, the experimental data confirm the general statement that at relatively low pressures, ($p_R \ll 1$), and relatively high temperatures ($T_R \gg 1$), the compressibility factor of any substance is equal to about unity, that is, the substance behaves as an ideal gas, whereas outside these ranges $\mathcal{Z}$ differs from unity, and the equation of state is not as simple as that of an ideal gas.

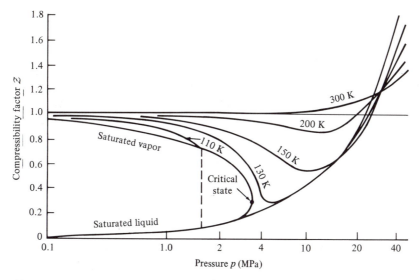

Figure 21.1 Compressibility of nitrogen (from G. J. Van Wylen and R. E. Sonntag, *Fundamentals of Classical Thermodynamics*, 2nd ed., Wiley, New York, 1976).

As stated earlier, it has not been possible to write an explicit p–v–T analytical relation for all the stable equilibrium states of a pure substance. Instead, various approximate equations of state have been proposed. A few of these equations are discussed in the sections that follow.

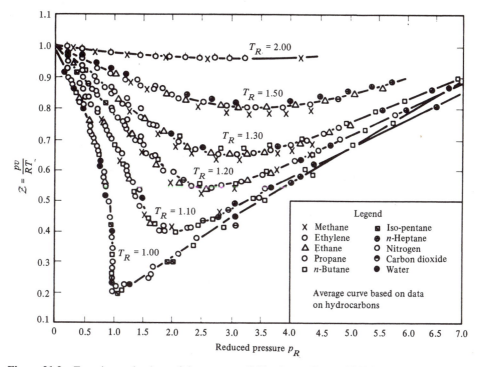

Figure 21.2 Experimental values of the compressibility factor $\mathcal{Z} = pv/RT$ for various substances graphed as a function of reduced pressure p_R and reduced temperature T_R. For all substances, the data fall on a single set of curves [from Gour-Jen Su, *Modified Law of Corresponding States*, *Ind. Eng. Chem.*, **38**, 803(1946)].

21.2   Van der Waals Equation

In 1873, Johannes van der Waals proposed an equation of state that represents in a qualitative way both liquid and vapor homogeneous states. For example, it yields the same value of the compressibility factor for all substances at the critical point, $Z_c = p_c v_c / RT_c = 0.375$. By comparison with the results listed in Table 19.1, we see that this value differs from the experimental values obtained for various substances but is of the same order of magnitude.

The van der Waals equation[1] is

$$p = \frac{RT}{v - b} - \frac{a}{v^2} \tag{21.2a}$$

or, equivalently,

$$\left(p + \frac{a}{v^2} \right)(v - b) = RT \tag{21.2b}$$

where R is the gas constant, and a and b are constants for each substance. Typical isotherms described by the van der Waals equation are shown in Figure 21.3.

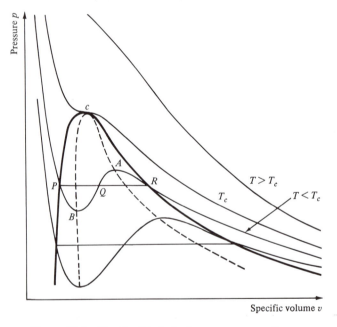

Figure 21.3   Van der Waals isotherms on a p–v diagram.

Below the critical temperature T_c each isotherm has a local maximum and a local minimum. We find each pair of extreme points by equating the derivative $(\partial p / \partial v)_T$ to zero, namely,

$$\left(\frac{\partial p}{\partial v} \right)_T = -\frac{RT}{(v - b)^2} + \frac{2a}{v^3} = 0 \tag{21.3}$$

---

[1]The equation is named in honor of Johannes Diderik van der Waals (1837–1923), Dutch physicist, who conceived it, and for which he received the Nobel Prize for physics in 1910.

and solving (21.2) and (21.3) for the values of p and v in terms of T. Moreover, we find the locus of the maxima and the minima, shown as the dashed curve in Figure 21.3, by eliminating T from (21.2) and (21.3). The result is the relation

$$p = \frac{a\,(v - 2\,b)}{v^3} \tag{21.4}$$

The critical isotherm is tangent to the dashed curve at the highest point, that is, the critical point, where the maximum and the minimum points coincide in a point of inflection. We find the location of this point as a solution of the set of equations consisting of (21.2) and the results of equating to zero either $(\partial p/\partial v)_T$ and $(\partial^2 p/\partial v^2)_T$ with $p = p(v, T)$ given by (21.2) or dp/dv with $p = p(v)$ given by (21.4). From the latter choice, we find that

$$\frac{dp}{dv} = \frac{6ab - 2av}{v^4} = 0 \quad \text{or} \quad v_c = 3b \tag{21.5}$$

and upon substituting this result in (21.2) and (21.4), that

$$v_c = 3\,b \qquad p_c = \frac{a}{27\,b^2} \qquad T_c = \frac{8\,a}{27R\,b} \tag{21.6}$$

In particular, we find that the compressibility factor $\mathcal{Z}_c$ at the critical point is independent of the magnitudes of the constants a and b, that is,

$$\mathcal{Z}_c = \frac{p_c v_c}{RT_c} = \frac{(a/27b^2)(3b)}{8a/27b} = \frac{3}{8} = 0.375 \tag{21.7}$$

Upon rewriting equation (21.2) as

$$\mathcal{Z} = \frac{3\,v_R}{3\,v_R - 1} - \frac{9}{8\,T_R v_R} \tag{21.8}$$

we observe that, according to the van der Waals equation, the compressibility factor $\mathcal{Z}$ is independent of the values of the constants a and b, that is, for all substances it is the same function of the reduced volume v_R and the reduced temperature T_R.

The experimental values of $\mathcal{Z}_c$ listed in Table 21.1 for various substances show that $\mathcal{Z}_c$ is not in good accord with (21.7) and not the same for all substances. Nevertheless, for problems in which low accuracy in the values of the specific volume is acceptable, the van der Waals equation can be adapted to yield accurate results on most other properties by disregarding the first of equations (21.6) and using the second and the third to determine the constants a and b from the experimental values of T_c, p_c, and v_c of each given substance, that is, using the values $a = 27R^2T_c^2/64p_c$ and $b = RT_c/8p_c$ as listed in Table 21.1.

To describe the saturation dome that includes the two-phase region of liquid–vapor states, the van der Waals equation must be completed by the conditions of mutual stable equilibrium for the coexisting phases. The conditions of temperature and pressure equality imply that in the two-phase region any isotherm becomes horizontal on a p–v diagram. The condition of chemical potential equality implies that the saturated liquid state P (Figure 21.3) and the saturated vapor state R are determined by drawing a horizontal line PR such that the area PBQ is equal to the area QAR. Thus a vapor dome is obtained. This is known as the *Maxwell graphic rule*.

At any temperature $T < T_c$, the portion of the isotherm between the saturated liquid state P and the minimum point B on the dashed curve in Figure 21.3 is assumed to represent metastable-equilibrium liquid states, and the portion of the isotherm between the maximum point A on the dashed curve and the vapor saturation state R is assumed to represent metastable-equilibrium vapor states.

TABLE 21.1.   Constants for the van der Waals and Dieterici equations of state for various substances, determined from the critical data in Table 19.1 and equations (21.6) and (21.10).

Substance	Formula	$\mathcal{Z}_c = \dfrac{p_c v_c}{R T_c}$	$a = \dfrac{27 R^2 T_c^2}{64 p_c}$ $\dfrac{\text{MPa m}^6}{\text{kmol}^2}$	$b = \dfrac{R T_c}{8 p_c}$ $\dfrac{\text{m}^3}{\text{kmol}}$	$a = \dfrac{4 R^2 T_c^2}{e^2 p_c}$ $\dfrac{\text{MPa m}^6}{\text{kmol}^2}$	$b = \dfrac{R T_c}{e^2 p_c}$ $\dfrac{\text{m}^3}{\text{kmol}}$
Acetic acid	$C_2H_4O_2$	0.200	1.7812	0.1068	2.2856	0.1156
Acetone	CH_3COCH_3	0.233	1.6017	0.1123	2.0552	0.1216
Acetylene	C_2H_2	0.271	0.4515	0.0522	0.5794	0.0565
Ammonia	NH_3	0.242	0.4255	0.0374	0.5460	0.0405
Argon	Ar	0.291	0.1361	0.0322	0.1746	0.0348
Benzene	C_6H_6	0.271	1.8831	0.1194	2.4164	0.1293
Carbon dioxide	CO_2	0.274	0.3659	0.0429	0.4695	0.0464
Carbon monoxide	CO	0.294	0.1474	0.0395	0.1891	0.0428
Chlorine	Cl_2	0.275	0.6586	0.0563	0.8452	0.0609
Chloroform	$CHCl_3$	0.293	1.5338	0.1019	1.9682	0.1103
Ethane	C_2H_6	0.285	0.5570	0.0650	0.7148	0.0704
Ethanol	C_2H_5OH	0.248	1.2176	0.0841	1.5623	0.0910
Ethylene	C_2H_4	0.277	0.4619	0.0583	0.5927	0.0631
Fluorine	F_2	0.288	0.1164	0.0287	0.1494	0.0311
Freon 12	CCl_2F_2	0.280	1.0484	0.0970	1.3453	0.1051
Freon 13	$CClF_3$	0.281	0.6784	0.0801	0.8705	0.0867
Freon 21	$CHCl_2F$	0.271	1.1512	0.0908	1.4771	0.0984
Freon 22	$CHClF_2$	0.267	0.7992	0.0771	1.0255	0.0835
Freon 114	$C_2Cl_2F_4$	0.274	1.5688	0.1335	2.0130	0.1445
Genetron 100	CH_3CHF_2	0.253	0.9690	0.0893	1.2434	0.0967
Genetron 101	CH_3CClF_2	0.279	1.1901	0.1034	1.5271	0.1119
Hydrogen	H_2	0.305	0.0248	0.0266	0.0318	0.0288
Hydrogen chloride	HCl	0.249	0.3699	0.0406	0.4746	0.0440
Isooctane	C_8H_{18}	0.266	3.3698	0.2280	4.3240	0.2390
Methane	CH_4	0.287	0.2303	0.0431	0.2956	0.0466
Methanol	CH_3OH	0.224	0.9467	0.0658	1.2148	0.0713
Methylene chloride	CH_2Cl_2	0.277	1.2479	0.0872	1.6013	0.0944
Naphthalene	$C_{10}H_8$	0.267	4.0309	0.1919	5.1724	0.2078
Neon	Ne	0.311	0.0209	0.0167	0.0268	0.0181
Nitric oxide	NO	0.251	0.1457	0.0289	0.1870	0.0312
Nitrogen	N_2	0.289	0.1369	0.0386	0.1756	0.0418
Nitrogen dioxide	NO_2	0.480	0.5357	0.0443	0.6875	0.0479
Nitrous oxide	N_2O	0.274	0.3859	0.0444	0.4952	0.0481
n-Octane	C_8H_{18}	0.258	3.8014	0.2382	4.8779	0.2579
Oxygen	O_2	0.288	0.1382	0.0318	0.1773	0.0345
Ozone	O_3	0.228	0.3565	0.0487	0.4575	0.0527
Propane	C_3H_8	0.280	0.9395	0.0905	1.2056	0.0980
Propylene	CH_2CHCH_3	0.276	0.8410	0.0821	1.0792	0.0889
Water	H_2O	0.229	0.5543	0.0305	0.7113	0.0330

21.3   Dieterici Equation

The van der Waals equation gives a value for $\mathcal{Z}_c$ at the critical point which is higher than that of any substance. Largely to overcome this discrepancy, Dieterici[2] in 1899 proposed the equation of state·

---

[2]C. H. Dieterici, *Ann. Phys. Chem.*, *11*, 700 (1899).

$$p = \frac{RT}{v-b}\, e^{-a/vRT} \tag{21.9}$$

where, for a given substance, a and b are constants different from the van der Waals constants. For the Dieterici equation, the relations between the constants a and b and the critical constants are

$$v_c = 2b \qquad p_c = \frac{a}{4e^2 b^2} \qquad T_c = \frac{a}{4Rb} \tag{21.10}$$

Consequently,

$$Z_c = \frac{p_c v_c}{RT_c} = \frac{(a/4e^2 b^2)(3b)}{a/4b} = \frac{2}{e^2} \approx 0.27 \tag{21.11}$$

a value that is nearly the average of values of pure substances.

Also, according to the Dieterici equation, the compressibility factor is a universal function of reduced volume and reduced temperature; that is,

$$Z = \frac{2v_R}{2v_R - 1}\, \exp\!\left(-\frac{2}{v_R T_R}\right) \tag{21.12}$$

The experimental values of Z_c listed in Table 21.1 are not the same for all substances. Nevertheless, also the Dieterici equation can be adapted to yield accurate results by disregarding the first of equations (21.10) and using the second and the third to determine the constants a and b from the experimental values of T_c, p_c, and v_c of each given substance, that is, using the values $a = 4R^2 T_c^2/e^2 p_c$ and $b = RT_c/e^2 p_c$ as listed in Table 21.1.

The isotherms of the Dieterici equation resemble those of the van der Waals equation in most respects. However, the Dieterici equation has proved superior to the van der Waals equation, particularly in the neighborhood of the critical point.

21.4 Virial Equations

A general type of equation of state, which can be made to represent almost any substance in a gaseous phase at the expense of an indefinite increase in the number of constants, is the virial type. It may be written as

$$\frac{pv}{RT} = 1 + \frac{B(T)}{v} + \frac{C(T)}{v^2} + \frac{D(T)}{v^3} + \cdots \tag{21.13}$$

where the coefficients $B(T)$, $C(T)$, $D(T)$, ... called the second, third, fourth, ... *virial coefficients*, are functions of temperature only. Substantial progress has been made in determining the forms of the functions $B(T)$ and $C(T)$ for simple molecules by means of models of the structure of the molecule and of the force fields between molecules.

21.5 Beattie–Bridgeman Equation

The Beattie–Bridgeman[3] equation of state is given by the relation

$$p = \frac{RT}{v}\left(1 - \frac{c}{vT^3}\right)\left[1 + \frac{B_o}{v}\left(1 - \frac{b}{v}\right)\right] - \frac{A_o}{v^2}\left(1 - \frac{a}{v}\right) \tag{21.14}$$

---

[3] J. A. Beattie and O. C. Bridgeman, *Proc. Am. Acad. Sci.*, *63*, 229 (1928).

where the five parameters a, b, c, A_o, and B_o are adjustable constants which must be fitted to data for each particular substance.

Equation (21.14) can also be written in virial form as

$$\frac{pv}{RT} = 1 + \frac{B(T)}{v} + \frac{C(T)}{v^2} + \frac{D(T)}{v^3} \tag{21.15}$$

where the virial coefficients are given by $B(T) = B_o - A_o/RT - c/T^3$, $C(T) = -B_o b + A_0 a/RT - B_o c/T^3$, $D(T) = B_o bc/T^3$.

21.6   Benedict–Webb–Rubin Equation

The Benedict–Webb–Rubin[4] equation of state is given by the relation

$$p = \frac{RT}{v} + \left(B_o RT - A_o - \frac{C_o}{T^2} \right) \frac{1}{v^2} + (bRT - A)\frac{1}{v^3} + \frac{a\alpha}{v^6}$$

$$+ \frac{c\left[1 + (\gamma/v^2)\right]}{T^2} \frac{1}{v^3} e^{-\gamma/v^2} \tag{21.16}$$

For a number of substances, values of the constants A_o, B_o, C_o, a, b, c, α, and γ in this equation are listed in Table 21.2.

TABLE 21.2.   Empirical constants for the Benedict–Webb–Rubin equation for various substances.[a]

Gas	A_o	B_o	$C_o \times 10^{-6}$	a	b	$c \times 10^{-6}$	$\alpha \times 10^3$	$\gamma \times 10^2$
Nitrogen	1.19250	0.0458000	0.005889	0.014900	0.0019815	0.000548	0.291545	0.75000
Methane	1.85500	0.0426000	0.022570	0.494000	0.0033800	0.002545	0.124359	0.60000
Ethylene	3.33958	0.0556833	0.131140	0.259000	0.0086000	0.021120	0.178000	0.92300
Propylene	6.11220	0.0850647	0.439182	0.774056	0.0187059	0.102611	0.455696	1.82900
Popane	6.87225	0.0973130	0.508256	0.947700	0.0225000	0.129000	0.607175	2.20000
i-Butane	10.23264	0.137544	0.849943	1.93763	0.0424352	0.286010	1.07408	3.40000
i-Butylene	8.95325	0.116025	0.927280	1.69270	0.0348156	0.274920	0.910889	2.95945
n-Butane	10.0847	0.124361	0.992830	1.88231	0.0399983	0.316400	1.10132	3.40000
i-Pentane	12.7959	0.160053	1.74632	3.75620	0.0668120	0.695000	1.70000	4.63000
n-Pentane	12.1794	0.156751	2.12121	4.07480	0.0668120	0.824170	1.81000	4.75000
n-Hexane	14.4373	0.177813	3.31935	7.11671	0.109131	1.51276	2.81086	6.66849
n-Heptane	17.5206	0.199005	4.74574	10.36475	0.151954	2.470000	4.35611	9.00000

[a] Units: atmospheres, liters, moles, kelvin.

21.7   Principle of Corresponding States

The empirical observation that experimental p–v–T data for all substances fall on a single surface when graphed in terms of either the compressibility factor $\mathcal{Z}$ or the reduced volume v_R versus the reduced temperature T_R and the reduced pressure p_R (Figure 21.2) leads to the conclusion that for all substances to a high degree of approximation, we can assume that

$$\mathcal{Z} = \mathcal{Z}(T_R, p_R) \tag{21.17}$$

or

$$v_R = v_R(T_R, p_R) \tag{21.18}$$

[4]M. Benedict, G. B. Webb, and L. C. Rubin, *Chem. Eng. Prog.*, 47, 419 (1951).

where each of the functions $\mathcal{Z}$ and v_R is a universal function of T_R and p_R, that is, a function that does not change from substance to substance. This conclusion is known as the *principle of corresponding states*.

The principle of corresponding states was originally stated by van der Waals on the basis of equation (21.8). Although it is an approximation and not a necessary consequence of the laws of thermodynamics, it provides a useful guide for estimating values of properties of pure substances in certain limited ranges of states.

On the basis of this empirical principle, generalized compressibility charts have been developed by fitting experimental data, and can be used to provide estimates of p–v–T values for all substances to within a reasonable degree of accuracy. Such charts[5] are shown in Figures 21.4 to 21.6 (see pp. 353–355), for various ranges of values of p_R. If the critical values of temperature, pressure, and specific volume for any one substance are known, the entire equation of state at high temperatures and a broad range of pressures can be found from Figures 21.4 to 21.6 with errors of less than 5%. This accuracy is adequate for a number of engineering calculations.

At low pressures and high temperatures, however, the principle of corresponding states loses its usefulness. To see this clearly, we observe that if $\mathcal{Z}$ were a single function of T_R and p_R for all substances, $p_R v_R / T_R$ would satisfy the relations

$$\mathcal{Z}_R = \frac{p_R v_R}{T_R} = \frac{pv/RT}{p_c v_c/RT_c} = \frac{\mathcal{Z}(T_R, p_R)}{\mathcal{Z}_c} \qquad (21.19)$$

Because $\mathcal{Z}$ tends to unity in the limit of low pressures and high temperatures, equation (21.19) implies that in such a limit $\mathcal{Z}_R$ tends to $1/\mathcal{Z}_c$. But the value of $\mathcal{Z}_c$ varies from substance to substance. For example, from the experimental data listed in Table 19.1 we see that $\mathcal{Z}_c$ ranges from 0.23 to 0.31, and $1/\mathcal{Z}_c$ from 3.2 to 4.3. Therefore, in the limit of very low pressures and very high temperatures, the value of $\mathcal{Z}_R$ is not the same for all substances, and hence the principle of corresponding states ceases to yield accurate results. We reach the same conclusion by expressing v_R in the form

$$v_R = \frac{T_R \mathcal{Z}(T_R, p_R)}{p_R \mathcal{Z}_c} \qquad (21.20)$$

We see that in the limit of low pressures and high temperatures where $\mathcal{Z}(T_R, p_R) \to 1$, $v_R \neq v_R(T_R, p_R)$ because $\mathcal{Z}_c$ does not have the same value for all substances.

Problems

21.1 For a fluid behaving according to the van der Waals equation, prove the following relations:

(a)
$$\alpha_p = \frac{Rv^2(v-b)}{RTv^3 - 2a(v-b)^2}$$

$$= \frac{1}{T} \frac{1 - b/v + a/v^2 p - ab/v^3 p}{1 - a/v^2 p + 2ab/v^3 p}$$

$$\kappa_T = \frac{v^2(v-b)^2}{RTv^3 - 2a(v-b)^2}$$

$$= \frac{1}{p} \frac{1 - b/v}{1 - a/v^2 p + 2ab/v^3 p}$$

---

[5]Prepared by L. C. Nelson and E. F. Obert, *Trans., ASME*, **76**, 1057 (1954).

(b)
$$ds = \frac{c_v}{T}dT + \frac{R}{v-b}dv$$

$$\left(\frac{\partial s}{\partial v}\right)_T = \frac{R}{v-b}$$

$$\left(\frac{\partial u}{\partial v}\right)_T = \frac{a}{v^2}$$

$$du = c_v dT + \frac{a}{v^2}dv$$

21.2 (a) Show that the van der Waals equation may be written as a power series of b/v in the form

$$\frac{pv}{RT} = 1 + \left(b - \frac{a}{RT}\right)\frac{1}{v} + \frac{b^2}{v^2} + \frac{b^3}{v^3} + \cdots$$

(b) Compare this series with the virial form [equation (21.13)], and find expressions for the virial coefficients $B, C, D, \ldots$.

(c) For $v \gg b$, find an approximate expression for pv/RT.

(d) Use the data in Table 21.1 to verify that for most gaseous states $(b/v)^2 \ll 1$, $(a/v^2 p)^2 \ll 1$, and $(b/v)(a/v^2 p) \ll 1$.

21.3 Use the results of Problem 21.1 to show that the van der Waals equation implies that

$$\int_{v_f}^{v_g} \left[\frac{RT}{v-b} - \frac{a}{v^2} - p_{\text{sat}}(T)\right]_{T,p} dv = \mu_f(T, p_{\text{sat}}(T)) - \mu_g(T, p_{\text{sat}}(T))$$

where the subscripts f and g denote, respectively, the saturated liquid and saturated vapor states at T. Explain why this relation proves the Maxwell graphic rule cited in Section 21.2.

21.4 Use the van der Waals equation and the results of Problems 21.1 and 21.2 to prove that

$$T\alpha_p - 1 \approx \frac{2a}{v^2 p} - \frac{b}{v} \quad \text{and} \quad p\kappa_T - 1 \approx \frac{a}{v^2 p} - \frac{b}{v}$$

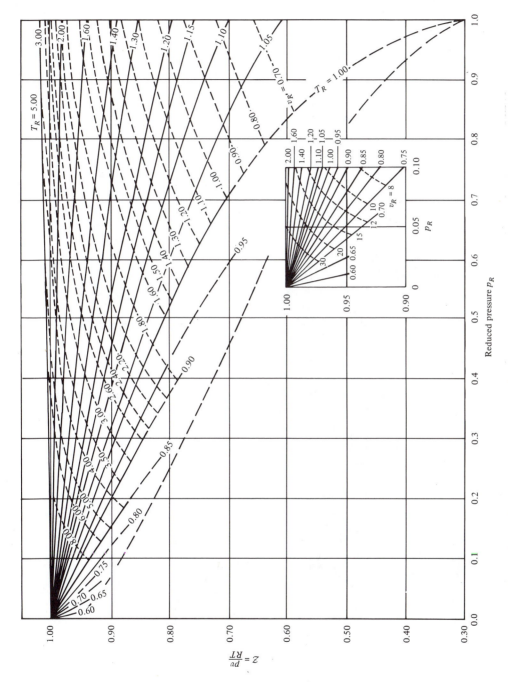

Figure 21.4 Generalized compressibility chart with p_R in the range from 0 to 1. *Source:* see text.

353

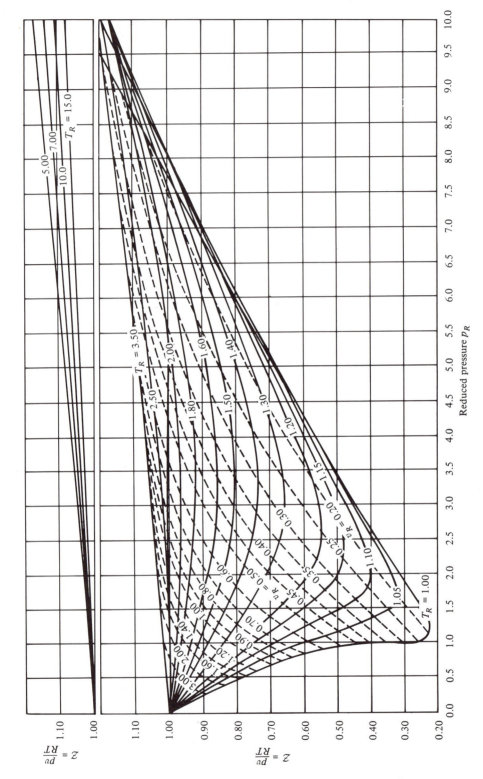

Figure 21.5 Generalized compressibility chart with p_R in the range from 0 to 10. *Source:* see text.

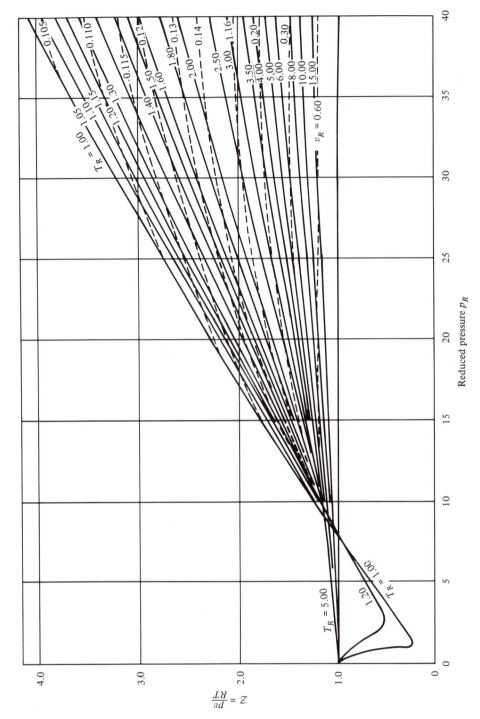

Figure 21.6 Generalized compressibility chart with p_R in the range from 0 to 40. *Source:* see text.

22   Bulk Flow

From the practical standpoint, the most important and interesting applications of the concepts, principles, and conclusions of thermodynamics involve systems in states that are not stable equilibrium, and interactions between systems that result in exchanges of energy, entropy, and amounts of constituents, and in changes in values of parameters. Typical examples are energy-processing devices, energy-conversion systems, chemical reactors, and materials-processing systems.

In general, the analysis of systems that are not in stable equilibrium states faces at least two major difficulties. The first is that we lack a practical equation of motion, an equation that predicts the evolution of states of interacting systems as a function of time and initial conditions. Because of this deficiency, we do not have the theoretical tools that are necessary to describe, for example, the generation of entropy that occurs during the evolution of nonequilibrium states, yet such a generation of entropy causes drastic deteriorations of the usefulness of the energies of the systems in question. The second difficulty is numerical and one that probably would persist even if a general equation of motion were known. It is that the number of independent properties required to characterize states that are not stable equilibrium is much larger than that for states that are. Accordingly, relations between properties of states that are not stable equilibrium are numerically much more complicated and therefore much more difficult to analyze than relations between properties of stable equilibrium states. Because of such difficulties, the theoretical understanding of problems involving states that are not stable equilibrium is limited and much more research work needs to be done.

However, one class of phenomena that can be analyzed successfully and is useful in the vast field of fluid-flow applications is one in which nonequilibrium states can be described in terms of a hybrid of properties of stable equilibrium and simple mechanical motion.

In this chapter, we define this special class of nonequilibrium states, and introduce another class of nonwork interactions that involves these states and that transfers energy, entropy, and amounts of constituents. Illustrative applications of these concepts are discussed in Chapter 23 as well as other chapters.

22.1   Bulk-Flow States

Up to this point we have been discussing properties of simple systems in stable equilibrium states. Here we relax this restriction. We consider a system with all the defining features of a simple system (Section 17.1) except that its constituents are also subject to an external gravitational force field of constant intensity $g \neq 0$. For any state of this system we denote by n and V the amounts of the r constituents and the volume, respectively, and by E, S, m, ξ, and z the properties energy, entropy, mass, velocity of the center of mass, and elevation of the center of mass in the gravity field, respectively. Each of these properties is well defined for all states, including nonequilibrium states.

Next, we consider a special class of states of the system just defined for which the values E of the energy, S of the entropy, n of the amounts of the r constituents, V of the volume, m of the total mass, ξ of the velocity of the center of mass, and z of the elevation of the center of mass in the gravity field are interrelated in such a way that

$$E = U(S, V, n) + m\frac{\xi^2}{2} + mgz \qquad (22.1)$$

where $U(S, V, n)$ is the stable equilibrium state relation between the internal energy U, the entropy S, the volume V, and the amounts of constituents n of a simple system consisting of the same r constituents as the given system.

When equation (22.1) is valid we say that the system is in a *bulk-flow state* or that the state is *bulk-flow*. In view of (22.1), we call U the *internal energy*, $m\xi^2/2$ the *kinetic energy*, and mgz the *potential energy* of the bulk-flow state, and we rewrite the relation in the form

$$e = \frac{E}{m} = u + \frac{\xi^2}{2} + gz \qquad (22.2a)$$

$$u = \frac{U}{m} = \frac{1}{m}U(S, V, n) \qquad (22.2b)$$

where u, $\xi^2/2$, and gz are the mass-specific internal, kinetic, and potential energies, respectively.

We conclude that a bulk-flow state is a particularly simple kind of nonequilibrium state that requires for its characterization a hybrid of two purely mechanical properties, namely, bodily velocity ξ and elevation z in the gravity field, plus the set of properties and variables that characterize the stable equilibrium states of a simple system, namely, the set S, V, n_1, n_2, ..., n_r. Because it is a nonequilibrium state, temperature, pressure, chemical potentials, and therefore enthalpy, Helmholtz free energy, and Gibbs free energy cannot be defined for a bulk-flow state as such. These properties, however, are defined for the stable equilibrium state of the simple system that has $U(S, V, n)$ as a characteristic function. In practice, for brevity we do refer to the temperature, pressure, chemical potentials, enthalpy, Helmholtz free energy, and Gibbs free energy of a bulk-flow state but what we mean are the respective properties that are derived from $U(S, V, n)$.

It is noteworthy that in the absence of the gravity field, that is, for $g = 0$, the system defined at the beginning of this section is simple. Then a bulk-flow state with center-of-mass velocity $\xi = 0$ is stable equilibrium. In other words, in the absence of gravity, any given bulk-flow state becomes a stable equilibrium state when viewed from a reference frame that moves with the center-of-mass velocity ξ.

It is clear that bulk-flow states form a very special and restrictive class, because not all nonequilibrium states satisfy the demanding requirement that $E - m\xi^2/2 - mgz$ be an energy related to S, V, and n in accordance with the state principle. Nevertheless, many practical situations conform to this requirement.

22.2 Bulk-Flow Interactions

An elementary process that constitutes the basis of many industrial applications and that involves bulk-flow states is shown schematically in Figure 22.1a. System A interacts

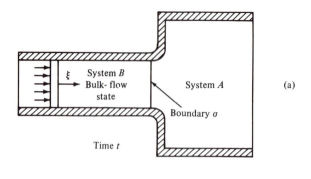

(a)

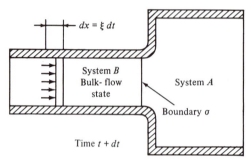

(b)

Figure 22.1 Schematic of a process experienced by system B during the time interval between t and $t + dt$ involving a bulk-flow interaction with system A across the boundary σ.

across the fixed boundary surface σ with a system B. The latter consists of matter confined in a pipe of cross-sectional area a, pushed by a piston moving at speed ξ and exerting a force per unit area p. The state of B is bulk-flow with speed ξ, elevation z in the gravity field, internal energy $U = mu$, entropy $S = ms$, volume $V = mv$, mass m, amounts of constituents $n_1 = y_1 n$, $n_2 = y_2 n$, ..., $n_r = y_r n$, temperature T, and pressure p equal to the force per unit area on the piston. We assume that the interaction between A and B at the boundary is such that as the piston is pushed over an infinitesimal distance dx (Figure 22.1b), matter flows between A and B across the boundary σ, the values ξ, z, u, s, v, y_1, y_2, ..., y_r, T, p remain invariant, and the process for system B is reversible.

To calculate how much energy, entropy, and amounts of constituents are transferred across the boundary σ, we consider the process experienced by system B during the time interval between t and $t + dt$. Because the intensive properties of B remain invariant, the changes in energy, entropy, volume, and amounts of constituents of B can be written, respectively, as $dE = d(U + m\xi^2/2 + mgz) = (u + \xi^2/2 + gz) dm$, $dS = s dm$, $dV = v dm$, $dn_i = y_i dm/M$, for $i = 1, 2, ..., r$, where M is the mean molecular weight, $M = m/n = \sum_i y_i M_i$, and M_i the molecular weight of the ith constituent. Moreover, $dV = -a dx$, where dx is the displacement of the piston. Denoting by $\delta E^{A\leftarrow}$, $\delta S^{A\leftarrow}$, and $\delta m^{A\leftarrow}$ the energy, entropy, and mass transferred across the boundary σ into system A, we write the energy, entropy, and mass balances for system B during the elementary reversible process in the forms

$$dE = -\delta E^{A\leftarrow} + p\,a\,dx = -\delta E^{A\leftarrow} - p\,dV \qquad (22.3)$$

$$dS = -\delta S^{A\leftarrow} \qquad (22.4)$$

$$dm = -\delta m^{A\leftarrow} \qquad (22.5)$$

where $p\,a\,dx = -p\,dV$ is the work done by the piston on system B as a result of the displacement of the point of application of the force pa by dx. Consistently with our notation, if mass flows into A, $\delta m^{A\leftarrow} > 0$, and if it flows out of A, $\delta m^{A\leftarrow} < 0$.

In terms of specific properties, equations (22.3) and (22.4) may be rewritten as

$$\delta E^{A\leftarrow} = -dE - p\,dV = -\left(u + \frac{\xi^2}{2} + gz\right)dm - pv\,dm$$

$$= \left(u + pv + \frac{\xi^2}{2} + gz\right)\delta m^{A\leftarrow} \tag{22.6}$$

$$\delta S^{A\leftarrow} = -dS = -s\,dm = s\,\delta m^{A\leftarrow} \tag{22.7}$$

where in writing the last of equations (22.6) and (22.7) we use (22.5). We see that as a result of the interaction across the boundary σ, the energy transferred per unit of mass transferred equals the specific energy $u + \xi^2/2 + gz$ of the bulk-flow state of the matter in system B plus an additional amount pv arising from the work done on system B as matter is displaced against its own pressure.

In summary, the elementary interaction just considered transfers energy, entropy, and mass so that the ratios of the energy and entropy transferred to the mass transferred are

$$\frac{\delta E^{A\leftarrow}}{\delta m^{A\leftarrow}} = h + \frac{\xi^2}{2} + gz \tag{22.8}$$

$$\frac{\delta S^{A\leftarrow}}{\delta m^{A\leftarrow}} = s \tag{22.9}$$

where the specific enthalpy $h = u + pv$. In a similar way, we find that the amount of the ith constituent transferred across the boundary σ is given by the relation

$$\frac{\delta n_i^{A\leftarrow}}{\delta m^{A\leftarrow}} = \frac{y_i}{M} = \frac{y_i}{\sum_{j=1}^{r} y_j M_j} \qquad \text{for } i = 1,\ 2,\ \ldots,\ r \tag{22.10}$$

where M_j is the molecular weight of constituent j.

We call *bulk-flow interaction* any interaction that transfers energy, entropy, and mass across a boundary between systems so that the amounts of energy transferred, entropy transferred, mass transferred, and amounts of constituents transferred are related to the bulk-flow state of the transferred matter by equations (22.8) to (22.10).

We can express a bulk-flow interaction also in terms of rates, that is, amounts transferred per unit of time interval. Indeed, upon denoting by $\dot{m}^\leftarrow$, $\dot{E}^\leftarrow$, $\dot{S}^\leftarrow$, and $\dot{n}_i^\leftarrow$ the flow rates into system A of mass, energy, entropy, and ith amount, respectively, we have the relations

$$\dot{E}^\leftarrow = \dot{m}^\leftarrow\left(h + \frac{\xi^2}{2} + gz\right) \tag{22.11}$$

$$\dot{S}^\leftarrow = \dot{m}^\leftarrow s \tag{22.12}$$

$$\dot{n}_i^\leftarrow = \dot{m}^\leftarrow\left(\frac{y_i}{M}\right) \tag{22.13}$$

So to characterize a bulk-flow interaction, we must specify the mass flow rate $\dot{m}^\leftarrow$ and the bulk-flow state of the matter that flows across the boundary where the interaction occurs. A positive value of $\dot{m}^\leftarrow$ signifies flow into A and a negative value flow out of A.

22.3   Work, Heat, and Bulk Flow

We consider a system A that has many ports and that experiences work, heat, and bulk-flow interactions either successively or simultaneously as it goes from state A_1 at time t to state A_2 at time $t + dt$. In following chapters we will see that such a system is representative of many devices and engines used in applications.[1]

System A is shown schematically in Figure 22.2. Its interactions are as follows: a work interaction that transfers energy at the rate $\dot{W}_s^{\rightarrow}$ by means such as shafts, cables, and electrical connections; another work interaction that transfers energy at the rate $p_o \dot{V}$ to an environment at pressure p_o by means of a moving piston separating system A from the environment, where $\dot{V} = dV/dt$ is the rate of change of the volume of A; a heat interaction that transfers energy at the rate $\dot{Q}_o^{\leftarrow}$ at the temperature T_o of the environment; other heat interactions each of which transfers energy at the rate $\dot{Q}_k^{\leftarrow}$ at temperature T_k, for $k = 1, 2, \ldots$; and bulk-flow interactions each of which transfers mass across a port q at the rate $\dot{m}_q^{\leftarrow}$, and is such that the bulk-flow state has specific enthalpy h_q, specific entropy s_q, speed ξ_q, elevation z_q in the gravity field, mole fractions $y_q = \{y_{q1}, y_{q2}, \ldots, y_{qr}\}$, and mean molecular weight M_q, for $q = 1, 2, \ldots$.

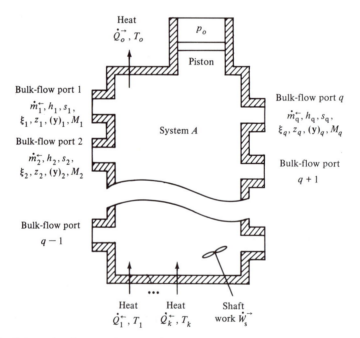

Figure 22.2   Schematic of a general system A experiencing work, heat, and bulk-flow interactions. This system is general enough to reduce in different special cases to several useful models of systems and devices used for many engineering applications.

In the time interval between t and $t + dt$, the energy of system A changes from E to $E + dE$, the entropy from S to $S + dS$, the volume from V to $V + dV$, the mass from m to $m + dm$, and the amounts of constituents from n_i to $n_i + dn_i$, for $i = 1, 2, \ldots, r$.

---

[1] According to a long-standing tradition, a system that experiences interactions, such as bulk-flow, capable of causing exchanges of amounts of constituents with other systems, is called an *open system*. Otherwise, the system is called a *closed system*. In this book, we do not use this terminology.

For such a process, the mass, energy, and entropy balances for system A are

$$m + dm - m = \sum_q \dot{m}_q^{\leftarrow} \, dt \tag{22.14}$$

$$E + dE - E = -\dot{W}_s^{\rightarrow} \, dt - p_o \dot{V} \, dt + \dot{Q}_o^{\leftarrow} \, dt$$

$$+ \sum_k \dot{Q}_k^{\leftarrow} \, dt + \sum_q \dot{m}_q^{\leftarrow} \left(h_q + \frac{\xi_q^2}{2} + gz_q \right) dt \tag{22.15}$$

$$S + dS - S = \frac{\dot{Q}_o^{\leftarrow}}{T_o} \, dt + \sum_k \frac{\dot{Q}_k^{\leftarrow}}{T_k} \, dt + \sum_q \dot{m}_q^{\leftarrow} s_q \, dt + \dot{S}_{\text{irr}} \, dt \tag{22.16}$$

where, of course, $\dot{S}_{\text{irr}}$ is the rate of entropy generation by irreversibility inside system A. Upon dividing each of these equations by dt, we find that

$$\frac{dm}{dt} = \sum_q \dot{m}_q^{\leftarrow} \tag{22.17}$$

$$\frac{dE}{dt} = -\dot{W}_s^{\rightarrow} - p_o \dot{V} + \dot{Q}_o^{\leftarrow} + \sum_k \dot{Q}_k^{\leftarrow} + \sum_q \dot{m}_q^{\leftarrow} \left(h_q + \frac{\xi_q^2}{2} + gz_q \right) \tag{22.18}$$

$$\frac{dS}{dt} = \frac{\dot{Q}_o^{\leftarrow}}{T_o} + \sum_k \frac{\dot{Q}_k^{\leftarrow}}{T_k} + \sum_q \dot{m}_q^{\leftarrow} s_q + \dot{S}_{\text{irr}} \tag{22.19}$$

Equation (22.17) shows that the rate of change dm/dt of the mass of system A equals the sum of the mass flow rates into all the bulk-flow ports. Some of these flow rates are positive—flow into A—and the others are negative—flow out of A. Equation (22.18) shows that the rate of change dE/dt of the energy of A equals the sum of the energy flow rates into the system associated with all the work and heat interactions, and the energy flow rates into all the bulk-flow ports. Equation (22.19) shows that the rate of change dS/dt of the entropy of A equals the entropy flow rates into the system associated with all the heat interactions, plus the entropy flow rates into all the bulk-flow ports, plus the rate of entropy that may be generated internally to system A if the process is irreversible.

If system A is in a steady state, namely, a state such that the value of no property changes with time, then $dm/dt = 0$, $dE/dt = 0$, $dS/dt = 0$, and $dV/dt = \dot{V} = 0$, and equations (22.17) to (22.19) become

$$\sum_q \dot{m}_q^{\leftarrow} = 0 \tag{22.20}$$

$$-\dot{W}_s^{\rightarrow} + \dot{Q}_o^{\leftarrow} + \sum_k \dot{Q}_k^{\leftarrow} + \sum_q \dot{m}_q^{\leftarrow} \left(h_q + \frac{\xi_q^2}{2} + gz_q \right) = 0 \tag{22.21}$$

$$\frac{\dot{Q}_o^{\leftarrow}}{T_o} + \sum_k \frac{\dot{Q}_k^{\leftarrow}}{T_k} + \sum_q \dot{m}_q^{\leftarrow} s_q + \dot{S}_{\text{irr}} = 0 \tag{22.22}$$

If system A has only two ports, 1 and 2, fixed volume, and a heat interaction only with the environment (Figure 22.3), the rate balance equations become

$$\frac{dm}{dt} = \dot{m}_1^{\leftarrow} + \dot{m}_2^{\leftarrow} \tag{22.23}$$

$$\frac{dE}{dt} = -\dot{W}_s^{\rightarrow} + \dot{Q}_o^{\leftarrow}$$
$$+ \dot{m}_1^{\leftarrow}\left(h_1 + \frac{\xi_1^2}{2} + gz_1\right) + \dot{m}_2^{\leftarrow}\left(h_2 + \frac{\xi_2^2}{2} + gz_2\right) \tag{22.24}$$

$$\frac{dS}{dt} = \frac{\dot{Q}_o^{\leftarrow}}{T_o} + \dot{m}_1^{\leftarrow}s_1 + \dot{m}_2^{\leftarrow}s_2 + \dot{S}_{\text{irr}} \tag{22.25}$$

Moreover, if the two-port system is in a steady state, and if the mass flow is into port 1 and out of port 2, we define the positive rate $\dot{m}$ so that

$$\dot{m} = \dot{m}_1^{\leftarrow} = -\dot{m}_2^{\leftarrow} \tag{22.26}$$

Then $\dot{m}_1^{\leftarrow} + \dot{m}_2^{\leftarrow} = 0$, as it should, and the energy and entropy rate balance equations become

$$0 = -\dot{W}_s^{\rightarrow} + \dot{Q}_o^{\leftarrow} + \dot{m}\left(h_1 + \frac{\xi_1^2}{2} + gz_1\right) - \dot{m}\left(h_2 + \frac{\xi_2^2}{2} + gz_2\right) \tag{22.27}$$

$$0 = \frac{\dot{Q}_o^{\leftarrow}}{T_o} + \dot{m}\,s_1 - \dot{m}\,s_2 + \dot{S}_{\text{irr}} \tag{22.28}$$

or, equivalently,

$$\dot{m}\left[(h_2 - h_1) + \frac{\xi_2^2 - \xi_1^2}{2} + g(z_2 - z_1)\right] = \dot{Q}_o^{\leftarrow} - \dot{W}_s^{\rightarrow} \tag{22.29}$$

$$\dot{m}\,(s_2 - s_1) = \frac{\dot{Q}_o^{\leftarrow}}{T_o} + \dot{S}_{\text{irr}} \tag{22.30}$$

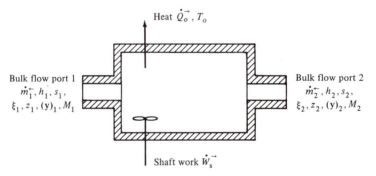

Figure 22.3 Schematic of a two-port flow device with two bulk-flow interactions, a shaft-work interaction, and a heat interaction at temperature T_o

In some applications, the mass rate balance may be more conveniently expressed in terms of velocities rather than mass flow rates. Specifically, for the bulk-flow interaction at port q we write

$$\dot{m}_q^{\rightarrow} = \boldsymbol{\xi}_q \cdot \mathbf{N}_q \frac{a_q}{v_q} = \rho_q \boldsymbol{\xi}_q \cdot \mathbf{N}_q a_q \tag{22.31}$$

where $\boldsymbol{\xi}_q$ is the velocity vector, $\boldsymbol{N}_q$ an outward unit vector normal to the interaction surface with area a_q, $\boldsymbol{\xi}_q \cdot \boldsymbol{N}_q$ the dot product of the vectors $\boldsymbol{\xi}_q$ and $\boldsymbol{N}_q$, v_q the mass-specific volume, and ρ_q the density. For example, when the velocity $\boldsymbol{\xi}_q$ and the unit vector $\boldsymbol{N}_q$ are parallel, then $\boldsymbol{\xi}_q \cdot \boldsymbol{N}_q = \pm \xi_q$, where the plus sign applies when $\boldsymbol{\xi}_q$ points outward, and the minus sign when $\boldsymbol{\xi}_q$ points inward. Using this notation, equation (22.17) becomes

$$\frac{dm}{dt} = \sum_q \rho_q \boldsymbol{\xi}_q \cdot \boldsymbol{N}_q \, a_q \qquad (22.32)$$

22.4 Combined Rate Balance

In the analysis of processes involving a given set of interactions, such as for the system A shown in Figure 22.2, we are always interested in finding the condition under which the best thermodynamic performance is achieved, that is, either the most benefit from specific changes of states is derived or the least effort to achieve desired changes of states is expended. To address this question here, we consider the system A in Figure 22.2 and combine the energy and entropy rate balances into one relation as follows.

Upon multiplying equation (22.19) by the environmental temperature T_o, subtracting the result from equation (22.18), and rearranging terms, we find that

$$\frac{d(E - T_o S + p_o V)}{dt} = -\dot{W}_s^{\rightarrow} + \sum_k \dot{Q}_k^{\leftarrow} \left(1 - \frac{T_o}{T_k} \right)$$

$$+ \sum_q \dot{m}_q^{\leftarrow} \left(h_q - T_o s_q + \frac{\xi_q^2}{2} + g z_q \right) - T_o \dot{S}_{\text{irr}} \qquad (22.33)$$

Equation (22.33) is called the *combined energy and entropy rate balance* or simply the *combined rate balance*.

If system A is in a steady state, its energy E, entropy S, and volume V are time invariant and (22.33) becomes

$$\dot{W}_s^{\rightarrow} - \sum_k \dot{Q}_k^{\leftarrow} \left(1 - \frac{T_o}{T_k} \right) = \sum_q \dot{m}_q^{\leftarrow} \left(h_q - T_o s_q + \frac{\xi_q^2}{2} + g z_q \right) - T_o \dot{S}_{\text{irr}} \qquad (22.34)$$

Each of the terms $\dot{Q}_k^{\leftarrow} (1 - T_o/T_k)$, for $k = 1, 2, \ldots$, can be interpreted as the *work rate equivalent with respect to a reservoir at temperature T_o* of the heat rate $\dot{Q}_k^{\leftarrow}$ from the energy source at temperature T_k [see equation (15.5)]. Therefore, the left-hand side of equation (22.34) is the work rate equivalent done by the system as a result of the work and heat interactions with other systems, excluding the bulk-flow interactions. It is fully accounted for by the changes in the bulk-flow states, and the irreversibility inside A.

Indeed, the sum over q in the right-hand side of (22.34) is specified by the bulk-flow states of the input and output streams. It may be negative or positive. If it is negative, it signifies the least rate at which work must be done on the system in order to achieve the specified changes from the input bulk-flow states to the output bulk-flow states. It differs from the work rate done on the system by the amount $T_o \dot{S}_{\text{irr}}$, which is the work rate equivalent, with respect to the reservoir at T_o, of the rate of entropy generated by irreversibility. If the sum over q is positive, it signifies the largest rate at which work can be done by the system while the streams change from the input bulk-flow states to the output bulk-flow states. It too differs from the work rate done by the system by the amount $T_o \dot{S}_{\text{irr}}$. Each of the two differences just cited is the smaller the smaller the value of the positive rate $\dot{S}_{\text{irr}}$. It follows that optimum performance—either least work rate into the system or largest work rate out of the system—is achieved if the process is reversible, that is, if $\dot{S}_{\text{irr}} = 0$.

It is noteworthy that equation (22.34) does not include the heat rate $\dot{Q}_o^{\leftarrow}$ at temperature T_o because the work rate equivalent of such a heat interaction with respect to a reservoir at T_o is null. Nevertheless, this heat plays an important role in determining the optimum thermodynamic performance. In general, the overall rate of entropy change from the inlet to the outlet bulk-flow streams differs from zero. To achieve optimum performance at steady state, this rate of entropy change must be balanced by the heat interactions [equation (22.19) with $dS/dt = 0$ and $\dot{S}_{\text{irr}} = 0$]. If a heat interaction occurs at temperature $T_k \neq T_o$, it has a work equivalent different from zero. On the other hand, the heat interaction at T_o has zero work equivalent and therefore can exchange any amount of entropy at the expense of a corresponding amount of valueless energy—energy with zero work equivalent. We can see this clearly if every $\dot{Q}_k = 0$, for $k = 1, 2, \ldots$, and the steady-state process is optimum, namely, $\dot{S}_{\text{irr}} = 0$, because then the entropy and the combined rate balances become

$$\frac{\dot{Q}_o^{\leftarrow}}{T_o} + \sum_q \dot{m}_q^{\leftarrow} s_q = 0 \qquad \text{or} \qquad \dot{Q}_o^{\leftarrow} = -T_o \sum_q \dot{m}_q^{\leftarrow} s_q \qquad (22.35)$$

$$\dot{W}_{\text{optimum}}^{\rightarrow} = \sum_q \dot{m}_q^{\leftarrow} \left(h_q - T_o s_q + \frac{\xi_q^2}{2} + g z_q \right) \qquad (22.36)$$

When the right-hand side of equation (22.36) is positive, it represents the largest rate at which shaft work can be done by exploiting the changes in bulk-flow states from the inlet to the outlet streams, and when it is negative the least rate of shaft work necessary to achieve the change in bulk-flow states from inflow to outflow.

For steady-state bulk flow processes such that changes in kinetic and potential energies are not important, equation (22.36) reduces to

$$\dot{W}_{\text{optimum}}^{\rightarrow} = \sum_q \dot{m}_q^{\leftarrow} (h_q - T_o s_q) \qquad (22.37)$$

Moreover, upon defining

$$\dot{H}_{\text{in}} = \sum_q \dot{m}_q^{\leftarrow} h_q \qquad \text{and} \qquad \dot{S}_{\text{in}} = \sum_q \dot{m}_q^{\leftarrow} s_q \qquad (22.38)$$

where the sums are taken only over the ports for which $\dot{m}_q^{\leftarrow} > 0$, and

$$\dot{H}_{\text{out}} = -\sum_q \dot{m}_q^{\leftarrow} h_q \qquad \text{and} \qquad \dot{S}_{\text{out}} = -\sum_q \dot{m}_q^{\leftarrow} s_q \qquad (22.39)$$

where the sums are taken only over the ports for which $\dot{m}_q^{\leftarrow} < 0$, we find that

$$\dot{W}_{\text{optimum}}^{\rightarrow} = \dot{H}_{\text{in}} - \dot{H}_{\text{out}} - T_o (\dot{S}_{\text{in}} - \dot{S}_{\text{out}}) \qquad (22.40)$$

In words, the optimum work rate depends not only on the rate of overall enthalpy change of the bulk-flow streams from inlet to outlet but also on the overall rate of entropy change of the bulk-flow streams from inlet to outlet.

Equation (22.40) may also be written in the form

$$\dot{W}_{\text{optimum}}^{\rightarrow} = (\dot{H}_{\text{in}} - T_o \dot{S}_{\text{in}}) - (\dot{H}_{\text{out}} - T_o \dot{S}_{\text{out}}) \qquad (22.41)$$

The expression $\dot{H} - T_o \dot{S}$ represents one of the availability rate functions discussed later in Chapter 24.

Clearly, if the steady-state process is not reversible, namely, $\dot{S}_{irr} \neq 0$, then, using equation (22.36), we can rewrite the combined rate balance equation (22.34) in the form

$$\dot{W}_s^{\rightarrow} - \sum_k \dot{Q}_k^{\leftarrow} \left(1 - \frac{T_0}{T_k}\right) = \dot{W}_{optimum}^{\rightarrow} - T_0 \dot{S}_{irr} \qquad (22.42)$$

where we see that the sum of the shaft work rate and the work rate equivalents of the heat rates differs from the optimum work rate because of the rate of entropy generation by irreversibility.

Problems

22.1 A stream of gases at $T_1 = 650°C$ and $p_1 = 1$ atm is available from a high-temperature industrial process at a rate $\dot{m} = 10$ kg/h. The stream can be used to raise steam in a waste-heat boiler and the steam, in turn, can power a turbogenerator to produce electricity. The stream gases can be modeled as a perfect gas with $c_p = 1.4$ kJ/kg K and $c_v = 1$ kJ/kg K.

(a) What is the largest rate of work that can be done using the stream in an environment at $T_0 = 20°C$ and $p_0 = 1$ atm?

(b) Under the conditions specified in Figure P22.1, at what rates are energy and entropy transferred out of the stream?

(c) If the power output $\dot{W}^{\rightarrow}$ of the turbogenerator is only 70% of the largest rate of work found in part (a), what is the rate of entropy generation by irreversibility? What is the total rate of entropy rejection into the atmosphere?

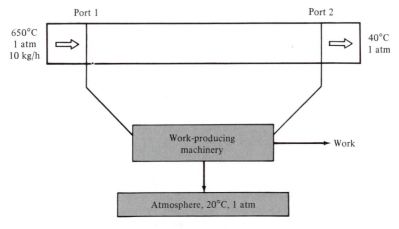

Figure P22.1

22.2 A rigid, insulated, empty vessel is to be filled with steam from a steam line at 7 bar and 320°C. Steam flows into the vessel until the pressure in the vessel is 7 bar. At this instant the valve on the charging line is closed.

(a) What is the temperature of the steam in the vessel at the instant the valve is closed?

(b) What is the final pressure of the steam in the vessel after it has reached mutual stable equilibrium with the environment at 20°C?

(c) How much energy is transferred per kilogram of steam as the steam and the environment come to mutual stable equilibrium?

22.3 A well-insulated container is connected to a vacuum pump. Initially, 1 kg of saturated triple-point liquid nitrogen, a negligible amount of saturated triple-point vapor, and no triple-point solid are present in the container. The pump is started and vapor is removed from the container until all of the liquid either has boiled away or is frozen to a saturated triple-point solid. In the final state, only saturated triple-point solid and saturated triple-point vapor remain in the container and no liquid is present. Triple-point data for nitrogen are given in the following table, where $T_{tp} = 63.15$ K and $p_{tp} = 12.5$ kPa.

	Enthalpy	Specific Volume
Solid	-25.78 kJ/kg	1.130×10^{-3} m^3/kg
Liquid	0 kJ/kg	1.153×10^{-3} m^3/kg
Vapor	216.3 kJ/kg	1.488 m^3/kg

(a) Determine the mass of the triple-point solid that remains in the container. For this calculation, assume that the mass of triple-point vapor remaining in the container is negligible.

(b) Estimate the mass of triple-point vapor remaining in the container at the final state so as to show that the assumption of negligible vapor mass used in part (a) is justified.

22.4 Geothermal saturated steam at 100°C is being considered for heating a building. A number of heating schemes are under consideration. In all these schemes, only the latent heat of the geothermal steam will be used. The temperature of the building is to be maintained at 20°C, and the average outdoor temperature is 5°C. The grid of an electric utility is also available for connection to any necessary machinery. The value of electricity is 10 times that of geothermal steam per unit of energy delivered, that is, electricity is valued at 10 cents/kWh, and energy from steam at 1 cent/kWh. These values include all capital and other charges for any type of equipment that might be used.

(a) What is the smallest energy cost per unit of energy delivered to the building? How is the smallest energy cost achieved? Make a sketch of the interacting systems involved.

(b) How much electricity is transferred per unit of energy delivered to the building under the conditions specified in part (a)? Is this electricity transferred to the building or to the grid?

(c) If 90% of the geothermal energy is used directly to heat the building, what is the cost per unit of energy delivered? How much entropy is generated by irreversibility during this process?

(d) If the geothermal energy is not used and the building is heated by a heat pump with a coefficient of performance $(COP)_{h.p.} = 2$, what is the cost of electricity per unit of energy delivered to the building? How much entropy is generated by irreversibility when the heat pump is used?

22.5 A chamber contains 1 mol of air at temperature T_i and pressure p_i. It is connected through a valve to a vertical cylinder equipped with a frictionless nonleaking piston. The piston is loaded by a weight that can be supported by a pressure p_f ($p_f < p_i$). Initially, the piston is at the bottom of the cylinder. The valve is opened slightly, and air flows into the cylinder slowly until the pressures on the two sides of the valve are equalized. Assume that the process involves no heat interactions and that the air behaves as a perfect gas.

(a) Show that the final air temperature in the chamber $T_f = T_i (p_f/p_i)^{(\gamma-1)/\gamma}$, where $\gamma = c_p/c_v$.

(b) Show that the number of moles of air left in the chamber $n_f = n_i (p_f/p_i)^{1/\gamma}$.

(c) Show that the final air temperature of the cylinder $T'_f = T_i[1 - (p_f/p_i)]/[1 - (p_f/p_i)^{1/\gamma}]$.

22.6 It has been proposed that icebergs be towed from the arctic region and used to provide heat and electricity for coastal cities. The temperature of seawater $T_s = 10°C$. Pure liquid water at $T_w = 10°C$ is available from the environment for the production of steam at 180°C and 200 kPa to be used for district heating. A city requires 500 MW of electric power for motors and lighting and 600 kg/s of steam for heating purposes.

(a) What is the least rate of ice consumption needed to satisfy the city's electricity and heating needs?

(b) What is the answer to part (a) if only the constant-temperature melting of the ice is exploited?

(c) What is the actual rate of ice consumption if only the constant-temperature melting of the ice is exploited and the entropy generation by irreversibility is four times the entropy supplied to the water stream at 10°C to produce the heating steam?

22.7 An engine has been devised to supply heat to an oven at 1000 K using as sources of energy saturated geothermal steam at 500 K and 1.8 MPa and a water pond at 293 K.

(a) What is the optimum amount of energy exchanged between the engine and the oven per kilogram of steam processed by the engine?

(b) What amount of energy is exchanged between the engine and the oven per kilogram of steam processed by the engine if the entropy generated by irreversibility per kilogram of steam is 1 kJ/kg K?

22.8 In a chemical plant, ethyl alcohol vapor, C_2H_5OH, is produced at the top of a rectification column, at a pressure of 1 atm and saturation temperature $T_{sat}(1 \text{ atm}) = 60.7°C$ as shown schematically in Figure P22.8. The vapor is condensed for further processing and shipment. The condensation is achieved by transferring energy to boiling water at a temperature close to the saturation temperature of ethyl alcohol $T_{sat}(1 \text{ atm})$. At this temperature, the pressure of boiling water is less than 1 atm and is achieved by a pumping device powered by high-pressure steam as shown in the figure. Under steady-state, bulk-flow conditions, and for the data listed on the figure, answer the following questions.

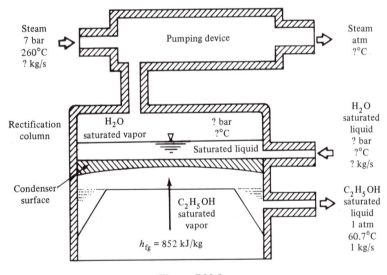

Figure P22.8

(a) If no temperature difference is necessary for the operation of the condenser, what are the temperature and pressure of the boiling water?

(b) What is the rate of heat across the condenser surface?

(c) What are the vapor flow rate and the liquid flow rate on the water side of the condenser?

(d) What are the smallest flow rate of 7-bar steam through the pumping device and the temperature of the corresponding discharge stream?

22.9 In open-cycle gas turbines used for peak electricity load, exhaust gases (products of combustion) are usually wasted even though they are hot. For an installation rated 50 MW electric, the gases are exhausted to the atmosphere at a temperature of 370°C and pressure of 1 atm, where the environmental temperature is 20°C. The difference between the rate of enthalpy flow of the exhaust gases at 370°C and the rate of enthalpy flow that the same gases would have at 20°C is 146 MW.

(a) In the temperature range between 15 and 600°C the products of combustion can be treated as a perfect gas. What is the difference between the rate of entropy flow of the exhaust gases at 370°C and 1 atm and the rate of entropy flow that the same gases would have at 20°C and 1 atm?

(b) What is the largest work that can be done by machinery that exploits the exhaust gases?

(c) What temperature should an energy source have in order for a perfect heat engine to yield the same work as in part (b), of course, in the same environment?

(d) If the exhaust gases are not utilized to produce work, what is the rate of entropy generation due to their discharge into the environment?

(e) The exhaust gases can be used in a waste-heat-recovery unit. One of these units (A) generates entropy because of irreversibility at the rate of 4 kW/K, and another (B) at the rate 24 kW/K. What is the power delivered by each unit?

(f) If work can be sold at 10 cents/kWh, what is the cost of 1 Wh/K of entropy generated by irreversibility in an environment at 20°C?

(g) Gas turbines operate for about 2000 h/yr. If the annual cost of capital, including interest, depreciation, profit, taxes, and maintenance and operating costs is 30% of the investment, how much would you be willing to pay to avoid generating 1 Wh/K of entropy?

22.10 A rigid metal tank of volume V connects to the atmosphere through a valve. Atmospheric air is at pressure p_0 and temperature T_0. Initially, the tank is fully evacuated and the valve is closed.

The valve is then opened and air flows into the tank. Two phases of this airflow are observed. In the first phase, which lasts a few seconds, the air flows rapidly into the tank at a decreasing rate until the pressure increases from zero to p_0. In the second phase, which lasts about 1 h, air continues to flow into the tank at a very low rate while the pressure remains constant at p_0. Eventually, this second phase ends when the air flow goes to zero.

Assume that air behaves as a perfect gas with constant R and specific heat ratio γ and that the metal tank has such a large heat capacity that its temperature remains essentially constant and equal to T_0 during the whole process. This problem is not a repetition of Problem 20.14 if you model the interaction between the contents of the tank and the atmosphere as bulk flow.

(a) Find the temperature T_1 of the air in the tank at the end of the first phase, and show that $T_1 = \gamma T_0$.

(b) Calculate the entropy created by irreversibility during the first phase of the filling process.

(c) Calculate the entropy created by irreversibility during the second phase of the filling process.

(d) Show that the overall entropy created by irreversibility during both filling phases is equal to $p_0 V / T_0$.

(e) If, instead of opening the valve, the empty tank is connected to appropriate cyclic machinery interacting with a weight and the environmental air, what is the largest work that can be done on the weight?

22.11 Ice is to be made from a water supply at 10°C by the process shown in Figure P22.11. The temperature of the ice is $-12°C$, and the temperature of the water exiting from the condenser is 27°C. All streams are at atmospheric pressure. What is the least work required to produce 1 ton of ice?

22.12 A waste-heat source from a chemical plant consists of 250 kg/s of N_2 at 700°C and 8 atm. The environment of the plant is at $T_0 = 25°C$. The specific energy and specific entropy of this stream are given by the relations $u = 0.46(T - T_1)$ kJ/kg and $s = 0.61 \ln(T/T_2) - 0.15 \ln(p/p_2)$ kJ/kg K, where T_1, T_2 are arbitrary reference temperatures in kelvin, and p_2 is an arbitrary reference pressure.

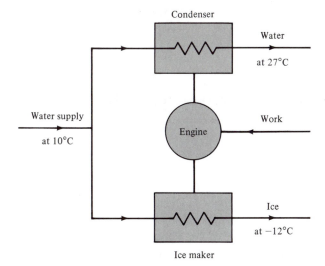

Figure P22.11

(a) If the stream is used in perfect equipment, what is the largest work that can be extracted?

(b) The stream is used in practical equipment that yields 70% of the answer in part (a). If the value of the work produced is 10 cents/kWh, what is an upper limit to the investment that you would be willing to make to recover this work? Assume that the equipment works for 7000 h/yr, and that the annual cost of the equipment that accounts for depreciation, profit, maintenance, and taxes is about 25% of the initial investment.

22.13 In a manufacturing plant operating in an environment at constant temperature T_0 and constant pressure p_0, certain raw materials are being gasified. As a result of the gasification, the energy, entropy, and volume of the raw materials change by ΔU, ΔS, and ΔV, respectively. The only source of energy for the plant is saturated steam that can be condensed at fixed temperature T_s. In addition to expanding against the atmosphere, the raw materials that are being gasified can exchange heat with the environment. In terms of the given quantities T_0, p_0, T_s, ΔU, ΔS, and ΔV, make the following calculations.

(a) Find an expression that gives the least amount of steam energy, $(\Delta H_s)_{least}$, that is required to gasify the raw materials.

(b) Find the entropy exchanged with the environment when the gasification proceeds as in part (a).

(c) If the energy supplied by the condensing steam is $2(\Delta H_s)_{least}$, find the amount of entropy generated by irreversibility.

22.14 A steel tank is to be filled with steam from a very large steam reservoir containing 100% dry saturated H_2O at 70 bar. The tank is spherical and has an internal diameter of 60 cm. The steel tank wall weighs 40 kg, has a specific heat $c = 0.42$ kJ/kg K, and is well insulated from the environment. Initially, the tank contains 18 kg of H_2O at 0.07 bar, and the tank and H_2O are in mutual stable equilibrium.

(a) How much H_2O must flow from the steam reservoir into the tank to bring the pressure in the tank to 70 bar with the tank and H_2O in mutual stable equilibrium? The insulation around the tank is perfect.

(b) How much entropy is generated by irreversibility?

23   Conversion Devices

In practical processes involving either conversion of one form of energy flow to another—energy-conversion systems—or conversion of one form of material flow to another—materials-conversion systems—or both, we use different devices to accomplish the desired tasks. For example, in converting energy from flowing steam into shaft work we use a steam turbine. Each of the devices is so designed as to achieve a process—a specified change of state of the substances and specified interactions with other systems.

In this chapter we consider a number of simplified models of such devices and derive interrelations between end states, states of inlet and outlet bulk-flow streams, interactions, irreversibility, and empirical characteristics of the equipment. Typical interconnections of these devices in energy-conversion systems are discussed in later chapters.

It is noteworthy that each of most of the devices discussed here is the subject of a separate and extensive discipline concerned with the description, design, and optimization of the device. Our purpose is not to summarize all these disciplines but to provide some overall operational characteristics. Moreover, our discussion is far from being exhaustive in covering the great variety of conversion devices that are available to engineers for different purposes. It covers only a few representative examples.

23.1   Steady Flow Through a Pipe

For many applications, steady flow through a pipe (Figure 23.1) is successfully modeled in terms of a two-port device experiencing bulk-flow interactions at the inlet and outlet ports, and a heat interaction at a rate $\dot{Q}^{\leftarrow}$ and a constant temperature T_{w} which is assumed to be the temperature of the inside wall of the pipe all along its length between the inlet and outlet ports.

Using the terminology listed in Figure 23.1, the mass flow rates at ports 1 and 2 are $\dot{m}_1^{\leftarrow} = \rho_1 \xi_1 a_1$ and $\dot{m}_2^{\rightarrow} = \rho_2 \xi_2 a_2$, respectively. At steady state, the mass rate balance yields $\dot{m}_1^{\leftarrow} = \dot{m}_2^{\rightarrow} = \dot{m}$, that is,

$$\rho_1 \xi_1 a_1 = \rho_2 \xi_2 a_2 = \dot{m} \tag{23.1}$$

and so the energy and entropy rate balances [equations (22.27) and (22.28)] are

$$\dot{m}(h_1 - h_2) + \dot{m}\left(\frac{\xi_1^2}{2} - \frac{\xi_2^2}{2}\right) + \dot{m}g(z_1 - z_2) + \dot{Q}^{\leftarrow} = 0 \tag{23.2}$$

$$\dot{m}(s_1 - s_2) + \frac{\dot{Q}^{\leftarrow}}{T_{\mathrm{w}}} + \dot{S}_{\mathrm{irr}} = 0 \tag{23.3}$$

Depending on the application, some of the terms on the left-hand side of (23.2) may be zero or negligible with respect to the others. For example, for a horizontal pipe, $z_1 = z_2$. Again, even if $z_1 \neq z_2$ and $\xi_1 \neq \xi_2$, we may find (Figure 23.2) that

370

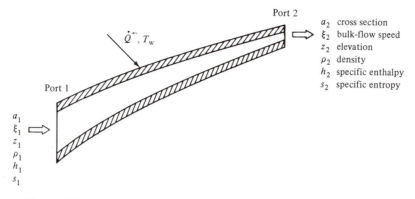

Figure 23.1   Flow through a pipe segment with uniform wall temperature.

$|g(z_1 - z_2)| \ll |h_1 - h_2|$ and $|\xi_1^2/2 - \xi_2^2/2| \ll |h_1 - h_2|$ so that the energy balance reduces to

$$\dot{m}(h_2 - h_1) = \dot{Q}^{\leftarrow} \tag{23.4}$$

For such a pipe, the entropy generation by irreversibility is

$$\dot{S}_{\text{irr}} = \dot{m}(s_2 - s_1) - \frac{\dot{m}(h_2 - h_1)}{T_w}$$

$$= \dot{m}(h_2 - h_1)\left(\frac{s_2 - s_1}{h_2 - h_1} - \frac{1}{T_w}\right)$$

$$= \dot{Q}^{\leftarrow}\left(\frac{s_2 - s_1}{h_2 - h_1} - \frac{1}{T_w}\right) \tag{23.5}$$

where in writing the last of equations (23.5) we use (23.4). So we see that we can relate the irreversibility to the inlet and outlet states of the stream and to the energy source.

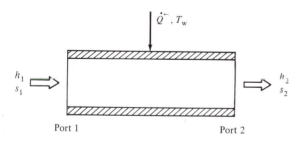

Figure 23.2   Flow through a pipe with uniform wall temperature and negligible changes in kinetic and potential energies of the flow.

For example, if the stream behaves as a perfect gas and there is no pressure drop between the inlet and outlet states, that is, $p_1 = p_2$, then

$$s_2 - s_1 = c_p \ln \frac{T_2}{T_1} \qquad h_2 - h_1 = c_p(T_2 - T_1)$$

and

$$\dot{S}_{\text{irr}} = \dot{Q}^{\leftarrow}\left(\frac{1}{T_{\text{lm}}} - \frac{1}{T_w}\right) \tag{23.6}$$

where

$$T_{\mathrm{lm}} = \frac{h_2 - h_1}{s_2 - s_1} = \frac{c_p\,(T_2 - T_1)}{c_p\,\ln(T_2/T_1)} = \frac{T_2 - T_1}{\ln(T_2/T_1)} \qquad (23.7)$$

The quantity T_{lm} is called the *logarithmic mean temperature* between T_1 and T_2. Because $\dot{S}_{\mathrm{irr}} \geq 0$, for energy to be transferred into the stream, that is, for $\dot{Q}^{\leftarrow} > 0$, we must have $T_{\mathrm{w}} \geq T_{\mathrm{lm}}$. Equations (23.6) and (23.7) also apply if the stream behaves as a perfect incompressible liquid and there is no pressure drop between the inlet and outlet bulk-flow states, that is, $p_1 = p_2$, because then

$$s_2 - s_1 = c\ln\frac{T_2}{T_1} \qquad h_2 - h_1 = c(T_2 - T_1)$$

Example 23.1. A bath tub is being filled with hot water at 50°C outflowing from a pipe at the rate of 1.3 kg/s. The water enters the pipe at 10°C, flows first through a pipe segment heated by a gas burner, and then through a long, well-insulated pipe segment before outflowing into the bath tub. The pipe cross-sectional area $a = 75$ mm². The bath tub has a volume of 0.5 m³ and is at an elevation 10 floors (30 m) lower than the gas burner. Find the rate $\dot{Q}^{\leftarrow}$ at which energy is received by the pipe in the heated segment. Find the cost of the energy if gas costs 3 cents/kWh.

Solution. Modeling water as a perfect incompressible fluid with constant specific heat $c = 4.2$ kJ/kg K and density $\rho = 10^3$ kg/m³, equation (23.4) yields $\dot{Q}^{\leftarrow} = \dot{m}c(T_2 - T_1) = 1.3 \times 4.2 \times (50 - 10) = 218.4$ kW. The bath tub takes $0.5 \times 10^3/1.3 = 385$ s $= 0.11$ h to be filled. So the cost of the energy is $218.4 \times 0.11 \times 0.03 = \$0.70$. We can also verify that indeed equation (23.4) applies. To this end we note that $h_2 - h_1 = c(T_2 - T_1) = 4.2 \times 40 = 168$ kJ/kg, $g(z_2 - z_1) = -9.8 \times 30 = -294$ m²/s² $= -0.294$ kJ/kg, and $\xi_2 = \xi_1$ because of (23.1). So we conclude that (23.4) is valid because $|g(z_2 - z_1)|$ is about three orders of magnitude smaller than $h_2 - h_1$. Assuming that the pipe wall temperature in the heated segment $T_{\mathrm{w}} = 600$ K, we find that $T_{\mathrm{lm}} = (T_2 - T_1)/\ln(T_2/T_1) = 40/\ln(323/283) = 302.6$ K, and $\dot{S}_{\mathrm{irr}} = \dot{Q}^{\leftarrow}(1/T_{\mathrm{lm}} - 1/T_{\mathrm{w}}) = 218.4(1/302.6 - 1/600) = 0.36$ kW/K.

For some applications it is convenient to write the mass, energy, and entropy rate balances for an infinitesimal length of pipe (Figure 23.3). For example, this would be required if the wall temperature of the pipe varies along its length. In steady state, the rate balances for an infinitesimal length are

$$d(\rho\xi a) = 0 \qquad (23.8)$$

$$-\dot{m}\,(dh + \xi\,d\xi + g\,dz) + \delta\dot{Q}^{\leftarrow} = 0 \qquad (23.9)$$

$$-\dot{m}\,ds + \frac{\delta\dot{Q}^{\leftarrow}}{T_{\mathrm{w}}} + \delta\dot{S}_{\mathrm{irr}} = 0 \qquad (23.10)$$

where the terminology is listed in Figure 23.3. The differentials of the specific enthalpy dh and the specific entropy ds satisfy the relation

$$dh = T\,ds + v\,dp = T\,ds + \frac{dp}{\rho} \qquad (23.11)$$

because the changes between the inlet and outlet bulk-flow states are also infinitesimal.

If the process is reversible ($\delta\dot{S}_{\mathrm{irr}} = 0$), and either the pipe is perfectly insulated ($\delta\dot{Q}^{\leftarrow} = 0$) or the wall temperature equals the fluid temperature ($T_w = T$), equation (23.9) reduces to

$$dp + \rho\xi\,d\xi + \rho g\,dz = 0 \qquad (23.12)$$

because then $dh = T\,ds + dp/\rho = \delta\dot{Q}^{\leftarrow} + dp/\rho$. Moreover, if $\rho = $ constant, that is, the fluid is incompressible, equation (23.12) may be written in the form

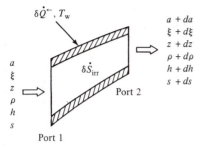

Figure 23.3 Flow through a pipe segment of infinitesimal length.

$$d\left(p + \frac{1}{2}\rho\xi^2 + \rho g z\right) = 0 \tag{23.13}$$

which is known as the *Bernoulli equation* for steady, reversible flow of an incompressible fluid in a perfectly insulated pipe segment.

For a process in a short pipe segment ($z_1 = z_2$) with rapidly varying cross-sectional area, and with negligible energy and entropy rate transfers from the wall, that is, such that $|\delta\dot{Q}^{\leftarrow}| \ll \dot{m}\,|d(h + \xi^2/2)|$, and $|\delta\dot{Q}^{\leftarrow}/T_w| \ll \dot{m}\,|ds|$, equations (23.9) and (23.10) yield that

$$d\left(h + \frac{\xi^2}{2}\right) = 0 \tag{23.14}$$

$$ds = \frac{\delta\dot{S}_{\text{irr}}}{\dot{m}} \tag{23.15}$$

and, upon integration of these results, that

$$h_2 + \frac{\xi_2^2}{2} = h_1 + \frac{\xi_1^2}{2} \tag{23.16}$$

$$s_2 = s_1 + \frac{\dot{S}_{\text{irr}}}{\dot{m}} \tag{23.17}$$

Under the conditions specified for this process, we see from (23.16) that the sum of the specific enthalpy and the specific kinetic energy of the flow at any cross section is a constant. The sum

$$h_{\text{o}} = h + \frac{\xi^2}{2} \tag{23.18}$$

is called the *specific stagnation enthalpy* of the bulk-flow state. In subsequent sections we discuss several practical devices that operate under conditions of constant specific stagnation enthalpy.

The enthalpy h_{o} is used to define two concepts of wide application in engineering practice. The *stagnation temperature* T_{o}, and the *stagnation pressure* p_{o} of a bulk-flow state are defined as the temperature and pressure that the substance would have in a stable equilibrium state with specific enthalpy h_{o} equal to the specific stagnation enthalpy $h + \xi^2/2$, and specific entropy s_{o} equal to the specific entropy s of the bulk-flow state. For example, for an incompressible fluid we readily verify that

$$T_{\text{o}} = T \tag{23.19}$$

$$p_{\text{o}} = p + \frac{\rho\xi^2}{2} \tag{23.20}$$

where T, p, and ξ are, respectively, the temperature, pressure, and speed of the bulk-flow state, and ρ the constant density (mass per unit volume) of the fluid. Again, for a perfect gas we readily verify that

$$T_o = T + \frac{\xi^2}{2c_p} \tag{23.21}$$

$$p_o = p \left(\frac{T_o}{T}\right)^{\gamma/(\gamma-1)} = p \left(1 + \frac{\xi^2}{2c_p T}\right)^{\gamma/(\gamma-1)} \tag{23.22}$$

where c_p is the specific heat at constant pressure of the perfect gas, and $\gamma = c_p/c_v$, that is, the ratio of the specific heats.

If the process in a pipe segment is reversible, the flow occurs at constant stagnation temperature and constant stagnation pressure.

Example 23.2. A stream with speed $\xi_1 = 10.2$ m/s flows from a pipe into a large tank (Figure 23.4). At port 1, the temperature $T_1 = 25°C$ and the pressure $p_1 = 200$ kPa. If there is no energy exchange and no entropy production by irreversibility in the pipe segment between port 1 and port 2, and the speed $\xi_2 = 0$, find the temperature T_2 and pressure p_2 in the tank: (a) if the fluid is water; and (b) if the fluid is air.

Solution. (a) Modeling water as an incompressible liquid, with $\rho = 10^3$ kg/m³ and $c = 4.2$ kJ/kg K, we find that $s_2 - s_1 = c \ln(T_2/T_1)$ and $h_2 - h_1 = c (T_2 - T_1) + (p_2 - p_1)/\rho$. Because there is no irreversibility, $s_2 = s_1$ and, therefore, $T_2 = T_1 = 25°C$. Moreover, equation (23.16) yields that $h_2 = h_{o1} = h_1 + \xi_1^2/2$ and, therefore, that $p_2 = p_{o1} = p_1 + \rho\xi_1^2/2 = 200 \times 10^3 + 10^3 \times 10.2^2/2 = 252 \times 10^3$ Pa = 252 kPa.

(b) Modeling air as a perfect gas with $\gamma = 1.4$ and $c_p = 10^3$ J/kg K, we find that $s_2 - s_1 = c_p \ln(T_2/T_1) - R \ln(p_2/p_1)$ and $h_2 - h_1 = c_p (T_2 - T_1)$. Accordingly, equation (23.16) implies that $T_2 = T_{o1} = T_1 + \xi_1^2/2c_p = 25 + 10.2^2/(2 \times 10^3) = 25.05°C$, and therefore $p_2 = p_{o1} = p_1(T_2/T_1)^{\gamma/(\gamma-1)} = 200 \times (298.20/298.15)^{1.4/0.4} = 200.12$ kPa.

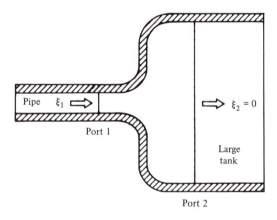

Figure 23.4 A short pipe segment with a very large change in cross-sectional area brings the flow to a condition of almost zero bulk velocity at port 2 where the enthalpy equals the stagnation enthalpy at port 1. If the process in the pipe segment is also reversible, then the pressure and temperature at port 2 equal, respectively, the stagnation pressure and stagnation temperature at port 1.

23.2 Diffusers

A *diffuser* is a short segment of pipe with increasing cross-sectional area along the direction of the flow (Figure 23.5). For given inlet and outlet flow speeds, it converts part of the kinetic energy of the incoming fluid into higher pressure of the outgoing fluid without any heat interactions. Typically, a diffuser process is irreversible, and the larger the irreversibility the smaller the conversion of kinetic energy into high pressure. In practice, the *efficiency of a diffuser* is defined as the ratio

$$\eta_D = \frac{p_{o2}}{p_{o1}} \tag{23.23}$$

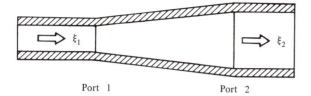

Figure 23.5 A diffuser is a short pipe segment with a cross-sectional area which increases in the direction of the flow. Its purpose is to convert part of the kinetic energy of the incoming fluid into higher pressure of the outgoing fluid.

where p_{o1} and p_{o2} are the stagnation pressures of the inlet and outlet bulk-flow states, respectively. The characterization of a diffuser by means of a value of the diffuser efficiency is equivalent to specifying the rate of entropy generation by irreversibility per unit mass flow rate and, therefore, η_D and the input bulk-flow state determine the effect of the diffuser, that is, the output bulk-flow state. We illustrate this point by the following example.

Example 23.3. A stream of $\dot{m} = 0.8$ kg/s of water flows into a cylindrical diffuser with inlet diameter $d_1 = 10$ mm, and outlet diameter $d_2 = 20$ mm. If the diffuser efficiency $\eta_D = 0.8$, the inlet temperature $T_1 = 20°C$, and the inlet pressure $p_1 = 150$ kPa, find the outlet temperature T_2 and the outlet pressure p_2.

Solution. The inlet and outlet cross-sectional areas are $a_1 = \pi d_1^2/4 = 78.5$ mm^2 and $a_2 = \pi d_2^2/4 = 314$ mm^2. The water density $\rho = 10^3$ kg/m^3, and therefore the inlet and outlet speeds are $\xi_1 = \dot{m}/\rho a_1 = 10.2$ m/s and $\xi_2 = \dot{m}/\rho a_2 = 2.55$ m/s. The inlet stagnation pressure $p_{o1} = p_1 + \rho \xi_1^2/2 = 202$ kPa, the outlet stagnation pressure $p_{o2} = \eta_D p_{o1} = 161.6$ kPa $= p_2 + \rho \xi_2^2/2$, and therefore $p_2 = p_{o2} - \rho \xi_2^2/2 = 158.3$ kPa.

From equation (23.16) we find that $h_2 - h_1 = \xi_1^2/2 - \xi_2^2/2$. On the other hand, for an incompressible fluid $h_2 - h_1 = c\,(T_2 - T_1) + (p_2 - p_1)/\rho$ so that $T_2 - T_1 = (\xi_1^2 - \xi_2^2)/2c - (p_2 - p_1)/\rho c = 0.0116 - 0.0002 = 0.0114°C$ for $c = 4.2$ kJ/kg K. Thus this diffuser achieves a pressure increase of 8.3 kPa with a negligible increase in temperature and a reduction of flow speed from 10.2 m/s to 2.55 m/s.

It is noteworthy that if the process were reversible, the change in pressure for the same change in kinetic energy would be $p_2 - p_1 = \rho \xi_1^2/2 - \rho \xi_2^2/2 = 48.8$ kPa and, therefore, $p_2 = 198.8$ kPa.

23.3 Nozzles

A *nozzle* is a short segment of pipe with rapidly decreasing cross-sectional area along the direction of the flow (Figure 23.6). For given inlet and outlet pressures, it converts part of the enthalpy of the incoming stream into kinetic energy of the outgoing stream without any heat interactions.

In practice, the *efficiency of a nozzle* is defined as the ratio

$$\eta_N = \frac{\xi_2^2}{\xi_{2s}^2} \tag{23.24}$$

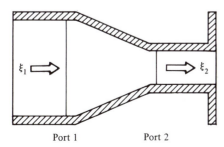

Figure 23.6 A nozzle is a short segment of a pipe with cross-sectional area which decreases rapidly in the direction of the flow. Its purpose is to convert part of the enthalpy of the incoming fluid into kinetic energy of the outgoing fluid.

where ξ_{2s} is the outlet flow speed that would be achieved for the same values of inlet and outlet pressures, p_1 and p_2, and the same inlet speed ξ_1 if the process were reversible, that is, $\dot{S}_{\text{irr}} = 0$ and $s_2 = s_1$. Equivalently, ξ_{2s} is the outlet speed when the flow occurs at constant stagnation temperature. For example, for an incompressible fluid we would find that $\xi_{2s}^2 = \xi_1^2 + 2(p_1 - p_2)/\rho$, and for a perfect gas $\xi_{2s}^2 = \xi_1^2 + 2c_pT_1\left[1 - (p_2/p_1)^{(\gamma-1)/\gamma}\right]$.

Example 23.4. In the Geneva Lake, in Geneva, Switzerland, a cylindrical nozzle jets a fountain of water about 70 m high. The nozzle inlet and outlet diameters are about 0.24 m and 0.07 m, respectively. Assuming that the nozzle efficiency $\eta_N = 0.95$, find the pressure that the pump exerts at the inlet in order to maintain the fountain.

Solution. At its high elevation, z, the specific potential energy of the water, gz, equals the specific kinetic energy at the outlet of the nozzle, $\xi_2^2/2$, and hence $\xi_2 = (2gz)^{1/2} = (2 \times 9.8 \times 70)^{1/2} = 37$ m/s, and the mass flow rate $\rho\xi_2 a_2 = 142$ kg/s $= 511$ t/h. The outlet pressure is atmospheric, that is, $p_2 = 101.3$ kPa, and the inlet speed is such that $\rho\xi_1 a_1 = \rho\xi_2 a_2$ or $\xi_1 = \xi_2 a_2/a_1 = 37 \times 0.07^2/0.24^2 = 3.1$ m/s. If the nozzle were reversible, the outlet speed $\xi_2 = \xi_{2s}$, $\xi_{2s}^2 = \xi_1^2 + 2(p_1 - p_2)/\rho$, and hence $p_1 = p_2 + \rho(\xi_{2s}^2 - \xi_1^2)/2 = 780.8$ kPa. For a nozzle efficiency $\eta_N = 0.95$, $\xi_{2s} = (\xi_2^2/\eta_N)^{1/2} = 38$ m/s, and therefore the pump must maintain a pressure $p_1 = 817$ kPa.

Example 23.5. Air flows from a pressurized tank into the atmosphere through an orifice that acts like a nozzle with a nozzle efficiency $\eta_N = 0.3$. If the air in the tank is at temperature $T_1 = 20°C$ and pressure $p_1 = 200$ kPa, find the nozzle outflow speed ξ_2 and temperature T_2. Repeat the problem for helium instead of air.

Solution. For air, $\gamma = 1.4$ and $c_p = 1$ kJ/kg K. The relation $\xi_{2s}^2 = \xi_1^2 + 2c_pT_1\left[1 - (p_2/p_1)^{(\gamma-1)/\gamma}\right]$ for $\xi_1 = 0$ and $p_2 = 101.3$ kPa yields $\xi_{2s} = 322$ m/s, and therefore $\xi_2 = \xi_{2s}(\eta_N)^{1/2} = 176$ m/s. Next, the relation $c_p(T_2 - T_1) + \xi_2^2/2 - \xi_1^2/2 = 0$ for $\xi_1 = 0$ yields $T_2 = T_1 - \xi_2^2/2c_p = 5°C$. For helium, $\gamma = 1.67$ and $c_p = 5.2$ kJ/kg K, and following the same procedure, we find that $\xi_{2s} = 853$ m/s, $\xi_2 = 467$ m/s, and $T_2 = -1°C$.

23.4 Throttles and Valves

A *throttle* is a short segment of pipe with a cross-sectional area that first decreases rapidly and then increases suddenly to its initial value (Figure 23.7). The sudden increase causes the flow streamlines to separate from the wall of the pipe and then experience turbulent mixing. The purpose of a throttle is to reduce the bulk-flow pressure from a value p_1 to a lower value p_2 by forcing the fluid through an abrupt restriction or *orifice*. If the cross-sectional area of the orifice is adjustable by means of an external mechanism (Figure 23.8), the device is called a *throttling valve* or, simply, a *valve*.

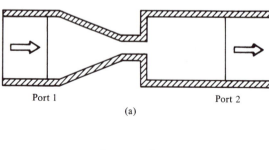

Port 1   Port 2

(a)

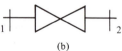

(b)

Figure 23.7 (a) A throttle is a short segment of pipe in which the cross-sectional area decreases rapidly and then returns suddenly to its initial value. Its purpose is to reduce the pressure of the fluid. (b) Symbol representing a throttle.

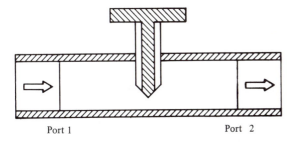

Figure 23.8 A valve is a throttling device with a restriction of variable cross-sectional area that can be controlled to regulate the pressure drop from inlet to outlet.

The fluid in a throttle experiences no interactions other than the two bulk flows at the inlet and outlet, no potential energy change because the pipe segment is short, $z_2 \approx z_1$, and no appreciable kinetic energy change because the cross-sectional area $a_2 \approx a_1$. In fact, if $a_2 = a_1$ and the fluid is incompressible, $\rho\xi_1 a_1 = \rho\xi_2 a_2$ and, therefore, $\xi_1 = \xi_2$.

In steady state, the energy and entropy rate balances are

$$\dot{m}(h_1 - h_2) = 0 \qquad \text{or} \qquad h_1 = h_2 \tag{23.25}$$

$$\dot{m}(s_1 - s_2) + \dot{S}_{\text{irr}} = 0 \tag{23.26}$$

We see that the process occurs at constant enthalpy, that is, the expansion is isenthalpic. Moreover, for a given pressure drop, such a process is irreversible as we can appreciate by observing the relative values of two isobars on an h versus s diagram (Figure 23.9). The relative positions and shapes of the isobars are as shown in the figure because $(\partial h/\partial p)_s = v > 0$ [equation (16.9)], $(\partial h/\partial s)_p = T > 0$ [equation (16.8)], and either $(\partial T/\partial s)_p = 0$ in a two phase region, or $(\partial T/\partial s)_p = T/c_p > 0$ in a vapor phase region.

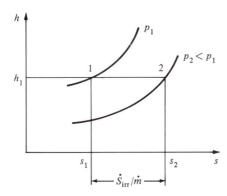

Figure 23.9 Representation of the end states of a throttle on an h versus s diagram. A specified reduction in pressure by a throttle or a valve is always accompanied by entropy generation by irreversibility.

The pressure drop caused by a throttle and the consequent entropy generation by irreversibility depend on the geometry of the orifice, the flow speed, and the type of fluid. For a given pressure drop, we can evaluate the entropy generation by irreversibility as follows. We recall that $dh = T\,ds + v\,dp$ and, therefore,

$$\left(\frac{\partial s}{\partial p}\right)_h = -\frac{v}{T} \tag{23.27}$$

Upon integrating (23.27) along a constant enthalpy line, we find that

$$s_2 - s_1 = -\int_{p_1}^{p_2} \left(\frac{v}{T}\right)_h dp \tag{23.28}$$

and using the entropy rate balance, that

$$\dot{S}_{\text{irr}} = \dot{m}(s_2 - s_1) = \dot{m}\int_{p_2}^{p_1} \left(\frac{v}{T}\right)_h dp \tag{23.29}$$

If the $\dot{m}$ moles per unit time behave as an ideal gas, then $v/T = R/p$ and

$$\dot{S}_{\text{irr}} = \dot{m}R \ln\frac{p_1}{p_2} \tag{23.30}$$

We see that the larger the ratio p_1/p_2, the larger is the entropy generation by irreversibility. The irreversibility can be reduced if the pressure lowering occurs while work is done by the stream, for example, by means of a turbine. Whether such work is advisable depends on the benefit from the work versus the cost of the turbine.

In contrast to the pressure and entropy changes, the temperature difference $T_2 - T_1$ across a throttle does not always have the same sign. Indeed, we can express this difference in the form

$$T_2 - T_1 = \int_{p_1}^{p_2} \left(\frac{\partial T}{\partial p}\right)_h dp \tag{23.31}$$

where the partial derivative $(\partial T/\partial p)_h$ is called the *Joule–Thomson coefficient* and denoted by μ_{JT}, that is,

$$\mu_{\text{JT}} = \left(\frac{\partial T}{\partial p}\right)_h \tag{23.32}$$

It is clear that the sign of $T_2 - T_1$ is determined by the sign of μ_{JT}, which turns out not to be definite. For example, for ideal gases $h = $ constant implies that $T = $ constant, and therefore $\mu_{\text{JT}} = 0$ and $T_2 = T_1$. Again, for a fluid obeying the van der Waals equation of state it can be shown that $\mu_{\text{JT}} \approx (2a - bRT)/c_p RT$. Again, for ideal incompressible behavior, $dh = c\,dT + v\,dp$ and hence $\mu_{\text{JT}} = -v/c < 0$ and $T_2 > T_1$ for $p_1 > p_2$. Again, for two-phase liquid–vapor flow, the Clausius–Clapeyron relation implies that $dT/dp = Tv_{\text{fg}}/h_{\text{fg}} > 0$, and therefore $\mu_{\text{JT}} > 0$ and $T_2 < T_1$ for $p_1 > p_2$.

For a general substance, the qualitative behavior of the constant-enthalpy lines on a temperature versus pressure diagram is shown in Figure 23.10. We see that at sufficiently low temperatures, in the vapor phase each constant-enthalpy line exhibits a maximum called the *inversion point*. The locus of inversion points, called the *inversion curve*, separates the region of states with positive values of the Joule–Thomson coefficient μ_{JT} from the region with negative μ_{JT}.

The existence of vapor states with positive Joule–Thomson coefficients is exploited in industrial gas liquefaction plants where a sequence of compressions and subsequent throttling expansions is used to cool a vapor to very low temperatures.

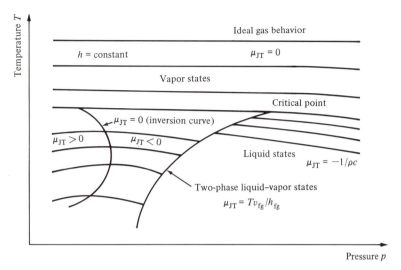

Figure 23.10 Constant enthalpy curves on a temperature versus pressure diagram. At each point (T, p), the slope of each $h = $ constant curve is the Joule-Thomson coefficient. The graph includes the inversion curve.

23.5 Compressors and Pumps

A *compressor* is a two-port device designed to increase the pressure of a compressible fluid, such as a gas (Figure 23.11). A *pump* is a two-port device designed to increase the pressure of an incompressible fluid such as a liquid (Figure 23.12). In either a compressor or a pump, the increase in pressure is achieved at the expense of work provided by means of a rotating shaft, the reciprocating motion of a piston, or a membrane.

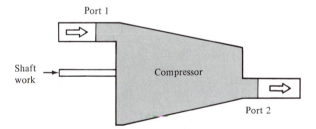

Figure 23.11 A compressor is a two-port device designed to increase the pressure of a compressible fluid, such as a gas.

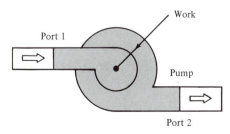

Figure 23.12 A pump is a two-port device designed to increase the pressure of an incompressible fluid, such as a liquid

In practice, heat interactions through the walls of a compressor or a pump are neglected. This is an excellent approximation because here both the path of the fluid particles through and their residence time in the device are short, and the characteristic time for energy transfer by means of heat interactions is generally much longer than that by means of the work interaction.[1] Accordingly, the steady-state energy and entropy rate balances are

$$\dot{m}(h_2 - h_1) = \dot{W}^{\leftarrow} \qquad \text{or} \qquad h_2 - h_1 = \frac{\dot{W}^{\leftarrow}}{\dot{m}} = w^{\leftarrow} \qquad (23.33)$$

$$\dot{m}(s_2 - s_1) = \dot{S}_{irr} \qquad \text{or} \qquad s_2 - s_1 = \frac{\dot{S}_{irr}}{\dot{m}} = s_{irr} \qquad (23.34)$$

For given pressures p_1 and p_2, the changes in specific enthalpy and specific entropy are shown on an h versus s diagram in Figure 23.13. We see from this diagram that the work per unit mass flow rate, $w^{\leftarrow} = h_2 - h_1$, is larger the larger the amount of irreversibility $s_{irr} = s_2 - s_1$. We also see that under the specified conditions, the smallest amount of work, $w^{\leftarrow}_{rev} = h_{2s} - h_1$, is required when the process is reversible, that is, when the outlet bulk-flow state $2s$ has a pressure equal to the specified outlet pressure, $p_{2s} = p_2$, and an entropy equal to that of the specified inlet state, $s_{2s} = s_1$.

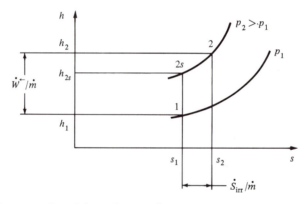

Figure 23.13 Representation of the end states of a compressor or a pump on a specific enthalpy versus specific entropy diagram.

For given pressures p_1 and p_2, it is customary to define an *efficiency of a compressor*, η_c, or an *efficiency of a pump*, η_p, as the ratio of the work $w^{\leftarrow}_{rev}$ that would be required if the process were reversible to the work $w^{\leftarrow}$ done on the stream in the course of the actual process, that is,

$$\eta_c \text{ (or } \eta_p) = \frac{w^{\leftarrow}_{rev}}{w^{\leftarrow}} = \frac{h_{2s} - h_1}{h_2 - h_1} \qquad (23.35)$$

The value of this efficiency ranges from zero to unity and is smaller the larger the irreversibility in the device.

For a pump operating on a perfect incompressible fluid for which $dh = c\,dT + v\,dp$, $ds = c\,dT/T$, v = constant, and c = constant, we find that

---

[1] Because the only interaction experienced by a compressor or a pump, in addition to the two bulk-flow interactions at the inlet and outlet ports, is work, it is customary to say that such devices are adiabatic. Here, however, this terminology is improper because the term adiabatic excludes not only heat interactions but also any other kind of nonwork interaction, including bulk flow.

$$w_{rev}^{\leftarrow} = v\,(p_2 - p_1) \tag{23.36}$$

$$w^{\leftarrow} = c\,(T_2 - T_1) + v\,(p_2 - p_1) = c\,(T_2 - T_1) + w_{rev}^{\leftarrow} \tag{23.37}$$

$$\eta_p = \frac{w_{rev}^{\leftarrow}}{w^{\leftarrow}} = \frac{v\,(p_2 - p_1)}{c\,(T_2 - T_1) + v\,(p_2 - p_1)} \tag{23.38}$$

The pressure difference $p_2 - p_1$ is called the *head of the pump*.

Example 23.6.   An irrigation system pumps water from a river and supplies a pipe at a mass flow rate $\dot m = 40$ kg/s, with a head $p_2 - p_1 = 300$ kPa. If the pump is powered by an electric motor and has an efficiency $\eta_p = 0.8$, find the electric power consumption.

Solution.   The work $w_{rev}^{\leftarrow} = v\,(p_2 - p_1) = 10^{-3} \times 300 = 300$ J/kg, and the work $w^{\leftarrow} = w_{rev}^{\leftarrow}/\eta_p = 300/0.8 = 375$ J/kg. Therefore, the electric power consumption $\dot W^{\leftarrow} = \dot m w^{\leftarrow} = 40 \times 375 = 15 \times 10^3$ W $= 15$ kW. At an electricity rate of 10 cents/kWh, the power cost is \$1.50 per hour. This cost is much less than the cost of labor performing the same task.

For a compressor operating on a perfect gas we find that

$$T_{2s} = T_1 \left(\frac{p_2}{p_1}\right)^{(\gamma-1)/\gamma} \tag{23.39}$$

$$w_{rev}^{\leftarrow} = h_{2s} - h_1 = c_p\,(T_{2s} - T_1) = c_p T_1 \left[\left(\frac{p_2}{p_1}\right)^{(\gamma-1)/\gamma} - 1\right] \tag{23.40}$$

$$w^{\leftarrow} = h_2 - h_1 = c_p\,(T_2 - T_1) \tag{23.41}$$

$$\eta_c = \frac{w_{rev}^{\leftarrow}}{w^{\leftarrow}} = \frac{T_{2s} - T_1}{T_2 - T_1} = \frac{T_1 \left[(p_2/p_1)^{(\gamma-1)/\gamma} - 1\right]}{T_2 - T_1} \tag{23.42}$$

Example 23.7.   An electrically driven compressor with efficiency $\eta_c = 0.8$ compresses $\dot m = 40$ kg/s of nitrogen and increases the pressure by $p_2 - p_1 = 300$ kPa. The initial gas temperature $T_1 = 25°C$, and pressure $p_1 = 101.3$ kPa. Find the power consumption and compare it to the result of Example 23.6.

Solution.   For nitrogen gas, $c_p = 1$ kJ/kg K and $\gamma = 1.4$. Therefore, $T_{2s} = T_1\,(p_2/p_1)^{(\gamma-1)/\gamma} = 298 \times (401.3/101.3)^{0.4/1.4} = 442$ K $= 169°C$, $T_2 = T_1 + (T_{2s} - T_1)/\eta_c = 298 + (442 - 298)/0.8 = 478$ K $= 205°C$, $w^{\leftarrow} = c_p\,(T_2 - T_1) = 1 \times 180 = 180$ kJ/kg, and $\dot W^{\leftarrow} = \dot m w^{\leftarrow} = 40 \times 180 = 7200$ kW. We see that the power consumption per unit mass flow rate required to compress nitrogen is much larger than that to compress a liquid between the same pressures. On the other hand, the power consumption per unit volume flow rate of nitrogen and that of a typical liquid are comparable. Indeed, the inlet specific volume of the nitrogen flow $v_1 = R_M T_1/p_1 = 0.286 \times 298/101.3 = 0.84$ m³/kg and therefore $w^{\leftarrow}/v_1 = 180/0.84 = 214$ kJ/m³, whereas for the water flow $v_1 = 10^{-3}$ m³/kg and $w^{\leftarrow}/v_1 = 0.375/10^{-3} = 375$ kJ/m³.

Example 23.8.   Repeat Example 23.7 using data from Table 20.4 instead of the perfect-gas model.

Solution.   Interpolating data from the table for $T_1 = 25°C$, we find that $h_1 = 8670$ J/mol and $p_{r1} = 1.003$. Therefore, equation (20.17) yields $p_{r2} = p_{r1}(p_2/p_1) = 1.003 \times (401.3/101.3) = 3.974$ and, interpolating again from the table, we find that $h_{2s} = 12{,}851$ J/mol and $T_{2s} = 168.1°C$. From equation (23.35), $h_2 = h_1 + (h_{2s} - h_1)/\eta_c = 8670 + (12{,}851 - 8670)/0.8 = 13{,}896$ J/mol or $w^{\leftarrow} = (h_2 - h_1)/M = (13{,}896 - 8670)/28.02 = 186.5$ kJ/kg, and from the table $T_2 = 213.7°C$. The reason for the differences in the values of $w^{\leftarrow}$ and T_2 between the present solution and that in Example 23.7 is that the perfect-gas model is inappropriate for nitrogen in the given range of temperatures.

23.6 Turbines

A *turbine* is a device designed to do work at the expense of a decrease in either the enthalpy of a fluid, or the kinetic energy of a fluid, or both (Figure 23.14). Heat interactions through the walls of a turbine are usually negligible for the reasons we discuss in connection with the operation of compressors and pumps.

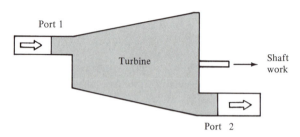

Figure 23.14 A turbine is a two-port device designed to do work at the expense of a decrease in either the enthalpy of a fluid, or the kinetic energy of a fluid, or both.

In *hydraulic turbines* the work is accounted for primarily by the change in kinetic energy of an incompressible fluid, in *gas turbines* by the change in enthalpy and pressure of a gaseous stream in states far from saturation, and in *steam turbines* or *vapor turbines* by the change in enthalpy and pressure of a steam or vapor stream in states close to saturation conditions.

For a gas or a steam turbine in a steady state, the energy and entropy rate balances are similar to those for a compressor, that is,

$$\dot{m}(h_1 - h_2) = \dot{W}^{\rightarrow} \quad \text{or} \quad h_1 - h_2 = \frac{\dot{W}^{\rightarrow}}{\dot{m}} = w^{\rightarrow} \qquad (23.43)$$

$$\dot{m}(s_2 - s_1) = \dot{S}_{\text{irr}} \quad \text{or} \quad s_2 - s_1 = \frac{\dot{S}_{\text{irr}}}{\dot{m}} = s_{\text{irr}} \qquad (23.44)$$

The difference, of course, is that work is done by the turbine ($\dot{W}^{\rightarrow} > 0$), whereas work is done on the compressor ($\dot{W}^{\leftarrow} > 0$). For a given pressure drop, the changes in specific enthalpy and specific entropy of a gas or steam turbine are shown on an h versus s diagram in Figures 23.15 and 23.16. We see from either of these diagrams that the work done per unit mass flow rate, $w^{\rightarrow} = h_1 - h_2$, is smaller the larger the amount of irreversibility, $s_{\text{irr}} = s_2 - s_1$. We also see that, under the specified conditions, the largest amount of work, $w^{\rightarrow}_{\text{rev}} = h_1 - h_{2s}$, for the specified outlet pressure, $p_{2s} = p_2$, would be done when the outlet bulk-flow state $2s$ has an entropy equal to that of the specified inlet state, $s_{2s} = s_1$, that is, when the process is reversible.

For a given pressure drop, it is common engineering practice to define an *efficiency of a turbine* η_t as the ratio of the work per unit flow rate $w^{\rightarrow}$ done in the actual process to the work $w^{\rightarrow}_{\text{rev}}$ that would be done if the process were reversible, that is,

$$\eta_t = \frac{w^{\rightarrow}}{w^{\rightarrow}_{\text{rev}}} = \frac{h_1 - h_2}{h_1 - h_{2s}} \qquad (23.45)$$

The value of this efficiency ranges from zero to unity, and it is smaller the larger the irreversibility in the turbine.

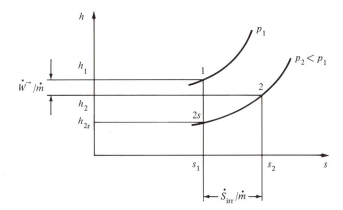

Figure 23.15 Representation of the end states of a fluid flowing through a turbine on a specific enthalpy versus specific entropy diagram.

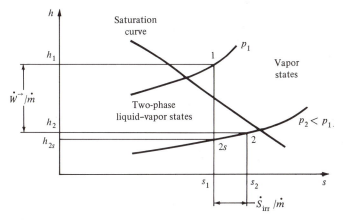

Figure 23.16 Representation of the end states of steam flowing through a steam turbine on a specific enthalpy versus specific entropy diagram. Here the steam is partially condensed prior to exiting the turbine.

For a gas turbine powered by a stream of a perfect gas, we have the following relations:

$$T_{2s} = T_1 \left(\frac{p_2}{p_1}\right)^{(\gamma-1)/\gamma} \tag{23.46}$$

$$w^{\rightarrow}_{rev} = c_p \left(T_1 - T_{2s}\right) = c_p T_1 \left[1 - \left(\frac{p_2}{p_1}\right)^{(\gamma-1)/\gamma}\right] \tag{23.47}$$

$$w^{\rightarrow} = c_p \left(T_1 - T_2\right) \tag{23.48}$$

$$\eta_t = \frac{w^{\rightarrow}}{w^{\rightarrow}_{rev}} = \frac{T_1 - T_2}{T_1 - T_{2s}} \tag{23.49}$$

Example 23.9. Find the work rate produced by a gas turbine with efficiency $\eta_t = 0.9$, and powered by a flow of nitrogen with $\dot{m} = 40$ kg/s, $T_1 = 900°C$, $p_1 = 1.013$ MPa, and $p_2 = 101.3$ kPa.

Solution. For nitrogen, $\gamma = 1.4$ and $c_p = 1$ kJ/kg K. Therefore, $T_{2s} = 1173 \times (101.3/1013)^{0.4/1.4} = 608$ K $= 335°C$, $T_2 = T_1 - \eta_t (T_1 - T_{2s}) = 664$ K $= 391°C$. So $\dot{W}^{\rightarrow} = \dot{m}(h_1 - h_2) = \dot{m} c_p (T_1 - T_2) = 40 \times 1 \times (900 - 391) = 20.4 \times 10^3$ kW $= 20.4$ MW.

Example 23.10. Repeat Example 23.9 using data from Table 20.4 instead of the perfect-gas model.

Solution. For $T_1 = 900°C$, by interpolation of data from the table, we find that $h_1 = 35,875$ J/mol and $p_{r1} = 153.92$, so that (20.17) yields $p_{r2} = p_{r1}(p_2/p_1) = 153.92 (101.3/1013) = 15.39$ and, again from the table, $h_{2s} = 18,895$ J/mol and $T_{2s} = 370.9°C$. From equation (23.45), $h_2 = h_1 - \eta_t (h_1 - h_{2s}) = 35,875 - 0.9 \times (35,875 - 18,895) = 20,593$ J/mol or $w^{\rightarrow} = (h_1 - h_2)/M = (35,875 - 20,593)/28.02 = 545.4$ kJ/kg and, from the table, $T_2 = 426.4°C$. Accordingly, the work rate is $\dot{W}^{\rightarrow} = \dot{m}w^{\rightarrow} = 21.8$ MW. The differences between the values of $\dot{W}^{\rightarrow}$ and T_2 here and in the solution of Example 23.9 are due to the inadequacy of the perfect-gas model for nitrogen in this range of temperatures.

Example 23.11. Find the rate of work done by a steam turbine having an efficiency $\eta_t = 0.9$, and operated by an inlet steam flow with $\dot{m} = 40$ kg/s, $T_1 = 580°C$, $p_1 = 4$ MPa, and outlet pressure p_2 equal to the saturation pressure at $45°C$. Evaluate the rate of entropy generation by irreversibility in the turbine.

Solution. To proceed with the solution, we find the properties of the bulk-flow states from the steam tables. In this connection it is noteworthy that some states may be two-phase and therefore their properties must be evaluated accordingly. A graphical illustration of this point is shown on the h versus s diagram in Figure 23.16. For $p_1 = 4$ MPa and $T_1 = 580°C$, we find that $h_1 = 3628.5$ kJ/kg and $s_1 = 7.3156$ kJ/kg K. Next, for $p_2 = p_{sat}(45°C) = 9.593$ kPa we find that $h_f = h_f(45°C) = 188.45$ kJ/kg, $h_g = h_g(45°C) = 2583.2$ kJ/kg, $h_{fg} = h_{fg}(45°C) = 2394.8$ kJ/kg, $s_f = s_f(45°C) = 0.6387$ kJ/kg K, $s_g = s_g(45°C) = 8.1648$ kJ/kg K, and $s_{fg} = s_{fg}(45°C) = 7.5261$ kJ/kg K. By comparing s_1 with s_f and s_g, we conclude that state $2s$ must be two-phase because $s_{2s} = s_1$ and $s_f < s_1 < s_g$. Accordingly, we compute the vapor fraction x_{2s} from the relation $s_{2s} = s_f + x_{2s}s_{fg} = s_1$ and find that $x_{2s} = (s_1 - s_f)/s_{fg} = 0.887$. Hence $h_{2s} = h_f + x_{2s}h_{fg} = 2313.1$ kJ/kg, $h_2 = h_1 - \eta_t (h_1 - h_{2s}) = 2444.5$ kJ/kg, and $\dot{W}^{\rightarrow} = \dot{m}(h_1 - h_2) = 47.4$ MW. State 2 is also two-phase because here $h_f < h_2 < h_g$. Specifically, $x_2 = (h_2 - h_f)/h_{fg} = 0.942$, $s_2 = s_g + x_2 s_{fg} = 7.7288$ kJ/kg K, and the rate of entropy generated by irreversibility $\dot{S}_{irr} = \dot{m}(s_2 - s_1) = 16.5$ kW/K.

Example 23.12. Repeat Example 23.11 for $p_1 = 1$ MPa.

Solution. For $p_1 = 1$ MPa and $T_1 = 580°C$ we find that $h_1 = 3653.5$ kJ/kg, $s_1 = 7.9776$ kJ/kg K, and the properties for the saturated states at $45°C$ listed in Example 23.11. Proceeding in the same manner as for Example 23.11, we calculate $x_{2s} = (s_1 - s_f)/s_{fg} = 0.975$, $h_{2s} = h_f + x_{2s}h_{fg} = 2523.7$ kJ/kg, $h_2 = h_1 - \eta_t (h_1 - h_{2s}) = 2636.7$ kJ/kg, and $\dot{W}^{\rightarrow} = \dot{m}(h_1 - h_2) = 40.7$ MW. Next we observe that $h_2 > h_g$, so that state 2 is single-phase. For $h_2 = 2636.7$ kJ/kg and $p_2 = p_{sat}(45°C) = 9.593$ kPa, we find that $T_2 = 73.2°C$, $s_2 = 8.3544$ kJ/kg K, and $\dot{S}_{irr} = \dot{m}(s_2 - s_1) = 15.1$ kW/K.

Example 23.13. Repeat Example 23.11 for $p_1 = 600$ kPa.

Solution. For $p_1 = 600$ kPa and $T_1 = 580°C$ we find that $h_1 = 3656.8$ kJ/kg, and $s_1 = 8.2162$ kJ/kg K; the properties for the saturated states at $45°C$ are as listed in Example 23.11. Here, both states $2s$ and 2 are single-phase superheated vapor states because $s_{2s} = s_1 > s_g(45°C)$. For $p_2 = p_{sat}(45°C) = 9.593$ kPa and $s_{2s} = s_1 = 8.2162$ kJ/kg K, we find that $T_{2s} = 48.8°C$, $h_{2s} = 2590.5$ kJ/kg, $h_2 = h_1 - \eta_t (h_1 - h_{2s}) = 2697.1$ kJ/kg, and $\dot{W}^{\rightarrow} = \dot{m}(h_1 - h_2) = 38.4$ MW. Moreover, $s_2 = 8.5211$ kJ/kg K, and $\dot{S}_{irr} = \dot{m}(s_2 - s_1) = 12.1$ kW/K.

23.7 Heat Exchangers

A *heat exchanger* is a device in which energy is transferred from a hot stream, gaseous or liquid, to a colder stream, gaseous or liquid. In the course of such a transfer, the hot stream is cooled, and the cold stream is heated (Figure 23.17).

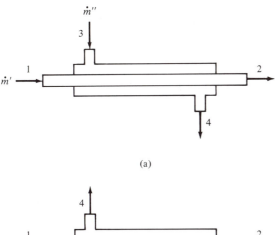

(a)

(b)

Figure 23.17 A heat exchanger is a four-port flow device designed to transfer energy and entropy from a hot or primary stream to a colder or secondary stream, without mixing of the streams. (a) Parallel flow heat exchanger. (b) Counterflow heat exchanger.

Different types of heat exchangers are distinguished depending on the flow patterns and the geometric arrangements of the flow streams, such as parallel-flow, counterflow, cross-flow, concentric-pipe, and shell-and-tube.

Upon modeling the inlet and outlet stream states as bulk-flow, disregarding speed and elevation changes, and using the terminology listed in the figure, we find that the steady-state energy and entropy rate balances are

$$\dot{m}'\,(h_1 - h_2) + \dot{m}''\,(h_3 - h_4) = 0 \tag{23.50}$$

$$\dot{m}'\,(s_1 - s_2) + \dot{m}''\,(s_3 - s_4) + \dot{S}_{\text{irr}} = 0 \tag{23.51}$$

These balances may be used to evaluate up to two unknown properties or to investigate the realizability of specified bulk-flow conditions. In particular, they may be combined to yield

$$\dot{S}_{\text{irr}} = \dot{Q}\left(\frac{s_4 - s_3}{h_4 - h_3} - \frac{s_2 - s_1}{h_2 - h_1}\right) \tag{23.52}$$

where

$$\dot{Q} = \dot{m}'\,(h_1 - h_2) = \dot{m}''\,(h_4 - h_3) \tag{23.53}$$

Equation (23.52) indicates that in steady state, the rate of entropy generated by irreversibility in a heat exchanger is proportional to the rate of energy $\dot{Q}$ transferred from the primary to the secondary stream and to the difference of the inverses of the two ratios $(h_4 - h_3)/(s_4 - s_3)$ and $(h_2 - h_1)/(s_2 - s_1)$. Because $\dot{S}_{\text{irr}}$ must be nonnegative, we conclude that

$$\frac{h_2 - h_1}{s_2 - s_1} \geq \frac{h_4 - h_3}{s_4 - s_3} \qquad \text{for} \quad \dot{Q} > 0 \tag{23.54}$$

Equation 23.54 has an important implication. It must be valid for all bulk-flow conditions of a heat exchanger. In particular, it must hold for any infinitesimal segment of the heat exchanger in which energy is transferred at the rate $\delta \dot{Q}$ between two streams each under isobaric conditions. To pursue the consequences of this implication, we subdivide each stream into thin consecutive slices and consider the energy transfer from a slice of the hot stream to an adjacent slice of the cold stream. By using the rate balances for these two slices, we can readily verify that

$$\delta \dot{S}_{\text{irr}} = \delta \dot{Q} \left[\left(\frac{\partial s}{\partial h} \right)_p'' - \left(\frac{\partial s}{\partial h} \right)_p' \right]$$

$$= \delta \dot{Q} \left(\frac{1}{T''} - \frac{1}{T'} \right) \tag{23.55}$$

where superscripts $'$ and $''$ refer to the primary and the secondary streams, respectively, and each of the temperatures $T' = (\partial h / \partial s)_p'$ and $T'' = (\partial h / \partial s)_p''$ is associated with the entering bulk-flow state of the corresponding slice. Because $\delta \dot{S}_{\text{irr}}$ must be nonnegative, we conclude from (23.55) that wherever energy is transferred from the primary stream to the secondary stream, the local bulk-flow state temperature T' must be greater than T'' or at least equal to T''. The irreversibility is larger the larger the difference between T' and T'' at each pair of adjacent slices.

If each of the two streams behaves either as a perfect gas with specific heat at constant pressure c_p, or as an incompressible fluid with specific heat c, it is convenient to denote the product of the mass flow rate and the specific heat by Γ, so that for the perfect gas stream $\Gamma = \dot{m} c_p$ and for the incompressible fluid stream $\Gamma = \dot{m} c$. If, in addition, the pressure drop from inlet to outlet is negligible for both streams, the following relations hold for the heat exchanger:

$$\dot{m}'(h_2 - h_1) = \Gamma'(T_2 - T_1) \qquad \dot{m}'(s_2 - s_1) = \Gamma' \ln(T_2/T_1)$$

$$\dot{m}''(h_4 - h_3) = \Gamma''(T_4 - T_3) \qquad \dot{m}''(s_4 - s_3) = \Gamma'' \ln(T_4/T_3)$$

$$T_{\text{lm}}' = \frac{T_2 - T_1}{\ln(T_2/T_1)} \qquad T_{\text{lm}}'' = \frac{T_4 - T_3}{\ln(T_4/T_3)}$$

$$\dot{Q} = \dot{m}' c_p' (T_1 - T_2) = \dot{m}'' c_p'' (T_4 - T_3) \tag{23.56}$$

$$\dot{Q} = \Gamma'(T_1 - T_2) = \Gamma''(T_4 - T_3)$$

$$\dot{S}_{\text{irr}} = \dot{Q} \left(\frac{1}{T_{\text{lm}}''} - \frac{1}{T_{\text{lm}}'} \right)$$

23.8 Heat Conductors and Heat Convectors

A *heat conductor* is a slab of stationary material, such as a solid, that can be modeled as a system experiencing a heat interaction at a rate $\dot{Q}_1^{\leftarrow}$ at temperature T_1, and another heat interaction at a rate $\dot{Q}_2^{\rightarrow}$ at temperature T_2 (Figure 23.18). A *heat convector* is a rigid container filled with a fluid in motion that can be modeled in the same way (Figure 23.19). The fluid motion occurs either spontaneously, such as in natural convection devices and heat pipes, or as a result of external forces, such as in forced convection devices. The energy and entropy rate balances for a heat conductor or a heat convector in steady state are

$$\dot{Q}_1^{\leftarrow} - \dot{Q}_2^{\rightarrow} = 0 \tag{23.57}$$

$$\frac{\dot{Q}_1^{\leftarrow}}{T_1} - \frac{\dot{Q}_2^{\rightarrow}}{T_2} + \dot{S}_{\text{irr}} = 0 \tag{23.58}$$

and upon eliminating $\dot{Q}_2^{\rightarrow}$ and defining $\dot{Q} = \dot{Q}_1^{\leftarrow} = \dot{Q}_2^{\rightarrow}$, we find that

$$\dot{S}_{\text{irr}} = \dot{Q}\left(\frac{1}{T_2} - \frac{1}{T_1}\right) \tag{23.59}$$

For a heat conductor with constant cross-sectional area a and length l, it is customary to express $\dot{Q}$ in the form

$$\dot{Q} = \frac{a\lambda\,(T_1 - T_2)}{l} \tag{23.60}$$

where λ is an empirically determined quantity called the *thermal conductivity* of the heat conductor. Using equation (23.60) in equation (23.59), we find that

$$\dot{S}_{\text{irr}} = \frac{a\lambda\,(T_1 - T_2)^2}{lT_1T_2} = \frac{l\dot{Q}^2}{a\lambda T_1 T_2} \tag{23.61}$$

Similarly, for a heat convector it is customary to express $\dot{Q}$ in the form

$$\dot{Q} = a\alpha\,(T_1 - T_2) \tag{23.62}$$

where a is the area of the surface where one of the heat interactions occurs, and α is an empirically determined quantity called the *heat transfer coefficient* at that surface. Using equation (23.62) in equation (23.59), we find that

$$\dot{S}_{\text{irr}} = \frac{a\alpha\,(T_1 - T_2)^2}{T_1T_2} = \frac{\dot{Q}^2}{a\alpha T_1 T_2} \tag{23.63}$$

We see that the rate of entropy generation by irreversibility is positive, as it should, regardless of the sign of the difference $T_1 - T_2$ or $\dot{Q}$. Moreover, for a given value either of $a\lambda/l$ for a heat conductor or of $a\alpha$ for a heat convector, the irreversibility is larger the

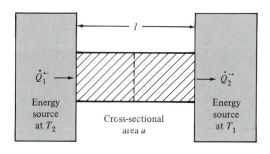

Figure 23.18 A heat conductor is a slab of stationary material that can be modeled as a system experiencing two heat interactions with energy sources at different temperatures.

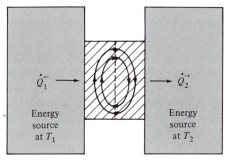

Figure 23.19 A heat convector is a rigid container filled with a fluid in motion that can be modeled as a system experiencing two heat interactions with energy sources at different temperatures.

larger the temperature difference $T_1 - T_2$, and for given values of $\dot{Q}$, and either λ, l, or α, the irreversibility $\dot{S}_{irr}$ is smaller the larger the area a.

It is noteworthy that the larger the area a, the more expensive is the slab. So in attempting to improve a heat conduction or heat convection design, we must balance the gain from the reduced irreversibility versus the higher cost of the larger cross-sectional area of either the conductor or the convector.

Problems

23.1 Consider a perfectly insulated, steady-state, horizontal segment of a pipe with variable cross-sectional area. Assume that the interactions at ports 1 and 2 satisfy the bulk-flow conditions. As usually, denote pressure by p, temperature by T, bulk-flow speed by ξ, and specific volume by v. Assume that the flowing substance behaves as a perfect incompressible fluid with specific volume v and specific heat c.

(a) Write a relation between the temperature change $T_2 - T_1$, the pressure change $p_2 - p_1$, the constants v and c, and the speeds ξ_1 and ξ_2 at the ports of the pipe.

(b) Write a condition, in terms of no other variables than p_1, p_2, ξ_1, ξ_2, and v, for the direction of the flow to be from port 1 to port 2.

Now assume that the flowing substance behaves as a perfect gas with specific heats c_v and c_p, specific heat ratio γ, and gas constant R.

(c) Write a condition, in terms of no other variables than p_1, p_2, ξ_1, ξ_2, T_1, c_p, and γ, for the direction of the flow to be from port 1 to port 2.

23.2 Liquid water enters a nozzle at 130 kPa, 30°C, and negligible speed, and exits at a pressure of 110 kPa. Model water as a perfect incompressible fluid with $c = 4.18$ kJ/kg K and $v = 10^{-3}$ m^3/kg.

(a) Find the largest possible exit speed from the nozzle under the specified conditions.

(b) If the nozzle efficiency is $\eta_N = 0.85$, find the temperature rise of the water and the exit speed.

(c) Find the entropy generated by irreversibility per unit flow rate.

23.3 Superheated steam at 260°C and 13 bar is supplied with negligible velocity to a well-insulated nozzle. The steam expands through the nozzle to a pressure of 1.4 bar and, because of irreversibility, achieves an exit speed 0.05 that achieved under isentropic expansion.

(a) What is the exit speed?

(b) How much entropy is generated by irreversibility per kilogram of steam flowing through the nozzle?

(c) What is the largest work (per kilogram of steam) that machinery can do using the superheated steam in an environment at $T_0 = 20$°C and $p_0 = 1$ bar?

(d) How much of the work potential in part (c) is lost as the steam expands through the nozzle?

23.4 It is estimated that a new natural gas well can deliver 3×10^4 m^3/h of almost pure methane, CH$_4$, at 20°C and 1 atm. The gas exits from the wellhead at 27°C and 7 atm. To be transported to the demand centers, the gas must be supplied to the pipeline at 37°C and 34 atm. This change in state is accomplished by using a well-insulated compressor that compresses the gas to 34 atm, and then a water-cooled exchanger that cools the gas at constant pressure to 37°C. The design of the compressor is such that the work required is 1.25 times that of the isentropic compressor, that' is, the efficiency is 0.8. Model the gas behavior of methane as ideal but not perfect, and use data from either Table 20.1 or 20.2.

(a) What is the power rating of the required compressor?

(b) What is the energy rate to the cooling water in the exchanger?

23.5 The water supply of a skyscraper is contained in a 10-m^3 tank at the top of the building about 200 m from the ground floor. Water is available at the ground level at $T_o = 7°C$. A pump consumes about twice the least work required to fill the tank.

(a) What is the least work required to fill the tank?

(b) If the pump is perfectly insulated, estimate the temperature change of the water from the inlet to the outlet of the pump.

(c) If the average water consumption of the building is 100 m^3/day, what is the power rating of the pump?

23.6 A stream of a perfect gas ($\dot{m} = 5$ kg/s, $c_p = 1$ kJ/kg K, $R = 0.286$ kJ/kg K) flows steadily through the power plant shown schematically in Figure P23.6. The stream enters turbine A in state 1, at temperature $T_1 = 750$ K and pressure $p_1 = 6$ MPa, and exits at an intermediate pressure $p_2 = 1.2$ MPa. The turbine is thermally insulated and has a turbine efficiency $\eta_t = 0.85$.

Upon exiting turbine A in state 2, the stream enters a waste-heat-recovery unit B that does work. The stream leaves the unit at pressure $p_3 = p_o = 1$ atm. In addition to the stream and the shaft, unit B interacts with the environmental reservoir at $T_o = 300$ K.

(a) Find the power $\dot{W}_A^{\rightarrow}$ produced by turbine A.

(b) Find the rate of entropy generation by irreversibility in turbine A.

(c) Find the largest shaft power $\dot{W}_B^{\rightarrow}$ that can be produced by unit B.

(d) If instead of being used in unit B, the stream in state 2 is fed to a heating network that does no work but utilizes all the energy of the stream, what would you use as a measure of how efficiently the stream in state 1 is used in turbine A?

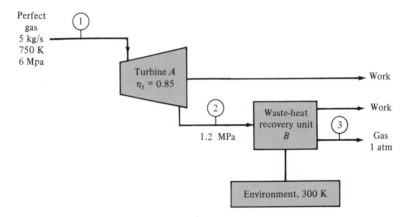

Figure P23.6

23.7 A steam turbogenerator and a nitrogen compressor are coupled together as shown schematically in Figure P23.7. The inlet conditions to the steam turbine are 100 kg/h, 40 bar, and 370°C, and the exit conditions 0.2 bar and $x = 0.95$. Nitrogen enters the compressor at 15 kg/h, 1 bar, and 15°C, and exits the aftercooler at 100 bar, and 40°C. The steam turbine delivers 5.5 hp to the compressor and the balance to the generator.

(a) Determine the power supplied from the turbine to the generator. The turbine and the compressor are very well insulated.

(b) Determine the rate of energy loss from the nitrogen as it flows through the compressor and aftercooler.

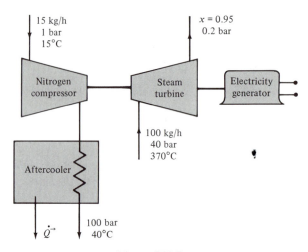

15 kg/h
1 bar
15°C

x = 0.95
0.2 bar

Nitrogen compressor

Steam turbine

Electricity generator

100 kg/h
40 bar
370°C

Aftercooler

$\dot{Q}$

100 bar
40°C

Figure P23.7

23.8 Water is used as the working fluid of a pressurized-water nuclear-power plant. Figure P23.8 shows a schematic of the plant. Saturated liquid exiting the reactor (state 1, pressure $p_1 = 10$ MPa) enters a flash evaporator at a lower pressure ($p_2 = 2.2$ MPa). The pressure reduction is achieved by a throttle. Thus, a fraction of the water flashes into saturated steam (state 2), and the remainder flows as liquid into the mixing chamber (state 3, $p_3 = p_2$). The steam enters the turbine, generates shaft work, and exits in state 4 (pressure $p_4 = 6.5$ kPa). After state 4, the working fluid enters the condenser, exits the condenser as saturated liquid at 35°C, enters the pump, and then the mixing chamber. The liquid exiting the mixing chamber is pumped to the pressure of the reactor and reenters the reactor in state 5 (pressure $p_5 = 10$ MPa). Assume that the pumps are 100% efficient and that liquid water is accurately modeled as a perfect incompressible fluid.

(a) How much steam exits the flash evaporator per kilogram of saturated liquid from the reactor?

(b) What are the specific enthalpy and the specific entropy of the water in state 5?

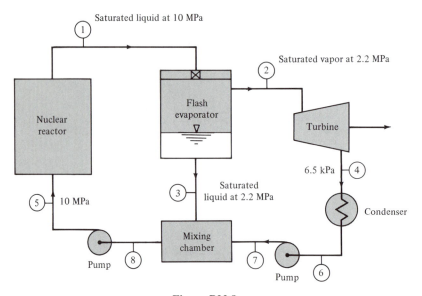

Saturated liquid at 10 MPa

Saturated vapor at 2.2 MPa

Nuclear reactor

Flash evaporator

Turbine

6.5 kPa

Saturated liquid at 2.2 MPa

10 MPa

Condenser

Pump

Mixing chamber

Pump

Figure P23.8

(c) What is the shaft work done by the turbine per kilogram of water leaving the reactor? The turbine efficiency $\eta_t = 0.95$.

(d) What are the energy and the entropy gained by the water in the nuclear reactor per kilogram of water leaving the reactor?

(e) What are the rates of entropy generation by irreversibility in the flash evaporator, the mixing chamber, and the turbine?

(f) Assuming an environmental temperature of $T_o = 25°C$, what is the rate of entropy generation by irreversibility in the condenser?

(g) Energy from nuclear fission in the reactor can be regarded as having approximately zero entropy. If so, what is the rate of entropy generation by irreversibility in the reactor?

(h) What fraction of the power from nuclear fission, $\dot{E}_{\text{fission}}$, is converted into net shaft power, that is, turbine power minus pump power? Verify that your answer is equal to $[\dot{E}_{\text{fission}} - T_o(\dot{S}_{\text{irr}})_{\text{overall}}]/\dot{E}_{\text{fission}}$, where $(\dot{S}_{\text{irr}})_{\text{overall}}$ is the overall rate of entropy generation by irreversibility in the entire plant.

23.9 The structure of a typical centrifugal compressor is shown in Figure P23.9. Air enters the compressor at plane 0 at 1 atm and 300 K at negligible bulk-flow velocity. The flow accelerates to velocity ξ_1 as it passes from plane 0 to plane 1, where it enters the impeller. In the impeller, a work per unit mass $w_c^{\leftarrow}$ is done on the airflow and the (absolute) velocity increases from ξ_1 to ξ_2. The flow is then decelerated in the diffuser to a lower velocity ξ_3. Finally, in the collector, the flow further decelerates to negligible velocity and exits the compressor at plane 4.

(a) Derive expressions for $(h_0 - h_1)$, $(h_2 - h_1)$, $(h_3 - h_2)$, $(h_4 - h_3)$, and $(h_4 - h_0)$ in terms of the variables just cited. Heat interactions are negligible at any stage of the process.

(b) Locate states 0, 1, 2, 3, and 4 on a carefully proportioned qualitative h versus s diagram for this real compressor in relation to the constant-pressure curves corresponding to $p_{\text{atm}} = p_0$, p_1, p_2, p_3, and p_4.

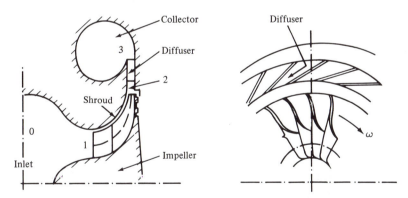

Figure P23.9

23.10 The turbocharger of a diesel engine is shown schematically in Figure P23.10. The compressor draws in atmospheric air at 1 atm and 300 K and delivers air at 2 atm to the engine intake manifold. The exhaust gases exit the engine exhaust manifold at 850 K and enter the turbine, which drives the compressor. The manifold conditions and the turbocharger speed can be assumed to be steady. The exhaust-gases mass flow rate is 4% higher than the intake-air mass flow rate because of fuel injection in the engine.

Both the inlet air and the exhaust gases may be modeled as perfect, with $R_a = 0.287$ kJ/kg K and $\gamma_a = 1.4$ for air, and $R_e = 0.290$ kJ/kg K and $\gamma_e = 1.32$ for the exhaust products of combustion.

(a) If both the compressor and the turbine are isentropic, what is the pressure in the exhaust manifold between the engine and the turbine?

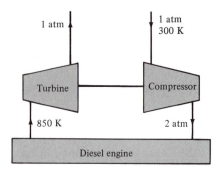

Figure P23.10

(b) If the compressor efficiency is 0.7 and the turbine efficiency 0.8, what is the pressure in the exhaust manifold?

23.11 A two-stage compressor with intercooler and postcooler is sketched in Figure P23.11. The first compressor stage, A, draws in 1 kg/s of air at atmospheric conditions ($p_1 = 101$ kPa and $T_1 = 300$ K) and delivers air at pressure p_2. This stream is fed to the intercooler, where it exchanges energy and entropy with the environment at $T_o = 300$ K and exits at pressure $p_3 = p_2$ and temperature $T_3 = 350$ K. The stream is then drawn into the second compressor stage, B, and is delivered at pressure $p_4 = 10$ MPa. Finally, the stream is fed to the postcooler and cooled down to atmospheric temperature $T_5 = 300$ K.

Assume that air behaves as a perfect gas with $R = 0.287$ kJ/kg K and $\gamma = 1.4$, and that the compressor stages have efficiencies $\eta_A = \eta_B = 0.8$.

(a) Find the intermediate pressure $p_2 = p_3$ for which the total power input, $\dot{W}_A^- + \dot{W}_B^-$, is the least. What is the outlet temperature T_4? Sketch the bulk-flow states of the stream on an h versus s diagram.

(b) Find the rates of entropy generation due to irreversibility in each of the two compressor stages, in the intercooler, and the postcooler.

(c) Compare the total power input $\dot{W}_A^- + \dot{W}_B^-$ and the total entropy generation by irreversibility in the two-stage compressor with intercooler and postcooler with the power input $\dot{W}_C$ and total entropy generation in a single-stage compressor C followed by a postcooler, assuming a compressor efficiency $\eta_C = 0.8$, the same inlet conditions as in part (a), and the same outlet conditions of 10 MPa and 300 K.

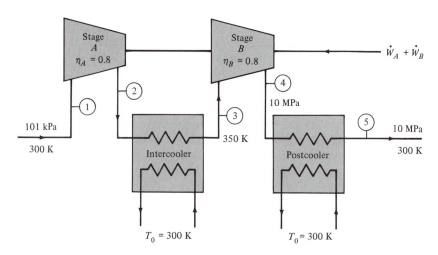

Figure P23.11

23.12 Consider the exchanges of energy and entropy between two perfect-gas streams in a counterflow heat exchanger in which pressure drops are not negligible (Figure P23.12). Assume that the two streams have the same heat capacity rate $\dot{n}c_p$, that the hot stream enters at temperature T_1 and pressure p_1 and exits at $T_1 - \Delta T$ and $p_1 - \Delta p_1$, and that the cold stream enters at temperature T_2 and pressure p_2 and exits at $p_2 - \Delta p_2$.

(a) Write an expression for the ratio of the rate of entropy generation by irreversibility and the heat capacity rate, $\dot{S}_{irr}/\dot{n}c_p$, in terms of the given quantities and the specific-heat ratio $\gamma = c_p/c_v$. Rewrite this expression in terms of $\Delta p_1/p_1$, $\Delta p_2/p_2$, $(T_1 - T_2)^2/T_1T_2$, and the so-called "heat-exchanger effectiveness" $\epsilon = \Delta T/(T_1 - T_2)$.

(b) Linearize the expression found in part (a) in the limit as $\Delta p_1/p_1 \ll 1$, $\Delta p_2/p_2 \ll 1$, and $\epsilon(1 - \epsilon)(T_1 - T_2)^2/T_1T_2 \ll 1$, and comment on the effects of the various terms on irreversibility and on methods to reduce such effects.

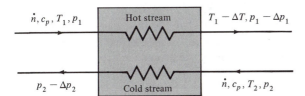

Figure P23.12

23.13 Combustion gases leave a kerosene-fueled burner at 1350 K and 1 MPa. The gases enter a gas turbine, do work, and exit the turbine at 890 K and atmospheric pressure. Energy losses from the turbine are negligible. The gases can be modeled as perfect with molecular mass $M = 29$ kg/kmol and specific-heat ratio $\gamma = 1.32$.

(a) What is the work per unit mass flowing through the gas turbine?

(b) What is the ratio of the work in part (a) to the work per unit mass that would be produced by a reversible turbine (with no energy losses) with the same inlet stream and the same outlet pressure as those of the gas turbine?

(c) What is the ratio of the work in part (a) to the largest work that could be produced by any device that can interact with the environment at 25°C and has the same inlet stream as the gas turbine?

(d) Comment on the meaning of the ratio of your answer in part (c) and your answer in part (b).

23.14 A cost-effective method of waste-energy recovery is to use the waste energy in products of combustion to preheat—in a counter-flow heat exchanger or "preheater"—the air used for combustion of a fuel. In such an application, the products of combustion flow at a rate of 14 kg/s, and are cooled from 320°C to 200°C. The air stream is heated from 35°C to 175°C. Assume perfect-gas behavior for both the products of combustion ($c_{ppc} = 1.088$ kJ/kg K) and the air ($c_{pa} = 1.005$ kJ/kg K), and neglect pressure drops in the preheater.

(a) At what rate is the air stream flowing in the preheater, and at what rate is energy transferred from the stream of products of combustion to the air stream?

(b) At what rate is entropy removed from the stream of products of combustion, and at what rate is entropy received by the air stream?

(c) If the energy transfer in part (a) is achieved reversibly by machinery interacting solely with the two streams, what is the rate of work done by this machinery for the given inlet and outlet temperatures of the two streams?

(d) If the energy transfer in part (a) is achieved reversibly by machinery with no net work done and no interactions other than with the two streams, what is the outlet air-stream temperature for the given inlet and outlet temperatures of the stream of products of combustion and the given inlet air-stream temperature?

23.15 A steady flow of air at 1000 K and 10 atm is available at the end of an industrial process. For air, $c_p = 1$ kJ/kg K and $\gamma = 1.4$. Different work-producing devices that could use this air and could interact with the environment at 300 K and 1 atm are being evaluated.

(a) What is the largest work per unit mass of air that can be done by any such device?

(b) What is the largest work per unit mass of air that can be done by any perfectly insulated gas turbine?

(c) What is the largest work per unit mass of air that can be done if the air is contaminated and can interact with the devices only via a constant-pressure heat exchanger?

23.16 In the hydroelectric power plant sketched in Figure P23.16, water flows through the turbine from a large reservoir at a rate of 10^5 kg/s. The difference in elevation between the water reservoir and the turbine exit plane is 180 m. The turbine is well insulated and its efficiency is 97%. The water velocity at the exit plane of the turbine is 6 m/s. The water reservoir temperature $T_o = 280$ K.

(a) What is the water temperature at the turbine exit plane?

(b) How much entropy is generated by irreversibility in the turbine?

(c) If the difference in elevation between the turbine exit plane and the discharge pond is 3 m, what is the water temperature of the discharge pond? How much entropy is generated by irreversibility in the discharge?

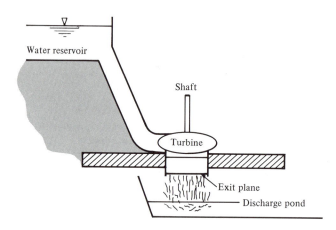

Figure P23.16

23.17 A source of high-pressure nitrogen is used to produce liquid nitrogen as shown in Figure P23.17. High-pressure gas ($p_1 = 2$ MPa, $T_1 = 27°C$) is cooled (state 2) and then expanded through a throttle to a two-phase mixture at atmospheric pressure. Saturated liquid (state 4) is extracted, and the saturated vapor (state 3) is passed through the heat exchanger to cool the incoming stream. The heated gas stream exits at $T_5 = 27°C$. Use tabulated data for nitrogen, and assume no pressure drops in the heat exchanger.

(a) What is the ratio $\dot{m}_4/\dot{m}_1$, namely, the mass flow rate of liquid N$_2$ per unit flow rate of high-pressure nitrogen entering the heat exchanger?

(b) What is the temperature T_2 of the nitrogen exiting the heat exchanger in state 2?

(c) What is the rate of entropy generation by irreversibility per unit flow rate of high-pressure nitrogen?

(d) How is the entropy generation by irreversibility distributed between the heat exchanger and the throttle?

23.18 In the Claude liquefaction system shown in Figure P23.18, nitrogen gas at 27°C and 1 atm (state 1) is compressed reversibly and isothermally to 5 MPa (state 2). The gas is then cooled to

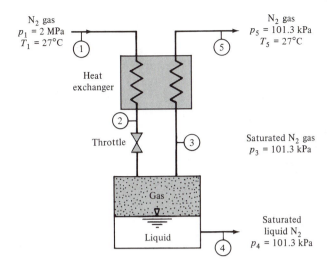

Figure P23.17

0°C (state 3) by passing through a heat exchanger. At the outlet of this heat exchanger, 60% of the total flow is diverted and expanded to 1 atm (state 4) through a well-insulated expander with efficiency of 75%. Two more energy exchanges occur (states 5 and 6) and, after throttling to 1 atm (state 7), saturated liquid nitrogen is withdrawn (state 8). Makeup nitrogen gas is supplied at the conditions of state 1.

(a) How much work is done by the expander per kilogram of nitrogen gas flowing in stream 2?

(b) What is the liquid nitrogen yield per kilogram of nitrogen gas flowing in stream 2?

(c) What is the net work required to produce 1 kg of liquefied nitrogen?

(d) What is the least work required by any machinery to produce liquid nitrogen at the conditions of state 8 starting from the environmental conditions of state 1?

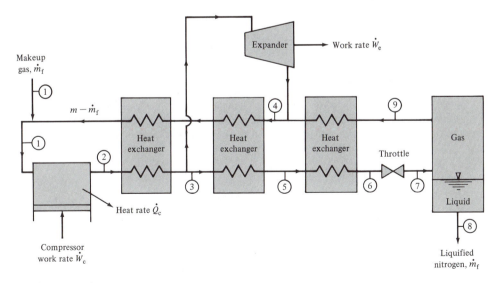

Figure P23.18

23.19 Fresh water may be produced from seawater by recovering energy from the exhaust of a steam turbine as shown in Figure P23.19. The two-phase mixture exiting from the turbine is

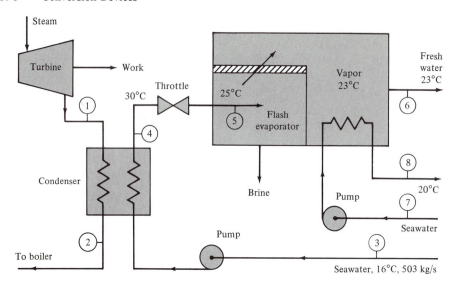

Figure P23.19

used to heat seawater from 16°C (state 3) to 30°C (state 4) at a rate of 503 kg/s. Then the heated seawater enters a flash evaporator, where the pressure is reduced to that corresponding to a saturation temperature $T_5 = 25$°C. During this expansion, some of the seawater flashes into vapor and the remainder is pumped back into the sea as brine. The vapor is condensed at $T_6 = 23$°C to form fresh water by utilizing seawater that enters the flash evaporator at 16°C (state 7) and leaves it at 20°C (state 8). For the purposes of this problem, model seawater as pure water.

(a) How much pure water is produced per hour?

(b) What is the flow rate of seawater at state 7?

(c) At what rate is entropy generated by irreversibility in the condenser if the minimum temperature difference between the condensing steam and the seawater is 10°C?

(d) What is the total rate of entropy generation by irreversibility per kilogram of pure water produced in the overall desalination process?

(e) What is the rate at which energy is supplied by the condensing steam to the desalination process per kilogram of pure water produced?

(f) Of the energy found in part (e), how much work can be done on a weight by cyclic machinery interacting with the environmental reservoir—the seawater—at 16°C? Compare this work with the least work of separation of 1 kg of pure water from seawater at 16°C, which is about 1.43 kJ/kg.

23.20 Use the results of Problems 19.24 and 21.4 to show that the Joule–Thomson coefficient $(\partial T/\partial p)_h$ may be approximated by the relation

$$\mu_{JT} = \left(\frac{\partial T}{\partial p}\right)_h \approx \frac{1}{c_p}\left(\frac{2a}{RT} - b\right)$$

and therefore that it changes sign at a temperature

$$T_{inversion} = \frac{2a}{bR}$$

where the inversion temperature $T_{inversion}$ is such that $\mu_{JT} < 0$ for $T > T_{inversion}$, and $\mu_{JT} > 0$ for $T < T_{inversion}$.

23.21 Use the results of Problem 23.20 and the van der Waals constants in Table 21.1 to estimate the Joule–Thomson coefficient (in K/MPa) at 400°C and 1 atm for water. Compare the result with an estimate using data from the steam tables.

23.22 Methane from a gas well is transported in a pipeline and reaches a demand center at 20°C and 30 atm. Before entering the local distribution network, the pressure of the methane must be decreased to 3 atm. The expansion occurs in a throttle and is immediately followed by heating in a heat exchanger, where the heating is usually provided by the combustion of a small fraction of the methane (which provides about 50 MJ/kg).

(a) Use the results of Problem 23.20 and the van der Waals constants in Table 21.1 to estimate the Joule–Thomson coefficient (in K/MPa) at 20°C and 1 atm for methane, and the temperature drop of the methane stream upon expansion in the throttle.

(b) Estimate the fraction of methane that must be burned to provide sufficient heating so that the methane stream remains at 20°C after the expansion.

23.23 An inventor claims to have built a machine that can compress a gaseous stream while interacting only with a reservoir at 180 K. Consider a stream of CO_2 and a stream of H_2 with the end states as listed in the following table.

	CO$_2$		H$_2$	
	State 1	State 2	State 1	State 2
T [K]	322	267	600	400
p [atm]	13.6	20.4	1	10
h [kJ/kg]	152.6	137.8	8881	5983
s [kJ/kg K]	1.133	1.068	69.84	54.507
v [m^3/kg]	0.04415	0.02441	24.43	1.658

(a) Does the device work for either of the two streams?

(b) For what ranges of reservoir temperatures can the machine operate for each of the two streams?

(c) What are the principal devices of the machine?

23.24 We need to supply 10^5 kJ of energy to an oven kept at fixed temperature $T = 1000$ K. We have three things at our disposal: (1) a waste stream of steam at $T_1 = 500$ K, $p_1 = 18$ bar, $u_1 = 3117.9$ kJ/kg, $h_1 = 3469.8$ kJ/kg, $s_1 = 7.4825$ kJ/kg K, $v_1 = 195.5$ cm^3/g; (2) plenty of environmental water at $T_2 = 293$ K, $p_2 = 1$ bar, $u_2 = 83.95$ kJ/kg, $h_2 = 83.95$ kJ/kg, $s_2 = 0.2966$ kJ/kg K, $v_2 = 1.0018$ cm^3/g; and (3) equipment of any kind.

(a) What is the least amount of steam that might be used to supply the energy to the oven?

(b) What kind of equipment would you use, and how would you arrange it?

(c) Depending on your answer in part (b), make an estimate of how much more steam will be used than your answer in part (a).

23.25 An arrangement for waste-heat recovery from steel slabs is shown in Figure P23.25. Steel slabs, 500 kg each, specific heat $c_s = 0.473$ kJ/kg K, enter the heat exchanger at 1000°C and at a rate 10 slab/min, and exit at 380°C. Eighty percent of the energy given up by the slabs in the heat exchanger is transferred to a stream of compressed air. The air behaves as a perfect gas with $c_p = 1$ kJ/kg K, $R = 0.286$ kJ/kg K. It powers a turbine-compressor unit. The air leaving the turbine has a pressure $p_4 = 1$ atm and is used to produce process steam.

(a) If the processes in the turbine and the compressor are reversible and isentropic, the inlet temperature to the turbine $T_3 = 1173$ K, and values of other parameters are as given in the figure, find the pressure $p_2 = p_3$ that yields the largest power input into the electricity generator.

(b) For the pressure found in part (a) and the values of parameters given in the figure but a turbine efficiency $\eta_t = 0.9$ and a compressor efficiency $\eta_c = 0.86$, find the airflow rate $\dot{m}_1$, the net power output $\dot{W}^{\rightarrow}$ from the generator, and the steam flow rate $\dot{m}_6$.

(c) How much entropy is generated by irreversiblity in the overall process in part (b) as the slabs pass through the heat exchanger and then cool to 300 K?

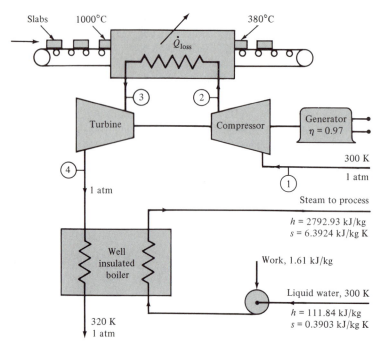

Figure P23.25

23.26 In a typical home heating system, high-temperature products of combustion at atmospheric pressure heat an air stream (Figure P23.26a). The air stream then maintains a living space at a fixed temperature $T_3 = 20°C$. As shown in the figure, in the course of heating, the products of combustion change their temperature from $T_1 = 560°C$ to $T_2 = 180°C$. This system is very simple and requires little capital investment. However, the process is highly irreversible.

An alternative system is shown in Figure P23.26b. The products of combustion interact with machinery that acts as a heat pump operating between the atmosphere at $T_0 = -18°C$ and the living space at $T_3 = 20°C$.

(a) If all processes in the alternative system are reversible, derive an expression for the energy transfer q_L to the living space per unit of energy extracted from the stream of the products of combustion. In the temperature range T_1 to T_2, treat the products of combustion as a perfect gas.

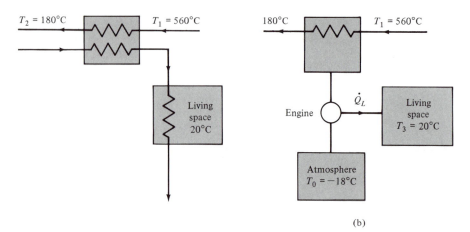

(b)

Figure P23.26

The machinery of the alternative system may be regarded as consisting of a work-producing engine operating between the hot stream and a reservoir at T_0, and a heat pump operating between T_3 and T_0 and powered by the work of the engine.

(b) If the engine does two-thirds of-the largest work possible for the given T_1, T_2, and T_0, and if the heat pump has one-half the coefficient of performance of a reversible heat pump operating between T_3 and T_0, how much energy is delivered to the living space?

(c) If the average heating load is 10 kW, the capital cost of machinery is about $700 per kilowatt of heating load, the capital cost of the heat exchanger for the products of combustion is $200 per kilowatt of heating load, the annual maintainance and depreciation costs are 20% of the capital costs, the fuel cost is $10 per gigajoule in the products of combustion, the products of combustion behave as a perfect gas with $R = 0.290$ kJ/kg K and $\gamma = 1.32$, and the equipment is used 4000 h/yr, which of the two heating systems is preferable?

23.27 The condenser of a steam electric power plant is sketched in Figure P23.27. Saturated water vapor enters at 30°C and, by fully condensing to saturated liquid, transfers 2 GW of energy to the cooling water pumped from a nearby river at 15°C to a pressure of 66 kPa higher that atmospheric. The efficiency of the pump is 80%. The cooling water exits the condenser at 25°C and atmospheric pressure and is then dumped into the river.

(a) What are the electric power consumption of the pump and the rate of entropy generation by irreversibility in the pump?

(b) What is the rate of entropy generation by irreversibility due to dumping the cooling water into the river?

(c) Assuming no pressure drop on the steam side of the condenser, what is the rate of entropy generation by irreversibility in the condenser?

(d) If there is a pressure drop of 1.077 kPa on the steam side of the condenser, what is the rate of entropy generation by irreversibility in the condenser? Compare your result with that found in part (c).

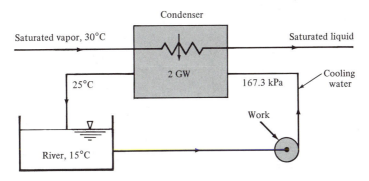

Figure P23.27

24   Availability Functions

In Chapter 14 we discuss briefly the concept of available energy and indicate that the difference between the available energies of two states of a system is equal to the optimum work that can be done in a weight process of the composite of the system and a reservoir. In Chapters 15 and 23 we introduce a number of efficiencies, each tailor-made for a specific application, such as the efficiency of a heat engine, the coefficient of performance of a heat pump, the efficiency of a compressor, and the efficiency of a turbine.

In this chapter we discuss some general features of the expressions for the optimum work in various processes and introduce a concept of thermodynamic efficiency that is universally applicable to all processes.

24.1   General Remarks

To accomplish almost every practical task, we exploit resources in our natural environment. Some resources are used as energy sources, others as raw materials. *Energy sources* are substances not in mutual stable equilibrium with the environment that can be used to power the energy conversion systems required by the various tasks. Typical sources are coal, oil, natural gas, uranium, and solar energy. Typical tasks are locomotion; motive power and process heat for manufacturing; space conditioning, such as heating, cooling, and ventilation; and electric power for communication devices, computers, industrial machines, home appliances, and lighting. *Raw materials* are substances used as feedstocks in manufacturing tasks—in materials-processing installations that produce different products. Examples of manufacturing tasks are the making of steel out of iron ore, and the making of aluminum out of bauxite. Most raw materials are in mutual stable equilibrium with the environment. They are reduced to desired products at the expense of energy sources. Other raw materials are only in partial mutual stable equilibrium with the environment and remain so if prevented from chemical (or nuclear) interactions with other environmental materials. Also, these raw materials are reduced to desired products at the expense of energy sources, such as in a refinery where crude oil is processed to yield petroleum products that are subsequently used as energy sources, or in an enrichment plant where natural uranium is processed to yield fissile uranium that is subsequently used as an energy source.

Each task is accomplished by means of an arrangement of devices, materials-processing systems, and energy-conversion systems interacting with each other, with resources, and with the natural environment. Examples of energy-conversion systems are discussed in Chapter 25. The selection, evaluation, and adoption of a particular arrangement involves the resolution and reconciliation of many complex and conflicting scientific, technical, economical, environmental, social, and safety questions. The complete discussion of these questions is of decisive importance but beyond the scope of this book, except for the questions that are related to thermodynamics, and are as follows.

1. What are the actual inlet, outlet, and end states of both the task and the energy sources used in a particular arrangement?

2. What are the optimum interactions required by the specified inlet, outlet, and end states of the task?

3. What are the optimum interactions that could be supplied by the energy sources employed if these sources were used in the best way physically possible?

4. If the answers to questions 1 and 2 differ, what aspects of the arrangement are the causes of the difference?

5. What can be done to change the difference between the answers to questions 1 and 2?

6. What is a universal measure of such a difference that characterizes how effectively the task is accomplished by a given arrangement?

It is clear that the concept of optimum that we use here is delimited only by the laws of physics, not by restrictions imposed either by economic, social, and environmental considerations, or by current technology. As such, it may well be secondary to all the other concerns. Nevertheless, it does provide a limit that cannot be exceeded under any circumstances.

24.2 The Environment as a Reservoir

In Chapters 15 and 23 we show that for given inlet, outlet, and end states of a system, a process is optimum when it is reversible. Accordingly, if the inlet, outlet, and end states are not matched so as to yield zero net differences in entropy, the entropy balance must be achieved by exchange of entropy with another system; otherwise, the process cannot be carried out reversibly. The only system that is readily available and can exchange large amounts of entropy at no cost is the environment. Similarly, the environment is a readily available no-cost[1] source of certain substances, such as the air we breathe, the water we drink, and the air intake of our automobile engines. It is also an easy-access sink of substances, such as the carbon dioxide we exhale, the liquid wastes we disperse in the rivers and the ocean, the products of combustion from automobile engines and energy-conversion systems, and many wastes from residential, commercial, and industrial activities.

For analyses of optimum processes, we model all substances in our natural environment, except energy sources, as a simple system behaving as a reservoir. We call it the *environmental reservoir* and denote it by R^*. Depending on the application, in order to focus our attention on the phenomena that are most prevalent, we find it convenient to impose different restrictions on the values of the amounts of constituents and the volume of the environmental reservoir. Whether we restrict either the values of the amounts of constituents, or the value of the volume, or both is evident from the context.

For example, in applications in which the system and the environment exchange entropy and energy but neither amounts of constituents nor volume, we model the environment as a reservoir R^* with fixed values of amounts of constituents and volume, and denote its constant temperature by T_{R^*}. Under these restrictions, the relation between entropy and energy differences [equation (14.40) for $E = U$] becomes

---

[1]The recent decades of heavy exploitation of our natural environment show that an unregulated use of the environment may cause a variety of serious alterations that result in enormous costs to our society and impacts on the quality of our lives. Thus the cost-free use of the environment should be allowed only for purposes that are unavoidable.

$$S_2^{R^*} - S_1^{R^*} = \frac{1}{T_{R^*}} (U_2^{R^*} - U_1^{R^*}) \tag{24.1}$$

where R_1^* and R_2^* are any two stable equilibrium states of R^*.

Again, in applications in which the system and the environment exchange entropy, energy, and volume but no amounts of constituents, we model the environment as a reservoir R^* with variable volume and fixed values of the amounts of constituents, and denote its constant temperature by T_{R^*} and its constant pressure by p_{R^*}. Under these restrictions, the relation between differences in values of energy, entropy, and volume of any two stable equilibrium states R_1^* and R_2^* [equation (14.40) for $E = U$] becomes

$$S_2^{R^*} - S_1^{R^*} = \frac{1}{T_{R^*}} (U_2^{R^*} - U_1^{R^*}) + \frac{p_{R^*}}{T_{R^*}} (V_2^{R^*} - V_1^{R^*}) \tag{24.2}$$

Finally, in applications in which the system and the environment exchange entropy, energy, volume, and amounts of constituents, we model the environment as a reservoir R^* with variable volume, and variable amounts of constituents, and denote its constant temperature by T_{R^*}, its constant pressure by p_{R^*}, and the constant chemical potentials by $\mu_{1R^*}, \mu_{2R^*}, \ldots, \mu_{rR^*}$. Here the relation between differences in values of energy, entropy, volume, and amounts of constituents of any two stable equilibrium states R_1^* and R_2^* [equation (14.40) for $E = U$] becomes

$$S_2^{R^*} - S_1^{R^*} = \frac{1}{T_{R^*}} (U_2^{R^*} - U_1^{R^*}) + \frac{p_{R^*}}{T_{R^*}} (V_2^{R^*} - V_1^{R^*}) - \sum_{i=1}^{r} \frac{\mu_{iR^*}}{T_{R^*}} \left[(n_i)_2^{R^*} - (n_i)_1^{R^*} \right] \tag{24.3}$$

Any state of the environmental reservoir is sometimes called a *passive* or *dead state* because starting from such a state and using no energy sources we can accomplish no useful task. Indeed, we can build neither a perpetual motion machine of the first kind (PMM1) nor a perpetual motion machine of the second kind (PMM2) using the environmental reservoir as a system.

In addition, the state of any system A in mutual stable equilibrium with R^* is sometimes called a passive or dead state and denoted by A_{0^*} because, once in such a state, system A is useless as well. In particular, if the environmental reservoir R^* is modeled as having variable volume and variable amounts of constituents, the dead state A_{0^*} of system A has the same values of temperature, pressure, and chemical potentials as the respective values of R^*, that is, $T_{0^*} = T_{R^*}$, $p_{0^*} = p_{R^*}$, and $\mu_{i0^*} = \mu_{iR^*}$ for $i = 1, 2, \ldots, r$. Again, if R^* is modeled as having fixed values of the volume and the amounts of constituents, the dead state A_{0^*} of A has temperature $T_{0^*} = T_{R^*}$, but values of pressure and chemical potentials not necessarily equal to the corresponding values of R^*.

Given a composite of a system A and the environmental reservoir R^*, spontaneous changes of state can occur only until A reaches mutual stable equilibrium with R^*, that is, only until A is in state A_{0^*}. After state A_{0^*} is reached, no further change in the state of the composite of A and R^* is possible without expenditure of an energy source because A_{0^*} has null adiabatic availability, null available energy with respect to R^*, and no reservoir other than the environmental is readily available.

In principle, it is always possible to create a reservoir at conditions of temperature, pressure, and chemical potentials different from those of the natural environment. But the creation of such a reservoir requires the expenditure of energy sources, and any benefit that could result would be at best equal to but usually less than the expenditure. For example, we recall from Chapter 15 that the gain in the work equivalent of a heat source with respect to a reservoir at temperature $T_R < T_{R^*}$ is at most equal to the amount of work required to cool a part of the natural environment from T_{R^*} to T_R.

24.3   Availability or Exergy

In previous discussions, we encounter some answers to questions related to optimum interactions. For example, in Chapter 14 we indicate that in changing the state of system A from state A_1 to state A_2 while the system is in combination with a reservoir R that has fixed amounts of constituents and parameters, and the composite AR experiences a weight process, the optimum work done on the weight is

$$(W_{12}^{AR\rightarrow})_{\text{optimum}} = \Omega_1^R - \Omega_2^R \qquad (24.4)$$

where Ω_1^R and Ω_2^R are the generalized available energies of the two states of A with respect to R, and to some reference values n and β of the amounts of constituents and the parameters of A. Hence if $\Omega_1^R > \Omega_2^R$, then $(W_{12}^{AR\rightarrow})_{\text{optimum}}$ is the largest work that the composite of A and R could do in a weight process under the specified conditions, whereas if $\Omega_2^R > \Omega_1^R$, then $-(W_{12}^{AR\rightarrow})_{\text{optimum}} = (W_{12}^{AR\leftarrow})_{\text{optimum}}$ is the least work required in a weight process for the composite of A and R to change the state of A from A_1 to A_2, again under the specified conditions.

If the system is simple and the reservoir environmental, equation (24.4) may be expressed in terms of energy and entropy in the form

$$(W_{12}^{AR^*\rightarrow})_{\text{optimum}} = (U_1 - T_{R^*}S_1) - (U_2 - T_{R^*}S_2) \qquad (24.5)$$

where U and S represent the internal energy and the entropy of system A, respectively, and in writing the equation we use the relation between energy, generalized available energy, and entropy introduced in Section 14.6. Moreover, under the specified conditions, we recall that the available energy is zero when A and R^* have the same temperature and conclude that

$$\Omega_1^{R^*} = (W_{10^*}^{AR^*\rightarrow})_{\text{rev}} = (U_1 - T_{R^*}S_1) - (U_{0^*} - T_{R^*}S_{0^*}) \qquad (24.6)$$

where U_{0^*} and S_{0^*} are the energy and entropy of system A in the dead state A_{0^*}, respectively, with temperature $T_{0^*} = T_{R^*}$, and values of the amounts of constituents and the parameters equal to the respective reference values n and β.

The expression $U - T_{R^*}S$ is called an *availability function* or *exergy function*. As equation (24.5) indicates, for the specified conditions the difference in the values of this function at two states yields the optimum work in a weight process for the composite of A and R^*.

The expression $(U_1 - T_{R^*}S_1) - (U_{0^*} - T_{R^*}S_{0^*})$, that is, the generalized available energy of state A_1 is also called the *availability* or *exergy* of state A_1. Under the specified conditions, it represents the optimum work that can be done as a result of the state of system A changing from A_1 to the state A_{0^*} with the reference values n and β, and temperature T_{0^*} equal to that of the environmental reservoir. It turns out that this work is not sign definite, namely, it can be either positive or negative.

Expressions analogous to equations (24.5) and (24.6) can be derived for conditions other than those involved in the definition of generalized available energy. Ideally, we should define a distinct name for each set of conditions, and the corresponding function and its differences. Because there are innumerable conditions that we must examine, we would then have so many names that it would be questionable whether the richness of the vocabulary would be of any help. To avoid this linguistic pile up, we proceed as follows.

First, we consider a system A and the environmental reservoir R^* with given specifications regarding whether the values of their respective amounts of constituents and volume are variable or fixed. We define as the *availability function* or *exergy function* corresponding to the given specifications that expression the differences of which yield the optimum work in a weight process for the composite of A and R^* as system A changes from a given state A_1 to another given state A_2. Moreover, we define as the *availability* or *exergy* corresponding to the given specifications and to state A_1 that expression which yields the optimum work in a weight process for the composite of A and R^* as system A changes from state A_1 to a state A_{0*} in which A and R^* are in mutual stable equilibrium. For example, for the conditions discussed at the beginning of this section, we summarize the results by writing

$$(W_{12}^{AR^* \rightarrow})_{\text{optimum}} = (U_1 - T_{R^*}S_1) - (U_2 - T_{R^*}S_2) \tag{24.7}$$

$$\text{availability function} = U - T_{R^*}S \tag{24.8}$$

$$\text{availability} = (U - U_{0*}) - T_{R^*}(S - S_{0*})$$

$$= (U - T_{R^*}S) - (U_{0*} - T_{R^*}S_{0*}) \tag{24.9}$$

Other examples are discussed in the following section.

Next, we consider a given type of interaction, such as work, heat, or bulk flow, and the environmental reservoir R^* with given specifications regarding whether the values of its amounts of constituents and volume are variable or fixed. We define as the *availability rate function* or *exergy rate function* corresponding to the given specifications that expression the differences of which yield the optimum work rate in a process for the composite of reservoir R^* and a system A maintained in a steady state by two given interactions of the same type. Moreover, we define as the *availability rate* or *exergy rate* corresponding to the given specifications and associated with a given interaction that expression which yields the optimum work rate in a process for the composite of reservoir R^* and a system A maintained in a steady state by the given interaction and an interaction of the same type with the reservoir R^*. An example of an availability rate is the work rate equivalent of a heat rate that we discuss in Section 22.4. In particular, using (22.34), with the right-hand-side set equal to zero, that is, assuming no bulk-flow interactions and no entropy generated by irreversibility, and considering the composite of R^* and a system A maintained at steady state by two heat interactions at temperatures T_{Q1} and T_{Q2}, respectively, we find that

$$(\dot{W}^{AR^* \rightarrow})_{\text{optimum}} = \dot{Q}_1^{\leftarrow}\left(1 - \frac{T_{R^*}}{T_{Q1}}\right) - \dot{Q}_2^{\rightarrow}\left(1 - \frac{T_{R^*}}{T_{Q2}}\right) \tag{24.10}$$

$$\text{availability rate function} = \dot{Q}\left(1 - \frac{T_{R^*}}{T_Q}\right) \tag{24.11}$$

$$\text{availability rate} = \dot{Q}_1^{\leftarrow}\left(1 - \frac{T_{R^*}}{T_{Q1}}\right) \tag{24.12}$$

where, of course, equation (24.12) represents the availability rate associated with the heat interaction at the rate $\dot{Q}_1^{\leftarrow}$ and temperature T_{Q1}.

24.4   Different Availabilities

24.4.1

Here we consider a simple system A, changing from a state A_1 with volume V_1 to a state A_2 with a different volume V_2, and surrounded by the environmental reservoir R^*, modeled as having variable volume but fixed values of the amounts of constituents, so that the reservoir experiences an equal and opposite change in volume.

At the moving boundary between A and R^* a volume exchange occurs according to the equation

$$V_2^{R^*} - V_1^{R^*} = -(V_2 - V_1) \tag{24.13}$$

The motion of the boundary against the constant reservoir pressure p_{R^*} results in a work interaction between A and R^*, and the work done by A on R^* is $p_{R^*}(V_2 - V_1)$. In a weight process for the composite of A and R^*, the work $p_{R^*}(V_2 - V_1)$ represents just an internal exchange between A and R^* and not work $W_{12}^{AR^* \rightarrow}$ done on the weight. To evaluate the optimum work $(W_{12}^{AR^* \rightarrow})_{\text{optimum}}$ done on the weight under the specified conditions, we begin by writing the energy and entropy balances

$$(U_2 + U_2^{R^*}) - (U_1 + U_1^{R^*}) = -(W_{12}^{AR^* \rightarrow}) \tag{24.14}$$

$$(S_2 + S_2^{R^*}) - (S_1 + S_1^{R^*}) = S_{\text{irr}} \tag{24.15}$$

Upon combining (24.13) to (24.15) with (24.2), and setting $S_{\text{irr}} = 0$ for optimality, we find that

$$(W_{12}^{AR^* \rightarrow})_{\text{optimum}} = (U_1 - T_{R^*}S_1 + p_{R^*}V_1) - (U_2 - T_{R^*}S_2 + p_{R^*}V_2) \tag{24.16}$$

$$\text{availability function} = U - T_{R^*}S + p_{R^*}V \tag{24.17}$$

$$\text{availability} = (U - U_{0^*}) - T_{R^*}(S - S_{0^*}) + p_{R^*}(V - V_{0^*})$$

$$= (U - T_{R^*}S + p_{R^*}V) - (U_{0^*} - T_{R^*}S_{0^*} + p_{R^*}V_{0^*}) \tag{24.18}$$

where U_{0^*}, S_{0^*}, and V_{0^*} are the energy, entropy, and volume of A in mutual stable equilibrium with the reservoir and, therefore, in the state A_{0^*} with temperature $T_{0^*} = T_{R^*}$ and pressure $p_{0^*} = p_{R^*}$. Although the value of $U_{0^*} + p_{R^*}V_{0^*} - T_{R^*}S_{0^*}$ equals that of the Gibbs free energy of state A_{0^*} because $T_{0^*} = T_{R^*}$ and $p_{0^*} = p_{R^*}$, it is noteworthy that $U + p_{0^*}V - T_{0^*}S$ is not a Gibbs free energy because U, S, V, T_{0^*}, and p_{0^*} are not all associated with the same state of system A.

24.4.2

Here we consider a simple system A, changing from a state A_1 with values V_1 and $(n)_1$ of the volume and the amounts of constituents to a state A_2 with values V_2 and $(n)_2$, and surrounded by the environmental reservoir R^*. The reservoir is modeled as having variable values of volume and amounts of constituents, so that it experiences changes in values of volume and in each of the amounts of constituents equal and opposite to the respective changes in values of A. Thus equation (24.13) holds also here and, in addition, we have

$$(n_i)_2^{R^*} - (n_i)_1^{R^*} = -\left[(n_i)_2 - (n_i)_1\right] \qquad \text{for } i = 1, 2, \ldots, r \tag{24.19}$$

where $(n_i)_1^{R^*}$ and $(n_i)_2^{R^*}$ are the values of the amount of the ith constituent of R^* at states R_1^* and R_2^*, respectively, and $(n_i)_1$ and $(n_i)_2$ the values of the amount of the same constituent at states A_1 and A_2 of A, respectively.

The energy and entropy balances for a weight process of the composite of A and R^* under the specified conditions are still given by (24.14) and (24.15). Combining these with (24.13), (24.19), and (24.3), and setting $S_{\text{irr}} = 0$ for optimality, we find that

$$(W_{12}^{AR^* \rightarrow})_{\text{optimum}} = \left[U_1 - T_{R^*}S_1 + p_{R^*}V_1 - \sum_{i=1}^{r} \mu_{iR^*}(n_i)_1\right]$$

$$- \left[U_2 - T_{R^*}S_2 + p_{R^*}V_2 - \sum_{i=1}^{r} \mu_{iR^*}(n_i)_2\right] \tag{24.20}$$

$$\text{availability function} = U - T_{R*}S + p_{R*}V - \sum_{i=1}^{r} \mu_{iR*}\, n_i \qquad (24.21)$$

$$\text{availability} = (U - U_{0*}) - T_{R*}(S - S_{0*}) + p_{R*}(V - V_{0*})$$

$$- \sum_{i=1}^{r} \mu_{iR*}\left[n_i - (n_i)_{0*} \right]$$

$$= \left[U - T_{R*}S + p_{R*}V - \sum_{i=1}^{r} \mu_{iR*}\, n_i \right]$$

$$- \left[U_{0*} - T_{R*}S_{0*} + p_{R*}V_{0*} - \sum_{i=1}^{r} \mu_{iR*}\, (n_i)_{0*} \right] \qquad (24.22)$$

24.4.3

As a third example, we consider a system A maintained at steady state by two bulk-flow interactions, and the environmental reservoir R^* modeled as having variable volume and amounts of constituents. We assume no changes in mass flow rate and composition, and negligible changes in kinetic and potential energies between the bulk-flow states of the inlet and outlet streams. For example, for the arrangement shown in Figure 24.1, bulk-flow states 1 and 2 are the states of the inlet and outlet streams of a steady-state device A which, in addition to these two bulk-flow interactions, is surrounded by the environmental reservoir and connected to a weight.

Adapting equation (22.34) to the conditions just specified for the composite of the device A and the reservoir R^*, we find that

$$(\dot{W}^{AR^* \rightarrow})_{\text{optimum}} = \dot{m}\,(h_1 - T_{R*}s_1) - \dot{m}\,(h_2 - T_{R*}s_2) \qquad (24.23)$$

$$\text{availability rate function} = \dot{m}\,(h - T_{R*}s) \qquad (24.24)$$

$$\text{availability rate} = \dot{m}\left[(h - h_{0*}) - T_{R*}(s - s_{0*}) \right]$$

$$= \dot{m}\,(h - T_{R*}s) - \dot{m}\,(h_{0*} - T_{R*}s_{0*}) \qquad (24.25)$$

where h_{0*} and s_{0*} are the specific enthalpy and the specific entropy of a bulk-flow state O^* at temperature $T_{0*} = T_{R*}$ and pressure $p_{0*} = p_{R*}$. It is noteworthy that the stream in bulk-flow state O^* is not in mutual stable equilibrium with R^*. The reason is that a bulk-flow state is not stable equilibrium unless the kinetic and potential energies are zero, and even if these energies are zero, the condition of chemical potential equality cannot be met because of the specification that no changes in compositions can occur.

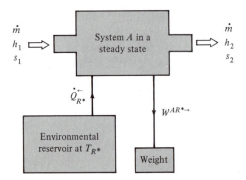

Figure 24.1 Schematic of a system A maintained in a steady state by two bulk-flow interactions, shaft work, and heat with the environmental reservoir.

In other words, the bulk-flow interaction at state O^* introduces into the environment substances that do not correspond to the environmental composition and therefore cause a subsequent irreversible mixing. This irreversibility is built into the system specifications we are considering. If the specifications are different, such as when chemical reactions are allowed, the availability rate has a different expression.

In processes involving many streams, the availability rate function and the availability rate are given by expressions similar to equations (24.24) and (24.25) except that here each rate is a sum over many streams. Specifically, recalling the notation introduced in Section 22.4, we have

$$(\dot{W}^{AR^* \rightarrow})_{\text{optimum}} = (\dot{H}_{\text{in}} - T_{R^*}\dot{S}_{\text{in}}) - (\dot{H}_{\text{out}} - T_{R^*}\dot{S}_{\text{out}}) \tag{24.26}$$

$$\text{availability rate function} = \dot{H} - T_{R^*}\dot{S} \tag{24.27}$$

$$\text{availability rate} = (\dot{H} - T_{R^*}\dot{S}) - \left[\dot{H}(T_{R^*}, p_{R^*}) - T_{R^*}\dot{S}(T_{R^*}, p_{R^*})\right] \tag{24.28}$$

where $\dot{H}$ and $\dot{S}$ represent net flow rates over many streams.

24.4.4

Under the same conditions as specified in Section 24.4.3, except that the changes in kinetic and potential energies of the bulk-flow streams are not negligible, equations (24.23) to (24.25) become

$$(\dot{W}^{AR^* \rightarrow})_{\text{optimum}} = \dot{m}\left(h_1 - T_{R^*}s_1 + \frac{\xi_1^2}{2} + g z_1\right)$$
$$- \dot{m}\left(h_2 - T_{R^*}s_2 + \frac{\xi_2^2}{2} + g z_2\right) \tag{24.29}$$

$$\text{availability rate function} = \dot{m}\left(h - T_{R^*}s + \frac{\xi^2}{2} + g z\right) \tag{24.30}$$

$$\text{availability rate} = \dot{m}\left[(h - h_{0^*}) - T_{R^*}(s - s_{0^*}) + \frac{\xi^2}{2} + g(z - z_{0^*})\right]$$
$$= \dot{m}\left(h - T_{R^*}s + \frac{\xi^2}{2} + g z\right)$$
$$- \dot{m}(h_{0^*} - T_{R^*}s_{0^*} + g z_{0^*}) \tag{24.31}$$

where h_{0^*}, s_{0^*}, and z_{0^*} refer to a bulk-flow state with temperature $T_{0^*} = T_{R^*}$, pressure $p_{0^*} = p_{R^*}$, bulk-flow speed $\xi_{0^*} = 0$, and the lowest elevation z_{0^*} in the environment.

It is clear that many more availability (availability rate) or exergy (exergy rate) functions can be defined, each associated with a particular set of conditions.

24.5 Availability or Exergy Analyses

An illustration of the usefulness of equations (24.26) and (24.28) is provided by the bulk-flow processes in Figure 24.2. Various substances enter a materials-processing system A in various bulk-flow streams with overall enthalpy rate $\dot{H}_1^{\text{materials}}$ and entropy rate $\dot{S}_1^{\text{materials}}$. The plant is designed to operate in steady state and to transform the entering streams into products having overall enthalpy rate $\dot{H}_2^{\text{materials}}$ and overall entropy rate $\dot{S}_2^{\text{materials}}$. The transformation requires shaft work from a power plant at a rate $\dot{W}_s^{A\leftarrow}$ and heat from the natural environment at temperature T_{R^*} at a rate $\dot{Q}_{R^*}^{A\leftarrow}$. Adapting the equations in Section 22.4 to the specifications of the materials-processing system, we find that

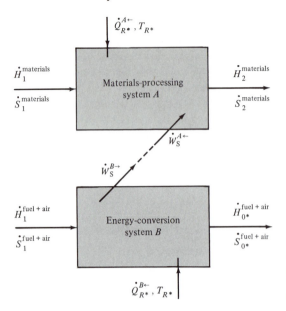

Figure 24.2 The burning of a fuel-air mixture in the energy-conversion system provides the work needed to process the bulk-flow stream through the materials-processing system.

$$\dot{W}_s^{A\leftarrow} = \left[\dot{H}_2^{\text{materials}} - \dot{H}_1^{\text{materials}} - T_{R*}(\dot{S}_2^{\text{materials}} - \dot{S}_1^{\text{materials}})\right] + T_{R*}\dot{S}_{\text{irr}}^A \qquad (24.32)$$

where $\dot{S}_{\text{irr}}^A$ is the rate of entropy generation by irreversibility in the materials-processing system A. It is noteworthy that $\dot{W}_s^{A\leftarrow}$ is optimum if $\dot{S}_{\text{irr}}^A = 0$, and that to each different value of $\dot{W}_s^{A\leftarrow}$ there corresponds a different value of $\dot{Q}_{R*}^{A\leftarrow}$.

The shaft work is provided by the energy-conversion system B (Figure 24.2), which converts a fuel and air stream into products of combustion. The fuel and air enter the system as bulk-flow streams with overall enthalpy rate $\dot{H}_1^{\text{fuel+air}}$ and overall entropy rate $\dot{S}_1^{\text{fuel+air}}$. The energy-conversion system does shaft work at a rate $\dot{W}_s^{B\rightarrow}$, and interacts with the natural environment at T_{R*} with heat at a rate $\dot{Q}_{R*}^{B\leftarrow}$. Moreover, we assume that the products of combustion exit in bulk-flow streams at temperature $T_{0*} = T_{R*}$ and pressure $p_{0*} = p_{R*}$. For these streams, we denote the overall enthalpy rate by $\dot{H}_{0*}^{\text{fuel+air}}$ and the overall entropy rate by $\dot{S}_{0*}^{\text{fuel+air}}$. If no entropy is generated by irreversibility in the energy-conversion system, the work rate is given by the availability rate of the fuel–air mixture, equation (24.28). However, if entropy is generated by irreversibility, $\dot{W}_s^{B\rightarrow}$ satisfies the relation

$$\dot{W}_s^{B\rightarrow} = \left[\dot{H}_1^{\text{fuel+air}} - \dot{H}_{0*}^{\text{fuel+air}} - T_{R*}(\dot{S}_1^{\text{fuel+air}} - \dot{S}_{0*}^{\text{fuel+air}})\right] - T_{R*}\dot{S}_{\text{irr}}^B \qquad (24.33)$$

where $\dot{S}_{\text{irr}}^B$ is the rate of entropy generation by irreversibility in the energy-conversion system.

Upon subtracting equation (24.32) from equation (24.33), recognizing that $\dot{W}_s^{B\rightarrow} = \dot{W}_s^{A\leftarrow}$, and rearranging terms we find that

$$\left[\dot{H}_1^{\text{fuel+air}} - \dot{H}_{0*}^{\text{fuel+air}} - T_{R*}(\dot{S}_1^{\text{fuel+air}} - \dot{S}_{0*}^{\text{fuel+air}})\right]$$
$$= \left[\dot{H}_2^{\text{materials}} - \dot{H}_1^{\text{materials}} - T_{R*}(\dot{S}_2^{\text{materials}} - \dot{S}_1^{\text{materials}})\right] + T_{R*}\dot{S}_{\text{irr}} \qquad (24.34)$$

where $\dot{S}_{\text{irr}} = \dot{S}_{\text{irr}}^A + \dot{S}_{\text{irr}}^B$, that is, the rate of entropy generation by irreversibility in both the materials-conversion and the energy-conversion systems.

The left-hand side of equation (24.34) is the availability rate of the fuel–air mixture, that is, the largest rate at which work could possibly be done by processing the mixture in the natural environment under the specified conditions on the types of interactions between the energy-conversion system and the environment. The bracketed term in the right-hand side is the least work rate required to achieve the change of state of the bulk-flow streams processed by the materials-processing system under the specified types of interactions with the environment. The term $T_{R*}\dot{S}_{\mathrm{irr}}$ is the work rate equivalent of the rate of entropy generation by irreversibility. It represents a loss of availability. It is a partial loss of the ability of the fuel source to perform a useful task. This loss has been incurred because the processes in the materials-processing and the energy-conversion systems are not the best achievable under the specified conditions, that is, the processes are not reversible. So, in contrast to energy, availability is not conserved. It is destroyed or consumed by the generation of entropy due to irreversibility.

For the arrangement in Figure 24.2, equation (24.34) provides answers to questions raised in Section 24.1. Specifically, it includes the inlet and outlet states of the task and the energy sources and therefore answers question 1: "What are the actual inlet, outlet, and end states of both the task and the energy sources used in the arrangement?" It specifies the optimum interactions required by the task and, therefore, answers question 2: "What are the optimum interactions required by the specified inlet, outlet, and end states of the task?" And it specifies the optimum interactions that could be supplied by the energy sources and thus provides an answer to question 3.

To answer question 4, "What aspects of the arrangement are the causes of the difference between the answers to questions 2 and 3?" we must look into the detailed design characteristics of the equipment used in the process. We illustrate the procedure by the following example.

Example 24.1.   We consider the refrigeration arrangement shown schematically in Figure 24.3. Saturated propane vapor (state 1) is compressed to superheated vapor (state 2) by means of the compressor. The shaft-work rate input to the compressor $\dot{W}^{\leftarrow} = 110$ kW. The superheated vapor is condensed to saturated liquid (state 3) in the refrigerant condenser, which is cooled by environmental water at almost fixed temperature $T_{R*} = 298$ K. Upon exiting the condenser, the propane is first partially vaporized (state 4) by means of a throttle, and then fully vaporized (state 1) by passing through the refrigerant evaporator. In the evaporator, the propane receives energy and entropy from the space B, which is being refrigerated and kept at fixed temperature

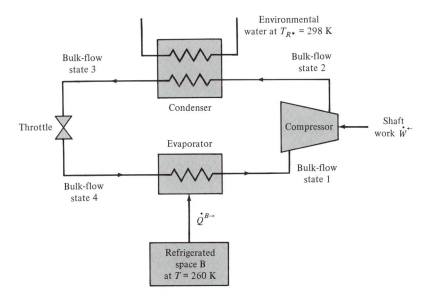

Figure 24.3   Schematic of a refrigeration system.

$T_B = 260$ K. The propane flow rate $\dot{m} = 1$ kg/s, and other properties of propane are listed in Table 24.1.

TABLE 24.1.   Data for Example 24.1.

State	T [K]	p [MPa]	h [kJ/kg]	s [kJ/kg K]
1	250	0.2161	514.80	2.1088
2	330	1.2690	620.00	2.1600
3	310	1.2690	263.18	1.0305
4	250	0.2161	263.18	1.0700

We neglect the small pressure drops in the evaporator and the condenser, and calculate

(a) the least work rate required to achieve the cooling,

(b) the overall rate of loss of availability, and

(c) the rate of loss of availability in each of the components of the refrigeration unit.

Solution.   **(a)** We find the heat rate $\dot{Q}^{B\rightarrow}$ from the refrigerated space from the energy rate balance of the evaporator, that is, $\dot{Q}^{B\rightarrow} = \dot{m}(h_1 - h_4) = 1 \times (514.80 - 263.18) = 251.62$ kW. The task of the refrigeration arrangement is to pump $\dot{Q}^{B\rightarrow}$ from the refrigerated space at temperature T_B to the environmental water at temperature T_{R*}. The least work rate required for the task is given by the relation

$$\dot{W}_{\text{least}}^{\leftarrow} = \left(\frac{T_{R*}}{T_B} - 1\right)\dot{Q}^{B\rightarrow} = \left(\frac{298}{260} - 1\right) \times 251.62 = 36.8 \text{ kW}$$

This relation is obtained by writing (15.14) in terms of rates and assuming that no entropy is generated by irreversibility.

(b) The input power is $\dot{m}(h_2 - h_1) = 105.2$ kW, and therefore the overall rate of loss of availability due to irreversibility $T_{R*}\dot{S}_{\text{irr}} = \dot{W}^{\leftarrow} - \dot{W}_{\text{least}}^{\leftarrow} = 105.2 - 36.8 = 68.4$ kW. The overall entropy generation rate is the sum of the entropy generation rates of all the components.

(c) We find the rate of loss of availability from the combined rate balance for each of the components of the refrigeration system.

Compressor: $T_{R*}\dot{S}_{\text{irr}} = \dot{W}^{\leftarrow} - \dot{m}[(h_2 - h_1) - T_{R*}(s_2 - s_1)] = 105.2 - (620 - 514.8) + 298(2.16 - 2.1088) = 15.3$ kW or $15.3/68.4 = 22.3\%$ of the total.

Condenser: $T_{R*}\dot{S}_{\text{irr}} = \dot{m}[(h_2 - h_3) - T_{R*}(s_2 - s_3)] = 20.2$ kW or 29.7% of the total.

Expansion valve: $T_{R*}\dot{S}_{\text{irr}} = \dot{m}T_{R*}(s_4 - s_3) = 11.8$ kW or 17.2% of the total.

Evaporator: $T_{R*}\dot{S}_{\text{irr}} = -\dot{W}_{\text{least}}^{\leftarrow} - \dot{m}[(h_1 - h_4) - T_{R*}(s_1 - s_4)] = 21.1$ kW or 30.8% of the total.

Example 24.1 is representative of the answer to question 4 in Section 24.1. It indicates the difference between actual and optimum performance, which results from the irreversibility occurring in the various devices of the system, and the degree to which each component contributes to the difference. Having this information, we can proceed to change the difference by concentrating our efforts on those components that hold the highest promise to yield better results.

In the refrigeration arrangement, the difference between the actual work rate done on the system and the least work rate required can be reduced by decreasing the temperature differences between the propane stream and the refrigerated space in the evaporator, and the propane stream and the environmental water in the condenser, by improving the performance of the motor and the compressor, and by replacing the expansion valve by a turbine. Such modifications are illustrative of the answers to question 5 in

Section 24.1, that is, "What can be done to change the difference between the optimum interactions required by the task and the optimum interactions that can be supplied by the energy sources?"

An analysis of a system based on considering the energy, entropy, and combined balances for each component of the system, and computing the availability or exergy consumption, that is, the entropy generation by irreversibility, is called an *availability analysis* or *exergy analysis*.

Availability analyses, as well as energy and other analyses, require that the inlet, outlet, and end states of the task be specified, and that the changes in availability and availability rates of feedstocks, products, and energy sources be evaluated. Because of practical considerations related to existing knowledge and technology, the specification of a desired task is very often relative to existing knowledge and technology and not absolute and, therefore, availability and other analyses yield results that are relative to existing knowledge and technology.

For example, a common process encountered in industry is the heat treating of alloy steel parts to produce a locally hard surface, such as the surface of a steel ball for a bearing, or the surface of the teeth of a gear. Although only a very small fraction of the material of each part needs to be hardened, conventional technology has required that the entire part be heated to about 900°C. So the task is defined according to this requirement. Another way to specify the task, however, is to say that only a small fraction near the surface of the material need be hardened. The availability change required by the first task is much larger than that required by the second. Moreover, the results of the two availability analyses are not comparable to each other, just as the task of making pig iron in a blast furnace is not comparable to that of making aluminum in an electrolytic cell.

In the example of steel hardening, the second specification of the task has, of course, little practical significance if we do not know how to treat just the surface without affecting the bulk of the processed piece. However, the lower availability change required by this specification in the framework of the conventional technology of the task can provide useful guidance for innovative approaches to the problem of metal hardening. In fact, recent developments in high-power lasers and electron-beam accelerators have led to the development of practical processes for localized heat treating. In one carburizing application, for example, electron-beam heat treating has reduced the energy needed for a particular part from 1 kWh to only 2 Wh. Thus, by redefining the task, the required availability was lowered well below the level that was previously thought to be optimum.

24.6   Thermodynamic Efficiency or Effectiveness

Associated with each task, such as heating a room or making a specified amount of steel out of iron ore, is a least work that must be done to accomplish the task. Such least work is equal to the change in availability of the substances processed to achieve the task, and is independent of any details of the arrangement of devices and engines used in the task.

In practice, however, each specific arrangement consumes a certain, not necessarily optimum, amount of fuel or energy source to accomplish the task. Associated with this amount of fuel or energy source is a largest work that can be delivered to a weight. Such largest work is equal to the availability of the fuel or energy source consumed, and is independent of any details of the energy-conversion systems and devices used to convert the fuel or energy source to work.

For emphasis, we denote the least work rate required by a specified task production rate as $\dot{W}_{\text{least}}^{\leftarrow}$ required by the actual task production rate, and the availability rate of the actual rate of fuel or energy source consumption as $\dot{W}_{\text{largest}}^{\rightarrow}$ of the actual energy source consumption rate, and define the *thermodynamic efficiency* or *effectiveness*[2] ϵ of the actual arrangement as the ratio of these two rates, that is,

$$\epsilon = \frac{\dot{W}_{\text{least}}^{\leftarrow} \text{ required by the actual task production rate}}{\dot{W}_{\text{largest}}^{\rightarrow} \text{ of the actual energy source consumption rate}} \tag{24.35}$$

The effectiveness is a measure of the degree to which the processes involved in carrying out the task and in converting the energy source are reversible. If the processes are reversible, $\epsilon = 1$. If the processes are irreversible, $\epsilon < 1$.

The concept of effectiveness is applicable to any process and can always be expressed in the form

$$\epsilon = 1 - \frac{T_{R*}\dot{S}_{\text{irr}}}{\dot{W}_{\text{largest}}^{\rightarrow} \text{ of the actual energy source consumption rate}} \tag{24.36}$$

because the difference between the denominator and the numerator in (24.35) is always $T_{R*}\dot{S}_{\text{irr}}$, where $\dot{S}_{\text{irr}}$ is the total rate of entropy generation by irreversibility in the process. For example, the effectiveness of the materials-processing plant discussed in Section 24.5 is of the form of (24.36) as we can readily verify by using (24.34).

The effectiveness may assume even negative values. A negative value signifies that ideally the task can be accomplished while the processed streams transfer energy to a weight rather than consume energy sources, that is, the processed streams can be used as energy sources themselves. Instead, because of large irreversibilities in the actual materials-processing and energy-conversion systems, not only is the contribution from the processed streams wasted but other energy sources are consumed.

The term $T_{R*}\dot{S}_{\text{irr}}$ is sometimes called *lost work rate*. It represents the work that could be produced in the absence of irreversibility, but is not produced because of irreversibility.

We can express the thermodynamic efficiency or effectiveness also in terms of batch quantities rather than rates. Then

$$\epsilon = \frac{W_{\text{least}}^{\leftarrow} \text{ required by the actual task production}}{W_{\text{largest}}^{\rightarrow} \text{ of the actual energy source consumption}}$$

$$= 1 - \frac{T_{R*}S_{\text{irr}}}{W_{\text{largest}}^{\rightarrow} \text{ of the actual energy source consumption}} \tag{24.37}$$

This effectiveness behaves exactly in the same way as the one defined by (24.35).

Subject to the qualifications discussed in the next section, the concept of thermodynamic efficiency or effectiveness is the answer to question 6 posed in Section 24.1, namely, the universal measure of how effectively the task is accomplished by a given arrangement.

24.7 Practical Limitations

The construction of each machine, engine, and device is in itself a task that involves materials-processing and energy-conversion systems and therefore the consumption of energy sources. When it is sizable, this consumption must be accounted for. An important

---

[2]In some literature on this subject, the concept of effectiveness defined here is called second-law efficiency. Such terminology, however, may be misleading because the concept is based not just on the second law but on the first law as well, and on many other concepts, such as work, heat, and bulk-flow interactions, and energy and entropy balances. All these concepts are certainly related to but not derivable solely from the second law.

requirement of any installation used in primary energy processing, such as the production of electricity from various energy sources, is that the installation be capable of extracting more availability from the sources than the availability consumed for the construction of the machinery.

In many applications it may be technically impossible to take full advantage of the availability of the energy sources utilized. This may happen because some of the availability is either lost in processes outside the application, or remains intact for use in subsequent applications. When this occurs, defining the effectiveness of the application in terms of the availability of the energy sources is misleading.

For example, if the only known method to carry out a process is by means of electrolysis, and electricity is generated from coal, it is hopeless to expect to improve the electrolytic process so as to take full advantage of the fuel availability. Electricity is not available in nature. Its generation entails losses. About two-thirds of the availability of the fuel is lost because of the irreversibilities inherent in modern coal-fired power plants. This loss should not be charged to the imperfections of the electrolytic process because it requires electricity to operate and the loss cannot be recovered no matter how perfect the electrolytic process is. To avoid this difficulty, the reasonable thing to do is to consider the availability of electricity as a source of input, and evaluate the effectiveness of the electrolytic process with respect to electricity rather than with respect to coal.

Again, in each stage of a steam turbine only some of the availability of the flowing steam is consumed. The remaining availability is ready for use in subsequent stages. Hence it is misleading to compute the effectiveness of one stage of the turbine with respect to the full availability of the steam flow. In fact, what is done in practice is to define the effectiveness between any two states of the fluid as follows. The stage task is defined as that of achieving the change from state 1 to state 2. The work done per unit flow rate is $h_1 - h_2$. The largest work per unit flow rate that could be done for this change of state by combining the stage with the environment is $h_1 - h_2 - T_{R^*}(s_1 - s_2)$. Hence the effectiveness of the stage is

$$\epsilon = \frac{h_1 - h_2}{h_1 - h_2 - T_{R^*}(s_1 - s_2)} = 1 + \frac{T_{R^*}(s_1 - s_2)}{h_1 - h_2 - T_{R^*}(s_1 - s_2)}$$

$$= 1 - \frac{T_{R^*} s_{\text{irr}}}{h_1 - h_2 - T_{R^*}(s_1 - s_2)} \tag{24.38}$$

where the entropy generation by irreversibility in the stage $s_{\text{irr}} = s_2 - s_1$. In contrast to an effectiveness evaluated with respect to the full availability of the steam, the effectiveness given by (24.38) provides a better measure of the margin of improvement of the stage.

Again, for a heat engine operating between two temperatures T_A and T_B, both different from T_{R^*}, it is appropriate to define the effectiveness by means of (15.7) rather than in terms of T_{R^*}. The reason is that the energy of the low-temperature heat reservoir can be used in combination with the environmental reservoir to produce more work.

In some applications, the properties of the materials of the equipment do not permit the full utilization of the fuel availability. For example, in oil-fired power plants, the exhaust combustion gases contain water vapor. If cooled to environmental temperature, the vapor condenses and corrodes the equipment. So exhaust gases are not cooled to such low temperatures. Correspondingly, the availability of the fuel should be evaluated with respect to a final state not in temperature equilibrium with the environment but at a temperature such that vapor condensation cannot occur.

24.8 Comments

In contrast to other measures of efficiency, each specifically defined for a class of applications, the concept of thermodynamic efficiency or effectiveness is applicable to any task without conceptual modifications. For example, miles per gallon of gasoline is a measure of thermodynamic performance of a transportation task by an automobile, and the larger the value of this measure, the better the performance. Again, equivalent barrels of oil per ton of steel is a measure of thermodynamic performance of a steelmaking task at a steel plant, and the smaller the value of this measure, the better the performance. Clearly, these two measures are not interchangeable and have different limiting values. In contrast, the concept of effectiveness can be applied to both an automobile and a steelmaking plant. The result for each of these two tasks would be a number less than unity, with an upper limit equal to unity. The upper limit of unity corresponds to perfect thermodynamic performance, namely, to all processes involved in the task being reversible.

Being directly related to irreversibility, the thermodynamic efficiency or effectiveness provides a realistic measure of the degree to which the performance of a task can be improved. Other measures of efficiency may be misleading. To illustrate the last assertion, we consider a perfectly insulated heat exchanger in which all the energy change of the primary stream is transferred to the secondary stream. Upon defining efficiency as the energy increase of the secondary stream divided by the energy decrease of the primary stream, we would find that this heat exchanger is 100% efficient. Such a result is correct but misleading. It implies that the heat exchanger is perfect and cannot be improved. However, if we define the effectiveness as the availability increase of the secondary stream divided by the availability decrease of the primary stream, we find that the best exchanger is less than 100% efficient, and subject to improvement by reduction of the temperature differences between the two streams. Clearly, the second answer is realistic and, more important, relevant to our concerns about efficient use of resources.

Another important characteristic of the thermodynamic efficiency or effectiveness is that it provides a realistic evaluation of tasks with dissimilar outputs. To see this point, we consider a cyclic device that produces work, $W^{\rightarrow}$, and heat, $Q^{\rightarrow}$, at temperature T_Q, while using heat, $Q_1^{\leftarrow}$, from a source at temperature $T_{Q1} > T_Q$. If these are the only interactions, the energy and entropy balances are

$$Q_1^{\leftarrow} = W^{\rightarrow} + Q^{\rightarrow} \tag{24.39}$$

$$\frac{Q_1^{\leftarrow}}{T_{Q1}} + S_{\text{irr}} = \frac{Q^{\rightarrow}}{T_Q} \tag{24.40}$$

If efficiency were defined as the energy out divided by the energy in, then this efficiency would be unity here, regardless of whether most of $Q_1^{\leftarrow}$ is provided as work or low-temperature heat $Q^{\rightarrow}$. We know, however, that heat is not equally valuable as work. For example, if T_Q were equal to the environmental temperature T_{R^*}, then $Q^{\rightarrow}$ would be entirely useless and yet the energy ratio would count it as equally useful as work.

These difficulties are eliminated if we compare the availability of the two outputs to the availability of the input, because then all interactions are evaluated on a comparable basis. Specifically, the effectiveness of the cyclic device is

$$\epsilon = \frac{W^{\rightarrow} + Q^{\rightarrow}(1 - T_{R^*}/T_Q)}{Q_1^{\leftarrow}(1 - T_{R^*}/T_{Q1})} = 1 - \frac{T_{R^*} S_{\text{irr}}}{Q_1^{\leftarrow}(1 - T_{R^*}/T_{Q1})} \tag{24.41}$$

where in writing the second of equations (24.41) we use (24.39) and (24.40).

Problems

24.1 A solar energy collector is shown schematically in Figure P24.1. A fraction of the incident radiation I (W/m^2) is reflected back into space, and some is lost by heat conduction to the structural materials. Thus the amount of energy per unit area, ΔE, delivered to the working fluid is appoximately $\Delta E = A I - B (T_1 - T_0)$, where T_0 and T_1 are the inlet and outlet temperatures of the working fluid and, for a typical collector design, $A = 0.5$, $B = 2$ W/m^2 K, and $I = 600$ W/m^2. The working fluid is incompressible.

(a) Write an expression for the ratio of the energy change, ΔE, and the entropy change, ΔS, of the working fluid in terms of T_1 and T_0,

(b) Write an expression for the availability of the fluid exiting the collector in terms of A, B, T_1, T_0, and I. The environment is at temperature $T_{R*} = T_0$.

(c) Find an expression that determines the temperature T_1 that yields the largest availability of the fluid. Find the value of T_1 for the given collector design, and $T_0 = 25°C$.

(d) Find the temperature at which the collected energy is the largest. What is the availability of this energy?

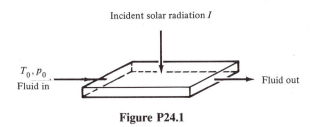

Figure P24.1

24.2 In an industrial process, steam at a pressure of 1 atm and 90% quality is vented to the atmosphere. The atmosphere is at $20°C$.

(a) How much availability is lost by venting 1 kg of steam?

(b) What is the entropy leaving the vent per kilogram of steam? What is the entropy increase of the atmosphere?

(c) Find the temperature of an equivalent energy reservoir having the same availability as the steam.

24.3 A significant amount of availability is contained in molten blast furnace slag, a by-product of the iron-making process. Typically, after being tapped from the blast furnace, the slag is poured into pits, where it is sprayed with large quantities of water, and the availability is dissipated. The system shown in Figure P24.3 has been proposed as a possible means of recovering some of the availability of molten blast furnace slag. Assume that the direct contact heat exchanger, the steam generator, condenser, turbine, pump, blower, and all connecting pipes are perfectly insulated. The generator efficiency is 95%. Other data are given in the following table.

State	Temperature	Substance	Energy/unit mass or energy flow	Entropy/unit mass or entropy flow
1	1700 K	Slag	1750 kJ/kg	3.831 kJ/kg K
2	450 K	Slag	242 kJ/kg	2.1617 kJ/kg K
3	—	Water	284 kW	—
4	—	Steam	9496 kW	—
5	300 K	Slag	200 kJ/kg	1.915 kJ/kg K

(a) Find the rate at which energy is rejected to the cooling water in the condenser.

(b) Find the power required to operate the pump and blower.

(c) What is the effectiveness of this system?

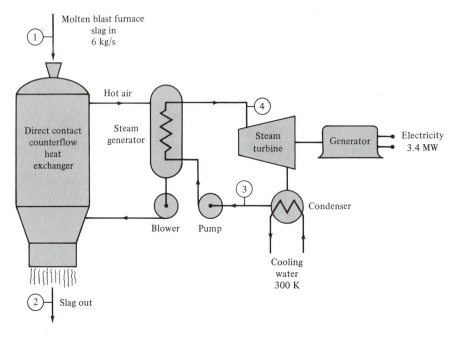

Figure P24.3

24.4 A separation plant of a propylene-propane mixture is shown in Figure P24.4. Electricity supplied to the three compressors 2, 5, and 6 and heat supplied by steam at 105°C (state 3) are used to separate the saturated liquid feed (state 1) into saturated vapor distillate (state 7) and saturated liquid bottoms (state 4). The condenser rejects heat at 25°C (state 8). Properties of the streams at each entry and exit point are listed in the following table. Other data are given in the figure.

State	Stream	T °C	p atm	$\dot{m}$ kmol/h	h kJ/kmol	s kJ/kmol K
1	Feed	52	20.0	273	−9,593.6	212.73
4	Saturated liquid	57.7	20.4	113.2	−72,458.8	213.74
7	Saturated vapor	46.7	19.0	159.5	46,976.3	241.52

(a) What is the least rate of work required for the separation?

(b) What is the rate of availability consumed in the process?

(c) What is the rate of entropy generation by irreversibility?

(d) What is the rate of entropy rejection at the condenser?

24.5 A cogeneration design consists of an indirect-fired air turbomachinery (turbine plus electricity generator) with a waste-heat-recovery steam generator (Figure P24.5a). The design is suitable for use with coal rather than oil or natural gas. The coal is burned in a fluidized-bed combustor and used to heat the working fluid indirectly. The working fluid is air (state 1) compressed to high pressure (state 2) and heated in an air heater. The hot compressed air (state 3) is expanded in a turbine, and the exhaust of the turbine (state 4) used to raise saturated process steam. The temperature profiles of the air and water in the waste-heat boiler are shown in Figure P24.5b. Note that the smallest difference between the two profiles is called the *pinch temperature*. In addition,

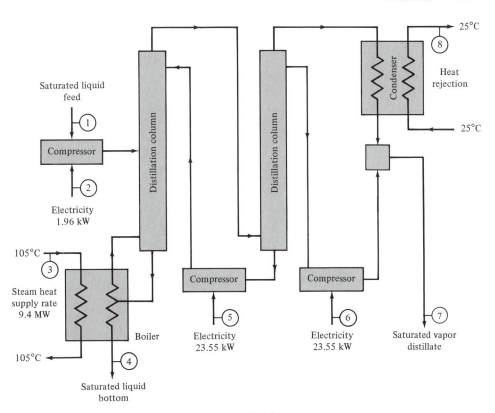

Figure P24.4

the section of the waste-heat boiler where the water is heated to saturation is called the *economizer*, and the section where steam is raised the *evaporator*. Typical values of some characteristics of this design are given in the following table.

Air heater efficiency = energy into air stream/energy from fuel	$\eta_H = 0.85$
Compressor efficiency	$\eta_c = 0.90$
Turbine efficiency	$\eta_t = 0.90$
Electric generator efficiency = power in/power out	$\eta_g = 0.96$
Compressor pressure ratio	$r_c = p_2/p_1$.
Turbine expansion ratio	$r_t = p_3/p_4 = 0.9r_c$
Environmental pressure	$p_{R^*} = p_1 = p_5 = 100 \text{ kPa}$
Air and environmental temperature	$T_{R^*} = T_1 = 15°C$
Condensate return temperature	$T_6 = 100°C$
Pinch-point temperature difference	$\Delta T_p = 20°C$
Temperature ratio	$\theta = T_3/T_1 = 4$
Air specific-heat ratio	$\gamma = c_p/c_v = 1.4$

(a) Find an expression for the exit temperature T_5 in terms of the properties of water, the steam pressure p_6, and the symbols defined in the table. Assume no pressure drop in the steam loop, i.e., $p_6 = p_7$, and no pressure drop in the air stream of the waste-heat boiler, i.e., $p_4 = p_5$.

(b) Find expressions for the rate of energy $\Delta \dot{H}_s$ required to transform the condensate into saturated steam, the availability rate $\Delta \dot{B}_s$—with respect to the environment—of the steam raised, the net electrical power output $\dot{W}_{el}$, the rate of energy input into the air stream $\dot{Q}_{in}$, and the steam mass flow rate $\dot{m}_s$ in terms of the air flow rate $\dot{m}_a$, the inlet temperature T_1, and the symbols defined in the table.

(c) Find an expression for the effectiveness of the cogeneration plant. Assume that the availability of the fuel is approximately equal to the energy supplied to the air heater (not to the air stream).

(d) Evaluate your answer in part (c) for the four combinations of conditions resulting from steam pressures $p_6 = 1$ MPa and $p_6 = 3$ MPa, and compressor pressure ratios $r_c = 4$ and $r_c = 12$.

(e) How would you allocate the fuel cost between the electricity and the process steam for the four combinations in part (d)?

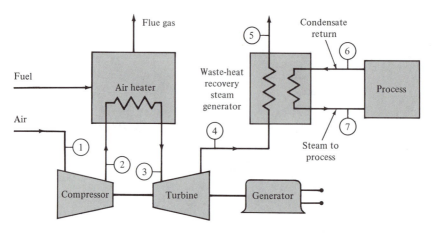

Figure P24.5a

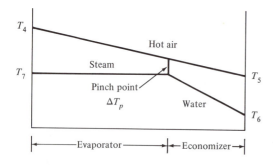

Figure P24.5b

24.6 A steel billet (mass 10^4 kg and specific heat 0.4 kJ/kg K) is to be cooled from 1000 K to 400 K by means of a stream of 10 kg/s of liquid water which, at atmospheric pressure, changes its temperature from 20°C to 80°C.

(a) What is the shortest lapse of time that machinery must operate in order to achieve the desired cooling effect?

(b) What is the effectiveness of machinery that performs the desired task in three times as much time as found in part (a)?

24.7 This problem compares two means of transportation: the bicycle and the Greyhound bus. A cyclist at the "cruise speed" of 30 km/h consumes the equivalent of 10 g of glucose per kilometer traveled. In the human body, the source of motive power is provided by the oxidation of glucose, $C_6H_{12}O_6$. The availability of glucose in an environment at 25°C is 15.6 MJ/kg. At the cruise speed, the total drag force acting on the bicycle and the rider is about 15 N. One of the pleasant aspects of cycling is that the removal of waste energy is very efficient due to the air stream in which the rider is immersed. Essentially all of the cooling—which maintains the body temperature at 37°C—is accomplished by evaporation of water from the surface of the skin.

A fully loaded 120-passengers Greyhound bus cruising at 30 km/h experiences a total drag force of about 2.7 kN and consumes fuel at a rate of about 0.25 kg/km. The availability of diesel fuel in an environment at 25°C is about 44 MJ/kg.

(a) How much availability is consumed by the cyclist per kilometer traveled at the cruise speed? Express the result in kcal/km.

(b) What is the effectiveness of bicycle riding at the given conditions?

(c) How much water is evaporated from the surface of the cyclist's skin per kilometer traveled at cruise speed in a cloudy day? Assume that glucose provides negligible entropy, and that the entropy created by irreversibilities in the human body is entirely removed by evaporation.

(d) How much availability is consumed by the Greyhound bus per passenger and per kilometer traveled at the cruise speed of 30 km/h?

(e) What is the effectiveness of bus transportation at this cruise speed?

24.8 In a given counterflow heat exchanger, each of the two streams behaves either as a perfect gas or as an incompressible fluid, and the pressure drop from inlet to outlet is negligible for both streams. Denote by Γ' and Γ'' the product of the mass flow rate and the specific heat of the hot and the cold streams, respectively, by T_1 and T_2 the inlet and outlet temperatures of the hot stream, and by T_3 the inlet temperature of the cold stream.

(a) Give an expression for the availability effectiveness of the heat exchanger, assuming that it has no energy losses, that is, that all the energy from the hot stream is transferred to the cold stream.

(b) Assume that cyclic machinery interacting with the two streams and a weight (but no other systems in the environment) produces work at the rate $\dot{W}$. What is the outlet temperature T_4 and the largest work rate $\dot{W}_{\text{largest}}$ if the machinery is perfect?

(c) Give an expression for the availability effectiveness of the heat exchanger of part (b) when $\dot{W}$ is not necessarily equal to $\dot{W}_{\text{largest}}$.

24.9 A utility purchases fuel oil at \$3 per 10^6 kJ and transforms it into electricity and process steam for manufacturing plants. It does so in a cogeneration plant (Figure P24.9), namely, by burning fuel in a steam boiler, supplying steam to process and to a back-pressure turbine, and getting more process steam from the exit of the back-pressure turbine. The electricity and other rates are shown in the figure. One-tenth of the energy of the fuel–air mixture is lost up the stack of the boiler and through the insulation of the installation. The fuel–air mixture availability is 42,000 kJ/kg of fuel.

(a) How much fuel is consumed in a year if the plant is operated 7000 h/yr, and used at an average of 80% of its capacity for all its products.

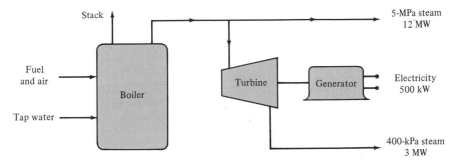

Figure P24.9

(b) The capital, maintenance and operation costs, and profit are about $2 per 10^6 kJ irrespective of the form of the product. What should be the sale prices of electricity, high-pressure steam, and low-pressure steam?

24.10 Diesel engines are used by many communities for the generation of electricity. One such installation is shown schematically in Figure P24.10a. Its operating characteristics are as follows. Number of diesel engines: 3; make: Pielstick; model: 12-cylinder PC4; power rating: 11.45 MW/engine; exhaust gas temperature: 425°C; exhaust gas mass flow rate (three engines): 249,000 kg/h; cooling-water temperature: 29°C; fuel: bunker C oil; fuel sulfur content: 3.5 to 4.5%. Depending on the cost of fuel relative to the cost of capital equipment, opportunities exist for recovering availability from either the exhaust gases, or the water cooling of the jackets of the engines, or both and thus increasing the energy productivity of the installation. One method to increase energy productivity is through a so-called *bottoming unit*. The bottoming unit consists of a waste-heat boiler and a steam turbogenerator (Figure 24.10b). The boiler is heated by the exhaust gases from the three diesels. To avoid condensation of sulfur oxide in the boiler, we must keep the exit temperature of the gases at 180°C. To avoid influences on the performance of the diesels, we must keep the pressure drop of exhaust gases in the waste-heat boiler, the so-called back pressure, at 4 in. of H_2O.

(a) What is the rate of enthalpy supplied to the waste-heat boiler by the exhaust gases? Assume that the exhaust gases are perfect.

(b) What is the rate of entropy supplied to the waste-heat boiler by the exhaust gases?

(c) What is the rate of availability supplied to the boiler?

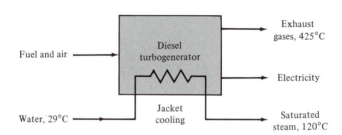

Figure P24.10a

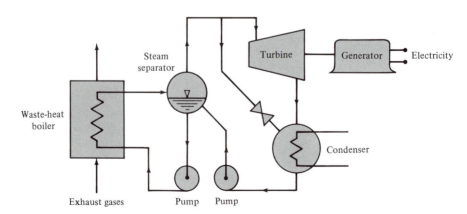

Figure P24.10b

(d) If the output of the bottoming unit is 3.7 MW, what is the entropy generated by irreversibility in the bottoming unit?

24.11 A special biological process can be modeled as shown in Figure P24.11. The environmental temperature $T_{R*} = T_0 = 20°C$. As a result of caloric consumption, and upon command from the central nervous system, cells c are set in a steady state in which they receive an influx of molecules, induce chemical reactions, and reject the products of the reactions at environmental conditions. The input enthalpy and entropy rates are $\dot{H}_c$ and $\dot{S}_c$, respectively. The output enthalpy and entropy rates are $\dot{H}_0$ and $\dot{S}_0$, respectively. During the course of the chemical reactions, energy is transferred to cells d that are at constant temperature $T_d = 36.7°C$. This transfer can occur by the engine-like arrangement shown in either Figure P24.11a or P24.11b.

(a) What is the availability received by cells c?

(b) What is the largest rate of availability transfer from cells c to cells d by means of the arrangement in Figure P24.11a?

(c) When the result in part (b) occurs, what is the rate at which cells d receive energy from cells c?

(d) Answer parts (b) and (c) for the arrangement in Figure P24.11b.

(e) Assume that both engines are totally irreversible, that is, engine 1 cannot do work and engine 2 cannot pump any energy in each of the two arrangements. What are the rates at which energy and availability is received by cells d?

(f) What are the rates of entropy generation by irreversibility in the two cases examined in part (e)?

(g) Assume that the effectiveness of engine 1 is η_1 and that of engine 2 η_2, and find an expression for the energy rate received by cells d in the arrangements in Figures P24.11a and P24.11b in terms of the specified enthalpies, entropies, temperatures, and η_1 and η_2.

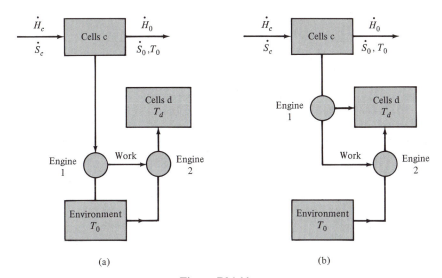

(a)   (b)

Figure P24.11

24.12 In a high-performance gas turbine for research or military applications, the blades in the first expansion stages are made of special materials and are water cooled so as to allow a very high inlet gas temperature $T_1 = 1400°C$. The inlet pressure is $p_1 = 1$ MPa. The blades are water cooled until the gas temperature $T_2 = 800°C$. Thereafter standard blade materials can operate without cooling. The turbine produces 10 MW of power. The low-temperature stages of the turbine (see Figure P24.12) have an efficiency $\eta = 0.9$. The air stream may be treated as a perfect gas with $\gamma = 1.4$. The gas exits the turbine at $T_3 = 400°C$ and $p_3 = 100$ kPa and is dumped into the atmosphere, which is at $T_{R^*} = 20°C$. The mass flow rate of cooling water is $\dot{m}_w = 1$ kg/s. It enters as saturated liquid at 200 kPa and exits as saturated vapor at 200 kPa.

(a) Find the intermediate pressure p_2 and the mass flow rate of air.

(b) Find the rates of entropy generation by irreversibility in the cooled stages and in the low-temperature stages of the turbine.

(c) At what rate is availability transferred to the cooling water?

(d) What is the availability effectiveness of the gas turbine if the availability transferred to the cooling water is dissipated into the atmosphere? What is the availability effectiveness if the cooling-water stream is considered as a useful by-product of the gas turbine?

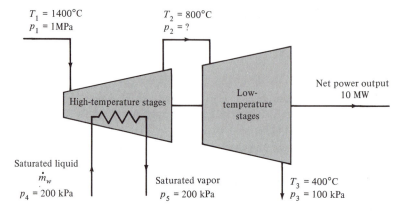

Figure P24.12

24.13 In a gas-turbine plant with air as the working fluid, the availability of the exhaust gases may be used in either of two different modes as shown in Figure P24.13. In mode 1 the exhaust gases are used as process heat in a nearby chemical plant. In mode 2 they are used in an engine that generates more power and then are discharged to the atmosphere. This engine is called a *bottoming unit* and often uses an organic fluid to carry energy from one part of the engine to another. The bottoming unit recovers 40% of the availability of the exhaust gases. The conditions that characterize the various streams are as follows: fuel availability rate 10 MW; compressor: efficiency 0.8, $T_{R^*} = 300$ K, $p_{R^*} = 101$ kPa, $p_1 = 1010$ kPa, $\dot{m} = 6.5$ kg/s, $c_p = 1$ kJ/kg K, $R = 0.29$ kJ/kg K; turbine: efficiency 0.9, $T_2 = 2000$ K, $p_2 = 1010$ kPa, $p_3 = 112$ kPa, $\dot{m} = 6.75$ kg/s, $c_p = 1.2$ kJ/kg K, $R = 0.27$ kJ/kg K.

(a) What is the net rate of work output by the gas-turbine plant?

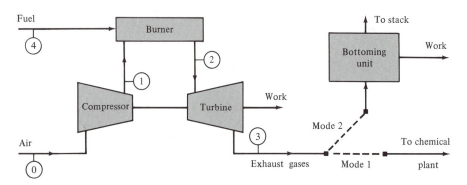

Figure P24.13

(b) What is the largest work rate that could be extracted from the exhaust stream 3?

(c) In mode 1 how would you allocate fuel cost between work output and process heat?

(d) In mode 2 how would you allocate fuel cost between the work output of the gas-turbine plant and the work output of the bottoming unit?

24.14 Machinery has been devised to heat air continuously from 20°C and 1 atm to 260°C and 5 atm. Air can be treated as a perfect gas with $c_p = 1.003$ kJ/kg K and $R = 0.287$ kJ/kg K. The only source of availability is saturated steam at 17 bar. Cooling water is available in large supply at 20°C.

(a) What is the largest amount of air, $m_{largest}$, that can be heated per kilogram of steam condensed to environmental conditions? What is the corresponding heat, q_s, to the environment?

(b) Per kilogram of condensing steam, find the amount of air, m, that is heated, and the heat, q, to the cooling water when the entropy generated by irreversibility in the process $s_{irr} = 2$ kJ/kg K.

(c) What amount of entropy generated by irreversibility would make the process incapable of heating any air?

24.15 A chemical plant has a waste stream of nitrogen, N_2, at 700°C and 8 atm, and a flow rate of 1.7×10^6 kg/h. An energy conservation proposal has been made to use this stream in a turbine-generator unit and a waste-heat boiler to produce electricity and saturated process steam, as shown in Figure P24.15. Design specifications are listed on the figure.

(a) What is the availability of the given waste stream of N_2?

(b) What are the electrical power generated and the temperature of the N_2 at the exit of the turbine?

(c) What is the rate of saturated process steam produced from water at 10 atm? Assume that 85% of the energy change of N_2 in the heat exchanger appears in the steam.

(d) What is the power required by the water pump?

(e) What is the availability of the process steam?

(f) What is the energy efficiency of the overall process?

(g) What is the availability of the outputs of the plant, and how does it compare to your answers in parts (a) and (f)?

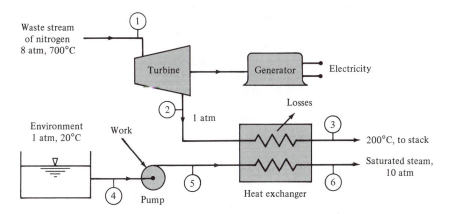

Figure P24.15

24.16 In open-cycle gas turbines used for peak electricity load, exhaust gases are usually wasted even though they are hot. For a 20-MWe installation, the gases are exhausted to the atmosphere and their power is about 2×10^8 kJ/h at a temperature of about 430°C. The turbine is in an environment at 15°C.

(a) Make an estimate of the largest rate at which work can be done using the exhaust gases in the given environment.

(b) If the exhaust gases were to be considered as an equivalent constant-temperature energy source, having the same availability as that found in part (a), what would be the temperature of this source?

(c) What is the rate of entropy flow from the source in part (b)?

(d) If the exhaust gases are not utilized to do work, what are the rates of entropy flow into the environment and entropy generation by irreversibility?

(e) The plant operators consider two types of engines that could utilize this energy source. Both are irreversible. Due to irreversibility, engine A generates 18,000 kJ/h K of entropy, while engine B generates 60,000 kJ/h K. What is the power delivered by each engine? What is the rate of entropy flow into the environment in each case?

(f) Part (e) should lead us to think that we really should be concerned about the penalty we pay for generating entropy rather than the "cost of energy" per se. If work is valued at 10 cents/kWh, what is the cost of 1 unit of entropy (1 kJ/K) generated by irreversibility in an environment at 15°C?

(g) If the annualized cost of capital equipment, including interest, depreciation, profit, and taxes is 25% of its purchase price, how much would you be willing to pay to avoid generating a unit of entropy per hour? Include in your estimate the effect of utilization of the machinery (hours per year).

(h) If unit A costs 12×10^6 and unit B 8×10^6, which would you choose? Assume that each operates 4000 h/yr.

24.17 Many industrial plants require both electricity and process steam for their operation. Widespread practice is to purchase electricity from a utility and raise process steam by burning fuel in a boiler at the plant. The raising of process steam results in waste of availability because the temperature of the products of combustion is much higher than the temperature of the required process steam. One approach to remedy this situation is to build a cogeneration plant. The fuel is used in an engine that produces electricity and the waste from the engine is used to produce process steam. If the needs of the plant are not in balance with those achieved by the cogenerator, either extra electricity or extra steam have to be produced or purchased separately, or one of these two forms of energy has to be sold to another user. An energy flowchart of a cogeneration plant is shown in Figure P24.17. All the electricity required is produced, but supplemental fuel firing is needed to produce the required process steam. As shown in the figure, the fuel consumption is as follows: diesel engine, 70 MW; boiler supplement, 22 MW. The energy produced and needed in the plant is as follows: electricity, 28 MW; 16-bar steam, 40 MW; 82°C water, 9.4 MW. At the time it was built, the cost of the plant was about 20×10^6, and the estimated time availability 8300 h/yr. For the purposes of this problem, assume that the maintenance, labor, taxes and insurance costs are $850,000 per year. Also assume that if no cogeneration plant were built, electricity would have to be purchased at 6 cents/kWh, and process steam would have to be generated in boilers that transfer 80% of the fuel enthalpy to the process stream. Fuel to run the diesel engine and the boilers would have been purchased at $6 per 10^6 kJ.

(a) What is the annual fuel consumption to produce the required hot water and steam in separate boilers (no cogeneration)?

(b) If the fuel can produce combustion gases at 1400°C, what is the availability of these gases? Assume an average atmospheric temperature of 10°C. Treat the gases as perfect at atmospheric pressure.

(c) If the availability of the fuel is equal to its energy, how does the answer in part (b) compare with the fuel availability?

(d) How much of the availability of the combustion gases is wasted per year by burning the fuel just to produce the steam and hot water? What is the entropy generated by irreversibility by such wastefulness?

(e) What would be the annual cost of purchasing fuel and electricity if the cogeneration plant were not built?

(f) If the local public utility delivers electricity with an overall efficiency of 30%, how much fuel is ultimately consumed to provide all the energy needs for the plant without cogeneration?

(g) What is the total fuel energy consumed in the cogeneration plant? What is its annual cost?

(h) How much money does the company save annually by the cogeneration plant? Does this justify the $20 million investment?

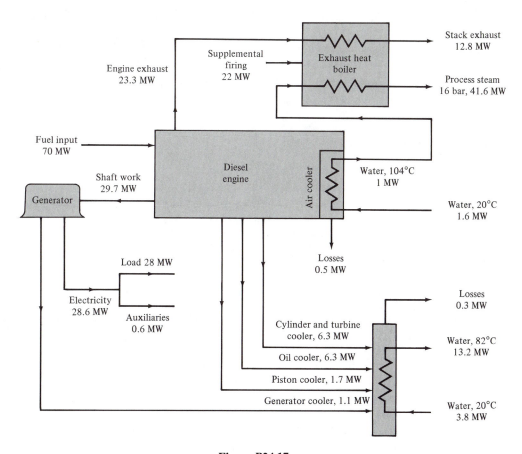

Figure P24.17

24.18 A feedwater heater used in a steam power plant is shown in Figure P24.18.

(a) Determine the amount of entropy generated by irreversibility per kilogram of feedwater leaving the heater at 26 bar and 53°C. Assume that the pump is perfectly insulated and reversible.

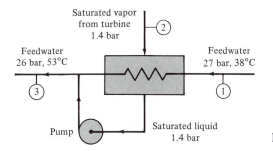

Figure P24.18

(b) Determine the amount of entropy generated by irreversibility if 1 kg of feedwater is to be changed from 27 bar and 38°C to 26 bar and 53°C in a counterflow heat exchanger in which the energy source is products of combustion that must exit the heat exchanger at 180°C. Assume that the products of combustion behave as a perfect gas. The gas flow rate is such that if the water were to be heated to its saturation temperature for $p = 26$ bar, the difference in temperature between gas and saturated water would be 10°C.

24.19 Assume that plenty of atmospheric air is available at 10°C and that a comfortable room temperature is 20°C.

(a) How much energy is required to heat 1 kg of atmospheric air to the required 20°C? Treat air as a perfect gas with $c_p = 1$ kJ/kg K and $R = 0.286$ kJ/kg K.

(b) What is the smallest amount of work required to accomplish the same task as in part (a)?

(c) If part (a) is accomplished either in a furnace that transfers 60% of the fuel enthalpy to the air, or by a heat pump that uses two times as much electricity as the minimum work required, compare the fuel consumed in the two methods. The electricity power plant uses the same fuel as the air furnace and transforms about one-third of the fuel enthalpy into electricity.

(d) Discuss whether the direct heating or the heat pump is more appropriate from the point of view of fuel conservation. How is your conclusion affected by economics?

24.20 A builder proposes three different systems to provide space heating and hot water to a two-apartment building: (1) electric resistance radiators and hot water heaters (capital cost of $100 per kilowatt of installed electric power); (2) a gas burner (capital cost of $200 per kilowatt of installed power); (3) a heat pump system (capital cost of $1000 per kilowatt of installed electric power).

Electric resistances transform electricity (purchased at 15 cents/kWh) directly into heating. The gas burner transforms into heating about 80% of the energy of a combustible gas (purchased at 6 cents/kWh). The heat pump consumes electricity (purchased at 15 cents/kWh) to drive an engine that pumps energy from the cold water in the environment to the space and the water to be heated, so that each kilowatthour of electricity consumed produces a useful heating of 2.5 kWh at the expense, of course, of a cooling (free of charge) of 1.5 kWh of the environmental water. The cost of capital, including interest rates, maintenance, insurance, and everything else, results in a yearly cost of 25% of the capital investment during the entire life of the equipment.

(a) For each of the three systems, express the yearly cost of the capital investment in terms of the required heating load $\dot{Q}$ (in kilowatts).

(b) For each of the three systems, express the yearly cost of fuel or electricity in terms of the heating load $\dot{Q}$ and the number of hours of operation n_h.

(c) Use the given data and compare the total yearly costs of the three systems. Show that the most economic solution is electric heating for $n_h < 500$ h, the gas burner for 500 h $< n_h <$ 2500 h, and the heat pump for $n_h > 2500$ h.

24.21 The heating and electric power loads of a food-processing complex, respectively, $\dot{Q} = 1200$ kW and $\dot{W} = 300$ kW, are to be supplied by a battery of so-called total energy systems or cogeneration units. In each unit, a derated automobile-type diesel engine drives an electricity generator, and the waste energy in the engine cooling water and exhaust gases is used for heating. The engine transfers 25% of the fuel energy rate $\dot{H}_f$ to the generator through a rotating shaft, and the generator transforms 90% of this rate into electricity. Of the fuel energy rate not transferred to the generator, 10% is lost to the atmosphere with the exhaust gases, and the remainder is collected by heat exchangers to provide the heating load.

Fuel is purchased at 6 cents/kWh. The cogeneration units are connected to the local electricity network. Any surplus of electric power is bought by the local utility at 10 cents/kWh. If necessary, to meet the electrical load, the generators output can be supplemented by electricity purchased from the local utility at 15 cents/kWh. If necessary to meet the heating load, fuel can be burned directly in a burner. The investment to purchase the cogeneration equipment is $1000 per kilowatt of electricty produced. The yearly cost of this investment, including interest rate, maintenance, insurance, and everything else, is 25% of the capital investment for the entire life of the equipment.

(a) Find the total yearly cost in terms of the number of hours of operation n_h if the cogeneration units are designed to meet exactly the heating load $\dot{Q}$.

(b) Find the total yearly cost in terms of n_h if the cogeneration units are designed to meet exactly the electricity load $\dot{W}$.

25   Energy Conversion Systems

As we discuss in Chapters 23 and 24, the accomplishment of almost every practical task requires the transfer of energy—or, more appropriately, availability—from an energy source to the task. The source may be a chemical fuel, such as coal, oil, or natural gas; a nuclear fuel, such as uranium; solar radiation, such as the input to photovoltaic cells; or an energy reservoir, such as a geothermal deposit, water at a high elevation, and other forms of stored solar energy (winds, and temperature differences in oceans and lakes).

The transfer of energy from the energy source to the task is achieved by means of a combination of devices. The devices experience interactions, such as work, heat, and bulk-flow interactions. Each combination or arrangement of devices is called an *energy conversion system*. Some of the interconnections between devices involve one or more fluids each of which circulates within the system from device to device. We call each such fluid a *working fluid*.

Energy-conversion systems are classified into various types, depending on the nature of the energy source, the working fluids, the sequence of states assumed by the system or a dominant working fluid, and the main purpose of the system. For example, in a gas-fired, organic-fluid, Rankine-cycle refrigeration system, the energy source is natural gas, the working fluid an organic compound that follows a particular cyclical change of state, and the system is used as a refrigerating unit.

In this chapter we discuss some typical energy-conversion systems first from the standpoint of the energy source and then from the standpoint of the sequence of states assumed by either the system or the dominant working fluid. Our discussions are limited to overall thermodynamic characteristics only. Many important aspects of detailed designs and analyses, as well as economics, environmental impact, and safety, are beyond the scope of this book.

25.1   Combustion Systems

To date, we get most of our energy from the combustion of coal, oil, and natural gas. Energy-conversion systems based on combustion are distinguished into two broad categories: *external combustion*, in which the fuel stream is separate from the working fluid that carries the energy to the task, and *internal combustion*, in which the fuel is part of the working fluid that supplies energy to the task.

25.1.1   External Combustion
A rudimentary schematic of an external combustion system is shown in Figure 25.1. Coal, oil, or natural gas and atmospheric air at environmental conditions are fed into a combustion chamber, where they combine chemically and transform into hot gaseous products of combustion. Energy and entropy from the hot gases are transferred to a

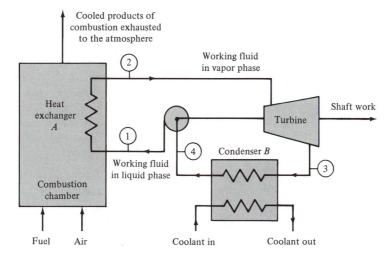

Figure 25.1 Schematic of an external combustion energy conversion system.

pressurized working fluid, such as water, as the hot gases pass through heat exchanger *A*. The cooled products of combustion are exhausted into the atmosphere. The working fluid changes from liquid state 1 to vapor state 2 in heat exchanger or boiler *A*, enters the turbine, does shaft work by expanding to state 3, and is restored to state 1 by passing through another heat exchanger or condenser *B*, and a pump.

Several features of an external combustion system in a steady state deserve to be emphasized. First, the working fluid undergoes fixed cyclic changes of state or cycles. The role of the working fluid is essential in that it carries availability from the fuel to the shaft, but auxiliary in that it is neither a source nor a sink of energy and entropy. Different sequences of states, and therefore different cycles are possible, and are discussed later.

Second, the work done by the turbine is much larger than the work done on the pump so that net availability is transferred from the fuel to the task.

Third, whereas the purpose of heat exchanger *A* is to transfer energy (and entropy) from the hot products of combustion to the working fluid, the purpose of heat exchanger or condenser *B* is to reject entropy (and energy) to the environment. This rejection of entropy is needed to maintain the energy-conversion system in a steady state. Indeed, entropy is transferred from the hot products of combustion to the working fluid and generated by irreversibility within the other devices of the system. Yet no entropy is transferred to the task.

The main sources of entropy generation are the processes of combustion in the burner and energy transfer from the hot products of combustion to the working fluid in heat exchanger *A*. In fact, these two sources of irreversibility are so large that they account for about a 50% loss of the fuel availability.

Example 25.1. A stream of hot products of combustion carries 70% of the availability rate of the fuel burned to generate it. The stream is at atmospheric pressure and temperature $T_1 = 1400$ K, and is cooled to $T_2 = 450$ K in a well-insulated boiler where saturated liquid water enters at $p_3 = 8$ MPa and exits as saturated vapor at $p_4 = 8$ MPa. Per unit of fuel availability rate, estimate the losses of availability in the boiler and in the discharge of the products of combustion into the atmosphere, assuming an environmental reservoir at $T_{R^*} = 290$ K.

Solution. The availability rate of the hot products of combustion, modeled as perfect gases, is
$$\dot{m}c_p\left[(T_1 - T_{R^*}) - T_{R^*}\ln(T_1/T_{R^*})\right] = \dot{m}c_p\left[(1400 - 290) - 290\ln(1400/290)\right] = 653\dot{m}c_p.$$
Thus the fuel availability rate is $653\dot{m}c_p/0.7 = 933\dot{m}c_p$. The energy transfer rate from

the hot products of combustion to the boiling water is $\dot{Q} = \dot{m}c_p(T_1 - T_2) = \dot{m}_w(h_4 - h_3) = \dot{m}_w h_{fg}(8\text{ MPa}) = 950\dot{m}c_p$. The entropy transfer rate from the products of combustion $\dot{m}c_p \ln(T_1/T_2) = 1.135\dot{m}c_p$, and to the water $\dot{m}_w s_{fg}(8\text{ MPa}) = \dot{m}_w h_{fg}(8\text{ MPa})/T_{sat}(8\text{ MPa}) = \dot{Q}/568\text{ K} = 1.673\,\dot{m}c_p$. Thus the rate of entropy generation by irreversibility in the boiler $\dot{S}_{irr} = (1.673 - 1.135)\dot{m}c_p = 0.538\,\dot{m}c_p$, and the rate of loss of availability per unit fuel availability rate $290 \times 0.538\dot{m}c_p/933\dot{m}c_p = 0.167$ or 16.7%. The availability rate of the outlet gases per unit of fuel availability rate $\dot{m}c_p[(450 - 290) - 290\ln(450/290)]/933\dot{m}c_p = 0.035$, that is, 3.5% of the fuel availability is lost in the discharge to the atmosphere. Thus, of the fuel availability rate, 30% is lost in the burner, 16.7% in the heat exchanger, and 3.5% in the discharge, for an overall loss of about 50%.

The fourth feature deserving comment is that most devices used as components in external combustion systems are designed so as to involve bulk-flow interactions and either shaft work only, such as in turbines and pumps, or heat only, such as in heat exchangers. The reason is that with today's materials and technology, work interactions can be achieved on a different time scale than heat interactions. For a given amount of energy transfer, the required residence time of the working fluid flowing through a turbine, pump, or compressor is much shorter than that through a heat exchanger. As a result, the combination of these two types of energy transfer in a single device creates technical conflicts and is usually avoided.

In some applications, upon exiting the turbine, the working fluid may be either rejected to the environment (Figure 25.2a), or used as a supply of relatively low temperature and low pressure process energy (Figure 25.2b). In the former case we say that we have an *open-cycle system* and in the latter that we have a *cogeneration system*, that is, a system that generates concurrently both work and nonwork.

25.1.2 Internal Combustion

A schematic of an internal-combustion system or *internal-combustion engine* is shown in Figure 25.3. It is a reciprocating engine in which a piston moves back and forth in a valved cylinder and does shaft work via a connecting rod and crank mechanism. The rotation of the crank produces a cyclical motion of the piston.

The majority of reciprocating engines operate on what is known as the *four-stroke cycle*, that is, the piston requires four sweeps of the cylinder—two revolutions of the crankshaft—to complete the sequence of events that produces one power cycle.

1. An intake stroke, which starts with the piston at the top position and ends with the piston at the bottom position. In spark-ignition engines, this stroke draws a mixture of air and gasoline or other fuel into the cylinder. In compression-ignition engines, this stroke draws only air into the cylinder. In either case, only the inlet valve is open.

2. A compression stroke during which both valves are closed and the contents of the cylinder are compressed. At the end of compression, combustion is initiated. In spark-ignition engines, the temperature of the fuel–air mixture at the end of the compression stroke is below the ignition temperature. A spark triggers the combustion by igniting a small amount of mixture near the spark plug. The flame then propagates very rapidly throughout the cylinder until all the initial fuel–air mixture is combusted. In compression-ignition engines, the temperature of the air at the end of the compression stroke is above the ignition temperature of a fuel–air mixture. Then liquid fuel is sprayed (injected) into the cylinder and the resulting fuel–air mixture auto-ignites. In either case, the temperature and pressure of the cylinder contents rises rapidly.

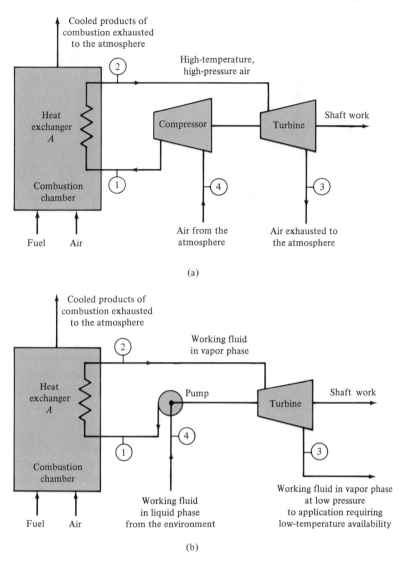

Figure 25.2 Schematics of: (a) an open-cycle system; and (b) a cogeneration system.

3. A power stroke that occurs with both valves closed. It starts with the piston at the top position and ends with the piston at the bottom position, as high-temperature, high-pressure gases push the piston down, and force the crank to rotate. About five times as much work is done by the gases in the course of a power stroke than is done on the gases in the course of the compression stroke.

4. An exhaust stroke during which the cylinder is purged of the gases through the exhaust valve as the piston moves from the bottom to the top position.

Here we usually identify the contents of the cylinder as the system. While the piston undergoes its cyclic change of position, the state of this system varies along the typical sequences of states that are discussed later.

As in the case of external-combustion systems, entropy must be discharged to the environment. The exhaust gases carry entropy, which accounts not only for the entropy of the inlet mixture but also for the entropy generated by irreversibility in the course of

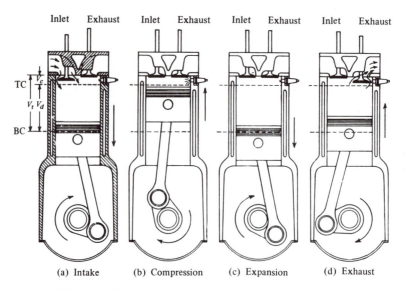

Inlet Exhaust Inlet Exhaust Inlet Exhaust Inlet Exhaust

(a) Intake (b) Compression (c) Expansion (d) Exhaust

Figure 25.3 Schematic of an internal combustion engine.

the four strokes. The primary cause of loss of fuel availability is the irreversibility of the process of combustion. Two other causes, however, contribute to losses. One is energy flow from the hot gases to the walls of the engine, and the other the fact that the exhaust gases are at relatively high temperature, much higher than in the case of external combustion. These three causes contribute an overall loss of over 50% of the fuel availability.

Example 25.2. For an automobile internal combustion, spark-ignition engine, assume that only 75% of the fuel availability is transformed by the combustion reaction in availability of the hot and compressed combustion products at $T_1 = 1500$ K and $p_1 = 4$ MPa. Assume also that: (1) 10% of the energy of the hot gases is transferred to the cylinder walls and, through the cooling system, to the environment; (2) the ratio between the energy and the entropy rates of this rejection from the hot gases is about 900 K; (3) the combustion products are exhausted at atmospheric pressure and $T_2 = 650$ K; and (4) the combustion products have a ratio of specific heats $\gamma = 1.33$. Per unit of availability in the fuel, estimate the losses of availability resulting from the energy and entropy rejection into the environment through the cooling water, and from the exhaust gases discharged into the atmosphere, assuming an environmental reservoir at $T_{R^*} = 290$ K and $p_{R^*} = 101$ kPa.

Solution. The availability of the hot products of combustion, modeled as a perfect gas, is $mc_v(T_1 - T_{R^*}) - T_{R^*}[mc_p \ln(T_1/T_{R^*}) - mR \ln(p_1/p_{R^*})] = mc_p[(1500 - 290)/1.33 - 290 \ln(1500/290) + 290(0.33/1.33) \ln(4000/101)] = 698mc_p$, where we use the relation $R/c_p = (\gamma - 1)/\gamma$. It follows that the fuel availability is $698mc_p/0.75 = 931mc_p$. The energy of the hot gases is $mc_v(T_1 - T_{R^*}) = mc_p(T_1 - T_{R^*})/\gamma = mc_p(1500 - 290)/1.33 = 910mc_p$. Of this energy, 10% is rejected to the environment through the cooling system, that is, $91mc_p$. The entropy transferred out of the products of combustion is $(91/900)mc_p = 0.101mc_p$, whereas that received by the environment is $(91/290)mc_p = 0.314mc_p$ and, therefore, the entropy generation by irreversibility in these transfers is $(0.314 - 0.101)mc_p = 0.213mc_p$, and the corresponding loss of availability per unit of fuel availability is $290 \times 0.213mc_p/931mc_p = 0.066$ or 6.6%. The availability of the exhaust gas stream discharged to the atmosphere, assuming that its pressure is close to atmospheric, is $mc_p[(T_2 - T_{R^*}) - T_{R^*} \ln(T_2/T_{R^*})] = mc_p[(650 - 290) - 290 \ln(650/290)] = 126mc_p$, and the corresponding loss per unit fuel availability is $126mc_p/931mc_p = 0.135$ or 13.5%. As a result, only $75\% - 6.6\% - 13.5\% \approx 55\%$ of the fuel availability remains for transfer as work through the piston during the power stroke. About one-fifth of this work, that is, 11% of the fuel availability, is used in the compression stroke, whereas a net of 44% remains for powering the transmission system and the auxiliaries.

25.2 Nuclear Fission Reactors

A schematic of a nuclear fission power reactor is shown in Figure 25.4. The fissionable material, usually the isotope of uranium that has a molecular weight of about 235, is dispersed in each of the fuel rods. Each fuel rod consists of other materials in addition to uranium 235 and, in water reactors, is surrounded by water.

Fission of a uranium 235 atom is induced by a relatively slowly moving neutron, with kinetic energy much less than 1 eV or 1.6×10^{-19} J. Each fission results in the production of two highly electrically charged, heavy nuclei, each having an extremely high momentum, as well as some other very high momentum particles, such as neutrons and electrons, and photons. Collectively, these particles have about 200 MeV of kinetic energy per fission (8.64×10^{10} kJ/kg of fissioned uranium, or 2×10^{13} kJ/kmol of fissioned uranium) and negligible entropy. The entropy of each fission particle is negligible because its kinetic energy is almost equal to the difference between its energy and the energy of its ground state. In other words, each fission particle is in a nonequilibrium state with an adiabatic availability almost equal to its energy above the ground-state energy.[1]

The fission fragments and the other particles produced by fission deposit their energy primarily in the fuel rods. For example, each heavily charged fission fragment is brought to a standstill within an infinitesimal distance from the location of its birth and therefore transfers almost all its kinetic energy to the constituents of the solid fuel rod. Much entropy is generated by irreversibility as a result of this transfer because the entropy per unit energy of the substances in the rod is much higher than the entropy per unit energy of the fission fragments. We can readily appreciate the magnitude of the availability loss

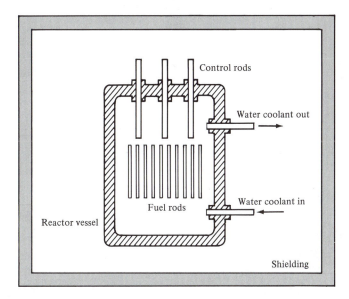

Figure 25.4 Schematic of a nuclear-fission reactor.

---

[1]It is noteworthy to compare the availability of a nuclear fuel to the availability of a chemical fuel such as coal, which is about 5×10^6 kJ/kmol. The ratio of the two availabilities is 4×10^6.

caused by this irreversibility by modeling the fuel rod as a system in an almost stable equilibrium state at an average temperature of about 1500 K. This is an oversimplified but reasonable representation of the fuel rods of a light-water reactor. With respect to an environment at about 300 K, each unit of energy in the fuel rod has an availability of $1 - (300/1500) = 0.8$. It follows that the loss of availability in transferring energy from the fission fragments to the fuel rods is about 20%. If all the energy of the fission particles were deposited in the fuel rods, the entropy generated by irreversibility per kilogram of fissioned uranium would be $0.2 \times 8.64 \times 10^{10}/300 = 5.76 \times 10^7$ kJ/kg K.

From each fuel rod, energy and entropy are transferred to the surrounding water, which may either boil in the reactor, as in a boiling-water reactor (Figure 25.5a), or remain liquid, as in a highly pressurized water reactor (Figure 25.5b). This transfer involves entropy generated by irreversibility as we can readily verify by modeling each fuel rod as a reservoir at an average temperature of about 1500 K, and the water as a reservoir at an average temperature of about 600 K. This modeling is again an oversimplification of the actual conditions and interactions of the fuel rods and the water. Nevertheless, it is

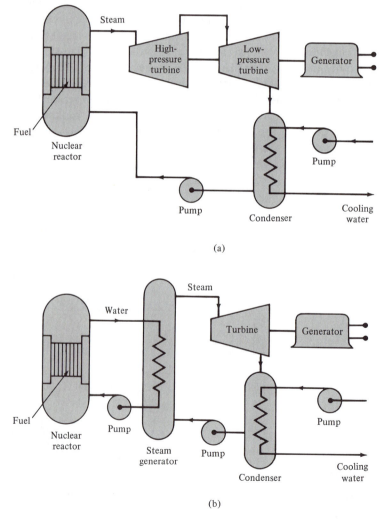

(a)

(b)

Figure 25.5 Schematics of: (a) a boiling-water-reactor (BWR) nuclear-fission energy conversion system; and (b) a pressurized-water-reactor (PWR) nuclear-fission energy conversion system.

sufficient for a good estimate of the availability. The availability of 1 unit of energy of the water at 600 K is $1 - (300/600) = 0.5$. Each such unit of energy is coming from the fuel rods where it has an availability of 0.8. It follows that the loss of availability resulting from the transfer of energy from the fuel rods to the circulating water is $0.8 - 0.5 = 0.3$ or about 30% of the fuel availability. The entropy generated by irreversibility per kilogram of fissioned uranium is $0.3 \times 8.64 \times 10^{10}/300 = 8.64 \times 10^7$ kJ/kg K. So without any loss of energy, the transfer of availability from the fission fragments to the water is accompanied by a loss of about one-half the availability of these fragments.

In a boiling-water reactor (Figure 25.5a), the water boils in the reactor core, and the generated steam may be fed directly into a turbine. Upon exiting the turbine, the water passes through a condenser and a pump, and returns to the inlet of the reactor ready to start another cyclic change of state. In a pressurized-water reactor (Figure 25.5b), the pressurized water is heated in the reactor core, enters a steam generator, and returns to the reactor via a pump, thus completing a cycle. In the heat exchanger, the secondary coolant or working fluid, which may also be water, begins a power-producing cyclic change of state such as that described in connection with external combustion systems. Details about cycles are discussed later.

In each of the systems in Figure 25.5, a condenser is necessary to dispose of the entropy generated by irreversibility not only in the course of the energy transfer processes in the reactor core but also those in the other components of the plant. The latter account for another 10 to 20% loss of availability.

25.3 A Solar System: Hydropower

The conversion of solar energy into work by a hydropower plant provides an example of an energy-conversion system involving an energy source in the form of electromagnetic radiation and a working fluid undergoing a special cycle. A schematic of the system is shown in Figure 25.6. It is a hybrid system, partly natural and partly manufactured.

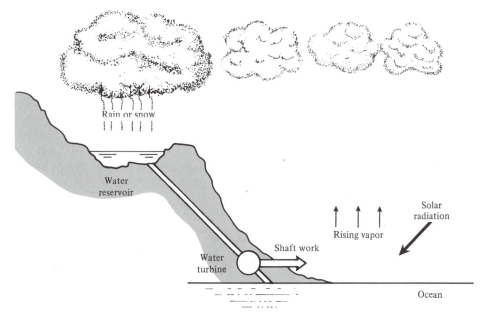

Figure 25.6 Schematic of the evaporation/precipitation system which converts solar energy into work in a hydropower plant.

At standard pressure of 1 atm, solar energy maintains the surface layers of the oceans and lakes—away from the poles—at temperatures of about 10 to 40°C, and vaporizes some of the water.[2] The vapor rises spontaneously to high altitudes against the force of gravity up to an altitude where it condenses into clouds of minute water droplets. When the clouds move into colder regions, the droplets coalesce and water from the condensation precipitates in the form of rain or snow, and some of it collects in reservoirs at high elevation, such as glaciers and lakes. From these reservoirs the water may eventually fall freely down to sea level in rivers and underground ducts, driven by the spontaneous trasformation of its potential energy into kinetic energy. Shaft work can be produced by interrupting the natural course and channeling water through a turbine.

The naturally occurring processes of a hydropower system are extremely inefficient. Energy is radiated by the sun at the rate of about 3.8×10^{22} MW. It results from nuclear fusion reactions occurring in a very small central core of the sun at temperatures around 20 million degrees. It is conducted to the outer surface, which is at a temperature of about 5760 K, and then radiated to the universe. Only a small part, 1.75×10^{11} MW, of the solar power is intercepted by the earth, and of this part 0.53×10^{11} MW is reflected away. The 1.22×10^{11} MW that reaches the surface contributes to the earth's activity as follows: 8.17×10^{10} MW to heating materials; 4.04×10^{10} MW to water evaporation and the metereological cycle; 3.68×10^{8} MW to convective currents in the oceans, winds, and waves; and 9.83×10^{7} MW to photosynthesis on the surface and in the oceans.

Associated with the solar radiation are an energy flow rate per unit surface area, J_e, and an entropy flow rate per unit surface area, J_s, given by the relations

$$J_e = \sigma\,(T^4 - T_{R*}^4) \qquad \text{and} \qquad J_s = \frac{4}{3}\,\sigma\,(T^3 - T_{R*}^3)$$

where $T = 5760$ K, T_{R*} is the temperature of the surface of the earth, and the Stefan-Boltzmann constant $\sigma = 5.6705 \times 10^{-8}$ W/m² K⁴. So with respect to an environmental temperature $T_{R*} = 300$ K, the availability rate per unit energy rate is $(J_e - T_{R*} J_s)/J_e = (T^4 - 4\,T_{R*}\,T^3/3 + T_{R*}^4/3)/(T^4 - T_{R*}^4) = 0.93$. However, most of the solar radiation availability used for natural evaporation is wasted. Rainfall from very high altitude to the surface of the earth, and subsequent evaporation or unexploited water flow to the oceans occur without any work production. Moreover, only a small fraction of the energy used in vaporization ends in water in high-elevation reservoirs. It has been estimated that of the 4.04×10^{10} MW used in vaporization of water from oceans and lakes, only about 10^7 MW may be usable in hydropower installations. The remainder is lost because of direct return to the oceans, loss of altitude in the course of rainfall, and a variety of reevaporation processes. Of course, all these processes occur spontaneously, have many beneficial effects, such as irrigation and replenishment of water reservoirs, and are cost free. Nevertheless, thermodynamically they are extremely inefficient.

Losses occurring in a constructed water fall and in a water turbine amount to only about 5% of the availability of the water stored in the reservoir at high elevation.

25.4 Working-Fluid Cycles

As we discuss in Sections 25.1 to 25.3, a characteristic feature of an energy-conversion system is that it either undergoes cyclic changes of state or includes a working fluid that undergoes cyclic changes of state. In the course of these changes, energy and entropy are carried from the energy source to the task and the environment. Some energy-conversion systems may include more than one fluid, each experiencing its own cycle. In this and

---

[2]We study the properties of the mixture of water vapor and atmospheric air in Chapter 27.

subsequent sections, we discuss the main thermodynamic characteristics of a number of cycles that are used in practice.

Strictly speaking, the term cycle means a sequence of states in which the final state of a system concides with the initial state. In many applications, such as external combustion systems, the working fluid follows such a sequence of states, and the use of the term cycle is appropriate. In other applications, such as internal combustion engines, the system does not experience a cycle. Nevertheless, it is part of the traditional terminology to characterize such changes also as cycles or open cycles because some aspects of the state of the system change cyclically. For example, in an internal combustion engine the configuration of the cylinder, piston, and valves changes cyclically, but the contents of the cylinder do not, even though their sequence of states is repeated by each subsequent cylinder charge.

Cycles are classified according to various schemes related to the prevailing form of aggregation of the working fluid, such as in gas cycles and vapor cycles; the purpose of the energy-conversion system, such as in power cycles and refrigeration cycles; the component or process that characterizes the energy-conversion system, such as in gas-turbine cycles or absorption cycles; and the names of pioneers who contributed to the development of the corresponding energy conversion systems, such as in Rankine cycles, Otto cycles, Diesel cycles, Joule–Brayton cycles, and Linde–Hampson cycles. We discuss various cycles in the following sections.

The sequence of states of a cycle is typically depicted on one or more diagrams, such as temperature–entropy, pressure–volume, enthalpy–entropy, and enthalpy–pressure diagrams. These diagrams represent properties of the stable equilibrium states that either approximate the states of a system or are associated with the bulk-flow states of a working fluid. For example, we recall from Chapter 22 that though a bulk-flow state in general is not a stable equilibrium state, a part of a bulk-flow state is uniquely related to a stable equilibrium state. Thus each point on a temperature versus specific entropy diagram characterizes only a part of the bulk-flow state of the working fluid, namely, the stable-equilibrium-state part. Moreover, the cycles on such diagrams are usually shown as continuous lines interconnecting the points on the diagram that represent the bulk-flow states at the inlet and outlet bulk-flow interactions of the various component devices. The implication of such a graph is that the working fluid is modeled as passing through a continuous sequence of bulk-flow states as it proceeds through each device. This implication may not be valid. However, in applications in which the interactions are specified, the implication is inconsequential.

Finally, we remark on some unfortunate, but commonly used expressions which are meaningless if not taken *cum grano salis*. Two such expressions are "efficiency of a cycle" and "reversible or irreversible cycle." A cycle by itself is neither efficient nor inefficient, and neither reversible nor irreversible. The concept of efficiency is related to the loss of the potential ability to produce a desired task, that is, to the loss of availability caused by the generation of entropy by irreversibility. The entropy generation is determined not only by the changes of state of the system or the working fluid, that is, the cycle, but also by the entropy exchanges with other systems, that is, by the particular arrangement of devices adopted to realize a given cycle. Only when the entropy generation by irreversibility is specified, the associated loss of availability and the efficiency of the particular arrangement are determined. Said differently, a given cycle may in principle be obtained by means of several different arrangements. Some of these arrangements, and not the cycle, may be reversible and therefore 100% efficient. Other arrangements may be irreversible, each with an efficiency more or less different from 100%. So the expression "efficiency of a given cycle" means "efficiency of an energy conversion system which results in the given cycle for either the system or the working fluid." Similarly, the expression "reversible (or irreversible) cycle" means "cycle of either the system or the working fluid of an energy conversion system operating reversibly (or irreversibly)."

25.5 Carnot Cycle

A cyclic sequence of states often used as a reference in discussions of the foundations of thermodynamics, and in analyses of energy-conversion systems is shown on a temperature versus specific entropy diagram in Figure 25.7a and on a pressure versus specific volume diagram in Figure 25.7b. The first segment of the sequence is isothermal from a state 1 with temperature T_1 to a state 2 with temperature $T_2 = T_1$. The second segment is isentropic from state 2 to a state 3 with specific entropy $s_3 = s_2$ and temperature T_3 smaller than T_1. The third is again isothermal from state 3 to state 4 so that $T_3 = T_4$. The fourth is isentropic from state 4 back to state 1, so that $s_4 = s_1$.

This cycle, with two isothermal segments and two isentropic segments, is called a *Carnot cycle*.

A system that may be experiencing a Carnot cycle is a reciprocating engine such as the one shown schematically in Figure 25.8, assumed to operate reversibly. It is called a reversible Carnot-cycle engine. A perfect gas, initially at temperature T_1, pressure p_1, and specific volume v_1 (state 1), is heated isothermally and reversibly by means of a work interaction and a heat interaction at temperature T_1 so that the gas expands to a larger specific volume v_2, at pressure p_2 and temperature $T_2 = T_1$ (state 2). At state 2 the heat interaction is interrupted, and the gas continues its expansion adiabatically and reversibly, and therefore isentropically, to an even larger specific volume v_3 at T_3, p_3, and $s_3 = s_2$ (state 3). After state 3, the gas is compressed isothermally and reversibly by means of a work interaction and a heat interaction at temperature T_3. The compression is carried up to state 4, at which $s_4 = s_1$, the specific volume is v_4, the pressure p_4, and, of course, the temperature $T_4 = T_3$. From state 4, the heat interaction is interrupted and the gas is compressed adiabatically and reversibly, and therefore isentropically, back to state 1.

Because the process is reversible, the heat to the gas during the isothermal expansion from state 1 to state 2 per unit mass of gas is $q_{12}^{\leftarrow} = T_1 (s_2 - s_1)$. It is represented by the area of the rectangle $122'1'$ in Figure 25.7a. Similarly, the heat from the gas during the isothermal compression from state 3 to state 4 per unit mass of gas is $q_{34}^{\rightarrow} = T_3 (s_3 - s_4) =$

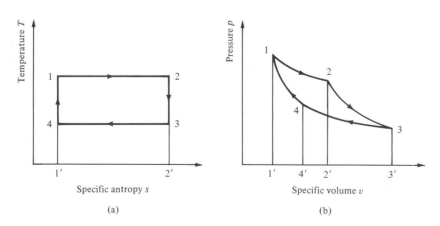

Figure 25.7 Representations of a Carnot cycle on a T–s and a p–v diagram.

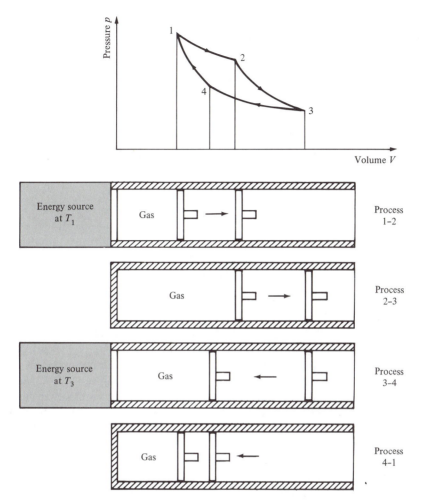

Figure 25.8   Schematic of a reciprocating piston-cylinder engine in which the working fluid may be following a Carnot cycle.

$T_3 (s_2 - s_1)$, where the last equation obtains because $s_3 = s_2$ and $s_4 = s_1$. This heat is represented by the area of the rectangle $432'1'$ in Figure 25.7a. We note that the ratio

$$\frac{q_{\overrightarrow{34}}}{q_{\overleftarrow{12}}} = \frac{T_3 (s_2 - s_1)}{T_1 (s_2 - s_1)} = \frac{T_3}{T_1} \tag{25.1}$$

and the net heat to the gas per unit mass

$$q_{\text{net}}^{\leftarrow} = \oint \delta q^{\rightarrow} = q_{\overleftarrow{12}} - q_{\overrightarrow{34}} \tag{25.2}$$

where $\oint$ denotes integration over the cycle 12341.

Again because the process is reversible, the work done by the gas during the expansion from state 1 to state 3 per unit mass of gas is $w_{\overrightarrow{123}} = \int_{v_1}^{v_3} p \, dv$, where the integral is taken along the line 123 and is represented by the area $1233'1'$ in Figure 25.7b. Similarly, the work done on the gas during the compression form state 3 to state 1 is $w_{\overleftarrow{341}} = \int_{v_1}^{v_3} p \, dv$, where the integral is taken along the line 143 and is represented by the area $1433'1'$ in Figure 25.7b. Each of the two integrals, $w_{\overrightarrow{123}}$ and $w_{\overleftarrow{341}}$, can readily be resolved by using the equation of state $pv = R_M T$ of the perfect gas and the specifications of the lines 123 and 143. We leave these resolutions as simple exercises for the reader. We note that the net work done by the gas per unit mass

$$w_{\text{net}}^{\rightarrow} = \oint p\,dv = w_{123}^{\rightarrow} - w_{341}^{\leftarrow} \tag{25.3}$$

Of course, the energy balance requires that the net work and the net heat be equal whenever the system undergoes a cycle, that is,

$$w_{\text{net}}^{\rightarrow} = q_{\text{net}}^{\leftarrow} \tag{25.4}$$

Formally, we could obtain (25.4) by integrating the energy balance $du = \delta q^{\leftarrow} - \delta w^{\rightarrow}$ along the cyclic process 12341 so that

$$\oint du = 0 = \oint \delta q^{\leftarrow} - \oint \delta w^{\rightarrow} = q_{\text{net}}^{\leftarrow} - w_{\text{net}}^{\rightarrow} \tag{25.5}$$

Upon combining equations (25.1) to (25.4), we find that

$$w_{\text{net}}^{\rightarrow} = w_{123}^{\rightarrow} - w_{341}^{\leftarrow} = q_{\text{net}}^{\leftarrow} = q_{12}^{\leftarrow} - q_{34}^{\rightarrow}$$

$$= q_{12}^{\leftarrow}\left(1 - \frac{q_{34}^{\rightarrow}}{q_{12}^{\leftarrow}}\right) = q_{12}^{\leftarrow}\left(1 - \frac{T_3}{T_1}\right)$$

$$= q_{12}^{\leftarrow}\,\eta_{\text{Carnot}} \tag{25.6}$$

where in writing the last of (25.6) we use the definition of the Carnot coefficient η_{Carnot} [equation (15.6)].

If the reciprocating piston-cylinder engine experiences a Carnot cycle but operates irreversibly, the net work is smaller than that given by (25.6) for given $q_{12}^{\leftarrow}$, T_1 and T_3. To illustrate this point, we assume that entropy is generated by irreversibility in the course of the isentropic expansion from state 2 to state 3, and in the course of the isentropic compression from state 4 to state 1. To maintain the expansion and the compression isentropic, this entropy must be transferred out of the system to an entropy sink, and such a transfer is accompanied by an energy transfer $q_{\text{sink}}^{\rightarrow}$, that is, the expansion and the compression are not adiabatic. As a result, the expansion yields less work and the compression requires more work than in the reversible operation. More precisely, we can readily show that the net work is

$$w_{\text{net}}^{\rightarrow} = q_{\text{net}}^{\leftarrow} = q_{12}^{\leftarrow} - q_{34}^{\rightarrow} - q_{\text{sink}}^{\rightarrow}$$

$$< q_{12}^{\leftarrow} - q_{34}^{\rightarrow} = q_{12}^{\leftarrow}\,\eta_{\text{Carnot}} \tag{25.7}$$

An example of an energy-conversion system in a steady state, in which the working fluid may be experiencing a Carnot cycle, is shown in Figure 25.9. The conditions of operation are such that all the states of the working fluid are under its saturation dome, and we assume that each component device of the system operates reversibly. Saturated liquid at temperature T_1 and pressure p_1 (state 1) enters a boiler and exits as saturated vapor at the same T_1 and p_1 (state 2). Upon exiting the boiler, the vapor enters a turbine, expands reversibly, does shaft work, and exits in a two-phase state 3 at temperature T_3 and pressure p_3. After the turbine, the two-phase fluid passes through a condenser where part of the vapor condenses at fixed T_3 and p_3 until the fluid reaches the two-phase state 4. From state 4, the two-phase fluid is compressed reversibly, back to state 1. As we discuss later, this scheme is not practical. Nevertheless, its study is informative.

In the boiler, the working fluid receives energy and entropy from an energy source at temperature T_1. The specific entropy changes from s_1 to $s_2 = s_1 + s_{\text{fg}}(T_1)$, and the specific enthalpy from h_1 to $h_2 = h_1 + h_{\text{fg}}(T_1) = h_1 + T_1\,s_{\text{fg}}(T_1) = h_1 + T_1\,(s_2 - s_1)$. So, in the boiler the heat rate to the working fluid per unit mass flow rate $q_{12}^{\leftarrow} = h_2 - h_1 = T_1\,s_{\text{fg}}(T_1) = T_1\,(s_2 - s_1)$. In the condenser, the working fluid rejects energy and entropy

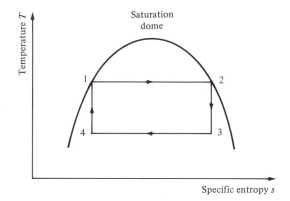

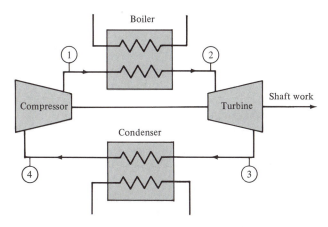

Figure 25.9 Schematic arrangement of a steady-state conversion system that could result in a Carnot cycle under the saturation dome for the working fluid.

to an energy sink at T_3. So in the condenser the heat rate from the working fluid per unit mass flow rate $\overrightarrow{q_{34}} = h_3 - h_4 = T_3 (s_3 - s_4)$. Because $s_3 = s_2$ and $s_4 = s_1$, we find that $s_2 - s_1 = s_3 - s_4$ and therefore $\overleftarrow{q_{12}}/T_1 = s_2 - s_1 = s_3 - s_4 = \overrightarrow{q_{34}}/T_3$ or

$$\frac{\overrightarrow{q_{34}}}{\overleftarrow{q_{12}}} = \frac{T_3}{T_1} \tag{25.8}$$

Because the turbine and the compressor operate reversibly, the isentropic changes from state 2 to state 3 ($s_3 = s_2$) in the turbine and from state 4 to state 1 ($s_4 = s_1$) in the compressor are achieved while the only interactions other than bulk flow are shaft work. Accordingly, the energy rate balance for the conversion system yields

$$\overrightarrow{w_{\text{net}}} = \overleftarrow{q_{12}} - \overrightarrow{q_{34}} = \overleftarrow{q_{12}} \left(1 - \frac{\overrightarrow{q_{34}}}{\overleftarrow{q_{12}}} \right)$$

$$= \overleftarrow{q_{12}} \left(1 - \frac{T_3}{T_1} \right) = \overleftarrow{q_{12}} \, \eta_{\text{Carnot}} \tag{25.9}$$

where $\overrightarrow{w_{\text{net}}}$ is the net shaft work rate per unit mass flow rate for turbine and compressor, that is, the difference of the rate of work done by the turbine and the rate of work done on the compressor divided by the mass flow rate of the working fluid.

If entropy were generated by irreversibility in the turbine and the compressor, the isentropic changes from state 2 to state 3 ($s_3 = s_2$) in the turbine and from state 4 to state 1 ($s_4 = s_1$) in the compressor would require the removal of this entropy by means of additional interactions such as heat to a sink. Then the energy rate balance per unit mass flow rate would be

$$w_{net}^{\rightarrow} = q_{\overline{12}}^{\leftarrow} - q_{\overline{34}}^{\rightarrow} - q_{sink}^{\rightarrow}$$

$$< q_{\overline{12}}^{\leftarrow} - q_{\overline{34}}^{\rightarrow} = q_{\overline{12}}^{\leftarrow}\,\eta_{Carnot} \tag{25.10}$$

or, equivalently, in terms of rates,

$$\dot{W}_{net}^{\rightarrow} \leq \dot{Q}_{\overline{12}}^{\leftarrow} - \dot{Q}_{\overline{34}}^{\rightarrow} = \dot{Q}_{\overline{12}}^{\leftarrow}\,\eta_{Carnot} \tag{25.11}$$

where $\dot{W} = \dot{m}\,w$, $\dot{Q} = \dot{m}\,q$, and $\dot{m}$ is the mass flow rate of the working fluid.

Traditionally, the ratio $w_{net}^{\rightarrow}/q_{\overline{12}}^{\leftarrow}$ [equation (25.6) or (25.9)], or $\dot{W}_{net}^{\rightarrow}/\dot{Q}_{\overline{12}}^{\leftarrow}$ [equation (25.11)] is called the *thermal efficiency of a reversible Carnot-cycle engine*. However, this terminology is improper. For example, for $T_1 = 600$ K, $T_3 = 300$ K, and reversible turbine and compressor operation, (25.11) yields $w_{net}^{\rightarrow}/q_{\overline{12}}^{\leftarrow} = \dot{W}_{net}^{\rightarrow}/\dot{Q}_{\overline{12}}^{\leftarrow} = 1 - T_3/T_1 = 0.5$. This ratio is smaller than unity not because of any imperfections related to the cycle, but because the availability of the energy source at 600 K with respect to an environment at $T_3 = 300$ K is only 50% of the energy. In a conversion system operating reversibly, this availability is received by the working fluid in the boiler and is carried to the turbine and the compressor, where it is all converted into shaft work, with a conversion efficiency of 100%.

When the turbine and the compressor operate irreversibly, it is impractical to remove the entropy generated by irreversibility directly from each of these devices. Instead, the removal is achieved in the condenser. Under such conditions, the cycle is no longer a Carnot cycle. On the T–s diagram, the cycle looks like that depicted in Figure 25.10, where the differences $s_3 - s_2$ and $s_1 - s_4$ represent the entropies generated by irreversibility in the turbine and the compressor, respectively. Per unit flow rate of the working fluid and given T_1 and T_3, the entropy difference $s_3 - s_4$ is greater than the difference $s_{3'} - s_{4'}$ of the Carnot cycle, the heat out of the condenser

$$q_{\overline{34}}^{\rightarrow} = T_3\,(s_3 - s_4) > T_3\,(s_{3'} - s_{4'}) = q_{\overline{3'4'}}^{\rightarrow} \tag{25.12}$$

and the net work

$$w_{net}^{\rightarrow} = q_{\overline{12}}^{\leftarrow} - q_{\overline{34}}^{\rightarrow} < q_{\overline{12}}^{\leftarrow} - q_{\overline{3'4'}}^{\rightarrow} = q_{\overline{12}}^{\leftarrow}\,\eta_{Carnot} \tag{25.13}$$

The availability of the heat $q_{\overline{12}}^{\leftarrow}$, that is, $q_{\overline{12}}^{\leftarrow}\,(1 - T_3/T_1)$, is an increasing function of T_1 and a decreasing function of T_3. Usually, the value of T_3 is dictated by the environment of the energy-conversion system. One method of increasing the temperature T_1 is by designing a conversion system with a working fluid that follows a Carnot cycle around the critical point (Figure 25.11a). At supercritical conditions, isothermal heating is not isobaric. It entails a pressure decrease that may be achieved by means of a turbine that is heated while the working fluid is expanding and doing shaft work (Figure 25.11b). Such a turbine is nearly impossible to build at a reasonable cost, mostly because of

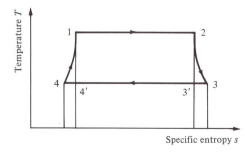

Figure 25.10 Representation on a T–s diagram of a cycle resulting from the arrangement in Figure 25.9 with irreversibilities in the turbine and the compressor.

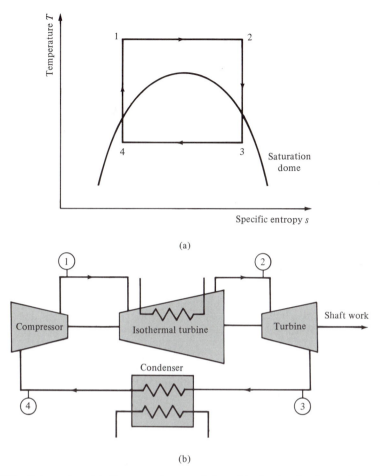

Figure 25.11   Schematics: (a) Carnot cycle around the critical point for the working fluid; and (b) arrangement of devices that may achieve this cycle.

the technical difficulty to subject the working fluid to both work and heat interactions simultaneously in one device.

In general, the Carnot cycle is not a good limiting cycle to strive for in an energy-conversion system using a two-phase working fluid such as water because of three important practical restrictions. First, the energy rate $\dot{Q}_{12}^{\leftarrow}$ comes from a source, such as products of combustion or nuclear fuel rods, that has much more availability per unit energy with respect to T_3 than the largest availability fraction $\dot{W}^{\rightarrow}/\dot{Q}_{12}^{\leftarrow} = (T_1 - T_3)/T_1$ [equation (25.11)] delivered to the working fluid. As we discuss earlier, the difference is due to entropy generated by irreversibility while energy is transferred from the energy source to the working fluid. To decrease this irreversibility and increase $\dot{W}^{\rightarrow}/\dot{Q}_{12}^{\leftarrow}$, we must increase T_1, the temperature at which the working fluid receives energy from the source. But increasing T_1 in the Carnot cycle in Figure 25.11 implies that the pressure must also be increased. This, in turn, makes the cost of the turbine prohibitively high. Second, the fraction of the working fluid that is in the form of liquid at the turbine outlet is relatively high. The presence of more than 10% liquid in the two-phase flow causes serious erosion of the blades of the turbine. This reduces both the performance and the life of the turbine. Third, both the expansion and compression of a two-phase fluid without heat involve unavoidable and relatively large amounts of entropy generation by irreversibility. This causes the rate of net work done by the working fluid in the course of a cycle to be much smaller than the largest possible.

All three restrictions are to a large degree bypassed by striving to operate the working fluid along the cycle described in the next section.

25.6   Rankine Cycle

A cyclic sequence of the stable equilibrium state parts of steady bulk-flow states of a simple system—working fluid of an energy-conversion system—is shown on a tempera-versus specific entropy diagram in Figure 25.12a. It may occur in an arrangement of devices, called a *reversible Rankine-cycle engine*, such as that shown in Figure 25.12b, where all devices operate reversibly. The sequence consists of four segments. The first segment is isobaric. It begins from state 1 at temperature T_1 and pressure p_1 of a compressed liquid which is transformed to a high temperature T_4 vapor at constant pressure p_1. Specifically, the liquid is heated first to the liquid saturation state 2 at T_2, p_1, next to the vapor saturation state 3 at $T_3 = T_2$, p_1, and then to the superheated vapor state 4 at T_4, p_1. During this heating the energy is supplied from a source at a rate $\dot{Q}_{14}^{\leftarrow}$. The second segment is isentropic. The superheated vapor enters a turbine, expands isentropically to a two-phase, high-quality state 5 at T_5, p_5, and does only shaft work at the rate $\dot{W}_t^{\rightarrow}$. The third segment is isobaric. The two-phase mixture passes through

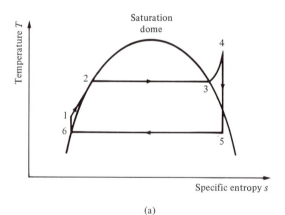

(a)

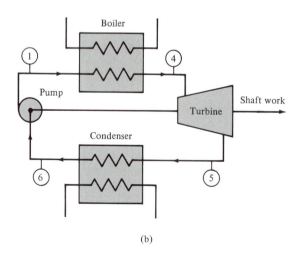

(b)

Figure 25.12   (a) Representation of a Rankine cycle on a T–s diagram; (b) Schematic of a conversion system in which the working fluid may follow a Rankine cycle.

a condenser, where it condenses to a saturated liquid state 6 at constant T_5 and p_5, as it rejects heat at the rate $\dot{Q}_{56}^{\rightarrow}$ and temperature T_5. The fourth segment is isentropic. The liquid from state 6 is compressed isentropically back to state 1 at T_1, p_1. The compression involves only shaft work done on the system at the rate $\dot{W}_c^{\leftarrow}$. This cycle, with two isobaric segments and two isentropic segments, is called a *Rankine cycle*.

Per unit mass flow rate, the energy input $q_{14}^{\leftarrow}$, the turbine work $w_t^{\rightarrow}$, the condenser heat $q_{56}^{\rightarrow}$, and the work done on the compressor $w_c^{\leftarrow}$ satisfy the relations

$$q_{14}^{\leftarrow} = h_4 - h_1 \qquad w_t^{\rightarrow} = h_4 - h_5 \qquad q_{56}^{\rightarrow} = h_5 - h_6 \qquad w_c^{\leftarrow} = h_1 - h_6 = v_6\,(p_1 - p_6)$$

where h denotes specific enthalpy, v specific volume, and in writing the last equation we assume that the fluid is incompressible. Accordingly, the net work done per unit flow rate of a fluid undergoing a Rankine cycle is

$$w_{net}^{\rightarrow} = w_t^{\rightarrow} - w_c^{\leftarrow} = h_4 - h_5 - h_1 + h_6 \tag{25.14}$$

For given T_1, $\dot{Q}_{14}^{\leftarrow}$, and T_5, $w_{net}^{\rightarrow}$ is the largest shaft work done per unit flow rate as the working fluid undergoes the specified reversible cyclic changes of steady bulk-flow states. It can be expressed as a fraction of the heat input $q_{14}^{\leftarrow}$ in the form

$$\eta = \frac{w_{net}^{\rightarrow}}{q_{14}^{\leftarrow}} = \frac{h_4 - h_5 - h_1 + h_6}{h_4 - h_1} = 1 - \frac{h_5 - h_6}{h_4 - h_1} \tag{25.15}$$

This fraction is called the *thermal efficiency of a reversible Rankine-cycle engine*.

The thermal efficiency of the reversible Rankine-cycle engine is an increasing function of the superheat temperature T_4 and the boiler pressure p_1, and a decreasing function of the condenser pressure p_5. We can verify these dependences by following the Rankine cycle on a Mollier diagram—specific enthalpy versus specific entropy diagram.

For example, for p_1 fixed we consider the two Rankine cycles 1234561 and 1234'5'61, corresponding to superheat temperatures T_4 and $T_{4'}$, respectively, as shown in Figure 25.13. For $T_{4'} > T_4$ we see that $h_{4'} > h_4$ and $h_{5'} > h_5$. However, $h_{4'} - h_4$ is larger than $h_{5'} - h_5$ because $(\partial h/\partial s)_p = T$ and the specific enthalpy increases faster along the

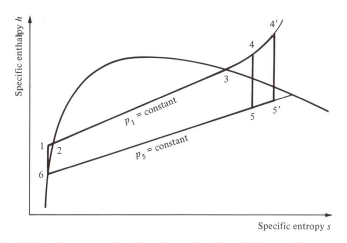

Figure 25.13 Mollier diagram used to investigate the effect of the superheat temperature on the thermal efficiency of a reversible Rankine-cycle engine.

isobar p_1 = constant with slope greater than T_4 than along the isobar p_5 = constant with slope $T_5 < T_4$. Accordingly, as T_4 increases, $(h_5 - h_6)/(h_4 - h_1)$ decreases, and the thermal efficiency [equation (25.15)] increases.

Again, for T_4 fixed, we consider the two Rankine cycles 1234561 and 1'2'3'4'5'61', corresponding to the two isobars p_1 and $p_{1'}$, respectively, and shown in Figure 25.14. For $p_{1'} > p_1$, $h_{5'}$ is less than h_5 by a larger amount than $h_{4'}$ is smaller than h_4 because in the vapor region $(\partial h/\partial s)_T \approx 0$. Accordingly, as p_1 increases, the thermal efficiency increases.

Regarding the dependence on p_5, we consider the two Rankine cycles 1234561 and 1'2345'6'1', corresponding to the two isobars p_5 and $p_{5'}$, respectively, and shown in Figure 25.15. For $p_{5'} < p_5$ we note that $h_{5'} - h_{6'}$ is smaller than $h_5 - h_6$ because the slope of the isobar $p_{5'}$ = constant is smaller than the slope of the isobar p_5 = constant, and that $h_4 - h_{1'}$ is larger than $h_4 - h_1$ because the temperature $T_{1'} < T_1$. Accordingly, the thermal efficiency increases as the condenser pressure p_5 decreases.

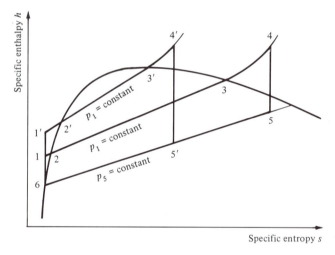

Figure 25.14 Mollier diagram used to investigate the effect of the boiler pressure on the thermal efficiency of a reversible Rankine-cycle engine.

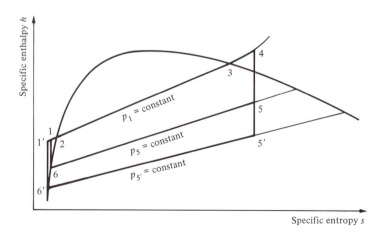

Figure 25.15 Mollier diagram used to investigate the effect of the condenser pressure on the thermal efficiency of a reversible Rankine-cycle engine.

It is noteworthy that heating the liquid from state 1 to state 2, and superheating the vapor from state 3 to state 4 permits transfer of more availability from the energy source to the working fluid than heating at constant temperature T_2 only. Also, expansion of the working fluid mostly as a vapor and compression as a pure liquid alleviate the problem of erosion. Finally, the shaft work done on the compressor is only a small fraction of the shaft work done by the turbine.

The actual cycle of a working fluid designed to operate under conditions that approximate a Rankine cycle deviates from it because of irreversibilities. Entropy is generated by irreversibility in each of the devices, boiler, turbine, condenser, and pump, as well as the piping that interconnects these devices. In detailed analyses, the irreversibility of each device is accounted for by means of a device efficiency, as we discuss in Chapter 23. Nevertheless, because the working fluid expands mostly as a vapor and is compressed as a liquid, the amount of entropy generated by irreversibility in the turbine and the compressor is reduced.

A modification of the Rankine-cycle engine that assures high quality, that is, high vapor fraction, of the working fluid exiting the turbine without undue increase of the superheat temperature is shown in Figure 25.16. The turbine consists of two stages, high pressure and low pressure. Upon exiting the high-pressure stage, the working fluid is

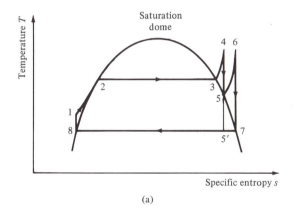

(a)

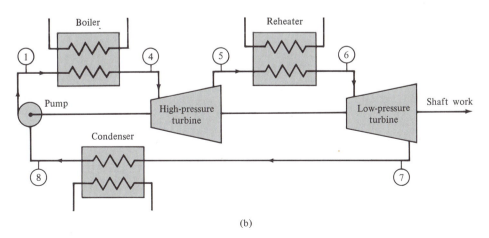

(b)

Figure 25.16 (a) Representation of a Rankine cycle with reheat on a T–s diagram; and (b) schematic of a conversion system in which the working fluid may follow a Rankine cycle with reheat.

returned to the superheater or reheater section of the boiler, reheated to another superheat temperature T_6, and then fed into the low-pressure turbine. As can readily be appreciated by inspection of the T–s diagram, the quality at the exit of the low-pressure turbine (state 7) is higher than is achieved (state $5'$) if the expansion continues isentropically without reheating. The modified cycle is called a *Rankine cycle with reheat*. For given high and low temperatures, the thermal efficiency of a properly designed reversible Rankine-cycle engine with reheat is generally higher than that of a Rankine-cycle engine without reheat.

25.7 Otto Cycle

A cyclic sequence of states used as a reference in discussions of spark-ignition, internal-combustion engines—engines in which the combustion of the fuel is ignited by an electric spark—is shown on a temperature versus specific entropy diagram and on a pressure versus specific volume diagram in Figure 25.17a. The first segment of the sequence is isochoric from state 1 with specific volume v_1 to a state 2 with specific volume $v_2 = v_1$. The second segment is an isentropic expansion from state 2 to a state 3 with specific entropy $s_3 = s_2$, and specific volume v_3 greater than v_2. The third segment is

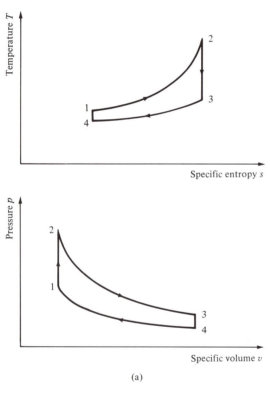

(a)

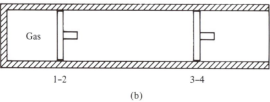

(b)

Figure 25.17 (a) Representations of an Otto cycle on T–s and p–v diagrams; and (b) schematic of a reciprocating piston-cyclinder engine which may follow an Otto cycle.

again isochoric from state 3 to a state 4, so that $v_4 = v_3$. The fourth segment is an isentropic compression from state 4 back to state 1. This cycle, with two isochoric and two isentropic segments, is called an *Otto cycle*.

A system that may be experiencing an Otto cycle is a reversible reciprocating piston–cylinder engine, as shown in Figure 25.17b. It is called a *reversible Otto-cycle engine*. A perfect gas, initially at pressure p_1 and specific volume v_1 (state 1), is heated reversibly at constant v_1 until the pressure becomes p_2 (state 2). The heat per unit amount of the gas is $q_{12}^{\leftarrow}$. At state 2, the heating is terminated, and the gas expands adiabatically and reversibly, that is, isentropically, to state 3, with $s_3 = s_2$. During the expansion, the work done by the gas per unit amount of gas is $w_{23}^{\rightarrow}$. After state 3, the gas is cooled reversibly at constant volume to state 4. The heat per unit amount of gas is $q_{34}^{\rightarrow}$. Finally, the gas is compressed adiabatically and reversibly back to state 1, and the work done on the gas per unit amount of gas is $w_{41}^{\leftarrow}$.

For a perfect gas, the heats $q_{12}^{\leftarrow}$ and $q_{34}^{\rightarrow}$, and the net work $w_{net}^{\rightarrow} = w_{23}^{\rightarrow} - w_{41}^{\leftarrow}$ satisfy the relations

$$q_{12}^{\leftarrow} = c_v\,(T_2 - T_1) \qquad q_{34}^{\rightarrow} = c_v\,(T_3 - T_4)$$

$$w_{net}^{\rightarrow} = w_{23}^{\rightarrow} - w_{41}^{\leftarrow} = q_{12}^{\leftarrow} - q_{34}^{\rightarrow} = q_{12}^{\leftarrow}\left(1 - \frac{q_{34}^{\rightarrow}}{q_{12}^{\leftarrow}}\right)$$

$$= q_{12}^{\leftarrow}\left(1 - \frac{T_3 - T_4}{T_2 - T_1}\right) = q_{12}^{\leftarrow}\left\{1 - \frac{[(T_3/T_4) - 1]\,T_4}{[(T_2/T_1) - 1]\,T_1}\right\} \qquad (25.16)$$

Because both the change from state 1 to state 4 and the change from state 2 to state 3 are isentropic, we have

$$T_1\,v_1^{\gamma-1} = T_4\,v_4^{\gamma-1} \qquad \text{and} \qquad T_2\,v_2^{\gamma-1} = T_3\,v_3^{\gamma-1}$$

where $\gamma = c_p/c_v$, the ratio of specific heats of the perfect gas. Upon combining these relations with $v_1 = v_2$, $v_3 = v_4$, and equation (25.16), we find that

$$\frac{T_3}{T_4} = \frac{T_2}{T_1} \qquad \frac{T_4}{T_1} = \left(\frac{v_1}{v_4}\right)^{\gamma-1} = \frac{1}{r^{\gamma-1}}$$

and

$$\frac{w_{net}^{\rightarrow}}{q_{12}^{\leftarrow}} = 1 - \frac{1}{r^{\gamma-1}} \qquad (25.17)$$

where $r = v_4/v_1 = v_3/v_2$, and is called the *compression ratio*. The ratio $w_{net}^{\rightarrow}/q_{12}^{\leftarrow}$ is called the *thermal efficiency of a reversible Otto-cycle engine*.

The working fluid of a spark-ignition, internal-combustion engine follows a sequence of states that resembles the Otto cycle. The processes of the cycle are shown schematically in Figure 25.18. A piston reciprocates in a cylinder fitted with an inlet and an outlet valve. Starting with the outlet valve closed and the inlet valve open, the piston moves from the left to the right, and a mixture of vaporized fuel (gasoline) and air at environmental conditions is drawn into the cylinder—the inlet stroke. When the piston reaches its extreme right position, the inlet valve closes and the mixture is in state 4. Starting from state 4, the piston moves from right to left and compresses the mixture to a state 1 at high pressure p_1, and temperature T_1—the compression stroke. At this state, the fuel–air mixture is ignited by means of an electric spark, and begins a spontaneous transformation into products of combustion—the flame propagation and combustion processes. This transformation is so fast that, to a good degree of approximation, can be

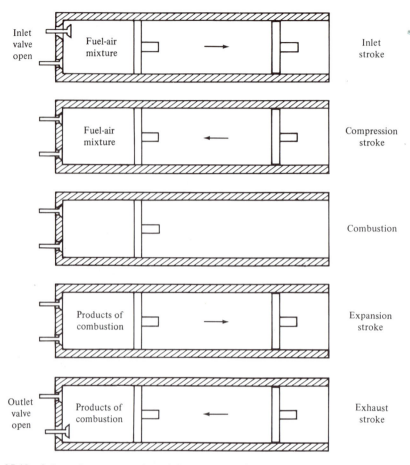

Figure 25.18 Schematic representation of the sequence of processes of a spark-ignition, internal-combustion engine following a sequence of states approximated by an Otto cycle.

regarded as occurring instantly, at constant volume and constant energy. It results in an almost instantaneous increase in pressure and temperature from p_1 and T_1 to p_2 and T_2, respectively. We discuss combustion processes in Chapter 31. The high pressure p_2 forces the piston to move again from left to right and to do work until the products of combustion reach state 3—the expansion stroke. At this state, the outlet valve opens and the pressure drops abruptly to a value slightly higher than atmospheric—cylinder decompression. Then the piston moves from right to left, the cylinder is cleared of most of the products of combustion—the exhaust stroke—and the engine is ready to repeat the sequence just described.

It is noteworthy that the compression and expansion processes can be regarded as adiabatic because they occur rather fast, leaving little time for heat interactions to take place. On the other hand, the combustion process differs from the heating process of an Otto-cycle engine in that it does not involve any heat interaction, and the decompression–exhaust–intake sequence of processes differs from the cooling process of an Otto-cycle engine in that it involves bulk-flow interactions instead of a heat interaction.

We see from equation (25.17) that the thermal efficiency of the reversible Otto-cycle engine is an increasing function of the compression ratio. The reason is that the availability associated with the energy $q_{12}^{\leftarrow}$ received by the gas in the process from state

1 to state 2 is an increasing function of the temperatures at which the energy is supplied, and such temperatures are increasing functions of the compression ratio.

In spark-ignition engines, the thermodynamic efficiency or effectiveness (Section 24.6) is defined as the net work divided by the availability of the fuel consumed, both evaluated over the same number of cycles. This efficiency is affected by irreversibilities. Nevertheless, many experimental and analytical studies indicate that also this efficiency is an increasing function of the compression ratio.

In practice, the increase of the compression ratio faces some limitations. When increased, it causes the temperature T_2 and the pressure p_2 to increase. But above a certain compression ratio, the fuel–air mixture reaches the spontaneous-ignition temperature before the piston has reached its extreme position in the cylinder. This causes the instantaneous ignition of the whole amount of fuel in the cylinder, and therefore its explosive combustion, at a rate much faster than when the combustion is initiated by a localized spark and continued by flame propagation through the cylinder. The explosive combustion and sudden pressure increase occur while the piston is still moving toward decreasing volumes. This occurrence, called *detonation* or *knock*, causes a much larger compression work and hence a much smaller net work than under normal operation. Moreover, it causes serious damages of the piston and the other mechanical components of the engine. So the gasoline is mixed with volatile compounds of lead, such as lead tetraethyl, that increase the auto-ignition temperature and allow the knock-free operation of engines with compression ratios up to about 12. But lead, which is poisonous, does not participate in the combustion process and is discharged to the atmosphere together with the exhaust combustion products. With about half a billion vehicles circulating throughout the world, lead pollution has become a severe environmental problem. Most industrialized countries are taking steps to eliminate completely the use of lead compounds. Various alcohols can be used as alternative antiknock compounds and are currently favored. However, these compounds participate in the combustion process and result in the formation of aldeheids, which are cancerous and mutageneous.

The compression ratio can be raised if the fuel is added not at the beginning of the compression process but at the end. The reference cycle and the internal combustion engines that result from this mode of operation are discussed in the next section.

25.8 Diesel Cycle

A cyclic sequence of states used as a reference in discussions of compression-ignition, internal-combustion engines—engines in which the fuel is injected in and ignited by high-pressure and high-temperature air—is shown on a temperature versus specific entropy diagram and on a pressure versus specific volume diagram in Figure 25.19a. The first segment of the sequence is an isobaric heating and expansion from a state 1 with pressure p_1 to a state 2 with pressure $p_2 = p_1$ and specific volume $v_2 > v_1$. The second segment is an isentropic expansion from state 2 to state 3 with specific entropy $s_3 = s_2$, and specific volume $v_3 > v_2$. The third segment is an isochoric cooling so that $v_4 = v_3$. The fourth segment is an isentropic compression from state 4 back to state 1. This cycle, with an isobaric segment, two isentropic segments, and an isochoric segment, is called a *Diesel cycle*.

A system that may be experiencing a Diesel cycle is a reciprocating piston-cylinder engine as shown in Figure 25.19b, provided that it is operated reversibly. It is called a *reversible Diesel-cycle engine*. A perfect gas, initially at pressure p_1 and specific volume v_1 (state 1), is heated reversibly at constant pressure p_1 until the specific volume becomes v_2 (state 2). The heat per unit amount of gas is $q_{12}^{\leftarrow}$. At state 2 the heating is terminated,

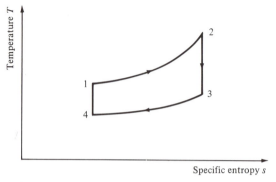

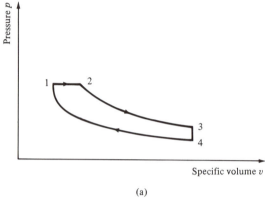

(a)

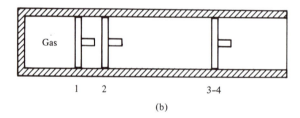

(b)

Figure 25.19 (a) Representations of a Diesel cycle on T–s and p–v diagrams; and (b) schematic of a reciprocating piston-cyclinder engine which may follow the cycle.

and the gas continues its expansion adiabatically and reversibly, that is, isentropically, to state 3 with $s_3 = s_2$. During the expansion from state 1 to state 3 the work done by the gas per unit amount of gas is $w_{23}^{\rightarrow}$. After state 3, the gas is cooled reversibly at constant volume to state 4. The heat per unit amount of gas is $q_{34}^{\rightarrow}$. Finally, the gas is compressed adiabatically and reversibly to state 1 and the work done on the gas per unit amount of gas is $w_{41}^{\leftarrow}$.

For a perfect gas, the heats $q_{12}^{\leftarrow}$ and $q_{34}^{\rightarrow}$, and the net work $w_{\text{net}}^{\rightarrow} = w_{23}^{\rightarrow} - w_{41}^{\leftarrow}$ satisfy the relations

$$q_{12}^{\leftarrow} = c_p\,(T_2 - T_1) \qquad q_{34}^{\rightarrow} = c_v\,(T_3 - T_4)$$

$$w_{\text{net}}^{\rightarrow} = w_{23}^{\rightarrow} - w_{41}^{\leftarrow} = q_{12}^{\leftarrow} - q_{34}^{\rightarrow} = q_{12}^{\leftarrow}\left[1 - \frac{c_v\,(T_3 - T_4)}{c_p\,(T_2 - T_1)}\right]$$

$$= q_{12}^{\leftarrow}\left\{1 - \frac{1}{\gamma}\frac{[(T_3/T_4) - 1]}{[(T_2/T_1) - 1]}\frac{T_4}{T_1}\right\} \tag{25.18}$$

In addition, for the isobaric change from state 1 to state 4, and the isentropic changes from state 2 to state 3, and from state 1 to state 4, we have the perfect gas relations

$$\frac{T_2}{T_1} = \frac{v_2}{v_1} = r_p \qquad T_1 v_1^{\gamma-1} = T_4 v_4^{\gamma-1} \qquad T_2 v_2^{\gamma-1} = T_3 v_3^{\gamma-1}$$

where we define the isobaric expansion ratio $r_p = v_2/v_1$. Upon combining these relations with $v_3 = v_4$ and equation (25.18), we find that

$$\frac{T_3}{T_4} = \frac{T_2}{T_1} r_p^{\gamma-1} = r_p^{\gamma} \qquad \frac{T_4}{T_1} = \left(\frac{v_1}{v_4}\right)^{\gamma-1} = \frac{1}{r^{\gamma-1}}$$

and

$$\frac{w_{\text{net}}^{\rightarrow}}{q_{12}^{\leftarrow}} = 1 - \frac{1}{\gamma}\left(\frac{r_p^{\gamma} - 1}{r_p - 1}\right)\frac{1}{r^{\gamma-1}} \tag{25.19}$$

where the compression ratio $r = v_4/v_1$. The ratio $w_{\text{net}}^{\rightarrow}/q_{12}^{\leftarrow}$ given by (25.19) is called the *thermal efficiency of a reversible Diesel-cycle engine*.

The working fluid of a compression-ignition, internal-combustion engine follows a sequence of states that resembles the Diesel cycle. The processes of the cycle are shown schematically in Figure 25.20. A piston reciprocates in a cylinder fitted with an inlet valve, an outlet valve, and an injector. Starting with the outlet valve closed and the inlet valve open, the piston moves from the left to the right, and atmospheric air at environmental conditions is drawn into the cylinder—the inlet stroke. When the piston reaches its extreme right position, the inlet valve closes, and the air is compressed from state 4 to a state 1 at high pressure p_1, and temperature T_1—the compression stroke. The compression ratio of diesel engines can be as high as 16. At state 1, the fuel injector begins to spray liquid fuel into the hot air—the fuel injection process. Part of the fuel evaporates from the spray of liquid droplets, mixes with the hot air, and ignites spontaneously because of the high temperature of the compressed air. Combustion increases the temperature and the rate of fuel evaporation from the spray droplets. By controlling the fuel injection, the combustion process proceeds simultaneously with the expansion against the piston so that the pressure remains approximately constant—the simultaneous combustion and expansion isobaric stroke—until the volume is $v_2 = r_p v_1$. Then combustion terminates and the products of combustion continue to do work until they reach state 3—the isentropic expansion stroke. At this state, the outlet valve opens, the pressure drops abruptly to a value slightly higher than atmospheric, the piston moves from right to left clearing the cylinder of most of the products of combustion—the exhaust stroke—and the engine is ready to repeat the sequence just described.

For the same high temperature T_2, the thermal efficiency of a reversible Diesel-cycle engine is larger than that of a reversible Otto-cycle engine, that is, the ratio of the availability and the energy received by the gas from the energy source is larger for a Diesel-cycle engine than for an Otto-cycle engine. To verify this assertion, we consider each of these cycles on a T–s diagram as shown in Figure 25.21. The relative position of the $p = $ constant and $v = $ constant lines passing trough the high temperature state 2 is as shown in the figure because $c_p > c_v$ and, therefore, $T(\partial s/\partial T)_p > T(\partial s/\partial T)_v$, that is, the slope of an isobar on the T–s diagram is smaller than the slope of an isochor. It follows that in the Diesel-cycle engine the temperature of the heat interaction with the source, and therefore the availability received by the gas per unit energy received, is higher than in the Otto-cycle engine.

25.9 Joule–Brayton Cycle

A cyclic sequence of stable-equilibrium-state parts of bulk-flow states used as a reference for gas-turbine energy conversion systems is shown in the temperature versus specific

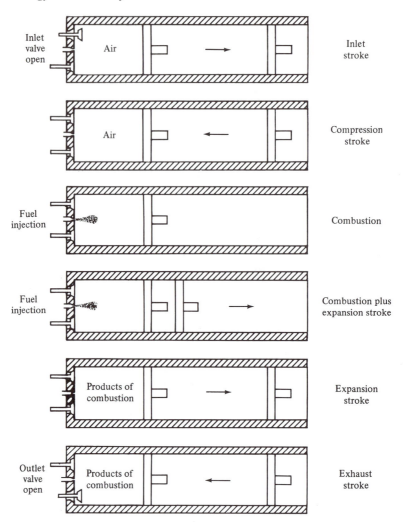

Figure 25.20 Schematic representation of the sequence of processes of a compression-ignition, internal-combustion engine following a sequence of states approximated by a Diesel cycle.

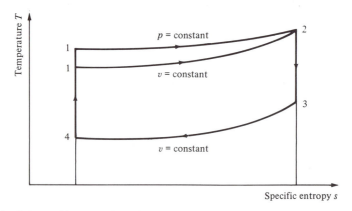

Figure 25.21 Superposition on a T–s diagram of a Diesel cycle and an Otto cycle with the same high temperature T_2.

entropy and the pressure versus specific volume diagrams in Figure 25.22. It is composed of two isobaric and two isentropic segments and is called the *Joule–Brayton cycle*.

A working fluid that may be experiencing a Joule–Brayton cycle is that of a reversible gas turbine power plant such as shown schematically in Figure 25.23a. It is called a *reversible Joule–Brayton-cycle engine*. Compressed air is heated at constant pressure p_1

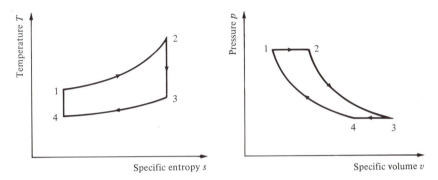

Figure 25.22 Representations of a Joule–Brayton cycle on T–s and p–v diagrams.

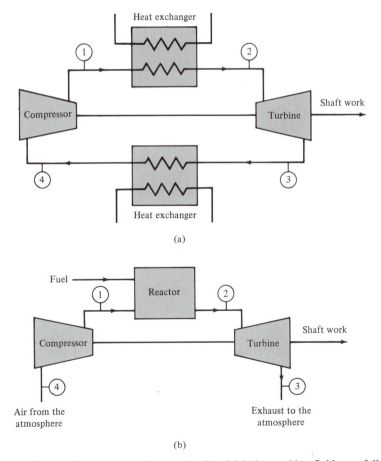

Figure 25.23 Schematics of a conversion system in which the working fluid may follow: (a) a reversible Joule–Brayton cycle; and (b) an open Joule–Brayton cycle.

from state 1 at temperature T_1 and specific volume v_1 to state 2 at temperature T_2 and specific volume v_2. The heat input rate per unit mass flow rate $q_{12}^{\rightarrow} = h_2 - h_1$, where h is the specific enthalpy. From state 2, the gas expands isentropically in a turbine to a state 3 at $p_3 < p_2$, $v_3 > v_2$, and with specific entropy $s_3 = s_2$. The turbine does shaft work only, at a rate per unit mass flow rate $w_{23}^{\rightarrow} = h_2 - h_3$. After state 3, the gas is cooled to state 4 at constant pressure so that $p_4 = p_3$, and until $s_4 = s_1$. The heat is $q_{34}^{\rightarrow} = h_3 - h_4$. From state 4, the gas is compressed isentropically back to state 1. The compression is done by a compressor that receives only shaft work at the rate per unit mass flow rate $w_{41}^{\leftarrow} = h_1 - h_4$. This work is a fraction of the work of the turbine.

The *thermal efficiency of a reversible Joule–Brayton-cycle engine* is defined by the relation

$$\frac{w_{net}^{\rightarrow}}{q_{12}^{\leftarrow}} = \frac{q_{12}^{\leftarrow} - q_{34}^{\rightarrow}}{q_{12}^{\leftarrow}} = 1 - \frac{q_{34}^{\rightarrow}}{q_{12}^{\leftarrow}} = 1 - \frac{h_3 - h_4}{h_2 - h_1} \tag{25.20}$$

If the working fluid is a perfect gas, then

$$\frac{w_{net}^{\rightarrow}}{q_{12}^{\leftarrow}} = 1 - \frac{c_p(T_3 - T_4)}{c_p(T_2 - T_1)} = 1 - \left[\frac{(T_3/T_4) - 1}{(T_2/T_1) - 1}\right]\frac{T_4}{T_1}$$

Because the expansion from state 2 to state 3 is between the same pressures as those of states 1 and 4, we have $T_3/T_4 = T_2/T_1$, $T_4/T_1 = (p_1/p_3)^{-(\gamma-1)/\gamma}$ and

$$\frac{w_{net}^{\rightarrow}}{q_{12}^{\leftarrow}} = 1 - \frac{1}{(p_1/p_3)^{(\gamma-1)/\gamma}} \tag{25.21}$$

We see that the thermal efficiency of the reversible Joule–Brayton-cycle engine is an increasing function of the pressure ratio p_1/p_3.

Figure 25.23b shows an arrangement of devices in which the working fluid follows a sequence of bulk-flow states close but not identical to the states of a Joule–Brayton cycle. Atmospheric air is drawn in from the environment at pressure p_4 and temperature T_4, compressed to state 1, mixed with a stream of fuel so that the products of the combustion in the well-insulated burner exit in state 2, expand in the turbine to state 3, and are discharged to the atmosphere. Even though the working fluid does not follow a cyclic sequence of states, it is customary to characterize the sequence just described as an *open Joule–Brayton cycle*. Because in most applications the temperature T_3 of the products of combustion at state 3 is higher than the temperature T_1, before being discharged to the atmosphere, these products can be used in a heat exchanger to preheat the compressed air as shown schematically in Figure 25.24.

25.10 Stirling Cycle

A cyclic sequence of states of an ideal gas, used as a reference in discussions of external-combustion reciprocating engines, is shown on a temperature versus specific entropy diagram, and on a pressure versus specific volume diagram in Figure 25.25. Its four segments are: an isothermal expansion from state 1, with temperature T_1 and specific volume v_1, to a state 2 with temperature $T_2 = T_1$ and specific volume $v_2 > v_1$; an isochoric cooling from state 2 to state 3 so that $v_3 = v_2$ and $T_3 < T_1$; an isothermal compression from state 3 to a state 4 with $T_4 = T_3$ and $v_4 < v_1$; and an isochoric heating from state 4 back to state 1. This cycle, with two isothermal and two isochoric segments, is called a *Stirling cycle*.

A schematic of a reversible reciprocating piston–cylinder engine in which the working fluid may follow a Stirling cycle is sketched in Figure 25.26. It is called a *reversible*

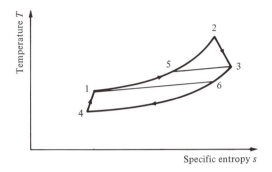

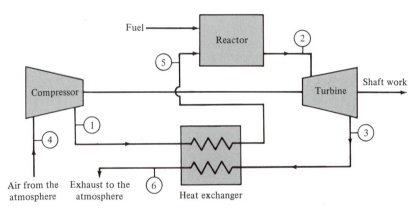

Figure 25.24   Schematic of a conversion system in which the working fluid follows an open Joule–Brayton cycle with preheating of the compressed air.

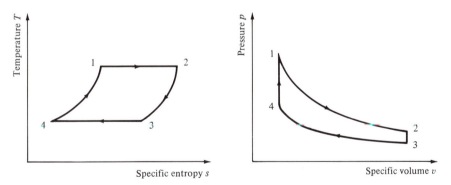

Figure 25.25   Representations of a Stirling cycle on T–s and p–v diagrams.

Stirling-cycle engine. It has two outer pistons and a displacer piston. The displacer piston does not change the total volume of the working fluid at all. It simply moves the contents of the cylinder either into the chamber heated by the hot external stream or into the chamber cooled by the cold external stream. Moreover, it is subjected to about the same pressures on both sides. Volume is affected only by the motion of the two outer pistons. During the isochoric process from state 2 to state 3, these two pistons move simultaneously to the right, thus without affecting the total volume.

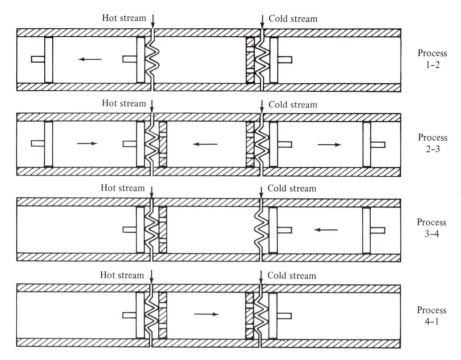

Figure 25.26   Schematic of a reciprocating piston-cyclinder engine in which the working fluid may follow a Stirling cycle.

For an ideal gas and all reversible processes, the net work per unit amount of the working fluid and per cycle, $w_{net}^{\rightarrow}$, is given by the cyclic integral $\oint p\,dv$, that is, the area of the cycle on the p–v diagram. It is the difference between the work done by the system during expansion, $w_{12}^{\rightarrow}$, and the work done on the system during compression, $w_{34}^{\rightarrow}$, that is, $w_{net}^{\rightarrow} = w_{12}^{\rightarrow} - w_{34}^{\rightarrow}$. Similarly, the net heat per unit amount of the working fluid and per cycle, $q_{net}^{\leftarrow}$, is given by the cyclic integral $\oint T\,ds$, that is, the area of the cycle on the T–s diagram. Because for an ideal gas the specific heat $c_v(T)$ is a function of temperature only, $q_{41}^{\leftarrow} = q_{23}^{\rightarrow} = \int_{T_3}^{T_1} c_v(T')\,dT'$ and therefore $q_{net}^{\leftarrow} = q_{12}^{\leftarrow} - q_{34}^{\rightarrow}$. Moreover, $s_1 - s_4 = s_2 - s_3 = \int_{T_3}^{T_1} c_v(T')\,dT'/T'$ and, therefore, $s_2 - s_1 = s_3 - s_4$, $q_{12}^{\leftarrow} = T_1(s_2 - s_1)$ and $q_{34}^{\rightarrow} = T_3(s_3 - s_4) = T_3(s_2 - s_1)$. Accordingly, $w_{net}^{\rightarrow} = q_{net}^{\leftarrow} = (T_1 - T_3)(s_2 - s_1)$, and

$$\frac{w_{net}^{\rightarrow}}{q_{12}^{\leftarrow}} = 1 - \frac{T_3}{T_1} = \eta_{\text{Carnot}} \tag{25.22}$$

The ratio $w_{net}^{\rightarrow}/q_{12}^{\leftarrow}$ is called the *thermal efficiency of a reversible Stirling-cycle engine*, and is identical to the Carnot coefficient corresponding to the same temperatures T_1 and T_3.

The pressure–volume diagram of an actual external combustion engine differs greatly from the Stirling cycle. Nevertheless, the corresponding efficiency $w_{net}^{\rightarrow}/q_{12}^{\leftarrow}$ may be larger than those of comparable spark-ignition and compression-ignition internal-combustion engines.

25.11   Refrigeration and Heating Cycles

Each discussion in Sections 25.5 to 25.10 refers to a *power cycle*, namely, a cyclic change of states of a working fluid used to transform an energy input from high-temperature sources to a work output.

By reversing the order of the sequence of states, each power cycle becomes either a *refrigeration cycle*, namely, a cyclic change of states of a working fluid used to transform a work input to cooling a low temperature system, or a *heating cycle*, namely, a cyclic change of states of a working fluid used to transform a work input to heating a high-temperature system. The reverse of any power cycle is also called a *reversed cycle*.

A cyclic sequence of states often used as a reference in discussions of the stable-equilibrium-state parts of the bulk-flow states of a refrigeration unit is shown on T–s and p–h diagrams in Figure 25.27. It occurs approximately in the arrangement of devices shown in Figure 25.28. Beginning in state 1, a compressed liquid at temperature T_1 and pressure p_1 is transformed isenthalpically to a state 2 at a lower temperature $T_2 < T_1$, and lower pressure $p_2 < p_1$. State 2 is a liquid–vapor mixture. The mixture is then heated isothermobarically to a saturated vapor state 3. The saturated vapor is compressed isentropically to a superheated state 4, such that $p_4 = p_1$ and $T_4 > T_1$. Following state 4, the superheated vapor is condensed isobarically at pressure $p_4 = p_1$ to the saturated liquid state 1. This cycle, with one isenthalpic, one isentropic, and two isobaric segments, is called a *vapor-compression refrigeration cycle* or a *reversed Rankine cycle*.

Per unit flow rate of the working fluid, the *coefficient of performance of a reversed-Rankine-cycle refrigeration unit*, $(COP)_{r.u.}$, is defined as the ratio of the heat out of

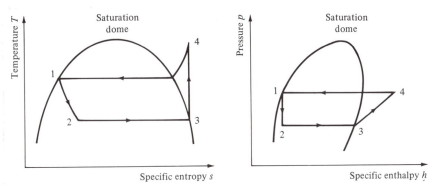

Figure 25.27   Representations of a reversed-Rankine refrigeration cycle on T–s and p–h diagrams.

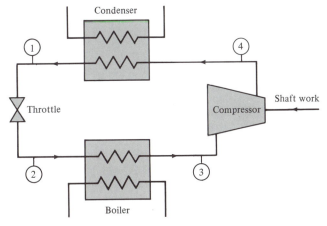

Figure 25.28   Schematic of a refrigeration system in which the working fluid may follow a reversed-Rankine cycle.

the refrigerated system, $q_{23}^{\leftarrow} = h_3 - h_2$, and the work done on the refrigeration unit, $w_{34}^{\leftarrow} = h_4 - h_3$, that is,

$$(\text{COP})_{\text{r.u.}} = \frac{q_{23}^{\leftarrow}}{w_{34}^{\leftarrow}} = \frac{h_3 - h_2}{h_4 - h_3} \tag{25.23}$$

The cyclic sequence of states in Figure 25.27 is also used as a reference in discussions of heat pumps. Then, per unit flow rate of the working fluid, a *coefficient of performance of a reversed-Rankine-cycle heat pump*, $(\text{COP})_{\text{h.p.}}$, is defined as the ratio of the heat to the heated system, $q_{41}^{\rightarrow} = h_4 - h_1$, and the work done on the heat pump, $w_{34}^{\leftarrow} = h_4 - h_3$, that is,

$$(\text{COP})_{\text{h.p.}} = \frac{q_{41}^{\rightarrow}}{w_{34}^{\leftarrow}} = \frac{h_4 - h_1}{h_4 - h_3} \tag{25.24}$$

From the pressure versus specific enthalpy diagram in Figure 25.27, we see that the $(\text{COP})_{\text{h.p.}}$ is greater than unity and that it is larger the smaller is the difference between the pressures p_1 and p_2.

In practice, because the compression from state 3 to state 4 is irreversible, the coefficient of performance of either the refrigeration unit or the heat pump is smaller than for a reversible compression. We can verify this fact by inspection of the p–h diagram in Figure 25.29. A given amount of entropy generated by irreversibility in the compressor causes an increase of the work $h_4 - h_3$ and therefore a decrease in the ratio $q_{23}^{\leftarrow}/w_{34}^{\leftarrow} = (h_3 - h_2)/(h_4 - h_3)$. It causes a decrease also in the ratio $q_{41}^{\rightarrow}/w_{34}^{\leftarrow} = (h_4 - h_1)/(h_4 - h_3)$ because the fractional increase of $h_4 - h_1$ is smaller than that of $h_4 - h_3$. The smallest value of $(\text{COP})_{\text{r.u.}}$ is almost zero. It corresponds to a refrigeration unit that is so irreversible that it hardly pumps any energy out of the low-temperature refrigerated system. The smallest value of $(\text{COP})_{\text{h.p.}}$ is almost unity. It corresponds to a heat pump that is so irreversible that it hardly pumps any energy to the high-temperature heated system.

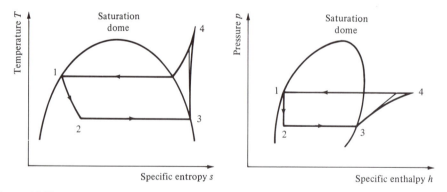

Figure 25.29 Representations on T–s and p–h diagrams of the working-fluid cycle of a refrigeration system with irreversible compression.

25.12 Linde–Hampson Liquefaction Cycle

A reversed cycle, used as a reference in discussions of liquefaction of constituents that at environmental conditions are in the gaseous phase, is shown on a temperature versus specific entropy diagram in Figure 25.30. Its four segments are an isothermal compression of the gas from state 1, with temperature T_1 and pressure p_1 larger than the corresponding critical values, to a state 2 in which the pressure of the gas is p_2; an isobaric cooling from state 2 to a state 3 in which the gas temperature is T_3 and the specific entropy s_3; an isenthalpic expansion from state 3 to a state 4, which consists of a two-phase mixture

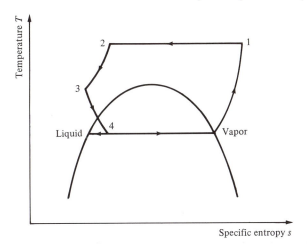

Figure 25.30   Representation of a Linde-Hampson refrigeration cycle on a T–s diagram.

at pressure $p_4 = p_1$; an isentropic separation of the liquid phase from the vapor phase in the mixture; and an isobaric compression from state 4 to state 1 at the constant pressure p_1. This reversed cycle, with one isothermal, two isobaric, and one isenthalpic segments, is called a *Linde–Hampson liquefaction cycle*. It may occur in the arrangement shown in Figure 25.31. In practice, the isothermal compression can be obtained by means of a two-stage compressor with an intercooler and an aftercooler. For the gas to cool upon the isenthalpic expansion through the throttle, the Joule–Thomson coefficient $(\partial T/\partial p)_h$ must be positive (Section 23.4).

A characteristic of a liquefaction system is the yield $Y = \dot{m}_L/\dot{m}_1$, that is, the ratio of the flow rate of liquid withdrawn, $\dot{m}_L$, and the flow rate of gas flowing through the compressor, $\dot{m}_1$. We may find this ratio by writing the mass and energy rate balances for the combined system consisting of the heat exchanger and the liquid receiver. Specifically, $\dot{m}_1 = \dot{m}_L + \dot{m}_V$ and $\dot{m}_1 h_1 = \dot{m}_L h_L + \dot{m}_V h_V$ or, eliminating $\dot{m}_V$ from these two equations, we find that

$$Y = \frac{\dot{m}_L}{\dot{m}_1} = \frac{h_V - h_1}{h_V - h_L} \tag{25.25}$$

The coefficient of performance of the liquefaction system can be defined as the ratio of the cooling effect per unit amount of gas through the compressor, $Y(h_1 - h_L)$, and the

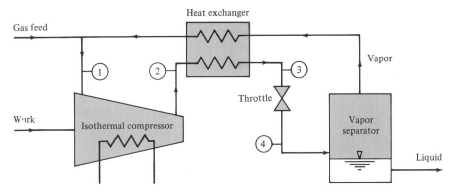

Figure 25.31   Schematic of a refrigeration system in which the working fluid may follow the Linde–Hampson cycle.

work done on the compressor, $w_{12}^{\leftarrow}$, that is,

$$\text{COP} = \frac{Y(h_1 - h_L)}{w_{12}^{\leftarrow}} \tag{25.26}$$

Problems

25.1 An advanced gas turbine, used as a peaking unit by a utility, exhausts hot gases at a rate $\dot{m} = 10^6$ kg/h, at temperature $T_{gi} = 260°C$, and pressure $p_{gi} = 1$ atm. To increase the fuel utilization, it is proposed to recover energy from the exhaust gases to power a Rankine cycle that generates additional power. The cycle uses water as the working fluid. The water is transformed into steam in a steam generator. Exhaust gases enter the steam generator at temperature $T_{gi} = 260°C$ and exit at T_{go}. Compressed water at pressure p enters the steam generator (state 1), is preheated to the boiling point (state 2), boils to saturation (state 3), and then is superheated to temperature T_{so} (state 4). The water flow is such that the minimum temperature difference at states 2 and 4 is (pinch temperature) $\Delta T_{min} = 10°C$.

(a) Draw a schematic of the temperature profiles of the exhaust gases and the water in the steam generator.

(b) What is the availability rate of the exhaust gases? Assume that the exhaust gases are perfect ($c_p = 1$ kJ/kg K), and that the environment is at $15°C$.

(c) What is the largest amount of power that can be obtained from the exhaust gases if the $10°C$ temperature difference between primary and secondary streams in the heat exchanger cannot be avoided?

(d) Draw a schematic of the power plant arrangement of devices, and representations of the Rankine cycle on a T–s diagram and a h–s diagram.

(e) Calculate the power generated by the Rankine cycle if the steam pressure is 3.4 bar. Assume that the expansion and compression of the cycle are isentropic and that the pressure drops in the steam generator and the condenser are zero. The condenser is cooled by environmental water at $15°C$. The temperature difference between condensing steam and environmental water is $15°C$. Neglect the power of the water pump.

25.2 A. I. Kalina has proposed a new cycle ("Combined Cycle System with Novel Bottoming Cycle," ASME Paper-84-GT-173 1984) for recovery of waste heat and conversion into work. A schematic of the cycle is shown in Figure P25.2. The working fluid is a water–ammonia solution. The composition of this fluid in the turbine is different from that in the condenser. The details of the cycle are as follows. A completely condensed working fluid of so-called "basic composition" leaves the condenser in state 1 (data in the accompanying table). It is then pumped to an intermediate pressure at state 2 and heated in a counterflow economizer to state 3 and in a counterflow heater to state 4. As a result, the fluid is partially evaporated. The released vapor is significantly enriched by the light component of the mixture, that is, ammonia. The vapor is separated from the liquid in the flashtank and then mixed with part of the remaining liquid to produce a so-called "working composition" (state 8). This newly created fluid, as well as the remaining liquid, is sent to a counterflow economizer to preheat the incoming fluid described above and to recuperate energy. After precooling in the economizer (change from state 8 to state 10), the working fluid is completely condensed by the cooling water (from state 10 to state 13), then pumped to a high pressure (state 14), and sent into the evaporator-boiler, where it is completely evaporated and superheated to state 15 in the counterflow waste-heat-recovery exchanger by the exhaust gases from a gas turbine. The superheated fluid enters a six-stage turbine, from which it exits in state 16. Subsequently, it goes through the heater and the economizer. After the economizer, it is mixed with liquid from the flashtank and exits the condenser in state 1 to repeat the cycle. Assume steady-state, bulk-flow conditions. The mass flow rate of the stream at state 15 is 2.72 kg/s.

(a) How much energy is transferred from the exhaust gases to the working fluid in the waste-heat-recovery exchanger per kilogram of working fluid in state 15?

State	T °C	p kPa	h kJ/kg	Ammonia mass fraction	Mass flow rate relative to stream 15
1	12.8	48.3	−151.4	0.288	4.1642
2	12.8	206.8	−151.2	0.288	4.1642
3	56.4	206.8	146.5	0.288	4.1642
4	62.8	206.8	241.1	0.288	4.1642
5	62.8	206.8	1511.6	0.912	0.4038
6	62.8	206.8	104.7	0.221	3.7604
7	62.8	206.8	104.7	0.221	0.5962
8	62.8	206.8	672.8	0.500	1.0000
9	62.8	206.8	104.7	0.221	3.1642
10	41.4	206.8	354.1	0.500	1.0000
11	41.4	206.8	4.7	0.221	3.1642
12	32.5	51.7	4.7	0.221	3.1642
13	12.8	206.8	−205.8	0.500	1.0000
14	12.8	8273.7	−197.7	0.500	1.0000
15	532.2	8273.7	2800.0	0.500	1.0000
16	68.4	55.2	1716.9	0.500	1.0000
17	62.8	54.5	1327.9	0.500	1.0000
18	41.4	51.7	718.6	0.500	1.0000
19	33.3	51.7	176.0	0.288	4.1642
20	560.0	—	—	Gas	5.79315
21	65.6	—	—	Gas	5.79315
22	10.0	—	—	Water	32.9300
23	23.9	—	—	Water	32.9300

(b) Assume the exhaust from the gas turbine behaves as a perfect gas. What is the entropy change of this exhaust stream as it goes through the waste-heat-recovery exchanger per kilogram of working fluid in state 15?

(c) What is the change in availability of the exhaust gases as they go through the waste-heat-recovery exchanger per kilogram of working fluid in state 15?

(d) What is the effectiveness of the bottoming unit?

(e) How much energy is carried away by the water coolant?

25.3 A bulk-flow stream undergoes a Rankine cycle with superheat. Steam at 30 bar and 350°C enters a well-insulated turbine, where it expands isentropically to the condenser pressure of 0.02 bar. The saturated condensate is pumped isentropically back to 30 bar by a well-insulated pump and then enters the boiler, where it is transformed to steam at the constant pressure of 30 bar.

(a) What is the thermal efficiency of the cycle?

(b) What is the ratio of the pump to the turbine power?

(c) What is the effectiveness of the net work production if the energy source is a perfect gas stream cooled from a temperature $T_1 = 400°C$ to a temperature $T_2 = 250°C$, and the environment is at temperature $T_{R^*} = 15°C$?

25.4 A bulk-flow stream undergoes a Carnot cycle. Saturated steam at 30 bar enters a well-insulated turbine, where it expands isentropically to the condenser pressure of 0.02 bar. Saturated liquid at 30 bar enters the boiler following isentropic compression in a well-insulated compressor.

(a) What is the thermal efficiency of the cycle?

(b) What is the ratio of the compressor to the turbine power?

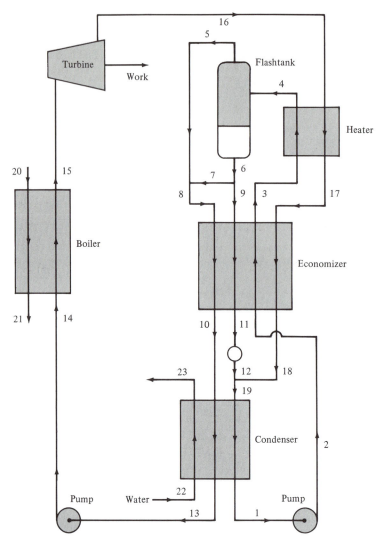

Figure P25.2

(c) What is the effectiveness of the net work production if the energy source is a perfect gas stream cooled from a temperature $T_1 = 400°C$ to a temperature $T_2 = 250°C$, and the environment is at temperature $T_{R*} = 15°C$?

25.5 A given amount of water is confined in a cylinder with a piston and undergoes the following cycle. Initially, the water is at temperature $T_1 = 100°C$ and specific volume $v_1 = 100$ cm^3/g. Beginning from that state, the water is: (1) heated to $T_2 = 260°C$ at constant volume by heat from a reservoir at $T_h = 280°C$; (2) expanded adiabatically and isentropically to $T_3 = T_1 = 100°C$; and (3) compressed isothermally to the initial state while interacting with a reservoir at $T_c = 80°C$.

   (a) What is the ratio of the net work to the heat supply?

   (b) What is the ratio of the compression to the expansion work?

   (c) What is the effectiveness of the net work production?

25.6 A schematic of a combined gas-turbine, steam-turbine power plant is shown in Figure P25.6. Compressed air is fed into the combustor, and the products of combustion are fed into the gas turbine. The exhaust from the gas turbine is passed through a waste-heat boiler, and the steam from the boiler through a steam turbine, and the other devices shown in the figure.

Air enters the compressor at $p_1 = 101$ kPa and $T_1 = 16°C$, and is discharged into the combustor at $p_2 = 1$ MPa. The combustion gases enter the gas turbine at $p_3 = p_2$ and $T_3 = 1360°C$ and exit the turbine at $p_4 = 120$ kPa. The net gas turbine power output $\dot{W}_t = 10$ MW. Steam is raised in the waste-heat boiler at $p_6 = 6$ MPa and $400°C$, and the steam turbine power output $\dot{W}_s = 10$ MW. The condenser pressure $p_7 = 10$ kPa. In addition to the energy provided by the waste gas stream, energy is added to the waste-heat boiler by burning fuel—for this reason this boiler is also called *supplementary fired boiler*—and the stack temperature $T_5 = 150°C$. The efficiency of the compressor is 86%, of the gas turbine 88%, and of the steam turbine 80%. The feed pump is 100% efficient. Model the various streams as you deem appropriate.

(a) What is the airflow rate through the gas turbine?

(b) What is the steam flow rate through the steam turbine?

(c) What is the supplementary energy rate to the boiler?

(d) What is the availability rate of the exhaust stream from the gas turbine? Assuming that the availability of the fuel is approximately equal to the energy it supplies, what is the availability rate of the exhaust plus the supplementary fuel?

(e) What is the availability effectiveness of the steam-turbine part of the plant? Of the overall combined plant?

(f) How would you allocate the fuel cost between the gas-turbine part and the steam-turbine part of the combined plant?

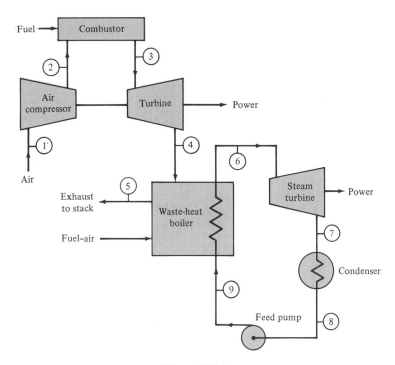

Figure P25.6

25.7 A combined gas and steam turbine power plant—a cogeneration installation—with extraction of process steam is shown schematically in Figure P25.7. Air at atmospheric conditions (state 1) is compressed (state 2) and fed to the burner heat exchanger, where it is heated by the hot products of combustion. The hot and compressed air is fed to the gas turbine (state 3), exits at atmospheric

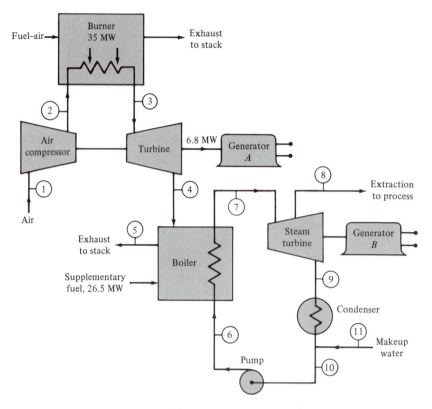

Figure P25.7

pressure (state 4), and passes through the supplementary fired boiler–heat–exchanger before going up the stack (state 5). The compressor, combustor, and gas turbine are all well insulated. Model the air as a perfect gas with specific heat ratio $\gamma_a = 1.4$ and molecular weight $M_a = 28.9$ kg/kmol. Pressurized water in state 6 enters the boiler, exits as steam in state 7, and powers a two-stage steam turbine. The steam exiting the first stage of the two-stage turbine in state 8 is fed partly to the second stage and partly to a process. The fluid exiting the second stage of the turbine in state 9 returns to the boiler in state 6 via a condenser and a pump. Conditions at various states are listed in the accompanying table.

State	T °C	p bar	$\dot{m}$ kg/s
1	20	1	?
2	?	?	?
3	?	?	?
4	428	1	?
5	204	1	?
6	?	41.8	13.3
7	441	41.8	13.3
8	335	18	12.5
9	?	0.05	?
10	Saturated liquid	0.05	13.3
11	20	1	?

(a) If the efficiency of the compressor is 0.9, that of the gas turbine 0.85, and the power input to generator A is 6800 kW, what are the air mass flow rate and the pressure to which the gas is compressed by the compressor? The energy and availability in the fuel input to the burner are each equal to 35 MW.

(b) What is the rate of entropy generation by irreversibility in the compressor–burner–gas turbine composite? What is the effectiveness of this composite with respect to the environment?

(c) If the efficiency of the second stage of the steam turbine is 0.95, what is the power output of the turbine? What is the power input to the pump?

(d) What is the effectiveness of the steam plant? The fuel input energy and availability to the boiler are equal to each other.

(e) What is the effectiveness of the overall cogeneration installation?

25.8 The solar thermal power plant Eurelios, located at Adrano, Italy, is rated to produce 1 MW of electricity. About 6200 m^2 of reflectors track the solar light and concentrate it onto a central boiler which produces superheated steam at 510°C in an environment at $T_{R*} = 300$ K.

The solar radiation incident on the outside of the earth's atmosphere can be modeled as the emission from the electromagnetic field in stable equilibrium at the surface of the sun at about $T_{sun} = 5760$ K, that is, with energy and entropy fluxes $J_E = 1.4$ kW/m^2 and $J_S = 0.324$ W/m^2 K, respectively. The atmosphere absorbs and diffuses a good part of this radiation, so that the average energy flux on the earth's surface at Adrano on a clear day is about 770 W/m^2.

(a) Assuming that the entropy flux is attenuated by the earth's atmosphere by the same factor as the energy flux, what is the average entropy flux of solar radiation at Adrano on a clear day?

(b) Assuming that cyclic machinery could do work on a weight while interacting only with the incident solar radiation and a reservoir at the environmental temperature T_{R*}, what is the largest fraction of the incident energy that could be transferred to the weight? Express your answer in terms of J_E, J_S, and T_{R*} only. Noticing that $J_E/J_S = (3/4)T_{sun}$, rewrite your answer in terms of T_{sun} and T_{R*} only. Compare your result with the Carnot coefficient $1 - T_{R*}/T_{sun}$. Is the Carnot coefficient relevant here?

(c) What fraction of the incident solar energy is transferred to the electricity network by the Adrano plant? What is the ratio of this fraction to the largest fraction that could have been transferred?

(d) Assume that the boiler at Adrano is a sphere with radius $r = 2$ m with its steel wall maintained at $T_W = 510$°C as a result of the following three effects: (1) incident solar radiation concentrated by the reflectors; (2) heat interaction on the inside with the boiling water and superheated steam; and (3) radiation emitted from the surface toward the environment at T_{R*}, with energy and entropy fluxes $J_E = \sigma (T_W^4 - T_{R*}^4)$ and $J_S = (4/3)\sigma (T_W^3 - T_{R*}^3)$, where $\sigma = 5.67 \times 10^{-8}$ W/m^2 K^4. Find the rates at which energy and entropy are supplied to the boiling water and steam in the boiler. What is the rate of entropy generation by irreversibility in the radiation-concentration part of Eurelios?

(e) What is the rate of entropy generation by irreversibility in the parts of the Eurelios plant not included in your answer in part (d)?

(f) What fraction of your answer in part (b) is lost in the radiation-concentration components of Eurelios? And what fraction in the remaining components of the plant?

26 Thermophysical Properties of Mixtures

In this chapter we return to the consideration of simple systems that consist of two or more different constituents, that are in stable equilibrium states, and that have constraints inhibiting all chemical and nuclear reaction mechanisms. The experimental and theoretical results on such systems are often referred to as the thermophysical properties of mixtures.

We encounter mixtures in almost all aspects of our daily experience. For example, the air we breathe, the beverages we drink, the food we eat, the fuels we burn in our automobiles, furnaces, and stoves, and the materials we process in our manufacturing industries all involve mixtures. Many of our physiological, homelife, and industrial activities involve the transfer of constituents from one mixture to another. For example, oxygen is dissolved into the blood flowing in our lungs while carbon dioxide is tranferred from the bloodstream to the air we exhale. Again, water vapor is extracted from an air stream in an air-conditioning system. Despite the great variety of conditions, many general features can be established from the study of stable equilibrium states. In this chapter we discuss experimental and theoretical results on the thermophysical properties of mixtures.

26.1 Specific Properties

Prior to discussing experimental results, we briefly review some of the relevant terminology. We consider amounts n_1, n_2, ..., n_r of r different types of molecules or atoms confined in a volume V under conditions that can be modeled by the stable equilibrium states of a simple system.

From the discussion in Section 17.5, we recall that each specific property of such a state depends at most on $r + 1$ independent variables. For example, the mole-specific internal energy $u = U/n$ depends at most on the mole-specific entropy $s = S/n$, the mole-specific volume $v = V/n$, and the mole fractions y_1, y_2, ..., y_r subject to the relation $y_1 + y_2 + \cdots + y_r = 1$, so that

$$u = u(s, v, y_1, y_2, \ldots, y_r) \tag{26.1}$$

where

$$y_i = \frac{n_i}{n} = \frac{n_i}{\sum_{j=1}^{r} n_j} \tag{26.2}$$

and the total amount of constituents

$$n = \sum_{j=1}^{r} n_j \tag{26.3}$$

The relative amounts of the constituents specified by the values of either the mole fractions y_1, y_2, ..., y_r or the mass fractions x_1, x_2, ..., x_r determine the *composition of the mixture*. Each *mass fraction* x_i of the ith constituent is defined by any of the relations

$$x_i = \frac{m_i}{m} = \frac{n_i M_i}{m} = \frac{n_i M_i}{\sum_{j=1}^r n_j M_j} = \frac{y_i M_i}{\sum_{j=1}^r y_j M_j} = \frac{y_i M_i}{M} \tag{26.4}$$

where m_i and M_i are the mass and the molecular weight of the ith constituent, respectively, m is the *mass of the mixture*, that is,

$$m = \sum_{j=1}^r m_j = \sum_{j=1}^r n_j M_j \tag{26.5}$$

M the *mean molecular weight of the mixture*, that is,

$$M = \frac{m}{n} = \sum_{j=1}^r y_j M_j = \frac{1}{\sum_{j=1}^r x_j/M_j} \tag{26.6}$$

and we use (26.2) and (26.3). Of course, the sum of the mass fractions equals unity, that is,

$$x_1 + x_2 + \cdots + x_r = 1 \tag{26.7}$$

Given the mole fractions and the molecular weights of all the constituents, equations (26.4) show that we can evaluate the mass fractions. Conversely, given the mass fractions and the molecular weights, we can evaluate the mole fractions because

$$y_i = \frac{n_i}{n} = \frac{x_i/M_i}{\sum_{j=1}^r x_j/M_j} \tag{26.8}$$

By virtue of equations (26.4) to (26.6), we can rewrite (26.1) in terms of mass-specific properties, that is, mass-specific internal energy $u = U/m$, mass-specific entropy $s = S/m$, mass-specific volume $v = V/m$, and mass fractions x_1, x_2, ..., x_r. The result is of the form

$$u = u(s, v, x_1, x_2, \ldots, x_r) \tag{26.9}$$

where we use the same symbols for mass-specific properties $u = U/m$, $s = S/m$, and $v = V/m$ as for the corresponding mole-specific ones, and recognize that, by virtue of (26.7), mass-specific properties of the stable equilibrium states of a simple system depend at most on $r + 1$ independent variables. Of course, the functional form expressed by equation (26.9) differs from that of equation (26.1), because it is obtained from (26.1) by replacing each y_i not simply by x_i but by (26.8).

Example 26.1.   Find the mass fractions and the mean molecular weight of a mixture which on a mole basis consists of 50% ammonia, NH_3, and 50% water, H_2O.

Solution.   Denoting ammonia as constituent 1 and water as constituent 2, we have $y_1 = 0.5$, $y_2 = 0.5$, $M_1 = 17.031$ kg/kmol and $M_2 = 18.015$ kg/kmol (molecular weights from Table 19.1). Using equations (26.6) and (26.4) we find that $M = y_1 M_1 + y_2 M_2 = 0.5 \times 17.031 + 0.5 \times 18.015 = 17.523$ kg/kmol, $x_1 = y_1 M_1/M = 0.5 \times 17.031/17.523 = 0.486$, $x_2 = y_2 M_2/M = 0.5 \times 18.015/17.523 = 0.514$, and $x_1 + x_2 = 0.486 + 0.514 = 1$.

26.2   Experimental Results on Binary Mixtures

As an illustration of the richness of interrelations between stable-equilibrium-state properties of simple systems with many constituents, we begin our discussions by presenting some typical results about two-constituent or *binary mixtures*. Here we describe the patterns of coexisting phases and forms of aggregation that are common to all binary mixtures.

According to the Gibbs phase rule, a nonreacting binary mixture consisting of 1, 2, 3, or 4 coexisting phases has, respectively, 3, 2, 1, or 0 of the intensive properties T, p, μ_1, μ_2 that can be varied independently. Hence graphs of interrelations between properties are more involved than those of pure substances.

For example, whereas the single-phase stable equilibrium states of a pure substance can be represented by points in a two-dimensional region of a $T - p$ diagram (Figure 19.7), the single-phase stable equilibrium states of a binary mixture are represented by points in a three-dimensional region of a $T - p - \mu_1$ diagram, where μ_1 is the chemical potential of constituent 1. Alternatively, because μ_1 is a function of either T, p, and y_1 or T, p, and x_1, the states may be represented on either a $T - p - y_1$ or a $T - p - x_1$ diagram, where y_1 and x_1 are the mole fraction and mass fraction of constituent 1, respectively. So for every combination of values of T and p, a binary mixture has an infinite number of states as either y_1 or x_1 varies from 0 to 1.

Again, whereas the two-phase states of a pure substance are represented on a $T - p$ diagram by points on a line that separates two single-phase regions (Figure 19.7), the two-phase states of a binary mixture are represented on a $T - p - \mu_1$ diagram by points on a surface that separates two single-phase regions. On a $T - p - y_1$ diagram (Figure 26.1), two-phase states are represented by points in a region delimited by two saturation surfaces that separate a single-phase region from a two-phase region. For either $y_1 = 0$ or $y_1 = 1$ the two surfaces coincide on a curve that represents the corresponding saturation relation for pure substance 2 or pure substance 1, respectively.

Each of the saturation surfaces represents a definite relation between T, p, and y_1. We can illustrate how such a relation arises by considering as an example liquid-vapor two-phase states. The equalities of chemical potentials for mutual stable equilibrium of

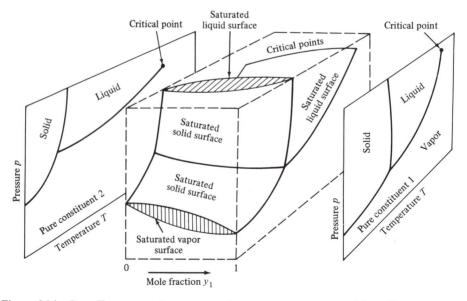

Figure 26.1 On a $T - p - y_1$ diagram, two-phase states are represented by points in a region delimited by two saturation surfaces. The diagrams in Figures 26.1 to 26.5 are based on C. E. Wales, *Chem. Eng.*, May 27, p. 120, June 24, p. 111, July 22, p. 141, Aug. 19, p. 167, Sept. 16, p. 187 (1983).

the coexisting phases are

$$\mu_{1g}(T, p, y_{1g}) = \mu_{1f}(T, p, y_{1f}) \tag{26.10}$$

$$\mu_{2g}(T, p, y_{1g}) = \mu_{2f}(T, p, y_{1f}) \tag{26.11}$$

where the subscripts f and g denote liquid and vapor phases, respectively. Upon solving equations (26.10) and (26.11), we find two functions, $y_{1f} = y_{1f}(T, p)$ and $y_{1g} = y_{1g}(T, p)$, representing the saturated-liquid and the saturated-vapor surfaces, respectively.

In practice, experimental data are presented by means of two-dimensional either $T - y_1$ or $p - y_1$ diagrams as sketched in Figures 26.2 and 26.3. Conceptually, each

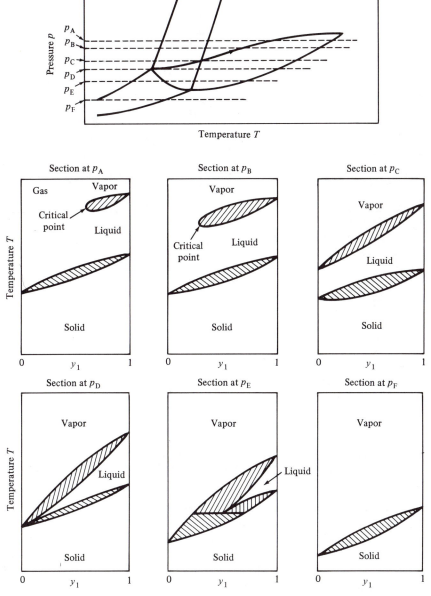

Figure 26.2 Typical $T - y_1$ diagrams obtained by cutting the $T - p - y_1$ diagram in Figure 26.1 at the values of pressure indicated on the $T - p$ projection of the same diagram sketched at the top.

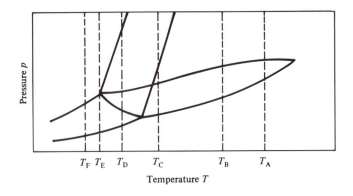

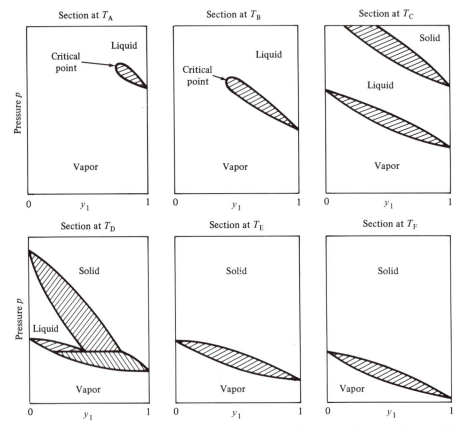

Figure 26.3 Typical $p - y_1$ diagrams obtained by cutting the $T - p - y_1$ diagram in Figure 26.1 at the values of temperature indicated on the $T - p$ projection of the same diagram sketched at the top.

of these diagrams is a cut of the $T - p - y_1$ diagram at a particular value of either the pressure or the temperature. At the top of each of Figures 26.2 and 26.3 a projection of the $T - p - y_1$ diagram of Figure 26.1 onto the $p - T$ plane indicates the values of pressure and temperature, respectively, at which the $T - p - y_1$ diagram is cut to yield the two-dimensional diagrams below. The $T - y_1$ and $p - y_1$ diagrams in Figures 26.2 and 26.3 show a variety of possible patterns. They all refer to a mixture in which constituent 1 is less volatile than constituent 2, that is, is such that at each value of the pressure the saturation temperature $T_{\text{sat},11}(p)$ of pure substance 1 is greater than the saturation temperature $T_{\text{sat},22}(p)$ of pure substance 2, $T_{\text{sat},11}(p) > T_{\text{sat},22}(p)$. Figure 26.4 presents a cut of the $T - p - y_1$ diagram at a particular value of y_1 yielding a two-dimensional $p - T$ diagram. A great variety of different shapes can be found for the $T - y_1$, $p - y_1$, and $p - T$ diagrams, depending on the details of the constituents in the mixture and on the nature of their intermolecular forces. Each of these diagrams is called a *phase diagram.*

The region of space enclosed within the saturated-liquid and the saturated-vapor surfaces corresponds to liquid–vapor two-phase states of the binary mixture. The two surfaces may be tangent to each other in the $T - p - y_1$ diagram for some values of T, p, and y_1. If at such values the two phases are still different from each other, for example, in the values of the density or other specific properties, the state of the mixture is called *azeotropic*. [1] Figure 26.5 shows $T - y_1$ and $p - y_1$ diagrams for mixtures with azeotropic liquid–vapor states. If, instead, the two phases are not different from each other, the values of T, p, and y_1 correspond to a critical point and belong to a curve in the $T - p - y_1$ diagram called the *critical-point curve* (Figure 26.1). This curve ends for

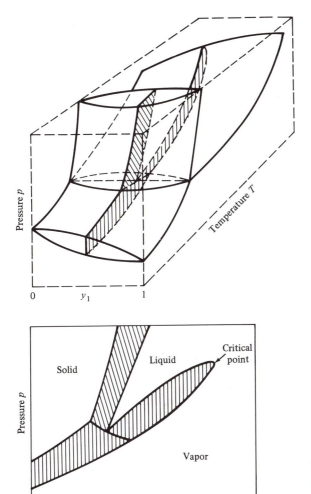

Figure 26.4   Typical $p - T$ diagram obtained by cutting the $T - p - y_1$ diagram in Figure 26.1 at a particular value of y_1.

---

[1]The term *azeotropic* means that the composition of the mixture does not change upon boiling at constant temperature and pressure.

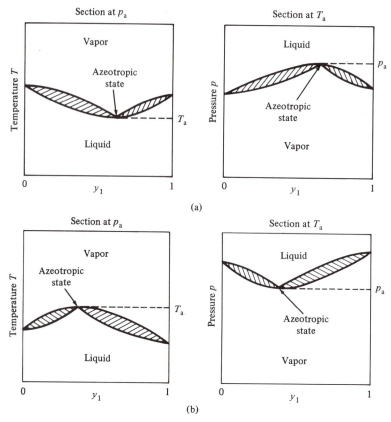

Figure 26.5 (a) Typical $T - y_1$ and $p - y_1$ diagrams for a mixture with low boiling-temperature azeotropic states. (b) Typical $T - y_1$ and $p - y_1$ diagrams for a mixture with high boiling-temperature azeotropic states.

$y_1 = 1$ at the critical point of pure substance 1 and for $y_1 = 0$ at the critical point of pure substance 2.

26.3 Partial Pressures

In Section 17.7 we express properties of mixtures in terms of the Gibbs free energy and the chemical potentials, and we conclude [equations (17.33) to (17.37)] that we can evaluate all the stable-equilibrium-state properties of a simple multiconstituent system if we know the functional dependence of each chemical potential μ_i on temperature, pressure and mole fractions, that is, if we have the functions $\mu_i = \mu_i(T, p, y_1, y_2, \ldots, y_r)$ for $i = 1, 2, \ldots, r$. In this section we discuss a method of measuring the chemical potentials of the r constituents of a mixture by performing measurements on r judiciously chosen one-constituent systems.

To this end, we consider $r + 1$ simple systems (Figure 26.6). System A consists of r constituents with amounts $n_1, n_2, \ldots, n_r$ confined in a volume V at temperature T and pressure p. Each of the other r simple systems, $11, 22, \ldots, ii, \ldots, rr$, consists of one of the constituents of A. We denote the system consisting of pure substance i by the double symbol ii, and any of its characteristics by the corresponding symbol with a double subscript ii. Accordingly, system ii consists of pure substance i, with amount n_{ii}, confined in a volume V_{ii}, at temperature T_{ii}, pressure p_{ii}, and with chemical

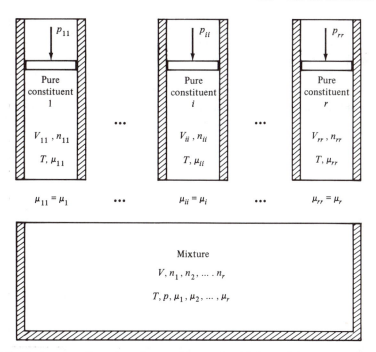

Figure 26.6   Each system ii consists of pure substance i and is in partial mutual stable equilibrium with system A through a rigid membrane permeable only to the ith constituent of A. The pressure p_{ii} at which this occurs is called the partial pressure of constituent i in system A.

potential μ_{ii}. The pressure p_{ii} is adjustable by a weighted piston, as shown schematically in Figure 26.6.

We select each system ii to be in partial mutual stable equilibrium with respect to the ith constituent of system A. This means that if system ii is placed in communication with A via a rigid *semipermeable membrane*, a membrane permeable only to constituent i and not to any other constituent, no energy, no entropy, and no amount of constituent i can be exchanged between systems ii and A without leaving net effects in their environment.

Because the only exchanges allowed by a rigid semipermeable membrane between systems ii and A are of amounts of constituent i but not of volume and amounts of any other constituent, it follows that the conditions that are necessary for partial mutual stable equilibrium are temperature equality and equality of the chemical potential of the only constituent that can be exchanged. Accordingly, we must have

$$T_{ii} = T \tag{26.12}$$

and

$$\mu_{ii}(T, p_{ii}) = \mu_i(T, p, y_1, y_2, \ldots, y_r) \tag{26.13}$$

for $i = 1, 2, \ldots, r$, where in writing $\mu_{ii}(T, p_{ii})$ instead of $\mu_{ii}(T_{ii}, p_{ii})$ in (26.13) we use (26.12).

Equation (26.13) shows that the μ_i of the ith constituent of system A can be measured as a function of temperature, pressure, and mole fractions by finding the pressure p_{ii} at which system ii is in partial mutual stable equilibrium with A and then, evaluating $\mu_{ii}(T, p_{ii})$, for example, from tabulated data on pure constituent i and the relation [equation (17.23)]

$$\mu_{ii}(T, p_{ii}) = h_{ii}(T, p_{ii}) - T s_{ii}(T, p_{ii})$$

where h_{ii} and s_{ii} denote specific enthalpy and specific entropy of system ii, respectively. Thus we have related properties of A to properties of one-constituent systems.

Each of equations (26.13) has an interesting implication. Because μ_{ii} is a function of temperature T and pressure p_{ii} only, it follows that p_{ii} is the pressure to which system ii must be brought in order to inhibit any flow of constituent i across a rigid semipermeable membrane that separates system ii from system A. The pressure p_{ii} is called the *partial pressure of constituent i*. Clearly, it is a property of system A.

Alternatively, we can think of (26.13) as defining p_{ii} as a function of T, p, y_1, y_2, $\ldots$, y_r, that is,

$$p_{ii} = p_{ii}(T, p, y_1, y_2, \ldots, y_r) \tag{26.14}$$

By measuring T, p, y_1, y_2, $\ldots$, y_r, and p_{ii} by means of the systems in Figure 26.6, we can disclose the functions $p_{ii}(T, p, y_1, y_2, \ldots, y_r)$ and, therefore, write each chemical potential as

$$\mu_i = \mu_{ii}(T, p_{ii}(T, p, y_1, y_2, \ldots, y_r))$$

$$= \mu_{ii}(T, p^*) + \int_{p^*}^{p_{ii}(T,p,y_1,y_2,\ldots,y_r)} v_{ii}(T, p')\, dp' \tag{26.15}$$

for $i = 1, 2, \ldots, r$, where v_{ii} denotes specific volume of system ii and, in writing the second of equations (26.15), we integrate from a reference value p^* to p_{ii} the general relation $(\partial\mu/\partial p)_T = v(T, p)$. The latter relation holds for any single-constituent simple system by virtue of (17.25). Once the functions $\mu_i(T, p, y_1, y_2, \ldots, y_r)$ are known, we can determine all the properties of the mixture as functions of temperature, pressure, and mole fractions by using equations (17.34) to (17.37), which we repeat here for convenience:

$$S = -\sum_{i=1}^{r} n_i \left(\frac{\partial\mu_i}{\partial T}\right)_{p,y} \tag{26.16}$$

$$V = \sum_{i=1}^{r} n_i \left(\frac{\partial\mu_i}{\partial p}\right)_{T,y} \tag{26.17}$$

$$H = \sum_{i=1}^{r} n_i \left[\frac{\partial(\mu_i/T)}{\partial(1/T)}\right]_{p,y} \tag{26.18}$$

$$U = H - pV = \sum_{i=1}^{r} n_i \left\{\left[\frac{\partial(\mu_i/T)}{\partial(1/T)}\right]_{p,y} - p\left(\frac{\partial\mu_i}{\partial p}\right)_{T,y}\right\} \tag{26.19}$$

It appears that equations (26.15) are mathematically just as complicated as the implicit relations for the $\mu_i(T, p, y_1, y_2, \ldots, y_r)$'s. And indeed they are. However, under certain practical conditions of temperature and pressure, these equations assume simpler forms when expressed in terms of partial pressures.

For example, if in the ranges of temperature T and pressure p of interest the pure substance ii behaves as an ideal gas, then $v_{ii}(T, p') = RT/p'$, and the integral in (26.15) can be resolved to yield

$$\mu_i = \mu_{ii}(T, p^*) + RT \ln \frac{p_{ii}(T, p, y_1, y_2, \ldots, y_r)}{p^*} \tag{26.20}$$

Again, if in the ranges of T and p of interest the pure substance ii behaves as ideal incompressible, then $v_{ii}(T, p') \approx v_{ii} \approx$ constant, and (26.15) yields

$$\mu_i = \mu_{ii}(T, p^*) - v_{ii} \left[p^* - p_{ii}(T, p, y_1, y_2, \ldots, y_r) \right] \tag{26.21}$$

For the systems in Figure 26.6, it is noteworthy that the temperature T and the partial pressures p_{ii} specify the pressure p of system A. To see this more explicitly, we write the Gibbs–Duhem relation for each of systems A, 11, 22, $\ldots$, ii, $\ldots$, rr, that is,

$$S \, dT - V \, dp + \sum_{i=1}^{r} n_i \, d\mu_i = 0 \tag{26.22}$$

$$s_{ii} \, dT_{ii} - v_{ii} \, dp_{ii} + d\mu_{ii} = 0 \qquad \text{for } i = 1, 2, \ldots, r \tag{26.23}$$

where we use equations (17.9) and (17.25). When equations (26.22) and (26.23) are applied to two neighboring states, in both of which the systems in Figure 26.6 are in partial mutual stable equilibrium, conditions (26.12) and (26.13) require that $dT = dT_{ii}$ and $d\mu_i = d\mu_{ii}$ for each $i = 1, 2, \ldots, r$. Upon substituting each $d\mu_i$ in (26.22) with the corresponding $d\mu_{ii}$ from (26.23), after some rearrangements we find that

$$dp = \frac{1}{V} \left(S - \sum_{i=1}^{r} n_i s_{ii} \right) dT + \frac{1}{V} \sum_{i=1}^{r} n_i v_{ii} \, dp_{ii} \tag{26.24}$$

which shows that the pressure p of system A can be regarded as a function of the temperature T and the partial pressures p_{ii}, that is,

$$p = p(T, p_{11}, p_{22}, \ldots, p_{rr}) \tag{26.25}$$

In particular,

$$\left(\frac{\partial p}{\partial T} \right)_{p_{11}, p_{22}, \ldots, p_{rr}} = \frac{1}{V} \left[S - \sum_{i=1}^{r} n_i \, s_{ii}(T, p_{ii}) \right] \tag{26.26}$$

$$\left(\frac{\partial p}{\partial p_{ii}} \right)_{T, p_{jj} \neq ii} = \frac{n_i \, v_{ii}(T, p_{ii})}{V} \qquad \text{for } i = 1, 2, \ldots, r \tag{26.27}$$

26.4   Equation of State in Terms of Partial Pressures

Because the specific Gibbs free energy g of a multiconstituent simple system is a function of T, p, and y only (Section 17.5), that is, $g = g(T, p, y_1, y_2, \ldots, y_r)$ subject to the relation $y_1 + y_2 + \cdots + y_r = 1$, it follows that also the specific volume $v = V/n = (\partial g/\partial p)_{T, y}$ [equation (17.22)] is a function of T, p, and y only, that is,

$$v = v(T, p, y_1, y_2, \ldots, y_r) \tag{26.28}$$

subject to the relation $y_1 + y_2 + \cdots + y_r = 1$. This, or an equivalent relation, is called an *equation of state* of the multiconstituent system, and is especially useful because measurements of temperature, pressure, specific volume, and mole fractions are relatively easy to perform.

For single-phase states, T, p, and $y_1, y_2, \ldots, y_r$ are independent of each other, as it is clear also from Figure 26.1. For q-phase states, instead, the conditions of chemical potential equality for mutual stable equilibrium of the q coexisting phases impose $q - 1$ relations between T, p, and y.

Regardless of the number of coexisting phases, we can substitute (26.14) into (26.25), and find the identity

$$p = p \left(T, p_{11}(T, p, y), p_{22}(T, p, y), \ldots, p_{rr}(T, p, y) \right) \tag{26.29}$$

subject to the relation $\sum_{i=1}^{r} y_i = 1$. For single-phase states, T, p, and $\boldsymbol{y}$ can be varied independently and, therefore, we can differentiate both sides of (26.29) with respect to p at constant T and $\boldsymbol{y}$ to yield the relation

$$1 = \sum_{i=1}^{r} \left(\frac{\partial p}{\partial p_{ii}}\right)_{T, p_{jj \neq ii}} \left(\frac{\partial p_{ii}}{\partial p}\right)_{T, \boldsymbol{y}} \tag{26.30}$$

or, equivalently,

$$1 = \sum_{i=1}^{r} \frac{n_i \, v_{ii}(T, p_{ii})}{V} \left(\frac{\partial p_{ii}}{\partial p}\right)_{T, \boldsymbol{y}} \tag{26.31}$$

where we use equations (26.27). Using the relations $V = nv$ and $y_i = n_i/n$, and writing explicitly the dependences, equation (26.31) becomes

$$v(T, p, \boldsymbol{y}) = \sum_{i=1}^{r} y_i \, v_{ii}(T, p_{ii}) \left(\frac{\partial p_{ii}}{\partial p}\right)_{T, \boldsymbol{y}} \tag{26.32}$$

This equation shows that the equation of state of system A in Figure 26.6 can be obtained from the knowledge of the equations of state of each of the single-constituent systems $11, 22, \ldots, rr$, and the dependences of each of the partial pressures p_{ii} on the pressure p at constant temperature T and mole fractions $\boldsymbol{y}$.

The equation of state is not a characteristic function and hence is not sufficient to determine all the stable-equilibrium-state properties. Nevertheless, equation (26.32) relates many properties of a multiconstituent system to the corresponding properties of single-constituent systems. It can also be rewritten as

$$\mathcal{Z}(T, p, \boldsymbol{y}) = \sum_{i=1}^{r} y_i \, \mathcal{Z}_{ii}(T, p_{ii}) \left(\frac{\partial \ln p_{ii}}{\partial \ln p}\right)_{T, \boldsymbol{y}} \tag{26.33}$$

where $\mathcal{Z} = pv/RT$ and $\mathcal{Z}_{ii} = p_{ii}v_{ii}/RT$ for $i = 1, 2, \ldots, r$ are the compressibility factors (Section 21.1) of the systems A, 11, 22, $\ldots$, rr (Figure 26.6), respectively.

26.5 Partial Properties in Terms of Partial Pressures

It is noteworthy that the meaning of the term "partial" in the expression "partial pressure" is not the same as in the definition of partial properties (Section 17.8), that is, partial pressure is not defined by a partial derivative with respect to some amount at constant T, p, and $\boldsymbol{n}$. Nevertheless, partial properties can be expressed in terms of partial pressures. Substituting the first of equations (26.15) into equations (17.47) to (17.50) and using the relation $(\partial \mu_{ii}/\partial p_{ii})_T = v_{ii}(T, p_{ii})$, we find that

$$s_i(T, p, \boldsymbol{y}) = -\left(\frac{\partial \mu_i}{\partial T}\right)_{p, \boldsymbol{y}} = -\left(\frac{\partial \mu_{ii}}{\partial T}\right)_{p_{ii}} - \left(\frac{\partial \mu_{ii}}{\partial p_{ii}}\right)_T \left(\frac{\partial p_{ii}}{\partial T}\right)_{p, \boldsymbol{y}}$$

$$= s_{ii}(T, p_{ii}) - v_{ii}(T, p_{ii}) \left(\frac{\partial p_{ii}}{\partial T}\right)_{p, \boldsymbol{y}} \tag{26.34}$$

$$v_i(T, p, \boldsymbol{y}) = \left(\frac{\partial \mu_i}{\partial p}\right)_{T, \boldsymbol{y}} = \left(\frac{\partial \mu_{ii}}{\partial p_{ii}}\right)_T \left(\frac{\partial p_{ii}}{\partial p}\right)_{T, \boldsymbol{y}}$$

$$= v_{ii}(T, p_{ii}) \left(\frac{\partial p_{ii}}{\partial p}\right)_{T, \boldsymbol{y}} \tag{26.35}$$

$$h_i(T, p, \boldsymbol{y}) = \mu_i + T\,s_i = \mu_{ii}(T, p_{ii}) + T\,s_{ii}(T, p_{ii}) - T\,v_{ii}(T, p_{ii})\left(\frac{\partial p_{ii}}{\partial T}\right)_{p,\boldsymbol{y}}$$

$$= h_{ii}(T, p_{ii}) - T\,v_{ii}(T, p_{ii})\left(\frac{\partial p_{ii}}{\partial T}\right)_{p,\boldsymbol{y}} \tag{26.36}$$

$$u_i(T, p, \boldsymbol{y}) = h_i - p\,v_i$$

$$= h_{ii}(T, p_{ii}) - T\,v_{ii}(T, p_{ii})\left(\frac{\partial p_{ii}}{\partial T}\right)_{p,\boldsymbol{y}} - p\,v_{ii}(T, p_{ii})\left(\frac{\partial p_{ii}}{\partial p}\right)_{T,\boldsymbol{y}}$$

$$= u_{ii}(T, p_{ii}) + v_{ii}(T, p_{ii})\left[p_{ii} - p\left(\frac{\partial p_{ii}}{\partial p}\right)_{T,\boldsymbol{y}} - T\left(\frac{\partial p_{ii}}{\partial T}\right)_{p,\boldsymbol{y}}\right] \tag{26.37}$$

Using these equations in (17.51) to (17.55), that is, in the equations

$$S(T, p, \boldsymbol{n}) = \sum_{i=1}^{r} n_i\,s_i(T, p, \boldsymbol{y}) \tag{26.38}$$

$$V(T, p, \boldsymbol{n}) = \sum_{i=1}^{r} n_i\,v_i(T, p, \boldsymbol{y}) \tag{26.39}$$

$$H(T, p, \boldsymbol{n}) = \sum_{i=1}^{r} n_i\,h_i(T, p, \boldsymbol{y}) \tag{26.40}$$

$$U(T, p, \boldsymbol{n}) = \sum_{i=1}^{r} n_i\,u_i(T, p, \boldsymbol{y}) \tag{26.41}$$

we see that we can indeed express all properties of an r-constituent simple system in terms of specific properties of single constituent systems and the partial pressures. In the next chapter we discuss ranges of conditions under which $p_{ii} = y_i p$ so that (26.34) to (26.37) are greatly simplified.

Problems

26.1 Consider Problem 10.5. The total potentials of electrons, ions, and atoms may also be expressed in the forms

$$\mu_e = kT \ln \frac{h^3 p_e(x)}{g_e\,(2\pi m_e)^{3/2}(kT)^{5/2}} + \epsilon_{oe} - q_e \psi(x)$$

$$\mu_i = kT \ln \frac{h^3 p_i(x)}{g_i\,(2\pi m_i)^{3/2}(kT)^{5/2}} + \epsilon_{oi} + q_e \psi(x)$$

$$\mu_a = kT \ln \frac{h^3 p_a(x)}{g_a\,(2\pi m_a)^{3/2}(kT)^{5/2}} + \epsilon_{oa}$$

where $p_e(x)$, $p_i(x)$, and $p_a(x)$ are, respectively, the partial pressures of electrons, ions, and atoms at location x, and all other symbols have the meanings specified in Problem 10.5. The values of the temperature and each total potential are independent of x.

(a) How does each of the partial pressures, $p_e(x)$, $p_i(x)$, and $p_a(x)$ vary as a function of location x?

(b) If $\mu_a = \mu_e + \mu_i$, and $\epsilon_e + \epsilon_i - \epsilon_a = V_i$, find an expression for the ratio $p_e(x)\,p_i(x)/p_a(x)$. This expression is also known as the *Saha equation*.

26.2 For a simple system consisting of two constituents, the Gibbs free energy may be expressed in the form

$$G = \frac{5}{2} n RT - \frac{3}{2} n RT \ln \frac{3 (RT)^{5/3}}{2 U_o (pV_o)^{2/3}} - n T S_o$$

$$+ n RT (y_1 \ln y_1 + y_2 \ln y_2 + A y_1 y_2)$$

where R, U_o, V_o, S_o, A are constants, T is the temperature, p the pressure, $n = n_1 + n_2$, n_1 the amount of constituent 1, n_2 the amount of constituent 2, y_1 the mole fraction of constituent 1, and y_2 the mole fraction of constituent 2.

(a) Find expressions for the chemical potentials μ_1 and μ_2 as functions of T, p, y_1, y_2, and the given constants.

(b) Find expressions for the chemical potentials μ_{11} and μ_{22} of the pure constituents as functions of T, p, and the given constants.

(c) Evaluate the partial pressures p_{11} and p_{22} as functions of T, p, y_1, y_2, and the given constants.

(d) Find expressions for the specific volumes v_{11} and v_{22} of the pure constituents. Do the pure constituents behave as ideal gases? If so, verify equation (26.20).

(e) Verify equations (26.26) and (26.27).

(f) Find expressions of the derivatives $(\partial p_{ii}/\partial p)_{T,y}$ for $i = 1$, 2, and write the equation of state $v = v(T, p, y)$ [equation (26.32)], and the compressibility factor $\mathcal{Z} = \mathcal{Z}(T, p, y)$ [equation (26.33)] as functions of T, p, y.

26.3 Dry air consists on a mole basis of 78.08% of nitrogen, N_2, 20.95% of oxygen, O_2, 0.934% of argon, Ar, 0.033% of carbon dioxide, CO_2, 18.18 ppm of neon, Ne, 5.24 ppm of helium, He, 2 ppm of methane, CH_4, 1.14 ppm of krypton, Kr, 0.5 ppm of nitrous oxide, N_2O, 0.5 ppm of hydrogen, H_2, and 0.087 ppm of xenon, Xe, where ppm = parts per million. The molecular weight (in kg/kmol) is 39.948 for argon, 20.183 for neon, 4.0026 for helium, 83.80 for krypton, and 131.30 for xenon. Values for the other constituents are listed in Table 19.1.

(a) What is the composition on a mass basis?

(b) What is the mean molecular weight of dry air?

27 Ideal-Gas Mixtures and Solutions

In Chapter 20 we approximate the behavior of a pure substance in certain ranges of temperatures and pressures by explicit and readily understandable mathematical relations among properties. Under certain conditions of temperature, pressure, and composition, it turns out that we may approximate also the behavior of a mixture by explicit and readily understandable mathematical forms. When such forms are valid, we say that they describe an ideal behavior of the mixture or that the mixture is ideal.

In this chapter we define ideal-gas mixtures and ideal solutions, and derive a number of practical relations between properties for each of these ideal behaviors. Methods for the description of mixtures and solutions in regions where the conditions of ideality do not apply are discussed in Chapter 28.

27.1 Gibbs–Dalton Mixtures

We consider the systems in Figure 26.6 under conditions of sufficiently high temperatures and low pressures so that system A is in the gaseous form of aggregation, that is, A is a *gas mixture*. Under such conditions, depending on the nature of the intermolecular forces, the different constituents may be so weakly coupled to one another that each constituent behaves as if the others were not present. Specifically, at a given temperature T and for a given amount n_i of constituent i in system A, the pressure p_{ii} at which system ii is in partial mutual stable equilibrium with A may be independent of the amounts of the other constituents and may remain unchanged even if all the other constituents are removed. If all constituents are removed except the ith, system A would consist of constituent i only, with amount n_i and volume V, at temperature T, and pressure p_{ii} because of the required temperature and chemical potential equalities of the two systems A and ii. These results would hold for $i = 1, 2, \ldots, r$ and, therefore, the specific volume $v_{ii}(T, p_{ii})$ is such that

$$n_i\, v_{ii}(T, p_{ii}) = V \qquad \text{for } i = 1,\, 2,\, \ldots,\, r \tag{27.1}$$

In addition, due to the absence of interactions between different constituents, the gas mixture in A with amounts $n_1, n_2, \ldots, n_r$, volume V, temperature T, and pressure p has the same energy and the same entropy as the composite of r single-constituent systems each of which has amount n_i, volume V, temperature T, and pressure p_{ii} for $i = 1, 2, \ldots, r$, so that

$$U(T, p, \boldsymbol{n}) = \sum_{i=1}^{r} n_i\, u_{ii}(T, p_{ii}) \tag{27.2}$$

$$S(T, p, \boldsymbol{n}) = \sum_{i=1}^{r} n_i\, s_{ii}(T, p_{ii}) \tag{27.3}$$

where in writing equations (27.2) and (27.3) we use the fact that energy and entropy are additive.

When equations (27.1) to (27.3) apply, we say that the system behaves as or is a *Gibbs–Dalton mixture*. It is noteworthy that (27.1) to (27.3) represent the condition of negligible interactions between molecules of different type, but are less restrictive about internal forces between molecules of the same type. For example, for a mixture consisting of oxygen and water, these equations imply negligible internal forces between the O_2 and the H_2O molecules, but do not require negligible forces between either O_2 molecules, or H_2O molecules. Said differently, a mixture may be Gibbs–Dalton even if at temperature T and partial pressure p_{ii}, constituent i for $i = 1, 2, \ldots, r$ does not behave as an ideal gas.

A feature of a Gibbs–Dalton mixture is that the pressure p equals the sum of the partial pressures of all the contituents, that is,

$$p = \sum_{i=1}^{r} p_{ii}(T, p, \boldsymbol{y}) \tag{27.4}$$

for all values of T and $y_1, y_2, \ldots, y_r$ within the range of Gibbs–Dalton behavior. This result is known as *Dalton's law of partial pressures* or, simply, as *Dalton's law*.[1]

Example 27.1. Prove equation (27.4).

Solution. Subtracting equation (27.3) from equation (26.38) and, substituting $(s_i - s_{ii})$ from (26.34), we find that

$$0 = \sum_{i=1}^{r} n_i (s_i - s_{ii}) = -\sum_{i=1}^{r} n_i v_{ii} \left(\frac{\partial p_{ii}}{\partial T} \right)_{p,y} = -V \sum_{i=1}^{r} \left(\frac{\partial p_{ii}}{\partial T} \right)_{p,y} \tag{27.5}$$

where in writing the last equation we use (27.1). Similarly, subtracting equation (27.2) from equation (26.41) and substituting $(u_i - u_{ii})$ from (26.37), we find that

$$0 = \sum_{i=1}^{r} n_i (u_i - u_{ii}) = \sum_{i=1}^{r} n_i v_{ii} \left[p_{ii} - p \left(\frac{\partial p_{ii}}{\partial p} \right)_{T,y} - T \left(\frac{\partial p_{ii}}{\partial T} \right)_{p,y} \right]$$

$$= V \sum_{i=1}^{r} p_{ii} - pV \sum_{i=1}^{r} \left(\frac{\partial p_{ii}}{\partial p} \right)_{T,y} \tag{27.6}$$

where in writing the last equation we use (27.1) and (27.5). By virtue of equations (26.31) and (27.1), $\sum_{i=1}^{r} (\partial p_{ii} / \partial p)_{T,y} = 1$, and thus Dalton's law is proved.

For a Gibbs–Dalton mixture, the enthalpy, the volume, and the compressibility factor satisfy the relations

$$H(T, p, \boldsymbol{n}) = \sum_{i=1}^{r} n_i \, h_{ii}(T, p_{ii}) \tag{27.7}$$

$$\frac{n}{V(T, p, \boldsymbol{n})} = \frac{1}{v(T, p, \boldsymbol{y})} = \sum_{i=1}^{r} \frac{1}{v_{ii}(T, p_{ii})} \tag{27.8}$$

$$V(T, p, \boldsymbol{n}) = \sum_{i=1}^{r} n_i \frac{p_{ii}}{p} \, v_{ii}(T, p_{ii}) \tag{27.9}$$

---

[1]Named in honor of John Dalton (1766–1844), English scientist, who discovered the validity of (27.4) experimentally.

$$\mathcal{Z}(T,p,y) = \sum_{i=1}^{r} n_i \, \mathcal{Z}_{ii}(T,p_{ii}) \tag{27.10}$$

Indeed, we can readily verify that equation (27.7) follows from $H = U + pV = \sum_{i=1}^{r} n_i u_{ii} + \sum_{i=1}^{r} p_{ii} V = \sum_{i=1}^{r} n_i u_{ii} + \sum_{i=1}^{r} p_{ii} n_i v_{ii} = \sum_{i=1}^{r} n_i (u_{ii} + p_{ii} v_{ii}) = \sum_{i=1}^{r} n_i h_{ii}$, where we use (27.1) and (27.4); equation (27.8) follows from $n/V = \sum_{i=1}^{r} y_i n/V = \sum_{i=1}^{r} y_i n/n_i v_{ii} = \sum_{i=1}^{r} y_i/y_i v_{ii} = \sum_{i=1}^{r} 1/v_{ii}$, where we use (27.1), $\sum_{i=1}^{r} y_i = 1$, and $y_i = n_i/n$; equation (27.9) follows from $V = (H - U)/p = \sum_{i=1}^{r} n_i (h_{ii} - u_{ii})/p = \sum_{i=1}^{r} n_i (p_{ii}/p) v_{ii}$; and equation (27.10) follows from $\mathcal{Z} = pV/nRT = (p/nRT) \sum_{i=1}^{r} n_i p_{ii} v_{ii}/p = \sum_{i=1}^{r} n_i p_{ii} v_{ii}/nRT = \sum_{i=1}^{r} y_i p_{ii} v_{ii}/RT = \sum_{i=1}^{r} y_i \mathcal{Z}_{ii}$.

From equation (27.8) we see that the number of molecules per unit volume in the mixture—the inverse of the mole-specific volume—equals the sum of the numbers of molecules per unit volume that each pure constituent would have at the same temperature as that of the mixture, and at a pressure equal to its partial pressure in the mixture.

If the equation of state of the ith constituent, $v_{ii} = v_{ii}(T, p_{ii})$, can be solved for p_{ii} to yield $p_{ii} = p_{ii}(T, v_{ii})$, then equation (27.1) yields

$$p_{ii} = p_{ii}\!\left(T, \frac{v}{y_i}\right) \tag{27.11}$$

If this can be done for every constituent, for $i = 1, 2, \ldots, r$, the Dalton law (equation 27.4) yields

$$p = \sum_{i=1}^{r} p_{ii}\!\left(T, \frac{v}{y_i}\right) \tag{27.12}$$

which is the equation of state of the Gibbs–Dalton mixture. In the next section we evaluate properties of the mixture when each constituent behaves as an ideal gas.

27.2 Ideal-Gas Mixtures

If constituent i of a Gibbs–Dalton mixture behaves as an ideal gas in a temperature range about T, and a pressure range about the partial pressure p_{ii}, and this holds for all $i = 1, 2, \ldots, r$, we say that we have a *Gibbs–Dalton mixture of ideal gases* or, simply, an *ideal-gas mixture.*[2] The equation of state of each constituent is

$$v_{ii}(T, p_{ii}) = \frac{RT}{p_{ii}} \qquad \text{for } i = 1, 2, \ldots, r \tag{27.13}$$

and, combining equations (27.1) and (27.13), we find that

$$p_{ii} = \frac{RT}{v_{ii}(T, p_{ii})} = \frac{RT}{V/n_i} = \frac{RT}{v/y_i} = y_i \frac{RT}{v} \tag{27.14}$$

where the specific volume of the mixture $v = V/n$. Then the Dalton law becomes

$$p = \sum_{i=1}^{r} p_{ii} = \sum_{i=1}^{r} y_i \frac{RT}{v} = \frac{RT}{v} \tag{27.15}$$

---

[2]The term ideal in the expression ideal gas mixture refers to the validity of equations (27.13) and their consequences, not to any aspect of optimum thermodynamic performance.

a result which indicates that the mixture itself behaves as an ideal gas. In addition, if the range of pressures in which each constituent behaves as an ideal gas includes the pressure p of the mixture, then we have $v_{ii}(T, p) = RT/p$ and

$$V(T, p, \boldsymbol{n}) = \frac{nRT}{p} = \sum_{i=1}^{r} n_i \frac{RT}{p} = \sum_{i=1}^{r} n_i v_{ii}(T, p) \tag{27.16}$$

Equation (27.16) indicates that the volume of an ideal-gas mixture equals the sum of the volumes that each constituent would occupy if the values of the amount n_i, the temperature T, and the pressure p were the same as the corresponding values of the mixture. This result is known as *Amagat's law of additive volumes* or, simply, *Amagat's law*.[3]

Combining equations (27.14) and (27.15), we find that

$$p_{ii} = y_i p \tag{27.17}$$

and conclude that the partial pressure of the ith constituent of an ideal-gas mixture equals the product of its mole fraction and the pressure of the mixture.

If the ideal-gas behavior of each constituent extends over the entire ranges of values of temperature and pressure of the mixture, then using equation (20.12a) we find that

$$s_{ii}(T, p_{ii}) = s_{ii}(T, p) - R \ln y_i \tag{27.18}$$

and therefore, equation (27.3) becomes

$$S(T, p, \boldsymbol{n}) = \sum_{i=1}^{r} n_i s_{ii}(T, p) - R \sum_{i=1}^{r} n_i \ln y_i \tag{27.19}$$

For an ideal gas, the internal energy and the enthalpy are independent of pressure. It follows that the same is true for an ideal-gas mixture. Indeed, equations (27.2) and (27.7) become

$$U(T, p, \boldsymbol{n}) = \sum_{i=1}^{r} n_i u_{ii}(T) \tag{27.20}$$

$$H(T, p, \boldsymbol{n}) = \sum_{i=1}^{r} n_i h_{ii}(T) \tag{27.21}$$

In contrast to the ideal-gas behavior of a system with one constituent, here the specific internal energy and specific enthalpy depend not only on temperature but also on composition. In particular, the mole-specific heats c_p and c_v depend on temperature and composition, as indicated by the relations

$$c_p(T, \boldsymbol{y}) = \sum_{i=1}^{r} y_i c_{pii}(T) \tag{27.22}$$

$$c_v(T, \boldsymbol{y}) = \sum_{i=1}^{r} y_i c_{vii}(T) \tag{27.23}$$

where $c_{pii}(T)$ and $c_{vii}(T)$ are the mole-specific heats of constituent i, and $c_p - c_v = R$ because $c_{pii} - c_{vii} = R$ for $i = 1, 2, \ldots, r$, and $\sum_{i=1}^{r} y_i = 1$.

---

[3]Named in honor of Emile Amagat (1841–1915), French physicist, who discovered the validity of (27.16) experimentally.

27.3   Entropy of Mixing

As an illustration of the usefulness of the results on ideal gas mixtures, we evaluate the difference in entropy between two states defined as follows. We consider a container of fixed volume V partitioned into r compartments by means of thin pistons that can move freely (Figure 27.1a). Compartment i confines an amount n_i of constituent i at temperature T and pressure p, for $i = 1, 2, \ldots, r$. Over some ranges around T and p, each constituent behaves as an ideal gas. Under these conditions, it follows that the constituents are in partial mutual stable equilibrium because they satisfy the temperature and pressure equalities.

Next, we remove the thin partitions and consider the stable equilibrium state that results from the mixing of the r constituents within the fixed volume V without any interactions with systems in the environment (Figure 27.1b). In that state, we assume that the mixture behaves as an ideal-gas mixture. We denote its temperature by T' and its pressure by p'.

To proceed with the evaluation of the entropy difference, we first show that under the conditions specified, the temperature T' and the pressure p' are equal to the temperature T and the pressure p, respectively, of each constituent in Figure 27.1a.

To show that $T' = T$, we write the energy balance for the change between the two states. Because no interactions with the environment occur, we have

$$U(T', p', \boldsymbol{n}) - \sum_{i=1}^{r} n_i\, u_{ii}(T, p) = 0 \qquad (27.24)$$

where the first term represents the energy of the mixture and the second term the sum of the energies of the r compartments in the initial state. Using equation (27.20) and the fact that the specific energy u of any ideal gas depends on T only, we find that equation (27.24) may be rewritten as

$$\sum_{i=1}^{r} n_i\, u_{ii}(T') - \sum_{i=1}^{r} n_i\, u_{ii}(T) = \sum_{i=1}^{r} n_i \int_{T}^{T'} c_{vii}(t)\, dt = 0 \qquad (27.25)$$

Each term in the last sum is nonnegative and therefore must be zero for the sum to be zero. This can be true only if $T' = T$.

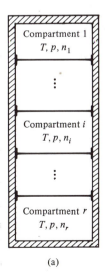

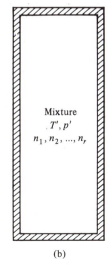

(a)

(b)

Figure 27.1   (a) A container of fixed volume V partitioned into r compartments by means of thin partitions that can move freely but confine in each compartment a different constituent at temperature T and pressure p; (b) Stable equilibrium state of the mixture that results upon removal of the thin partitions from (a).

To show that $p' = p$, we express the volume $V(T', p', n)$ in the form

$$V(T', p', n) = \sum_{i=1}^{r} n_i v_{ii}(T', p') = \sum_{i=1}^{r} n_i v_{ii}(T, p) \tag{27.26}$$

where in writing the first equality we use equation (27.16), and the second the fact that the volume of the mixture has been specified as equal to the sum of the volumes of the r compartments. Next, using the equation of state of an ideal gas, we express $v_{ii}(T, p) = RT/p$, and rewrite (27.26) in the form

$$\sum_{i=1}^{r} n_i \left[v_{ii}(T', p') - v_{ii}(T, p) \right] = \frac{RT'}{p'} \sum_{i=1}^{r} n_i - \frac{RT}{p} \sum_{i=1}^{r} n_i = \frac{nRT'}{p'} - \frac{nRT}{p} = 0 \tag{27.27}$$

which implies that $p' = p$ because $T' = T$.

The entropy balance during the spontaneous mixing process is

$$S(T', p', n) - \sum_{i=1}^{r} n_i s_{ii}(T, p) = S_{\text{irr}} \tag{27.28}$$

where the first term in the left-hand side is the entropy of the stable equilibrium state of the mixture, the second term the sum of the entropies of the r compartments in the initial state, and S_{irr} is the entropy generated by irreversibility during the spontaneous mixing process. Using the results $T' = T$ and $p' = p$, the final entropy is given by (27.19), so equation (27.28) becomes

$$S_{\text{irr}} = \sum_{i=1}^{r} n_i \left[s_{ii}(T, p) - R \ln y_i \right] - \sum_{i=1}^{r} n_i s_{ii}(T, p)$$

$$= -R \sum_{i=1}^{r} n_i \ln y_i = -nR \sum_{i=1}^{r} y_i \ln y_i \tag{27.29}$$

It is noteworthy that each term in the sum $\sum_i y_i \ln y_i$ is negative because the function $y \ln y$ is negative for any y with $0 < y < 1$.

The entropy generated by irreversibility, S_{irr}, as a result of mixing ideal gases under the conditions discussed in this section is sometimes referred to as the *entropy of mixing*. Clearly, mixing ideal gases that are initially not in temperature and pressure equality would result in amounts of entropy generation that differ from the entropy of mixing.

Just after removal of the thin partitions of the system in Figure 27.1a, the state is not stable equilibrium, and has nonzero adiabatic availability Ψ. To evaluate Ψ, we recall that it is equal to the difference between the energy of the given state, and the energy of the stable equilibrium state with the same values of $n_1, n_2, \ldots, n_r$, V, and S as the corresponding values of the given state. The latter energy is that of a mixture of the r constituents at temperature T' and pressure p' determined by the conditions that the volume and entropy of the stable equilibrium state are the same as the corresponding values of the given states. Using equation (27.28) with $S_{\text{irr}} = 0$, equation (27.19), and assuming that each constituent behaves as a perfect gas, we find that the condition of entropy equality implies the relation

$$0 = \sum_{i=1}^{r} n_i \left[s_{ii}(T', p') - R \ln y_i - s_{ii}(T, p) \right]$$

$$= \sum_{i=1}^{r} n_i \left[c_{pii} \ln \frac{T'}{T} - R \ln \frac{p'}{p} - R \ln y_i \right]$$

$$= nc_p \ln \frac{T'}{T} - nR \ln \frac{p'}{p} - nR \sum_{i=1}^{r} y_i \ln y_i$$

$$= nc_v \ln \frac{T'}{T} - nR \sum_{i=1}^{r} y_i \ln y_i \qquad (27.30)$$

where in writing the last of (27.30) we use the condition of volume equality [equation (27.27)], that is, the relation $p'/p = T'/T$. From equation (27.30) we find that

$$\frac{T'}{T} = \frac{p'}{p} = \exp \left(\frac{R}{c_v} \sum_{i=1}^{r} y_i \ln y_i \right) \qquad (27.31)$$

and, therefore, that

$$\Psi = \sum_{i=1}^{r} n_i \, u_{ii}(T, p) - U(T', p', \boldsymbol{n})$$

$$= \sum_{i=1}^{r} n_i \left[u_{ii}(T, p) - u_{ii}(T', p') \right] = \sum_{i=1}^{r} n_i c_{vii} (T - T')$$

$$= nc_v (T - T') = nc_v T \left[1 - \exp \left(\frac{R}{c_v} \sum_{i=1}^{r} y_i \ln y_i \right) \right] \qquad (27.32)$$

where in writing the various forms of this equation we use the perfect-gas relation between energy and temperature, and equation (27.31).

27.4 Psychrometry

To a good degree of approximation, *moist air*, the mixture of air and water vapor under conditions usually prevailing in our environment, may be regarded as an ideal-gas mixture, provided that the pressure is not extremely high and the temperature not extremely low. Mixtures of air and water vapor are important in many industrial applications, such as drying of paper and foodstuffs, air conditioning, and cooling towers of power plants. The study of the properties of moist air is known as *hydrometry* or *psychrometry*. In this section we discuss some of the terminology and the relations used in engineering applications involving mixtures of air and water vapor.

The term *dry air* is used to denote the usual mixture of gases that constitute the atmosphere at sea level, exclusive of water vapor. The distinction between water vapor and dry air is useful because the amount of water vapor contained in a given amount of air is extremely variable compared with the amounts of the other constituents, and because water vapor alone can often be condensed out of the mixture without unusual changes in the temperature and the volume of the mixture.

Dry air, corresponding to atmospheric air at sea level exclusive of water vapor, is a mixture with fixed composition containing, on a mole basis, 78.08% of nitrogen, N_2, 20.95% of oxygen, O_2, 0.934% of argon, Ar, 0.033% of carbon dioxide, CO_2, and traces, accounting for less than 0.01%, of other constituents, such as neon, Ne, helium, He, methane, CH_4, krypton, Kr, nitrous oxide, N_2O, hydrogen, H_2, and xenon, Xe. The resulting mean molecular weight (Section 26.1) $M_a = 28.964$ kg/kmol. Because it behaves as an ideal-gas mixture and the composition is fixed, we can think of dry air as a pure substance consisting of a fictitious single constituent with molecular weight M_a. Moreover, for applications that involve a restricted range of temperatures about the standard environmental temperature, dry air may be modeled as a perfect gas with specific heat ratio $\gamma_a = 1.4$ and, therefore, specific heat under constant pressure $c_{pa} = \gamma_a R/M_a(\gamma_a - 1) = 1.4 \times 8.314/28.964 \times 0.4 = 1$ kJ/kg K.

For moist air, that is, a binary mixture of dry air considered as a pure substance and water vapor, the conditions in typical engineering applications that involve temperatures in the range from a few degrees below 0°C to about 50°C are such that the mixture behaves as an ideal-gas mixture. Indeed, the mole fraction y_v for water vapor under such conditions is always so much smaller than unity that the partial pressure $p_v = y_v p$, for $p = 1$ atm, is much smaller than atmospheric. At such temperatures and partial pressures, we see in Figures 19.10, 19.12, 19.13, and 19.15 that the water vapor behaves as a perfect gas even when the conditions approach the saturation curve. For example, we see that the constant-enthalpy curves coincide with the constant-temperature curves, showing that enthalpy is a function of temperature only, as it should be for an ideal-gas. Moreover, Figure 19.15 shows that at low pressures and not too high temperatures, the specific heat of the vapor under constant pressure is approximately constant and equal to 1.82 kJ/kg K, as it should be for a perfect gas.

With the approximations just discussed, we can estimate the properties of moist air for all the conditions of practical interest. The composition may be specified by either the mole fraction y_v or the mass fraction x_v of vapor in the binary mixture of dry air and water vapor. The fractions x_v and y_v satisfy the relations [equations (26.4) and (26.8)]

$$x_v = \frac{y_v M_v}{y_v M_v + (1 - y_v)M_a} = \frac{1}{1 + (1 - y_v)/0.622\, y_v} \tag{27.33}$$

$$y_v = \frac{x_v/M_v}{x_v/M_v + (1 - x_v)/M_a} = \frac{1}{1 + 0.622(1 - x_v)/x_v} \tag{27.34}$$

where we use the ratio $M_v/M_a = 18.015/28.964 = 0.622$.

Because the amount of dry air is fixed, a practical alternative to describing the composition of moist air in terms of either x_v or y_v is to use the *specific humidity* or *humidity ratio* ω defined as

$$\omega = \frac{m_v}{m_a} = \frac{x_v}{1 - x_v} \tag{27.35}$$

where m_v is the mass of water vapor in the mixture and m_a the mass of dry air. Alternatively, the specific humidity may be expressed in terms of the mole fraction y_v, that is,

$$\omega = \frac{y_v M_v}{(1 - y_v)M_a} = \frac{0.622\, y_v}{1 - y_v} \tag{27.36}$$

At a given pressure p and a given temperature T, moist air behaves as an ideal-gas mixture at compositions y_v in the range between $y_v = 0$ (dry air) and a maximum value $y_{vs}(T, p)$ at which the mixture is saturated, that is, in mutual stable equilibrium with the liquid water phase. At mole fractions higher than $y_{vs}(T, p)$ the mixture is two-phase, one phase consisting of saturated moist air with mole fraction $y_v = y_{vs}(T, p)$, and the other phase consisting of saturated liquid water with a negligible amount of dry air dissolved in it. Equating the chemical potentials of the water vapor in the mixture and the almost pure liquid water in the condensed phase, we can readily show that

$$y_{vs}(T, p) = \frac{p_{sat}(T)}{p} \tag{27.37}$$

where $p_{sat}(T)$ is the saturation pressure of pure water at the temperature T of the mixture.

Example 27.1. Verify (27.37).

Solution. The amount of dry air dissolved in the liquid phase is negligible because the typical temperatures we are considering are much higher than the critical temperature of dry air, and therefore we can treat dry air as a noncondensable gas. Modeling the condensed phase as pure liquid water, for mutual stable equilibrium we must require that the chemical potential $\mu_v(T, p, y_v)$ of the water vapor in the mixture be equal to the chemical potential $\mu_{ww}(T, p)$ of the pure liquid water in the condensed phase. Modeling the moist air as an ideal-gas mixture, and the pure liquid as an ideal incompressible fluid, we express the chemical potentials in the forms

$$\mu_v(T, p, y_v) = \mu_{vv}(T, p_v) = \mu_{vv}(T, y_v p) = \mu_g(T, p_{sat}(T)) + RT \ln \frac{y_v p}{p_{sat}(T)}$$

$$\mu_{ww}(T, p) = \mu_f(T, p_{sat}(T)) + v_w[p - p_{sat}(T)]$$

where $p_{sat}(T)$ is the saturation pressure of pure water at temperature T, we use equations (26.20) and (26.21), and subscripts f and g stand for saturated liquid and saturated vapor, respectively. Because by definition of the saturation pressure $\mu_f(T, p_{sat}(T)) = \mu_g(T, p_{sat}(T))$, the chemical potential equality $\mu_{ww}(T, p) = \mu_v(T, p, y_v)$ yields

$$y_v = \frac{p_{sat}(T)}{p} \exp\left\{ \frac{v_w}{RT} [p - p_{sat}(T)] \right\} \approx \frac{p_{sat}(T)}{p}$$

where the approximation is very accurate in the range of conditions under consideration. For example, for $T = 300$ K and $p = 100$ kPa, the argument of the exponential is about $10^{-3} \times 100/(8.314 \times 300) = 4 \times 10^{-5}$, and hence the exponential is almost equal to unity (see also Section 20.5).

Because the mole fraction y_v of vapor in moist air can take values only between 0 and $y_{vs}(T, p)$, a practical alternative to describing the composition in terms of y_v, x_v, or ω is to use the *relative humidity* ϕ, defined as

$$\phi = \frac{y_v}{y_{vs}(T, p)} = \frac{y_v p}{p_{sat}(T)} \tag{27.38}$$

where in writing the second of equations (27.38) we use (27.37). Other useful relations between relative humidity ϕ, specific humidity ω, mole fraction y_v, and mass fraction x_v are as follows

$$y_v = \frac{\omega}{\omega + 0.622} = \phi \frac{p_{sat}(T)}{p} \tag{27.39}$$

$$\omega = \frac{0.622 \, \phi \, p_{sat}(T)}{p - \phi \, p_{sat}(T)} \tag{27.40}$$

$$\phi = \frac{\omega p}{(\omega + 0.622) \, p_{sat}(T)} \tag{27.41}$$

$$x_v = \frac{\omega}{\omega + 1} = \frac{0.622 \, \phi \, p_{sat}(T)}{p + (1 - 0.622) \, \phi \, p_{sat}(T)} \tag{27.42}$$

Equation (27.40) is often presented in the form of an ω versus T diagram for a fixed value of p and various values of ϕ. Such a chart, which can readily be constructed from the knowledge of $p_{sat}(T)$ for pure water, is called the *psychrometric chart*. For atmospheric pressure, $p = 101.3$ kPa, graphs of ω versus T for $\phi = 10\%$, 20%, ..., 100% are shown in Figure 27.2.

For moist air with given values of T, p, and y_v, the temperature T_{dp} that satisfies the relation

$$p_{sat}(T_{dp}) = y_v p \tag{27.43}$$

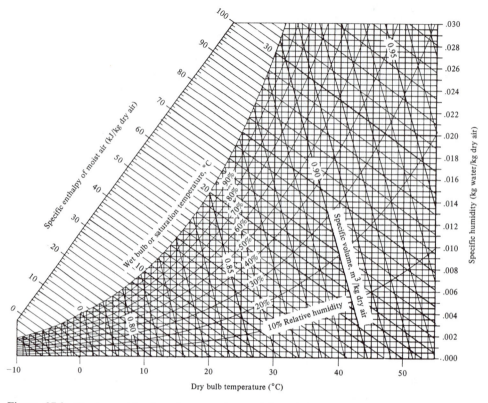

Figure 27.2 Psychrometric chart for $p = 1$ atm (adapted from Z. Zhang, and M. Pate, "A Methodology for Implementing a Psychrometric Chart in a Computer Graphics System," ASHRAE Transactions, 94, 1, 1988).

is called the *dew-point temperature*. Upon cooling the given moist air at constant pressure, the dew-point temperature is the temperature at which the mixture reaches saturation. Any further cooling at constant pressure results in condensation of part of the water vapor in the air. On the psychrometric chart, the dew-point temperature of a mixture in any state is found by reading the temperature corresponding to the intersection of the $\phi = 100\%$ curve and the constant-ω line passing through that state.

The psychrometric chart also shows curves at constant enthalpy per unit mass of dry air. Such curves are useful in analyses of bulk-flow devices involving the flow of moist air. In general, it is convenient to express the extensive properties of the mixture per unit mass of dry air in the mixture. For example, expressing the relations developed for ideal-gas mixtures on a mass basis, and using the specific humidity ω instead of the mass fraction x_v, we readily obtain the following relation

$$h_{\mathrm{m}}(T, \omega) = \frac{H}{m_{\mathrm{a}}} = h_{\mathrm{a}}(T) + \omega\, h_{\mathrm{v}}(T)$$

$$= h_{\mathrm{a}}(T_{\mathrm{o}}) + \omega\, h_{\mathrm{g}}(T_{\mathrm{o}}) + (c_{\mathrm{pa}} + \omega\, c_{\mathrm{pv}})(T - T_{\mathrm{o}}) \qquad (27.44)$$

where T_{o} is an arbitrary reference temperature, $h_{\mathrm{g}}(T_{\mathrm{o}})$ is the value of the saturated-vapor specific enthalpy taken from the steam tables, and the specific heats under constant pressure $c_{\mathrm{pa}} = 1$ kJ/kg K for dry air and $c_{\mathrm{pv}} = 1.82$ kJ/kg K for water vapor. Clearly, the value of $h_{\mathrm{g}}(T_{\mathrm{o}})$ must be taken from the same steam tables from which we take the other properties of water in the given application. From equation 27.44, we can evaluate ω as a function of T and h_m, that is,

$$\omega = \frac{h_m - h_a(T)}{h_g(T)} \tag{27.45}$$

and setting $T_o = T$, we find that

$$\left(\frac{\partial \omega}{\partial T}\right)_{h_m} = -\frac{c_{pa} + \omega c_{pv}}{h_g(T)} \tag{27.46}$$

Because ω is a small number—between 0 and 0.03—and $h_g(T)$ varies from 2501.4 to 2592.1 kJ/kg in the range 0 to 50°C, we conclude that the right-hand side of equation (27.46) is approximately constant—it ranges from -4.0×10^{-4} to -3.8×10^{-4} K^{-1}. Thus the constant-enthalpy curves on the ω–T psychrometric chart are approximately straight lines, all with approximately the same slope.

Yet another term used in psychrometry is that of *wet-bulb temperature*. It refers to the temperature reading of a thermometer with its bulb covered by a wick wetted with liquid water and placed in a stream of the moist air. As it passes through the wick, the moist air stream becomes saturated ($\phi = 100\%$) at the same temperature as the liquid water in the wick, that is, at the wet-bulb temperature T_{wb} registered on the thermometer. The relation between the wet-bulb temperature T_{wb} and the specific humidity ω for moist air at temperature T and pressure p is given by

$$\omega = \frac{c_{pa}(T_{wb} - T) + 0.622\, p_{sat} h_{fg}(T_{wb})/[p - p_{sat}(T_{wb})]}{h_{fg}(T_{wb}) + c_{pv}(T - T_{wb})} \tag{27.47}$$

From this relation we find that

$$\left(\frac{\partial \omega}{\partial T}\right)_{T_{wb}} = -\frac{c_{pa} + \omega c_{pv}}{h_{fg}(T_{wb}) + c_{pv}(T - T_{wb})} \tag{27.48}$$

Because ω is small and $h_{fg}(T_{wb}) \gg c_{pv}(T - T_{wb})$, the right-hand side of equation (27.48) is approximately equal to the ratio $c_{pa}/h_{fg}(T_{wb})$. Thus the constant-wet-bulb-temperature curves on the ω–T psychrometric chart are approximately straight lines with slopes slightly changing from about -4.0×10^{-4} K^{-1} to about -4.2×10^{-4} K^{-1} as $h_{fg}(T_{wb})$ varies from 2501.3 kJ/kg to 2373.1 kJ/kg in the range 0 to 50°C. For contrast with the wet-bulb temperature T_{wb}, the temperature T of the moist air is sometimes called the *dry-bulb temperature*.

Example 27.2. Prove equation (27.47).

Solution. As a result of passing through the wet wick, the moist air is saturated with water vapor. Denoting by ω_{wb} the specific humidity of the saturated moist air leaving the wick at the wet-bulb temperature T_{wb}, and using equation (27.40) with $\phi = 1$, we find that $\omega_{wb} = 0.622\, p_{sat}(T_{wb})/[p - p_{sat}(T_{wb})]$. Per unit of mass of dry air in the flow through the wick, the amount of water evaporated from the wick to saturate the moist air is $\omega_{wb} - \omega$, where ω is the inlet specific humidity. The energy balance of the wick at steady state yields

$$h_a(T) + \omega h_g(T) + (\omega_{wb} - \omega) h_f(T_{wb}) - h_a(T_{wb}) - \omega_{wb} h_g(T_{wb}) = 0$$

or

$$\omega [h_g(T) - h_f(T_{wb})] = h_a(T_{wb}) - h_a(T) + \omega_{wb} [h_g(T_{wb}) - h_f(T_{wb})]$$

Recalling that here both water vapor and dry air behave as perfect gases, and using equation (27.40) with $\phi = 1$ for ω_{wb}, we find equation (27.47).

27.5   Ideal Solutions

A mixture in the liquid form of aggregation in which the amount of one constituent is larger than the amounts of all the other constituents is called a *solution*. The prevailing constituent is called the *solvent*, and each of the other constituents a *solute*. A solution is called *dilute* if the mole fraction of the solvent is much greater than the mole fraction of any solute.

At any conditions of temperature and pressure, if a solution is sufficiently dilute, the nearest neighbors of the molecules of the solvent are almost always molecules of the solvent. Under such conditions it can be shown with the tools of quantum thermodynamics, and it is verified experimentally, that the chemical potential of the solvent is given by an expression of the form

$$\mu_i = \mu_{ii}(T,p) + RT \ln y_i \tag{27.49}$$

where $\mu_{ii}(T,p)$ is the chemical potential of the pure solvent at the temperature T and pressure p of the mixture. When equation (27.49) is satisfied, we say that constituent i follows an *ideal-solution behavior*.[4]

In general, the ideal solution behavior is exhibited by any constituent i of a liquid mixture in the limit as y_i tends to unity, namely, as the ith constituent becomes the solvent of a dilute solution. However, there are examples of special mixtures consisting of constituents that present very similar molecular structures so that each molecule experiences the same interactions whether its nearest neighbors are of the same or of another constituent. For such special liquid mixtures, equation (27.49) holds for every constituent at all compositions, and we say that the mixture is an *ideal solution*. Often, the behavior of a solution is described in terms of deviations from the ideal-solution behavior. Thus the study of ideal solutions constitutes a useful starting point also for the study of solutions that do not conform to the requirements of ideality.

Upon using equation (27.49) for a constituent with ideal-solution behavior and equations (26.34) to (26.37), we find that

$$s_i = s_{ii}(T,p) - R \ln y_i \tag{27.50}$$

$$v_i = v_{ii}(T,p) \tag{27.51}$$

$$h_i = h_{ii}(T,p) \tag{27.52}$$

$$u_i = u_{ii}(T,p) \tag{27.53}$$

Moreover, for an ideal solution, substituting equations (27.50) to (27.53) for each $i = 1, 2, \ldots, r$ into (26.38) to (26.41), we find that

$$S(T,p,n_1,n_2,\ldots,n_r) = \sum_{i=1}^{r} n_i s_{ii}(T,p) - R \sum_{i=1}^{r} n_i \ln y_i \tag{27.54}$$

$$V(T,p,n_1,n_2,\ldots,n_r) = \sum_{i=1}^{r} n_i v_{ii}(T,p) \tag{27.55}$$

$$H(T,p,n_1,n_2,\ldots,n_r) = \sum_{i=1}^{r} n_i h_{ii}(T,p) \tag{27.56}$$

---

[4]Here, too, the term "ideal" in the expression ideal solution refers to the validity of (27.49) and its consequences, not to any aspect of optimum thermodynamic performance.

$$U(T, p, n_1, n_2, \ldots, n_r) = \sum_{i=1}^{r} n_i u_{ii}(T, p) \tag{27.57}$$

We notice that some of these results are formally identical to the corresponding results for an ideal-gas mixture, such as equations (27.54) and (27.55) versus equations (27.19) and (27.16). The difference, of course, is that here we discuss liquid forms of aggregation.

If a constituent with ideal-solution behavior exhibits ideal incompressible behavior as a pure substance at the same temperature and pressure as those of the solution, then combining equation (26.21) with $p^* = p$ and equation (27.49), we find that

$$RT \ln y_i = -v_{ii}(p - p_{ii}) \tag{27.58}$$

where v_{ii} is the constant mole-specific volume of the pure incompressible liquid, and p_{ii} the partial pressure of the constituent in the solution. The difference $p - p_{ii}$ is called the *osmotic pressure of constituent i* in the solution. It is the pressure difference that a rigid semipermeable membrane, permeable only to the ith constituent, must sustain in order to maintain partial mutual stable equilibrium between the solution and the pure constituent, that is, to prevent the migration of molecules of the constituent through the membrane that would be driven by a chemical potential difference.

Using the identity $y_i = 1 - \sum_{j \neq i} y_j$, the fact that any constituent of any solution exhibits ideal-solution behavior in the limit as $y_i \to 1$, and equation (27.58), we conclude that the osmotic pressure of the solvent of a dilute solution is given by the relation

$$p - p_{ii} = -\frac{RT}{v_{ii}} \ln y_i = -\frac{RT}{v_{ii}} \ln \left(1 - \sum_{j \neq i} y_j \right) = \frac{RT}{v_{ii}} \sum_{j \neq i} y_j \tag{27.59}$$

where we use the approximation $\ln(1 - \epsilon) \approx -\epsilon$ valid in the limit as $\epsilon \to 0$. Equation (27.59) is known as the *van't Hoff relation for osmotic pressure*.[5] It indicates that the osmotic pressure of the solvent of a dilute solution depends only on the temperature, the constant specific volume of the pure solvent, and the sum of the mole fractions of the solutes, and that it is independent of the nature and properties of the solutes.

Example 27.3.   The molecular weight of the solute in a binary solution may be determined from measurements of osmotic pressure of the solvent. Show how this may be done.

Solution.   Denoting the solvent as constituent 1 and the mass fraction of the solute by x_2, and using equations (26.4) and (26.6), we find that $x_2 = y_2 M_2 n/m$, and rewrite (27.59) as

$$M_2 \frac{n v_{11}}{V} = \frac{RT}{p - p_{11}} \frac{m x_2}{V}$$

where we divide both sides of the equation by V to put in evidence the term $m x_2/V$, that is, the mass of solute per unit volume of solution, which is readily measurable. By measuring temperature T, osmotic pressure $p - p_{11}$, and $m x_2/V$ for increasingly dilute solutions, and extrapolating to zero x_2, the term $M_2 n v_{11}/V$ tends to M_2 because $n v_{11}$ tends to V, and hence we obtain a measurement of the molecular weight of the solute.

Example 27.4.   Evaluate the osmotic pressure of water in a solution which on a mass basis consists of 3.5% of salt (sodium cloride), NaCl, and 96.5% of water, H_2O, at a temperature $T = 10°C$.

Solution.   Denoting water as constituent 1 and salt as constituent 2, and using (26.8), we find that $M_1 = 18.015$ kg/kmol and $M_2 = 58.5$ kg/kmol, $y_1 = (0.965/18.015)/(0.965/18.015 +$

---

[5]This relation is named in honor of Jacobus Hendricus van't Hoff (1852–1911), a Dutch chemist, for his contributions to chemical thermodynamics, for which he earned the first Nobel Prize for chemistry in 1901.

$0.035/58.5) = 0.989 = 98.9\%$, and $y_2 = 1 - y_1 = 0.011 = 1.1\%$. This is a dilute solution. Upon assuming ideal-solution behavior for water, the solvent, and using equation (27.59), we find that $p - p_{11} = RTy_2/v_{11} = 8.314 \times 283 \times 0.011/(18.015 \times 10^{-3}) = 1437$ kPa, where for the mole-specific volume of water we use $v_{11} = 18.015 \times 10^{-3}$ m^3/kg. It is noteworthy that the osmotic pressure is very large, 1.4 MPa or about 14 atm.

Osmotic pressure plays an important role in many biological phenomena. Indeed, many biological membranes are semipermeable, that is, permeable only to some constituents and not to others. For example, semipermeable membranes in the roots of a tree maintain a high osmotic pressure. This pressure sustains at equilibrium the high columns of fluid that feed all the vital parts of the tree. Again, the permeability of biological membranes in a living cell, controlled by enzymes and electric signals, regulates the activities and the metabolism of the cell.

27.6   Entropy of Dilution and Work of Separation

As an example of application of the results just developed, we consider the system in Figure 27.3a. It consists of a container partitioned into two compartments by means of a movable piston. One compartment contains a dilute solution at temperature T, pressure p, and solvent mole fraction y_i. The other compartment contains an infinitesimal amount of pure constituent i, the solvent, at the same temperature and pressure as the solution. A piston at the top of the container maintains a constant pressure.

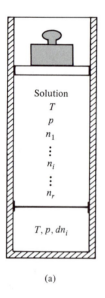

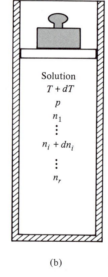

Figure 27.3   (a) A container maintained at a fixed pressure and partitioned into two compartments in partial mutual stable equilibrium, one containing a dilute solution and the other an infinitesimal amount of solvent. (b) The solution in the stable equilibrium state that results upon removal of the partition in (a).

Upon removal of the partition, the system evolves spontaneously, that is, without any interactions with the environment except possibly a volume change, to a new stable equilibrium state (Figure 27.3b), in which the solution has temperature $T + dT$, pressure p, and a composition infinitesimally more dilute. It is interesting to show that under the specified conditions, the dilution process is also isothermal, that is, $dT = 0$, and isochoric, that is, $dV = 0$.

To show that $dT = 0$, we write the energy balance for the system during the process. Because it is isobaric and involves no interactions other than the work due to volume change, the process is isoenthalpic, and the energy balance is

$$H(T + dT, p, n_1, \ldots, n_i + dn_i, \ldots, n_r)$$

$$-H(T, p, n_1, \ldots, n_i, \ldots, n_r) - h_{ii}(T, p)\, dn_i = 0 \qquad (27.60)$$

or, equivalently,

$$\left(\frac{\partial H}{\partial T}\right)_{p,n} dT + \left(\frac{\partial H}{\partial n_i}\right)_{T,p,n} dn_i - h_{ii}(T, p)\, dn_i = 0 \qquad (27.61)$$

Recalling the definition of partial enthalpy h_i [equations (17.42) and (27.52)] for the ideal-solution behavior of the solvent, we find that the second and third terms in (27.61) annul each other, and hence $dT = 0$.

To show that $dV = 0$, we write

$$dV = V(T, p, n_1, \ldots, n_i + dn_i, \ldots, n_r) - V(T, p, n_1, \ldots, n_i, \ldots, n_r)$$

$$- v_{ii}(T, p)\, dn_i = \left(\frac{\partial V}{\partial n_i}\right)_{T,p,n} dn_i - v_{ii}(T, p)\, dn_i = 0 \qquad (27.62)$$

where in writing the last equality we use (27.51).

Now the entropy balance for this process yields

$$S(T, p, n_1, \ldots, n_i + dn_i, \ldots, n_r)$$

$$-S(T, p, n_1, \ldots, n_i, \ldots, n_r) - s_{ii}(T, p)\, dn_i = \delta S_{\text{irr}} \qquad (27.63)$$

and, therefore,

$$\delta S_{\text{irr}} = \left(\frac{\partial S}{\partial n_i}\right)_{T,p,n} dn_i - s_{ii}(T, p)\, dn_i = -R \ln y_i\, dn_i \approx R \left(\sum_{j \neq i} y_j\right) dn_i \qquad (27.64)$$

where in the last expression we use the approximation $\ln(1 - \epsilon) \approx -\epsilon$. This entropy, generated by irreversibility under the conditions just discussed, is sometimes called the *entropy of dilution*. It is independent of T, p, and the nature and properties of the solutes and the solvent.

It is noteworthy that the difference in available energy with respect to a given reservoir at temperature T_0 between the initial and the final states of the system in Figure 27.3 is equal to $T_0\, \delta S_{\text{irr}}$, where δS_{irr} is given by (27.64). It follows that $T_0\, \delta S_{\text{irr}}$ equals the least *work of separation* of an infinitesimal amount of solvent with respect to a reservoir at T_0.

Example 27.5.   Find the least work of separation of 1 kg of pure water out of a large amount of seawater at environmental temperature $T_0 = 10°\text{C}$. The mole fraction of salt in seawater is $y_2 = 1.1\%$.

Solution.   Using equation (27.64) for $i = 1$ and assuming that $dn_1 = 1/18.015 = 0.0555$ kmol is infinitesimal with respect to the overall amount of seawater treated, we find that $T_0\, \delta S_{\text{irr}} = RT_0\, dn_1\, y_2 = 8.314 \times 283.15 \times 1 \times 0.011/18.015 = 1.4$ kJ.

Example 27.6.   Assuming that seawater is an ideal solution namely, that (27.49) holds for both water and salt, estimate the least work of separation of 0.035 *kg* of salt out of a large amount of seawater at $T_0 = 10°\text{C}$.

Solution.   Using equation (27.64) for $i = 2$, without the approximation in the last equality, assuming that $dn_2 = 0.035/58.5 = 0.6 \times 10^{-3}$ kmol is infinitesimal, and $y_2 = 0.011$, we find that $T_0\, \delta S_{\text{irr}} = -RT_0\, dn_2 \ln y_2 = -8.314 \times 283 \times 0.035 \times \ln(0.011)/58.5 = 6.4$ kJ. Noting that 0.035 kg is the amount of salt in 1 kg of seawater with mass fraction $x_2 = 3.5\%$, and comparing the results of this and Example 27.5, we see that separating the solute out of a dilute solution requires much more work expenditure than separating the solvent.

27.7 Two-Phase Liquid–Vapor States

In this section we discuss the two-phase liquid–vapor states of a mixture at conditions such that the vapor phase behaves as an ideal-gas mixture, and the liquid phase exhibits ideal-solution behavior at least for constituent 1. This model is fairly accurate as long as the liquid phase is a dilute solution with constituent 1 as the solvent, and the gas phase is not extremely compressed. For example, many dilute aqueous solutions—dilute solutions of various constituents in water—at ordinary temperatures and pressures are well represented by the model just defined.

Moreover, we consider only two-phase states at temperatures T that are below the critical temperature $T_{c,11}$ of constituent 1, so that we can use the pressure–temperature saturation relation for pure constituent 1, that is, the relation

$$p_{sat,11} = p_{sat,11}(T) \tag{27.65}$$

This equation results from the condition of chemical potential equality between liquid and vapor phases of pure constituent 1 at temperature T, that is, the condition

$$\mu_{11g}(T, p_{sat,11}) = \mu_{11f}(T, p_{sat,11}) \tag{27.66}$$

Under the assumptions cited, we can write the chemical potential of constituent 1 in the vapor phase of the mixture as

$$\mu_{1g} = \mu_{11g}(T, p_{11g}) = \mu_{11g}(T, y_{1g}p)$$
$$= \mu_{11g}(T, p_{sat,11}(T)) + RT \ln \frac{y_{1g}p}{p_{sat,11}(T)} \tag{27.67}$$

where we use equation (27.17) ($p_{11g} = y_{1g}p$) and the ideal-gas relation $\mu(T, p'') = \mu(T, p') + RT \ln(p''/p')$ with $p'' = y_{1g}p$ and $p' = p_{sat,11}(T)$. Moreover, we can write the chemical potential of constituent 1—the solvent—in the liquid phase of the mixture—dilute solution—as

$$\mu_{1f} = \mu_{11f}(T, p) + RT \ln y_{1f}$$
$$= \mu_{11f}(T, p_{sat,11}(T)) + v_{11f}\left[p - p_{sat,11}(T)\right] + RT \ln y_{1f} \tag{27.68}$$

where we use the ideal incompressible-fluid relation $\mu(T, p'') = \mu(T, p') + v(p'' - p')$ with $p'' = p$ and $p' = p_{sat,11}(T)$.

Thus, equating (27.67) and (27.68), namely, imposing the chemical potential equality for constituent 1, $\mu_{1g} = \mu_{1f}$, after a few rearrangements we find that

$$\frac{y_{1g}}{y_{1f}} = \frac{p_{sat,11}(T)}{p} \exp\left\{\frac{v_{11f}}{RT}\left[p - p_{sat,11}(T)\right]\right\} \tag{27.69}$$

If T is well below $T_{c,11}$ and p is not much larger than $p_{sat,11}(T)$, the exponential term in the right-hand side of (27.69) is approximately equal to unity, and therefore,

$$y_{1g}p = y_{1f}p_{sat,11}(T) \tag{27.70}$$

The exponential term is usually referred to as the Poynting correction (Section 20.6). Equation (27.70) is known as *Raoult's law*. In the study of psychrometry in Section 27.4 we derive a special form of Raoult's law [equation (27.37)] which applies to mutual stable equilibrium between pure liquid water—a limiting case of a dilute solution—and moist air at temperature T, pressure p, and water–vapor mole fraction y_{vs}.

Of course, Raoult's law would apply to all constituents if the liquid phase is an ideal solution, that is, if every constituent has ideal-solution behavior at all compositions.

Such a behavior occurs only very rarely in practice. Examples are the binary mixture of ethylene bromide and propylene bromide, and the binary mixture of benzene and bromobenzene in restricted ranges of T and p. For this reason we continue our discussion of a dilute solution without assuming that it is ideal. To simplify the presentation, we consider only binary mixtures.

Ideal-solution behavior of one constituent in a liquid phase of a binary mixture has an important implication on the behavior of the other constituents. Indeed, upon writing the Gibbs–Duhem relation [equation (26.22) divided by n] for the liquid phase of a binary mixture, we find that

$$s_f \, dT - v_f \, dp + y_{1f} \, d\mu_{1f} + y_{2f} \, d\mu_{2f} = 0 \tag{27.71}$$

Upon differentiating the first equation (27.68), substituting the result in equation (27.71), and setting $dT = 0$ and $dp = 0$, we find that

$$RT + y_{2f} \left(\frac{\partial \mu_{2f}}{\partial y_{1f}} \right)_{T,p} = 0 \tag{27.72}$$

or, equivalently,

$$\left(\frac{\partial \mu_{2f}}{\partial \ln y_{2f}} \right)_{T,p} = RT \tag{27.73}$$

where in writing equation (27.73) we use the relation $y_{1f} = 1 - y_{2f}$. Integration of equation (27.73) yields

$$\mu_{2f} = \lambda_{2f}(T, p) + RT \ln y_{2f} \tag{27.74}$$

where the constant of integration $\lambda_{2f}(T, p)$ is a function of T and p only. We see that even if the behavior of constituent 2 is not ideal, the ideal-solution behavior of constituent 1 implies a definite dependence of μ_{2f} on the mole fraction y_{2f}. This is a striking example of the many interrelations among properties that exist at stable equilibrium by virtue of the laws of thermodynamics. It is noteworthy that if constituent 2 exhibited ideal-solution behavior, $\lambda_{2f}(T, p)$ would be equal to $\mu_{22f}(T, p)$.

For the vapor phase, using equation (27.67) we find that

$$\mu_{2g} = \mu_{22g}(T, p_{\text{sat},22}(T)) + RT \ln \frac{y_{2g} p}{p_{\text{sat},22}(T)} \tag{27.75}$$

Substituting equations (27.74) and (27.75) into the chemical potential equality condition, $\mu_{2g} = \mu_{2f}$ yields

$$y_{2g} p = y_{2f} \, \mathcal{H}_{2f}(T, p) \tag{27.76}$$

where

$$\mathcal{H}_{2f}(T, p) = p_{\text{sat},22}(T) \exp \left[\frac{\lambda_{2f}(T, p) - \mu_{22f}(T, p_{\text{sat},22}(T))}{RT} \right] \tag{27.77}$$

and in writing (27.77) we use the relation $\mu_{22g}(T, p_{\text{sat},22}(T)) = \mu_{22f}(T, p_{\text{sat},22}(T))$. Equation (27.76) is known as *Henry's law* and the function $\mathcal{H}_{2f}(T, p)$ as *Henry's constant* for solute 2 in the liquid phase. Henry's constant must be determined empirically for each mixture. For ideal-solution behavior also of constituent 2, and negligible Poynting correction (Section 20.6), Henry's law reduces to Raoult's law, that is, $\mathcal{H}_{2f}(T, p)$ reduces to the vapor pressure $p_{\text{sat},22}(T)$.

Because Raoult's law [equation (27.70)] for the solvent is always satisfied in the limit of very dilute solutions ($y_{1f} \to 1$), it follows that Henry's law [equation (27.76)] is always satisfied in the limit as $y_{2f} \to 0$. For a given value of T, Figure 27.4 shows graphs of

$y_{1g}p$, $y_{2g}p$, and p versus y_{1f} for a mixture of nitromethane, CH_3NO_2 (constituent 1), and carbon tetrachloride, CCl_4 (constituent 2).[6] The three straight lines in Figure 27.4 would represent the graphs of $y_{1g}p$, $y_{2g}p$, and p versus y_{1f} if the liquid phase behaved as an ideal solution, that is, for ideal-solution behavior of each constituent at all compositions. From the figure we see that constituent 1 exhibits ideal-solution behavior only for y_{1f} between about 0.9 and 1, whereas constituent 2 does the same only for $y_{2f} = 1 - y_{1f}$ between about 0.95 and 1. Therefore, Raoult's and Henry's laws hold for this mixture only within these restricted ranges of compositions. The slope of the relation between $y_{1g}p$ and y_{1f} equals $p_{sat,11}(T)$ in the limit as $y_{1f} \to 1$, and $\mathcal{H}_{1f}(T, p_{sat,22}(T))$ in the limit as $y_{1f} \to 0$. Similarly, the slope of the relation between $y_{2g}p$ and $y_{2f} = 1 - y_{1f}$ equals $p_{sat,22}(T)$ in the limit as $y_{2f} \to 1$ ($y_{1f} \to 0$), and $\mathcal{H}_{2f}(T, p_{sat,11}(T))$ in the limit as $y_{2f} \to 0$ ($y_{1f} \to 1$).

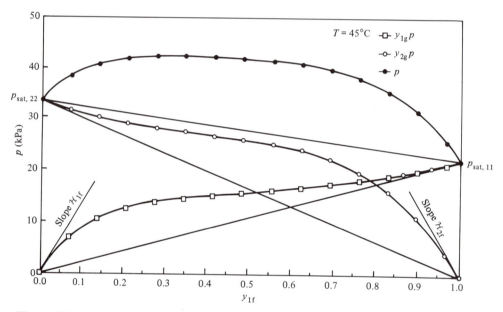

Figure 27.4 Graphs of $y_{1g}p$, $y_{2g}p$, and p versus y_{1f} for a mixture of nitromethane, CH_3NO_2 (constituent 1), and carbon tetrachloride, CCl_4 (constituent 2), at $T = 45°C$ illustrating Raoult's and Henry's laws.

In summary, for a binary mixture in a two-phase liquid–vapor state with ideal-solution behavior of constituent 1, the six variables y_{1g}, y_{2g}, y_{1f}, y_{2f}, T, and p are related by four relations, consistently with the phase rule (Chapter 18) for $r = 2$ and $q = 2$. These relations are

$$y_{1g} + y_{2g} = 1 \quad y_{1f} + y_{2f} = 1 \quad y_{2g}p = y_{2f}\,\mathcal{H}_{2f}(T, p) \quad y_{1g}p = y_{1f}\,p_{sat,11}(T) \quad (27.78)$$

For given values of T and p, the mole fractions are uniquely determined by these equations and are called *solubilities*. For example, y_{1g} is the solubility of liquid constituent 1 in gaseous constituent 2. Again, the value of y_{2f} is the solubility of gaseous constituent 2 in liquid constituent 1. The ratio

---

[6]For illustration purposes, we use this mixture throughout this section and Section 28.4.

$$\alpha_{2,1} = \frac{y_{2g}/y_{2f}}{y_{1g}/y_{1f}} = \frac{1}{\alpha_{1,2}} \tag{27.79}$$

is called the *relative volatility* of constituent 2 to constituent 1 and, for the conditions we are studying, equations (27.78) imply that $\alpha_{2,1} = \mathcal{H}_{2f}(T,p)/p_{sat,11}(T)$.

If a gas phase and a liquid phase that are not in mutual stable equilibrium are placed in contact, the spontaneous process by which the two phases exchange constituents so as to reach mutual stable equilibrium is called *absorption*. Absorption is used in a variety of industrial and laboratory applications that require separation of constituents in a mixture. Another widely used process of separation that we discuss briefly below is *distillation*.

For ideal-solution behavior of both constituents, equations (27.78) hold for all compositions, with $\mathcal{H}_{2f}(T,p) = p_{sat,22}(T)$. Upon combining these equations, we find that

$$p = y_{1f}\,p_{sat,11}(T) + (1 - y_{1f})\,p_{sat,22}(T) \tag{27.80}$$

$$\frac{1}{p} = \frac{y_{1g}}{p_{sat,11}(T)} + \frac{1 - y_{1g}}{p_{sat,22}(T)} \tag{27.81}$$

Using (27.80) and (27.81) for $p_{sat,11}(T) = 22.67$ kPa and $p_{sat,22}(T) = 33.32$ kPa, we obtain the graphs of p versus y_{1f} and p versus y_{1g} in Figure 27.5. Equation (27.80) yields a straight line from $p_{sat,11}(T)$ to $p_{sat,22}(T)$ with slope $dp/dy_1 = p_{sat,11} - p_{sat,22}$. It is called the *bubble line*. Equation (27.81) yields a curve that starts at $y_1 = 0$ with slope $dp/dy_1 = (p_{sat,11} - p_{sat,22})\,p_{sat,22}/p_{sat,11}$, and ends at $y_1 = 1$ with slope $dp/dy_1 = (p_{sat,11} - p_{sat,22})\,p_{sat,11}/p_{sat,22}$. It is called the *dew line*.

Combining equations (27.80) and (27.81) so as to eliminate the variable p, and again using the values of $p_{sat,11}(T) = 22.67$ kPa and $p_{sat,22}(T) = 33.32$ kPa, we find a relation between y_{1g} and y_{1f} that can be solved numerically to yield the graph in Figure 27.6.

Finally, equations (27.80) and (27.81) can be combined with the relations $p_{sat,11} = p_{sat,11}(T)$ and $p_{sat,22} = p_{sat,22}(T)$, and solved for a given value of the pressure p. For $p = 1$ atm and using the relations $p_{sat,11}(T) = \exp(17.08 - 4439/T)$ for nitromethane

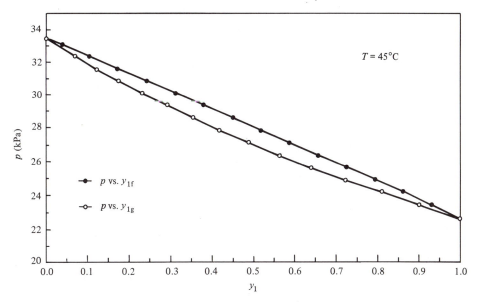

Figure 27.5 Graphs of p versus y_{1f} and p versus y_{1g} based on equations (27.80) and (27.81) with $p_{sat,11}(T) = 22.67$ kPa and $p_{sat,22}(T) = 33.32$ kPa.

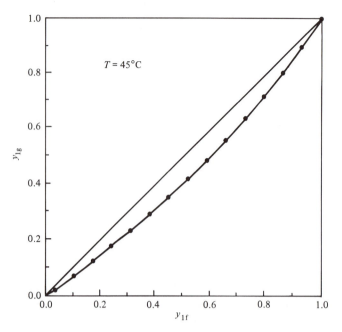

Figure 27.6 Graph of y_{1g} versus y_{1f} based on equations (27.80) and (27.81) with $p_{sat,11}(T) = 22.67$ kPa and $p_{sat,22}(T) = 33.32$ kPa.

and $p_{sat,22}(T) = \exp(16.33 - 4078/T)$ for carbon tetrachloride, we find the graphs of T versus y_{1f} and T versus y_{1g} shown in Figure 27.7. Again, the T versus y_{1f} curve is called the bubble line, and the T versus y_{1g} curve the dew line.

The graphs in Figure 27.7 have the following interpretation. We start with the liquid mixture at low temperature T and a given composition y_{1f} represented by point M. We

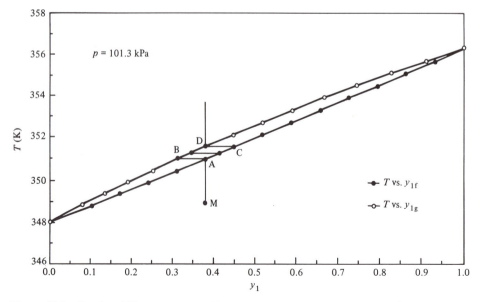

Figure 27.7 Graphs of T versus y_{1f} and T versus y_{1g} for $p = 1$ atm based on equations (27.80) and (27.81), with $p_{sat,11}(T) = \exp(17.08 - 4,439/T)$ and $p_{sat,22}(T) = \exp(16.33 - 4,078/T)$.

increase the temperature while maintaining the pressure fixed. As soon as we reach the temperature corresponding to point A on the bubble line, some vapor begins to form and its mole fraction y_{1g} corresponds to that of point B on the dew line. At point B, the vapor phase is richer in constituent 2 than the liquid phase. As temperature is increased further, the mole fraction y_{1f} increases due to the higher volatility of constituent 2 than of constituent 1, and the liquid phase follows the bubble line up to point C. Correspondingly, the vapor phase follows the dew line up to point D. At point D no liquid is left, and the vapor mole fraction y_{1g} has reached the starting liquid mole fraction. Further increase of temperature is depicted by the vertical line passing through D.

In each stage of a distillation column (Figure 27.8) the liquid to be separated or enriched in one of the constituents is fed to a midstage of the column and brought to mutual stable equilibrium with its vapor phase. The vapor phase, richer in the more volatile constituent, bubbles upward to the upper stage, and the liquid phase, depleted of the more volatile constituent, drains down to the lower stage. When liquid–vapor mutual stable equilibrium, or almost mutual stable equilibrium, is reached at every stage of the column, the vapor outflow from the top stage is richer in the more volatile constituent than the liquid outflow from the bottom stage.

More specifically, for the kth stage we have $(y_{1g}/y_{2g})_k = (\alpha_{1,2})_k\,(y_{1f}/y_{2f})_k$, where $(\alpha_{1,2})_k$ is the relative volatility at the conditions of stage k. Assuming that no vapor

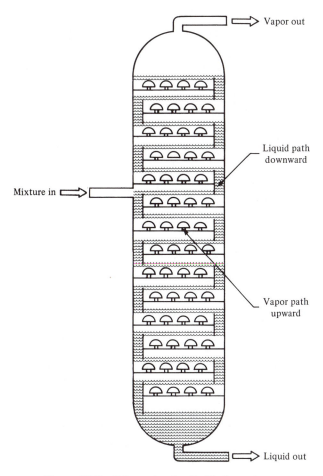

Figure 27.8 Schematic of a distillation column.

is extracted from the top and no liquid from the bottom of the distillation column—a condition called *total reflux operation*—a mass balance for constituent 1 at each stage yields that $(y_{1f})_k = (y_{1g})_{k-1}$ or $(y_{1f}/y_{2f})_k = (y_{1g}/y_{2g})_{k-1}$, so that $(y_{1g}/y_{2g})_k = (\alpha_{1,2})_k (y_{1g}/y_{2g})_{k-1} = (\alpha_{1,2})_k (\alpha_{1,2})_{k-1} (y_{1f}/y_{2f})_{k-1} = \cdots$, and therefore, $(y_{1f}/y_{2f})_{top} = \prod_{k=1}^{K}(\alpha_{1,2})_k (y_{1f}/y_{2f})_{bottom}$, where K is the number of stages. This result is known as the *Fenske relation*. Assuming that $\alpha_{1,2}$ is the same for all stages, the Fenske relation yields the number K of distillation plates necessary to achieve the desired separation. As the condition of total reflux is not satisfied because some distillate is withdrawn from the top and from the bottom of the column, the number of plates that are actually needed for a given separation is higher.

Example 27.7. Estimate the relative volatility of benzene to bromobenzene and the solubilities of benzene in gaseous and liquid bromobenzene at 80°C and 50 kPa, assuming that the saturated-vapor pressures of pure benzene and pure bromobenzene are, respectively, 100 and 10 kPa.

Solution. Denoting benzene as constituent 1 and assuming ideal-solution behavior, the relative volatility $\alpha_{1,2} = p_{sat,11}/p_{sat,22} = 100/10 = 10$. Equation 27.80 yields the solubility of benzene in liquid bromobenzene, $y_{1f} = (p - p_{sat,22})/(p_{sat,11} - p_{sat,22}) = (50 - 10)/(100 - 10) = 0.444$, and (27.81) or (27.70) the solubility of benzene in gaseous bromobenzene $y_{1g} = y_{1f}p_{sat,11}/p = 0.444 \times 100/50 = 0.888$.

The boiling temperature of a solution depends on the mole fraction of the solute. To see this clearly, we consider a binary mixture at conditions such that both the pressure p and the saturated-vapor pressure $p_{sat,11}$ of constituent 1 are much higher than the saturated-vapor pressure $p_{sat,22}$ of constituent 2 and hence than Henry's constant $\mathcal{H}_{2f}$. For example, this condition is valid for a dilute mixture of water and salt at ordinary pressures and temperatures. Under these conditions, equation (27.80) reduces to

$$p = y_{1f} p_{sat,11}(T) \tag{27.82}$$

and equation (27.81) implies that $y_{1g} \approx 1$, that is, $y_{2g} \ll 1$. We conclude that constituent 2 has a negligible relative volatility in constituent 1, and hence the vapor phase consists of pure constituent 1. Thus, for given p and y_{1f}, equation (27.82) defines the boiling temperature T of the solution.

Upon differentiating equation (27.82) and using the Clausius–Clapeyron relation [equation (19.10)] for pure constituent 1, $d \ln p_{sat,11}/dT = h_{fg,11}/Tp_{sat,11}v_{fg,11} \approx h_{fg,11}/RT^2$, we find that

$$\frac{dp}{p} = \frac{dy_{1f}}{y_{1f}} + \frac{d \ln p_{sat,11}}{dT} dT = -\frac{dy_{2f}}{y_{1f}} + \frac{h_{fg,11}}{RT^2} dT \tag{27.83}$$

where we use the relations $p_{sat,11}v_{fg,11} \approx RT$, $dy_{1f} = -dy_{2f}$. Thus we have

$$-\left(\frac{\partial T}{\partial y_{1f}}\right)_p = \left(\frac{\partial T}{\partial y_{2f}}\right)_p = \frac{RT^2}{y_{1f} h_{fg,11}(T)} \tag{27.84}$$

From equation (27.84) we see that an increase in the mole fraction of the nonvolatile solute in the liquid phase causes the boiling temperature of the solution to rise by an amount related to the enthalpy of vaporization of the volatile solvent. This result, usually referred to as *boiling-point raising*, has been used as the basis of experimental methods to find the enthalpy of vaporization of the solvent, and the molecular mass of the solute from relatively simple measurements of boiling temperatures. It is noteworthy that (27.84) is independent of the properties of the solute and hence can be applied as well to any dilute solution of solvent 1 and $r - 1$ different solutes.

Example 27.8.   Find the boiling temperature of a solution of 150 g of NaCl ($M = 58.5$ kg/kmol) in 1 kg of H_2O at atmospheric pressure.

Solution.   At atmospheric pressure, the boiling temperature of pure water ($y_{1f} = 1$) is $T_{b0} = 100°C$. Assuming that the right-hand side remains approximately constant, we can integrate equation (27.84) to its value for pure water, that is, $T_b - T_{b0} \approx RT_{b0}^2 y_{2f}/h_{fg,11}(T_{b0}) \approx 0.461 \times (373)^2 y_{2f}/2257 \approx y_{1f} \times 28°C$. For the given conditions, $y_{2f} = 150/58.5/(1000/18+150/58.5) = 0.044$ and hence $T_b = 100 + 0.044 \times 28 = 101.2°C$.

27.8   Two-Phase Liquid–Solid States

Solid solutions are of interest in metallurgy and other fields. Here, however, we restrict the discussion to a pure solid phase in mutual stable equilibrium with a dilute liquid solution. Assuming ideal-solution behavior for constituent 1, we rewrite (27.68) as

$$\mu_{1f} = \mu_{11f}(T, p_{sat,11}^s(T)) + v_{11f}\left[p - p_{sat,11}^s(T)\right] + RT \ln y_{1f} \tag{27.85}$$

where $p_{sat,11}^s(T)$ denotes the saturated-solid pressure at temperature T as defined by the chemical potential equality condition

$$\mu_{11f}(T, p_{sat,11}^s(T)) = \mu_{11s}(T, p_{sat,11}^s(T)) \tag{27.86}$$

for temperatures T greater than the triple-point temperature $T_{tp,11}$ of constituent 1.

For the pure solid phase at T and p, we write

$$\mu_{11s} = \mu_{11s}(T, p_{sat,11}^s(T)) + v_{11s}\left[p - p_{sat,11}^s(T)\right] \tag{27.87}$$

Thus the chemical potential equality $\mu_{1f} = \mu_{11s}$ yields the relation

$$RT \ln y_{1f} = (v_{11s} - v_{11f})\left[p - p_{sat,11}^s(T)\right] = -v_{fs,11}\left[p - p_{sat,11}^s(T)\right] \tag{27.88}$$

which defines the freezing temperature T of the solution. Differentiating this expression, assuming that $v_{fs,11} = $ constant, and using the Clausius–Clapeyron relation, we find that

$$R \ln y_{1f}\, dT + RT\, \frac{dy_{1f}}{y_{1f}} = -v_{fs,11}\, dp + v_{fs,11}\, \frac{dp_{sat,11}^s}{dT}\, dT$$

$$= -v_{fs,11}\, dp + \frac{h_{sf,11}}{T}\, dT \tag{27.89}$$

Because the solution is dilute, we can neglect $R \ln y_{1f}$ compared to $h_{sf,11}/T$ and therefore find that

$$-\left(\frac{\partial T}{\partial y_{1f}}\right)_p = \left(\frac{\partial T}{\partial y_{2f}}\right)_p = -\frac{RT^2}{y_{1f}\, h_{sf,11}(T)} \tag{27.90}$$

where we use the relation $dy_{1f} = dy_{2f}$. From equation (27.90) we see that an increase in the mole fraction of solute 2 causes a depression of the freezing temperature of the solution. This result is referred to as the *freezing-point depression* and is used as the basis of experimental methods to determine the enthalpy of solidification of the solvent and the molecular mass of the solute. The freezing-point depression is exploited in our everyday life when salts are spread on the roads in wintertime to reduce the freezing temperature of the resulting aqueous solution.

Example 27.9.   Find the freezing temperature of a solution of 150 g of NaCl ($M = 58.5$ kg/kmol) in 1 kg of H_2O at atmospheric pressure.

Solution.   At atmospheric pressure, the freezing or melting temperature of pure water ($y_{1f} = 1$) is $T_{m0} = 0°C$. Assuming that the right-hand side remains approximately constant, we can integrate equation (27.90) to its value for pure water, that is, $T_m - T_{m0} \approx -RT_{m0}^2 y_{2f}/h_{sf,11}(T_{m0}) \approx 0.461 \times (273)^2 y_{2f}/334 \approx -y_{2f} \times 103°C$. For the given conditions, $y_{2f} = 0.044$ and hence $T_m = 0 - 0.044 \times 103 = -4.5°C$.

Problems

27.1 A mixture consists of nitrogen, N_2, argon, Ar, and hydrogen, H_2. The partial pressures are: $p(N_2) = 0.5$ bar, $p(Ar) = 1$ bar, and $p(H_2) = 0.5$ bar. If it behaves as an ideal-gas mixture of perfect gases, determine: **(a)** the molecular weight of the mixture; and **(b)** the volume of 50 kg of the mixture at 25°C.

27.2 On a mole basis, air may be regarded as consisting of 21% O_2, and 79% N_2.

(a) Assuming that air behaves as an ideal-gas mixture of perfect gases at atmospheric temperature (25°C) and pressure (1 atm), calculate the least work required to separate 1 kg of pure O_2 from the mixture at atmospheric temperature and pressure.

(b) On a mole volume basis, if the fraction of oxygen is y_1, make a graph of the least work required to obtain 1 kg of pure O_2 from the mixture at atmospheric conditions as a function of y_1. From this graph, what conclusion can you draw concerning the extraction of rare gases from the atmosphere?

27.3 Consider two different, chemically inert gases A and B each in a separate container. The two gases are at the same temperature T_1 and pressure p_1, but have volumes V_1^A and V_1^B and numbers of moles n^A and n^B, respectively. The behavior of each gas is perfect.

(a) If an energy source at temperature T_1 is available, what is the largest work that can be done using the isothermal expansions of gas A from volume V_1^A to a volume $V_2^A = V_1^A + V_1^B$, and of gas B from volume V_1^B to a volume $V_2^B = V_1^A + V_1^B$?

(b) How much energy and entropy must be supplied to the two gases to bring about the changes in part (a)?

(c) With respect to an environment at temperature T_{R*}, what is the availability associated with the energy used in part (a)?

(d) What is the reduction in available energy of the two gases as a result of the two processes in part (a)?

(e) If a mixture consists of the two amounts of gases just specified, and has a volume $V = V_1^A + V_1^B$, temperature T_1, and pressure p_1, what is the difference between the available energy of the mixture and that of the two gases at the end of the processes in part (a)?

27.4 A rigid, perfectly insulated tank with volume $V = 1$ m^3 is connected, via two valves, to two separate supply lines, A and B. Line A supplies gaseous isooctane (C_8H_{18}) at constant pressure $p_A = 500$ kPa and constant temperature $T_A = 70$°C. Line B supplies oxygen (O_2) at constant pressure $p_B = 200$ kPa and constant temperature $T_B = 15$°C.

Initially, the tank is entirely evacuated and the two valves are closed. A controller operates the valves on lines A and B so that they open and close simultaneously. The valves are opened for a short time and then closed when the pressure and the temperature in the tank are $p_F = 100$ kPa and $T_F = 107$°C. No chemical reaction mechanism is active between the two gases.

Assume that both isooctane and oxygen behave as perfect gases in the range of conditions that apply to this problem, and that their mixture behaves as an ideal-gas mixture. For isooctane, the specific heat ratio $\gamma = 1.1$. For oxygen, $\gamma = 1.4$.

(a) After the valves are closed, how many moles, n, of mixture are in the tank?

(b) After the valves are closed, what are the mole fractions of isooctane and oxygen in the tank?

(c) How much entropy is generated by irreversibility during the tank-filling process?

27.5 A mixture of water vapor and other substances is at 1000 K and 50 MPa. Through a semipermeable membrane that is permeable only to water molecules, the mixture is in partial mutual stable equilibrium with pure water vapor at 5 MPa. Suddenly, the pressure of the pure water vapor outside the membrane is increased to 6 MPa.

(a) In which direction can water vapor flow spontaneously across the semipermeable membrane? Justify your answer quantitatively.

(b) If the mixture behaves as an ideal-gas mixture and consists of 5 kg of water vapor and 40 kg of other substances, what is the value of the volume of the mixture? What are the partial pressure and the specific volume (per unit mass) of the other substances?

27.6 The atmosphere is not a simple system because of the presence of the gravitational field. However, within each layer of atmosphere, between elevations z and $z + dz$, the gravitational potential may be considered as uniform. Thus we can model the atmosphere as a stratification of many thin layers, and each layer as a simple multiconstituent system with a superimposed uniform gravity field. The fundamental relation between energy E_z, entropy S_z, volume V_z, elevation z, and amounts of constituents $n_{z1}, n_{z2}, \ldots, n_{zr}$ of the layer between z and $z + dz$ can be written as

$$E_z = U(S_z, V_z, n_{z1}, n_{z2}, \ldots, n_{zr}) + m_z g z$$

where $U(S, V, n_1, n_2, \ldots, n_r)$ is the fundamental relation of a simple multiconstituent system in the absence of gravity.

(a) Show that the total potential of the ith constituent in the layer between z and $z + dz$ is given by

$$\mu_{iz} = \mu_i(T_z, p_z, y_{z1}, y_{z2}, \ldots, y_{zr}) + M_i g z$$

where M_i is the molecular weight of the ith constituent, and $\mu_i(T, p, y_1, y_2, \ldots, y_r)$ is the chemical potential of the ith constituent in the absence of gravity.

The so-called "isothermal atmosphere" consists of a stratification of thin layers, each modeled as just described and all in mutual stable equilibrium. Moreover, in the absence of gravity, each layer is assumed to behave as a Gibbs–Dalton mixture of perfect gases. Clearly, this model neglects all the temperature nonuniformities and the flows that actually occur in the real atmosphere.

(b) Show that for an isothermal atmosphere the partial pressure of the ith constituent as a function of the elevation z is given by the relation

$$p_{iiz} = p_{iio} \exp\left(-\frac{M_i g z}{RT}\right)$$

where p_{iio} is the partial pressure at sea level.

(c) For an isothermal atmosphere at 288.15 K with mole fractions at sea level ($z = 0$) given by $y_{10} = 0.21$ of oxygen and $y_{20} = 0.79$ of nitrogen, find expressions for the atmospheric pressure p_z and the oxygen mole fraction y_{1z} as functions of the elevation z. What is the pressure and oxygen mole fraction at the top of Mount Everest ($z = 8848$ m) according to this model?

A better model consists of considering the atmospheric temperature as a function of elevation, that is,

$$T = T_o - az$$

where $T_o = 288.15$ K and $a = 6.5$ K/km, as derived from worldwide-average temperature measurements.

(d) Show that the expression for the partial pressure of the ith constituent as a function of elevation becomes

$$p_{iiz} = p_{iio} \exp\left[-\frac{M_i g}{Ra} \ln \frac{1}{1 - (az/T_o)}\right]$$

(e) What is the pressure and oxygen mole fraction at the top of Mount Everest according to the better model?

(f) Compare the values of the pressure at $z = 8848$ m found in parts (c) and (e) with the measured worldwide-average atmospheric pressure as a function of elevation given by the relation $p = 101.325 (1 - 0.02256 z)^{5.256}$, where p is in kPa and z in km.

27.7 When an almost empty automobile gas tank is filled with gasoline, the air and fuel–vapor mixture in the tank is expelled into the atmosphere. The hydrocarbons thus expelled are a source of air pollution.

(a) If the pressure of saturated vapor of gasoline at the tank temperature of 27°C is 6.1 kPa, estimate the maximum mass of gasoline that would be emitted when filling an almost empty 75-liter tank. Assume that gasoline has a formula C_8H_{15}.

(b) With an appropriate assumption about fuel economy, find the equivalent hydrocarbon emissions (in grams per mile traveled) due to filling the tank. The U.S. Environmental Protection Agency (EPA) has set exhaust emission standards for hydrocarbons at 0.41 g/mile. How do these two emission rates compare?

27.8 A steady flow of a gas at 900°C and 1 atm is cooled by a spray of liquid water at 10°C. The result of this constant-pressure evaporation and mixing process is the steady flow of a mixture of the gas and water vapor at 650°C and 1 atm. This mixture behaves as a Gibbs–Dalton mixture of ideal gases, the gas is perfect with $R = 0.3$ kJ/kg K and $c_p = 1.1$ kJ/kg K, and water vapor at 650°C and 1 atm behaves as an ideal (but not perfect) gas.

(a) What is the mass flow rate of the liquid water spray as a fraction of the mass flow rate of gas?

(b) What is the partial pressure of steam in the resulting mixture?

(c) How much entropy is generated by irreversibility per kilogram of gas entering the spray cooler?

27.9 For the purposes of this problem, air can be modeled as a Gibbs–Dalton mixture of perfect gases consisting of oxygen and nitrogen with mole fractions, molecular weights, and specific-heat ratios, respectively, $y_1 = 0.21$, $y_2 = 0.79$, $M_1 = 32$ kg/kmol, $M_2 = 28$ kg/kmol, and $\gamma_1 = \gamma_2 = 1.4$.

(a) What is the least work required for completely separating 1 kg of air at 25°C and 1 atm into pure oxygen and pure nitrogen each at 25°C and 1 atm?

(b) What is the least work required for separating 0.233 kg of pure oxygen (at 25°C and 1 atm) out of a very large amount of air at 25°C and 1 atm?

(c) What is the least work required for separating 0.767 kg of pure nitrogen (at 25°C and 1 atm) out of a very large amount of air at 25°C and 1 atm?

(d) What is the least work (per kg of air) required for separating a stream of air at 25°C and 1 atm into a stream of pure oxygen and a stream of pure nitrogen, each at 25°C and 1 atm?

(e) What is the least work (per 0.233 kg of pure oxygen) required for separating a small stream of pure oxygen at 25°C and 1 atm out of a very large stream of air at 25°C and 1 atm?

(f) What is the least work (per 0.767 kg of pure nitrogen) required for separating a small stream of pure nitrogen at 25°C and 1 atm out of a very large stream of air at 25°C and 1 atm?

(g) Compare all your answers in parts (a) to (e) and comment.

27.10 A steady-state air-conditioning device operates by drawing a stream of 20 kg/h of air at 35°C and 80% relative humidity into a heat exchanger where the air is cooled and dehumidified to 27°C and 100% relative humidity. Condensate at 27°C is collected at the bottom of the heat exchanger and discharged.

(a) At what rate is condensate water removed from the heat exchanger?

(b) At what rate is energy removed from the air stream in the heat exchanger?

(c) If cooling is achieved by water flowing in a serpentine of the heat exchanger, entering at 15°C and exiting at 20°C, at what rate is entropy created by irreversibility in the heat exchanger?

27.11 The concentration of helium, He, in the atmosphere is about 1 part in 1,400,000 by weight at sea level. To investigate how this varies with elevation, consider an isothermal atmosphere at $T = 290$ K, and assume that it consists of air and helium. Air may be regarded as a mixture of

N_2 and O_2, and be treated as a perfect gas with a molecular weight of 28.9. The molecular weight of helium is 4. Problem 27.6 provides useful ideas.

(a) Find an expression for the variation of He concentration with elevation.

(b) What is the concentration of He in the atmosphere at an elevation of 8848 m?

27.12 Two streams, one of hydrogen and another of nitrogen, are mixed in a steady-flow well-insulated mixer in the ratio 5 mol of hydrogen per mole of nitrogen. Hydrogen enters at 2 bar and 50°C, and nitrogen at 2 bar and 150° C. After mixing, the pressure in the outlet stream is 1.8 bar. Assume that all streams behave as perfect gases.

(a) What is the temperature of the exit stream?

(b) What is the entropy change between outlet and inlet streams per mole of mixture?

(c) Is the process reversible?

27.13 Volumetric analysis of an ideal-gas mixture yields 60% N_2, 22% CO_2, 11% CO, and 7% O_2. Data about these gases are listed in Table 20.1.

(a) Determine the composition of the mixture on a mass basis.

(b) What is the mass of 0.03 m^3 of the mixture when the pressure is 1 bar and the temperature is 25°C?

(c) The mixture in part (b) is heated in a bulk-flow process from an initial temperature of 40°C to 260°C. Determine the energy required for the heating and the entropy change of the mixture.

27.14 A stream of 4 mol of N_2, and a stream of 1 mol of O_2, each at standard conditions, $T_0 = 25°C$ and $p_0 = 1$ atm, are mixed and form a mixture at the same temperature and pressure.

(a) What is the entropy of mixing per mole of air formed? Assume that each component of the mixture is a perfect gas.

(b) What is the least work required to separate 1 mol of the mixture into N_2 and O_2?

(c) Calculate the cost of the work required to recover 10^3 kg of pure oxygen from the atmosphere ($T_0 = 300$ K, $p_0 = 1$ atm). The process is only 15% effective, and the electricity costs $0.15 per kilowatthour. On a mole basis, assume that the atmosphere consists of 79% N_2 and 21% O_2 only.

27.15 In areas where the temperature is high and humidity is low, some measure of air conditioning can be achieved by evaporative cooling. Water sprayed into the air evaporates and results in a decrease in the temperature of the mixture. Such a scheme is shown in Figure P27.15. Assume that the atmospheric air is at 38°C, 10% relative humidity, and 1 atm, and that cooling water at 10°C is sprayed into the air.

(a) If the air–water vapor mixture is cooled to 27°C, what is the relative humidity?

(b) What is the lowest temperature to which the air can be cooled?

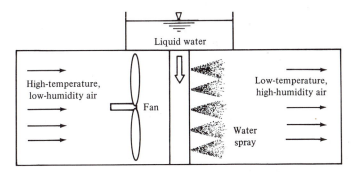

Figure P27.15

27.16 A steady-state air-conditioning device operates by spraying liquid water at atmospheric temperature and pressure onto a stream of atmospheric air. If the atmospheric air is not saturated, that is, the relative humidity is less than 100%, the evaporation of the water droplets draws energy from the air stream and reduces the temperature. If the atmospheric-air stream enters the spray at 35°C and leaves at 25°C and 90% relative humidity, what is the relative humidity of the atmospheric air?

27.17 A cooling tower of a nuclear power plant is shown schematically in Figure P27.17. It transfers energy at the rate of 2 GW from the condenser to the air in the atmosphere. Atmospheric air at 21°C and 60% relative humidity is drawn into the bottom of the tower by large fans. From the condenser, warm liquid water at 40°C and atmospheric pressure is sprayed from showers at the top of the tower and breaks up into small droplets that mix effectively with the upward stream of air, so that air leaves the top of the cooling tower at 35°C and 100% relative humidity. Cold water collects at the bottom of the tower at 20°C. Because some water is lost to the atmosphere, makeup water at 14°C from a nearby small river is added at the bottom of the tower. The cooled water is circulated back to the condenser.

(a) Find the mass flow rates of dry air in the air stream, of water from and to the condenser, and of water lost to the atmosphere and made up from the nearby river.

(b) At what rate is entropy generated by irreversibility within the cooling tower?

(c) At what rate is entropy generated by irreversibility in discharging the warm and saturated air from the tower into the atmosphere?

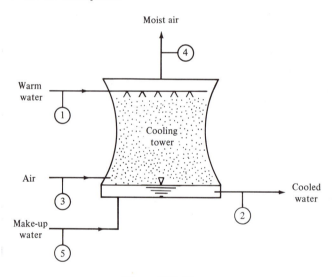

Figure P27.17

27.18 A stream of humid air at 50°C, 800 kPa, 10% relative humidity, and negligible speed is expanded to 100 kPa. To simplify the numerical solution of this problem, approximate the enthalpy of vaporization and the pressure–temperature saturation relation by the expressions

$$h_{\mathrm{fg}} = 3328.4 - 2.9\,T$$

$$p_{\mathrm{sat}}(T) = 101.3 \exp\left(56.54 - \frac{7212}{T} - 6.284 \ln T\right)$$

where h_{fg} is in kJ/kg, p is in kPa, and T in kelvin.

(a) If the expansion occurs in a throttle at constant enthalpy, what is the condition of the outlet stream? Is there any condensate?

(b) If the expansion occurs in a perfect nozzle at constant entropy, what is the condition of the outlet stream? Is there any condensate?

(c) If the expansion occurs in an isentropic turbine, what is the condition of the outlet stream, and how much work is done by the turbine per kilogram of dry air? Is there any condensate?

27.19 At each elevation, atmospheric air can be modeled as an ideal binary mixture in a gravity field with constant acceleration g (see Problem 27.6). Similarly, ocean water can be modeled as an ideal binary mixture of salt and water in a gravity field with constant acceleration g. At each depth the specific energy e_z of the mixture in the layer can be written as $e_z = u(s_z, v_z, y_z) + M_z g z$, where M_z is the mean molecular weight of the mixture within the layer and $u(s, v, y)$ the energy relation of a simple system in the absence of gravity. For each layer, we denote the temperature, pressure, and chemical potentials by the symbols T_z, p_z, and $\mu_{1,\text{chem}}(T_z, p_z, y_z)$ and $\mu_{2,\text{chem}}(T_z, p_z, 1 - y_z)$. For salt, assume that the density $\rho_1 = M_1/v_1 = 2.165\, M_2/v_2$ and the molecular weight $M_1 = 58.5$.

(a) Show that the total potential of salt at depth z is given by $\mu_{1,\text{tot}} = \mu_{1,\text{chem}}(T_z, p_z, y_z) + M_1 g z$, where M_1 is the molecular weight of salt.

(b) For a column of seawater at constant temperature T, show that the mole fraction of salt satisfies the relation $(y_z/y_0)\,[(1 - y_0)/(1 - y_z)]^{v_1/v_2} = \exp[(M_1 - M_2 v_1/v_2)(-z)g/RT]$.

(c) Estimate the concentration of salt at the bottom of the Mariana Trench ($z = -11{,}033$ m) assuming a constant temperature of $10°C$. At $z = 0$, assume that $y_0 = 0.011$ (3.5% by weight).

27.20 An oxygen manufacturing plant processes 10,000 mol of air per hour in an environment at $T_0 = 20°C$. Oxygen exits the plant at $T_1 = 27°C$ and $p_1 = 5$ atm, and nitrogen at $T_2 = 30°C$ and $p_2 = 2$ atm. On a mole basis, assume that air consists of 21% oxygen and 79% nitrogen only.

(a) What is the smallest amount of power required to accomplish this separation?

(b) What fraction of this power is due to pressure differences, what fraction to temperature differences, and what fraction to the unmixing process?

27.21 An oxygen separation plant handling 10,000 lbmol of air per hour (Figure P27.21) is operating in an environment at $25°C$ and 1 atm, under the conditions listed in the accompanying table. The air blower, the two-stage compressor, and the expander each has an efficiency of 0.75.

(a) What is the smallest work rate to carry out the process?

(b) What are the rates of work done by the air blower and the nitrogen two-stage compressor? What is the rate of work returned to the process by the expander (turbine)?

(c) What is the effectiveness of the process?

Stream	Description	y_{N_2} %	$\dot{n}$ lbmol/h	T °C	p atm	h kJ/lbmol	s kJ/lbmol K
1	Column feed	79	10,000	−189.7	1.27	1,053	72.033
2	Heated overhead	99	16,727	−190.9	1.14	1,041	69.321
3	Column bottom	10	2,247	−181.6	1.27	1,150	76.794
4	Remade nitrogen	99	6,638		5.1	1,082	64.784
5	Reflux	99	6,638	−178.8	5.03	−1,110	41.514
6	Cooled reflux	99	6,638	−183.9	4.96	−1,248	39.983
7	Expanded reflux	99	6,638	−194.6	1.2	−1,248	40.211
8	Column overhead	99	14,391	−194.4	1.2	986	68.408
9	Expanded nitrogen	99	2,336	−194.4	1.2	986	68.408
10	8 + 9	99	16,727	−194.4	1.2	986	68.408
11	Liquid from tower	10	4,813	−183.3	1.34	−1,869	43.536
12	Product nitrogen	99	7,753	22.2	1	3,895	87.050
13	Product oxygen	10	2,247	22.2	1.14	3,890	93.059
14	Initial feed	79	10,000	25	1	3,932	90.165
15	Exchanger feed	79	10,000	25	1.41	3,932	88.873
16	Compressed nitrogen	99	8,974	25	5.24	3,920	80.897
17	Recycle nitrogen	99	8,974	22.2	1	3,895	87.050
18	Feed to expander	99	2,336	−160.8	5.17	1,378	67.392
19	Overhead leak	99	16,727	−189.8	1.14	1,063	69.551

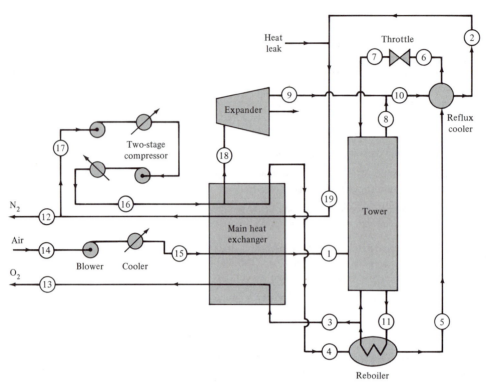

Figure P27.21

(d) What are the rates of entropy generated by irreversibility in the air blower, air cooler, nitrogen compressor, nitrogen cooler, main exchanger, expander, tower, reboiler, reflux cooler, valve, and heat leak?

(e) What is the work rate that must be done to overcome these irreversibilities?

(f) How does the sum of your answers in parts (a) and (e) compare with your answer in part (b)?

27.22 Repeat Problem 27.21 for the ideal oxygen separation plant shown in Figure P27.22 and the data listed in the accompanying table.

Stream	Description	y_{N_2} %	$\dot{n}$ lbmol/h	T °C	p atm	h kJ/lbmol	s kJ/lbmol K
1	Column feed	79	10,000	−191.1	1	1,036	72.756
2	Heated overhead	99	13,833	−191.7	1	1,032	69.691
3	Column bottom	10	2,247	−183.8	1	1,127	77.434
4	Recycle nitrogen	99	6,080	−183.8	3.26	1,060	66.015
5	Reflux	99	6,080	−185.6	3.26	−1,293	39.514
6	Cooled reflux	99	6,080	−190.4	3.26	−1,423	37.988
7	Expanded reflux	99	6,080	−195.7	1	−1,423	38.046
8	Column overhead	99	13,833	−195.6	1	975	68.979
9	Product nitrogen	99	7,753	24.8	1	3,930	87.194
10	Product oxygen	10	2,247	24.8	1	3,926	93.681
11	Initial feed	79	10,000	25	1	3,932	90.153
12	Compressed nitrogen	99	6,080	25	3.26	3,926	82.690
13	Recycle nitrogen	99	6,080	24.8	1	3,930	87.194
14	Liquid from tower	10	4,650	−185.6	1	−1,941	42.640

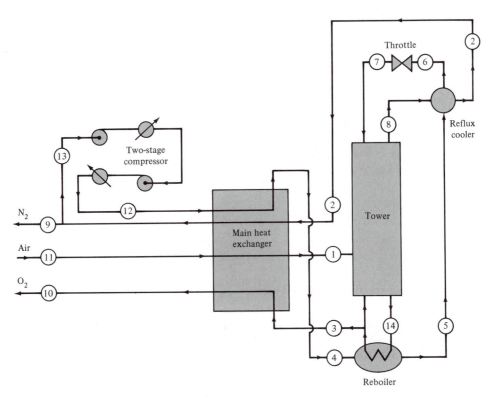

Figure P27.22

27.23 Calcium hypochlorite is produced by absorbing chlorine in milk of lime. A chlorine-rich gas mixture of inert gases and chlorine enters the absorption apparatus at 740 mm Hg and 24°C. The partial pressure of chlorine is 59 mm Hg. At the exit of the absorption apparatus the temperature is 26.7°C, the pressure 743 mm Hg, and the partial pressure of chlorine 0.5 mm Hg.

(a) Calculate the volume of gases leaving the apparatus per cubic meter entering.

(b) Calculate the weight of chlorine absorbed per cubic meter of gas mixture entering.

27.24 Industrial dryers use large amounts of energy to evaporate water or other solvents from products. In most cases, the vapor produced in the dryer goes to waste. In the United States, over 10^{15} kJ of energy is used each year for drying in the food, textile, and paper industries. In a typical dryer, shown schematically in Figure P27.24a, air flows across the product, picks up the vapor, and exhausts it. Before it is heated, this air is often mixed with recirculated exhaust vapor. Because the absolute humidity of the exhaust is low, ranging from 5 to 40%, the latent heat can be recovered at relatively low temperature, and therefore is not useful. By replacing the air with superheated steam as the drying medium, we can raise the temperature at which the latent heat can be recovered so that recovery becomes useful. A method for achieving this recovery, and therefore improving the energy use in drying, is shown schematically in Figure P27.24b. Superheated steam in state 1 at atmospheric pressure p_1 and temperature T_1, and rate $\dot{m}_s$ is fed into the dryer by a circulation blower. As it cools, the steam removes water from the product and the product exits almost free of water. The steam and evaporate leave the dryer at a slightly superheated state 2. The evaporate $\dot{m}_w$ is split off the main steam flow, compressed to pressure p_4 (state 4), condensed in the heat exchanger to state 5, and rejected to the atmosphere in state 6 after isenthalpic throttling. The main steam flow $\dot{m}_s$ is heated in the heat exchanger and recirculated back in the dryer. The power of the blower is negligible.

(a) As an aside that is not needed for this problem, answer the following questions: (1) What is the annual energy consumption in your country? (2) What are the percentages of the various

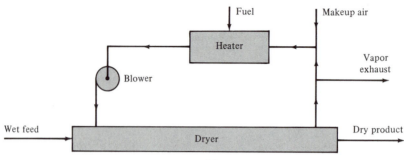

(a)
Figure P27.24a

energy sources used? (3) What fraction of energy is used for the generation of electricity? The answers help your understanding of the energy issues, including whether 10^{15} kJ is a large or small amount of energy.

(b) What is the energy rate supplied by the recirculating steam to the dryer under the conditions specified in Figure P27.24b?

(c) If the pinch temperature (Figure P27.24c) in the heat exchanger is 20°C, the efficiency of the compressor $\eta_c = 0.85$, and the pressure of the compressed evaporate $p_4 = 1$ MPa, what is the flow rate $\dot{m}_s$ of the recirculating steam? The pinch temperature is defined in Figure P27.24c.

(d) How much entropy is generated by irreversibility in the heat exchanger?

(e) What is the temperature T_1 of the recirculating steam? Neglect any effect of the blower.

(f) What is the power input to the compressor?

(g) Specify fully the condition of discharge (state 6).

(h) Verify that your calculations are correct by making an overall energy balance of the superheated-steam dryer.

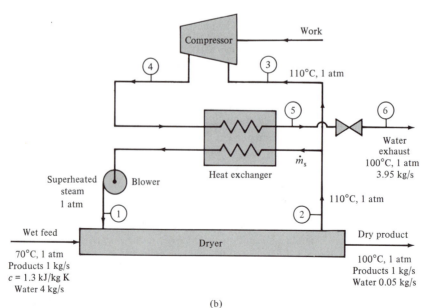

(b)
Figure P27.24b

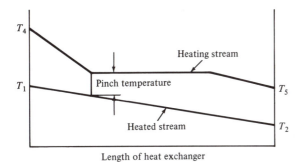

Figure P27.24c

(i) Assuming that drying with air is done with an overall energy efficiency of 0.5 (between process stream and fuel), and that power for the compressor of the superheated-steam dryer is generated with an effectiveness of 1/3, how much fuel is saved per kilogram of product being dried when superheated steam is used?

27.25 Tropical air at 40°C and 100% relative humidity is to be cooled and dehumidified.

(a) Find the mass of water per unit mass of air at these tropical ambient conditions.

(b) Find the least work per kilogram of dry air required in a steady-flow process to produce dry air at 20°C and liquid water at 40°C.

27.26 A mixture consists of 50 kmol of O_2 and 50 kmol of N_2 at environmental temperature $T_0 = 20°C$ and pressure $p_0 = 1$ atm. The only source of availability is a hot air stream at $T = 827°C$ and $p = 15$ atm.

(a) What is the smallest mass flow rate, $(\dot{m}_g)_s$, of the air stream that can be used in perfect equipment to separate 100 kmol/h of the mixture into its constituents each at $T_0 = 20°C$ and $p_0 = 1$ atm?

(b) If the available equipment generates entropy at the rate 2600 kJ/h K, what is the necessary mass flow rate, $\dot{m}_g$, of the hot gaseous stream to achieve the separation of the mixture?

(c) What is the effectiveness of the equipment in part (b)?

27.27 Many pairs of metals are miscible in the liquid phase but immiscible in the solid phase, that is, they form a single phase when mixed as liquids but separate in two different solid phases upon solidification of their liquid mixture.

(a) What is the variance F of a three-phase state in which the liquid-mixture phase coexists with both solid phases of a two-constituent simple system?

(b) Assuming ideal-solution behavior for the liquid phase at all compositions, show that at a given pressure p and a given temperature T the *solubility* y_{1f} of the solid phase of constituent 1 in the binary liquid mixture of the two constituents is given by

$$y_{1f} = \exp\left\{\frac{s_{sf,11}(p)}{R}\left[1 - \frac{T^s_{sat,11}(p)}{T}\right]\right\}$$

where $s_{sf,11}(p)$ is the specific entropy of melting of solid 1, and $T^s_{sat,11}(p)$ the melting temperature at pressure p. Write the analogous relation for constituent 2.

(c) For copper, Cu, the enthalpy of melting $h_{sf}(1 \text{ atm}) = 13$ kJ/mol and the melting temperature $T^s_{sat}(1 \text{ atm}) = 1083°C$. For aluminum, Al, $h_{sf}(1 \text{ atm}) = 10.7$ kJ/mol and the melting temperature $T^s_{sat}(1 \text{ atm}) = 660°C$. Estimate the solubility of solid aluminum in a liquid copper-aluminum mixture at 600°C and 1 atm. Does your answer depend on any of the properties of copper?

(d) The *eutectic composition* is the only composition at which the liquid phase can coexist at a given pressure with both the solid phase of constituent 1 and the solid phase of constituent 2. The temperature T_E of such a three-phase state is called the *eutectic temperature*. Show that the eutectic temperature satisfies the relation

$$\exp\left\{\frac{s_{sf,11}(p)}{R}\left[1-\frac{T_{sat,11}^s(p)}{T_E}\right]\right\} + \exp\left\{\frac{s_{sf,22}(p)}{R}\left[1-\frac{T_{sat,22}^s(p)}{T_E}\right]\right\} = 1$$

(e) Under the modeling assumptions in part (b), estimate the eutectic temperature and composition for a copper–aluminum mixture. Compare your results with the experimental eutectic temperature $T_E = 540°C$, and the eutectic composition on a mass basis of 60% Al and 40% Cu. Comment on the reason for the differences between your estimates and the experimental values.

28 Nonideal Mixtures

Outside the ranges of relatively low pressures and high temperatures in which a mixture behaves as an ideal-gas mixture, the effects of intermolecular forces play an important role in determining the relations between the properties of the mixture. Similar remarks can be made about solutions that are not dilute and therefore cannot be modeled as ideal.

In this chapter we discuss concepts and relations that apply to mixtures under conditions that cannot be accurately represented by the idealized models discussed in Chapter 27. Traditionally, such concepts and relations are said to represent nonideal mixtures.[1]

28.1 Fugacity and Activity of a Constituent in a Mixture

For the general treatment of a mixture, it is useful to extend to mixtures the concepts of fugacity and activity discussed in Section 20.5. Rather than working directly with the chemical potential of constituent i in the mixture, that is,

$$\mu_i = \mu_i(T, p, y_1, y_2, \ldots, y_r) \tag{28.1}$$

we define the *fugacity of constituent i in the mixture*, $\pi_i(T, p, y_1, y_2, \ldots, y_r; p_o, y_{1o}, y_{2o}, \ldots, y_{ro})$, by the relation

$$\pi_i(T, p, \boldsymbol{y}; p_o, \boldsymbol{y}_o) = \pi_i(T, p_o, \boldsymbol{y}_o; p_o, \boldsymbol{y}_o) \exp\left[\frac{\mu_i(T, p, \boldsymbol{y}) - \mu_i(T, p_o, \boldsymbol{y}_o)}{RT}\right] \tag{28.2}$$

together with the condition

$$\pi_i(T, p, \boldsymbol{y}; p_o, \boldsymbol{y}_o) = y_i p \quad \text{in the limits of } T \text{ high and } p \text{ low} \tag{28.3}$$

that is, in the limits in which the system behaves as an ideal-gas mixture, where p_o is a reference pressure and $\boldsymbol{y}_o$ a reference composition. Because the mixture approaches ideal-gas mixture behavior, and each pure constituent approaches ideal gas behavior at sufficiently low pressures and high temperatures, equation (28.3) and relation $\sum_{i=1}^{r} y_{io} = 1$ indicate that the reference values p_o and $\boldsymbol{y}_o$ are not entirely arbitrary.

The dimensionless ratio

$$\phi_i = \frac{\pi_i(T, p, \boldsymbol{y}; p_o, \boldsymbol{y}_o)}{y_i p} \tag{28.4}$$

is called the *fugacity coefficient of constituent i*, and indicates the departure from ideal mixture behavior. In terms of the fugacity and the fugacity coefficient, the chemical potential of constituent i can be expressed as

---

[1]It is noteworthy that the term nonideal does not represent any deficiency with respect to the performance characteristics of mixtures. It simply connotes that the relations between properties of a mixture are mathematically more complicated than are their limits in the region of ideal behavior.

$$\mu_i = \mu_{io} + RT \ln \frac{\pi_i}{\pi_{io}}$$

$$= \mu_{io} + RT \ln \frac{y_i p}{\pi_{io}} + RT \ln \phi_i \tag{28.5}$$

where $\mu_{io} = \mu_i(T, p_o, \boldsymbol{y}_o)$ and $\pi_{io} = \pi_i(T, p_o, \boldsymbol{y}_o; p_o, \boldsymbol{y}_o)$.

When the mixture is in a gaseous form of aggregation, p_o and $y_{1o}, y_{2o}, \ldots, y_{ro}$ may be conveniently chosen to correspond to ideal-gas behavior of pure constituent i, that is, $y_{io} = 1$ and p_o sufficiently low so that $\mu_{io} = \mu_{ii}(T, p_o)$, $\pi_{io} = p_o$, and

$$\mu_i = \mu_{ii}(T, p_o) + RT \ln \frac{y_i p}{p_o} + RT \ln \phi_i \tag{28.6}$$

When the mixture is in a condensed form of aggregation, it is more convenient to define the *activity of constituent i in the mixture* as the dimensionless function $a_i(T, p, y_1, y_2, \ldots, y_r; p_o, y_{1o}, y_{2o}, \ldots, y_{ro})$ given by the relation

$$a_i(T, p, \boldsymbol{y}; p_o, \boldsymbol{y}_o) = \exp \left[\frac{\mu_i(T, p, \boldsymbol{y}) - \mu_i(T, p_o, \boldsymbol{y}_o)}{RT} \right]$$

$$= \frac{\pi_i(T, p, \boldsymbol{y}; p_o, \boldsymbol{y}_o)}{\pi_i(T, p_o, \boldsymbol{y}_o; p_o, \boldsymbol{y}_o)} \tag{28.7}$$

Because for ideal behavior $a_i = y_i$, the *activity coefficient of constituent i*

$$\gamma_i = \frac{a_i(T, p, \boldsymbol{y}; p_o, \boldsymbol{y}_o)}{y_i} \tag{28.8}$$

is an indicator of departure from ideal behavior. In terms of the activity and the activity coefficient, the chemical potential may be expressed as

$$\mu_i = \mu_{io} + RT \ln a_i$$

$$= \mu_{io} + RT \ln y_i + RT \ln \gamma_i \tag{28.9}$$

where again $\mu_{io} = \mu_i(T, p_o, \boldsymbol{y}_o)$.

The choices of the reference values p_o and $\boldsymbol{y}_o$ depend on the application. For example, in applications involving two-phase liquid–vapor states it is convenient to choose $y_{io} = 1$ and $p_o = p_{\text{sat},ii}(T)$, so that $\mu_{io} = \mu_{ii}(T, p_{\text{sat},ii}(T))$ and

$$\mu_i = \mu_{ii}(T, p_{\text{sat},ii}(T)) + RT \ln a_i$$

$$= \mu_{ii}(T, p_{\text{sat},ii}(T)) + RT \ln \gamma_i y_i \tag{28.10}$$

Again, in applications involving just a liquid phase, it is convenient to choose $y_{io} = 1$ and $p_o = p$ so that

$$\mu_i = \mu_{ii}(T, p) + RT \ln a_i$$

$$= \mu_{ii}(T, p) + RT \ln \gamma_i y_i \tag{28.11}$$

It is noteworthy that when the dependences are not shown explicitly, much care must be taken at specifying the context so that the choices of values of p_o, and $y_{1o}, y_{2o}, \ldots, y_{ro}$ are unambiguous.

As an illustration of the use of fugacities and activities, we consider two coexisting phases, a liquid and a vapor. Using equation (28.6) for the vapor phase, equation (28.10) for the liquid phase, and the chemical potential equality for constituent i, we find that

$$\mu_{iig}(T,p_o) + RT \ln \frac{y_{ig}p}{p_o} + RT \ln \phi_{ig} = \mu_{iif}\left[T,p_{\text{sat},ii}(T)\right] + RT \ln \gamma_{if} y_{if} \qquad (28.12)$$

or

$$y_{ig}\phi_{ig}p = \gamma_{if}y_{if}p_o \exp\left[\frac{\mu_{iig}(T,p_{\text{sat},ii}(T)) - \mu_{iig}(T,p_o)}{RT}\right] \qquad (28.13)$$

where we use the fact that $\mu_{iif}(T,p_{\text{sat},ii}(T)) = \mu_{iig}(T,p_{\text{sat},ii}(T))$. Equation (28.13) is a generalization of Raoult's and Henry's laws.

If at $p_{\text{sat},ii}(T)$ pure constituent i behaves as an ideal gas, (28.13) becomes

$$y_{ig}\phi_{ig}p = \gamma_{if}\, y_{if}\, p_{\text{sat},ii}(T) \qquad (28.14a)$$

$$y_{ig}\phi_{ig}p = a_{if}\, p_{\text{sat},ii}(T) \qquad (28.14b)$$

Moreover, if both the gas phase and the liquid phase are ideal, then $\phi_{ig} = 1$ and $\gamma_{if} = 1$, and equation (28.14) reduces to Raoult's law, that is, $y_{ig}p = y_{if}\, p_{\text{sat},ii}(T)$.

28.2   Enthalpies, Entropies, and Volumes of Isothermobaric Mixing

We consider a system consisting of a container partitioned into two compartments by means of a movable partition (Figure 28.1a). One compartment contains a mixture at temperature T, pressure p, and amounts $n_1, n_2, \ldots, n_r$. The other compartment contains an infinitesimal amount dn_i of pure constituent i at temperature T and pressure p. A piston at the top of the container maintains the pressure p constant, and a contact with a reservoir at T maintains the temperature T constant. Upon removal of the partition, the system reaches a stable equilibrium state at T and p with amounts $n_1, \ldots, n_i + dn_i, \ldots, n_r$ (Figure 28.1b).

In general, the enthalpy, entropy, and volume of the system without the partition differ from the corresponding values of the system with the partition. In the limit of dn_i approaching zero, the differences in enthalpy, entropy, and volume, per unit amount

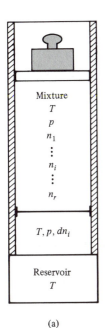
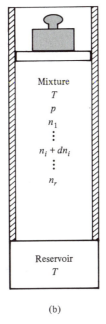

Figure 28.1 (a) A container maintained at fixed pressure and temperature, partitioned into two compartments in partial mutual stable equilibrium, one containing a mixture and the other an infinitesimal amount of one of the constituents. (b) Stable equilibrium state of the mixture that results upon removing the partition from (a).

dissolved, between the initial and the final states of this isothermobaric mixing process are called, respectively, *enthalpy*, *entropy*, and *volume of isothermobaric mixing of constituent* i in the initial mixture. We denote these quantities by the symbols $\Delta h_{i,\mathrm{mix}}$, $\Delta s_{i,\mathrm{mix}}$, and $\Delta v_{i,\mathrm{mix}}$.

For the enthaply of isothermobaric mixing, we find that

$$
\begin{aligned}
\Delta h_{i,\mathrm{mix}} &= \lim_{dn_i \to 0} \frac{H(T,p,n_1,\ldots,n_i+dn_i,\ldots,n_r) - \left[H(T,p,\boldsymbol{n}) + h_{ii}(T,p)\,dn_i\right]}{dn_i} \\
&= \left(\frac{\partial H}{\partial n_i}\right)_{T,p,n} - h_{ii}(T,p) = h_i(T,p,\boldsymbol{y}) - h_{ii}(T,p) \\
&= \left[\frac{\partial(\mu_i/T)}{\partial(1/T)}\right]_{p,y} - \left[\frac{\partial(\mu_{ii}/T)}{\partial(1/T)}\right]_{p,y} \\
&= \left[\frac{\partial(\mu_{io}/T)}{\partial(1/T)}\right]_{p,y} + R\left[\frac{\partial \ln a_i}{\partial(1/T)}\right]_{p,y} - \left[\frac{\partial(\mu_{ii}/T)}{\partial(1/T)}\right]_{p,y}
\end{aligned}
\tag{28.15}
$$

where in writing the third and the fourth of equations (28.15) we use equations (17.42) and (17.49), and in writing the last equation we use (28.9).

Similarly, for the entropy and the volume of isothermobaric mixing we find that

$$
\begin{aligned}
\Delta s_{i,\mathrm{mix}} &= \lim_{dn_i \to 0} \frac{S(T,p,n_1,\ldots,n_i+dn_i,\ldots,n_r) - \left[S(T,p,\boldsymbol{n}) + s_{ii}(T,p)\,dn_i\right]}{dn_i} \\
&= \left(\frac{\partial S}{\partial n_i}\right)_{T,p,n} - s_{ii}(T,p) = s_i(T,p,\boldsymbol{y}) - s_{ii}(T,p) \\
&= -\left(\frac{\partial \mu_i}{\partial T}\right)_{p,y} + \left(\frac{\partial \mu_{ii}}{\partial T}\right)_{p,y} \\
&= -\left(\frac{\partial \mu_{io}}{\partial T}\right)_{p,y} - R\ln a_i - RT\left(\frac{\partial \ln a_i}{\partial T}\right)_{p,y} + \left(\frac{\partial \mu_{ii}}{\partial T}\right)_{p,y}
\end{aligned}
\tag{28.16}
$$

$$
\begin{aligned}
\Delta v_{i,\mathrm{mix}} &= \lim_{dn_i \to 0} \frac{V(T,p,n_1,\ldots,n_i+dn_i,\ldots,n_r) - \left[V(T,p,\boldsymbol{n}) + v_{ii}(T,p)\,dn_i\right]}{dn_i} \\
&= \left(\frac{\partial V}{\partial n_i}\right)_{T,p,n} - v_{ii}(T,p) = v_i(T,p,\boldsymbol{y}) - v_{ii}(T,p) \\
&= \left(\frac{\partial \mu_i}{\partial p}\right)_{T,y} - \left(\frac{\partial \mu_{ii}}{\partial p}\right)_{T,y} \\
&= \left(\frac{\partial \mu_{io}}{\partial p}\right)_{T,y} + RT\left(\frac{\partial \ln a_i}{\partial p}\right)_{T,y} - \left(\frac{\partial \mu_{ii}}{\partial p}\right)_{T,y}
\end{aligned}
\tag{28.17}
$$

Now we recall from the preceding section that the most convenient choices of y_{io} and p_o in the definition of the activity of constituent i depend on whether it is present in the mixture only in the liquid phase, only in the gaseous phase, or in both phases of a liquid–vapor state of the mixture. Here we proceed by choosing $y_{io} = 1$ and $p_o = p$, which is most appropriate for mixtures in the liquid phase. Then $\mu_{io} = \mu_{ii}(T,p)$, a_i satisfies equation (28.11), and

$$\Delta h_{i,\text{mix}} = R\left(\frac{\partial \ln a_i}{\partial(1/T)}\right)_{p,y} = R\left(\frac{\partial \ln \gamma_i}{\partial(1/T)}\right)_{p,y} \qquad (28.18)$$

$$\Delta s_{i,\text{mix}} = -R\ln a_i - RT\left(\frac{\partial \ln a_i}{\partial T}\right)_{p,y} = -R\ln a_i + \frac{\Delta h_{i,\text{mix}}}{T} \qquad (28.19)$$

$$\Delta v_{i,\text{mix}} = RT\left(\frac{\partial \ln a_i}{\partial p}\right)_{T,y} \qquad (28.20)$$

The value of $\Delta v_{i,\text{mix}}$ is often very small, but that of $\Delta h_{i,\text{mix}}$ is significant for most mixtures. For example, for a dilute solution of NaCl in water at 25°C and 1 atm, $\Delta h_{\text{NaCl,mix}} \approx -3.9$ kJ/mol. The fact that the value of $\Delta h_{i,\text{mix}}$ is negative implies that upon isothermobaric mixing, the heat is from the mixture to the reservoir (Figure 28.1). In general, values of $\Delta h_{i,\text{mix}}$ may be determined by calorimetric measurements.

The *overall enthalpy, entropy,* and *volume of isothermobaric mixing* are defined by the relations

$$\Delta H_{\text{mix}} = H(T,p,\mathbf{n}) - \sum_{i=1}^{r} n_i\, h_{ii}(T,p) = \sum_{i=1}^{r} n_i\left[h_i(T,p,\mathbf{y}) - h_{ii}(T,p)\right]$$

$$= \sum_{i=1}^{r} n_i \Delta h_{i,\text{mix}} \qquad (28.21)$$

$$\Delta S_{\text{mix}} = S(T,p,\mathbf{n}) - \sum_{i=1}^{r} n_i\, s_{ii}(T,p) = \sum_{i=1}^{r} n_i\left[s_i(T,p,\mathbf{y}) - s_{ii}(T,p)\right]$$

$$= \sum_{i=1}^{r} n_i \Delta s_{i,\text{mix}} = \frac{\Delta H_{\text{mix}}}{T} - R\sum_{i=1}^{r} n_i \ln a_i \qquad (28.22)$$

$$\Delta V_{\text{mix}} = V(T,p,\mathbf{n}) - \sum_{i=1}^{r} n_i\, v_{ii}(T,p) = \sum_{i=1}^{r} n_i\left[v_i(T,p,\mathbf{y}) - v_{ii}(T,p)\right]$$

$$= \sum_{i=1}^{r} n_i \Delta v_{i,\text{mix}} \qquad (28.23)$$

where in writing the last of equations (28.22) we use (28.19).

28.3   Van Ness and Duhem–Margules Relations

By virtue of the Gibbs–Duhem relation, only $r-1$ activities of an r-constituent mixture need to be measured. To see this more explicitly, we write the Gibbs–Duhem relation for an r-constituent mixture at temperature T, pressure p, and given amounts n_1, n_2, $\ldots$, n_r, and for each of r one-constituent systems at the same T and p. Thus we have

$$S(T,p,\mathbf{n})\,dT - V(T,p,\mathbf{n})\,dp + \sum_{i=1}^{r} n_i\, d\mu_i(T,p,\mathbf{y}) = 0 \qquad (28.24)$$

$$s_{ii}(T,p)\,dT - v_{ii}(T,p)\,dp + d\mu_{ii}(T,p) = 0 \quad \text{for } i = 1,\ 2,\ \ldots,\ r \qquad (28.25)$$

Upon subtracting from equation (28.24) all equations (28.25), each multiplied by the corresponding amount n_i in the mixture, we find that

$$\Delta S_{\text{mix}}\,dT - \Delta V_{\text{mix}}\,dp + \sum_{i=1}^{r} n_i\, d\left[\mu_i(T,p,\mathbf{y}) - \mu_{ii}(T,p)\right] = 0 \qquad (28.26)$$

where we use (28.22) and (28.23). Upon expressing each difference $\mu_i - \mu_{ii}$ in terms of the activity of constituent i in the mixture [equations (28.11)], we find that

$$\left(\Delta S_{\text{mix}} + R \sum_{i=1}^{r} n_i \ln a_i\right) dT - \Delta V_{\text{mix}}\, dp + RT \sum_{i=1}^{r} n_i\, d\ln a_i$$

$$= \frac{\Delta H_{\text{mix}}}{T}\, dT - \Delta V_{\text{mix}}\, dp + RT \sum_{i=1}^{r} n_i\, d\ln a_i = 0 \qquad (28.27)$$

where in writing the second of equations (28.27) we use (28.22). Because $\sum_{i=1}^{r} y_i = 1$, the sum $\sum_{i=1}^{r} n_i\, d\ln a_i$ equals $\sum_{i=1}^{r} n_i\, d\ln \gamma_i$, and so equation (28.27) becomes

$$\sum_{i=1}^{r} n_i\, d\ln a_i = \sum_{i=1}^{r} n_i\, d\ln \gamma_i = \frac{\Delta V_{\text{mix}}}{RT}\, dp - \frac{\Delta H_{\text{mix}}}{RT^2}\, dT \qquad (28.28)$$

In this form, the equation is known as the *van Ness relation*. It is very useful to check the consistency of experimental data on activities and activity coefficients, and to interpolate and extrapolate between data points.

Upon recalling the dependences of the activity, that is, $a_i = a_i(T, p, \boldsymbol{y}; y_{io} = 1, p_o = p)$, and expanding the differential $d\ln a_i$, we can express the van Ness relation in the form

$$\sum_{i=1}^{r} n_i \left[\left(\frac{\partial \ln a_i}{\partial T}\right)_{p,y} + \frac{\Delta h_{i,\text{mix}}}{RT^2}\right] dT + \sum_{i=1}^{r} n_i \left[\left(\frac{\partial \ln a_i}{\partial p}\right)_{T,y} - \frac{\Delta v_{i,\text{mix}}}{RT}\right] dp$$

$$+ \sum_{i=1}^{r} n_i \sum_{j=1}^{r} \left(\frac{\partial \ln a_i}{\partial y_j}\right)_{T,p,y} dy_j = 0 \qquad (28.29)$$

But each of the terms between square brackets is equal to zero by virtue of (28.18) and (28.20), and hence we find the relation

$$\sum_{i=1}^{r} y_i \sum_{j=1}^{r} \left(\frac{\partial \ln a_i}{\partial y_j}\right)_{T,p,y} dy_j = 0 \qquad (28.30)$$

Equation (28.30) is known as the *Duhem–Margules relation*. Of course, we must recall that the mole fractions are not all independent and, therefore, equation (28.30) must be used in conjunction with the relation $\sum_{i=1}^{r} y_i = 1$. In practice, we substitute this relation in the expression of the activity so as to eliminate the mole fraction y_r, and thus we introduce another function a_i' for the activity of the ith constituent given by the relation

$$a_i'(T, p, y_1, y_2, \ldots, y_{r-1}; y_{io} = 1, p_o = p)$$

$$= a_i\left(T, p, y_1, y_2, \ldots, y_{r-1}, y_r = \left(1 - \sum_{j=1}^{r-1} y_j\right); y_{io} = 1, p_o = p\right) \qquad (28.31)$$

for $i = 1, 2, \ldots, r$. We can express the Duhem–Margules relation in terms of the activities a_i' as follows. Upon substituting $dy_r = -\sum_{j=1}^{r-1} dy_j$ in equation (28.30), we find that

$$\sum_{i=1}^{r} y_i \sum_{j=1}^{r} \left(\frac{\partial \ln a_i}{\partial y_j} \right)_{T,p,y} dy_j$$

$$= \sum_{i=1}^{r} y_i \sum_{j=1}^{r-1} \left(\frac{\partial \ln a_i}{\partial y_j} \right)_{T,p,y} dy_j + \sum_{i=1}^{r} y_i \left(\frac{\partial \ln a_i}{\partial y_r} \right)_{T,p,y} dy_r$$

$$= \sum_{i=1}^{r} y_i \sum_{j=1}^{r-1} \left[\left(\frac{\partial \ln a_i}{\partial y_j} \right)_{T,p,y} - \left(\frac{\partial \ln a_i}{\partial y_r} \right)_{T,p,y} \right] dy_j \qquad (28.32)$$

Next, upon differentiating a_i' with respect to y_j for $j \neq r$, we find that

$$\left(\frac{\partial \ln a_i'}{\partial y_j} \right)_{T,p,y} = \left(\frac{\partial \ln a_i}{\partial y_j} \right)_{T,p,y} - \left(\frac{\partial \ln a_i}{\partial y_r} \right)_{T,p,y} \qquad (28.33)$$

and, upon substituting equation (28.33) into (28.32), that

$$\sum_{i=1}^{r} y_i \sum_{j=1}^{r-1} \left(\frac{\partial \ln a_i'}{\partial y_j} \right)_{T,p,y} dy_j = 0 \qquad (28.34)$$

For a binary mixture, $r = 2$, $y_2 = 1 - y_1$, and the Duhem–Margules relation becomes

$$y_1 \left(\frac{\partial \ln a_1'}{\partial y_1} \right)_{T,p} + (1 - y_1) \left(\frac{\partial \ln a_2'}{\partial y_1} \right)_{T,p} = 0 \qquad (28.35a)$$

or, equivalently,

$$\left(\frac{\partial \ln a_1'}{\partial \ln y_1} \right)_{T,p} = \left(\frac{\partial \ln a_2'}{\partial \ln y_2} \right)_{T,p} \qquad (28.35b)$$

where we reemphasize that $a_1'(T, p, y_1; y_{1o} = 1, p_o = p) = a_1(T, p, y_1, y_2 = 1 - y_1; y_{1o} = 1, p_o = p)$ and $a_2'(T, p, y_1; y_{2o} = 1, p_o = p) = a_2(T, p, y_1, y_2 = 1 - y_1; y_{2o} = 1, p_o = p)$ or, equivalently,

$$\left(\frac{\partial \ln a_1'}{\partial \ln y_1} \right)_{T,p} = \left(\frac{\partial \ln a_2''}{\partial \ln y_2} \right)_{T,p} \qquad (28.35c)$$

where $a_2''(T, p, y_2; y_{2o} = 1, p_o = p) = a_2(T, p, y_1 = 1 - y_2, y_2; y_{2o} = 1, p_o = p)$.

Equation (28.35) can be used to find the activity of constituent 2 when that of constituent 1 has been measured. The equation also implies that if a_1' and a_2' are graphed versus y_1 at given values of T and p (Figure 28.2), the shape of either graph determines that of the other. In the limit as $y_1 \rightarrow 1$, the behavior of constituent 1 becomes ideal, $a_1' = y_1$, $(\partial \ln a_1'/\partial \ln y_1)_{T,p} = 1$ and equation (28.35b) requires that $(\partial \ln a_2'/\partial \ln y_2)_{T,p} = 1$ or, integrating, $a_2' = y_2 C_2(T,p)$, where $C_2(T,p)$ is the constant of integration that depends on T and p. Similarly, in the limit as $y_1 \rightarrow 0$, the behavior of constituent 2 becomes ideal, $a_2'' = y_2$, $(\partial \ln a_2''/\partial \ln y_2)_{T,p} = 1$, (28.35c) requires that $(\partial \ln a_1'/\partial \ln y_1)_{T,p} = 1$, and $a_1' = y_1 C_1(T,p)$, where $C_1(T,p)$ is the constant of integration that depends on T and p. Only for ideal mixture behavior $C_1(T,p) = C_2(T,p) = 1$. For the liquid phase of a two-phase mixture with ideal-gas-mixture behavior of the vapor phase and negligible Poynting correction, $C_1(T,p)$ and $C_2(T,p)$ are related to Henry's constant [equation (27.77)], namely, $C_1(T,p) = \mathcal{H}_{1f}(T,p)/p_{\text{sat},11}(T)$ and $C_2(T,p) = \mathcal{H}_{2f}(T,p)/p_{\text{sat},22}(T)$.

Example 28.1. Prove the last assertion.

Solution. For ideal-gas-mixture behavior of the vapor phase, equation (28.14b) for $i = 2$ becomes $y_{2g}p = a_{2f}\, p_{\text{sat},22}(T)$. For ideal-solution behavior of constituent 1 in the liquid phase, in the

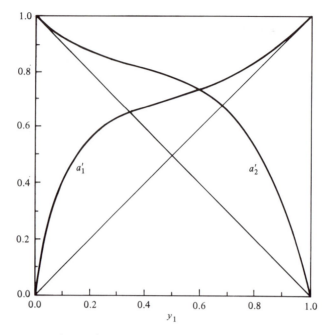

Figure 28.2 Graphs of a_1' and a_2' versus y_1 at given values of T and p for a liquid mixture of nitromethane and carbon tetrachloride.

limit as $y_1 \rightarrow 1$, Henry's law [equation (27.76)] implies that $y_2\, p = y_{2f}\, \mathcal{H}_{2f}(T, p)$. Therefore, $a_{2f}\, p_{\mathrm{sat},22}(T) = y_{2f}\, \mathcal{H}_{2f}(T, p)$. Now we recall that a_{2f} in equation (28.14b) is the activity defined by equation (28.11) (with $p_o = p$) whereas a_2'' is the activity defined by equation 28.10 [with $p_o = p_{\mathrm{sat},22}(T)$]. For a liquid phase, using equations (28.10) and (28.11) we find that

$$a_{2f}'' = a_{2f} \exp\left[\mu_{22f}(T, p_{\mathrm{sat},22}(T)) - \mu_{22f}(T, p)\right] \approx a_{2f}$$

where we neglect the Poynting correction. Thus we verify that $a_{2f}''\, p_{\mathrm{sat},22}(T) = y_{2f}\, \mathcal{H}_{2f}(T, p)$.

28.4 Van Laar and Margules Equations for Binary Mixtures

Several semiempirical relations have been used in the literature to correlate experimental data on activities for mixtures, that is, explicit forms of the relations $a_i' = a_i'(T, p, y_1, y_2, \ldots, y_{r-1}; y_{io} = 1, p_o = p)$ defined in the preceding section. The simplest of such relations are the *two-suffix Margules equations*

$$\ln a_1' = \ln y_1 + \frac{A}{RT}\,(1 - y_1)^2 \tag{28.36a}$$

$$\ln a_2' = \ln(1 - y_1) + \frac{A}{RT}\,(y_1)^2 \tag{28.36b}$$

where $A = A(T, p)$ is at most a function of temperature and pressure but not of composition. Of course, we can readily check that equations (28.36) satisfy the Duhem–Margules relation [equation (28.35)]. These equations represent well only simple liquid mixtures of constituents that have similar molecular structures. It is noteworthy that equations (28.36b) and (28.18) for $i = 1$ in the limit as $y_1 \rightarrow 0$ imply that $A = \Delta h_{1,\mathrm{mix}}$, and equations (28.36a) and (28.18) for $i = 2$ in the limit as $y_1 \rightarrow 1$ that $A = \Delta h_{2,\mathrm{mix}}$. It follows that equations (28.36) are good approximations only for mixtures for which $\Delta h_{1,\mathrm{mix}} \approx \Delta h_{2,\mathrm{mix}}$.

The *three-suffix Margules equations* are

$$\ln a_1' = \ln y_1 + \frac{2B - A}{RT}(1 - y_1)^2 + \frac{2(A - B)}{RT}(1 - y_1)^3 \tag{28.37a}$$

$$\ln a_2' = \ln(1 - y_1) + \frac{2A - B}{RT}(y_1)^2 + \frac{2(B - A)}{RT}(y_1)^3 \tag{28.37b}$$

where again $A = A(T, p)$ and $B = B(T, p)$ are functions of T and p only, and the Duhem–Margules relation is satisfied. Here, according to equation (28.18) applied both in the limit as $y_1 \rightarrow 0$ and in the limit as $y_1 \rightarrow 1$, we find that $A = \Delta h_{1,\text{mix}}$ and $B = \Delta h_{2,\text{mix}}$. Clearly, equations (28.37) reduce to equations (28.36) if $A = B$. Often the Margules equations are written in terms of the dimensionless functions $A' = A/RT$ and $B' = B/RT$. We can readily check that (28.37) satisfy the Duhem–Margules relation [equation (28.35)].

Finally, the *van Laar equations* are

$$\ln a_1' = \ln y_1 + \frac{A/RT}{\left[1 + Ay_1/B(1 - y_1)\right]^2} \tag{28.38a}$$

$$\ln a_2' = \ln(1 - y_1) + \frac{B/RT}{\left[1 + B(1 - y_1)/Ay_1\right]^2} \tag{28.38b}$$

where again we can verify that $A = \Delta h_{1,\text{mix}}$ and $B = \Delta h_{2,\text{mix}}$, and that equations (28.38) reduce to equations (28.36) if $A = B$. The van Laar equations are often given in terms of the dimensionless functions $A' = A/RT$ and $B' = B/RT$. Although successful in correlating activity data for several binary mixtures, it is noteworthy that these equations do not satisfy exactly the Duhem–Margules relation.

As an application of the empirical equations just presented, we consider the two-phase liquid–vapor states of a binary mixture at conditions such that the vapor phase behaves as an ideal-gas mixture, whereas the liquid phase does not exhibit ideal-solution behavior. The conditions of mutual stable equilibrium of the two coexisting phases [equation (28.14b) for $i = 1$ and $i = 2$] become

$$y_{1g}p = a_{1f}\,p_{\text{sat},11}(T) \tag{28.39}$$

$$y_{2g}p = a_{2f}\,p_{\text{sat},22}(T) \tag{28.40}$$

Expressing the activities in terms of equations (28.38) and using the relations $y_{1g} + y_{2g} = 1$ and $y_{1f} + y_{2f} = 1$, after a few rearrangements we find that

$$p = y_{1f}\exp\left\{\frac{A/RT}{\left[1 + Ay_{1f}/B(1 - y_{1f})\right]^2}\right\}p_{\text{sat},11}(T)$$

$$+ (1 - y_{1f})\exp\left\{\frac{B/RT}{\left[1 + B(1 - y_1)/Ay_1\right]^2}\right\}p_{\text{sat},22}(T) \tag{28.41}$$

$$\frac{1}{p} = \frac{y_{1g}}{p_{\text{sat},11}(T)}\exp\left\{-\frac{A/RT}{\left[1 + Ay_{1f}/B(1 - y_{1f})\right]^2}\right\}$$

$$+ \frac{1 - y_{1g}}{p_{\text{sat},22}(T)}\exp\left\{-\frac{B/RT}{\left[1 + B(1 - y_1)/Ay_1\right]^2}\right\} \tag{28.42}$$

These equations are analogous to equations (27.80) and (27.81) for ideal-solution behavior of the liquid phase. For a mixture of nitromethane and carbon tetrachloride experimental data are reasonably approximated by the van Laar equations assuming the constants $A' =$

$A/RT = 1.8$ and $B' = B/RT = 1.4$, and using the relations $p_{\text{sat},11}(T) = \exp(17.08 - 4439/T)$ for nitromethane and $p_{\text{sat},22}(T) = \exp(16.33 - 4078/T)$ for carbon tetrachloride. Thus, equations (27.80) and (27.81) can be solved numerically to yields the graphs in Figures 28.3 to 28.6, where we report also the results corresponding to ideal-solution behavior ($A' = 0$ and $B' = 0$).

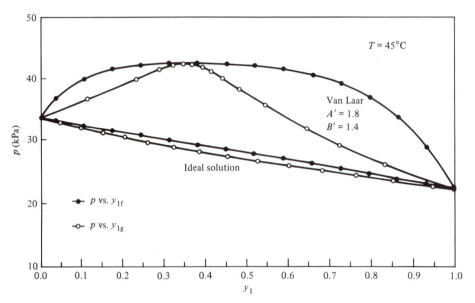

Figure 28.3 Graphs of p versus y_{1f} and p versus y_{1g} based on equations (28.41) and (28.42) with $p_{\text{sat},11}(T) = 22.67$ kPa and $p_{\text{sat},22}(T) = 33.32$ kPa.

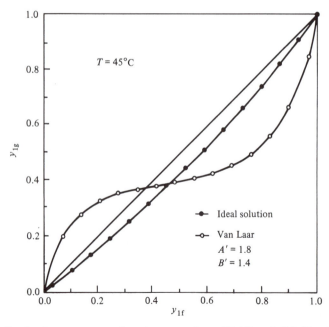

Figure 28.4 Graph of y_{1g} versus y_{1f} based on equations (28.41) and (28.42) with $p_{\text{sat},11}(T) = 22.67$ kPa and $p_{\text{sat},22}(T) = 33.32$ kPa.

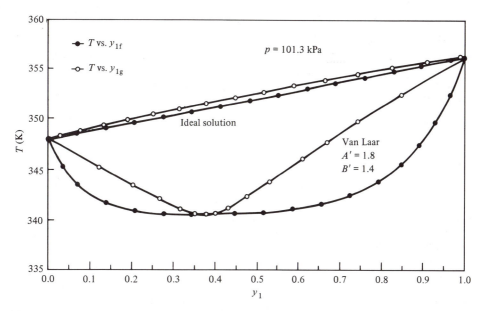

Figure 28.5 Graphs of T versus y_{1f} and T versus y_{1g} for $p = 1$ atm based on equations (28.41) and (28.42), with $p_{sat,11}(T) = \exp(17.08 - 4439/T)$ and $p_{sat,22}(T) = \exp(16.33 - 4078/T)$.

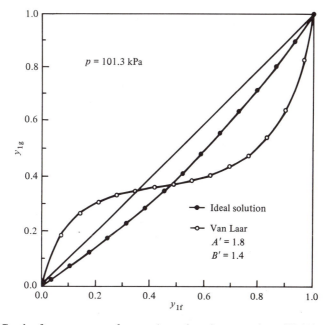

Figure 28.6 Graph of y_{1g} versus y_{1f} for $p = 1$ atm based on equations (28.41) and (28.42) with $p_{sat,11}(T) = 22.67$ kPa and $p_{sat,22}(T) = 33.32$ kPa.

Problems

28.1 Consider a binary solution of salt (NaCl, constituent 1, $M_1 = 58.5$ kg/kmol) and water (H_2O, constituent 2, $M_2 = 18$ kg/kmol) at temperature T and pressure p, and with mole fraction y_1 of salt.

The nonideal behavior of this solution can be approximated by the two-suffix Margules equations with $A = -3385$ kJ/kmol.

(a) If the average mass fraction of salt in seawater is $x_1 = 3.5\%$, what is the osmotic pressure of water in seawater at $5°C$?

(b) What is the boiling temperature of a solution of 50 g of NaCl in 1 kg of H_2O at atmospheric pressure?

(c) What is the freezing temperature of a solution of 50 g of NaCl in 1 kg of H_2O at atmospheric pressure?

(d) What is the least work required to separate 1 kg of pure water at $10°C$ out of a very large amount of seawater at the environmental temperature of $10°C$?

(e) Boil 1 kg water on your stove and, while it is boiling, throw in 20 g of salt. Due to effective mixing in the pot, the salt dissolves very quickly and you observe a small burst of water vapor bubbles. To explain this, you need to account for: the nonideal behavior of the saltwater solution, the boiling temperature-raising effect, and the heating of the salt initially at room temperature. Assuming that the salt is at $80°C$ when it reaches the liquid in the pot and that the specific heat of salt is 0.8 kJ/kg K, evaluate the volume of the water vapor bubbles that are formed due to dissolving the salt in the initially saturated liquid water. Notice that if the salt were not preheated, the effect would not occur in the bulk but only locally, that is, the salt is preheated by the liquid water and liberates the enthalpy of dilution while dissolving, which creates vapor at temperature slightly higher than the bulk of the still-pure liquid water, so the bubbles collapse.

(f) Repeat parts (a) to (e) for ideal solution behavior, obtained by setting $A = 0$ in the two-suffix Margules equations.

29   Chemical Reactions

Up to this point we have not discussed the effects of chemical reactions. In this chapter and the next we define the concepts and terminology used in analyses of chemical reactions and evaluate the effects of such reactions on stable equilibrium states.

29.1   Conservation of Atomic Nuclei

We consider a system composed of different constituents, such as oxygen molecules, hydrogen molecules, water molecules, solid carbon, and carbon dioxide, that are chemically active. Each constituent has an amount that takes a nonzero value in at least some states of the system. Each amount can change due to internal forces that result in chemical reaction mechanisms by which molecules of some constituents may disappear to give rise to molecules of other constituents.

In some discussions, we denote each constituent by its chemical symbol, such as O_2, H_2, H_2O, C, and CO_2. On other occasions, however, we may find it more convenient to denote each constituent by a common symbol with a subscript that represents the constituent and an argument that stands for the phase of the constituent. Specifically, if there are r constituents, we denote each by $A_i(p)$, where $i = 1, 2, \ldots, r$ and p stands for s, l, or g for a solid, liquid, or gaseous form of aggregation, respectively. In particular, for a system consisting of gaseous hydrogen molecules, gaseous oxygen molecules, and liquid water molecules, we can use the symbols $A_1(g)$ for gaseous H_2, $A_2(g)$ for gaseous O_2, and $A_3(l)$ for liquid H_2O.

Associated with each state of an r-constituent system are the r amounts n_1, n_2, $\ldots$, n_r, and the composition $y_1 = n_1/n$, $y_2 = n_2/n$, $\ldots$, $y_r = n_r/n$, where $n = \sum_{i=1}^{r} n_i$.

For a system with no chemical reactions, no nuclear reactions, no creation and no annihilation reactions, and no interactions that exchange matter with other systems, the amounts and the composition are the same for all the states. If chemical reactions are part of the system specification, however, the amounts of constituents and the composition of the system may change from state to state even in the absence of interactions with other systems. For example, C and O_2 in a system may combine spontaneously to form CO_2, so the amounts of C, O_2, and CO_2 may change even in the absence of transfers of these constituents to or from other systems.

Despite the changes in amounts of constituents due to chemical reactions, the number of atomic nuclei of each kind is conserved because chemical reactions affect the arrangements of nuclei in molecules but not the nuclei themselves. We refer to this characteristic feature of chemical reactions as the *principle of conservation of atomic nuclei*. For example, if the initial state of an isolated system consists of 3 mol of H_2, plus 2 mol of O_2, plus 1 mol of H_2O or, equivalently, 8 mol of hydrogen nuclei, and 5 mol of oxygen nuclei, then any other state of this isolated system consists of 8 mol of hydrogen nuclei and 5 mol of oxygen nuclei, regardless of the number of mol of H_2, O_2, and H_2O, which may change according to the chemical reaction mechanism $2H_2 + O_2 = 2H_2O$.

In general, in a given system many reactions may occur between the various constituents. Depending on the conditions under which the system is studied, however, some of these reactions may be entirely negligible. For example, a reaction may be proceeding at a rate much smaller than any of the rates of the other reactions. Although operative in principle, the slow reaction has no discernible effect on the behavior of the system over a short period of time and therefore can be disregarded. Formally, problems of this sort are analyzed by defining a system in which all the chemical reactions except those that have been selected as dominant are forbidden by *passive resistances*, *anticatalysts*, or *reaction inhibitors*. An effective selection of the dominant reactions in a given system is a difficult and important task of chemical kinetics. It is not addressed in this book. Here we assume that a system has been defined together with a set of dominant reaction mechanisms, and proceed from there.

29.2 Stoichiometry

Each chemical reaction mechanism is specified by a relation that determines which constituents change from one type to another. To illustrate this point, we consider a system in which only one chemical reaction mechanism is dominant, such as the water–gas shift reaction

$$C(s) + H_2O(g) = CO(g) + H_2(g) \tag{29.1}$$

In this reaction solid carbon, $C(s)$, reacts with water vapor, $H_2O(g)$, to form gaseous carbon monoxide, $CO(g)$, and gaseous hydrogen, $H_2(g)$. The meaning of relation (29.1) is solely that in the given system one carbon atom and one water molecule may disappear to give rise to one carbon monoxide molecule and one hydrogen molecule, regardless of how many molecules of each of these constituents are present. We can interchange the left-hand side and the right-hand side of relation (29.1), and then the meaning of the result is that one carbon monoxide molecule and one hydrogen molecule may disappear to give rise to one carbon atom and one water molecule, and again this meaning is independent of the number of molecules present. Occasionally, we may associate both meanings with relation (29.1) as is.

The specification of a chemical reaction mechanism is not arbitrary. The multipliers in front of each chemical compound symbol must be consistent with the principle of conservation of atomic nuclei in chemical reactions. For example, the number of carbon nuclei in the right-hand side of relation (29.1) must be equal to the number of carbon nuclei in the left-hand side, that is, unity. When the principle of conservation of atomic nuclei is satisfied, we say that the reaction mechanism is *stoichiometrically balanced*.

For convenience it is customary to call *reactants* the constituents in the left-hand side of a reaction mechanism, and *products* those in the right-hand side. If the order of the two sides of a reaction mechanism is inverted, the reactants become products and the products reactants. For example, if we write

$$CO(g) + H_2(g) = C(s) + H_2O(g) \tag{29.2}$$

then CO and H_2 are the reactants, and C and H_2O the products. But it is clear that both relations (29.1) and (29.2) describe exactly the same reaction mechanism, and that the reaction occurs while, in general, both reactants and products are present.

The water–gas shift reaction [relation (29.1)] may also be expressed in the form

$$A_1(s) + A_2(g) = A_3(g) + A_4(g) \tag{29.3}$$

where $A_1(s)$ denotes solid C, $A_2(g)$ gaseous H_2O, $A_3(g)$ gaseous CO, and $A_4(g)$ gaseous H_2. This form is convenient for discussions of chemical reactions in general. When each constituent is denoted by the same symbol A with a subscript i that differs from constituent to constituent, any chemical reaction mechanism in a system with r constituents can be expressed in the form

$$\sum_{i=1}^{r} \nu_i A_i(\text{p}) = 0 \qquad (29.4)$$

where ν_i is the *stoichiometric coefficient* or *stoichiometric number of the ith constituent in the reaction*. If any constituent $A_i(\text{p})$ does not participate in the chemical reaction mechanism either as a reactant or as a product, its stoichiometric number ν_i equals zero.

It is customary to assign positive values to the stoichiometric numbers of products, and negative values to the stoichiometric numbers of reactants. For example, if the constituents of a system are C, H_2O, CO, and H_2, and they can react by the mechanism described by relation (29.1), then we have $r = 4$, and may write the reaction in the form of relation (29.4) with the following assignments:

$$
\begin{array}{lll}
A_1(s) & \text{stands for solid C} & \nu_1 = -1 \\
A_2(g) & \text{stands for gaseous } H_2O & \nu_2 = -1 \\
A_3(g) & \text{stands for gaseous CO} & \nu_3 = 1 \\
A_4(g) & \text{stands for gaseous } H_2 & \nu_4 = 1
\end{array}
$$

It is evident that the stoichiometric coefficients ν_i describe only the reaction mechanism and do not represent the amounts of constituents. The former are fixed for each reaction mechanism, whereas the latter may vary as the state changes as a result of the reaction. We will see, however, that the stoichiometric coefficients are related to the changes in values of the amounts n_i that occur as a result of the reaction described by the ν_i's.

The stoichiometric coefficients of a chemical reaction mechanism are specified only up to a common factor. For example, the expressions

$$4\,C(s) + 4\,H_2O(g) = 4\,CO(g) + 4\,H_2(g)$$

$$2\,C(s) + 2\,H_2O(g) = 2\,CO(g) + 2\,H_2(g)$$

$$\tfrac{1}{2}\,C(s) + \tfrac{1}{2}\,H_2O(g) = \tfrac{1}{2}\,CO(g) + \tfrac{1}{2}\,H_2(g)$$

all specify the same chemical reaction mechanism as relation (29.1). This is so because a reaction mechanism is fully specified by the relative proportions by which the constituents can combine. In general, it is convenient to select the values of the stoichiometric coefficients so that they are the smallest integers satisfying the relative proportions characteristic of the reaction. In reactions that describe the formation of compound molecules out of simpler molecules, the so-called formation reactions (see Section 29.10), the values of the stoichiometric coefficients are selected so that the coefficient of the molecule of interest is equal to unity. For example, if CO(g) is the molecule of interest, the stoichiometric coefficient of this molecule is selected equal to unity rather than any other acceptable value, and then the other coefficients of the formation-reaction mechanism are adjusted accordingly.

To simplify the notation, in what follows we omit the form-of-aggregation argument from the symbolic representation of each molecule except where the designation is needed for clarity.

Example 29.1. Nitrogen, N_2, and water, H_2O, may combine to produce ammonia, NH_3, and oxygen O_2. Assuming that ammonia is the molecule of interest, write the chemical reaction mechanism and determine its stoichiometric coefficients.

Solution. The chemical reaction mechanism must be $a\,N_2 + b\,H_2O = NH_3 + c\,O_2$, where we select the stoichiometric coefficient of NH_3 as unity because it is the product of interest. Using the principle of conservation of atomic nuclei of nitrogen, hydrogen, and oxygen we find that $2\,a = 1$, $2\,b = 3$, and $b = 2\,c$, respectively, or $a = \frac{1}{2}$, $b = \frac{3}{2}$, and $c = \frac{3}{4}$. Moreover, if A_1 stands for N_2, A_2 for H_2O, A_3 for NH_3, and A_4 for O_2, then $\nu_1 = -\frac{1}{2}$, $\nu_2 = -\frac{3}{2}$, $\nu_3 = 1$, and $\nu_4 = \frac{3}{4}$.

29.3   Reaction Coordinate

In a system consisting of H_2O, H_2, and O_2 molecules, and in which the only chemical reaction mechanism is

$$2\,H_2 + O_2 = 2\,H_2O \tag{29.5}$$

we can define A_1 as H_2O, A_2 as H_2, and A_3 as O_2, $\nu_1 = 2$, $\nu_2 = -2$, $\nu_3 = -1$, and rewrite the reaction mechanism as

$$\nu_1 A_1 + \nu_2 A_2 + \nu_3 A_3 = \sum_{i=1}^{3} \nu_i A_i = 0 \tag{29.6}$$

With this choice of values of the ν_i's, we regard H_2O as the product ($\nu_1 > 0$) and H_2 and O_2 as the reactants ($\nu_2 < 0$, $\nu_3 < 0$).

Now we consider two states with different amounts. In one state, the amounts of water, hydrogen, and oxygen molecules are n_1, n_2, and n_3, respectively, and in the other state $n_1 + dn_1$, $n_2 + dn_2$, and $n_3 + dn_3$, all expressed in moles. We assume that the change in amounts occurred only because of the reaction and not as a result of either exchanges of matter with other systems, or nonchemical creation and annihilation processes. Accordingly, the values dn_1, dn_2, and dn_3 cannot be arbitrary. For example, if the number of water molecules has increased by 2, that is, $dn_1 = 2$, the numbers of hydrogen and oxygen molecules must have decreased by 2 and 1, that is, $dn_2 = -2$ and $dn_3 = -1$, respectively, because then and only then the change obeys the chemical reaction mechanism and the principle of conservation of atomic nuclei. More generally, if $d\epsilon$ is an arbitrary number and $dn_1 = 2\,d\epsilon$, we must have $dn_2 = -2\,d\epsilon$ and $dn_3 = -d\epsilon$. These conditions are called *proportionality relations* because, for a general reaction mechanism, they establish the proportions between the changes dn_i that are consistent with stoichiometric coefficients ν_i. For our example, the proportionality relations can be rewritten in the compact forms

$$\frac{dn_1}{2} = \frac{dn_2}{-2} = \frac{dn_3}{-1} = d\epsilon \tag{29.7a}$$

or

$$\frac{dn_1}{\nu_1} = \frac{dn_2}{\nu_2} = \frac{dn_3}{\nu_3} = d\epsilon \tag{29.7b}$$

More generally, for a system in which the chemical reaction mechanism is given by

$$\sum_{i=1}^{r} \nu_i A_i = 0 \tag{29.8}$$

the proportionality relations are

$$dn_i = \nu_i\,d\epsilon \qquad \text{for } i = 1, 2, \ldots, r \tag{29.9a}$$

or, equivalently,

$$\frac{dn_1}{\nu_1} = \frac{dn_2}{\nu_2} = \cdots = \frac{dn_r}{\nu_r} = d\epsilon \tag{29.9b}$$

Associated with each chemical reaction mechanism is a variable ϵ, called the *reaction coordinate*, such that its variations satisfy equations (29.9). The absolute value of ϵ may be determined by associating the value $\epsilon = 0$ with the values of the amounts of constituents of the system at some arbitrarily selected state, and then varying both the amounts of constituents and ϵ in conformity with equations (29.9). We discuss this point in the next section.

29.4 Changes in Amounts and Composition

We examine changes in amounts and composition corresponding to two different processes. The first process involves a system subject to work and heat interactions only, and to one chemical reaction between the set of r constituents A_1, A_2, ..., A_r. The chemical reaction mechanism is characterized by the set of r stoichiometric coefficients ν_1, ν_2, ..., ν_r. In the initial state, the amounts of the constituents are specified by the set n_{1a}, n_{2a}, ..., n_{ra}. As a result of the interactions and the chemical reaction the state of the system changes. In the final state, the amounts of the constituents are n_{1b}, n_{2b}, ..., n_{rb} (Figure 29.1). We call this a *batch process*.

For the reaction coordinate, we select the value $\epsilon = 0$ to correspond to the initial amounts. Because the changes in amounts are determined only by the specified chemical reaction mechanism, the values of the final amounts and the reaction coordinate must be consistent with the proportionality relations [equations (29.9)]. Thus, upon integrating each of these relations from the initial amount to the final amount, we find that

$$\int_{n_{ia}}^{n_{ib}} dn_i = n_{ib} - n_{ia} = \int_0^{\epsilon} \nu_i \, d\epsilon = \nu_i \epsilon \qquad \text{for } i = 1, 2, \ldots, r$$

or, equivalently,

$$n_{ib} = n_{ia} + \nu_i \epsilon \qquad \text{for } i = 1, 2, \ldots, r \qquad (29.10)$$

The set of values n_{1b}, n_{2b}, ..., n_{rb} is compatible with the given set of values n_{1a}, n_{2a}, ..., n_{ra} (Section 2.2) because it conforms to the requirements imposed by the chemical reaction mechanism.

Initial state

$n_{1a}, n_{2a}, ..., n_{ra}$

Chemical reaction mechanism

$$\sum_{i=1}^{r} \nu_i A_i = 0$$

Final state

$n_{1b}, n_{2b}, ..., n_{rb}$

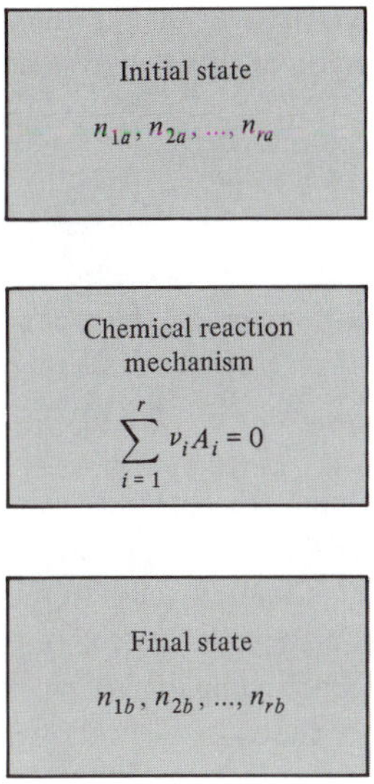

Figure 29.1 Schematics of end states and chemical reaction mechanism of a batch process.

It is noteworthy that one or more of the constituents of the system may not be participating in the chemical reaction. For each of these constituents the stoichiometric coefficient is zero and, therefore, the amount is not subject to change by the chemical reaction.

Equations (29.10) are very useful for analyses of batch processes. For example, given
the initial amounts $n_{1a}, n_{2a}, \ldots, \dot{n}_{ra}$, the stoichiometric coefficients $\nu_1, \nu_2, \ldots, \nu_r$, and
the final amount of constituent 1, n_{1b}, we can use (29.10) for $i = 1$ and determine the
value of ϵ, that is, $\epsilon = (n_{1b} - n_{1a})/\nu_1$. Then we can substitute this value of ϵ in equations
(29.10) for $i = 2, 3, \ldots, r$ and find the final amounts of all the other constituents. For
example, if $n_{1a} = 1$ mol, $\nu_1 = -1$, and $n_{1b} = 0$, then $\epsilon = (n_{1b} - n_{1a})/\nu_1 = 1$, and
$n_{ib} = n_{ia} + \nu_i$. Moreover, if $\nu_k = 0$, then $n_{kb} = n_{ka}$.

We can also use the reaction coordinate to express changes in composition. Upon
defining

$$n_a = \sum_{i=1}^{r} n_{ia} \quad n_b = \sum_{i=1}^{r} n_{ib} \quad y_{ia} = \frac{n_{ia}}{n_a} \quad y_{ib} = \frac{n_{ib}}{n_b} \quad \nu = \sum_{i=1}^{r} \nu_i \qquad (29.11)$$

and using equations (29.10), we find that the final composition $y_{1b}, y_{2b}, \ldots, y_{rb}$ is related
to the initial composition $y_{1a}, y_{2a}, \ldots, y_{ra}$ by means of the expressions

$$n_b = \sum_{i=1}^{r} n_{ib} = \sum_{i=1}^{r} (n_{ia} + \nu_i \epsilon) = n_a + \nu \epsilon \qquad (29.12)$$

$$y_{ib} = \frac{n_{ib}}{n_b} = \frac{n_{ia} + \nu_i \epsilon}{n_a + \nu \epsilon} = \frac{y_{ia} + \nu_i \epsilon/n_a}{1 + \nu \epsilon/n_a} \qquad \text{for } i = 1, 2, \ldots, r \qquad (29.13)$$

Clearly, equations (29.11) to (29.13) refer to all constituents, including those that do not
participate in the chemical reaction and for which the stoichiometric coefficients are zero.

Before proceeding with the discussion of the second process, we should clarify a
typical source of confusion. The reaction relation $\sum_{i=1}^{r} \nu_i A_i = 0$ specifies the mechanism by which the constituents of a system can react, but does not specify the amounts
with which the constituents are present before, during, and after the occurrence of the
chemical reaction.

We can use the chemical reaction mechanism and find a relation between the amounts
of constituents of two states. To this end, we multiply each stoichiometric coefficient by ϵ
and write the chemical reaction mechanism as $\sum_{i=1}^{r}(\nu_i \epsilon)A_i = 0$. Then we substitute for
the product $\nu_i \epsilon$ the expression $n_{ib} - n_{ia}$ [equation (29.10)], including all the constituents
for which $\nu_i = 0$, and find that

$$\sum_{i=1}^{r} n_{ia} A_i - \sum_{i=1}^{r} n_{ib} A_i = 0 \qquad \text{or} \qquad \sum_{i=1}^{r} n_{ia} A_i = \sum_{i=1}^{r} n_{ib} A_i \qquad (29.14)$$

This balance between initial and final amounts is useful in many analyses. However,
it is also sometimes a source of confusion because one might be tempted to identify
the stoichiometric coefficient ν_i with the amount either n_{ia} or n_{ib} and not, as correctly
specified by equations (29.10), with the difference $n_{ib} - n_{ia}$.

As we have already stated, each sum in relation (29.14) includes all constituents of
the system, regardless of whether they do or do not participate in the chemical reaction
mechanism, namely, regardless of whether their stoichiometric coefficients are different
from or equal to zero. Each constituent for which $\nu_i = 0$ is called an *inert substance* of
the chemical reaction. The inclusion of the inert substances in each sum in (29.14) is
necessary because these constituents determine the composition of the state even though
they do not participate in the reaction mechanism.

It is noteworthy that neither the chemical reaction mechanism [relation (29.4)] nor relation (29.14) is strictly an algebraic equation. Rather, each is a combination of symbols some with a quantitative measure, such as the ν_i's and the n_i's, and some without such a measure, such as the A_i's.

Example 29.2. In a chemical reactor, an amount $n_{1a} = 0.05$ mol of propane, C_3H_8, is transformed completely into CO_2 and H_2O by reacting with a mixture consisting of 0.315 mol of O_2 and 1.185 mol of N_2. Find the composition of the final state.

Solution. The chemical reaction mechanism is

$$C_3H_8(g) + 5\,O_2(g) = 3\,CO_2(g) + 4\,H_2O(g)$$

The constituents of the system are C_3H_8, O_2, N_2, CO_2, and H_2O. The stoichiometric coefficients, initial amounts, and initial composition are listed in Table 29.1.

TABLE 29.1

i	A_i	ν_i		n_{ia}	y_{ia}
1	C_3H_8	-1		0.050	0.032
2	O_2	-5	$0.21 \times 1.5 = 0.315$		0.203
3	N_2	0	$0.79 \times 1.5 = 1.185$		0.765
4	CO_2	3		0.000	0.000
5	H_2O	4		0.000	0.000
$\sum_{i=1}^{5}$		$\nu = 1$		$n_a = 1.550$	1.000

In the final state, the number of moles of propane $n_{1b} = 0$ because we assume that it is completely changed into CO_2 and H_2O. Therefore, the reaction coordinate $\epsilon = (n_{1b} - n_{1a})/\nu_1 = -0.05/(-1) = 0.05$ and, using equations (29.10) to (29.13), we find the results listed in Table. 29.2.

TABLE 29.2

i	A_i		n_{ib}	y_{ib}
1	C_3H_8	$n_{1a} + \nu_1\epsilon = 0.000$		0.000
2	O_2	$n_{2a} + \nu_2\epsilon = 0.315 + (-5) \times 0.05 = 0.065$		0.040
3	N_2	$n_{3a} + \nu_3\epsilon = 1.185 + 0 \times 0.05 = 1.185$		0.741
4	CO_2	$n_{4a} + \nu_4\epsilon = 0 + 3 \times 0.05 = 0.150$		0.094
5	H_2O	$n_{5a} + \nu_5\epsilon = 0 + 4 \times 0.05 = 0.200$		0.125
$\sum_{i=1}^{5}$			$n_b = 1.600$	1.000

The balance between initial and final amounts is

$$0.05\,C_3H_8 + 0.315\,O_2 + 1.185\,N_2 = 0.15\,CO_2 + 0.2\,H_2O + 0.065\,O_2 + 1.185\,N_2$$

where the coefficients on the left-hand side represent the respective amounts of the initial state, and the coefficients on the right-hand side the respective amounts of the final state, all in moles. Both the value of ϵ and the values of the final amounts are obtained by assuming that all the propane is exhausted.

As the second process, we examine that of a chemical reactor under steady-state bulk-flow conditions (Figure 29.2). We call this a *steady-state rate process*. The r

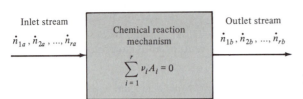

Figure 29.2   Schematic of a steady-state rate process.

constituents in the chemical reactor are A_1, A_2, ..., A_r. They enter the reactor as one stream or many streams, and their inflow rates are specified by the set $\dot{n}_{1a}$, $\dot{n}_{2a}$, ..., $\dot{n}_{ra}$, where the dot stands for a derivative with respect to time. As a result of a chemical reaction mechanism with stoichiometric coefficients ν_1, ν_2, ..., ν_r, the outlet stream is characterized by outflow rates specified by the set $\dot{n}_{1b}$, $\dot{n}_{2b}$, ..., $\dot{n}_{rb}$. In some reactors, the outlet may also consist of more than one stream. It is noteworthy that one or more of either the inlet rates or the outlet rates, as well as one or more of the stoichiometric coefficients may be null.

Changes in amounts may be due either to exchanges of constituents through the inlet and outlet ports, or to the chemical reaction mechanism inside the chemical reactor, or both. In a small interval of time dt, the change dn_i in the ith amount in the reactor is the sum of the contribution $(\dot{n}_{ia} - \dot{n}_{ib})\,dt$ due to the inlet and outlet streams, and the contribution $\nu_i\,d\epsilon$ due to the chemical reaction mechanism [equation (29.9a)], that is,

$$dn_i = (\dot{n}_{ia} - \dot{n}_{ib})\,dt + \nu_i\,d\epsilon \quad \text{or} \quad \frac{dn_i}{dt} = \dot{n}_{ia} - \dot{n}_{ib} + \nu_i\,\dot{\epsilon} \quad \text{for } i = 1,\,2,\,\ldots,\,r \quad (29.15)$$

where $\dot{\epsilon}$ denotes the *reaction coordinate rate*. At steady state, the amounts inside the reactor remain constant so that $dn_i/dt = 0$, and equations (29.15) become

$$\dot{n}_{ib} = \dot{n}_{ia} + \nu_i\,\dot{\epsilon} \qquad \text{for } i = 1,\,2,\,\ldots,\,r \qquad (29.16)$$

Equations (29.16) are analogous to (29.10) and are very useful for analyses of steady-state rate processes.

We can also use $\dot{\epsilon}$ to express changes in composition. Upon defining

$$\dot{n}_a = \sum_{i=1}^{r} \dot{n}_{ia} \quad \dot{n}_b = \sum_{i=1}^{r} \dot{n}_{ib} \quad y_{ia} = \frac{\dot{n}_{ia}}{\dot{n}_a} \quad y_{ib} = \frac{\dot{n}_{ib}}{\dot{n}_b} \quad \nu = \sum_{i=1}^{r} \nu_i \qquad (29.17)$$

and using equations (29.16), we find that the outlet composition y_{1b}, y_{2b}, ..., y_{rb} is related to the inlet composition y_{1a}, y_{2a}, ..., y_{ra} by means of the expressions

$$\dot{n}_b = \sum_{i=1}^{r} \dot{n}_{ib} = \sum_{i=1}^{r} (\dot{n}_{ia} + \nu_i\,\dot{\epsilon}) = \dot{n}_a + \nu\,\dot{\epsilon} \qquad (29.18)$$

$$y_{ib} = \frac{\dot{n}_{ib}}{\dot{n}_b} = \frac{\dot{n}_{ia} + \nu_i\,\dot{\epsilon}}{\dot{n}_a + \nu\,\dot{\epsilon}} = \frac{y_{ia} + \nu_i\,\dot{\epsilon}/\dot{n}_a}{1 + \nu\,\dot{\epsilon}/\dot{n}_a} \qquad \text{for } i = 1,\,2,\,\ldots,\,r \qquad (29.19)$$

In the next section we develop an approach that unifies the formal treatments of changes in compositions of both batch and steady-state rate processes.

29.5   Degree of Reaction

For the batch process in Figure 29.1, we define the ratio ξ by the relation

$$\xi = \frac{\epsilon}{n_a} \qquad (29.20)$$

where ϵ is the reaction coordinate, and n_a is the total amount of constituents in the initial state of the process. This ratio is called the *degree of reaction*.

Upon substituting (29.20) in equations (29.12) and (29.13), we find that

$$\frac{n_b}{n_a} = 1 + \nu\,\xi \tag{29.21}$$

$$y_{ib} = \frac{y_{ia} + \nu_i\,\xi}{1 + \nu\,\xi} \qquad \text{for } i = 1, 2, \ldots, r \tag{29.22}$$

For the steady-state rate process in Figure 29.2, we define the ratio ξ by the relation

$$\xi = \frac{\dot{\epsilon}}{\dot{n}_a} \tag{29.23}$$

where $\dot{\epsilon}$ is the reaction coordinate rate, and $\dot{n}_a$ the total inflow rate at the inlet port, expressed in moles per unit time. This ratio is also called the degree of reaction.

Upon substituting (29.23) in equations (29.18) and (29.19), we find that

$$\frac{\dot{n}_b}{\dot{n}_a} = 1 + \nu\,\xi \tag{29.24}$$

$$y_{ib} = \frac{y_{ia} + \nu_i\,\xi}{1 + \nu\,\xi} \qquad \text{for } i = 1, 2, \ldots, r \tag{29.25}$$

that is, two equations that are formally identical to equations (29.21) and (29.22). By introducing the degree of reaction, we see that both the batch-process and the rate-process compositions are described by the same expression.

Example 29.3. In a rate process, gaseous propane, C_3H_8, is supplied to a chemical reactor together with a mixture of oxygen, O_2, and nitrogen, N_2. As a result of a single chemical reaction mechanism, propane is transformed into carbon dioxide, CO_2, and water, H_2O. The inlet flow rates are 4 mol O_2/mol of C_3H_8, and 15.04 mol N_2/mol of C_3H_8. The outlet stream consists of CO_2, C_3H_8, H_2O, and N_2. Find the composition of the oulet stream and the degree of reaction.

Solution. The balance between input and output flow rates per unit flow rate of C_3H_8 is $C_3H_8 + 4\,O_2 + 15.04\,N_2 = \alpha\,CO_2 + \beta\,C_3H_8 + \gamma\,H_2O + 15.04\,N_2$, where the relative rates are determined by the principle of conservation of atomic nuclei. So, for carbon we find $\alpha + 3\beta = 3$, for hydrogen $8\beta + 2\gamma = 8$, and for oxygen $2\alpha + \gamma = 8$, or $\alpha = 2.4$, $\beta = 0.2$, and $\gamma = 3.2$. The total amount of the outlet stream $n_b = 2.4 + 0.2 + 3.2 + 15.04 = 20.84$. So, denoting C_3H_8, O_2, N_2, CO_2, and H_2O by the subscripts 1, 2, 3, 4, and 5, respectively, we find: $y_{1b} = \beta/n_b = 0.009$; $y_{2b} = 0$; $y_{3b} = 15.04/20.84 = 0.722$; $y_{4b} = \alpha/n_b = 0.115$; $y_{5b} = \gamma/n_b = 0.154$. Regarding the degree of reaction, we note that $n_{1a} = 1$, $n_{1b} = 0.2$, $\nu_1 = -1$, $\epsilon = (n_{1b} - n_{1a})/\nu_1 = 0.8$, $n_a = 1 + 4 + 15.04 = 20.04$, and $\xi = \epsilon/n_a = 0.8/20.04 = 0.04$. The stoichiometric coefficient $\nu_1 = -1$ because the chemical reaction mechanism is $C_3H_8 + 5\,O_2 = 3\,CO_2 + 4\,H_2O$.

29.6 Multiple Reactions

A chemical reaction can often be regarded as a combination of two or more reactions. For example, the reaction described by relation (29.1) can be regarded as the sum of the two reaction mechanisms

$$H_2O = H_2 + \tfrac{1}{2}\,O_2 \tag{29.26}$$

and

$$C + \tfrac{1}{2}\,O_2 = CO \tag{29.27}$$

because upon combining relations (29.26) and (29.27) we reproduce relation (29.1). In some applications, the representation of a reaction mechanism as a combination of other reaction mechanisms is very useful because the properties of the partial reaction mechanisms may be easier to study than those of the overall reaction mechanism. In other systems, the need to consider multiple reaction mechanisms may arise because different reaction mechanisms may have different time characteristics.

To analyze systems in which several reaction mechanisms are effective, we adopt the following terminology. We denote the stoichiometric coefficient of constituent i in mechanism j by $\nu_i^{(j)}$. The values of i range from 1 to r, and the values of j from 1 to τ, where r is the number of constituents of the system, and τ the number of reaction mechanisms that need to be considered. For example, for a system consisting of C, H_2O, CO, H_2, and O_2, denoting the reaction mechanism described by relation (29.26) by $j = 1$, and that by relation (92.27) by $j = 2$, we have the stoichiometric coefficients $\nu_i^{(1)}$ and $\nu_i^{(2)}$ listed in Table 29.3.

TABLE 29.3

i	A_i	$\nu_i^{(1)}$	$\nu_i^{(2)}$	ν_i
1	C	0.0	−1.0	−1.0
2	H_2O	−1.0	0.0	−1.0
3	CO	0.0	1.0	1.0
4	H_2	1.0	0.0	1.0
5	O_2	0.5	−0.5	0.0
$\sum_{i=1}^{5}$		$\nu^{(1)} = 0.5$	$\nu^{(2)} = -0.5$	$\nu = 0$

We can readily verify the values of these coefficients by writing the two reaction mechanisms in the forms

$$j = 1: \quad \sum_{i=1}^{5} \nu_i^{(1)} A_i = 0 \tag{29.28}$$

$$j = 2: \quad \sum_{i=1}^{5} \nu_i^{(2)} A_i = 0 \tag{29.29}$$

and then combining relations (29.28) and (29.29) to find that

$$\sum_{j=1}^{2} \sum_{i=1}^{5} \nu_i^{(j)} A_i = \sum_{i=1}^{5} \left(\sum_{j=1}^{2} \nu_i^{(j)} \right) A_i = \sum_{i=1}^{5} \nu_i A_i = 0 \tag{29.30}$$

where

$$\nu_i = \sum_{j=1}^{2} \nu_i^{(j)} \quad \text{for } i = 1, 2, \ldots, 5 \tag{29.31}$$

The values of the stoichiometric numbers ν_i resulting from equations (29.31) are those of relation (29.1), that is, of the overall reaction mechanism.

In analyzing systems with many chemical reactions, we need a reaction coordinate for each reaction. For example, if the amount of constituent i changes from n_i to $n_i + dn_i$ due to only two reactions, then

$$dn_i = dn_i^{(1)} + dn_i^{(2)} \qquad (29.32)$$

where $dn_i^{(1)}$ is the change due to the first reaction mechanism and $dn_i^{(2)}$ the change due to the second reaction mechanism. Each of the two partial changes obeys its own proportionality relations [equations (29.9)] so that

$$dn_i^{(1)} = \nu_i^{(1)} d\epsilon_1 \quad \text{and} \quad dn_i^{(2)} = \nu_i^{(2)} d\epsilon_2 \qquad \text{for } i = 1, 2, \ldots, r \qquad (29.33)$$

where ϵ_1 and ϵ_2 are the reaction coordinates of the first and second reactions, respectively. Thus, equation (29.32) becomes

$$dn_i = \nu_i^{(1)} d\epsilon_1 + \nu_i^{(2)} d\epsilon_2 \qquad \text{for } i = 1, 2, \ldots, r \qquad (29.34)$$

Equations (29.34) are the proportionality relations for a system with r constituents and two chemical reaction mechanisms.

More generally, for r constituents, τ concurrent chemical reaction mechanisms, and an overall reaction that is the sum of the τ concurrent reactions, we have the following relations:

<u>Partial reactions</u>

$$\sum_{i=1}^{r} \nu_i^{(j)} A_i = 0 \qquad \text{for } j = 1, 2, \ldots, \tau \qquad (29.35)$$

<u>Overall reaction</u>

$$\sum_{i=1}^{r} \nu_i A_i = 0 \qquad (29.36)$$

where

$$\nu_i = \sum_{j=1}^{\tau} \nu_i^{(j)} \qquad \text{for } i = 1, 2, \ldots, r \qquad (29.37)$$

and, for subsequent use, we define the following sums of stoichiometric numbers:

$$\nu^{(j)} = \sum_{i=1}^{r} \nu_i^{(j)} \qquad \text{for } j = 1, 2, \ldots, \tau \qquad (29.38)$$

$$\nu - \sum_{j=1}^{\tau} \sum_{i=1}^{r} \nu_i^{(i)} = \sum_{j=1}^{\tau} \nu^{(j)}$$

$$= \sum_{i=1}^{r} \sum_{j=1}^{\tau} \nu_i^{(j)} = \sum_{i=1}^{r} \nu_i \qquad (29.39)$$

For the overall reaction mechanism, the proportionality relations for batch and steady-state rate processes are, respectively,

$$dn_i = \sum_{j=1}^{\tau} \nu_i^{(j)} d\epsilon_j \qquad \text{for } i = 1, 2, \ldots, r \qquad (29.40)$$

and

$$\dot{n}_{ib} = \dot{n}_{ia} + \sum_{j=1}^{\tau} \nu_i^{(j)} \dot{\epsilon}_j \qquad \text{for } i = 1, 2, \ldots, r \qquad (29.41)$$

where ϵ_j and $\dot{\epsilon}_j$ are the reaction coordinate and the reaction coordinate rate of the jth partial reaction mechanism, respectively. Upon selecting the values $\epsilon_j = 0$ for $j = 1, 2, \ldots, \tau$ to correspond to the initial amounts of constituents, and integrating equations (29.40), we find that

$$n_{ib} = n_{ia} + \sum_{j=1}^{\tau} \nu_i^{(j)} \epsilon_j \qquad \text{for } i = 1, 2, \ldots, r \tag{29.42}$$

Equations (29.42) express the condition that the set of values $n_{1b}, n_{2b}, \ldots, n_{rb}$ is compatible with the given set of values $n_{1a}, n_{2a}, \ldots, n_{ra}$ because the two sets conform to the requirements imposed by the τ chemical reaction mechanisms.

As a result of these relations and definitions, we can readily verify that the initial and final amounts and compositions of either batch or steady-state rate processes with τ concurrent chemical reactions satisfy the following equalities:

$$\frac{n_b}{n_a} = 1 + \sum_{j=1}^{\tau} \nu^{(j)} \xi_j \tag{29.43}$$

$$\frac{\dot{n}_b}{\dot{n}_a} = 1 + \sum_{j=1}^{\tau} \nu^{(j)} \xi_j \tag{29.44}$$

$$y_{ib} = \frac{y_{ia} + \sum_{j=1}^{\tau} \nu_i^{(j)} \xi_j}{1 + \sum_{j=1}^{\tau} \nu^{(j)} \xi_j} \qquad \text{for } i = 1, 2, \ldots, r \tag{29.45}$$

where the degree of reaction of the jth reaction ξ_j equals either ϵ_j/n_a or $\dot{\epsilon}_j/\dot{n}_a$ for $j = 1, 2, \ldots, \tau$. Equations (29.43) to (29.45) reduce to equations (29.21), (29.24), and either (29.22) or (29.25) for a single reaction, that is, for $\tau = 1$.

The proportionality relations for each partial reaction mechanism imply that

$$\frac{1}{\nu_i} \sum_{j=1}^{\tau} \nu_i^{(j)} \xi_j = \frac{1}{\nu} \sum_{j=1}^{\tau} \nu^{(j)} \xi_j = \xi \qquad \text{for } i = 1, 2, \ldots, r \tag{29.46}$$

where ξ is the degree of reaction of the overall reaction mechanism.

Example 29.4.   Prove equations (29.46).

Solution.   We assume that the partial reactions and the overall reaction satisfy relations (29.35) to (29.39). In terms of the reaction coordinate ϵ and the degree of reaction ξ, we have the equalities

$$dn_i = \nu_i \, d\epsilon = \sum_{j=1}^{\tau} \nu_i^{(j)} d\epsilon_j \qquad \text{and} \qquad \nu_i \, d\xi = \sum_{j=1}^{\tau} \nu_i^{(j)} d\xi_j$$

Upon integrating the second equality from the initial zero values of the degrees of reaction to some final values, we find that

$$\xi = \frac{1}{\nu_i} \sum_{j=1}^{\tau} \nu_i^{(j)} \xi_j$$

Moreover, upon summing the results of the integrations over all i's, we find that

$$\left(\sum_{i=1}^{r} \nu_i \right) \xi = \nu \xi = \sum_{i=1}^{r} \sum_{j=1}^{\tau} \nu_i^{(j)} \xi_j = \sum_{j=1}^{\tau} \left(\sum_{i=1}^{r} \nu_i^{(j)} \right) \xi_j = \sum_{j=1}^{\tau} \nu^{(j)} \xi_j$$

So both of equations (29.46) are verified.

29.7 Energy and Entropy Balances for a Batch Process

We consider a simple system consisting of r constituents with constraints that inhibit all the chemical reaction mechanisms. Initially, the system is in a stable equilibrium state, denoted by subscript a and having the characteristics listed in Figure 29.3a. The constraint that inhibits the chemical reaction $\sum_{i=1}^{r} \nu_i A_i = 0$ is removed for a lapse of time and then imposed again. As a result of work and heat interactions plus the effects of the chemical reaction during the given lapse of time (Figure 29.3b), the final state of the system is again stable equilibrium, denoted by subscript b and having the characteristics listed in Figure 29.3c. In particular, the initial and final volumes are equal, $V_a = V_b$, the shaft work done on other systems is $W_s^{\rightarrow}$, and the heat interactions occur with q different sources, each supplying heat $Q_k^{\leftarrow}$ at temperature T_k for $k = 1, 2, \ldots, q$. It is noteworthy that the initial and final states are stable equilibrium states of a simple system with no reaction mechanisms, whereas during the given lapse of time the system is changed temporarily to one in which the given reaction mechanism is active. In other words, the given reaction mechanism is activated—"switched on"—temporarily and then deactivated again—"switched off." For many practical batch processes, this is a reasonable model of the end states of the process.[1]

For the batch process just defined, the energy and entropy balances are

$$U_b - U_a = \sum_{k=1}^{q} Q_k^{\leftarrow} - W_s^{\rightarrow} \tag{29.47}$$

$$S_b - S_a = \sum_{k=1}^{q} \frac{Q_k^{\leftarrow}}{T_k} + S_{\text{irr}} \tag{29.48}$$

For a stable equilibrium state of a simple system with no chemical reactions, the internal energy U and the entropy S are functions of the temperature T, the pressure

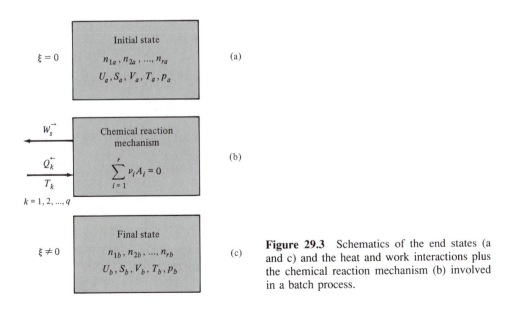

Figure 29.3 Schematics of the end states (a and c) and the heat and work interactions plus the chemical reaction mechanism (b) involved in a batch process.

[1]This model includes the possibility that one of the end states, or both, be chemical equilibrium. Chemical equilibrium states are discussed in Chapter 30.

p, and the amounts $n_1, n_2, \ldots, n_r$. In terms of partial properties and the composition $y_1, y_2, \ldots, y_r$, U and S can be expressed as [equations (26.41 and 26.38)]

$$U(T, p, n_1, n_2, \ldots, n_r) = \sum_{i=1}^{r} n_i \, u_i(T, p, y_1, y_2, \ldots, y_r) \qquad (29.49)$$

$$S(T, p, n_1, n_2, \ldots, n_r) = \sum_{i=1}^{r} n_i \, s_i(T, p, y_1, y_2, \ldots, y_r) \qquad (29.50)$$

where u_i is the partial internal energy, and s_i the partial entropy of constituent i.

Because the composition $y_{1a}, y_{2a}, \ldots, y_{ra}$ differs from the composition $y_{1b}, y_{2b}, \ldots, y_{rb}$, the evaluation of each of the differences $U_b - U_a$ and $S_b - S_a$ in terms of tabulated experimental data requires a special technique. We discuss this technique in Section 29.9.

29.8   Energy and Entropy Balances for a Steady-State Rate Process

We consider a system consisting of r constituents with constraints that inhibit all chemical reaction mechanisms except $\sum_{i=1}^{r} \nu_i A_i = 0$. The system is maintained in a steady state by shaft work interactions, heat interactions with q heat sources each at temperature T_k, for $k = 1, 2, \ldots, q$, and two bulk-flow interactions with streams of different but fixed compositions at the inlet and outlet ports (Figure 29.4). We denote the bulk-flow state of the inlet stream by subscript a, and that of the outlet stream by subscript b. We assume that these are bulk-flow states of a system subject to no reaction mechanisms. In other words, we assume that the given chemical reaction mechanism is active only within the system, not within either the inlet or the outlet stream. For many practical rate processes, this is a reasonable assumption that results in an acceptable model of the conditions prevailing at the inlet and outlet ports.[2]

Considering conditions under which kinetic and gravitational energy changes between the inlet and outlet streams with respect to changes in enthalpy are negligible, we have that the energy and entropy rate balances for the system are

$$\dot{H}_b - \dot{H}_a = \sum_{k=1}^{q} \dot{Q}_k^{\leftarrow} - \dot{W}_s^{\rightarrow} \qquad (29.51)$$

$$\dot{S}_b - \dot{S}_a = \sum_{k=1}^{q} \frac{\dot{Q}_k^{\leftarrow}}{T_k} + \dot{S}_{\mathrm{irr}} \qquad (29.52)$$

where H denotes enthalpy, S entropy, and a dot over a symbol a time rate.

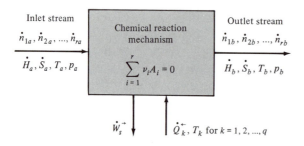

Figure 29.4 Bulk flows, heat and work interactions plus a chemical reaction mechanism involved in a steady-state rate process.

---

[2]Again, this model includes the possibility that the stable equilibrium part of either the inlet or the outlet state, or both, be a chemical equilibrium state. Chemical equilibrium states are discussed in Chapter 30.

For each bulk-flow state of given temperature, pressure, and composition of a stream subject to no reaction mechanisms, the rates of enthalpy flow and entropy flow can be expressed in terms of partial properties as

$$\dot{H} = \sum_{i=1}^{r} \dot{n}_i \, h_i(T, p, y_1, y_2, \ldots, y_r) \tag{29.53}$$

$$\dot{S} = \sum_{i=1}^{r} \dot{n}_i \, s_i(T, p, y_1, y_2, \ldots, y_r) \tag{29.54}$$

where $\dot{n}_i$, h_i, and s_i are the mole flow rate, the partial enthalpy, and the partial entropy of constituent i, respectively. The proof of equations (29.53) and (29.54) is left as an exercise for the reader.

If there are more than one inlet port, outlet port, or both, a bulk-flow state is defined for each port, and the rates of energy flow and entropy flow are sums over the various ports of expressions of the forms of equations (29.53) and (29.54), respectively.

In general, the inlet and outlet compositions differ and, for this reason, the evaluation of each of the differences $\dot{H}_b - \dot{H}_a$ and $\dot{S}_b - \dot{S}_a$ in terms of tabulated data requires the special technique discussed in the next section.

29.9 Mixture Properties at Different Compositions

For pure constituents, values of properties of stable equilibrium states are tabulated or may be evaluated by means of special explicit mathematical forms. We recall that for each constituent and each extensive property, a convenient reference state and reference value are selected. This value is arbitrary. Often, different investigators make different choices not only for different constituents but even for the same constituent. The arbitrary choice of the reference value is inconsequential because in most applications we evaluate differences in the values of the property of the constituent in two states, and thus the reference value cancels out.

For multiconstituent systems, however, some constituents may combine chemically to form other constituents, and the evaluation of extensive properties must be made on the basis of either tabulated properties of pure constituents, or of changes in values of properties between states of different compositions, or both. Under these circumstances, a normalization of the various reference values is required because the arbitrarily and independently selected reference values of the various constituents may no longer be compatible.

For example, in the tabulation of properties of stable equilibrium states of H_2 we can assign a zero value to the energy of an arbitrarily chosen state of H_2. The same arbitrariness applies to tabulated properties of O_2 and H_2O. However, when we consider a system in which H_2 and O_2 may combine to form H_2O, and wish to use the tabulated data on H_2, O_2, and H_2O, this arbitrariness is no longer acceptable because the energy of H_2O bears a definite relation to the energies of H_2 and O_2. To make ideas specific, if the zero energy of H_2 corresponds to the ground-state energy of free H_2, and the zero energy of O_2 to the ground-state energy of free O_2, we note that the zero energy of H_2O cannot correspond to the ground-state energy of free H_2O. The reason is that it takes a definite amount of energy to dissociate H_2O into H_2 and O_2, and therefore the ground-state energy of free H_2O is less than that of free H_2 plus one half that of free O_2.

We illustrate the technique used to normalize the various references for a system with r chemically reacting constituents by considering the expression for the difference in enthalpies of two states a and b. These states differ in composition as a result of an overall chemical reaction with stoichiometric coefficients $\nu_1, \nu_2, \ldots, \nu_r$. The result derived for enthalpy can easily be generalized to other extensive properties.

First, we express the enthalpies H_a and H_b in terms of partial enthalpies of the constituents in states a and b, so that

$$H_b - H_a = \sum_{i=1}^{r} n_{ib}\, h_{ib} - \sum_{i=1}^{r} n_{ia}\, h_{ia} \tag{29.55}$$

where the functional dependences of the partial enthalpies are omitted but should be kept in mind, that is, $h_{ia} = h_i(T_a, p_a, y_{ia}, y_{2a}, \ldots, y_{ra})$ and $h_{ib} = h_i(T_b, p_b, y_{1b}, y_{2b}, \ldots, y_{rb})$.

Next, we add and substract to each of the partial enthalpies h_{ia} and h_{ib} the quantity $h_{ii}^{\circ}(T) = h_{ii}(T, p_o)$, that is, the specific enthalpy of the pure constituent i at temperature T and pressure p_o. Thus we have the relations

$$h_{ia} = h_{ia} - h_{ii}^{\circ}(T) + h_{ii}^{\circ}(T) \qquad \text{and} \qquad h_{ib} = h_{ib} - h_{ii}^{\circ}(T) + h_{ii}^{\circ}(T) \tag{29.56}$$

and recall that the value of $h_{ii}^{\circ}(T)$ depends on the form of aggregation of pure constituent i at T and p_o, namely, on whether i is solid, liquid, or vapor. Upon combining equations (29.55) and (29.56), we find that

$$H_b - H_a = \sum_{i=1}^{r} n_{ib}\left[h_{ib} - h_{ii}^{\circ}(T)\right] - \sum_{i=1}^{r} n_{ia}\left[h_{ia} - h_{ii}^{\circ}(T)\right] + \sum_{i=1}^{r}(n_{ib} - n_{ia})\, h_{ii}^{\circ}(T)$$

$$= \sum_{i=1}^{r} n_{ib}\left[h_{ib} - h_{ii}^{\circ}(T)\right] - \sum_{i=1}^{r} n_{ia}\left[h_{ia} - h_{ii}^{\circ}(T)\right] + n_a\, \xi\, \Delta h^{\circ}(T) \tag{29.57}$$

where we use equations (29.10) to evaluate the differences $n_{ib} - n_{ia}$, and equation (29.20) to write $\epsilon = n_a\, \xi$, that is,

$$\sum_{i=1}^{r}(n_{ib} - n_{ia})\, h_{ii}^{\circ}(T) = \sum_{i=1}^{r} \nu_i\, \epsilon\, h_{ii}^{\circ}(T) = \epsilon\, \Delta h^{\circ}(T) = n_a\, \xi\, \Delta h^{\circ}(T) \tag{29.58}$$

and define

$$\Delta h^{\circ}(T) = \sum_{i=1}^{r} \nu_i\, h_{ii}^{\circ}(T) \tag{29.59}$$

The function $\Delta h^{\circ}(T)$ is called the *enthalpy of reaction at temperature T and pressure* p_o. When the temperature $T = T_0 = 25°C = 77°F$ and the pressure $p_o = 1$ atm, then $\Delta h^{\circ}(T_0)$ is denoted simply by Δh°, and is called the *enthalpy of reaction at standard conditions*. The value of either $\Delta h^{\circ}(T)$ or Δh° depends on the forms of aggregation of each pure product and each pure reactant considered in the chemical reaction mechanism because the value of $h_{ii}^{\circ}(T)$ depends on the form of aggregation of constituent i.

Using equations (29.21) and (29.22), we can rewrite equation (29.57) in the form

$$\frac{(H_b - H_a)}{n_a} = (1 + \nu\, \xi) \sum_{i=1}^{r} y_{ib}\left[h_{ib} - h_{ii}^{\circ}(T)\right]$$

$$- \sum_{i=1}^{r} y_{ia}\left[h_{ia} - h_{ii}^{\circ}(T)\right] + \xi\, \Delta h^{\circ}(T) \tag{29.60}$$

We see from this form that the value of $H_b - H_a$ can be found without concern about the arbitrary zero reference values for the different constituents, provided that the enthalpy of reaction $\Delta h^\circ(T)$ is known. Indeed, each of the two sums on the right-hand side of equation (29.60) represents a difference between the enthalpies of two states of the same composition. In particular, the first sum is the difference between the specific enthalpy of a nonreacting mixture in a stable equilibrium state at temperature T_b, pressure p_b, and composition y_{1b}, y_{2b}, ..., y_{rb}, and the specific enthalpy of a composite of r pure constituents, the ith pure constituent and its amount being the same as the respective constituent and amount in the mixture. Each pure constituent is in a stable equilibrium state at temperature T and pressure p_o and, therefore, in partial mutual stable equilibrium with each of the other pure constituents. Because the state of the mixture and the state of the composite have the same composition y_{1b}, y_{2b}, ..., y_{rb}, the difference between the specific enthalpies of the two states is independent of the reference values. Similar remarks apply to the second sum except that here the mixture is at temperature T_a, pressure p_a, and composition y_{1a}, y_{2a}, ..., y_{ra}, and the ith pure constituent and its amount are the same as the respective constituent and amount of the mixture at T_a, p_a, y_{1a}, y_{2a}, ..., y_{ra}. The third term on the right-hand side of equation (29.60), the enthalpy of reaction, is a weighted sum over the specific enthalpies of all the constituents. As such, it contains all the arbitrary reference values. Its determination is discussed in Section 29.10.

If the multiconstituent system behaves as a Gibbs–Dalton mixture (Section 27.1), then

$$H(T, p, n_1, n_2, \ldots, n_r) = \sum_{i=1}^{r} n_i \, h_{ii}(T, p_{ii}) \qquad (29.61)$$

where $h_{ii}(T, p_{ii})$ is the specific enthalpy of pure constituent i at temperature T and pressure p_{ii} equal to the partial pressure of constituent i in the mixture. Accordingly, equation (29.60) becomes

$$\frac{(H_b - H_a)}{n_a} = (1 + \nu \xi) \sum_{i=1}^{r} y_{ib} \left[h_{ii}(T_b, p_{iib}) - h_{ii}(T, p_o) \right]$$

$$- \sum_{i=1}^{r} y_{ia} \left[h_{ii}(T_a, p_{iia}) - h_{ii}(T, p_o) \right] + \xi \, \Delta h^\circ(T) \qquad (29.62)$$

that is, the two sums in the right-hand side of (29.60) are determined by data on pure constituents only. If the multiconstituent system does not behave as a Gibbs–Dalton mixture, these two sums must be evaluated by using either appropriate tabulated values or some analytical model for the mixture.

Using the notation introduced in Section 29.8, we can show that the ratio $(\dot{H}_b - \dot{H}_a)/\dot{n}_a$ that appears in the energy balance of a steady-state rate process is formally equal to the expression $(H_b - H_a)/n_a$ given by the right-hand side of either equation (29.62) or equation (29.60), depending on whether the system does or does not behave as a Gibbs–Dalton mixture, respectively. We do so by substituting $\dot{H}$ for H and $\dot{n}_i$ for n_i in the first of equations (29.57), and using the relations $\dot{n}_{ib} - \dot{n}_{ia} = \nu_i \dot{\epsilon}$ and $\dot{\epsilon} = \dot{n}_a \xi$.

All that we have done for differences in the enthalpies of two states of different compositions can be repeated for differences in internal energies, entropies, and Gibbs free energies. To show this formally, we introduce the symbol Z to denote any of the extensive properties internal energy U, enthalpy H, entropy S, or Gibbs free energy G, and the symbols z_{ia}, z_{ib}, $z_{ii}(T, p)$, z_{ii}°, $\Delta z^\circ(T)$, and Δz° to represent concepts analogous to those we have introduced for the case $z = h$. Thus we find that

$$\frac{Z_b - Z_a}{n_a} \quad \text{or} \quad \frac{\dot{Z}_b - \dot{Z}_a}{\dot{n}_a}$$

$$= (1 + \nu\,\xi) \sum_{i=1}^{r} y_{ib} \left[z_{ib} - z_{ii}^{o}(T) \right] - \sum_{i=1}^{r} y_{ia} \left[z_{ia} - z_{ii}^{o}(T) \right] + \xi\,\Delta z^{o}(T) \tag{29.63}$$

where $z_{ii}^{o}(T) = z_{ii}(T, p_0)$,

$$\Delta z^{o}(T) = \sum_{i=1}^{r} \nu_i\, z_{ii}^{o}(T) \tag{29.64}$$

and each of the partial properties z_{ia} and z_{ib}, in general, depends on temperature, pressure, and composition. Moreover, if in states a and b the system behaves as a Gibbs–Dalton mixture, equation (29.63) becomes

$$\frac{Z_b - Z_a}{n_a} \quad \text{or} \quad \frac{\dot{Z}_b - \dot{Z}_a}{\dot{n}_a} = (1 + \nu\,\xi) \sum_{i=1}^{r} y_{ib} \left[z_{ii}(T_b, p_{iib}) - z_{ii}(T, p_0) \right]$$

$$- \sum_{i=1}^{r} y_{ia} \left[z_{ii}(T_a, p_{iia}) - z_{ii}(T, p_0) \right] + \xi\,\Delta z^{o}(T) \tag{29.65}$$

As an example, we apply equation (29.65) to systems that behave as Gibbs–Dalton mixtures of perfect gases. Then the partial pressure p_{ii} of each constituent in the mixture is related to the pressure p of the mixture and the mole fraction y_i by the relations $p_{iia} = y_{ia}p_a$ and $p_{iib} = y_{ib}p_b$. For the entropy difference, that is, for $Z = S$, we find that

$$\frac{S_b - S_a}{n_a} \quad \text{or} \quad \frac{\dot{S}_b - \dot{S}_a}{\dot{n}_a} = (1 + \nu\,\xi) \left[\left(\sum_{i=1}^{r} y_{ib} c_{pii} \right) \ln \frac{T_b}{T} - R \ln \frac{p_b}{p_0} - R \sum_{i=1}^{r} y_{ib} \ln y_{ib} \right]$$

$$- \left[\left(\sum_{i=1}^{r} y_{ia} c_{pii} \right) \ln \frac{T_a}{T} - R \ln \frac{p_a}{p_0} - R \sum_{i=1}^{r} y_{ia} \ln y_{ia} \right] + \xi\,\Delta s^{o}(T) \tag{29.66}$$

where c_{pii} is the perfect-gas specific heat at constant pressure of constituent i, R is the universal gas constant, and we use the expression for the perfect-gas specific entropy as a function of temperature and pressure [equation (20.21)]. The quantity $\Delta s^{o}(T)$ is given by the relation

$$\Delta s^{o}(T) = \sum_{i=1}^{r} \nu_i\, s_{ii}^{o}(T) \tag{29.67}$$

and is called the *entropy of reaction at temperature T and pressure p_0*.

As other examples, we use equation (29.65) for $Z = H$ and $Z = U$, and for Gibbs–Dalton mixtures of perfect gases. Thus we find that

$$\frac{H_b - H_a}{n_a} \quad \text{or} \quad \frac{\dot{H}_b - \dot{H}_a}{\dot{n}_a}$$

$$= (1 + \nu\,\xi) \left(\sum_{i=1}^{r} y_{ib} c_{pii} \right)(T_b - T) - \left(\sum_{i=1}^{r} y_{ia} c_{pii} \right)(T_a - T) + \xi\,\Delta h^{o}(T) \tag{29.68}$$

and

$$\frac{U_b - U_a}{n_a} \quad \text{or} \quad \frac{\dot{U}_b - \dot{U}_a}{\dot{n}_a}$$

$$= (1 + \nu \xi)\left(\sum_{i=1}^{r} y_{ib} c_{vii}\right)(T_b - T) - \left(\sum_{i=1}^{r} y_{ia} c_{vii}\right)(T_a - T) + \xi \Delta u^\circ(T) \quad (29.69)$$

where c_{vii} is the perfect-gas specific heat at constant volume of constituent i. The quantity $\Delta h^\circ(T)$ is the enthalpy of reaction defined by equation (29.59), whereas $\Delta u^\circ(T)$ is given by the relation

$$\Delta u^\circ(T) = \sum_{i=1}^{r} \nu_i\, u_{ii}^\circ(T) \quad (29.70)$$

and is called the *internal energy of reaction at temperature T and pressure p_o*.

29.10   Reactions of Formation and Standard Properties of Formation

We now address the problem of constructing a data base in which the reference values of the properties of all constituents are compatible and consistent with experimental measurements. To this end, among all the conceivable chemical compounds, we select a minimal subset whose components are sufficient to form all other compounds by means of some chemical reactions. In other words, we select a *complete set of independent constituents* such that: (1) there exist chemical reaction mechanisms, consistent with the rule of conservation of atomic nuclei, by which all other constituents can be formed starting only with constituents in the set; and (2) there exist no chemical reaction mechanisms that involve only constituents in the set.

In practice, the complete set of independent constituents consists of the so-called *elemental species*. A chemical species is called elemental if: (1) it is formed by atomic nuclei of only one kind; and (2) its molecular structure and form of aggregation are the most stable (for any kind of atomic nucleus involved) at standard conditions of temperature and pressure, $T_o = 25°C$ and $p_o = 1$ atm. For example, gaseous H_2, O_2, N_2, liquid Hg, and solid Cu with the most stable crystallographic structure are elemental species. In contrast, H_2O, CO_2, and H_2SO_4 are not elemental species because they contain atomic nuclei of more than one kind. Again, H, O, and O_3 are not elemental species because their state of aggregation is not the most stable at standard conditions for the types of nuclei involved.

We arbitrarily assign zero values to the enthalpy and entropy of each elemental species at standard conditions. These arbitrary assignments introduce no inconsistencies because, by definition, there exists no chemical reaction mechanism which transforms a set of elemental species into a different set of elemental species, that is, the elemental species are chemically independent. Moreover, they imply that the Gibbs free energy of each elemental species under standard conditions is also zero.

We represent an elemental species j by the symbol A_j^e, where the superscript e denotes the fact that the species is elemental. By definition, given any arbitrary chemical constituent A_i there exists a chemical reaction mechanism in which A_i is the only product with stoichiometric coefficient equal to unity, and the reactants are all elemental species A_j^e. We call such a reaction a *reaction of formation of constituent i*, and describe it by the chemical reaction mechanism

$$\sum_{j} \alpha_{ij} A_j^e = A_i \quad (29.71)$$

where each α_{ij} is a positive number consistent with the rule of conservation of atomic nuclei. In terms of the earlier notation, if relation (29.71) were written in the form $\sum_k \nu_k A_k = 0$, the stoichiometric coefficients would be $\nu_i = 1$ and $\nu_j = -\alpha_{ij}$ for the index j running over the complete set of elemental species. Examples of reactions of formation of H_2O, CO_2, and H_2SO_4 are

$$H_2 + \tfrac{1}{2}O_2 = H_2O \qquad C + O_2 = CO_2 \qquad H_2 + S + 2O_2 = H_2SO_4$$

The enthalpy of reaction at standard conditions for relation (29.71) is called the *enthalpy of formation of constituent i*. It is denoted by $(\Delta h_f^o)_i$ and is given by equation (29.59) for $T = T_0$ and the values of the stoichiometric coefficients of relation (29.71), namely,

$$(\Delta h_f^o)_i = h_{ii}^o - \sum_j \alpha_{ij} h_{jj}^o \tag{29.72}$$

Because the elemental species are chemically independent, if constituent A_i is an elemental species, then the only possible but trivial mechanism of formation according to relation (29.71) is such that $\alpha_{ij} = \delta_{ij}$, where δ_{ij} is the Kronecker delta, that is, $\alpha_{ii} = 1$ and $\alpha_{ij} = 0$ for $j \neq i$. Accordingly, equation (29.72) indicates that the enthalpy of formation of an elemental species is always equal to zero, namely,

$$(\Delta h_f^o)_i = 0 \qquad \text{for } A_i \text{ an elemental species} \tag{29.73}$$

We can make similar remarks about *"property Z" of formation of constituent i* at standard conditions. Such a property is given by the relation

$$(\Delta z_f^o)_i = z_{ii}^o - \sum_j \alpha_{ij} z_{jj}^o \tag{29.74}$$

Again we observe that, if constituent A_i is an elemental species, the only possible but trivial mechanism of formation according to relation (29.71) is such that $\alpha_{ij} = \delta_{ij}$ and therefore that the "property Z" of formation of the ith elemental species is always equal to zero, that is,

$$(\Delta z_f^o)_i = 0 \qquad \text{for } A_i \text{ any elemental species} \tag{29.75}$$

The enthalpy of formation of a constituent other than an elemental species is measured and tabulated. The experimental setup is conceptually very simple. It is shown schematically in Figure 29.5. The inputs are heat and separate streams of the elemental species in the mole proportions required by the chemical reaction of formation, and each at temperature T_0 and pressure p_0. The output is 1 mol of the constituent, also at temperature T_0 and pressure p_0. The enthalpy of formation is given by the net heat per mole of the constituent that is formed because this heat equals the difference between

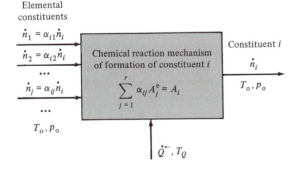

Elemental
constituents

$\dot{n}_1 = \alpha_{i1}\dot{n}_i$

$\dot{n}_2 = \alpha_{i2}\dot{n}_i$

...

$\dot{n}_j = \alpha_{ij}\dot{n}_i$

...

T_0, p_0

Chemical reaction mechanism
of formation of constituent i

$\sum_{j=1}^{r} \alpha_{ij} A_j^e = A_i$

Constituent i

$\dot{n}_i$

T_0, p_0

$\dot{Q}^\leftarrow, T_Q$

Figure 29.5 Schematic of an experimental set-up for measuring the enthalpy of formation.

the enthalpy of the constituent and the sum of the enthalpies of the input streams. It is measurable regardless of the assumed values of properties of the reference states.

The Gibbs free energy of formation of a constituent other than an elemental species is also measured and tabulated. The experimental setup is shown schematically in Figure 29.6. The inputs are heat $q_o^{\leftarrow} = \dot{Q}_o^{\leftarrow}/\dot{n}_i$ at temperature T_o, and separate streams of the elemental species in the mole proportions required by the chemical reaction for the formation of 1 mol of constituent, each at temperature T_o and pressure p_o. The outputs are shaft work $w_s^{\rightarrow} = \dot{W}_s^{\rightarrow}/\dot{n}_i$, and 1 mol of the constituent also at T_o and p_o. The combined energy and entropy balance for this process is readily found to be

$$w_s^{\rightarrow} = -(\Delta h_f^o)_i + T_o\,(\Delta s_f^o)_i - T_o s_{irr}$$
$$= -(\Delta g_f^o)_i - T_o s_{irr}$$

It follows that upon carrying the process reversibly ($s_{irr} = 0$) and measuring the largest shaft work $(w_s^{\rightarrow})_{largest}$, we obtain $-(\Delta g_f^o)_i$.

Elemental
constituents

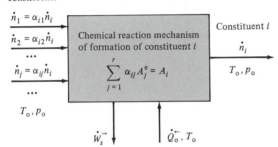

Figure 29.6 Schematic of an experimental set-up for measuring the Gibbs free energy of formation.

Properties of formation are used in evaluating properties of reaction. For example, solving equation (29.72) for h_{ii}^o, that is, the specific enthalpy of constituent i at standard temperature $T = T_o$ and pressure p_o, we find that

$$h_{ii}^o = (\Delta h_f^o)_i + \sum_j \alpha_{ij} h_{jj}^o \tag{29.76}$$

Substituting this result into the expression for the enthalpy of reaction at standard conditions [equation (29.59) for $T = T_o$], we also find that

$$\Delta h^o = \sum_{i=1}^r \nu_i h_{ii}^o = \sum_{i=1}^r \nu_i (\Delta h_f^o)_i + \sum_j h_{jj}^o \sum_{i=1}^r \nu_i \alpha_{ij} = \sum_{i=1}^r \nu_i (\Delta h_f^o)_i \tag{29.77}$$

where in writing the last of equations (29.77) we use the general relation

$$\sum_{i=1}^r \nu_i \alpha_{ij} = 0 \quad \text{for every elemental species } j \tag{29.78}$$

To prove the validity of equation (29.78), we substitute in the reaction mechanism, $\sum_{i=1}^r \nu_i A_i = 0$, the reaction of formation of each constituent i and find that

$$\sum_j \left(\sum_{i=1}^r \nu_i \alpha_{ij} \right) A_j^e = 0 \tag{29.79}$$

that is, a chemical reaction mechanism that involves only elemental species A_j^e. But elemental species are chemically independent and, therefore, each stoichiometric coefficient $\sum_{i=1}^r \nu_i \alpha_{ij}$, for all j's in relation (29.79), must be zero, that is, (29.78) is valid.

The expression $\Delta h^\circ = \sum_{i=1}^r \nu_i (\Delta h_f^\circ)_i$ is called a *Hess relation*. It relates the enthalpy of reaction at T_o, p_o to a weighted sum of the enthalpies of formation of the reactants and products partaking in the chemical reaction mechanism, the weights being the stoichiometric coefficients.

The Hess relation can be readily shown to hold for any of the extensive properties H, S, U, G, as well as the volume V. Denoting by Z any such property, the corresponding Hess relation reads

$$\Delta z^\circ = \sum_{i=1}^r \nu_i (\Delta z_f^\circ)_i \qquad (29.80)$$

and has an interpretation analogous to that of the enthalpy of reaction.

The usefulness of Hess relations [equation (29.80) for z either h, s, u, g, or v] follows from the fact that the values of the standard properties of formation $(\Delta z_f^\circ)_i$ for most chemical constituents are tabulated and, therefore, the properties Δz° of any reaction can be readily evaluated by using the appropriate Hess relations.

Typically, only values of the enthalpy of formation Δh_f° and the Gibbs free energy of formation Δg_f° are tabulated. Values of other properties of formation are evaluated by using appropriate property interrelations. For example, using the relation $g = h - Ts$, we verify that the entropy of formation Δs_f° is related to Δh_f° and Δg_f° by the expression

$$\Delta s_f^\circ = \frac{\Delta h_f^\circ - \Delta g_f^\circ}{T_o} \qquad (29.81)$$

Values of Δh_f°, Δg_f°, and Δs_f° are listed in Table 29.4 for various substances. The data in this table refer to the *ideal-gas state at standard conditions*, that is, the ideal-gas state at temperature $T_o = 25°C$ and pressure $p_o = 1$ atm. Such a state is a stable equilibrium state of the substance if at T_o and p_o the substance is an ideal gas. If the substance is not an ideal gas at T_o and p_o, the ideal-gas state is fictitious, and values of its properties are obtained by extrapolating the ideal-gas relations and specific-heat data (such as the data in Tables 20.1 and 20.2) from a region of ideal-gas behavior to T_o and p_o. For example, the specific enthalpy at T_o and p_o is given by the relation $h_{ii}(T_o, p_o) = h_{ii}(T, p_o) - \int_{T_o}^T c_{pii}(T') \, dT'$, where $c_{pii}(T)$ is obtained from either Table 20.1 or 20.2. Similarly, the specific entropy, $s_{ii}(T_o, p_o) = s_{ii}(T, p_o) - \int_{T_o}^T c_{pii}(T') \, dT'/T'$.

Tables of properties often include absolute values of specific entropies s_{ii}° of constituents at standard conditions. Each such value should not be confused with an entropy of formation Δs_f°. The value of s_{ii}° is absolute, that is, it is evaluated so that at absolute temperature $T = 0$, the value of the entropy of the constituent is equal to zero. On the other hand, Δs_f° is evaluated so that the entropy of formation of each elemental species is equal to zero at T_o and p_o.

Example 29.5.   Verify equation (29.81) for gaseous *n*-octane, C_8H_{18}, and for carbon dioxide, CO_2, using the data from Table 29.4.

Solution.   For n-octane, C_8H_{18}, we find that $\Delta h_f^\circ = -208.6$ MJ/kmol, $\Delta g_f^\circ = 16.4$ MJ/kmol and therefore $\Delta s_f^\circ = (\Delta h_f^\circ - \Delta g_f^\circ)/T_o = (-208,600 - 16,400)/298.15 = -754.6$ kJ/kmol K, which coincides with the tabulated value. For CO_2, we find $\Delta s_f^\circ = (-393,200 + 394,600)/298.15 = 2.7$ kJ/kmol K, which differs from the tabulated value 2.9 kJ/kmol K because of rounding off errors in the tabulated values of Δh_f° and Δg_f°.

Example 29.6.   For the reaction mechanism $CO + \frac{1}{2} O_2 = CO_2$, evaluate the enthalpy of reaction Δh°, the Gibbs free energy of reaction Δg°, and the entropy of reaction Δs° at standard conditions.

TABLE 29.4.   Values of the enthalpy of formation, the Gibbs free energy of formation, and the entropy of formation of various substances in the ideal-gas state at standard temperature, $T_o = 25°C$, and pressure, $p_o = 1$ atm.

Substance	Formula	Δh_f^o MJ/kmol	Δg_f^o MJ/kmol	Δs_f^o kJ/kmol K
Acetic acid	$C_2H_4O_2$	−336.5	−286.3	−168.4
Acetone	CH_3COCH_3	−217.7	−153.2	−216.5
Acetylene	C_2H_2	226.9	209.3	58.8
Ammonia	NH_3	−45.7	−16.2	−99.1
Argon	Ar	0	0	0
Benzene	C_6H_6	83.0	129.7	−156.9
Carbon	C	0	0	0
Carbon dioxide	CO_2	−393.8	−394.6	2.9
Carbon monoxide	CO	−110.6	−137.4	89.7
Chlorine	Cl_2	0	0	0
Chloroform	$CHCl_3$	−101.3	−68.6	−109.8
Ethane	C_2H_6	−84.7	−33.0	−173.7
Ethanol	C_2H_5OH	−235.0	−168.4	−223.3
Ethylene	C_2H_4	52.3	68.2	−53.1
Fluorine	F_2	0	0	0
Freon 12	CCl_2F_2	−481.5	−442.5	−130.6
Freon 13	$CClF_3$	−695.0	−654.4	−136.2
Freon 21	$CHCl_2F$	−298.9	−268.4	−102.5
Freon 22	$CHClF_2$	−502.0	−470.9	−104.3
Genetron 100	CH_3CHF_2	−494.0	−436.5	−192.9
Hydrogen	H_2	0	0	0
Hydrogen (atomic)	H	218.0	203.3	49.4
Hydrogen chloride	HCl	−92.4	−95.3	10.0
Hydroxyl	OH	39.5	34.3	17.4
Isooctane	C_8H_{18}	−224.3	13.7	−798.6
Methane	CH_4	−74.9	−50.9	−80.6
Methanol	CH_3OH	−201.3	−162.6	−129.8
Methylene chloride	CH_2Cl_2	−95.5	−68.9	−89.0
Naphthalene	$C_{10}H_8$	151.1	223.7	−243.8
Nitric oxide	NO	90.4	86.8	12.4
Nitrogen	N_2	0	0	0
Nitrogen (atomic)	N	472.8	455.6	57.6
Nitrogen dioxide	NO_2	33.9	52.0	−60.8
Nitrous oxide	N_2O	81.6	103.7	−74.1
n-Octane	C_8H_{18}	−208.6	16.4	−754.6
Oxygen	O_2	0	0	0
Oxygen (atomic)	O	249.2	231.8	58.3
Ozone	O_3	142.8	162.9	−67.5
Propane	C_3H_8	−103.9	−23.5	−269.8
Propylene	CH_2CHCH_3	20.4	62.8	−142.0
Water	H_2O	−242.0	−228.8	−44.4

Source: Data from R. C. Reid, J. M. Prausnitz, and T. K. Sherwood, *The Properties of Gases and Liquids*, McGraw-Hill, New York, 1977.

Solution.   Using the Hess relation [equation (29.80)] for $z = h$ and data from Table 29.4 we find
that $\Delta h^\circ = -(-110.6) - (\frac{1}{2})(0) + (-393.8) = -283.2$ MJ/kmol. Again, for $z = g$, we find
that $\Delta g^\circ = -(-137.4) - (\frac{1}{2})(0) + (-394.6) = -275.2$ MJ/kmol. Finally, we find Δs° either
as $\Delta s^\circ = -(89.7) - (\frac{1}{2})(0) + (2.9) = -86.8$ kJ/kmol K or, less accurately (see example 29.5),
as $\Delta s^\circ = (\Delta h^\circ - \Delta g^\circ)/T_o = (-238,200 + 257,200)/298.15 = -87.2$ kJ/kmol K.

Another example of the use of tabulated data is provided by the evaluation of the
energy of formation Δu_f°. Starting from the relation $u = h - pv$ for each pure constituent,
we find that

$$\Delta u_f^\circ = \Delta h_f^\circ - p_o \Delta v_f^\circ \tag{29.82}$$

where Δv_f° is the volume of formation. In general, the value of Δv_f° of a constituent
must be evaluated using equation (29.74) for $z = v$, and tabulated values of the specific
volumes at standard conditions for the constituent and the elemental species. However,
if at standard conditions the constituent and the elemental species in the reaction of
formation behave as ideal gases, each obeys the equation of state $pv = RT$, and

$$\Delta u_f^\circ = \Delta h_f^\circ - \nu_f RT_o \tag{29.83}$$

where ν_f is the sum of the stoichiometric coefficients of the reaction of formation. Properties of reaction at standard conditions can be evaluated by means of equation (29.80)
and the tabulated values of the corresponding properties of formation (Table 29.4).

Properties of reaction can also be measured directly. As an example, we consider
the measurement of the enthalpy of reaction Δh° at standard conditions. In general,
it involves three separate experiments. In the first, we use the steady-state chemical
reactor shown schematically in Figure 29.7a, and measure the heat rate $\dot{Q}_A^\leftarrow$ necessary to
convert the bulk-flow stream from the inlet to the outlet conditions. This heat rate may
be supplied at one or many different temperatures and satisfies the energy rate balance

$$\dot{H}_b - \dot{H}_a = \sum_{i=1}^{r} \dot{n}_{ib}(h_{ib} - h_{ii}^o) - \sum_{i=1}^{r} \dot{n}_{ia}(h_{ia} - h_{ii}^o) + \dot{n}_a \xi \Delta h^\circ = \dot{Q}_A^\leftarrow \tag{29.84}$$

where in writing the first of equations (29.84) we use (29.63) for $Z = H$ and $T = T_o$.

In the second and third experiments, we use two steady-state mixers B and C,
and measure the heat rates $\dot{Q}_B^\leftarrow$ and $\dot{Q}_C^\leftarrow$ necessary to convert two different r pure-
constituent streams into mixtures without chemical reactions, as shown schematically in
Figures 29.7b and 29.7c, respectively. The energy rate balances for the two mixers are

$$\sum_{i=1}^{r} \dot{n}_{ia}(h_{ia} - h_{ii}^o) = \dot{Q}_B^\leftarrow \tag{29.85}$$

$$\sum_{i=1}^{r} \dot{n}_{ib}(h_{ib} - h_{ii}^o) = \dot{Q}_C^\leftarrow \tag{29.86}$$

It is noteworthy that each of the heat rates $\dot{Q}_B^\leftarrow$ and $\dot{Q}_C^\leftarrow$ may also be supplied at one or
many different temperatures.

Upon combining the results of these three experiments, we find that

$$\Delta h^\circ = \frac{\dot{Q}_A^\leftarrow + \dot{Q}_B^\leftarrow - \dot{Q}_C^\leftarrow}{\dot{n}_a \xi} \tag{29.87}$$

that is, we verify that the enthalpy of reaction can be measured directly by carrying out
the three heat-rate measurements in Figure 29.7.

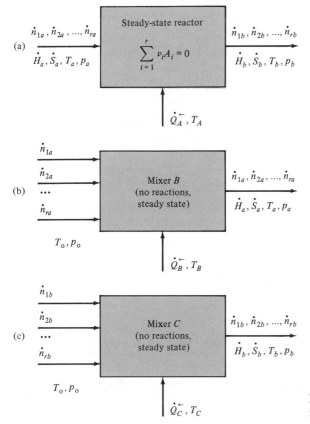

Figure 29.7 A scheme for measuring the enthalpy of reaction.

For the particular selection $T_a = T_b = T_0$ and $p_a = p_b = p_0$, the heat rate $\dot{Q}_A^{\leftarrow}$ necessary to carry out the chemical reaction at standard conditions of temperature and pressure is usually much larger than the heat rates $\dot{Q}_B^{\leftarrow}$ and $\dot{Q}_C^{\leftarrow}$ required by the two mixers. Both rates $\dot{Q}_B^{\leftarrow}$ and $\dot{Q}_C^{\leftarrow}$ are zero if the mixtures in the bulk-flow states a and b are Gibbs–Dalton.

The Hess relation [equation (29.80)] holds also for temperatures other than T_0. In general, we have

$$\Delta z^{\circ}(T) = \sum_{i=1}^{r} \nu_i \left[\Delta z_f^{\circ}(T) \right]_i \qquad (29.88)$$

Moreover, the dependence of a property of reaction on temperature can be readily evaluated from data on the pure constituents that enter the chemical reaction mechanism, and the relation

$$\Delta z^{\circ}(T) = \Delta z^{\circ} + \sum_{i=1}^{r} \nu_i \left[z_{ii}(T, p_0) - z_{ii}(T_0, p_0) \right] \qquad (29.89)$$

In particular, for the enthalpy of reaction

$$\Delta h^{\circ}(T) = \Delta h^{\circ} + \sum_{i=1}^{r} \nu_i \left[h_{ii}(T, p_0) - h_{ii}(T_0, p_0) \right] \qquad (29.90)$$

Example 29.7. For the reaction mechanism $CO + \frac{1}{2} O_2 = CO_2$, evaluate the enthalpy of reaction $\Delta h^{\circ}(T)$, the entropy of reaction $\Delta s^{\circ}(T)$, and the Gibbs free energy of reaction $\Delta g^{\circ}(T)$ at $T = 3200$ K.

Solution. Using equation (29.90), the ideal-gas relations, and the approximate expressions given in Table 20.2, we find that

$$\Delta h^\circ(T) - \Delta h^\circ$$

$$= \sum_{i=1}^{r} \nu_i \left[a_{ii}(T - T_0) + \tfrac{4}{5} b_{ii}(T^{5/4} - T_0^{5/4}) + \tfrac{2}{3} c_{ii}(T^{3/2} - T_0^{3/2}) + \tfrac{4}{7} d_{ii}(T^{7/4} - T_0^{7/4}) \right]$$

$$= \Delta a(T - T_0) + \tfrac{4}{5}\Delta b(T^{5/4} - T_0^{5/4}) + \tfrac{2}{3}\Delta c(T^{3/2} - T_0^{3/2}) + \tfrac{4}{7}\Delta d(T^{7/4} - T_0^{7/4})$$

$$= 11{,}637 \text{ kJ/kmol}$$

where $T = 3200$ K, $T_0 = 298.15$, $\Delta a = \sum_{i=1}^{r} \nu_i a_{ii}$, $\Delta b = \sum_{i=1}^{r} \nu_i b_{ii}$, $\Delta c = \sum_{i=1}^{r} \nu_i c_{ii}$, $\Delta d = \sum_{i=1}^{r} \nu_i d_{ii}$, and the values of a_{ii}, b_{ii}, c_{ii}, d_{ii} are taken from Table 20.2 and summarized in Table 29.5.

TABLE 29.5

i	A_i	ν_i	a_{ii}	b_{ii}	c_{ii}	d_{ii}
1	CO	-1	62.8	-22.6	4.60	-0.272
2	O_2	$-1/2$	10.3	5.4	-0.18	0
3	CO_2	1	-55.6	30.5	-1.96	0
			$\Delta a = -123.55$	$\Delta b = 50.4$	$\Delta c = -6.47$	$\Delta d = 0.272$

Using $\Delta h^\circ = -283{,}200$ kJ/kmol from Example 29.6, we find that $\Delta h^\circ(3200\text{ K}) = -283{,}200 + 11{,}637 = -271{,}563$ kJ/kmol. Similarly, for the entropy we find

$$\Delta s^\circ(T) - \Delta s^\circ$$

$$= \sum_{i=1}^{r} \nu_i \left[a_{ii}\ln\frac{T}{T_0} + 4 b_{ii}(T^{1/4} - T_0^{1/4}) + 2 c_{ii}(T^{1/2} - T_0^{1/2}) + \tfrac{4}{3} d_{ii}(T^{3/4} - T_0^{3/4}) \right]$$

$$= \Delta a \ln\frac{T}{T_0} + 4\Delta b(T^{1/4} - T_0^{1/4}) + 2\Delta c(T^{1/2} - T_0^{1/2}) + \tfrac{4}{3}\Delta d(T^{3/4} - T_0^{3/4})$$

$$= 5.05 \text{ kJ/kmol K}$$

Using $\Delta s^\circ = -86.8$ kJ/kmol K from Example 29.6, we find that $\Delta s^\circ(3200\text{ K}) = -86.8 + 5.05 = -81.75$ kJ/kmol K. Finally, $\Delta g^\circ(T) = \Delta h^\circ(T) - T\Delta s^\circ(T) = -271{,}563 - 3200(-81.75) = -9963$ kJ/kmol.

Problems

29.1 The industrial production of ammonia, NH_3, is carried out mainly by catalytic synthesis according to the reaction mechanism $N_2 + 3H_2 = 2NH_3$. The initial mixture of nitrogen, N_2, and hydrogen, H_2, is prepared with the water-gas process by reacting carbon, C, at high temperature with controlled amounts of combustion air and water vapor, H_2O, according to the reaction mechanisms $C + H_2O = CO + H_2$ and $2C + O_2 + 3.773 N_2 = 2CO + 3.773 N_2$. The mixture of CO, H_2, and N_2 is then "converted" according to the reaction mechanism $CO + H_2O = CO_2 + H_2$, and finally, the carbon dioxide, CO_2, is separated out. Consider an ammonia production plant in which each of the four reactions goes to completion. Find the number of moles of C, H_2O, and air to be fed to the plant in order to produce 1 mol of NH_3, and have CO_2 as the only other product in the plant.

29.2 An industrial process for manufacturing cyclohexane, C_6H_{12}, by catalytic hydrogenation of benzene, C_6H_6, is shown in Figure P29.2. Liquid benzene from storage (state 1) is pumped directly to the reactor with makeup hydrogen (state 3) and recycle hydrogen (state 4) for the exothermic reaction mechanism $C_6H_6 + 3\,H_2 = C_6H_{12}$. The heat of reaction is used partially to heat the reactants, and partially to produce saturated steam at 148°C from environmental water at 38°C. The gaseous reactor effluent (state 6) is partially condensed with cooling water at 38°C and flashed at high presssure to separate the hydrogen from the cyclohexane. The hydrogen is partially recycled (state 7) and partially purged (state 8). The liquid cyclohexane is recycled (state 9), vented (state 11), and the remainder collected as product (state 12). The conditions at each

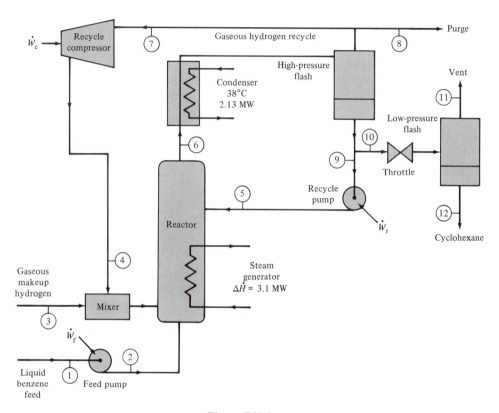

Figure P29.2

State	| Flow rate [kmol/h]					T	p	$\dot{H}$	$\dot{S}$
	H_2	N_2	C_6H_6	C_6H_{12}	Total	°C	bar	MJ/h	MJ/h K
1	0	0	92.14	0	92.14	38	1	7858.6	17.212
2	0	0	92.14	0	92.14	39.3	22.5	7895.14	17.271
3	282.96	0.84	0	0	283.80	49	22.5	2608.76	30.414
4	65.15	7.01	0	1.61	73.77	63	22.5	592.85	8.928
5	0.60	0.13	0.05	56.33	57.11	49	22	−5177.52	12.621
6	72.53	7.98	0.13	150.00	230.64	200	21	−5345.94	61.33
7	65.15	7.01	0	1.61	73.77	49	20	561.12	8.90
8	5.80	0.62	0	0.14	6.56	49	20	50.15	0.794
9	0.60	0.13	0.05	56.33	57.11	49	20	−5180.1	12.617
10	0.98	0.22	0.08	91.92	93.20	49	20	−8453.2	20.555
11	0.95	0.21	0	0.69	1.85	49	1	−33.12	0.393
12	0.03	0.01	0.08	91.23	91.35	49	1	−8420.08	20.304

state are listed in the accompanying table. At 38°C the Gibbs free energy of reaction is about −90 kcal/mol.

(a) What is the least rate of hydrogen required for the hydrogenation of the amount of benzene given in the table?

(b) What is the optimum rate of work required for the hydrogenation of 92.14 kmol/h of benzene in an environment at about 38°C? Should this work be done on the benzene or be done by the benzene?

(c) What is the smallest rate of work required to operate the plant as shown in the figure?

(d) What are the work rates to operate the recycle compressor, $\dot{W}_c$, the feed pump, $\dot{W}_f$, and the recycle pump, $\dot{W}_r$?

(e) At what rate is availability consumed in the process?

(f) Find the rates of availabilities lost in the various components, and compare their sum to your answer in part (e).

(g) What is the effectiveness of the process?

29.3 A gaseous mixture consists, on a mole basis, of 80% of methane, CH_4, and 20% of hydrogen, H_2.

(a) How many kilograms of O_2 are required for complete transformation of 1 kg of the mixture into carbon dioxide, CO_2, and water, H_2O?

(b) If the O_2 is supplied by atmospheric air, assuming that on a mole basis air consists of 21% of O_2 and 79% of N_2, how many kilograms of air are required for the reaction in part (a)?

(c) On a mole basis, find the composition of the result in part (b) assuming that all products are gases.

29.4 Hydrogen peroxide, H_2O_2, and the hydrocarbon $C_{16}H_{30}$ are reactants in a chemical reactor yielding CO_2 and H_2O only.

(a) How many kilograms of hydrogen peroxide are required per kilogram of $C_{16}H_{30}$?

(b) How many kilograms of H_2O are formed per kilogram of $C_{16}H_{30}$?

29.5 The enthalpy of reaction at 25°C and 1 atm of the oxidation reaction of gaseous acetylene, $C_2H_2 + \frac{5}{2}O_2 = 2CO_2 + H_2O$, is −48.3 MJ/kg of C_2H_2, assuming the H_2O in the gaseous form of aggregation.

(a) What is the internal energy of the reaction at 25°C and 1 atm?

(b) What is the enthalpy of reaction at 25°C and 1 atm with the H_2O in the liquid form of aggregation?

29.6 In an intermediate step of a coal gasification process, benzene, C_6H_6, is reacted with steam according to the mechanism

$$4\,C_6H_6 + 18\,H_2O = 9\,CO_2 + 15\,CH_4$$

(a) Evaluate the enthalpy, Gibbs free energy, and entropy of reaction at standard temperature and pressure.

(b) Evaluate the Gibbs free energy of reaction at 1000 K and 50 atm, assuming ideal-gas behavior for each of the constituents.

30   Chemical Equilibrium

In this chapter we discuss the effects of chemical reactions in determining the composition of either a stable equilibrium state or a bulk-flow state, and the dependences of these effects on temperature and pressure.

30.1 Necessary Conditions for Stable Equilibrium

Chemical reactions may cause spontaneous changes in the values of the amounts of constituents of a system. Such spontaneous changes may occur only until the system reaches a stable equilibrium state. For a simple system A with r constituents and τ chemical reactions in a stable equilibrium state, the following conditions are necessary:

$$\sum_{i=1}^{r} \nu_i^{(j)} \, \mu_i(T, p, y_{1o}, y_{2o}, \ldots, y_{ro})$$

$$= \sum_{i=1}^{r} \nu_i^{(j)} \, \mu_i \left(T, p, \frac{n_{1a} + \sum_{j=1}^{\tau} \nu_1^{(j)} \epsilon_{jo}}{n_a + \sum_{j=1}^{\tau} \nu^{(j)} \epsilon_{jo}}, \ldots, \frac{n_{ra} + \sum_{j=1}^{\tau} \nu_r^{(j)} \epsilon_{jo}}{n_a + \sum_{j=1}^{\tau} \nu^{(j)} \epsilon_{jo}} \right) = 0$$

$$\text{for } j = 1, 2, \ldots, \tau \qquad (30.1)$$

where T is the temperature, p the pressure, n_{ia}, for $i = 1, 2, \ldots, r$, the amount of the ith constituent for which the values of the τ reaction coordinates $\epsilon_1, \epsilon_2, \ldots, \epsilon_\tau$ are all equal to zero, $n_a = \sum_{i=1}^{r} n_{ia}$, the $\nu_i^{(j)}$'s are the stoichiometric coefficients that specify the τ chemical reaction mechanisms

$$\sum_{i=1}^{r} \nu_i^{(j)} A_i = 0 \qquad \text{for } j = 1, 2, \ldots, \tau \qquad (30.2)$$

$\nu^{(j)} = \sum_{i=1}^{r} \nu_i^{(j)}$ for $j = 1, 2, \ldots, \tau$, $\epsilon_{1o}, \epsilon_{2o}, \ldots, \epsilon_{\tau o}$ are the values of the reaction coordinates at the stable equilibrium state, $y_{1o}, y_{2o}, \ldots, y_{ro}$ the values of the mole fractions given by the relations

$$y_{io} = \frac{n_{ia} + \sum_{j=1}^{\tau} \nu_i^{(j)} \epsilon_{jo}}{n_a + \sum_{j=1}^{\tau} \nu^{(j)} \epsilon_{jo}} \qquad \text{for } i = 1, 2, \ldots, r \qquad (30.3)$$

and $\mu_i(T, p, y_{1o}, y_{2o}, \ldots, y_{ro})$, for $i = 1, 2, \ldots, r$, is the chemical potential of the ith constituent of a simple system B that we call the "surrogate" of system A and that consists of the same r constituents as A at temperature T, pressure p, and composition $y_{1o}, y_{2o}, \ldots, y_{ro}$ but with all the chemical reactions switched off—completely inhibited. In Sections 30.2 to 30.5 we show that conditions (30.1) are very useful to characterize the stable equilibrium states of system A. The derivation of conditions (30.1) is discussed in Sections 30.6 and 30.7.

To emphasize that system A is subject to a set of chemical reactions, we use the expression "chemical equilibrium" as synonymous with stable equilibrium. Accordingly, for each of the stable equilibrium states of system A, we say that the system is in a *chemical equilibrium state* or, simply, that the state is *chemical equilibrium*. Moreover, the jth of equations (30.1) is called the *chemical equilibrium equation* for the jth reaction mechanism $\sum_{i=1}^{r} \nu_i^{(j)} A_i = 0$.

Because each chemical equilibrium equation is in terms of the chemical potentials of a system with all the reaction mechanisms switched off—the surrogate system—we are able to capitalize on all the results discussed in Chapters 26 to 28. For given explicit functional forms of $\mu_i = \mu_i(T, p, \boldsymbol{y})$, for $i = 1, 2, \ldots, r$, and for given values of T, p, and $\boldsymbol{n}_a = n_{1a}, n_{2a}, \ldots, n_{ra}$, we can solve the τ equations (30.1) for the τ unknowns $\epsilon_{1o}, \epsilon_{2o}, \ldots, \epsilon_{\tau o}$ and hence determine the chemical equilibrium composition $y_{1o}, y_{2o}, \ldots, y_{ro}$ from equations (30.3), and the chemical equilibrium values of the amounts of constituents $n_{1o}, n_{2o}, \ldots, n_{ro}$ from the relations

$$n_{io} = n_{ia} + \sum_{j=1}^{\tau} \nu_i^{(j)} \epsilon_{jo} \qquad \text{for } i = 1, 2, \ldots, r \tag{30.4}$$

The procedure for determining the composition and the values of the amounts of constituents of a chemical equilibrium state is more fully developed in Section 30.2 for a simple system subject to a single chemical reaction mechanism, in Section 30.4 for a simple system subject to many chemical reaction mechanisms, and in Section 30.8 for a simple system subject to all conceivable chemical reaction mechanisms.

30.2   Chemical Equilibrium with Respect to a Single Reaction

For a simple system subject to a single chemical reaction mechanism, $\sum_{i=1}^{r} \nu_i A_i = 0$, there is only one reaction coordinate ϵ and only one chemical equilibrium equation [equations (30.1) for $\tau = 1$]

$$\sum_{i=1}^{r} \nu_i \mu_i(T, p, y_1, y_2, \ldots, y_r)$$

$$= \sum_{i=1}^{r} \nu_i \mu_i\left(T, p, \frac{n_{1a} + \nu_1 \epsilon}{n_a + \nu \epsilon}, \frac{n_{2a} + \nu_2 \epsilon}{n_a + \nu \epsilon}, \ldots, \frac{n_{ra} + \nu_r \epsilon}{n_a + \nu \epsilon}\right) = 0 \tag{30.5}$$

where for convenience we drop the subscript o.

As a first important application, we consider the chemical equilibrium states of gaseous mixtures. We recall that $\mu_i(T, p, y_1, y_2, \ldots, y_r) = \mu_{ii}(T, p_{ii})$ [equation (26.13)] and rewrite equation (30.5) as

$$\sum_{i=1}^{r} \nu_i \mu_{ii}(T, p_{ii}(T, p, y_1, y_2, \ldots, y_r)) = 0 \tag{30.6}$$

where μ_{ii} denotes the chemical potential of pure constituent i and p_{ii} the partial pressure of constituent i in the surrogate system.

If in the ranges of temperature T and pressure p of interest, each pure constituent i behaves as an ideal gas, that is, $v_{ii}(T, p') = RT/p'$, the chemical equilibrium equation may be expressed as an explicit form of the partial pressures. Specifically, we recall that [equation (26.20)]

$$\mu_{ii}(T, p_{ii}) = \mu_{ii}(T, p_o) + RT \ln \frac{p_{ii}(T, p, y_1, y_2, \ldots, y_r)}{p_o} \qquad \text{for } i = 1, 2, \ldots, r \tag{30.7}$$

where p_o is some reference pressure, such as the standard reference pressure $p_o = 1$ atm. Upon substituting equations (30.7) into equation (30.6) and rearranging terms, we find that

$$\sum_{i=1}^{r} \nu_i \mu_{ii}(T, p_{ii}) = \sum_{i=1}^{r} \nu_i \mu_{ii}(T, p_o) + RT \ln \prod_{i=1}^{r} \left(\frac{p_{ii}}{p_o}\right)^{\nu_i} = 0 \qquad (30.8)$$

or, equivalently,

$$\prod_{i=1}^{r} \left(\frac{p_{ii}}{p_o}\right)^{\nu_i} = \frac{1}{(p_o)^{\nu}} \prod_{i=1}^{r} (p_{ii})^{\nu_i}$$

$$= \exp\left[-\frac{1}{RT}\sum_{i=1}^{r} \nu_i \mu_{ii}(T, p_o)\right] = \exp\left[-\frac{\Delta g^{\circ}(T)}{RT}\right]$$

$$= K(T) \qquad (30.9)$$

where $\nu = \sum_{i=1}^{r} \nu_i$, and in writing the second of equations (30.9) we use the fact that $\mu_{ii}(T, p_o) = g_{ii}(T, p_o)$, that is, that the chemical potential of a pure constituent equals its specific Gibbs free energy, and the definition of the Gibbs free energy of reaction at temperature T and pressure p_o, that is, $\Delta g^{\circ}(T) = \sum_{i=1}^{r} \nu_i g_{ii}(T, p_o) = \sum_{i=1}^{r} \nu_i \mu_{ii}(T, p_o)$. The function of temperature $K(T)$ defined by the relation

$$K(T) = \exp\left[-\frac{1}{RT}\sum_{i=1}^{r} \nu_i \mu_{ii}(T, p_o)\right] = \exp\left[-\frac{\Delta g^{\circ}(T)}{RT}\right] \qquad (30.10)$$

is called the *equilibrium constant of the reaction at temperature T*, and equation (30.9) the *law of mass action of Guldberg, Waage, van't Hoff, and Horstmann*.[1]

For a given chemical reaction mechanism and a given value of T, the value of the equilibrium constant $K(T)$ is found by first evaluating $\Delta g^{\circ}(T)$ as done in Example 29.7, and then using equation (30.10).

Example 30.1. Evaluate the equilibrium constant at 3200 K for the reaction mechanism $CO + \frac{1}{2} O_2 = CO_2$.

Solution. From Example 29.7, $\Delta g^{\circ}(3200 \text{ K}) = -9963$ kJ/kmol, and therefore equation (30.10) yields $K(3200 \text{ K}) = \exp[(9963/(8.314 \times 3200))] = 1.454$. This value is within 6% of the value 1.545 obtained directly from the *JANAF Thermochemical Tables*.

In the handbooks of physical chemistry, values of the equilibrium constant $K(T)$ are tabulated or graphed versus temperature T for each of a large number of chemical reactions. Representative data are shown in Figure 30.1. It is noteworthy that the data in Figure 30.1 can be approximated fairly accurately by expressions of the form $K(T) \approx \exp(\Delta a - \Delta b/T)$, where Δa and Δb are constants. Values of the constants Δa and Δb for the chemical reaction mechanisms of formation of various constituents are given in Table 30.1.

For the reaction of formation of constituent i, we denote the equilibrium constant by $[K_f(T)]_i$, that is, $[K_f(T)]_i = \exp\{(-[\Delta g_f^{\circ}(T)]_i/RT)\}$. By virtue of the Hess relation $\Delta g^{\circ}(T) = \sum_{i=1}^{r} \nu_i [\Delta g_f^{\circ}(T)]_i$ [equation (29.88) for $z = g$], we can rewrite equation (30.10) as $\ln K(T) = \sum_{i=1}^{r} \nu_i \ln[K_f(T)]_i$. It follows that using the values of the equilibrium constants of the reactions of formation we can evaluate the equilibrium constant of any reaction mechanism. In particular, for a reaction with stoichiometric coefficients

---

[1] In honor of Cato Maximilian Guldberg (1836–1902), Norwegian mathematician and chemist.

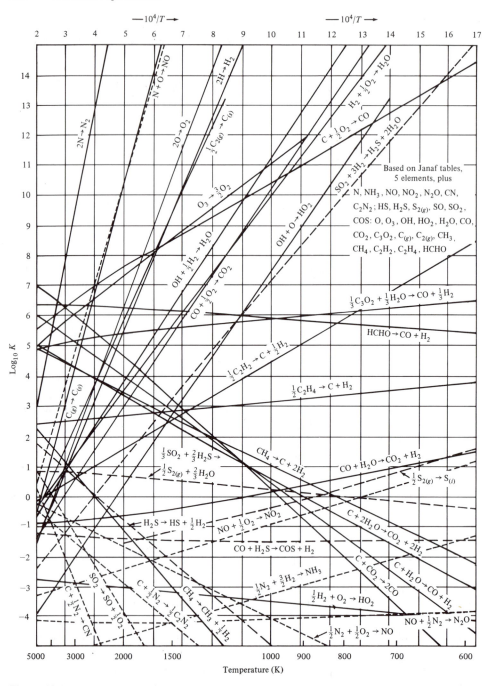

Figure 30.1 Graph of $\log_{10} K(T)$, the logarithm to the base 10 of the equilibrium constant $K(T)$, of various chemical reaction mechanisms versus temperature T on a $1/T$ scale (from M. Modell and R.C. Reid, *Thermodynamics and its Applications*, Prentice-Hall, 1974).

ν_i, for $i = 1, 2, \ldots, r$, the values of the constants Δa and Δb in the approximation $\ln K(T) \approx \Delta a - \Delta b/T$ are $\Delta a = \sum_{i=1}^{r} \nu_i (a_{\mathrm{f}})_i$ and $\Delta b = \sum_{i=1}^{r} \nu_i (b_{\mathrm{f}})_i$, where the constants $(a_{\mathrm{f}})_i$ and $(b_{\mathrm{f}})_i$ correspond to the reaction of formation of the ith constituent (see Table 30.1).

TABLE 30.1. Values of the constants a_f and b_f for the reaction mechanisms of formation of various substances in ideal-gas states at standard pressure, $p_o = 1$ atm, for use in the approximate expression

$$K(T) = \exp\left(\Delta a - \frac{\Delta b}{T}\right)$$

where $\Delta a = \sum_{i=1}^r \nu_i (a_f)_i$, $\Delta b = \sum_{i=1}^r \nu_i (b_f)_i$, and T is in kelvin within the range 298 to 5000 K.[a]

Substance	Formula	a_f	b_f [K]
Acetylene	C_2H_2	6.325	26,818
Ammonia	NH_3	−13.951	−6,462
Carbon	C	18.871	86,173
Carbon (diatomic)	C_2	22.870	100,582
Carbon dioxide	CO_2	−0.010	−47,575
Carbon monoxide	CO	10.098	−13,808
Carbon tetrafluoride	CF_4	−18.143	−112,213
Chlorine (atomic)	Cl	7.244	14,965
Chloroform	$CHCl_3$	−13.284	−12,327
Ethylene	C_2H_4	−9.827	4,635
Fluorine (atomic)	F	7.690	9,906
Freon 12	CCl_2F_2	−14.830	−58,585
Freon 21	$CHCl_2F$	−12.731	−34,190
Hydrogen (atomic)	H	7.104	26,885
Hydronium ion	H_3O^+	−8.312	71,295
Hydroxyl	OH	1.666	4,585
Hydroxyl ion	OH^-	−6.753	−20,168
Methane	CH_4	−13.213	−10,732
Nitric oxide	NO	1.504	10,863
Nitrogen (atomic)	N	7.966	57,442
Nitrogen dioxide	NO_2	−7.630	3,870
Nitrogen oxide	N_2O	−8.438	10,249
Oxygen (atomic)	O	7.963	30,471
Oxygen ion	O^-	0.528	10,048
Ozone	O_3	−8.107	17,307
Proton	H^+	13.437	188,141
Water	H_2O	−6.866	−29,911

Source: Regression of data from the *JANAF Thermochemical Tables*, 2nd ed., D. R. Stull and H. Prophet, project directors, NSRDS-NBS37. U.S. Department of Commerce National Bureau of Standards, Washington, D.C., 1971.

[a]For any elemental species, $a_f = 0$ and $b_f = 0$.

Example 30.2. Using data from Table 30.1 for the reaction mechanism $CO + \frac{1}{2} O_2 = CO_2$, evaluate the equilibrium constant $K(T)$ as a function of T, and at $T = 3200$ K.

Solution. Setting $i = 1$ for CO, $i = 2$ for O_2, and $i = 3$ for CO_2, from Table 30.1 we have $(a_f)_1 = 10.098$, $(b_f)_1 = -13,808$ K, $(a_f)_3 = -0.010$, and $(b_f)_3 = -47,575$ K. Moreover, $(a_f)_2 = 0$ and $(b_f)_2 = 0$ because O_2 is an elemental species and the reaction of formation of any elemental species has stoichiometric coefficients all equal to zero and, therefore, Gibbs free energy of reaction equal to zero, and equilibrium constant equal to unity. Thus we find that $\Delta a = \sum_{i=1}^r \nu_i (a_f)_i = -(10.098) - \frac{1}{2}(0) + (-0.010) = -10.108$, $\Delta b = \sum_{i=1}^r \nu_i (b_f)_i = -(13,808) - \frac{1}{2}(0) + (-47,575) = -33,767$, $K(T) = \exp(-10.108 + 33,767/T)$, and $K(3200\,\text{K})$

= 1.559. This value is within 1% of the value 1.545 obtained directly from the *JANAF Thermochemical Tables*, and is in good agreement with the value we obtain in Example 30.1 and the value we can estimate from Figure 30.1.

Each equilibrium constant is typically reported only for one choice of values of the stoichiometric coefficients of the corresponding chemical reaction mechanism. If we denote tabulated or graphical values by the subscript tab, we note that an arbitrary choice of stoichiometric coefficients ν_i must be such that $\nu_i = \lambda(\nu_i)_{tab}$, where the positive or negative multiplier λ is the same for all constituents, that is, for $i = 1, 2, \ldots, r$. Upon considering equation (30.10), we see that the equilibrium constant $K(T)$ for a choice of ν_i's different from the tabulated values is related to the tabulated equilibrium constant $K(T)_{tab}$ by the equation $K(T) = [K(T)_{tab}]^\lambda$ because $\Delta g^\circ(T) = \lambda \Delta g^\circ(T)_{tab}$.

In general, the partial pressure $p_{ii} = p_{ii}(T, p, y_1, y_2, \ldots, y_r)$. But if the surrogate simple system behaves as an ideal-gas mixture, $p_{ii} = y_i p$ for each $i = 1, 2, \ldots, r$ and, therefore,

$$\frac{p_{ii}}{p} = y_i = \frac{n_i}{n} = \frac{n_{ia} + \nu_i \epsilon}{n_a + \nu \epsilon} = \frac{y_{ia} + \nu_i \xi}{1 + \nu \xi} \qquad \text{for } i = 1, 2, \ldots, r \qquad (30.11)$$

where ξ is the degree of reaction. Upon substituting equations (30.11) into (30.9), and after some rearrangements we find that

$$\prod_{i=1}^{r} (y_i)^{\nu_i} = \prod_{i=1}^{r} \left(\frac{y_{ia} + \nu_i \xi}{1 + \nu \xi} \right)^{\nu_i} = \left(\frac{p_o}{p} \right)^\nu K(T) \qquad (30.12)$$

Thus, given the temperature T, the pressure p, the composition $y_{1a}, y_{2a}, \ldots, y_{ra}$ at which $\xi = 0$, and the chemical reaction mechanism $\sum_{i=1}^{r} \nu_i A_i = 0$, we can use the tabulated value of the equilibrium constant $K(T)$, and solve equation (30.12) for the value of the degree of reaction ξ at chemical equilibrium. Then, using equations (30.11) again, we determine the chemical equilibrium composition — the chemical equilibrium mole fractions — and the corresponding amounts of constituents.

Example 30.3. A system consists initially of 0.3 mol of H_2, and 0.7 mol of O_2. These two gases react according to the chemical reaction mechanism $2H_2 + O_2 = 2H_2O$. If the final state of the system is a chemical equilibrium state at $T = 4000$ K and $p = 2$ atm, find its composition.

Solution. The stoichiometric coefficients, the initial composition, and the composition in terms of the degree of reaction ξ are listed in Table 30.2.

TABLE 30.2

i	A_i	ν_i	y_{ia}	y_i
1	H_2O	2	0.0	$2\xi/(1 - \xi)$
2	H_2	-2	0.3	$(0.3 - 2\xi)/(1 - \xi)$
3	O_2	-1	0.7	$(0.7 - \xi)/(1 - \xi)$
$\sum_{i=1}^{3}$		-1	1.0	1.0

For the reaction mechanism $H_2 + \frac{1}{2}O_2 = H_2O$ and $T = 4000$ K, Table 30.1 indicates that the chemical equilibrium constant is approximately $[K(T)]_{tab} = \exp(-6.866 + 29,911/T)$, and therefore $[K(4000 \text{ K})]_{tab} = 1.844$. Because the stoichiometric coefficients of this reaction differ from those of our reaction by a factor of 2, we have $K(4000 \text{ K}) = (1.844)^2 = 3.4$. Moreover, at $T = 4000$ K, and $p = 2$ atm, H_2, O_2, and H_2O behave as ideal gases and, therefore, equation (30.12) applies. So we find that

$$\left(\frac{2\xi}{1-\xi}\right)^2\left(\frac{0.3-2\xi}{1-\xi}\right)^{-2}\left(\frac{0.7-\xi}{1-\xi}\right)^{-1} = \left(\frac{1}{2}\right)^{-1}\times 3.4 \quad \text{or} \quad \frac{\xi^2(1-\xi)}{(0.3-2\xi)^2(0.7-\xi)} = 1.7$$

The three solutions of this cubic equation are: $\xi = 0.1021$, $\xi = 0.3054$, and $\xi = 0.5926$. The second and third solutions are not acceptable because they yield negative values for y_2. The acceptable solution, $\xi_o = 0.1021$, yields the composition of the chemical equilibrium state as $y_{1o} = 0.2274$, $y_{2o} = 0.1067$, $y_{3o} = 0.6659$. It is noteworthy that, at the specified conditions of temperature and pressure, not all the hydrogen is converted into water vapor even though there is adequate oxygen present to do so. In other words, in the chemical equilibrium state at $T = 4000$ K and $p = 2$ atm, all the constituents are present with nonzero amounts.

Example 30.4.   Find the chemical equilibrium state of a gas mixture at 3200 K and 69 kPa initially consisting, on a mole basis, of 40% of CO_2 and 60% of O_2, assuming that the two gases are subject to the chemical reaction mechanism $CO_2 = CO + \frac{1}{2}O_2$. Assume that this is an ideal-gas mixture.

Solution.   For this mixture, the initial composition and the composition in terms of the degree of reaction ξ are listed in Table 30.3.

TABLE 30.3

i	A_i	ν_i	y_{ia}	y_i
1	CO_2	-1.0	0.4	$(0.4-\xi)/(1+0.5\xi)$
2	CO	1.0	0.0	$\xi/(1+0.5\xi)$
3	O_2	0.5	0.6	$(0.6+0.5\xi)/(1+0.5\xi)$
$\sum_{i=1}^{3}$		0.5	1.0	1.0

For the reaction mechanism $CO + \frac{1}{2}O_2 = CO_2$ and $T = 3200$ K, in Example 30.2 we find that $[K(3200 \text{ K})]_{\text{tab}} = 1.559$. Because the stoichiometric coefficients of this reaction differ from those of our reaction by a factor of -1, we have $K(3200 \text{ K}) = (1.559)^{-1} = 0.641$. So the chemical equilibrium equation is

$$\left(\frac{\xi}{1+0.5\xi}\right)\left(\frac{0.6+0.5\xi}{1+0.5\xi}\right)^{0.5}\left(\frac{1+0.5\xi}{0.4-\xi}\right) = \left(\frac{101.3}{69}\right)^{0.5}\times 0.641$$

Solving this equation for ξ, we find $\xi = 0.1973$, and then $y_{1o} = 0.1845$, $y_{2o} = 0.1796$, and $y_{3o} = 0.6359$.

If any one of the constituents of the system does not behave as an ideal gas, the equilibrium constant is still useful but the chemical equilibrium equation is not readily resolvable. To see this clearly, we express the chemical potential of each pure substance in terms of its fugacity (Section 20.5), that is,

$$\mu_{ii}(T,p_{ii}) = \mu_{ii}(T,p_o) + RT \ln \frac{\pi_{ii}(T,p_{ii})}{\pi_{ii}(T,p_o)} \qquad \text{for } i = 1, 2, \ldots, r \qquad (30.13)$$

Substituting equations (30.13) in (30.6), and after some rearrangements, we find that

$$\prod_{i=1}^{r}\left[\frac{\pi_{ii}(T,p_{ii})}{\pi_{ii}(T,p_o)}\right]^{\nu_i} = K(T) \qquad (30.14)$$

where again $K(T)$ is given by equation (30.10). Of course, under ideal-gas behavior, equation (30.14) reduces to equation (30.9) because then $\pi_{ii}(T,p_{ii}) = p_{ii}$ and $\pi_{ii}(T,p_o) = p_o$.

Another important class of applications involves chemical equilibrium states of ideal and nonideal solutions. For an ideal solution, we recall that [equation (27.49)]

$$\mu_i = \mu_{ii}(T,p) + RT \ln y_i \qquad \text{for } i = 1, 2, \ldots, r \tag{30.15}$$

Substituting these relations in equation (30.5), and after some rearrangements, we find that

$$\prod_{i=1}^{r} (y_i)^{\nu_i} = K(T,p) \tag{30.16}$$

where

$$K(T,p) = \exp\left[-\frac{1}{RT} \sum_{i=1}^{r} \nu_i \, \mu_{ii}(T,p)\right]$$

$$= K(T) \exp\left[-\frac{1}{RT} \sum_{i=1}^{r} \nu_i \int_{p_o}^{p} v_{ii}(T,p')\,dp'\right] \tag{30.17}$$

$v_{ii}(T,p)$ denotes the specific volume of pure constituent i, $K(T)$ is given by equation (30.10), and in writing the second of equations (30.17) we integrate the expression $(\partial \mu/\partial p)_T = v$ from p_o to p.

If at temperatures around T, and in the pressure range between p_o and p, each pure constituent of the solution behaves as an ideal incompressible liquid or solid, $v_{ii}(T,p) = v_{ii}^o(T) = $ constant or, at least, is independent of pressure, and therefore

$$K(T,p) = K(T) \exp\left[-\frac{p - p_o}{RT} \sum_{i=1}^{r} \nu_i \, v_{ii}^o(T)\right] = K(T) \exp\left[-\frac{p - p_o}{RT} \Delta v^o(T)\right] \tag{30.18}$$

where $\Delta v^o(T) = \sum_{i=1}^{r} \nu_i \, v_{ii}(T,p_o)$, that is, the volume of the reaction.

For a nonideal mixture or solution, we recall that [equation (28.9)]

$$\mu_i = \mu_{ii}(T,p) + RT \ln a_i \qquad \text{for } i = 1, 2, \ldots, r \tag{30.19}$$

where the activity of the ith constituent in the mixture $a_i = a_i(T,p,y_1,y_2,\ldots,y_r)$. Substituting these relations in equation (30.5), we find that

$$\prod_{i=1}^{r} (a_i)^{\nu_i} = K(T,p) \tag{30.20}$$

where $K(T,p)$ is given by equation (30.17).

All the results of this and the preceding section can readily be adapted to any bulk-flow state for which the stable equilibrium part is a chemical equilibrium state. Indeed, the mole fraction of each constituent in a bulk-flow state is related to the degree of reaction ξ by the same formal expression as that of a batch process (Section 29.5), and therefore the chemical equilibrium relations in terms of ξ remain intact.

Example 30.5.   A bulk-flow stream of carbon dioxide, CO_2, enters a steady-state chemical reactor at a flow rate of 1 kmol/s, temperature $T_o = 25°C$, and pressure $p_o = 1$ atm. A bulk-flow stream consisting of CO_2, CO, and O_2 exits the reactor at a constant flow rate, at temperature $T = 3200$ K and pressure $p = p_o = 1$ atm. In addition, the reactor interacts with a reservoir at temperature T_R. If the potential and kinetic energies of the bulk-flow streams are negligible, and the outlet stream is in chemical equilibrium with respect to the chemical reaction mechanism $CO_2 = CO + \frac{1}{2}O_2$, find: **(a)** the flow rate of the outlet stream; **(b)** the heat rate with the reservoir; and **(c)** the smallest temperature T_R for which the heat interaction can occur.

Solution. **(a)** The stoichiometric coefficients, and the flow rates and compositions of the inlet and outlet streams are listed in Table 30.4.

TABLE 30.4

i	A_i	ν_i	$\dot{n}_{ia}$	y_{ia}	$\dot{n}_{ib}$	y_{ib}
1	CO_2	-1.0	1	1	$1 - \xi$	$(1 - \xi)/(1 + 0.5\,\xi)$
2	CO	1.0	0	0	ξ	$\xi/(1 + 0.5\,\xi)$
3	O_2	0.5	0	0	$0.5\,\xi$	$0.5\,\xi/(1 + 0.5\,\xi)$
$\sum_{i=1}^{3}$		0.5	1	1	$1 + 0.5\,\xi$	1

The composition of the outlet stream is found from the balance between the inlet and outlet flow rates, which can be expressed in terms of the degree of reaction ξ as $CO_2 = (1 - \xi)\,CO_2 + \xi\,CO + (\xi/2)\,O_2$. For $T = 3200$ K, the equilibrium constant of the dissociation reaction mechanism, $CO_2 = CO + \frac{1}{2}\,O_2$, $K(3200\ \text{K}) = 0.641$ (Example 30.4). Assuming that the outlet stream behaves as an ideal-gas mixture, the chemical equilibrium equation becomes ($p = p_o$)

$$\left(\frac{1 + \xi/2}{1 - \xi}\right)\left(\frac{\xi}{1 + \xi/2}\right)\left(\frac{\xi/2}{1 + \xi/2}\right)^{1/2} = 0.641$$

and yields $\xi = 0.58$. Hence $\dot{n}_b = 1.29$ kmol/s per kilomole of CO_2 per second, $y_{1b} = 0.33$, $y_{2b} = 0.45$, $y_{3b} = 0.22$.

(b) The heat rate per unit flow rate of CO_2 is given by the energy rate balance

$$\frac{\dot{Q}^{\leftarrow}}{\dot{n}_a} = \frac{\dot{Q}^{\leftarrow}}{\dot{n}_{1a}} = \frac{\dot{H}_b - \dot{H}_a}{\dot{n}_a} = (1 + \nu\,\xi)\sum_{i=1}^{3} y_{ib}\,(h_{ib} - h_{ii}^o) - (h_{11a} - h_{11}^o) + \xi\,\Delta h^o$$

where in writing the last of these equations we use equation (29.63) for $\dot{Z} = \dot{H}$ and $T = T_o$. Assuming ideal-gas mixture behavior, we can substitute $h_{ii}(T_b)$ for h_{ib} in the sum, for $i = 1, 2, 3$. Using Table 20.2, after some computation we find that $h_{1b} - h_{11}^o = 172{,}804 - 6926 = 165{,}878$ kJ/kmol, $h_{2b} - h_{22}^o = 109{,}326 - 8787 = 100{,}539$ kJ/kmol, and $h_{3b} - h_{33}^o = 115{,}211 - 7805 = 107{,}406$ kJ/kmol. Of course, $h_{11a} - h_{11}^o = 0$ because $T_a = T_o$. Using the enthalpies of formation from Table 29.4 and the dissociation reaction mechanism, we have $\Delta h^o = -110{,}600 + (1/2)(0) - (-393{,}800) = 283{,}200$ kJ/kmol. So

$$\frac{\dot{Q}^{\leftarrow}}{\dot{n}_{1a}} = (1 + 0.5 \times 0.58)(0.33 \times 165{,}878 + 0.45 \times 100{,}539 + 0.22 \times 107{,}406)$$

$$+\, 0.58 \times 283{,}200 = 323{,}715 \text{ kJ/kmol of } CO_2$$

(c) The temperature T_R satisfies the entropy rate balance

$$\frac{\dot{Q}^{\leftarrow}}{\dot{n}_a T_R} = \frac{\dot{Q}^{\leftarrow}}{\dot{n}_{1a} T_R} = \frac{\dot{S}_b - \dot{S}_a}{\dot{n}_a} - \frac{\dot{S}_{\text{irr}}}{\dot{n}_a}$$

$$= (1 + \nu\,\xi)\sum_{i=1}^{3} y_{ib}\,(s_{ib} - s_{ii}^o) - (s_{11a} - s_{11}^o) + \xi\,\Delta s^o - \frac{\dot{S}_{\text{irr}}}{\dot{n}_a}$$

where in writing the last of these equations we use equation (29.63) for $\dot{Z} = \dot{S}$ and $T = T_o$. The smallest value of T_R obtains when $\dot{S}_{\text{irr}} = 0$. Assuming ideal-gas mixture behavior and noting that $p_b = p_o$, we can write $s_{ib} = s_{ii}(T_b, p_o) - R \ln y_{ib}$ for $i = 1, 2, 3$. Using Table 20.2, after some computation we find that $s_{1b} - s_{11}^o = 247.1 - 122.5 + 9.2 = 133.8$ kJ/kmol K, $s_{2b} - s_{22}^o = 193.1 - 115.0 + 6.6 = 84.7$ kJ/kmol K, and $s_{3b} - s_{33}^o = 225.2 - 142.2 + 12.6 =$

95.6 kJ/kmol K. In addition, $s_{11a} - s_{11}^{o} = 0$. Using Table 29.4 and proceeding as in Example 29.6, we have $\Delta s^o = 89.7 + (1/2)(0) - 2.9 = 86.8$ kJ/kmol K. So

$$\frac{\dot{S}_b - \dot{S}_a}{\dot{n}_a} = (1 + 0.5 \times 0.58)(0.33 \times 133.8 + 0.45 \times 84.7 + 0.22 \times 95.6)$$

$$+ 0.58 \times 86.8 = 183.6 \text{ kJ/kmol K}$$

and $(T_R)_{\text{smallest}} = \dot{Q}^{\leftarrow}/(\dot{S}_b - \dot{S}_a) = 323,715/183.6 = 1763$ K.

30.3   Temperature and Pressure Effects

It is clear from the chemical equilibrium equation [equation (30.5)] that the composition of the chemical equilibrium state depends on the temperature T and the pressure p. To examine explicitly these dependences, we begin by proving that

$$\frac{d\ln K(T)}{dT} = \frac{d}{dT}\left[-\frac{1}{RT} \sum_{i=1}^{r} \nu_i \mu_{ii}(T, p_0) \right] = \frac{\Delta h^o(T)}{RT^2} \tag{30.21a}$$

or, equivalently, that

$$\frac{d\ln K(T)}{d(1/T)} = -\frac{\Delta h^o(T)}{R} \tag{30.21b}$$

where $\Delta h^o(T)$ is the enthalpy of reaction at temperature T and pressure p_0 for the reaction mechanism $\sum_{i=1}^{r} \nu_i A_i = 0$. Equation 30.21 is known as the *van't Hoff relation for chemical equilibrium* and is useful to interpolate and extrapolate data on equilibrium constants.

To prove equation (30.21), we recall that for a pure constituent simple system $\mu = g = h - Ts$, and $(\partial\mu/\partial T)_p = -s$. Hence we find that

$$\left[\frac{\partial(\mu/T)}{\partial T}\right]_p = \frac{1}{T}\left(\frac{\partial\mu}{\partial T}\right)_p - \frac{\mu}{T^2} = -\frac{s}{T} - \frac{\mu}{T^2} = -\frac{h}{T^2} \tag{30.22}$$

$$\frac{d\ln K(T)}{dT} = -\frac{1}{R}\frac{d}{dT}\sum_{i=1}^{r}\frac{\nu_i \mu_{ii}(T, p_0)}{T} = -\frac{1}{R}\sum_{i=1}^{r}\nu_i \frac{\partial}{\partial T}\left[\frac{\mu_{ii}(T, p)}{T}\right]_{p=p_0}$$

$$= \frac{1}{R}\sum_{i=1}^{r}\frac{\nu_i h_{ii}(T, p_0)}{T^2} = \frac{1}{R}\sum_{i=1}^{r}\frac{\nu_i h_{ii}^o(T)}{T^2} = \frac{\Delta h^o(T)}{RT^2} \tag{30.23}$$

where in writing the last of equations (30.23) we use the definition of $\Delta h^o(T)$ [equation (29.59)].

At a given temperature, a chemical reaction is called either *exothermic* if the enthalpy of reaction $\Delta h^o(T)$ is negative or *endothermic* if $\Delta h^o(T)$ is positive. Because $\Delta h^o(T)$ is a very weak function of T, the vast majority of chemical reactions are either exothermic or endothermic at all values of temperature.

Exothermic means that energy must be transferred out of a steady-state chemical reactor (Figure 30.2) in order to maintain the outlet bulk-flow stream at the same temperature T and pressure p_0 as the inlet bulk-flow stream. If one of the reactants, say, constituent 1, has a stoichiometric coefficient equal to -1, the energy transferred out of the reactor per mole of that reactant is $e^{\rightarrow} = \dot{Q}^{\rightarrow}/\dot{n}_{1a} = -\Delta h^o(T)$.

Conversely, endothermic means that energy must be transferred into the chemical reactor to maintain the same inlet and outlet temperature T and pressure p_0. If one of

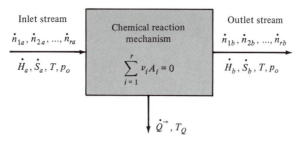

Figure 30.2 A steady-state chemical reactor with inlet and outlet bulk-flow streams maintained at the same temperature T and pressure p_o.

the reactants has a stoichiometric coefficient equal to -1, the energy transferred into the reactor per mole of that reactant is $e^{\leftarrow} = \Delta h^o(T)$.

Example 30.6. Consider the reaction mechanism $C + O_2 = CO_2$, and the steady state reactor shown in Figure 30.2. Assume that the inlet stream consists of C and O_2, each at T_o and p_o, that the reactor interacts with a reservoir at T_o, and that the outlet stream is pure CO_2. Neglecting kinetic and potential energies of the bulk-flow streams, evaluate the heat between the reactor and the reservoir, $q^{\rightarrow} = \dot{Q}^{\rightarrow}/\dot{n}_C$, and the amount of entropy generated by irreversibility, $s_{\mathrm{irr}} = \dot{S}_{\mathrm{irr}}/\dot{n}_C$.

Solution. Using subscripts 1, 2, and 3 to denote CO_2, C, and O_2, respectively, the energy balance per unit flow rate of C is

$$q^{\rightarrow} = h_a - h_b = h_{22}^o + h_{33}^o - h_{11}^o = -\Delta h^o$$

where Δh^o is the enthalpy of the reaction $C + O_2 = CO_2$ at standard conditions, that is, the enthalpy of formation of CO_2. From Table 29.4 we find $\Delta h^o = -393.8$ MJ/kmol. So the reaction is exothermic at temperature T_o, and $q^{\rightarrow} = 393.8$ MJ/kmol. Per mole of C, the entropy generated by irreversibility is

$$s_{\mathrm{irr}} = s_b - s_a + \frac{q^{\rightarrow}}{T_o} = s_{11}^o - s_{22}^o - s_{33}^o + \frac{q^{\rightarrow}}{T_o} = \Delta s^o + \frac{q^{\rightarrow}}{T_o}$$

where $\Delta s^o = s_{11}^o - s_{22}^o - s_{33}^o$, that is, the entropy of the reaction $C + O_2 = CO_2$ at standard conditions or, equivalently, the entropy of formation of CO_2. Again from Table 29.4 we find that $\Delta s^o = 2.9$ kJ/kmol K and, therefore, $s_{\mathrm{irr}} = 2.9 \times 10^{-3} + 393.8/298 = 1.325$ MJ/kmol K.

We see from (30.21a) that for an exothermic (endothermic) reaction the equilibrium constant $K(T)$ is a decreasing (increasing) function of T. Equivalently, a graph of $\ln K(T)$ versus $1/T$ has a slope $-\Delta h^o(T)/R$ [equation (30.21b)]. This slope is positive for exothermic reactions and negative for endothermic reactions. Figure 30.1 shows graphs of $\log_{10} K(T)$ versus $1/T$ for various chemical reactions. The slope of each graph, $-\Delta h^o(T)/2.3R$ [recall that $\ln x = (\ln 10) \log_{10} x = 2.3 \log_{10} x$], is approximately constant, thus verifying the statement that $\Delta h^o(T)$ is a very weak function of temperature. In view of this observation, for a range of temperatures $T_1 < T < T_2$ within which $\Delta h^o(T)$ is approximately constant, we can integrate the van't Hoff relation and obtain the expression

$$\Delta h^o(T) \approx R T_1 T_2 \frac{\ln K(T_2) - \ln K(T_1)}{T_2 - T_1} \tag{30.24}$$

and the interpolation formula

$$\ln K(T) \approx \ln K(T_1) + \frac{1 - T_1/T}{1 - T_1/T_2} \left[\ln K(T_2) - \ln K(T_1) \right] \tag{30.25}$$

Equation (30.25) also justifies the approximation $\ln K(T) \approx \Delta a - \Delta b/T$ in Table 30.1. This approximation is better the weaker the dependence of $\Delta h^\circ(T)$ on T. Equation (30.25) becomes exact if $\Delta h^\circ(T)$ is independent of T. Indeed, upon combining equations (29.59) and (29.67) with the relation $(\partial h/\partial T)_p = T\,(\partial s/\partial T)_p$ [equation (19.16)], we find that $d\,\Delta h^\circ(T)/dT = T\,d\,\Delta s^\circ(T)/dT$. Therefore, if $\Delta h^\circ(T)$ is a constant, then $\Delta s^\circ(T)$ is also a constant, and $\ln K(T) = -\Delta g^\circ(T)/RT = \Delta s^\circ(T)/R - \Delta h^\circ(T)/RT = \Delta a - \Delta b/T$, where $\Delta a = \Delta s^\circ(T)/R = $ constant and $\Delta b = \Delta h^\circ(T)/R = $ constant.

This completes our discussion of the dependence of the equilibrium constant on T. Next, we comment on the meaning of the degree of reaction ξ. In general, it is not advisable to assign a physical meaning to the value of ξ because it depends on the values of the stoichiometric coefficients, and we have seen that the choice of values of the stoichiometric coefficients of a chemical reaction mechanism has a degree of arbitrariness. Nevertheless, in special reactions, ξ is related to the yield of products.

For given initial amounts of constituents and a chemical reaction mechanism, upon differentiating equation (29.22) [or (29.25)] we find that

$$dy_i = \frac{\nu_i - \nu\, y_{ia}}{(1 + \nu\,\xi)^2}\, d\xi \qquad (30.26)$$

For a constituent i that is present as a product but not as a reactant, the stoichiometric coefficient $\nu_i > 0$, the initial mole fraction $y_{ia} = 0$, and equation (30.26) shows that $dy_i/d\xi > 0$. This means that y_i is an increasing function of ξ and therefore that the higher the value of ξ, the higher the value of y_i for that product. So we confirm the statement that ξ is related to the yield of products of the reaction.

In the sense just cited, the yield of products in a chemical equilibrium state corresponding to a given chemical reaction mechanism, namely, the value of ξ that satisfies the chemical equilibrium equation, depends on the temperature T and pressure p of that state. To investigate these dependences, we consider an ideal gas mixture, take the logarithm of equation (30.12), differentiate the result, and after some algebra find that

$$F_a(\xi)\, d\xi = \frac{\Delta h^\circ(T)}{RT^2}\, dT - \frac{\nu}{p}\, dp \qquad (30.27)$$

where

$$F_a(\xi) = \frac{1}{1 + \nu\,\xi} \sum_{i=1}^{r} \nu_i \frac{\nu_i - \nu\, y_{ia}}{y_{ia} + \nu_i\,\xi} = \frac{1}{1 + \nu\,\xi} \sum_{i=1}^{r} \left(\frac{\nu_i^2}{y_i} - \nu\,\nu_i \right)$$

$$= \frac{1}{1 + \nu\,\xi} \sum_{i=1}^{r} \left(\frac{\nu_i}{\sqrt{y_i}} - \nu\,\sqrt{y_i} \right)^2 \qquad (30.28)$$

and we use equation (30.22), and the identities $\sum_{i=1}^{r} \nu_i = \nu$ and $\sum_{i=1}^{r} y_i = 1$. The function $F_a(\xi)$ is positive definite because $1 + \nu\,\xi = \sum_{i=1}^{r} n_i / \sum_{i=1}^{r} n_{ia} > 0$. So, Equation 30.27 indicates that the degree of reaction ξ—the yield of a product of reaction—depends on T and p as follows. At constant pressure, that is, for $dp = 0$, ξ is an increasing $[(\partial \xi/\partial T)_p > 0]$ or decreasing $[(\partial \xi/\partial T)_p < 0]$ function of temperature T depending on whether the reaction is endothermic $[\Delta h^\circ(T) > 0]$ or exothermic $[\Delta h^\circ(T) < 0]$, respectively. At constant temperature, that is, for $dT = 0$, ξ is an increasing $[(\partial \xi/\partial p)_T > 0]$, invariant $[(\partial \xi/\partial p)_T = 0]$, or decreasing $[(\partial \xi/\partial p)_T < 0]$ function of pressure p depending on whether $\nu < 0$, $\nu = 0$, or $\nu > 0$, respectively.

Example 30.7.   Consider the outlet stream of the chemical reactor in Example 30.5. Estimate the change in the yield of carbon monoxide, CO, if **(a)** the pressure is doubled and the temperature is maintained at 3200 K and **(b)** the pressure is maintained at p_0 and the temperature is reduced at 2800 K.

Solution.   Rather than repeating the computations in Example 30.5 for the new conditions, we apply the foregoing discussion. Using the results of Example 30.5, we find that equation (30.26) for $i = 2$ becomes $dy_2 = (1 - 0.5 \times 0)/(1 + 0.5 \times 0.58)^2 \, d\xi = 0.60 \, d\xi$, $F_a(\xi) = [1/(1 + 0.5 \times 0.58)] \times \{[1/0.33 - 0.5 \times (-1)] + [1/0.45 - 0.5 \times 1] + [0.5^2/0.22 - 0.5 \times 0.5]\} = 4.76$, $\nu/p = 0.5/1$ atm $= 0.5$ atm^{-1}. Using the results of Example 29.7, we have $\Delta h^\circ(T)/RT^2 = 271{,}563/(8.314 \times 3200^2) = 3.19 \times 10^{-3}$ K^{-1}. Next, we integrate equation (30.27) approximately assuming that the coefficients of the differentials are constant, that is, we write $F_a(\xi) \, \Delta\xi \approx [\Delta h^\circ(T)/RT^2] \, \Delta T - (\nu/p) \, \Delta p$ or, substituting the values just found, $4.76 \, \Delta\xi \approx 3.19 \times 10^{-3} \, \Delta T - 0.5 \, \Delta p$. For condition (a), $\Delta p = 1$ atm, $\Delta T = 0$, and therefore $\Delta\xi \approx -0.5/4.76 = -0.105$, $\Delta y_2 \approx 0.60 \times (-0.105) = -0.06$, and the new yield of carbon monoxide $y_2 \approx 0.45 - 0.06 = 0.39$. For condition (b), $\Delta p = 0$, $\Delta T = 2800 - 3200 = -400$ K, and therefore $\Delta\xi \approx 3.19 \times 10^{-3} \times (-400)/4.76 = -0.27$, $\Delta y_2 \approx 0.60 \times (-0.27) = -0.16$, and the new yield of carbon monoxide $y_2 \approx 0.45 - 0.16 = 0.29$.

30.4   Chemical Equilibrium with Respect to Many Reactions

We consider a simple system with r constituents subject to τ different chemical reaction mechanisms. We assume that A behaves as a nonideal mixture or solution, so that

$$\mu_i = \mu_{ii}(T, p) + RT \ln a_i \qquad \text{for } i = 1, \, 2, \, \ldots, \, r \tag{30.29}$$

where the activity of the ith constituent in the mixture $a_i = a_i(T, p, y_1, y_2, \ldots, y_r)$. Substituting equations (30.29) in the chemical equilibrium equations [equations (30.1)] and rearranging terms, we find that

$$\prod_{i=1}^{r} (a_i)^{\nu_i^{(j)}} = K_j(T, p) \qquad \text{for } j = 1, \, 2, \, \ldots, \, \tau \tag{30.30}$$

where

$$K_j(T, p) = \exp\left[-\frac{1}{RT} \sum_{i=1}^{r} \nu_i^{(j)} \mu_{ii}(T, p)\right]$$

$$= K_j(T) \exp\left[-\frac{1}{RT} \sum_{i=1}^{r} \nu_i^{(j)} \int_{p_0}^{p} v_{ii}(T, p') \, dp'\right] \tag{30.31}$$

the equilibrium constant of the jth reaction

$$K_j(T) = \exp\left[-\frac{1}{RT} \sum_{i=1}^{r} \nu_i^{(j)} \mu_{ii}(T, p_0)\right] = \exp\left\{-\frac{1}{RT} \left[\Delta g^\circ(T)\right]_j\right\} \tag{30.32}$$

and $[\Delta g^\circ(T)]_j$ is the Gibbs free energy of the jth reaction at temperature T and standard pressure p_0.

For special behavior, the general equations (30.30) and (30.31) simplify as follows. For either ideal-gas mixture or ideal-solution behavior, we have

$$a_i = y_i = \frac{n_{ib}}{n_b} = \frac{y_{ia} + \sum_{j=1}^{\tau} \nu_i^{(j)} \xi_j}{1 + \sum_{j=1}^{\tau} \nu^{(j)} \xi_j} \qquad \text{for } i = 1, \, 2, \, \ldots, \, r \tag{30.33}$$

where ξ_j is the degree of reaction of the jth chemical reaction. For ideal-gas behavior of each of the constituents, we have

$$K_j(T, p) = \left(\frac{p_0}{p}\right)^{\nu^{(j)}} K_j(T) \qquad \text{for } j = 1, \, 2, \, \ldots, \, \tau \tag{30.34}$$

For ideal incompressible behavior of each of the constituents, we have

$$K_j(T,p) = K_j(T)\exp\left\{-\frac{p-p_0}{RT}\left[\Delta v^o(T)\right]_j\right\} \qquad \text{for } j = 1, 2, \ldots, \tau \qquad (30.35)$$

where $[\Delta v^o(T)]_j = \sum_{i=1}^{r} \nu_i^{(j)} v_{ii}(T, p_0)$, that is, the volume of the jth reaction at T and p_0.

As an illustration of these special results, we consider an ideal-gas mixture in chemical equilibrium with respect to τ chemical reaction mechanisms for which the chemical equilibrium equations become

$$\prod_{i=1}^{r}\left(\frac{n_{ib}}{n_b}\right)^{\nu_i^{(j)}} = \prod_{i=1}^{r}\left(\frac{y_{ia} + \sum_{j=1}^{\tau}\nu_i^{(j)}\xi_j}{1 + \sum_{j=1}^{\tau}\nu^{(j)}\xi_j}\right)^{\nu_i^{(j)}} = \left(\frac{p_0}{p}\right)^{\nu^{(j)}}K_j(T)$$

$$\text{for } j = 1, 2, \ldots, \tau \qquad (30.36)$$

These τ equations can in principle be solved for the τ degrees of reaction $\xi_1, \xi_2, \ldots, \xi_\tau$. So the composition and the values of the amounts of constituents of the chemical equilibrium state at T and p can be evaluated. However, it is often more convenient to evaluate the chemical equilibrium composition by considering the overall reaction mechanism (Section 29.6) and applying the methods discussed in the preceding sections.

Here, too, the results can readily be adapted to any bulk-flow state for which the stable equilibrium part is a chemical equilibrium state with respect to τ chemical reaction mechanisms.

Example 30.8. Pure water, as well as any aqueous solution, is always in chemical equilibrium with respect to the dissociation reaction $2\,H_2O = H_3O^+ + OH^-$. The equilibrium constant of this reaction at standard temperature and pressure $K_1(25°C) = 10^{-14}$. Thus the ions H_3O^+ and OH^- are always present with very small mole fractions, which nevertheless are important for many applications because ionized molecules are very reactive chemical compounds. It is standard practice to denote with the symbol pH the negative of the logarithm to the base 10 of the mole fraction of the H_3O^+ ions, that is,

$$pH = -\log_{10} y_{H_3O^+}$$

(a) Verify that for pure water at standard temperature and pressure pH $= 7$.

(b) A pH indicator is a molecule of the form RH, where R stands for some radical and H for hydrogen, which in aqueous solution dissociates according to the chemical reaction mechanism $RH + H_2O = R^- + H_3O^+$ with equilibrium constant at standard temperature and pressure $K_2(25°C)$, and such that the molecule RH has one color (e.g., red), whereas the ion R^- has another color (e.g., yellow). Added in minute quantities to an acqueous solution, the pH indicator colors the solution according to the ratio y_{R^-}/y_{RH} of, say, yellow to red constituents. The indicator is characterized by the value of its pK defined by the relation $pK = -\log_{10} K_2$. For a pH indicator with a given value of pK, show that the color of the solution is determined by its pH.

(c) Find the pH at standard temperature and pressure of a solution of acetic acid, CH_3COOH, in water with mole fraction $y_{CH_3COOH} = 10^{-5}$, given that the equilibrium constant of the dissociation reaction $CH_3COOH + H_2O = CH_3COO^- + H_3O^+$ is $K_3(25°C) = 1.8 \times 10^{-5}$.

(d) Find the pH at standard temperature and pressure of a solution of ammonia, NH_3, in water with mole fraction $y_{NH_3} = 10^{-5}$, given that the equilibrium constant of the dissociation reaction $NH_3 + H_2O = OH^- + NH_4^+$ is $K_4(25°C) = 1.8 \times 10^{-5}$.

Solution. All the aqueous solutions in this problem are so dilute that $y_{H_2O} \approx 1$.

(a) For chemically pure water, the initial mole fractions of H_3O^+ and OH^- are zero. Thus the chemical equilibrium equation is $\xi_1^2/(1 - 2\xi_1)^2 = 10^{-14}$ and yields $y_{H_3O^+} = \xi_1 \approx 10^{-7}$ and hence pH $= 7$.

(b) The chemical equilibrium equation of the dissociation reaction of the pH indicator can be written as $(y_{R^-})(y_{H_3O^+})/(y_{RH})(y_{H_2O}) = 10^{-pK}$. Taking logarithms to the base 10, recalling that $y_{H_2O} \approx 1$ and that $y_{H_3O^+} = 10^{-pH}$, we find that $\log_{10}[(y_{R^-})/(y_{RH})] = pH - pK$.

(c) Initially, the nonzero mole fractions are $y_{CH_3COOH} = 10^{-5}$ and $y_{H_2O} \approx 1$. The chemical equilibrium equations $\xi_1(\xi_1 + \xi_2)/(1 - 2\xi_1 - \xi_2)^2 = 10^{-14}$ and $\xi_2(\xi_1 + \xi_2)/(1 - 2\xi_1 - \xi_2)(10^{-5} - \xi_2) = 1.8 \times 10^{-5}$ yield $\xi_1 \approx 1.4 \times 10^{-9}$ and $\xi_2 \approx 7.15 \times 10^{-6}$. Hence, $y_{H_3O^+} = \xi_1 + \xi_2 \approx 7.15 \times 10^{-6} \approx 10^{-5.146}$ and $pH = 5.146$.

(d) Initially, the nonzero mole fractions are $y_{NH_3} = 10^{-5}$ and $y_{H_2O} \approx 1$. The chemical equilibrium equations [formally identical to those in part (c)] $\xi_1(\xi_1 + \xi_2)/(1 - 2\xi_1 - \xi_2)^2 = 10^{-14}$ and $\xi_2(\xi_1 + \xi_2)/(1 - 2\xi_1 - \xi_2)(10^{-5} - \xi_2) = 1.8 \times 10^{-5}$ yield $\xi_1 \approx 1.4 \times 10^{-9}$ and $\xi_2 \approx 7.15 \times 10^{-6}$. But here $y_{H_3O^+} = \xi_1 \approx 1.4 \times 10^{-9} \approx 10^{-8.854}$ and $pH = 8.854$.

Example 30.9. A bulk-flow stream of carbon dioxide, CO_2, oxygen, O_2, and nitrogen, N_2, enters a steady-state chemical reactor at 25°C, 1 atm, flow rate of 1 kmol/s, and mole fractions $y_1 = 0.5$ of CO_2, $y_2 = 0.105$ of O_2, and $y_4 = 0.395$ of N_2. A bulk-flow stream consisting of CO_2, CO, O_2, N_2, NO and NO_2 exits the reactor at a constant flow rate, at 4000 K and 1 atm. In addition, the reactor interacts with a reservoir at temperature T_R. If the potential and kinetic energies of the bulk-flow streams are negligible and the outlet stream is in chemical equilibrium with respect to the partial chemical reaction mechanisms $CO_2 = CO + \frac{1}{2}O_2$, $\frac{1}{2}N_2 + \frac{1}{2}O_2 = NO$, and $\frac{1}{2}N_2 + O_2 = NO_2$ find: **(a)** the composition of the outlet stream; **(b)** the heat rate with the reservoir; and **(c)** the smallest temperature T_R for which the heat interaction can occur.

Solution. **(a)** By combining the three partial chemical reaction mechanisms, we find the overall reaction mechanism

$$CO_2 + N_2 + O_2 = CO + NO + NO_2$$

The stoichiometric coefficients, compositions of the inlet and outlet streams, and coefficients a_f and b_f for the reaction mechanisms of formation of the various constituents (Table 30.1) are listed in Table 30.5.

TABLE 30.5

i	A_i	ν_i	y_{ia}	y_{ib}	$(a_f)_i$	$(b_f)_i$
1	CO_2	-1	0.5	$0.5 - \xi$	-0.010	$-47,575$
2	CO	1	0	ξ	10.098	$-13,808$
3	O_2	-1	0.105	$0.105 - \xi$	0	0
4	N_2	-1	0.395	$0.395 - \xi$	0	0
5	NO	1	0	ξ	1.504	10,863
6	NO_2	1	0	ξ	-7.630	3,870
$\sum_{i=1}^{6}$		0	1	1	$\Delta a = 3.982$	$\Delta b = 48,500$

The equilibrium constant of the overall reaction $K(T) = \exp(\Delta a - \Delta b/T) = \exp(3.982 - 48,500/4000) = 2.9 \times 10^{-4}$, and the chemical equilibrium equation becomes

$$\frac{\xi^3}{(0.5 - \xi)(0.105 - \xi)(0.395 - \xi)} = 2.9 \times 10^{-4}$$

and yields $\xi = 0.0167$. Hence $y_{1b} = 0.4833$, $y_{2b} = y_{5b} = y_{6b} = 0.0167$, $y_{3b} = 0.0883$, and $y_{4b} = 0.3783$.

(b) The heat rate per unit flow rate is given by the energy rate balance

$$\frac{\dot{Q}^{\leftarrow}}{\dot{n}_a} = \frac{\dot{H}_b - \dot{H}_a}{\dot{n}_a} = (1 + \nu\,\xi)\sum_{i=1}^{6} y_{ib}\,(h_{ib} - h_{ii}^{\circ}) - \sum_{i=1}^{6} y_{ia}\,(h_{ia} - h_{ii}^{\circ}) + \xi\,\Delta h^{\circ}$$

where in writing the last of these equations we use equation (29.63) for $\dot{Z} = \dot{H}$ and $T = T_o$. Assuming ideal-gas mixture behavior, we can substitute $h_{ii}(T_b)$ for h_{ib} and $h_{ii}(T_a)$ for h_{ia} for $i = 1, 2, \ldots, 6$. Using Table 20.2, after some computation we find that $h_{1b} - h_{11}^{\circ} = 223{,}221 - 6926 = 216{,}295$ kJ/kmol, $h_{2b} - h_{22}^{\circ} = 139{,}167 - 8787 = 131{,}380$ kJ/kmol, $h_{3b} - h_{33}^{\circ} = 148{,}265 - 7805 = 140{,}460$ kJ/kmol, $h_{4b} - h_{44}^{\circ} = 136{,}153 - 8975 = 127{,}178$ kJ/kmol, $h_{5b} - h_{55}^{\circ} = 143{,}104 - 8661 = 134{,}443$ kJ/kmol, and $h_{6b} - h_{66}^{\circ} = 207{,}212 - 6047 = 201{,}165$ kJ/kmol. Clearly, $h_{ia} - h_{ii}^{\circ} = h_{ii}(T_a) - h_{ii}^{\circ} = 0$ for $i = 1, 2, \ldots, 6$ because $T_a = T_o = 25°C$. Using the enthalpies of formation in Table 29.4 and the overall reaction mechanism, we have $\Delta h^{\circ} = 393{,}800 - 110{,}600 + 90{,}400 + 33{,}900 = 407{,}500$ kJ/kmol. So

$$\frac{\dot{Q}^{\leftarrow}}{\dot{n}_a} = (0.4833 \times 216{,}295 + 0.0167 \times 131{,}380 + 0.0883 \times 140{,}460$$

$$+ 0.3783 \times 127{,}178 + 0.0167 \times 134{,}443 + 0.0167 \times 201{,}165)$$

$$+ 0.0167 \times 407{,}500 = 179{,}653 \text{ kJ/kmol}$$

(c) The temperature T_R satisfies the entropy rate balance

$$\frac{\dot{Q}^{\leftarrow}}{\dot{n}_a T_R} = \frac{\dot{S}_b - \dot{S}_a}{\dot{n}_a} - \frac{\dot{S}_{irr}}{\dot{n}_a}$$

$$= (1 + \nu\,\xi)\sum_{i=1}^{6} y_{ib}\,(s_{ib} - s_{ii}^{\circ}) - \sum_{i=1}^{6} y_{ia}\,(s_{ia} - s_{ii}^{\circ}) + \xi\,\Delta s^{\circ} - \frac{\dot{S}_{irr}}{\dot{n}_a}$$

where in writing the last of these equations we use equation (29.63) for $\dot{Z} = \dot{S}$ and $T = T_o$. The smallest value of T_R obtains when $\dot{S}_{irr} = 0$. Assuming ideal-gas mixture behavior and noting that $p_b = p_o$, we can substitute $s_{ib} = s_{ii}(T_b, p_o) - R \ln y_{ib}$ for $i = 1, 2, \ldots, 6$. Using Table 20.2, after some computation we find that $s_{1b} - s_{11}^{\circ} = 261.1 - 122.5 + 6.0 = 144.6$ kJ/kmol K, $s_{2b} - s_{22}^{\circ} = 201.4 - 115.0 + 34.0 = 120.4$ kJ/kmol K, $s_{3b} - s_{33}^{\circ} = 234.4 - 142.2 + 20.2 = 112.4$ kJ/kmol K, $s_{4b} - s_{44}^{\circ} = 198.1 - 113.8 + 8.1 = 92.4$ kJ/kmol K, $s_{5b} - s_{55}^{\circ} = 211.9 - 122.6 + 34.0 = 123.3$ kJ/kmol K, $s_{6b} - s_{66}^{\circ} = 275.9 - 144.9 + 34.0 = 165.0$ kJ/kmol K. Moreover, $s_{ia} - s_{ii}^{\circ} = -R \ln y_{ia}$ for $i = 1, 2, \ldots, 6$, and therefore $\sum_{i=1}^{6} y_{ia}\,(s_{ia} - s_{ii}^{\circ}) = -R \sum_{i=1}^{6} y_{ia} \ln y_{ia} = 7.9$ kJ/kmol K. Using the entropies of formation in Table 29.4 and the overall reaction mechanism, we have $\Delta s^{\circ} = -2.9 + 89.7 + 12.4 - 60.8 = 38.4$ kJ/kmol K. So

$$\frac{\dot{S}_b - \dot{S}_a}{\dot{n}_a} = (0.4833 \times 144.6 + 0.0167 \times 120.4 + 0.0883 \times 112.4$$

$$+ 0.3783 \times 92.4 + 0.0167 \times 123.3 + 0.0167 \times 165.0)$$

$$+ 0.0167 \times 38.4 = 122.2 \text{ kJ/kmol K}$$

and $(T_R)_{\text{smallest}} = \dot{Q}^{\leftarrow}/(\dot{S}_b - \dot{S}_a) = 179{,}653/122.2 = 1470$ K.

30.5   Phase Rule for Chemically Reacting Systems

In Section 18.3 we establish the number of intensive properties T, p, μ_1, μ_2, $\ldots$, μ_r that can be varied independently in a heterogeneous state of q coexisting phases of a simple system with r constituents that do not react chemically. This number is called variance, is denoted by F, and equals $r + 2 - q$ (Gibbs phase rule).

We define coexistence as mutual stable equilibrium between the phases, and therefore as a condition in which each phase is in a stable equilibrium state having the common intensive properties T, p, μ_1, μ_2, ..., μ_r.

If τ chemical reaction mechanisms are effective among r constituents of a phase, the state of the phase must be chemical equilibrium, that is, the r chemical potentials of the phase must satisfy the τ chemical equilibrium equations [equations (30.1)]. Moreover, for q coexisting phases, the τ necessary conditions are common to all the phases because the values of the chemical potentials and the stoichiometric coefficients are common to all such phases. It follows that the chemical equilibrium equations impose τ restrictions on the number of intensive properties μ_1, μ_2, ..., μ_r in addition to those imposed by the requirements of mutual stable equilibrium between coexisting phases.

So, the number of intensive properties T, p, μ_1, μ_2, ..., μ_r that can be varied independently in a heterogeneous state of q coexisting phases of a simple system with r constituents that can react according to τ chemical reaction mechanisms is

$$F = r + 2 - q - \tau \tag{30.37}$$

Here too, as in Section 18.3, the number F is called *variance*.

For $r = 1$, namely, for systems consisting of only one constituent, there can be no chemical reaction mechanisms. Therefore, $\tau = 0$ and the implications of the phase rule are as discussed in Section 18.3. For $r = 2$, namely, for systems consisting of two constituents, there can be at most one chemical reaction mechanism. Therefore, either $\tau = 0$ or $\tau = 1$. For example, for a system consisting only of oxygen molecules, O_2, and nitrogen molecules, N_2, no chemical reaction is conceivable. Again, for a system consisting only of hydrogen molecules, H_2, and hydrogen atoms, H, the only conceivable chemical reaction mechanism is $2\,H = H_2$. If the mechanism is active, $\tau = 1$, and the number of phases that can coexist in a stable equilibrium state is 1, 2, or 3, and the variance is 2, 1, or 0, respectively. For a one-phase ideal-gas mixture of H_2 and H subject to the reaction mechanism $2\,H = H_2$, temperature and pressure may be regarded as the independent intensive properties, but then each chemical potential and each mole fraction is fixed by the values of T and p. Alternatively, a mole fraction of one constituent and the temperature (or the pressure) may be taken as independent, and then the pressure (or the temperature) is fixed by their values. Analogous conclusions can be reached for each possible combination of values of r, q, and τ.

30.6 Derivation of Conditions for Chemical Equilibrium

In this section we derive necessary conditions for a system subject to chemical reaction mechanisms to be in a stable equilibrium state. For a simple system A with r constituents and τ chemical reactions, and for given values U of the energy and V of the volume, and values n_1, n_2, ..., n_r of the amounts of constituents that are compatible with given values n_{1a}, n_{2a}, ..., n_{ra}, the system admits a very large number of states. But the second law requires that among all these states one and only one be stable equilibrium. As we discuss in Section 30.1, this state is called chemical equilibrium. At this state the values of the amounts of constituents, n_{1o}, n_{2o}, ..., n_{ro}, and the corresponding values of the reaction coordinates, ϵ_{1o}, ϵ_{2o}, ..., $\epsilon_{\tau o}$, satisfy equations (30.4).

The values U, V, $\boldsymbol{n}_a = \{n_{1a}, n_{2a}, ..., n_{ra}\}$, and the stoichiometric coefficients $\boldsymbol{\nu} = \{\nu_i^{(j)}$, for $i = 1, 2, ..., r$ and $j = 1, 2, ..., \tau\}$ determine uniquely the values of all the properties and quantities that characterize the chemical equilibrium state, including the values of the entropy, each ϵ_{jo}, and each n_{io}. We write the dependences of the latter quantities in the forms

$$S = S(U, V, \boldsymbol{n}_a; \boldsymbol{\nu}) \qquad\qquad (30.38)$$

$$\epsilon_{jo} = \epsilon_{jo}(U, V, \boldsymbol{n}_a; \boldsymbol{\nu}) \qquad \text{for } j = 1, 2, \ldots, \tau \qquad (30.39)$$

$$n_{io} = n_{io}(U, V, \boldsymbol{n}_a; \boldsymbol{\nu}) \qquad \text{for } i = 1, 2, \ldots, r \qquad (30.40)$$

In general, we cannot find the explicit functional forms of equations (30.38) to (30.40). For simple systems, however, the problem is somewhat less complicated because we can express chemical equilibrium properties in terms of stable equilibrium properties of a multiconstituent system in which all the chemical reaction mechanisms are inhibited—switched off. To see how this is done, we proceed as follows.

First, we consider a simple system B consisting of the same r types of constituents as system A but with all the chemical reaction mechanisms inhibited—switched off. Of course, A and B are different systems because they are subject to different internal forces and constraints. We assume that B is in a stable equilibrium state with values U of the energy, V of the volume, and $\boldsymbol{n} = \{n_1, n_2, \ldots, n_r\}$ of the amounts of constituents of its r constituents. We denote the entropy at that stable equilibrium state by the fundamental relation

$$S_{\text{off}} = S_{\text{off}}(U, V, \boldsymbol{n}) \qquad\qquad (30.41)$$

where we use the subscript off to emphasize that all the reaction mechanisms are switched off.

Next, we assume that the chemical reaction mechanisms are instantly switched on, that is, all the reactions defined by the stoichiometric coefficients $\boldsymbol{\nu}$ are no longer inhibited. As a result, we obtain again system A. Because in our discussion of chemical equilibrium we go back and forth between systems A and B by switching off and switching on the chemical reaction mechanisms, we call system B the *surrogate system* of A.

Because the surrogate system B is simple and initially in a stable equilibrium state,[2] immediately after switching on the reaction mechanisms the state of system A has the same values of S, U, V, n_1, n_2, $\ldots$, n_r as the corresponding values of the stable equilibrium state of B. In general, however, this state of A is not stable equilibrium. For example, if B is a quiescent mixture of gasoline vapor and air at room temperature and we activate the reaction mechanisms by a minute spark, we instantly produce a nonequilibrium state of a system A in which the reactions are no longer inhibited—the burning of the gasoline is proceeding—even though the instantaneous perturbations of the values of S, U, V, n_1, n_2, $\ldots$, n_r introduced by the spark are negligible.

Among all the states of A that may be obtained from B in the manner just cited, we consider the subset that has given values U and V of the energy and the volume, and amounts of constituents that are compatible with given values n_{1a}, n_{2a}, $\ldots$, n_{ra}. We

---

[2]In Section 17.1 the definition of a simple system includes the requirement that the instantaneous switching on or off of one or more chemical reaction mechanisms causes negligible instantaneous changes in the values of entropy, energy, volume, and amounts of constituents, provided that the system is initially in a stable equilibrium state. In general, the instantaneous switching on or off of chemical reaction mechanisms has definite effects on the system. For example, using the tools of quantum theory, we can show that the switching on of a reaction mechanism requires the switching on of an additional term in the Hamiltonian of the system which may affect the value of the energy. Again, using the tools of quantum theory, we can show that the switching off of a reaction mechanism requires the "destruction" of correlations among constituents which in general require a reduction of the value of the entropy. Nevertheless, we can also show that these effects become less and less important, and negligible for all practical purposes, as the values of each of the amounts of constituents increase beyond the order 10. Consistently with the discussion in Section 17.2, here we see another reason why the simple system model is highly accurate as long as the values of the amounts of constituents are relatively large.

denote each of these states by A_ϵ, and recognize that it corresponds to a set of values of the τ reaction coordinates $\epsilon = \{\epsilon_1, \epsilon_2, \ldots, \epsilon_\tau\}$ such that

$$n_i = n_{ia} + \sum_{j=1}^{\tau} \nu_i^{(j)} \epsilon_j \qquad \text{for } i = 1, 2, \ldots, r \tag{30.42}$$

and all the n_i's are nonnegative. Among all the states A_ϵ, the one with the highest entropy is the unique chemical equilibrium state with energy U, volume V, and amounts of constituents compatible with n_a. We denote it by A_{ϵ_0}.

Example 30.10. Prove that indeed A_{ϵ_0} is a chemical equilibrium state.

Solution. To proceed with the proof, we assume the contrary and reach a contradiction. We assume that A_0 is the chemical equilibrium state corresponding to the values U, V, n_a, ν, and that it differs from A_{ϵ_0}. Then $S_{\epsilon_0} < S_0$ because A_0 has the highest entropy. Now, starting from A_0 we switch off the chemical reaction mechanisms. Because system A is simple and A_0 is a stable equilibrium state, the resulting state B_0 of surrogate system B has the same values U, V, and n_0 as A_0 and, in particular, its entropy is S_0. If state B_0 were stable equilibrium, then upon switching the chemical reaction mechanisms back on, we would return to state A_0 and conclude that it belongs to the family of states A_ϵ and hence coincides with A_{ϵ_0}. If, instead, state B_0 were not stable equilibrium, the stable equilibrium state of B with values U, V, n_0 would have entropy $S > S_0$, and switching on the reactions beginning from this state would yield a state in the family of the states A_ϵ that has entropy $S > S_0 > S_{\epsilon_0}$. The conclusion that $S > S_{\epsilon_0}$ contradicts our stipulation that A_{ϵ_0} has the highest entropy and therefore is invalid. So A_{ϵ_0} must be a chemical equilibrium state.

As a result of these observations, we see that we can express the entropy S_ϵ of a state A_ϵ in terms of the entropy $S_{\text{off}}(U, V, n)$ of the state of the surrogate system to which A_ϵ corresponds. Moreover, we can determine the chemical-equilibrium entropy $S(U, V, n_a; \nu)$ [equation (30.38)] by a suitable maximization of the function $S_{\text{off}}(U, V, n)$.

To pursue this maximization, we first write the entropy S_ϵ of a state A_ϵ in the form

$$S_\epsilon = S_{\text{off}}(U, V, n_a + \nu \cdot \epsilon) \tag{30.43}$$

where in the fundamental relation $S_{\text{off}} = S_{\text{off}}(U, V, n)$ we use the shorthand notation $n_a + \nu \cdot \epsilon$ for the set $n = \{n_{1a} + \sum_{j=1}^{\tau} \nu_1^{(j)} \epsilon_j, n_{2a} + \sum_{j=1}^{\tau} \nu_2^{(j)} \epsilon_j, \ldots, n_{ra} + \sum_{j=1}^{\tau} \nu_r^{(j)} \epsilon_j\}$. Then we note that in order for A_{ϵ_0} to be the state of highest entropy among all the states A_ϵ with given U, V, and n_a, the values ϵ_0 must be such that

$$\left(\frac{\partial S_\epsilon}{\partial \epsilon_j} \right)_{U,V,n_a,\nu,\epsilon} = 0 \qquad \text{for } j = 1, 2, \ldots, \tau \tag{30.44}$$

where the subscripts n_a, ν, and ϵ denote, respectively, that each of the amounts n_{ia}, each of the stoichiometric coefficients $\nu_i^{(j)}$, and each of the reaction coordinates ϵ_i that do not appear in the derivative are kept fixed. Using equation (30.43) in equation (30.44), we find that

$$\left(\frac{\partial S_\epsilon}{\partial \epsilon_j} \right)_{U,V,n_a,\nu,\epsilon} = \sum_{i=1}^{r} \left(\frac{\partial S_{\text{off}}}{\partial n_i} \right)_{U,V,n} \left(\frac{\partial n_i}{\partial \epsilon_j} \right)_{n_a,\nu,\epsilon}$$

$$= -\sum_{i=1}^{r} \frac{\mu_i}{T} \nu_i^{(j)} = 0 \qquad \text{for } j = 1, 2, \ldots, \tau \tag{30.45}$$

where T is the temperature and μ_i the chemical potential of constituent i of the stable equilibrium state of the surrogate system B that corresponds to A_{ϵ_0}, and in writing the second of equations (30.45) we use the relation $(\partial S/\partial n_i)_{U,V,n} = -\mu_i/T$ [equation (10.7) for simple systems with $E = U$] and equations (30.42). For finite values of T, we see from equations (30.45) that a set of necessary conditions that relate U, V, n_a, ν, and ϵ_o at chemical equilibrium are

$$\sum_{i=1}^{r} \nu_i^{(j)} \mu_i(U, V, n_a + \nu \cdot \epsilon_o) = 0 \qquad \text{for } j = 1, 2, \ldots, \tau \qquad (30.46)$$

In Section 30.7 we show that T and μ_i for $i = 1, 2, \ldots, r$ are also equal to the temperature and chemical potentials of the chemical equilibrium state of system A. Each of equations (30.46) is called a *chemical equilibrium equation* for the jth reaction mechanism $\sum_{i=1}^{r} \nu_i^{(j)} A_i = 0$.

For each given set of values U, V, n_a, and ν, equations (30.46) are τ necessary conditions for chemical equilibrium. They may be solved to yield equations (30.38) to (30.40) and, therefore, all properties of the chemical equilbrium state. They confirm the statement made earlier to the effect that properties of chemical equilibrium may be expressed in terms of properties of a multiconstituent system with all chemical reaction mechanisms switched off.

For the extremum corresponding to equations (30.44) to be a maximum, it is also necessary that the second differential of S_ϵ with respect to the τ reaction coordinates $\epsilon_1, \epsilon_2, \ldots, \epsilon_\tau$ be negative. This turns out to be true because the fundamental relation of surrogate system B [equation (30.41)] is concave with respect to all the variables n_i for $i = 1, 2, \ldots, r$.

Example 30.11. Show that indeed the second differential of S_ϵ is negative.

Solution. Using equation (30.43), we find that the second differential of S_ϵ is given by the relation

$$(d^2 S_\epsilon)_{U,V,n_a,\nu} = \frac{1}{2!} \sum_{j=1}^{r} \sum_{k=1}^{r} \left(\frac{\partial^2 S_\epsilon}{\partial \epsilon_j \partial \epsilon_k} \right)_{U,V,n_a,\nu} d\epsilon_j \, d\epsilon_k$$

$$= \frac{1}{2!} \sum_{j=1}^{r} \sum_{k=1}^{r} \sum_{p=1}^{r} \sum_{q=1}^{r} \left(\frac{\partial^2 S_{\text{off}}}{\partial n_p \partial n_q} \right)_{U,V,n} \nu_p^{(j)} \nu_q^{(k)} d\epsilon_j \, d\epsilon_k$$

$$= \frac{1}{2!} \sum_{q=1}^{r} \sum_{p=1}^{r} \left(\frac{\partial^2 S_{\text{off}}}{\partial n_p \partial n_q} \right)_{U,V,n} \sum_{j=1}^{r} \nu_p^{(j)} d\epsilon_j \sum_{k=1}^{r} \nu_q^{(k)} d\epsilon_k$$

$$= \frac{1}{2!} \sum_{p=1}^{r} \sum_{q=1}^{r} \left(\frac{\partial^2 S_{\text{off}}}{\partial n_p \partial n_q} \right)_{U,V,n} dn_p \, dn_q < 0$$

where we use $dn_i = \sum_{j=1}^{r} \nu_i^{(j)} d\epsilon_j$ for $i = 1, 2, \ldots, r$. The inequality is always satisfied because the fundamental relation of the surrogate system B is concave with respect to every n_i, for $i = 1, 2, \ldots, r$ [relation (10.10)].

Each of the necessary conditions for chemical equilibrium—each of equations (30.46)—is expressed as a function of energy, volume, and amounts of constituents of the chemical equilibrium state. In the next section we rewrite these conditions in terms of temperature and pressure rather than energy and volume.

30.7 Conditions for Chemical Equilibrium in Terms of Temperature and Pressure

Rather than using energy, volume, and amounts of constituents as independent variables, we can more conveniently express each chemical equilibrium equation [equations

(30.46)] in terms of temperature, pressure, and mole fractions. To this end, we note that the stable equilibrium state of the surrogate system B obtained by switching off the reaction mechanisms at a chemical equilibrium state of a system A has not only the same values of energy, entropy, volume, and amounts of constituents as the chemical equilibrium state, but also the same values of temperature, pressure, and chemical potentials, $\mu_1, \mu_2, \ldots, \mu_r$.

Example 30.12. Prove the last assertion.

Solution. To prove the assertion, first we extend the discussion in Chapters 9 to 11 to simple systems with chemical reaction mechanisms, and define temperature, pressure, and chemical potentials, respectively, by the relations $1/(\partial S/\partial U)_{V,n_a,\nu}, (\partial S/\partial V)_{U,n_a,\nu}/(\partial S/\partial U)_{V,n_a,\nu}$, and $-(\partial S/\partial n_{ia})_{U,V,n_a,\nu}/(\partial S/\partial U)_{V,n_a,\nu}$ for $i = 1, 2, \ldots, r$, where the fundamental relation for the chemical equilibrium states is $S = S(U, V, n_a; \nu)$ [equation (30.38)].

Next, we express this fundamental relation in terms of that of the surrogate system B by evaluating equation (30.43) at ϵ_0 and using equation (30.39) so that

$$S = S_{\text{off}}(U, V, n_a + \nu \cdot \epsilon_0(U, V, n_a; \nu)) \tag{30.47}$$

Thus, for the inverse temperature of a chemical equilibrium state we find that

$$\left(\frac{\partial S}{\partial U}\right)_{V,n_a,\nu} = \left(\frac{\partial S_{\text{off}}}{\partial U}\right)_{V,n} + \sum_{i=1}^{r} \left(\frac{\partial S_{\text{off}}}{\partial n_i}\right)_{U,V,n} \sum_{j=1}^{\tau} \nu_i^{(j)} \left(\frac{\partial \epsilon_{jo}}{\partial U}\right)_{V,n_a,\nu}$$

$$= \left(\frac{\partial S_{\text{off}}}{\partial U}\right)_{V,n} \left[1 - \sum_{j=1}^{\tau} \left(\frac{\partial \epsilon_{jo}}{\partial U}\right)_{V,n_a,\nu} \sum_{i=1}^{r} \nu_i^{(j)} \mu_i\right]$$

$$= \left(\frac{\partial S_{\text{off}}}{\partial U}\right)_{V,n} = \frac{1}{T(U, V, n_a + \nu \cdot \epsilon_0(U, V, n_a; \nu))} \tag{30.48}$$

where in writing the first and the third of equations (30.48) we use equation (30.47) and the chemical equilibrium equations [equations (30.46)], respectively. So the temperature of a chemical equilibrium state equals the temperature of the corresponding state of the surrogate system B.

For the pressure of the chemical equilibrium state we find that

$$(\partial S/\partial V)_{U,n_a,\nu}/(\partial S/\partial U)_{V,n_a,\nu}$$

$$= T\left[\left(\frac{\partial S_{\text{off}}}{\partial V}\right)_{U,n} + \sum_{i=1}^{r} \left(\frac{\partial S_{\text{off}}}{\partial n_i}\right)_{U,V,n} \sum_{j=1}^{\tau} \nu_i^{(j)} \left(\frac{\partial \epsilon_{jo}}{\partial V}\right)_{U,n_a,\nu}\right]$$

$$= T\left(\frac{\partial S_{\text{off}}}{\partial V}\right)_{U,n} - \sum_{j=1}^{\tau} \left(\frac{\partial \epsilon_{jo}}{\partial V}\right)_{U,n_a,\nu} \sum_{i=1}^{r} \nu_i^{(j)} \mu_i$$

$$= T\left(\frac{\partial S_{\text{off}}}{\partial V}\right)_{U,n} = p(U, V, n_a + \nu \cdot \epsilon_0(U, V, n_a; \nu)) \tag{30.49}$$

where in writing the third of equations (30.49) we use equations (30.46). So the pressure of a chemical equilibrium state equals the pressure of the corresponding state of the surrogate system B.

For each chemical potential of the chemical equilibrium state we find

$$-(\partial S/\partial n_{ia})_{U,V,n_a,\nu}/(\partial S/\partial U)_{V,n_a,\nu}$$

$$= -T\left[\left(\frac{\partial S_{\text{off}}}{\partial n_i}\right)_{U,V,n} + \sum_{k=1}^{r} \left(\frac{\partial S_{\text{off}}}{\partial n_k}\right)_{U,V,n} \sum_{j=1}^{\tau} \nu_k^{(j)} \left(\frac{\partial \epsilon_{jo}}{\partial n_{ia}}\right)_{U,V,n_a,\nu}\right]$$

$$= -T\left(\frac{\partial S_{\text{off}}}{\partial n_i}\right)_{U,V,\boldsymbol{n}} + \sum_{j=1}^{\tau}\left(\frac{\partial \epsilon_{jo}}{\partial n_{ia}}\right)_{U,V,\boldsymbol{n}_a,\boldsymbol{\nu}}\sum_{k=1}^{r}\nu_k^{(j)}\mu_k$$

$$= -T\left(\frac{\partial S_{\text{off}}}{\partial n_i}\right)_{U,V,\boldsymbol{n}} = \mu_i(U,V,\boldsymbol{n}_a+\boldsymbol{\nu}\cdot\boldsymbol{\epsilon}_o(U,V,\boldsymbol{n}_a;\boldsymbol{\nu})) \tag{30.50}$$

where in writing the third of equations (30.50) we use equations (30.46). So the chemical potential of the ith constituent of a chemical equilibrium state of system A equals the chemical potential the ith constituent of the corresponding state of the surrogate system B.

It is worth emphasizing that the identity of values of temperature, pressure, and chemical potentials of a chemical equilibrium state with the values of the respective properties of a stable equilibrium state of the surrogate system obtains only at chemical equilibrium, because only then are the chemical equilibrium equations [equations (30.46)] satisfied. Away from chemical equilibrium states, temperature, pressure, and chemical potentials are not defined for system A because all such states are not stable equilibrium.

Finally, we note that equations (30.48) to (30.50) indicate that, geometrically, the surfaces represented by the functions $S = S(U, V, \boldsymbol{n}_a; \boldsymbol{\nu})$ and $S_{\text{off}} = S_{\text{off}}(U, V, \boldsymbol{n}_a+\boldsymbol{\nu}\cdot\boldsymbol{\epsilon})$ have a contact of first degree for each given set of values of U, V, and $\boldsymbol{n}_a$, at $\boldsymbol{\epsilon} = \boldsymbol{\epsilon}_o(U, V, \boldsymbol{n}_a)$, namely, at each chemical equilibrium state.

From Chapter 17 we recall that each chemical potential of a multiconstituent system in which all chemical reaction mechanisms are switched off may be expressed in the form $\mu_i = \mu_i(T, p, y_1, y_2, \dots, y_r)$. Using the stoichiometric relations developed in Chapter 29, we write the mole fractions and the chemical equilibrium equations in the forms

$$y_i = \frac{n_{ia} + \sum_{j=1}^{\tau}\nu_i^{(j)}\epsilon_j}{n_a + \sum_{j=1}^{\tau}\nu^{(j)}\epsilon_j} \tag{30.51}$$

and

$$\sum_{i=1}^{r}\nu_i^{(j)}\mu_i\left(T, p, \frac{n_{1a}+\sum_{j=1}^{\tau}\nu_1^{(j)}\epsilon_{jo}}{n_a+\sum_{j=1}^{\tau}\nu^{(j)}\epsilon_{jo}}, \dots, \frac{n_{ra}+\sum_{j=1}^{\tau}\nu_r^{(j)}\epsilon_{jo}}{n_a+\sum_{j=1}^{\tau}\nu^{(j)}\epsilon_{jo}}\right) = 0$$

$$\text{for } j = 1, 2, \dots, \tau \tag{30.52}$$

where $n_a = \sum_{i=1}^{r}n_{ia}$ and $\nu^{(j)} = \sum_{i=1}^{r}\nu_i^{(j)}$.

Equations (30.52) represent the chemical equilibrium equations as functions of T, p, and the mole fractions of the chemical equilibrium state. The chemical potentials are those of surrogate system B. As we discuss in Sections 30.2 to 30.5, for given values of T, p, and $\boldsymbol{n}_a$, we can solve the τ equations (30.52) for the τ unknowns $\epsilon_{1o}, \epsilon_{2o}, \dots, \epsilon_{\tau o}$ and hence determine the chemical equilibrium composition, $y_{1o}, y_{2o}, \dots, y_{ro}$, and the values of the corresponding amounts of constituents $n_{1o}, n_{2o}, \dots, n_{ro}$. Conversely, if the values of T, p, y_1, y_2, $\dots$, y_r are given but do not satisfy equations (30.52), we would conclude that the state is not chemical equilibrium but a state A_ϵ that can be described using the stable-equilibrium properties of the surrogate system. Then, of course, the values of T and p refer to the state of the surrogate system.

Example 30.13. Consider a chemical equilibrium state of a system A. For given values of temperature T and pressure p, and values of the amounts of constituents $\boldsymbol{n}$, compatible with given values $\boldsymbol{n}_a$ according to equations (30.42), show that the lowest value of the Gibbs free energy of the surrogate system B obtains at the state of B that corresponds to the chemical equilibrium state of A.

Solution. For the surrogate simple system, we recall from Chapter 17 that the Gibbs free energy $G_{\text{off}} = G_{\text{off}}(T, p, n_1, n_2, \dots, n_r)$, where the subscript off indicates the absence of all chemical reaction mechanisms. If the amounts of constituents are compatible with $\boldsymbol{n}_a$ and the τ chemical reaction mechanisms—conform to equations (30.42)—we may rewrite G_{off} in the form $G_{\text{off}} = G_{\text{off}}(T, p, \boldsymbol{n}_a + \boldsymbol{\nu} \cdot \boldsymbol{\epsilon})$. For given T, p, and $\boldsymbol{n}_a$, an extreme value of G_{off} obtains when

$$\left(\frac{\partial G_{\text{off}}}{\partial \epsilon_j}\right)_{T,p,\boldsymbol{n}_a,\boldsymbol{\nu},\boldsymbol{\epsilon}} = \sum_{i=1}^{r} \left(\frac{\partial G_{\text{off}}}{\partial n_i}\right)_{T,p,\boldsymbol{n}} \left(\frac{\partial n_i}{\partial \epsilon_j}\right)_{\boldsymbol{n}_a,\boldsymbol{\nu},\boldsymbol{\epsilon}} = \sum_{i=1}^{r} \mu_i \, \nu_i^{(j)} = 0$$

for $j = 1, 2, \ldots, \tau$, where we use the equation $(\partial G/\partial n_i)_{T,p,\boldsymbol{n}} = \mu_i$ for $i = 1, 2, \ldots, r$. These conditions are satisfied when the chemical potentials satisfy the chemical equilibrium equations [equations (30.46)], that is, when the stable equilibrium state of the surrogate system corresponds to the chemical equilibrium state of A. The extreme value of G_{off} is a minimum because the second differential of G_{off} with respect to all the ϵ_j's is positive. The proof of this statement is left to the reader.

We conclude our derivation of the chemical equilibrium equations by summarizing the results pictorially with the help of the energy versus entropy graphs introduced in Chapter 13. For given values of V and $\boldsymbol{n}_a$, three projections of states are superimposed on the single U versus S diagram shown in Figure 30.3: (1) the projection of the states of system A; (2) the projection of the states of the surrogate system B with values $\boldsymbol{n}_1 = \boldsymbol{n}_a + \boldsymbol{\nu} \cdot \epsilon_1$; and (3) the projection of the states of the surrogate system B with values $\boldsymbol{n}_2 = \boldsymbol{n}_a + \boldsymbol{\nu} \cdot \epsilon_2$, where ϵ_1 and ϵ_2 are two different sets of values of the reaction coordinates $\boldsymbol{\epsilon}$. The curves that represent the stable equilibrium states of the surrogate system B corresponding to the two sets of values $\boldsymbol{n}_1 = \boldsymbol{n}_a + \boldsymbol{\nu} \cdot \epsilon_1$ and $\boldsymbol{n}_2 = \boldsymbol{n}_a + \boldsymbol{\nu} \cdot \epsilon_2$ also represent the loci of all the states A_{ϵ_1} and A_{ϵ_2} of system A.

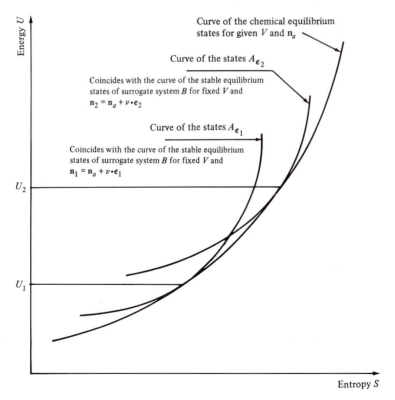

Figure 30.3   Pictorial representation of the chemical equilibrium states on an energy versus entropy diagram.

The set of values ϵ_1 is chosen so that $\epsilon_1 = \epsilon_o(U_1, V, n_a; \nu)$ and, therefore, at the energy U_1 the locus of states A_{ϵ_1} is tangent to the curve of the chemical equilibrium states of A. Similarly, the set ϵ_2 is such that $\epsilon_2 = \epsilon_o(U_2, V, n_a; \nu)$, and the locus of states A_{ϵ_2} is tangent to the chemical equilibrium curve at energy U_2. We see that the curve of the chemical equilibrium states is the envelope of the loci of states A_ϵ for ϵ taking on all possible values. We also see that for $U \neq U_1$ the states A_{ϵ_1} represent states of system A that are not stable equilibrium, yet can be described using the stable-equilibrium properties of the surrogate system.

30.8   Complete Chemical Equilibrium

Many problems cannot be adequately modeled by a limited set of chemical reaction mechanisms. Typically, this is the case when the number of relevant reaction mechanisms is very large or not well defined. A practical approach to these problems consists of considering all conceivable reaction mechanisms between the constituents, and using the largest entropy principle without specifying explicitly any of the reaction mechanisms. Under these conditions, we say that we find the *complete chemical equilibrium state*, and that the system is in *complete chemical equilibrium*.

To characterize the complete chemical equilibrium state, we use the result proved in Section 30.6 that the state of the surrogate system obtained by switching off the reaction mechanisms has the highest entropy among all the states with given energy U and volume V, and with amounts of constituents compatible with the given values n_a.

In Section 30.6, we maximize the stable-equilibrium entropy $S_{\text{off}}(U, V, n)$ of the surrogate system for given values of U and V and subject to the compatibility conditions given by equations (30.42), that is, in our shorthand notation, the condition $n = n_a + \nu \cdot \epsilon$. Here the compatibility conditions between n and n_a are different. They are obtained from the principle of conservation of atomic nuclei. Specifically, upon denoting the number of atomic nuclei of type j by c_j, and the number of atomic nuclei of type j that are present in one molecule of constituent i by a_{ji}, we find that the values $n_1, n_2, \ldots, n_r$ of the amounts of constituents that at any state are compatible with the given values $n_{1a}, n_{2a}, \ldots, n_{ra}$ satisfy the relations of conservation of atomic nuclei

$$c_j = \sum_{i=1}^{r} a_{ji} n_i = \sum_{i=1}^{r} a_{ji} n_{ia} \qquad \text{for } j = 1, 2, \ldots, k \qquad (30.53)$$

where k is the number of different types of nuclei present in the system. For given values n_a, equations (30.53) determine the values of the c_j's. For example, in a system consisting of the three constituents $A_1 = H_2O$, $A_2 = H_2$, and $A_3 = O_2$, we have two types of atomic nuclei, hydrogen corresponding to $j = 1$, and oxygen corresponding to $j = 2$. Accordingly, the values n_a, the coefficients a_{ji}, and the compatible values n satisfy the relations $a_{11} = 2$, $a_{12} = 2$, $a_{13} = 0$, $a_{21} = 1$, $a_{22} = 0$, $a_{23} = 2$, $c_1 = 2n_1 + 2n_2 + 0n_3 = 2n_1 + 2n_2 = 2n_{1a} + 2n_{2a}$, and $c_2 = 1n_1 + 0n_2 + 2n_3 = n_1 + 2n_3 = n_{1a} + 2n_{2a}$.

In view of these observations and for given values of U and V, we conclude that we can characterize the complete chemical equilibrium state by maximizing the stable-equilibrium-state entropy $S_{\text{off}}(U, V, n)$ of the surrogate system, subject to the compatibility conditions between n and n_a expressed by (30.53). We can solve this constrained maximization problem by introducing a Lagrange multiplier λ_j for each of the k constraining equations (30.53), and then, for given values of U, V, and n_a, finding the extremum of the expression[3]

$$L = S_{\text{off}}(U, V, n_1, n_2, \ldots, n_r) + \sum_{j=1}^{k} \lambda_j \left(\sum_{i=1}^{r} a_{ji} n_i - \sum_{i=1}^{r} a_{ji} n_{ia} \right) \tag{30.54}$$

with respect to $n_1, n_2, \ldots, n_r$. Accordingly, we must have

$$\frac{\partial L}{\partial n_i} = \left(\frac{\partial S_{\text{off}}}{\partial n_i} \right)_{U, V, n} + \sum_{j=1}^{k} \lambda_j a_{ji}$$

$$= -\frac{\mu_i}{T} + \sum_{j=1}^{k} \lambda_j a_{ji} = 0 \qquad \text{for } i = 1, 2, \ldots, r \tag{30.55}$$

where in writing the second of equations (30.55) we use the relation $(\partial S/\partial n_i)_{U, V, n} = -\mu_i/T$. Thus we find that

$$\mu_i = T \sum_{j=1}^{k} \lambda_j a_{ji} \qquad \text{for } i = 1, 2, \ldots, r \tag{30.56}$$

To proceed further we must know the explicit dependences of the μ_i's on $n_1, n_2, \ldots, n_r$ because then we can solve equations (30.56) for each n_i as a function of the Lagrange multipliers λ_j, substitute the results in equations (30.53), determine the values of the λ_j's in terms of the values of the c_j's and thus express the values of the n_i's of the complete chemical equilibrium state in terms of the c_j's or, equivalently, of n_a, and either U and V or T and p.

We can carry out this program explicitly if the surrogate system behaves as an ideal-gas mixture. Then $p_{ii} = y_i p$ and

$$\mu_i = \mu_{ii}(T, p_{ii}) = \mu_{ii}(T, p_o) + RT \ln \frac{y_i p}{p_o} = T \sum_{j=1}^{k} \lambda_j a_{ji} \tag{30.57}$$

where in writing the last of these equations we use equation (30.56). Upon solving equation (30.57) for y_i, we find that

---

[3]It is noteworthy that also the maximization in Section 30.6 can be performed by introducing a Lagrange multiplier λ_i for each of the r constraining equations (30.42) and then, for given values of U, V, and n_a, finding the extremum of the expression

$$L = S_{\text{off}}(U, V, n_1, n_2, \ldots, n_r) + \sum_{i=1}^{r} \lambda_i \left(n_i - n_{ia} - \sum_{j=1}^{\tau} \nu_i^{(j)} \epsilon_j \right)$$

The extremum occurs when the values of the partial derivatives $\partial L/\partial n_i$ for $i = 1, 2, \ldots, r$, and $\partial L/\partial \epsilon_j$ for $j = 1, 2, \ldots, \tau$ are equal to zero. Accordingly, we have

$$\frac{\partial L}{\partial n_i} = \left(\frac{\partial S_{\text{off}}}{\partial n_i} \right)_{U, V, n} + \lambda_i = -\frac{\mu_i}{T} + \lambda_i = 0 \qquad \text{for } i = 1, 2, \ldots, r$$

$$\frac{\partial L}{\partial \epsilon_j} = -\sum_{i=1}^{r} \lambda_i \nu_i^{(j)} = 0 \qquad \text{for } j = 1, 2, \ldots, \tau$$

Each equation in the first set implies that $\lambda_i = \mu_i/T$, for $i = 1, 2, \ldots, r$. Substituting these results in the equations in the second set, we find the chemical equilibrium equations $\sum_{i=1}^{r} \mu_i \nu_i^{(j)} = 0$ for $j = 1, 2, \ldots, \tau$.

$$y_i = \frac{p_o}{p} \exp\left[\frac{1}{R}\sum_{j=1}^{k}\lambda_j a_{ji} - \frac{\mu_{ii}(T,p_o)}{RT}\right] \tag{30.58}$$

$$n_i = ny_i = \frac{np_o}{p} \exp\left[\frac{1}{R}\sum_{j=1}^{k}\lambda_j a_{ji} - \frac{\mu_{ii}(T,p_o)}{RT}\right] \tag{30.59}$$

where $n = \sum_{i=1}^{r} n_i$. Thus the Lagrange multipliers λ_j satisfy the relations

$$1 = \frac{p_o}{p} \sum_{i=1}^{r}\exp\left[\frac{1}{R}\sum_{j=1}^{k}\lambda_j a_{ji} - \frac{\mu_{ii}(T,p_o)}{RT}\right] \tag{30.60}$$

$$c_q = \frac{np_o}{p} \sum_{i=1}^{r} a_{qi}\exp\left[\frac{1}{R}\sum_{j=1}^{k}\lambda_j a_{ji} - \frac{\mu_{ii}(T,p_o)}{RT}\right] \qquad \text{for } q = 1, 2, \ldots, k \quad (30.61)$$

where equation (30.60) follows from the condition $\sum_{i=1}^{r} y_i = 1$, and equations (30.61) from equations (30.53). Given the amounts of constituents n_a and, therefore, the k values of the c_q's, the values of T and p, and the expressions for $\mu_{ii}(T,p_o)$, we can solve the $k+1$ equations (30.60) and (30.61) for the $k+1$ unknown values n and λ_j for $j = 1, 2, \ldots, k$. This is done numerically with the help of a computer. The values of n and the λ_j's are then used in equations (30.58) and (30.59) to find the composition and the amounts of constituents of the complete chemical equilibrium state, i.e., the values of $y_{1o}, y_{2o}, \ldots, y_{ro}$ and $n_{1o}, n_{2o}, \ldots, n_{ro}$.

Finally, we note that equations (30.55) define an extremum that is a maximum. The proof of this assertion follows from the relation

$$\left(\frac{\partial^2 L}{\partial n_i^2}\right)_{U,V,n} = \left(\frac{\partial^2 S}{\partial n_i^2}\right)_{U,V,n} < 0 \qquad \text{for } i = 1, 2, \ldots, r \tag{30.62}$$

and the fact that S is concave with respect to $n_1, n_2, \ldots, n_r$.

Problems

30.1 A superheater is to be designed to heat 8000 kg/h of steam at 1 atm from a temperature of 500°C to 2500°C.

(a) Calculate the required energy rate neglecting dissociation.

(b) Calculate the fractions of the steam that dissociate into H_2 and O_2 at 1 atm, and temperatures of 500°C and 2500°C.

(c) What is the required energy rate when dissociation is included?

30.2 Consider the reaction mechanism $CH_4 + H_2O = CO + 3H_2$.

(a) Find the equilibrium constant at $T = 25°C$ and $p = 1$ atm.

(b) Assume that initially 1 mol of CH_4 and 1 mol of H_2O are present, and find the values of the reaction coordinate, ϵ, and the equilibrium mole fractions at chemical equilibrium at $T = 25°C$ and $p = 1$ atm.

(c) Assume that initially 1 mol of CO and 3 mol of H_2 are present, and repeat part (b).

30.3 A catalyst has been found which, at 500°C, enhances the reaction mechanism $CO + 2H_2 = CH_3OH$. Estimate the order of magnitude of the pressure that would be required to make this reaction feasible as an industrial process.

30.4 On a mole basis, the exhaust gas from a liquid hydrogen–liquid oxygen rocket is found to have the following composition: H_2O, 61.3%; H_2, 32.2%; H, 3.55%; OH, 2.63%; O_2, 0.13%; O, 0.17%. Verify that the gas temperature is about 3200 K, and estimate its pressure.

30.5 Gaseous hydrogen is to be manufactured by steam cracking of methane in a catalytic reactor according to the reaction mechanism $CH_4 + 2\,H_2O = CO_2 + 4\,H_2$ at 480°C and 1 atm.

(a) If 5 mol of steam are fed to the reactor for every mole of methane, what is the chemical equilibrium composition of the product stream?

(b) Is this reaction exothermic or endothermic? How much energy must be added or removed from the reactor per mole of methane fed?

(c) Will an increase in pressure result in an increased or decreased yield (moles of CH_4 reacted/moles of CH_4 fed)?

(d) It has been suggested that the chemical equilibrium yield of this reaction can be increased by diluting the feedstream with an inert gas. Do you agree with this suggestion?

30.6 It has been suggested that water is an extremely dangerous material to use for extinguishing hot fires where large amounts of iron (or other metals) are present. It is feared that the iron (or other metals) may decompose the water by forming a metal oxide and liberating hydrogen. The reaction mechanism is $Fe + H_2O = FeO + H_2$. The properties of formation of FeO are $\Delta h_f^o = -246$ MJ/kmol and $\Delta g_f^o = -266$ MJ/kmol.

(a) If the temperature in a certain fire is 1000°C, determine the ratio y_{H_2}/y_{H_2O} at chemical equilibrium.

(b) Does it appear that this reaction will proceed to any appreciable extent at 1000°C?

(c) Determine the amount of energy liberated per mole of iron oxidized according to the given reaction mechanism.

30.7 A stoichiometric mixture of hydrogen and nitrogen is delivered to the reaction chamber of an ammonia synthesis plant at 25°C. Nitrogen, hydrogen, and ammonia leave the reaction chamber at 500 K and 5 atm.

(a) What is the yield of ammonia from the reactor expressed as a mole fraction?

(b) How much energy must be supplied to or removed from the reaction chamber per mole of ammonia produced?

30.8 Carbon dioxide gas enters a steady-flow hot-gas generator at 298 K and 6 atm. Calculate the energy transfer required to generate an outlet stream at a temperature of 2800 K and a pressure of 5 atm. Assume that the outlet stream is in chemical equilibrium at the outlet conditions.

30.9 The reaction mechanism for ionization of cesium, Cs, is $Cs = Cs^+ + e^-$, where Cs^+ is the positive cesium ion, and e^- the electron. The equilibrium constant K for the reaction at 2000 K is 15.63.

(a) Determine the extent to which cesium is ionized at 2000 K and 1 atm.

(b) Determine the pressure at which the ionization is 0.9999 complete at 2000 K.

30.10 Consider a steady-state bulk-flow chemical reactor. The inlet stream is an ideal gas mixture of equal moles of SO_2 and H_2 at $T_1 = 700$ K, $p_1 = 1$ atm, a flow rate $\dot{m}_0 = 1$ kg/s, and subject to the chemical reaction mechanism $SO_2 + 3\,H_2 = H_2S + 2\,H_2O$. The reactor interacts with an energy source at appropriate temperature so that the products (exit stream) are also an ideal-gas mixture at $T_2 = T_1 = 700$ K, $p_2 = p_1 = 1$ atm, and in chemical equilibrium. All constituents behave as perfect gases.

(a) Determine the value $\Delta h^o(700\ K)$ of the enthalpy of reaction.

(b) Determine the mole fractions in the outlet stream.

(c) Is energy received from or rejected to the energy source? What is the value of this energy?

(d) How would you adjust the temperature, pressure, and inlet composition of the reactor to minimize the amount of SO_2 in the exit stream?

30.11 Sulfur dioxide, SO_2, is a product of combustion of sulfur-containing fuels and a major contributor to air pollution. A new idea for reducing SO_2 emissions to the atmosphere is to convert SO_2 into an aqueous solution of sulfuric acid and collect the sulfuric acid for disposal. Because the acid is in condensed form, it can be disposed of more easily than SO_2 gas. The overall conversion takes place in two steps: $SO_2(g) + \frac{1}{2}O_2(g) = SO_3(g)$ (gas phase reaction); $SO_3(g) + H_2O(l) = H_2SO_4$ (aqueous solution). Under typical conditions, the conversion of SO_2 to SO_3 occurs in the gas phase at $T = 610°C$ and $p = 1$ atm.

(a) If the mole fractions in the feedstream from the combustor are 12% SO_2, 8% O_2, and 80% inert constituents, what are the equilibrium mole fractions of SO_2, SO_3, and O_2 for the gas-phase reaction? For this reaction, $\Delta g°(610°C) = -3.75$ kcal/mol.

(b) For the conditions of part (a), what percent of the initial SO_2 is converted to SO_3?

(c) Assuming that the reaction temperature must remain at $610°C$, what should the reaction pressure be to increase the conversion percentage to 80%? What should the pressure be for 99% conversion?

30.12 Hydrocarbon reforming is one of the main processes used for the industrial production of hydrogen. It is based on the reaction of a hydrocarbon, typically methane, with water vapor: $CH_4 + H_2O = CO + 3H_2$. The reaction is endothermic, and the equilibrium constant at low temperature implies a very small yield of hydrogen. Thus the reforming of methane is performed at high temperature, about $1100°C$, maintained by combusting controlled amounts of methane according to the reaction mechanism $2CH_4 + 3O_2 = 2CO + 4H_2O$.

(a) Find the equilibrium constants of the hydrogen producing reaction at $25°C$ and at $1100°C$.

(b) At steady state, a well-insulated reactor is fed with separate streams of 1 mol/s methane, 1 mol/s water vapor, and $\dot{x}$ mol/s oxygen, each at standard pressure of 1 atm, and preheated to $1100°C$. For each molecule of oxygen, the burning of methane is complete. The outlet stream is a mixture of methane, water vapor, carbon monoxide, and hydrogen in chemical equilibrium at 1 atm and $1100°C$. Find the mole flow rates, and the mole fractions of each of the constituents in the outlet stream, as a function of the oxygen inflow rate $\dot{x}$, and the reaction coordinate rate $\dot{\epsilon}$ for the hydrogen-producing reaction.

(c) Write the two independent equations that define the values of $\dot{x}$ and $\dot{\epsilon}$ for the conditions specified. Estimate the mole flow rate of hydrogen in the outlet stream, and the mole flow rate of oxygen in the inlet stream that is necessary to maintain the specified steady state.

(d) Would an increase in the outlet pressure increase the yield of hydrogen?

30.13 A laboratory reactor for the production of small quantities of hydrochloric acid, HCl, processes a liquid solution of 2 g/s of sodium chloride, NaCl, and 1 g/s of sulfuric acid, H_2SO_4, in 100 g/s of water. The outlet liquid stream is a water solution in chemical equilibrium with respect to the reaction mechanism $NaCl + H_2SO_4 = NaHSO_4 + HCl$. The inlet and outlet streams are at standard conditions ($25°C$, 1 atm). The reactor experiences only a heat interaction with an energy source at $25°C$. Data for the reacting species in aqueous solution are given in the accompanying table. Assume that the behavior of each constituent is ideal, and that each of the mole specific volumes v_{ii} is constant.

i	Constituent	M_i g/mol	v_{ii} cm³/mol	$(\Delta h_f°)_i$ kcal/mol	$(\Delta g_f°)_i$ kcal/mol
1	H_2O	18			
2	NaCl	58.5	27	−97	−94
3	H_2SO_4	98.1	53	−217	−178
4	$NaHSO_4$	120.1	49	−269	−242
5	HCl	36.5	24	−40	−31

(a) Evaluate the equilibrium constant $K(25°C, 1 \text{ atm})$.

(b) Find the chemical equilibrium composition of the outlet stream and the rate of hydrochloric acid production.

(c) Find the direction and rate of the heat exchange needed to maintain the reactor at the specified steady state.

(d) Find the rate of entropy generation by irreversibility in the reactor.

30.14 A bulk-flow chemical reactor for the industrial production of hydrogen (Figure P30.14) is fed with a stream of 1 mol/s composed of 50% CO and 50% H_2O at temperature $T = 700$ K and pressure $p_0 = 1$ atm. The exit stream is at the same temperature T and pressure p_0, but it is a mixture with the chemical equilibrium composition according to the water-gas reaction mechanism $CO + H_2O = CO_2 + H_2$. The reactor exchanges energy with a source at $T = 700$ K. Treat the inlet and outlet streams as if each were an ideal gas mixture.

(a) Find the mole fraction y_i of each constituent in the outlet stream (Figure P30.14a).

(b) Find the rate $\dot{Q}$ at which energy is exchanged with the source. Is the reaction endothermic or exothermic?

(c) Find the rate of entropy generated by irreversibility.

(d) Find the smallest work rate required to separate H_2 from the other products in the outlet streams (Figure P30.14b).

(e) Now assume that hydrogen is produced in a reversible fuel cell operating between the same inlet and outlet conditions, in contact with the same energy source, but experiencing, in addition, an electrical work interaction (Figure P30.14c). Find the electrical power $\dot{W}_{rev}$ of the reversible fuel cell. Is the fuel cell producing or consuming electrical power?

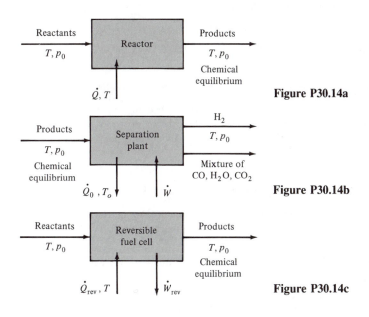

Figure P30.14a

Figure P30.14b

Figure P30.14c

30.15 A mixture of N_2O_4 and NO_2 at temperature $T_a = 80°C$ and high pressure p_a is throttled to temperature T_b and pressure $p_b = p_0 = 1$ atm. For all temperatures and pressures the mixture is in chemical equilibrium according to the reaction mechanism $N_2O_4 = 2 NO_2$. At $T_a = 80°C$ and p_a, the composition of the mixture is $y_a(N_2O_4) = 0.637$ and $y_a(NO_2) = 0.363$. At all T and p, the mixture behaves as a perfect gas with $c_p(NO_2) = c_p(N_2O_4) = 0.174$ cal/g K. For the given reaction mechanism, assume that $\Delta g°(T) = 13.8 - 0.042\,T$ in kcal/mol for the temperature T in kelvin, and $\Delta h° = 13.9$ kcal/mol.

(a) Calculate the value of the pressure p_a.

(b) Derive an expression for the chemical equilibrium mole fractions y_1 and y_2 ($y_1 + y_2 = 1$) at pressure $p_b = p_0$ as a function of the temperature T_b.

(c) Using the fact that the throttle is isenthalpic, derive another relation between the mole fractions y_1 and y_2 and the temperature T_b at the exit of the throttle.

(d) Calculate the values of y_1, y_2, and T_b.

30.16 Methanol gas, CH_3OH, can be synthesized from carbon monoxide, CO, and hydrogen H_2, in accordance with the chemical reaction mechanism $CO + 2H_2 = CH_3OH$. The equilibrium constant of this reaction $K(127°C) = 1.377$.

(a) Calculate the equilibrium yield of methanol when 1 kmol of CO and 2 kmol of hydrogen react at $127°C$ and a pressure of 5 atm.

(b) How many kilomoles of CO should be reacted with 2 kmol of hydrogen in order to obtain the maximum possible concentration of methanol in the equilibrium gas mixture at $127°C$ and 5 atm?

30.17 Hydrogen and oxygen are used in a fuel cell to produce electricity. The reaction mechanisms are $H_2 = 2H^+ + 2e^-$ at the anode, $\frac{1}{2}O_2 + 2H^+ + 2e^- = H_2O$ at the cathode, and $H_2 + \frac{1}{2}O_2 = H_2O$ overall. In a test at standard atmospheric conditions, a fuel cell produces a voltage of 0.9 volt and a current of 96 ampere.

(a) What is the largest voltage of a hydrogen–oxygen fuel cell? Use the fact that H^+ carries a charge of 96,500 coulomb per mole.

(b) What is the rate of entropy generated by irreversibility in the given cell?

(c) What is the rate of energy transfer to the environment?

(d) What is the rate of water production in the cell?

30.18 Consider the dissociation chemical reaction mechanism $H_2O = H_2 + \frac{1}{2}O_2$.

(a) What is the enthalpy of reaction (the *heat of dissociation*) of water vapor at 2000 K?

(b) At a pressure of 1 atm, what is the temperature at which 40% of the H_2O is dissociated?

(c) At what temperatures is 20% of the H_2O dissociated at 1 atm and at 10 atm?

30.19 Initially pure nitrogen reaches chemical equilibrium with respect to the reaction mechanism $N_2 = 2N$. Each constituent behaves as an ideal gas, and dissociation occurs at a fixed pressure $p = 1$ atm and a fixed temperature T. At chemical equilibrium, the volume has increased by 10% with respect to the initial gas (before dissociation).

(a) What is the value of the reaction coordinate at chemical equilibrium?

(b) What is the value of the temperature T at chemical equilibrium?

30.20 Consider a system initially charged with 1 kmol of pure hydrogen, H_2. The system is maintained at $1000°C$ and 1 atm, and the dissociation reaction mechanism $H_2 = 2H$ is effective. Assume that both H_2 and H behave as ideal gases.

(a) Express the enthalpy, entropy, and Gibbs free energy of the surrogate system at $1000°C$ and 1 atm in terms of the reaction coordinate ϵ and other data.

(b) Find the value of the reaction coordinate for which the Gibbs free energy has the smallest value. Verify that this value corresponds to the chemical equilibrium state for the given reaction mechanism.

30.21 Consider the ammonia synthesis reaction mechanism

$$N_2 + 3H_2 = 2NH_3$$

A fixed-volume reactor is filled with 500 mol of N_2 and 100 mol of H_2.

(a) Find the chemical equilibrium composition at $800°C$ and 100 atm.

(b) Starting from the chemical equilibrium state in part (a) the energy is suddenly increased in such a way that the new state is not chemical equilibrium, but the surrogate system (with the reaction mechanisms switched off) is at $1000°C$ and 100 atm and has the composition found in part (a). What is the change in enthalpy of the system?

(c) Find the temperature and the composition of the chemical equilibrium state reached by the system in a spontaneous process starting from the state in part (b) and such that the surrogate system follows a constant-pressure path at 100 atm. Compare the value of the temperature with the value 1000°C imposed by the sudden perturbation in part (b).

(d) Starting from the chemical equilibrium state in part (a), 200 mol of N_2 at 800°C is injected in the reactor in such a way that the new state is not chemical equilibrium but the surrogate system is at 800°C and 100 atm. What is the change in Gibbs free energy of the system?

(e) Find the composition of the chemical equilibrium state reached by the system in a spontaneous process starting from the state in part (b) and such that the surrogate system follows a constant-temperature and constant-pressure path at 800°C and 100 atm. Compare the value of the amount of N_2 with the value imposed by the sudden perturbation in part (d).

(f) Interpret your answers in parts (c) and part (e) in terms of the natural tendency of systems in stable equilibrium states to respond to a perturbation by moderating its effect. This moderation is a manifestation of the stability of equilibrium known as the *Le Châtelier–Braun principle* (see also Problem 16.10).

30.22 In terms of the entropy S_ϵ of a state A_ϵ, that is, $S_\epsilon = S_{\text{off}}(U, V, n_a + \nu \cdot \epsilon)$, the partial derivative

$$Y_j = \left(\frac{\partial S_\epsilon}{\partial \epsilon_j} \right)_{U,V,n_a,\nu,\epsilon} = -\sum_{i=1}^{r} \frac{\mu_i}{T} \nu_i^{(j)}$$

is called the *affinity* of the jth chemical reaction mechanism. At a chemical equilibrium state the affinity of each reaction mechanism is equal to zero [equation (30.45)].

If a system passes through a sequence of states A_ϵ, namely, states that are stable equilibrium for the surrogate system with all the reaction mechanisms switched off, show that in a spontaneous process at fixed values of U, V, n_a, ν, the reaction coordinates $\epsilon_1, \epsilon_2, \ldots, \epsilon_\tau$ can only change in such a way that

$$\sum_{j=1}^{\tau} Y_j \, d\epsilon_j \geqq 0$$

31 Combustion

In this chapter, we consider a special class of exothermic reactions in which the reactants are oxygen and certain substances called chemical fuels. The oxygen is supplied as part of the atmospheric air even though it could be supplied in pure form as well. Each such reaction is called a *combustion reaction* or *fuel-oxidation reaction*. Chemical fuels represent about 90% of the energy sources we use in energy conversion devices.

31.1 Fuels

The main *chemical fuels* are coal, consisting primarily of carbon, and liquid and gaseous hydrocarbons, consisting primarily of compounds of carbon and hydrogen. Coal exists in deposits in the solid form of aggregation, and its composition differs considerably from mine to mine. In addition to carbon, coal contains differing amounts of sulfur, hydrogen, nitrogen, and ash. Typical exothermic reaction mechanisms associated with the combustion or burning of coal are

$$C + O_2 = CO_2 \qquad C + \tfrac{1}{2}O_2 = CO \qquad S + O_2 = SO_2 \qquad H_2 + \tfrac{1}{2}O_2 = H_2O$$

Liquid hydrocarbon fuels are derived from crude oil by means of various refining processes, such as distillation and catalytic cracking. Examples are gasoline, diesel oil, and kerosene. Gaseous hydrocarbons are obtained from natural gas wells, from oil wells (associated gas), and from processing of other fuels, such as coal gasification.

Each hydrocarbon fuel consists of a mixture of compounds of the form C_kH_l, where k and l are integers. Some predominant compounds are isooctane, C_8H_{18}, in gasoline, and dodecane, $C_{12}H_{26}$, in diesel oil. The reaction mechanism of combustion of a hydrocarbon is

$$C_kH_l + \left(k + \frac{l}{4}\right)O_2 = k\,CO_2 + \frac{l}{2}H_2O \tag{31.1}$$

If all the carbon, C, and hydrogen, H_2, in a fuel are converted into carbon dioxide, CO_2, and water, H_2O, respectively, we say that the fuel is *completely oxidized* and that the *combustion is complete*. Otherwise, we say that the *combustion is incomplete*.

31.2 Combustion Stoichiometry

The oxygen required in each combustion reaction is obtained either from a pure oxygen source or from air. Pure oxygen is produced primarily by air liquefaction and subsequent separation by distillation of liquid air, and is used in special applications, such as basic oxygen furnaces in steelmaking and welding. In the vast majority of combustion reactions, atmospheric air provides the required oxygen.

As we discuss in Section 27.4, dry air is a mixture with fixed composition containing, on a mole basis, 20.95% of oxygen, 78.08% of nitrogen, and small quantities of several other constituents. It has a mean molecular weight $M_{air} = 28.96$ kg/kmol.

In most combustion studies, the nitrogen and the other constituents of dry atmospheric air are assumed to be inert, that is, not to participate in the prevailing chemical reaction mechanisms. So dry air is regarded as a binary mixture consisting on a mole basis of 20.95% of oxygen and 79.05% of a fictitious constituent with a molecular weight M such that $0.2095 \times 31.999 + 0.7905 \times M = 28.96$ kg/kmol, that is, $M = 28.16$ kg/kmol. This fictitious constituent is called *atmospheric nitrogen*. It differs so slightly from pure nitrogen, N_2, that for most applications its specific heat, specific enthalpy, and specific entropy are sufficiently approximated by the corresponding properties of N_2. In short, when atmospheric air provides the oxygen for a combustion reaction, we assume that every mole of oxygen is accompanied by $79.05/20.95 = 3.77$ mol of atmospheric nitrogen or simply nitrogen. Accordingly, each mole of O_2 corresponds to 4.77 mol of dry air.

On a mass basis, the mass fractions of dry air are [equation (26.4)] $20.95 \times 31.999/28.96 = 23.14\%$ oxygen and $79.05 \times 28.16/28.96 = 76.86\%$ nitrogen. So each kilogram of oxygen is accompanied by 3.32 kg of nitrogen or, equivalently, there is 1 kg of O_2 in every 4.32 kg of dry air.

Although it does not participate in the combustion reaction mechanism, nitrogen appears in both the reactants and the products and plays an important role in determining their compositions and states.

It is noteworthy that under certain conditions, it may be completely misleading to treat nitrogen as an inert substance. For example, if the products of combustion are at sufficiently high temperature, the reaction mechanisms of formation of nitric oxide and nitrogen dioxide become active and cannot be neglected. These oxides are potential causes of severe air pollution.

As we discuss in Section 27.4, atmospheric air is in general a mixture of dry air and water vapor. The typical values of the specific humidity ω range between 0 and 0.04, and the properties of moist air differ from those of an equal amount of dry air by a few percent. In many combustion calculations, the water vapor in the air is neglected. For more accurate results, however, the effect of the humidity of atmospheric air should be included.

A common way to specify the composition of the reactants in a process involving combustion is to specify the *air–fuel ratio*, that is, the ratio of the amount of dry air and the amount of fuel. Depending on whether the amounts are expressed on a mole or on a mass basis, we denote the air–fuel ratio by n_{air}/n_{fuel} or m_{air}/m_{fuel}. Alternatively, it is also common to use the reciprocal of the air–fuel ratio, that is, the *fuel–air ratio*, n_{fuel}/n_{air} or m_{fuel}/m_{air}. Often, the air–fuel mass ratio m_{air}/m_{fuel} is also denoted by the symbol A/F or, simply, AF.

The least amount of dry air that supplies sufficient oxygen for the complete combustion of a given fuel is called *stoichiometric* or *theoretical amount of air* for complete oxidation of the fuel. We denote the corresponding air–fuel ratio by $(n_{air}/n_{fuel})_s$ or $(m_{air}/m_{fuel})_s$ where the subscript s stands for stoichiometric.

According to relation (31.1), the least amount of oxygen, O_2, required to burn completely 1 mol, of a hydrocarbon $C_k H_l$ is $(k + l/4)$ mol, and therefore the least amount of dry air is $4.77(k + l/4)$ mol. So the stoichiometric air–fuel ratio is, on a mole basis,

$$\left(\frac{n_{air}}{n_{fuel}}\right)_s = 4.77 \left(k + \frac{l}{4}\right) \tag{31.2}$$

and on a mass basis,

$$\left(\frac{m_{air}}{m_{fuel}}\right)_s = 4.77\left(k + \frac{l}{4}\right)\frac{28.96}{M_{fuel}} \tag{31.3}$$

where M_{fuel} is the molecular weight of the fuel.

Relations (31.1) to (31.3) are valid also for the combustion stoichiometry of a fuel that consists of a mixture of hydrocarbons. For our purposes, such a fuel may be represented by the formula C_kH_l where the numbers of moles of carbon nuclei, k, and of hydrogen nuclei, l, per mole of fuel are not necessarily integers. For example, most gasolines are liquid mixtures of various hydrocarbons with a typical average value of k around 8 and a typical average value of l around 15. Values of k and l for a variety of fuels and the stoichiometric air–fuel ratio $(A/F)_s = (m_{air}/m_{fuel})_s$ are listed in Table 31.1.

Example 31.1. Evaluate the stoichiometric air–fuel ratio for the combustion of pure carbon, C, and express the balance between amounts of reactants and products as in equation (29.14), assuming that the reactants consist of 1 mol of C and stoichiometric air.

Solution. The chemical reaction mechanism is

$$C + O_2 = CO_2$$

This mechanism indicates that the oxidation of 1 mol of C into 1 mol of CO_2 requires 1 mol of O_2 and therefore 4.77 mol of dry air. So the stoichiometric air–fuel ratio $(n_{air}/n_{fuel})_s = 4.77$ and $(m_{air}/m_{fuel})_s = 4.77 \times 28.96/12.01 = 11.51$, where 12.01 is the molecular weight of C. Using equation (29.14), the balance between amounts of reactants and products on a mole basis is

$$1\ \text{mol C} + 1\ \text{mol O}_2 + 3.77\ \text{mol N}_2 = 1\ \text{mol CO}_2 + 3.77\ \text{mol N}_2$$

or, equivalently, on a mass basis,

$$12.01\ \text{kg C} + 32\ \text{kg O}_2 + 106.16\ \text{kg N}_2 = 44.01\ \text{kg CO}_2 + 106.16\ \text{kg N}_2$$

where we use the molecular weight 28.16 for atmospheric nitrogen.

Example 31.2. Repeat Example 31.1 for isooctane, C_8H_{18}, instead of carbon.

Solution. The reaction mechanism is

$$C_8H_{18} + 12.5\,O_2 = 8\,CO_2 + 9\,H_2O$$

Here, for complete combustion of 1 mol of C_8H_{18}, the minimum amount of O_2 is 12.5 mol, and therefore [equation (31.2)] the stoichiometric air–fuel ratio $(n_{air}/n_{fuel})_s = 4.77(8+18/4) = 59.63$ and $(m_{air}/m_{fuel})_s = 4.77\,(8 + 18/4) \times 28.96/114.232 = 15.12$. The balance between amounts of reactants and products on a mole basis is

$$1\ \text{mol}\,C_8H_{18} + 12.5\ \text{mol}\,O_2 + 47.13\ \text{mol}\,N_2 = 8\ \text{mol}\,CO_2 + 9\ \text{mol}\,H_2O + 47.13\ \text{mol}\,N_2$$

or, equivalently,

$$114.2\ \text{kg}\,C_8H_{18} + 400\ \text{kg}\,O_2 + 1327.2\ \text{kg}\,N_2 = 352.1\ \text{kg}\,CO_2 + 162.1\ \text{kg}\,H_2O + 1327.2\ \text{kg}\,N_2$$

In practice, the amount of dry air in the reactants may differ from the theoretical amount. It is expressed as *percent theoretical air* or by means of the *relative air–fuel ratio* λ defined by the relation

$$\lambda = \frac{n_{air}/n_{fuel}}{(n_{air}/n_{fuel})_s} = \frac{m_{air}/m_{fuel}}{(m_{air}/m_{fuel})_s} \tag{31.4}$$

For example, 120% theoretical air or $\lambda = 1.2$ means that the dry air is 1.2 times the theoretical air. Alternatively, the difference between the actual amount of air in the reactants and the theoretical air is expressed as a percentage of theoretical air, and we

TABLE 31.1. Typical values of molecular weight, M, density, ρ, enthalpy of vaporization, h_{fg}, specific heat at constant pressure, c_p, higher heating value, HHV, lower heating value, LHV, and stoichiometric air–fuel ratio $(A/F)_s$ of fuels.[a]

Fuel	Formula (Typical composition)	M $\dfrac{\text{kg}}{\text{kmol}}$	ρ $\dfrac{\text{kg}}{\text{liter}}$	h_{fg} $\dfrac{\text{kJ}}{\text{kg}}$	$(c_p)_f$ $\dfrac{\text{kJ}}{\text{kg K}}$	$(c_p)_g$ $\dfrac{\text{kJ}}{\text{kg K}}$	HHV $\dfrac{\text{MJ}}{\text{kg}}$	LHV $\dfrac{\text{MJ}}{\text{kg}}$	$(A/F)_s$
Practical fuels									
Gasoline	$C_{7.92}H_{14.8}$ (l)	110	0.75	305	2.4	1.7	47.3	44.0	14.6
Light diesel	$C_{12.3}H_{22.1}$ (l)	170	0.86	270	2.2	1.7	44.8	42.5	14.5
Heavy diesel	$C_{14.6}H_{24.8}$ (l)	200	0.89	230	1.9	1.7	43.8	41.4	14.4
Natural gas	$C_{1.1}H_{3.9}N_{0.1}$ (g)	18	—	—	—	2	50.0	45.0	14.5
Hydrocarbons									
Methane	CH_4 (g)	16.04	—	—	—	2.2	55.5	50.0	17.23
Propane	C_3H_8 (g)	44.10	—	—	—	1.6	50.4	46.4	15.67
Isooctane	C_8H_{18} (l)	114.23	0.692	308	2.1	1.6	47.8	44.3	15.13
Cetane	$C_{16}H_{34}$ (l)	226.44	0.773	358	—	1.6	47.3	44.0	14.82
Benzene	C_6H_6 (l)	78.11	0.879	433	1.7	1.1	41.9	40.2	13.27
Toluene	C_7H_8 (l)	92.14	0.867	412	1.7	1.1	42.5	40.6	13.50
Alcohols									
Methanol	CH_3OH (l)	32.04	0.792	1103	2.6	1.7	22.7	20.0	6.47
Ethanol	C_2H_5OH (l)	46.07	0.785	840	2.5	1.9	29.7	26.9	9.00
Other fuels									
Carbon	C (s)	12.01	2	—	—	—	33.8	33.8	11.51
Carbon monoxide	CO (g)	28.01	—	—	—	1.1	10.1	10.1	2.467
Hydrogen	H_2 (g)	2.015	—	—	—	1.4	142.0	120.0	34.3

Source: Data mainly from J. B. Heywood, *Internal Combustion Engine Fundamentals*, McGraw-Hill, New York, 1988.

[a] Values of h_{fg} at 25°C for liquid fuels. Values of ρ and $(c_p)_f$ at 25°C and 1 atm for liquid fuels. Values of $(c_p)_g$ at 25°C and saturation pressure for liquid fuels, and at 1 atm and 25°C for gaseous fuels.

talk of *percent excess air* or *percent air deficiency.* For example, 120% theoretical air is equivalent to 20% excess dry air, whereas 80% theoretical air is equivalent to 20% dry air deficiency. Yet another term used in internal combustion engine practice is the *fuel–air equivalence ratio,* defined as the inverse of λ.

Example 31.3. Evaluate the air–fuel ratio for the complete combustion of methane, CH_4, with 150% theoretical air.

Solution. The chemical reaction mechanism is

$$CH_4 + 2\,O_2 = CO_2 + 2\,H_2O$$

The stoichiometric air–fuel ratio $(n_{air}/n_{fuel})_s = 4.77(1 + 4/4) = 9.54$. Accordingly, for 150% theoretical air the air–fuel ratio $(n_{air}/n_{fuel}) = 1.5 \times 9.54 = 14.31$ and $m_{air}/m_{fuel} = 14.31 \times 28.96/16.04 = 25.83$. The balance between amounts of reactants and products on a mole basis is

$$1 \text{ mol } CH_4 + 3 \text{ mol } O_2 + 11.31 \text{ mol } N_2$$

$$= 1 \text{ mol } CO_2 + 2 \text{ mol } H_2O + 1 \text{ mol } O_2 + 11.31 \text{ mol } N_2$$

or, equivalently, on a mass basis,

$$16.04 \text{ kg } CH_4 + 96 \text{ kg } O_2 + 318.5 \text{ kg } N_2$$

$$= 44.01 \text{ kg } CO_2 + 36.03 \text{ kg } H_2O + 32 \text{ kg } O_2 + 318.5 \text{ kg } N_2$$

Example 31.4. Calculate the air–fuel ratio and the mole fractions of the products of the complete combustion of isooctane with 200% theoretical air.

Solution. The stoichiometric air–fuel ratio $(n_{air}/n_{fuel})_s = 4.77(1 + 18/4) = 59.63$ (Example 31.2). Accordingly, for 200% theoretical air the air–fuel ratio $n_{air}/n_{fuel} = 2 \times 59.63 = 119.26$ and $m_{air}/m_{fuel} = 119.26 \times 28.96/114.23 = 30.23$. The balance between amounts of reactants and products on a mole basis is

$$1 \text{ mol } C_8H_{18} + 25 \text{ mol } O_2 + 94.3 \text{ mol } N_2$$

$$= 8 \text{ mol } CO_2 + 9 \text{ mol } H_2O + 12.5 \text{ mol } O_2 + 94.3 \text{ mol } N_2$$

or, equivalently, on a mass basis,

$$114 \text{ kg } C_8H_{18} + 800 \text{ kg } O_2 + 2656 \text{ kg } N_2$$

$$= 352 \text{ kg } CO_2 + 162 \text{ kg } H_2O + 400 \text{ kg } O_2 + 2656 \text{ kg } N_2$$

The total amount of products per mole of fuel is $n_b = 8 + 9 + 12.5 + 94.3 = 123.8$ mol. So the composition of the products is: $y_{CO_2} = 8/123.8 = 0.0646$; $y_{H_2O} = 9/123.8 = 0.0727$; $y_{O_2} = 12.5/123.8 = 0.1010$; $y_{N_2} = 94.3/123.8 = 0.7617$.

For the complete combustion of a hydrocarbon, C_kH_l, according to the reaction mechanism represented by relation (31.1), and in air having a relative air–fuel ratio λ ($\lambda \geqq 1$) and a specific humidity ω, the stoichiometric coefficients and the amounts of constituents per unit amount of the hydrocarbon before and after combustion are listed in Table 31.2. All are expressed in terms of k, l, λ, and ω. Per unit amount of the hydrocarbon, the stoichiometric amount of oxygen is $(k + l/4)$, and the actual amount $\lambda(k + l/4)$. Therefore, the corresponding amount of dry air $n_{air} = 4.77\lambda(k + l/4)$, and for a specific humidity $\omega = m_v/m_{air}$, the amount of water vapor in the moist air $n_v = m_v/M_v = m_{air}\,\omega/M_v = n_{air}\,\omega\,M_{air}/M_v$, that is, $n_v = n_{air}\,\omega/0.622$.

The listings in Table 31.2 are valid for either a batch process or a steady-state rate process. They are helpful in subsequent calculations.

TABLE 31.2. Stoichiometry for the complete combustion of a hydrocarbon in excess moist air ($\lambda \geqq 1$) according to the reaction mechanism in relation (31.1), and amounts of constituents on a mole basis before and after combustion in terms of the relative air–fuel ratio, λ, and the specific humidity, ω, of the air.

i	A_i	ν_i	Amounts before combustion n_{ia}/n_{1a} or $\dot{n}_{ia}/\dot{n}_{1a}$	Amounts after combustion n_{ib}/n_{1a} or $\dot{n}_{ib}/\dot{n}_{1a}$
1	C_kH_l	-1	1	0
2	O_2	$-\left(k+\dfrac{l}{4}\right)$	$\lambda\left(k+\dfrac{l}{4}\right)$	$(\lambda-1)\left(k+\dfrac{l}{4}\right)$
3	N_2	0	$3.77\,\lambda\left(k+\dfrac{l}{4}\right)$	$3.77\,\lambda\left(k+\dfrac{l}{4}\right)$
4	H_2O	$\dfrac{l}{2}$	$\dfrac{4.77}{0.622}\,\omega\,\lambda\left(k+\dfrac{l}{4}\right)$	$\dfrac{l}{2}+\dfrac{4.77}{0.622}\,\omega\,\lambda\left(k+\dfrac{l}{4}\right)$
5	CO_2	k	0	k
$\sum_{i=1}^{5}$	$\dfrac{l}{4}-1$		$1+4.77\,\lambda\left(k+\dfrac{l}{4}\right)\left(1+\dfrac{\omega}{0.622}\right)$	$\dfrac{l}{4}+4.77\,\lambda\left(k+\dfrac{l}{4}\right)\left(1+\dfrac{\omega}{0.622}\right)$

31.3 Heating Values

We consider a combustion chamber in a steady state (Figure 31.1) such that: (1) the inlet bulk-flow stream of fuel is at standard temperature T_0 and pressure p_0; (2) the inlet bulk-flow stream[1] of dry air is also at standard temperature T_0 and pressure p_0, and provides the theoretical amount of oxygen for the complete combustion of the fuel, that is, $\lambda = 1$; (3) the outlet bulk flow is a mixture of atmospheric nitrogen and the products of complete combustion, that is, CO_2 and H_2O, also at temperature T_0 and pressure p_0; and (4) the only additional interaction is heat at a total rate $\dot{Q}^{\rightarrow}$ and at one or more unspecified temperatures.

Under the conditions just cited, the energy rate balance is

$$\dot{H}_a - \dot{H}_b = \dot{Q}^{\rightarrow} \tag{31.5}$$

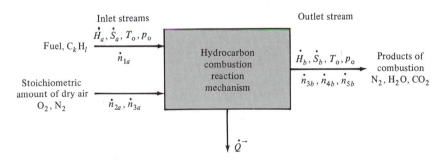

Figure 31.1 Sketch of the steady-state combustion chamber used in the definition of the heating value of a fuel at constant pressure, HV_p.

---

[1]In this chapter we consider only bulk-flow streams with negligible changes, between inlet and outlet, in potential and kinetic energies.

If the flow rate of the fuel is $\dot{n}_{1a}$ (mass flow rate $\dot{m}_f$), the ratio $\dot{Q}^{\rightarrow}/\dot{n}_{1a}$ ($\dot{Q}^{\rightarrow}/\dot{m}_f$) is defined as the *heating value of the fuel at constant pressure* or, simply, the *heating value of the fuel*. We denote it by the symbol HV_p.

In general, the heating value can be measured for each fuel by measuring $\dot{Q}^{\rightarrow}$ and $\dot{n}_{1a}$. If the chemical composition of the fuel is accurately known, the heating value may also be evaluated from the data on enthalpies of formation. In fact, for all practical purposes it is equal to the negative of the enthalpy of the (combustion) reaction.

To see this clearly, we note that

$$\dot{H}_a - \dot{H}_b = \dot{n}_{1a}h^{\circ}_{11a}(T_{\rm o}) + (\dot{n}_{2a}h^{\circ}_{2a} + \dot{n}_{3a}h^{\circ}_{3a}) - \sum_{i=3}^{5}\dot{n}_{ib}h^{\circ}_{ib} \qquad (31.6)$$

where the subscripts 1 to 5 correspond to the molecules listed in Table 31.2, and we use the fact that the inlet fuel and air streams are not mixed. For complete combustion, $0 = \dot{n}_{1b} = \dot{n}_{1a} + \nu_1\dot{\epsilon} = \dot{n}_{1a} - \dot{\epsilon}$ or $\dot{\epsilon} = \dot{n}_{1a}$ and, therefore, $\dot{n}_{ib} = \dot{n}_{ia} + \nu_i\dot{n}_{1a}$ for $i = 1, 2, \dots, r$.

The inlet air stream behaves as an ideal-gas mixture, and therefore, $h^{\circ}_{ia} = h^{\circ}_{iia}(T_{\rm o})$ for $i = 2, 3$. If the fuel behaves as an ideal gas at $T_{\rm o}$ and $p_{\rm o}$, then $h^{\circ}_{11a}(T_{\rm o})$ coincides with the specific enthalpy $h_{11}(T_{\rm o})$ of the ideal-gas state. If the fuel is liquid at $T_{\rm o}$ and $p_{\rm o}$, then $h^{\circ}_{11a}(T_{\rm o}) = h_{11}(T_{\rm o}) - h_{11,{\rm fg}}(T_{\rm o})$, where $h_{11,{\rm fg}}(T_{\rm o})$ is the enthalpy of vaporization of the fuel at $T_{\rm o}$.

Depending on the hydrogen-to-carbon ratio of the fuel, the outlet stream may consist of a single-phase gaseous mixture of CO_2, N_2, and water vapor, or a two-phase mixture of gaseous CO_2, N_2, water vapor, and liquid water, typically in the form of small droplets. The gaseous phase behaves as an ideal gas mixture so that $h_{ib} = h_{ii}(T_{\rm o})$ for $i = 3, 4, 5$, that is, each partial enthalpy equals the corresponding specific enthalpy at $T_{\rm o}$. It is noteworthy that $h_{44}(T_{\rm o})$, the specific enthalpy of the ideal-gas state (Section 29.10) of H_2O at standard temperature $T_{\rm o}$, is very well approximated by $h_{44,{\rm g}}(T_{\rm o})$, the specific enthalpy of saturated water vapor at $T_{\rm o}$, because at $T_{\rm o} = 25°C$ and $p_{\rm sat}(T_{\rm o}) = 3.197$ kPa water vapor behaves as an ideal gas. When the liquid phase is present, the specific enthalpy of the liquid water droplets is well approximated by $h_{44,{\rm f}}(T_{\rm o})$, the specific enthalpy of saturated liquid water at $T_{\rm o}$.

We recall that in an ideal gas mixture at $T_{\rm o}$ and $p_{\rm o}$ the mole fraction of water vapor [equation (27.37)] cannot exceed $p_{\rm sat}(T_{\rm o})/p_{\rm o} = 3.197/101.325 = 0.0316$. Therefore, if $\dot{n}_{4b}/\dot{n}_b \leqq 0.0316$, the products stream is single phase. If, instead, $\dot{n}_{4b}/\dot{n}_b > 0.0316$, the gaseous phase is saturated with water vapor, and coexists with the liquid phase consisting of water droplets. Denoting the water vapor rate by $\dot{n}_{4b,{\rm g}}$ and the liquid water rate by $\dot{n}_{4b,{\rm f}}$, where $\dot{n}_{4b,{\rm g}} + \dot{n}_{4b,{\rm f}} = \dot{n}_{4b}$, the condition that in the gaseous phase the water vapor mole fraction is 0.0316, namely, that $\dot{n}_{4b,{\rm g}}/(\dot{n}_b - \dot{n}_{4b,{\rm f}}) = 0.0316$, implies that

$$\dot{n}_{4b,{\rm f}} = \frac{\dot{n}_{4b} - 0.0316\,\dot{n}_b}{1 - 0.0316} \qquad (31.7)$$

Expressing $\dot{n}_{4b}$ and $\dot{n}_b$ in terms of relations listed in Table 31.2, we find that

$$\frac{\dot{n}_{4b,{\rm f}}}{\dot{n}_{1a}} = \frac{\dfrac{l}{2} - 0.0316\left[\dfrac{l}{2} + 4.77\left(k + \dfrac{l}{4}\right)\right]}{1 - 0.0316} = 0.4611\,l - 0.156\,k \qquad (31.8)$$

and that the condition $\dot{n}_{4b}/\dot{n}_b > 0.0316$ or, equivalently, $l/k > 0.337$ is satisfied by most common hydrocarbons.

As a result of the foregoing remarks, we conclude the following. For a gaseous hydrocarbon with $l/k \leqq 0.337$, equations (31.5) and (31.6) yield

$$HV_p = \frac{\dot{Q}^{\rightarrow}}{\dot{n}_{1a}} = \frac{\dot{H}_a - \dot{H}_b}{\dot{n}_{1a}} = -\sum_{i=1}^{5} \nu_i h_{ii}(T_o) = -\Delta h^{\circ} \qquad (31.9)$$

where $h_{ii}(T_o)$ is the enthalpy of the ideal-gas state of the ith constituent at T_o and p_o, and Δh° the enthalpy of reaction of the combustion reaction mechanism or, simply, the *enthalpy of combustion*. The latter enthalpy can be evaluated by means of the Hess relation [equation (29.80) with $z = h$] and the data in Table 29.4.

For a gaseous hydrocarbon and $l/k > 0.337$, using equation (31.7) and after a few rearrangements we find that

$$HV_p = \frac{\dot{Q}^{\rightarrow}}{\dot{n}_{1a}} = -\Delta h^{\circ} + (0.461\,l - 0.156\,k)\,h_{44,\text{fg}}(T_o) \qquad (31.10)$$

where the enthalpy of vaporization of water at 25°C on a mole basis $h_{44,\text{fg}}(T_o) = 44$ MJ/kmol.

For a liquid hydrocarbon and $l/k \leq 0.337$, we find that

$$HV_p = \frac{\dot{Q}^{\rightarrow}}{\dot{n}_{1a}} = -h_{11,\text{fg}}(T_o) - \Delta h^{\circ} \qquad (31.11)$$

where $h_{11,\text{fg}}(T_o)$ is the enthalpy of vaporization of the fuel at T_o.

For a liquid hydrocarbon with $l/k > 0.337$, we find that

$$HV_p = \frac{\dot{Q}^{\rightarrow}}{\dot{n}_{1a}} = -h_{11,\text{fg}}(T_o) - \Delta h^{\circ} + (0.461\,l - 0.15\,k)\,h_{44,\text{fg}}(T_o) \qquad (31.12)$$

For most hydrocarbons, the correction due to the enthalpy of vaporization of water is of the order of 10% of the predominant term $-\Delta h^{\circ}$. Rather than by the heating value HV_p, fuels are commonly characterized by the *lower heating value* LHV and the *higher heating value* HHV. These are defined like the heating value HV_p except that the H_2O in the outlet stream is assumed to be all in the vapor phase when computing the LHV, and all in the liquid phase when computing the HHV. Thus for a gaseous hydrocarbon, $C_kH_l(g)$, the lower heating value

$$LHV = -\Delta h^{\circ} \qquad (31.13)$$

for a liquid hydrocarbon, $C_kH_l(l)$,

$$LHV = -h_{11,\text{fg}}(T_o) - \Delta h^{\circ} \qquad (31.14)$$

and for either a gaseous or a liquid hydrocarbon, the higher heating value

$$HHV = LHV + \frac{l}{2}\,h_{44,\text{fg}}(T_o) \qquad (31.15)$$

where $h_{44,\text{fg}}(T_o) = 44$ MJ/kmol, $h_{11,\text{fg}}(T_o)$ is the enthalpy of vaporization of the fuel at 25°C, and the enthalpy of reaction Δh° is evaluated using the Hess relation data from Table 29.4.

Example 31.5. Evaluate the heating value at constant pressure HV_p, the higher heating value HHV, and the lower heating value LHV of methane, $CH_4(g)$, and benzene, $C_6H_6(l)$, and compare the results with the data in Table 31.1.

Solution. For methane, $k = 1$ and $l = 4$. From the data in Table 29.4 we find that $\Delta h^{\circ} = -(-74.9)+(4/2)(-242.0)+(-393.8) = -802.9$ MJ/kmol and hence [equation (31.13)] LHV $= 802.9$ MJ/kmol $= 50.056$ MJ/kg. Because $l/k = 4 > 0.337$, we use equation (31.10) and find that $HV_p = -(-802.9) + (0.461 \times 4 - 0.156)\,44 = 802.9 + 74.325 = 877.225$ MJ/kmol $= 54.690$ MJ/kg. Moreover, we use equation (31.15) and find HHV $= 802.900 + 2 \times 44 = 890.916$ MJ/kmol $= 55.543$ MJ/kg.

For benzene, $k = 6$ and $l = 6$. So $\Delta h^\circ = -(83.0) + (6/2)(-242.0) + 6(-393.8) = -3171.8$ MJ/kmol, and LHV $= -33.822 + 3171.8 = 3138$ MJ/kmol $= 40.174$ MJ/kg, where we use equation (31.14) and $h_{11,\text{fg}}(T_0) = 433$ kJ/kg $= 33.822$ MJ/kmol. Because $l/k = 1 > 0.337$, we use equation (31.12) to find $\text{HV}_p = 33.822 - (-3171.8) + (0.461 \times 6 - 0.156 \times 6)44 = -33.822 + 3171.8 + 80.693 = 3218.7$ MJ/kmol $= 41.207$ MJ/kg, and equation (31.15) to find HHV $= 3138 + 3 \times 44 = 3270$ MJ/kmol $= 41.864$ MJ/kg.

Another heating value sometimes used in internal combustion engine practice is the *heating value of the fuel at constant volume*. It is denoted by the symbol HV_V. To express HV_V in terms of other properties of the fuel, we consider a gastight, rigid combustion chamber of volume V (Figure 31.2) such that: (1) initially the chamber contains a mixture of a fuel and the stoichiometric amount of dry air at standard temperature T_0 and pressure p_0; (2) the final state is a mixture of atmospheric nitrogen and the products of complete combustion, that is, CO_2 and H_2O, also at temperature T_0; and (3) the only interaction is heat, $Q^\rightarrow$, at one or more unspecified temperatures.

Under these conditions, the energy balance is

$$U_a - U_b = Q^\rightarrow \tag{31.16}$$

and HV_V is defined as the ratio

$$\text{HV}_V = \frac{Q^\rightarrow}{n_{1a}} = \frac{U_a - U_b}{n_{1a}} \tag{31.17}$$

A first estimate of the heating value of the fuel at constant volume is given by the relation $\text{HV}_V \approx -\Delta u^\circ$, where the internal energy of reaction of the combustion reaction mechanism or, simply, the *internal energy of combustion* $\Delta u^\circ = \Delta h^\circ - \nu R T_0$. However, a more accurate calculation requires consideration of the liquid phase in the initial state if the hydrocarbon is normally liquid, the humidity in the air, and the liquid water phase in the final state, including the fact that the final pressure is not p_0.

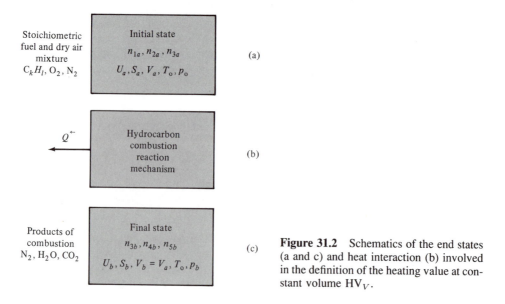

Figure 31.2   Schematics of the end states (a and c) and heat interaction (b) involved in the definition of the heating value at constant volume HV_V.

31.4   Adiabatic Flame Temperature

The burning of a fuel in a combustion chamber can be achieved under diverse conditions, depending on the amount of energy and entropy that is transferred out of the chamber during the combustion by means of interactions such as work and heat but not bulk flow. In general, the temperature of the products of combustion is higher the less the energy transferred out of the combustion chamber by work and heat, that is, the more the energy of the products of combustion, and the lower the dilution of the final composition by excess air. Traditionally, the temperature of the products of combustion that obtains under conditions of no net transfer of energy by means of work and heat into or out of the combustion chamber is called the *adiabatic flame temperature*. As we discuss in Section 31.6, the adiabatic flame temperature is useful because it is associated with an upper limit of the availability of the products of combustion in each combustion process.

In what follows, we evaluate adiabatic flame temperatures for two types of combustion processes, a batch process, and a rate process. First, we consider a batch process in which a fuel and air mixture with given values of relative air–fuel ratio λ, specific humidity ω, temperature T_a, and pressure p_a undergoes a combustion reaction in a fixed-volume chamber, and in the absence of any net interactions with systems in the environment (Figure 31.3). As a result, the products of combustion end in a stable equilibrium state at temperature T_b and pressure p_b. By definition, T_b is the adiabatic flame temperature.

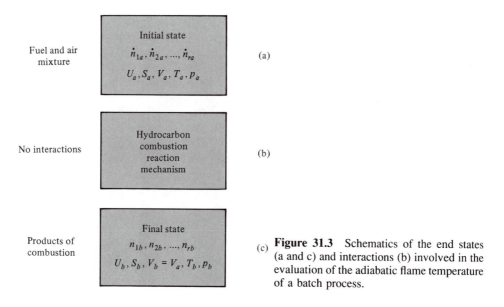

Figure 31.3 Schematics of the end states (a and c) and interactions (b) involved in the evaluation of the adiabatic flame temperature of a batch process.

The values of T_b and p_b can be found from the condition that $V_b = V_a = V$, the energy balance

$$U_b - U_a = 0 \tag{31.18}$$

and the equation of state of the products of combustion. Because the compositions of the initial and final states differ, the difference $U_b - U_a$ must be evaluated using the relations developed in Section 29.9.

For ideal-gas mixture behavior at both the initial and the final states, the energy balance becomes

$$\sum_{i=1}^{r} n_{ib}\left[u_{ii}(T_b) - u_{ii}^{o}\right] - \sum_{i=1}^{r} n_{ia}\left[u_{ii}(T_a) - u_{ii}^{o}\right] + \epsilon\,\Delta u^{o} = 0 \tag{31.19}$$

Using the ideal-gas relations $u_{ii}(T) = h_{ii}(T) - RT$ and $\Delta u^{o} = \Delta h^{o} - \nu RT_{o}$, the equalities $n_{ib} = n_{ia} + \nu_i\,\epsilon$ and $n_b = n_a + \nu\epsilon$, and the condition of complete combustion,

$\epsilon = n_{1a}$ in equation (31.19), and dividing the result by the amount of fuel n_{1a}, we find that

$$\sum_{i=1}^{r} \frac{n_{ib}}{n_{1a}} [h_{ii}(T_b) - h_{ii}^{\circ}] - \frac{n_b}{n_{1a}} R T_b = \sum_{i=1}^{r} \frac{n_{ia}}{n_{1a}} [h_{ii}(T_a) - h_{ii}^{\circ}] - \frac{n_a}{n_{1a}} R T_a - \Delta h^{\circ}$$

(31.20)

For a hydrocarbon, $C_k H_l$, each of the ratios n_{ib}/n_{1a} and n_{ia}/n_{1a} can be expressed in terms of k, l, λ, and ω (Table 31.2).

The temperature T_b is found from equation (31.20) by one of three methods. The first is a trial-and-error method. A trial value of T_b is assumed, and each of the differences $h_{ii}(T_b) - h_{ii}^{\circ}$ as well as $h_{ii}(T_a) - h_{ii}^{\circ}$ are evaluated from tabulated or graphical experimental data on constituent i. If the equality between the left-hand and right-hand sides of equation (31.20) is not satisfied, another trial value of T_b is assumed until the values of the two sides of the equation are matched. After two wrong trials, a linear interpolation (extrapolation) technique may be used to accelerate the convergence to the correct estimate of T_b.

The second method consists in using an explicit formula for each difference $h_{ii}(T) - h_{ii}^{\circ}$. For each constituent, except the fuel, we assume ideal-gas behavior in the range of temperatures between $T_0 = 300$ K and T_b (up to 4000 K), and express the temperature dependence of the specific heat at constant pressure in the form [equation (20.13b)]

$$c_p(T) = a + b T^{1/4} + c T^{1/2} + d T^{3/4}$$

(31.21)

to find that

$$h_{ii}(T) - h_{ii}^{\circ} = \int_{T_0}^{T} c_{pii}(T)\, dT$$

$$= a_{ii}(T - T_0) + \frac{4}{5} b_{ii}(T^{5/4} - T_0^{5/4}) + \frac{2}{3} c_{ii}(T^{3/2} - T_0^{3/2}) + \frac{4}{7} d_{ii}(T^{7/4} - T_0^{7/4}) \quad (31.22)$$

where the values of the coefficients a_{ii}, b_{ii}, c_{ii}, d_{ii} are listed in Table 20.2 for various constituents. Next, we substitute equation (31.22) for $T = T_a$ and $T = T_b$ for each constituent into equation (31.20), and find that

$$a_b(T_b - T_0) + \frac{4}{5} b_b(T_b^{5/4} - T_0^{5/4}) + \frac{2}{3} c_b(T_b^{3/2} - T_0^{3/2}) + \frac{4}{7} d_b(T_b^{7/4} - T_0^{7/4}) - \frac{n_b}{n_{1a}} R T_b$$

$$= a_a(T_a - T_0) + \frac{4}{5} b_a(T_a^{5/4} - T_0^{5/4}) + \frac{2}{3} c_a(T_a^{3/2} - T_0^{3/2})$$

$$+ \frac{4}{7} d_a(T_a^{7/4} - T_0^{7/4}) + [h_{11}(T_a) - h_{11}^{\circ}] - \frac{n_a}{n_{1a}} R T_a - \Delta h^{\circ} \quad (31.23a)$$

or, equivalently,

$$\frac{4}{7} d_b T_b^{7/4} + \frac{2}{3} c_b T_b^{3/2} + \frac{4}{5} b_b T_b^{5/4} + \left(a_b - \frac{n_b}{n_{1a}} R\right) T_b$$

$$= \frac{4}{7} d_a T_a^{7/4} + \frac{2}{3} c_a T_a^{3/2} + \frac{4}{5} b_a T_a^{5/4} + \left(a_a - \frac{n_a}{n_{1a}} R\right) T_a$$

$$+ [h_{11}(T_a) - h_{11}^{\circ}] + \frac{4}{7} (d_b - d_a) T_0^{7/4} + \frac{2}{3} (c_b - c_a) T_0^{3/2}$$

$$+ \frac{4}{5} (b_b - b_a) T_0^{5/4} + (a_b - a_a) T_0 - \Delta h^{\circ} \quad (31.23b)$$

where $a_x = \sum_{i=2}^{r}(n_{ix}/n_{1a})\,a_{ii}$, $b_x = \sum_{i=2}^{r}(n_{ix}/n_{1a})\,b_{ii}$, $c_x = \sum_{i=2}^{r}(n_{ix}/n_{1a})\,c_{ii}$, and $d_x = \sum_{i=2}^{r}(n_{ix}/n_{1a})\,d_{ii}$, for $x = a,\, b$. The difference $h_{11}(T_a) - h_{11}^{\circ}$ may be evaluated by using the data in Table 20.1, which includes several hydrocarbons. Equation (31.23) determines the value of the adiabatic flame temperature T_b because all other terms in the equation are specified. A numerical illustration of this method is given in Example 31.6.

In the third method, we assume that each constituent behaves as a perfect gas over the entire range of temperatures between T_0 and T_b, and assign to it a suitably selected constant value of the specific heat at constant pressure. For example, for the products of combustion of a given batch process, we make an educated guess of the value of T_b, and select a specific heat c_p for each constituent appropriate for the range between T_0 and T_b. Under these approximations, equation (31.20) yields

$$T_b = \frac{\sum_{i=1}^{r}\dfrac{n_{ib}}{n_{1a}}\,c_{pii}\,T_0 + \sum_{i=1}^{r}\dfrac{n_{ia}}{n_{1a}}\,c_{pii}\,(T_a - T_0) - \dfrac{n_a}{n_{1a}}\,RT_0 - \Delta h^{\circ}}{\sum_{i=1}^{r}\dfrac{n_{ib}}{n_{1a}}\,c_{pii} - \dfrac{n_b}{n_{1a}}\,R} \qquad (31.24a)$$

or, equivalently, using the ideal-gas relations $c_{pii} = c_{vii} + R$,

$$T_b = T_0 + \frac{\sum_{i=1}^{r}\dfrac{n_{ia}}{n_{1a}}\,c_{vii}\,(T_a - T_0) - \Delta h^{\circ} + \nu\,RT_0}{\sum_{i=1}^{r}\dfrac{n_{ib}}{n_{1a}}\,c_{vii}} \qquad (31.24b)$$

Example 31.6. A perfectly insulated, constant-volume combustion chamber of an internal combustion engine is initially filled with a homogeneous gas mixture of n-octane, C_8H_{18}, and 50% excess theoretical air ($\lambda = 1.5$) at temperature $T_a = 700$ K, pressure $p_a = 10$ atm, and specific humidity $\omega = 0.01$. Find the adiabatic flame temperature and the final pressure of this process assuming that complete combustion occurs according to the chemical reaction mechanism $C_8H_{18} + 12.5\,O_2 = 8\,CO_2 + 9\,H_2O$, that all other reaction mechanisms are inactive, and that there is ideal-gas mixture behavior before and after combustion.

Solution. For $k = 8$, $l = 18$, $\lambda = 1.5$, and $\omega = 0.01$, we use the general relations in Table 31.2 and find the values of the amounts before and after combustion normalized with respect to n_{1a}. They are listed in Table 31.3, together with the values of the coefficients a_{ii}, b_{ii}, c_{ii}, d_{ii} (in kJ/kmol) of the various constituents taken from Table 20.2.

TABLE 31.3

i	A_i	ν_i	$\dfrac{n_{ia}}{n_{1a}}$	$\dfrac{n_{ib}}{n_{1a}}$	a_{ii}	b_{ii}	c_{ii}	d_{ii}
1	C_8H_{18}	-1	1	0	—	—	—	—
2	O_2	-12.5	18.75	6.25	10.3	5.4	-0.18	0
3	N_2	0	70.74	70.74	72	-26.9	5.19	-0.298
4	H_2O	9	1.44	10.44	180	-85.4	15.6	-0.858
5	CO_2	8	0	8	-55.6	30.5	-1.96	0
$\sum_{i=1}^{5}$		$\nu = 3.5$	$\dfrac{n_a}{n_{1a}} = 91.93$	$\dfrac{n_b}{n_{1a}} = 95.43$				

From the values listed in the table, we find that

$$a_a = \sum_{i=2}^{5} \frac{n_{ia}}{n_{1a}} a_{ii} = 5546 \text{ kJ/kmol} \qquad b_a = \sum_{i=2}^{5} \frac{n_{ia}}{n_{1a}} b_{ii} = -1925 \text{ kJ/kmol}$$

$$c_a = \sum_{i=2}^{5} \frac{n_{ia}}{n_{1a}} c_{ii} = 386.2 \text{ kJ/kmol} \qquad d_a = \sum_{i=2}^{5} \frac{n_{ia}}{n_{1a}} d_{ii} = -22.3 \text{ kJ/kmol}$$

$$a_b = \sum_{i=2}^{5} \frac{n_{ib}}{n_{1a}} a_{ii} = 6592 \text{ kJ/kmol} \qquad b_b = \sum_{i=2}^{5} \frac{n_{ib}}{n_{1a}} b_{ii} = -2517 \text{ kJ/kmol}$$

$$c_b = \sum_{i=2}^{5} \frac{n_{ib}}{n_{1a}} c_{ii} = 513.2 \text{ kJ/kmol} \qquad d_b = \sum_{i=2}^{5} \frac{n_{ib}}{n_{1a}} d_{ii} = -30.0 \text{ kJ/kmol}$$

$$\frac{n_a}{n_{1a}} = 91.93 \qquad\qquad \frac{n_b}{n_{1a}} = 95.43$$

For C_8H_{18} we use data from Table 20.1, and for $T_a = 700$ K find $h_{11}(T_a) - h_{11}^o = 113,009$ kJ/kmol. From the enthalpies of formation (Table 29.4), we find $\Delta h^o = -(-208.6) + 9(-242.0) + 8(-393.8) = -5119.8$ MJ/kmol. Upon using all these values plus $T_a = 700$ K, $T_o = 298$ K, and $R = 8.314$ kJ/kmol in equation (31.23b), we find that

$$(6.592 - 0.793)T_b - 2.014 T_b^{5/4} + 0.342 T_b^{3/2} - 0.017 T_b^{7/4} = 6656.1 \qquad (31.25)$$

The valid solution of this equation is $T_b = 2551$ K. Moreover, the condition of constant volume, $V = V_a = V_b$, and the ideal gas relations $p_b V = n_b R T_b$ and $p_a V = n_a R T_a$ yield $p_b = p_a (n_b/n_a)(T_b/T_a) = 10(95.43/91.93)(2551/700) = 37.8$ atm. This completes the calculations of the adiabatic flame temperature and the pressure of a batch combustion process.

Before proceeding with the discussion of the second combustion process, it is noteworthy that the temperature T_b found in Example 31.6 may not be realistic. The reason is that, at such a high temperature, the assumption that only the combustion reaction mechanism is active and all the other reaction mechanisms are inactive is not valid. Said differently, the assumption that the products of combustion consist only of O_2, N_2, CO_2, and H_2O is not correct. We verify the assertion later in this section.

Next, we consider a rate process in a steady-state reactor with an inlet bulk-flow stream consiting of a fuel and air mixture with given values of relative air–fuel ratio λ, specific humidity ω, temperature T_a and pressure p_a, an outlet bulk-flow stream consisting of the products of combustion at a given pressure p_b and temperature T_b, and neither work nor heat interactions (Figure 31.4). By definition, for this process T_b is the adiabatic flame temperature.

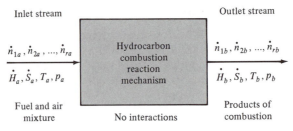

Figure 31.4 Schematic of the steady-state rate process used to evaluate the adiabatic flame temperature.

The value of T_b can be found from the energy rate balance

$$\dot{H}_b - \dot{H}_a = 0 \tag{31.26}$$

Because the compositions of the inlet and outlet bulk-flow states differ, the difference $\dot{H}_b - \dot{H}_a$ must be evaluated using the relations developed in Section 29.9. In particular, for ideal-gas mixture behavior of both the inlet and outlet streams, the energy rate balance becomes

$$\sum_{i=1}^{r} \dot{n}_{ib} \left[h_{ii}(T_b) - h_{ii}^{\circ} \right] - \sum_{i=1}^{r} \dot{n}_{ia} \left[h_{ii}(T_a) - h_{ii}^{\circ} \right] + \dot{\epsilon}\, \Delta h^{\circ} = 0 \tag{31.27}$$

or, using the condition of complete combustion, $\dot{\epsilon} = \dot{n}_{1a}$, and dividing through by the inlet fuel flow rate $\dot{n}_{1a}$,

$$\sum_{i=1}^{r} \frac{\dot{n}_{ib}}{\dot{n}_{1a}} \left[h_{ii}(T_b) - h_{ii}^{\circ} \right] = \sum_{i=1}^{r} \frac{\dot{n}_{ia}}{\dot{n}_{1a}} \left[h_{ii}(T_a) - h_{ii}^{\circ} \right] - \Delta h^{\circ} \tag{31.28}$$

For complete combustion of a hydrocarbon, $C_k H_l$, and given values of the relative air–fuel ratio $\lambda > 1$, the specific humidity ω, and the temperature T_a of the inlet stream, the values of the stoichiometric coefficients ν_i, and the ratios $\dot{n}_{ia}/\dot{n}_{1a}$ and $\dot{n}_{ib}/\dot{n}_{1a}$ are determined by expressions identical to those of the corresponding quantities ν_i, n_{ia}/n_{1a} and n_{ib}/n_{1a} of the batch process (Table 31.2).

It is clear that the temperature T_b can be found from equation (31.28) by means of one of the three methods discussed in connection with the batch combustion process. In particular, by means of the third method, equation (31.28) yields

$$T_b = T_o + \frac{\sum_{i=1}^{r} \frac{\dot{n}_{ia}}{\dot{n}_{1a}} c_{pii} (T_a - T_o) - \Delta h^{\circ}}{\sum_{i=1}^{r} \frac{\dot{n}_{ib}}{\dot{n}_{1a}} c_{pii}} \tag{31.29}$$

In general, for a given fuel and given values of λ, ω, and T_a, the value of the adiabatic flame temperature of the rate process, $(T_b)_{\text{rate}}$, differs from the value of the adiabatic flame temperature of the batch process, $(T_b)_{\text{batch}}$. Typically, $(T_b)_{\text{rate}} < (T_b)_{\text{batch}}$.

Example 31.7. Verify the last assertion.

Solution. We use the assumptions and approximations that lead to equation (31.24b) for $(T_b)_{\text{batch}}$ and equation (31.29) for $(T_b)_{\text{rate}}$. For a given fuel, and given values of λ and ω, each ratio n_{ia}/n_{1a} and n_{ib}/n_{1a} for the batch process is identical to the corresponding ratio $\dot{n}_{ia}/\dot{n}_{1a}$ and $\dot{n}_{ib}/\dot{n}_{1a}$ for the rate process (Table 31.2). Using these identities and the perfect gas relations $c_{pii} = c_{vii} + R$, and eliminating Δh° from equations (31.24b) and (31.29), after a few rearrangements we find that

$$(T_b)_{\text{batch}} - (T_b)_{\text{rate}} = \frac{\dfrac{n_b}{n_{1a}} R}{\sum_{i=1}^{r} \dfrac{n_{ib}}{n_{1a}} c_{vii}} \left[(T_b)_{\text{rate}} - \frac{n_a}{n_b} T_a \right] \tag{31.30}$$

For the complete combustion of a hydrocarbon, the value of the ratio $n_a/n_b = (n_a/n_{1a})/(n_b/n_{1a})$ is very close to unity. For example, for stoichiometric and dry air, $\lambda = 1$ and $\omega = 0$, and using Table 31.2, we find that the value of n_a/n_b is 1.17 for hydrogen, H_2, 1.04 for acetylene, C_2H_2, 1 for methane, CH_4, and ethylene, C_2H_4, 0.99 for benzene, C_6H_6, and 0.96 for propane, C_3H_8. Moreover, under typical combustion conditions, the temperature T_a of the initial state of either a batch process or the inlet stream of a rate process is well below 1000 K, whereas the adiabatic flame temperature T_b for a rate process is well above 1500 K. Thus the term $(T_b)_{\text{rate}} - (n_a/n_b) T_a$ in equation (31.30) is positive, and therefore $(T_b)_{\text{batch}} > (T_b)_{\text{rate}}$.

Example 31.8.   A steady-state burner is fed with a homogeneous gas mixture of n-octane, C_8H_{18}, and 50% excess theoretical air ($\lambda = 1.5$) at temperature $T_a = 700$ K and specific humidity $\omega = 0.01$. The burner experiences no other interactions, except the inlet and outlet bulk flows. Find the adiabatic flame temperature assuming that the combustion is complete and that all other reaction mechanisms are inactive.

Solution.   For $k = 8$, $l = 18$, $\lambda = 1.5$, and $\omega = 0.01$, the values of the ratios $\dot{n}_{ia}/\dot{n}_{1a}$ and $\dot{n}_{ib}/\dot{n}_{1a}$ are the same as those of the corresponding ratios n_{ia}/n_{1a} and n_{ib}/n_{1a} listed in Example 31.6. As a result, equation (31.28) is identical to equation (31.20), except for the absence of the terms $(n_b/n_{1a})RT_b$ and $(n_a/n_{1a})RT_a$. So we can readily verify that T_b satisfies the equation

$$6.592\,T_b - 2.014\,T_b^{5/4} + 0.342\,T_b^{3/2} - 0.017\,T_b^{7/4} = 7191.5$$

The valid solution of this equation is $T_b = 2160$ K. Just as we note in connection with Example 31.6, however, we should caution here that this value of T_b may not be realistic because of the omission of all chemical reaction mechanisms other than complete combustion. We investigate this point immediately below.

In addition to the reaction mechanism of combustion of a given fuel, many other reaction mechanisms are active at the conditions of high temperatures that are typically reached by the products of combustion. At temperatures above 1700 K, most of the conceivable reaction mechanisms involving the atomic nuclei C, O, H, and N are sufficiently active that the composition is almost that of complete chemical equilibrium (Section 30.8). For example, for the products of complete combustion of ordinary hydrocarbons under ordinary combustion conditions, experimental studies and numerical evaluation of the complete chemical equilibrium compositions show that the predominant species are N_2, CO_2, H_2O, CO, O_2, H_2, OH, NO, H, O, and NO_2. If the combustion occurs under conditions of air deficiency, then also a variety of unburned hydrocarbons, generally denoted as HC, become relevant. Many other species are present in traces.

A good estimate of the adiabatic flame temperature of a hydrocarbon, C_kH_l, is obtained by considering only two important reaction mechanisms, in addition to that describing the fuel combustion. They are the two dissociation mechanisms

$$j = 2: \qquad CO_2 = CO + \frac{1}{2}O_2 \qquad\qquad (31.31)$$

$$j = 3: \qquad H_2O = H_2 + \frac{1}{2}O_2 \qquad\qquad (31.32)$$

which have equilibrium constants

$$K_2(T) = \exp\left(10.108 - \frac{33{,}767}{T}\right) \qquad\qquad (31.33)$$

$$K_3(T) = \exp\left(6.866 - \frac{29{,}911}{T}\right) \qquad\qquad (31.34)$$

where the numerical values in the exponents are obtained from Table 30.1. The evaluation proceeds as follows. We assume that the fuel is completely oxidized according to the combustion reaction mechanism, and that the products of combustion consist of a mixture of N_2, CO_2, H_2O, CO, O_2, and H_2 at chemical equilibrium with respect to the dissociation reaction mechanisms represented by relations (31.31) and (31.32).

For complete combustion of a hydrocarbon in either a batch process or a rate process, and for given values of the relative air–fuel ratio $\lambda > 1$, and the specific humidity ω, the reaction coordinate of the mechanisms represented by relations (31.1), (31.31), and (31.32) are $\epsilon_1 = n_{1a}$, ϵ_2, and ϵ_3, respectively. Moreover, upon defining the *modified degrees of reaction* $\xi_1' = \epsilon_1/n_{1a} = 1$, $\xi_2' = \epsilon_2/n_{1a}$, and $\xi_3' = \epsilon_3/n_{1a}$, we can express the stoichiometric coefficients and the ratios n_{ia}/n_{1a}, $\dot{n}_{ia}/\dot{n}_{1a}$, n_{ib}/n_{1a}, $\dot{n}_{ib}/\dot{n}_{1a}$ as listed in Table 31.4, where we set $n_{1b} = 0$ and $\dot{n}_{1b} = 0$ because the combustion is complete.

TABLE 31.4.   **Stoichiometry for the complete combustion of a hydrocarbon according to the reaction mechanism given by relation (31.1) in terms of the relative air–fuel ratio, λ, the specific humidity, ω, and the modified degrees of reaction of the dissociation reaction mechanisms given by relations (31.31) and (31.32).**

i	A_i	ν_i	$\nu_i^{(2)}$	$\nu_i^{(3)}$	Amounts before combustion n_{ia}/n_{1a} or $\dot{n}_{ia}/\dot{n}_{1a}$	Amounts after combustion n_{ib}/n_{1a} or $\dot{n}_{ib}/\dot{n}_{1a}$
1	C_kH_l	-1	0	0	1	0
2	O_2	$-\left(k+\dfrac{l}{4}\right)$	$\dfrac{1}{2}$	$\dfrac{1}{2}$	$\lambda\left(k+\dfrac{l}{4}\right)$	$(\lambda-1)\left(k+\dfrac{l}{4}\right)+\dfrac{\xi_2'}{2}+\dfrac{\xi_3'}{2}$
3	N_2	0	0	0	$3.77\,\lambda\left(k+\dfrac{l}{4}\right)$	$3.77\,\lambda\left(k+\dfrac{l}{4}\right)$
4	H_2O	$\dfrac{l}{2}$	0	-1	$\dfrac{4.77}{0.622}\,\omega\,\lambda\left(k+\dfrac{l}{4}\right)$	$\dfrac{l}{2}+\dfrac{4.77}{0.622}\,\omega\,\lambda\left(k+\dfrac{l}{4}\right)-\xi_3'$
5	CO_2	k	-1	0	0	$k-\xi_2'$
6	CO	0	1	0	0	ξ_2'
7	H_2	0	0	1	0	ξ_3'

$$\sum_{i=1}^{7}$$

	ν	$\nu^{(2)}$	$\nu^{(3)}$	$1+$	$\dfrac{l}{4}+\dfrac{\xi_2'}{2}+\dfrac{\xi_3'}{2}+$
	$=\dfrac{l}{4}-1$	$=\dfrac{1}{2}$	$=\dfrac{1}{2}$	$4.77\,\lambda\left(k+\dfrac{l}{4}\right)\left(1+\dfrac{\omega}{0.622}\right)$	$4.77\,\lambda\left(k+\dfrac{l}{4}\right)\left(1+\dfrac{\omega}{0.622}\right)$

For an ideal gas mixture of products of combustion, the ratios either $n_{ib}/n_b = (n_{ib}/n_{1a})/(n_b/n_{1a})$ or $\dot{n}_{ib}/\dot{n}_b = (\dot{n}_{ib}/\dot{n}_{1a})/(\dot{n}_b/\dot{n}_{1a})$ satisfy the chemical equilibrium equations [equations (30.36)]

$$\prod_{i=2}^{7}\left(\frac{n_{ib}}{n_b}\right)^{\nu_i^{(2)}}=\left(\frac{p_o}{p_b}\right)^{\nu^{(2)}}K_2(T_b) \tag{31.35}$$

$$\prod_{i=2}^{7}\left(\frac{n_{ib}}{n_b}\right)^{\nu_i^{(3)}}=\left(\frac{p_o}{p_b}\right)^{\nu^{(3)}}K_3(T_b) \tag{31.36}$$

Moreover, the energy balance for the batch process can be expressed in the form

$$\sum_{i=1}^{r}\frac{n_{ib}}{n_{1a}}\left[h_{ii}(T_b)-h_{ii}^\circ\right]-\frac{n_b}{n_{1a}}RT_b$$

$$=\sum_{i=1}^{r}\frac{n_{ia}}{n_{1a}}\left[h_{ii}(T_a)-h_{ii}^\circ\right]-\frac{n_a}{n_{1a}}RT_a-\Delta h^\circ-\xi_2'(\Delta h^\circ)_2-\xi_3'(\Delta h^\circ)_3 \tag{31.37}$$

and for the rate process in the form

$$\sum_{i=1}^{r}\frac{n_{ib}}{n_{1a}}\left[h_{ii}(T_b)-h_{ii}^\circ\right]$$

$$=\sum_{i=1}^{r}\frac{n_{ia}}{n_{1a}}\left[h_{ii}(T_a)-h_{ii}^\circ\right]-\Delta h^\circ-\xi_2'(\Delta h^\circ)_2-\xi_3'(\Delta h^\circ)_3 \tag{31.38}$$

where $(\Delta h^\circ)_2$ and $(\Delta h^\circ)_3$ are the enthalpies of reaction of the dissociation mechanisms represented by relations (31.31) and (31.32), respectively. The proofs of equations (31.37) and (31.38) are left as exercises for the reader.

Example 31.9. Find the adiabatic flame temperature of the batch combustion process defined in Example 31.6, but subject to the dissociation mechanisms represented by relations (31.31) and (31.32). Repeat the calculations assuming a rate process and either $p_b = 10$ atm or $p_b = 1$ atm.

Solution. In Example 31.6, the hydrocarbon, C_kH_l, is n-octane, that is, $k = 8$ and $l = 18$, the relative air–fuel ratio $\lambda = 1.5$, the specific humidity $\omega = 0.01$, $T_a = 700$ K, and $p_a = 10$ atm. Using the expressions in Table 31.4, we find the initial and final amounts as functions of ξ_2' and ξ_3' as listed in Table 31.5, together with the values of the coefficients a_{ii}, b_{ii}, c_{ii}, d_{ii} of the various constituents taken from Table 20.2.

TABLE 31.5

i	A_i	$\dfrac{n_{ia}}{n_{1a}}$	$\dfrac{n_{ib}}{n_{1a}}$	a_{ii}	b_{ii}	c_{ii}	d_{ii}
1	C_8H_{18}	1	0	—	—	—	—
2	O_2	18.75	$6.25 + \dfrac{\xi_2'}{2} + \dfrac{\xi_3'}{2}$	10.3	5.4	−0.18	0
3	N_2	70.74	70.74	72	−26.9	5.19	−0.298
4	H_2O	1.44	$10.44 - \xi_3'$	180	−85.4	15.6	−0.858
5	CO_2	0	$8 - \xi_2'$	−55.6	30.5	−1.96	0
6	CO	0	ξ_2'	62.8	−22.6	4.6	−0.272
7	H_2	0	ξ_3'	79.5	−26.3	4.23	−0.197

$\displaystyle\sum_{i=1}^{7}$ $\dfrac{n_a}{n_{1a}} = 91.93$ $\dfrac{n_b}{n_{1a}} = 95.43 + \dfrac{\xi_2'}{2} + \dfrac{\xi_3'}{2}$

From these values we find that $n_b/n_{1a} = 95.43 + \xi_2'/2 + \xi_3'/2$, $a_b = \sum_{i=1}^{7}(n_{ib}/n_{1a})\,a_{ii} = 6592 + 124\,\xi_2' - 95\,\xi_3'$, $b_b = \sum_{i=1}^{7}(n_{ib}/n_{1a})\,b_{ii} = -2517 - 50.4\,\xi_2' + 61.8\,\xi_3'$, $c_b = \sum_{i=1}^{7}(n_{ib}/n_{1a})\,c_{ii} = 513.2 + 6.47\,\xi_2' - 11.5\,\xi_3'$, $d_b = \sum_{i=1}^{7}(n_{ib}/n_{1a})\,d_{ii} = -30.0 - 0.272\,\xi_2' + 0.66\,\xi_3'$, where the units of a_b, b_b, c_b, d_b are kJ/kmol. As in Example 31.6, $\Delta h^\circ = -5119.8$ MJ/kmol. Using Table 29.4, we evaluate $(\Delta h^\circ)_2 = -(-393.8) + (-110.6) = 283.2$ MJ/kmol and $(\Delta h^\circ)_3 = -(-242.0) = 242$ MJ/kmol. Upon using all these relations in equation (31.37), we find

$$(5.7986 + 0.12\,\xi_2' - 0.099\,\xi_3')\,T_b + (-2.014 - 0.04\,\xi_2' + 0.049\,\xi_3')\,T_b^{5/4}$$

$$+ (0.342 + 0.004\,\xi_2' - 0.008\,\xi_3')\,T_b^{3/2} + (-0.0171 - 0.0001\,\xi_2' + 0.0003\,\xi_3')\,T_b^{7/4}$$

$$- 6656.5 + 283.2\,\xi_2' + 242.0\,\xi_3' = 0 \tag{31.39}$$

The constant-volume condition and the ideal-gas relations $p_b V = n_b R T_b$ and $p_a V = n_a R T_a$ yield $p_b = p_a n_b T_b / n_a T_a = 10\,(95.43 + \xi_2'/2 + \xi_3'/2)\,T_b/(700 \times 91.93) = (95.43 + \xi_2'/2 + \xi_3'/2)(T_b/6435)$ atm. Therefore, the chemical equilibrium equations (31.35) and (31.36) become

$$\frac{\xi_2'\,(6.25 + \xi_2'/2 + \xi_3'/2)^{1/2}}{(8 - \xi_2')} = \left(\frac{6435}{T_b}\right)^{1/2} \exp\left(10.108 - \frac{33{,}767}{T_b}\right) \tag{31.40}$$

$$\frac{\xi_3'\,(6.25 + \xi_2'/2 + \xi_3'/2)^{1/2}}{(10.44 - \xi_3')} = \left(\frac{6435}{T_b}\right)^{1/2} \exp\left(6.866 - \frac{29{,}911}{T_b}\right) \tag{31.41}$$

Solving the three equations (31.39) to (31.41) by trial and error, we find $T_b = 2531$ K, $\xi_2' = 0.194$, and $\xi_3' = 0.046$. Therefore, $p_b = 37.6$ atm. The resulting final compositions are listed in the batch-process columns of Table 31.6.

TABLE 31.6

		Batch process		Rate process $p_b = 10$ atm		Rate process $p_b = 1$ atm	
i	A_i	n_{ib}/n_{1a}	y_{ib}	$\dot{n}_{ib}/\dot{n}_{1a}$	y_{ib}	$\dot{n}_{ib}/\dot{n}_{1a}$	y_{ib}
1	C_8H_{18}	0	0	0	0	0	0
2	O_2	6.37	0.0667	6.27	0.0657	6.32	0.0662
3	N_2	70.74	0.7404	70.74	0.7411	70.74	0.7407
4	H_2O	10.39	0.1088	10.43	0.1092	10.40	0.1089
5	CO_2	7.81	0.0817	7.96	0.0834	7.89	0.0826
6	CO	0.194	0.0020	0.038	0.0004	0.113	0.0012
7	H_2	0.046	0.0005	0.012	0.0001	0.035	0.0004

We note that about 2.5% of the CO_2, and about 0.5% of the H_2O are dissociated according to relations (31.31) and (31.32).

For the rate process, the energy balance equation and the chemical equilibrium equations become

$$(6.592 + 0.124\,\xi_2' - 0.095\,\xi_3')\,T_b + (-2.014 - 0.04\,\xi_2' + 0.049\,\xi_3')\,T_b^{5/4}$$
$$+ (0.342 + 0.004\,\xi_2' - 0.008\,\xi_3')\,T_b^{3/2} + (-0.0171 - 0.0001\,\xi_2' + 0.0003\,\xi_3')\,T_b^{7/4}$$
$$- 7191.5 + 283.2\,\xi_2' + 242.0\,\xi_3' = 0 \tag{31.42}$$

$$\frac{\xi_2'\,(6.25 + \xi_2'/2 + \xi_3'/2)^{1/2}}{(8 - \xi_2')\,(95.43 + \xi_2'/2 + \xi_3'/2)^{1/2}} = \left(\frac{p_o}{p_b}\right)^{1/2} \exp\left(10.108 - \frac{33,767}{T_b}\right) \tag{31.43}$$

$$\frac{\xi_3'\,(6.25 + \xi_2'/2 + \xi_3'/2)^{1/2}}{(10.44 - \xi_3')\,(95.43 + \xi_2'/2 + \xi_3'/2)^{1/2}} = \left(\frac{p_o}{p_b}\right)^{1/2} \exp\left(6.866 - \frac{29,911}{T_b}\right) \tag{31.44}$$

For $p_b = 10$ atm, the three equations yield $T_b = 2157$ K, $\xi_2' = 0.038$, and $\xi_3' = 0.012$, and for $p_b = 1$ atm, $T_b = 2150$ K, $\xi_2' = 0.113$, and $\xi_3' = 0.035$. The resulting compositions are listed in the rate-process columns of Table 31.6. We note that the amounts of CO_2 and H_2O that are dissociated are less than for the batch process because the temperature is lower. Because both $\nu^{(2)}$ and $\nu^{(3)}$ are positive, we also verify that the higher the pressure the lower the dissociation.

Even though they are present only in traces, some of the many constituents of the products of combustion are very important for considerations of environmental pollution. In particular, carbon monoxide, CO, nitric oxide, NO, nitrogen dioxide, NO_2, and unburned or partially burned hydrocarbons, denoted generically by HC, are important causes of urban air pollution. They are due primarily to the combustion of gasoline and diesel fuels in automotive engines. In addition, because gasoline and diesel fuels contain a small amount of sulfur, sulfur dioxide, SO_2, and sulfur trioxide, SO_3, are present in the products of combustion. Sulfur oxides combine with water vapor in the atmosphere, producing sulfuric acid and giving rise to the phenomenon known as "acid rain." The products of combustion of other fuels are characterized by the presence of other pollutants. For example, alcohol fuels, such as methanol and ethanol, produce aldehydes, such as formaldehyde. Aldehydes are eye and respiratory irritants, in addition to being cancerous and mutageneous. Estimates of the production of these pollutants in various combustion engines require consideration of the kinetics of their formation reaction mechanisms and are beyond our scope here.

Example 31.10.   Estimate the amount of NO in the products of combustion of the batch process considered in Example 31.9.

Solution.   We assume that the products of combustion are at chemical equilibrium also with respect to the reaction mechanism of formation of NO,

$$j = 4 : \qquad \frac{1}{2}N_2 + \frac{1}{2}O_2 = NO$$

Instead of repeating the lengthy calculations, here we assume that the values of the temperature and the pressure are sufficiently well approximated by the values found in Example 31.9, that is, $T_b = 2531$ K and $p_b = 37.6$ atm, and that the amounts of O_2, N_2, and NO are corrected according to the stoichiometric and chemical equilibrium equations of the mechanism of formation of NO. Thus, denoting NO by $i = 8$, $n_{8b}/n_{1a} = \xi_4'$, and using the results of Example 31.9, we find that $n_{2b}/n_{1a} = 6.37 - \xi_4'/2$, $n_{3b}/n_{1a} = 70.74 - \xi_4'/2$, and the chemical equilibrium equation

$$\frac{\xi_4'}{(6.37 - \xi_4'/2)^{1/2}(70.74 - \xi_4'/2)^{1/2}} = \exp\left(1.504 - \frac{10,863}{T_b}\right)$$

where we use the data for NO in Table 30.1. For $T_b = 2531$ K, this equation yields $n_{8b}/n_{1a} = \xi_4' = 1.24$, and therefore the NO mole fraction $y_{8b} \approx 0.013$. Of course, in combustion energy-conversion devices, the products of combustion are cooled down before discharge to the atmosphere, and therefore the chemical equilibrium mole fraction of NO decreases. However, the reaction mechanism by which this occurs becomes less and less active as the temperature decreases and, below a temperature around 1700 K and in the absence of special catalysts, the mole fraction of NO remains unchanged. For $T_b = 1700$ K and all other quantities unchanged, we find that $n_{8b}/n_{1a} = \xi_4' = 0.16$ and, therefore, $y_{8b} \approx 0.0017$. This value is indicative of the NO content of the exhaust gases of an automobile engine.

31.5   Availability of a Fuel

We consider a combustion chamber in a steady state (Figure 31.5) such that: (1) the inlet bulk-flow streams are a fuel and stoichiometric dry air for complete combustion of the fuel, each at temperature T_o and pressure p_o; (2) the outlet bulk flow stream is a mixture of the products of complete combustion plus nitrogen, also at T_o and p_o; and (3) the additional interactions are shaft work at the rate $\dot{W}_s^{\rightarrow}$, and heat at the rate $\dot{Q}_o^{\leftarrow}$ and temperature T_o with the environment modeled as a reservoir with variable volume and amounts of constituents as in Section 24.4.3.

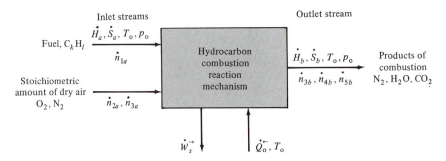

Figure 31.5   Schematic of a steady-state rate process used to evaluate the availability of a fuel.

Under these conditions, the energy and entropy rate balances are

$$\dot{H}_b - \dot{H}_a = \dot{Q}_o^{\leftarrow} - \dot{W}_s^{\rightarrow} \tag{31.45}$$

$$\dot{S}_b - \dot{S}_a = \frac{\dot{Q}_o^{\leftarrow}}{T_o} + \dot{S}_{irr} \tag{31.46}$$

Upon multiplying equation (31.46) by T_o and subtracting the result from equation (31.45), we find that

$$\dot{W}_s^{\rightarrow} = \dot{H}_a - \dot{H}_b - T_o(\dot{S}_a - \dot{S}_b) - T_o\dot{S}_{irr} \tag{31.47}$$

and the largest shaft work rate ($\dot{S}_{irr} = 0$)

$$(\dot{W}_s^{\rightarrow})_{largest} = \dot{H}_a - \dot{H}_b - T_o(\dot{S}_a - \dot{S}_b) \tag{31.48}$$

Because the outlet bulk flow stream is in temperature and pressure equality with an environment at the standard conditions T_o and p_o, $(\dot{W}_s^{\rightarrow})_{largest}$ is the availability rate of the fuel under the conditions specified at the beginning of this section.

With a reasoning analogous to that used in Section 31.3 to derive equation (31.9), we can evaluate the differences $\dot{H}_a - \dot{H}_b$ and $\dot{S}_a - \dot{S}_b$. For brevity, here we present only the results, that is,

$$\frac{\dot{H}_a - \dot{H}_b}{\dot{n}_{1a}} = -\Delta h^o \tag{31.49}$$

$$\frac{\dot{S}_a - \dot{S}_b}{\dot{n}_{1a}} = -\Delta s^o + R\sum_{i=3}^{5} \frac{\dot{n}_{ib}}{\dot{n}_{1a}} \ln y_{ib} - R\sum_{i=2}^{3} \frac{\dot{n}_{ia}}{\dot{n}_{1a}} \ln y'_{ia} \tag{31.50}$$

where Δs^o is the entropy of reaction of the combustion reaction mechanism or, simply, the *entropy of combustion* and, of course, we assume ideal-gas-mixture behavior for both the atmospheric air and the products of combustion, recall that here $T_a = T_b = T_o$, $p_a = p_b = p_o$, and $\dot{\epsilon} = \dot{n}_a\xi = \dot{n}_{1a}$ (complete combustion), use subscripts $i = 1, 2, \ldots, 5$ as in Table 31.2, and denote by y'_{ia} the composition of the inlet air stream. Substituting equations (31.49) and (31.50) in equation (31.48), we find that

$$\frac{(\dot{W}_s^{\rightarrow})_{largest}}{\dot{n}_{1a}} = -\Delta h^o + T_o\Delta s^o - RT_o\left(\sum_{i=3}^{5} \frac{\dot{n}_{ib}}{\dot{n}_{1a}} \ln y_{ib} - \sum_{i=2}^{3} \frac{\dot{n}_{ia}}{\dot{n}_{1a}} \ln y'_{ia}\right)$$

$$= -\Delta g^o - RT_o\left(\sum_{i=3}^{5} \frac{\dot{n}_{ib}}{\dot{n}_{1a}} \ln y_{ib} - \sum_{i=2}^{3} \frac{\dot{n}_{ia}}{\dot{n}_{1a}} \ln y'_{ia}\right) \approx -\Delta g^o \tag{31.51}$$

where Δg^o is the Gibbs free energy of the combustion reaction mechanism or, simply, the *Gibbs free energy of combustion*, and in the last of equations (31.51) we neglect the bracketed terms. These terms—the product of T_o and the difference between the entropy of mixing of the products of combustion and the entropy of mixing of the theoretical air—are negligibly smaller than the value of Δg^o for all fuels. So we conclude that the availability associated with the combustion of a fuel is about equal to the negative of the Gibbs free energy of the (combustion) reaction.

For various fuels, values of Δg^o and Δh^o for the combustion reaction evaluated with all the constituents in their ideal-gas state at standard T_o and p_o are given in Table 31.7.

TABLE 31.7. **Values of enthalpy, Gibbs free energy, and entropy of combustion of fuels at standard temperature, $T_o = 25°C$, and pressure, $p_o = 1$ atm.**[a]

Fuel	Formula	M $\dfrac{kg}{kmol}$	Δh^o $\dfrac{MJ}{kg}$	Δg^o $\dfrac{MJ}{kg}$	Δs^o $\dfrac{kJ}{kg\,K}$	$\dfrac{\Delta h^o - \Delta g^o}{\Delta g^o}$ $\%$
Hydrogen	H_2	2.016	−120.0	−113.5	−22.0	+5.8
Carbon (graphite)	C	12.011	−32.8	−32.9	0.2	−0.2
Methane	CH_4	16.043	−50.0	−49.9	−0.3	+0.2
Acetylene	C_2H_2	26.038	−48.3	−47.1	−3.7	+2.4
Ethylene	C_2H_4	28.054	−47.2	−46.9	−1.1	+0.7
Ethane	C_2H_6	30.07	−47.5	−48.0	1.5	−1.0
Propylene	C_3H_6	42.081	−45.8	−45.9	0.4	−0.3
Propane	C_3H_8	44.097	−46.4	−47.1	2.3	−1.5
n-Butane	C_4H_{10}	58.12	−45.8	−46.6	2.7	−1.7
n-Pentane	C_5H_{12}	72.15	−45.4	−46.3	2.9	−1.9
Benzene	C_6H_6	78.114	−40.6	−40.8	0.5	−0.4
n-Hexane	C_6H_{14}	86.18	−45.1	−46.1	3.1	−2.0
n-Heptane	C_7H_{16}	100.21	−45.0	−45.9	3.2	−2.1
n-Octane	C_8H_{18}	114.232	−44.8	−45.8	3.3	−2.2
Isooctane	C_8H_{18}	114.232	−44.7	−45.8	3.7	−2.4
n-Nonane	C_9H_{20}	128.26	−44.7	−45.7	3.4	−2.2
n-Decane	$C_{10}H_{22}$	142.29	−44.6	−45.7	3.5	−2.3
Carbon monoxide	CO	28.01	−10.1	−9.2	−3.1	+10.1
Methanol	CH_3OH	32.042	−21.1	−21.5	1.4	−1.9
Ethanol	C_2H_5OH	46.069	−27.8	−28.4	2.1	−2.2
Ethylene glycol	$(CH_2OH)_2$	62.07	−17.1	−18.6	5.1	−8.1
Sulfur	S	32.064	−9.2	−9.3	0.3	−0.9
Sulfur monoxide	SO	48.063	−6.3	−5.8	−1.6	+8.5

Source: Data from R. C. Weast, editor, *CRC Handbook of Chemistry and Physics*, 66th ed., CRC Press, Boca Raton, Fla., 1985.

[a] Each constituent before and after combustion is assumed to be in its ideal-gas state at T_o and p_o.

We note that for all hydrocarbons and most other fuels, $\Delta g^o \approx \Delta h^o$. So we conclude that the availability of a fuel is also about equal to its lower heating value.

Example 31.11. For the combustion of methane under the conditions defined at the beginning of this section, show that the bracketed terms in equation (31.51) are indeed negligible with respect to Δg^o.

Solution. The combustion reaction mechanism is $CH_4 + 2O_2 = CO_2 + 2H_2O$ and the balance between inlet and outlet flow rates per unit fuel flow rate is $CH_4 + 2O_2 + 7.54N_2 = CO_2 + 2H_2O + 7.54N_2$. So, denoting the constituents CH_4, O_2, N_2, CO_2, and H_2O with the indexes $i = 1, 2, \ldots, 5$, respectively, the atmospheric air composition is $y'_{2a} = 0.2095$, $y'_{3a} = 0.7905$, and $y'_{1a} = y'_{4a} = y'_{5a} = 0$, and the composition of the outlet products of combustion is $y'_{1b} = y'_{2b} = 0$, $y'_{3b} = 7.54/10.54 = 0.7156$, $y'_{4b} = 1/10.54 = 0.0948$, and $y'_{5b} = 2/10.54 = 0.1896$. Using these data in equation (31.51) we find $RT_o[\sum_{i=3}^{5}(\dot{n}_{ib}/\dot{n}_{1a}) \ln y_{ib} - \sum_{i=2}^{3}(\dot{n}_{ia}/\dot{n}_{1a}) \ln y'_{ia}] = 8.314 \times 298.15 \times (7.54 \ln 0.7156 + \ln 0.0948 + 2 \ln 0.1896 - 2 \ln 0.2095 - 7.54 \ln 0.7905) = -6.92$ MJ/kmol. This value is indeed small compared to $\Delta g^o = -49.9 \times 16.04 = -800.4$ MJ/kmol (Table 31.7).

Tables 31.8 to 31.12 list values of adiabatic flame temperature, pressure, entropy generation by irreversibility, and composition of some of the product gases, for the

TABLE 31.8. Values of adiabatic flame temperature, pressure, entropy generation by irreversibility times $T_o = 298$ K, and composition of some of the product gases for the complete combustion of various hydrocarbons in a perfectly insulated constant-volume combustion chamber.[a]

Fuel	Formula	T_b K	p_b atm	$\dfrac{T_o S_{irr}}{(-n_{1a}\Delta g^o)}$ %	CO_2 $\dfrac{kmol}{MJ}$	CO $\dfrac{mol}{MJ}$	H_2 $\dfrac{mol}{MJ}$	NO $\dfrac{mol}{MJ}$	NO_2 $\dfrac{mmol}{MJ}$	N_2O $\dfrac{mmol}{MJ}$
Hydrogen	H_2	2870.8	8.33	17.2	0	0	382	108	13.8	5.2
Carbon	C	2677.9	7.58	22.7	2.05	489	0	105	19.3	5.6
Methane	CH_4	2655.4	9.02	24.2	1.00	247	98.8	83.5	12.5	4.2
Acetylene	C_2H_2	3009.7	10.00	19.2	0.99	642	68.3	174	30.6	8.7
Ethylene	C_2H_4	2811.0	9.63	22.4	1.10	418	83.9	120	19.8	6.1
Ethane	C_2H_6	2691.1	9.42	24.9	1.09	293	87.3	91.1	14.0	4.6
Propylene	C_3H_6	2770.7	9.69	23.7	1.16	387	77.1	111	18.0	5.6
Propane	C_8H_8	2698.7	9.56	25.3	1.14	307	81.4	92.9	14.5	4.7
n-Butane	C_4H_{10}	2702.7	9.64	25.6	1.16	315	78.1	93.9	14.7	4.8
n-Pentane	C_5H_{12}	2704.0	9.68	25.7	1.18	320	76.0	94.3	14.8	4.8
Benzene	C_6H_6	2767.6	9.61	23.7	1.43	452	44.6	115	19.6	5.9
n-Hexane	C_6H_{14}	2703.9	9.71	25.8	1.19	322	74.4	94.3	14.9	4.8
n-Heptane	C_7H_{16}	2707.6	9.75	25.9	1.19	326	73.7	95.1	15.0	4.9
n-Octane	C_8H_{18}	2706.9	9.76	26.0	1.20	327	72.7	95.0	15.0	4.9
Isooctane	C_8H_{18}	2702.6	9.74	26.2	1.21	323	72.0	94.0	14.8	4.8
n-Nonane	C_9H_{19}	2707.1	9.77	26.0	1.21	328	72.0	95.1	15.0	4.9
n-Decane	$C_{10}H_{22}$	2707.2	9.78	26.1	1.21	329	71.5	95.1	15.0	4.9

[a]Initially the mixture of each hydrocarbon with the stoichiometric amount of dry air ($\lambda = 1$ and $\omega = 0$) is at $T_a = 298$ K and $p_a = 1$ atm. Values are determined assuming chemical equilibrium for the product gases with respect to the reaction mechanisms $CO_2 = CO + (1/2)O_2$, $H_2O = H_2 + (1/2)O_2$, $N_2 + O_2 = 2NO$, $(1/2)N_2 + O_2 = NO_2$ and $N_2 + (1/2)O_2 = N_2O$. The loss of availibility, $T_o S_{irr}$, and the amount of each product, n_{ib}, are expressed per unit of availibility of the hydrocarbon, that is, as ratios $T_o S_{irr}/(-n_{1a}\Delta g^o)$ and $n_{ib}/(-n_{1a}\Delta g^o)$, where Δg^o is the Gibbs free energy of the hydrocarbon.

complete combustion of various hydrocarbons under various conditions in either a perfectly insulated constant-volume combustion chamber or a perfectly insulated steady-state burner. The values are determined assuming chemical equilibrium for the product gases with respect to the reaction mechanisms $CO_2 = CO + (1/2)O_2$, $H_2O = H_2 + (1/2)O_2$, $N_2 + O_2 = 2NO$, $(1/2)N_2 + O_2 = NO_2$ and $N_2 + (1/2)O_2 = N_2O$. For each hydrocarbon, the entropy (rate) generated by irreversibility times the environmental temperature $T_o = 298$ K, and the composition of products are expressed per unit amount of availability rate of the hydrocarbon.

31.6 Loss of Availability Due to Combustion

The process of burning a fuel without any shaft work and heat interactions is irreversible. As a result, the availability of the products of combustion is appreciably less than that of the fuel. To illustrate this assertion, we consider complete combustion of a fuel in a batch process and in a steady-state rate process.

For the batch process (Figure 31.3), the energy and entropy balances are

$$U_b - U_a = 0 \tag{31.52}$$

$$S_b - S_a = S_{irr} \tag{31.53}$$

TABLE 31.9.   Values of adiabatic flame temperature, entropy generation by irreversibility times $T_o = 298$ K, and composition of some of the product gases for the complete combustion of various hydrocarbons in a perfectly insulated steady state burner.[a]

Fuel	Formula	T_b K	$\dfrac{T_o \dot{S}_{irr}}{(-\dot{n}_{1a}\Delta g^o)}$ %	CO_2 $\dfrac{kmol}{MJ}$	CO $\dfrac{mol}{MJ}$	H_2 $\dfrac{mol}{MJ}$	NO $\dfrac{mol}{MJ}$	NO_2 $\dfrac{mmol}{MJ}$	N_2O $\dfrac{mmol}{MJ}$
Hydrogen	H_2	2448.5	20.9	0	0	231	46.3	7.4	2.3
Carbon	C	2326.0	26.0	2.25	280	0	44.6	9.5	2.5
Methane	CH_4	2266.0	28.3	1.12	124	57.5	31.3	5.6	1.7
Acetylene	C_2H_2	2598.0	22.6	1.17	457	51.4	87.3	19.8	4.6
Ethylene	C_2H_4	2416.6	26.2	1.27	254	56.6	52.9	10.8	2.8
Ethane	C_2H_6	2300.5	29.0	1.23	155	53.2	35.6	6.6	1.9
Propylene	C_3H_6	2378.5	27.6	1.33	227	50.8	47.0	9.4	2.5
Propane	C_8H_8	2307.9	29.4	1.28	165	50.3	36.6	6.9	2.0
n-Butane	C_4H_{10}	2311.8	29.6	1.31	170	48.6	37.1	7.0	2.0
n-Pentane	C_5H_{12}	2313.2	29.7	1.32	173	47.4	37.4	7.1	2.0
Benzene	C_6H_6	2382.6	27.5	1.61	270	30.1	49.6	10.4	2.7
n-Hexane	C_6H_{14}	2313.2	29.8	1.34	175	46.4	37.4	7.1	2.0
n-Heptane	C_7H_{16}	2316.7	29.9	1.34	178	46.2	37.8	7.2	2.0
n-Octane	C_8H_{18}	2316.1	30.0	1.35	178	45.6	37.8	7.2	2.0
Isooctane	C_8H_{18}	2312.1	30.2	1.35	175	45.0	37.2	7.1	2.0
n-Nonane	C_9H_{19}	2316.3	30.0	1.36	179	45.2	37.8	7.3	2.0
n-Decane	$C_{10}H_{22}$	2316.4	30.1	1.36	179	44.9	37.8	7.3	2.0

[a]In the inlet stream the mixture of each hydrocarbon with the stoichiometric amount of dry air ($\lambda = 1$ and $\omega = 0$) is at $T_a = 298$ K and $p_a = 1$ atm. Values are determined assuming chemical equilibrium for the gases in the outlet stream with respect to the reaction mechanisms $CO_2 = CO + (1/2)O_2$, $H_2O = H_2 + (1/2)O_2$, $N_2 + O_2 = 2NO$, $(1/2)N_2 + O_2 = NO_2$ and $N_2 + (1/2)O_2 = N_2O$. The rate of loss of availability, $T_o \dot{S}_{irr}$, and the rate of the amount of each product, $\dot{n}_{ib}$, are expressed per unit of availability rate of the hydrocarbon, that is, as ratios $T_o \dot{S}_{irr}/(-\dot{n}_{1a}\Delta g^o)$ and $\dot{n}_{ib}/(-\dot{n}_{1a}\Delta g^o)$, where Δg^o is the Gibbs free energy of the hydrocarbon.

The energy balance determines the adiabatic flame temperature of the products of combustion (Section 31.4), and use of this temperature in equation (31.53) yields the entropy S_{irr} generated by irreversibility.

With respect to an environmental reservoir at standard temperature T_0 and with fixed amounts of constituents and parameters, the availability of the initial fuel–air mixture is $U_a - U_o - T_o(S_a - S_o)$, and that of the final mixture of products of combustion $U_b - U_o - T_o(S_b - S_o)$, where U_o and S_o refer to the state of the products of combustion at the standard temperature T_o in the initial volume. The difference between these two availabilities is the loss of availability as the fuel–air mixture is transformed into products of combustion at the adiabatic flame temperature T_b. Accordingly, using equations (31.52) and (31.53), we find that

$$\text{Loss of availability} = U_a - U_b - T_o(S_a - S_b) = T_o S_{irr} = T_o(S_b - S_a) \quad (31.54)$$

The entropy generated by irreversibility per unit amount of fuel, $(S_b - S_a)/n_{1a}$ is obtained by using equation (29.63) for $Z = S$. In particular, for ideal-gas-mixture behavior at both the initial and the final states of the batch process, we find that

TABLE 31.10.   Values of adiabatic flame temperature, pressure, entropy generation by irreversibility times $T_0 = 298$ K, and composition of some of the product gases for the complete combustion of isooctane, C_8H_{18}, in a perfectly insulated constant-volume combustion chamber, as functions of the relative air–fuel ratio λ.[a]

λ	T_b K	p_b atm	$\dfrac{T_0 S_{irr}}{(-n_{1a}\Delta g^o)}$ %	CO_2 $\dfrac{kmol}{MJ}$	CO $\dfrac{mol}{MJ}$	H_2 $\dfrac{mol}{MJ}$	NO $\dfrac{mol}{MJ}$	NO_2 $\dfrac{mmol}{MJ}$	N_2O $\dfrac{mmol}{MJ}$
1.0	2912	44.8	20.8	1.17	357	73.8	126	17	6.3
1.1	2843	43.2	21.5	1.30	230	44.9	167	31	8.5
1.2	2758	41.5	22.1	1.39	143	27.6	196	48	10.1
1.3	2667	39.9	22.8	1.44	86.8	17.0	213	65	11.1
1.4	2577	38.4	23.4	1.48	52.1	10.5	220	82	11.6
1.5	2489	37.0	23.9	1.50	31.4	6.60	220	98	11.7
1.6	2408	35.7	24.5	1.51	19.0	4.19	216	113	11.6
1.7	2332	34.5	25.0	1.52	11.7	2.70	209	126	11.3
1.8	2262	33.4	25.4	1.52	7.28	1.78	200	138	11.0
1.9	2198	32.4	25.9	1.53	4.61	1.19	190	149	10.5
2.0	2138	31.4	26.3	1.53	2.96	0.78	180	159	10.0
2.5	1901	27.8	28.2	1.53	0.40	0.13	132	195	7.7
3.0	1732	25.2	29.8	1.53	0.08	0.02	96	216	5.8
3.5	1605	23.3	31.1	1.53	0.00	0.02	71	227	4.4
4.0	1506	21.8	32.2	1.53	0.00	0.00	53	234	3.4

[a] Initially the mixture of isooctane and dry air ($\omega = 0$) is at $T_a = 700$ K and $p_a = 10$ atm. Values are determined under the same assumptions as those used in Table 31.8. The loss of availability, $T_0 S_{irr}$, and the amount of each product, n_{ib}, are expressed per unit availibility of C_8H_{18}, that is, $T_0 S_{irr}/(-n_{1a}\Delta g^o)$ and $n_{1b}/(-n_{1a}\Delta g^o)$, where Δg^o is the Gibbs free energy of combustion of C_8H_{18}.

$$\frac{S_{irr}}{n_{1a}} = \frac{S_b - S_a}{n_{1a}}$$

$$= \sum_{i=1}^{r} \frac{n_{ib}}{n_{1a}} \left[s_{ii}^o(T_b) - s_{ii}^o \right] - \frac{n_b}{n_{1a}} R \ln \frac{p_b}{p_0} - R \sum_{i=1}^{r} \frac{n_{ib}}{n_{1a}} \ln y_{ib}$$

$$- \sum_{i=1}^{r} \frac{n_{ia}}{n_{1a}} \left[s_{ii}^o(T_a) - s_{ii}^o \right] + \frac{n_a}{n_{1a}} R \ln \frac{p_a}{p_0} + R \sum_{i=1}^{r} \frac{n_{ia}}{n_{1a}} \ln y_{ia} + (\Delta s^o)' \quad (31.55)$$

where $s_{ii}^o(T_b) = s_{ii}(T_b, p_0)$,

$$(\Delta s^o)' = \Delta s^o + \xi_2' (\Delta s^o)_2 + \xi_3' (\Delta s^o)_3 \quad (31.56)$$

Δs^o is the entropy of reaction of the combustion mechanism, $(\Delta s^o)_2$ and ξ_2' are the entropy of reaction and the reaction coordinate, respectively, of the dissociation reaction mechanism in relation (31.31), and $(\Delta s^o)_3$ and ξ_3' of the dissociation reaction mechanism in relation (31.32).

For the steady-state rate process (Figure 31.4), the energy and entropy rate balances are

$$\dot{H}_b - \dot{H}_a = 0 \quad (31.57)$$

$$\dot{S}_b - \dot{S}_a = \dot{S}_{irr} \quad (31.58)$$

With respect to an environmental reservoir at standard temperature T_0 and pressure p_0 with variable amounts of constituents and volume, the availability rate of the inlet fuel

TABLE 31.11.   **Values of adiabatic flame temperature, pressure, entropy generation by irreversibility times $T_0 = 298$ K, and composition of some of the product gases for the complete combustion of isooctane, C_8H_{18}, in a perfectly insulated constant-volume combustion chamber, as functions of the specific humidity ω and the relative air–fuel ratio λ.^{a}**

λ	ω	T_b K	p_b atm	$\dfrac{T_0 S_{\mathrm{irr}}}{(-n_{1a}\Delta g)}$ %	CO_2 $\dfrac{\mathrm{kmol}}{\mathrm{MJ}}$	CO $\dfrac{\mathrm{mol}}{\mathrm{MJ}}$	H_2 $\dfrac{\mathrm{mol}}{\mathrm{MJ}}$	NO $\dfrac{\mathrm{mol}}{\mathrm{MJ}}$	NO_2 $\dfrac{\mathrm{mmol}}{\mathrm{MJ}}$	N_2O $\dfrac{\mathrm{mmol}}{\mathrm{MJ}}$
1.0	0.00	2911.7	44.8	20.8	1.17	357	74	126	17	6.3
1.0	0.01	2885.7	44.3	20.7	1.19	338	77	120	16	6.0
1.0	0.02	2860.5	43.8	20.8	1.21	319	79	114	15	5.7
1.0	0.03	2835.8	43.4	20.8	1.23	301	82	108	14	5.4
1.0	0.04	2811.7	42.9	20.9	1.25	284	83	103	13	5.1
1.5	0.00	2489.3	37.0	23.9	1.50	31	6.6	220	98	11.7
1.5	0.01	2460.8	36.5	23.9	1.50	27	6.7	210	96	11.1
1.5	0.02	2433.1	36.1	23.9	1.51	24	6.8	200	94	10.5
1.5	0.03	2406.3	35.7	24.0	1.51	21	6.7	191	91	10.0
1.5	0.04	2380.2	35.3	24.1	1.51	18	6.6	182	89	9.5
2.0	0.00	2138.2	31.4	26.3	1.53	3.0	0.8	180	159	10.0
2.0	0.01	2113.3	31.1	26.3	1.53	2.5	0.8	170	155	9.4
2.0	0.02	2089.4	30.7	26.3	1.53	2.1	0.8	160	151	8.9
2.0	0.03	2066.3	30.3	26.3	1.53	1.8	0.8	151	147	8.3
2.0	0.04	2044.0	30.0	26.4	1.53	1.5	0.8	143	143	7.8

^{a}Initially the mixture of isooctane and air is at $T_a = 700$ K and $p_a = 10$ atm. Values are determined under the same assumptions as those used in Table 31.8. The loss of availability, $T_0 S_{\mathrm{irr}}$, and the amount of each product, n_{ib}, are expressed per unit of availability of isooctane, that is, as ratios $T_0 S_{\mathrm{irr}}/(-n_{1a}\Delta g^\circ)$, and $n_{1b}/(-n_{1a}\Delta g^\circ)$, where Δg° is the Gibbs free energy combustion of C_8H_{18}.

and air streams is $\dot{H}_a - \dot{H}_0 - T_0(\dot{S}_a - \dot{S}_0)$, and that of the outlet mixture of products of combustion $\dot{H}_b - \dot{H}_0 - T_0(\dot{S}_b - \dot{S}_0)$, where $\dot{H}_0$ and $\dot{S}_0$ refer to the bulk-flow state of the products of combustion at the standard temperature T_0 and pressure p_0. The difference between these two availability rates is the rate of loss of availability as the fuel and air streams are transformed into products of combustion at the adiabatic flame temperature T_b. Accordingly, using equations (31.57) and (31.58), we find that

$$\text{Rate of loss of availability} = \dot{H}_a - \dot{H}_b - T_0(\dot{S}_a - \dot{S}_b)$$

$$= T_0 \dot{S}_{\mathrm{irr}} = T_0(\dot{S}_b - \dot{S}_a) \qquad (31.59)$$

The rate of entropy generation by irreversibility per unit fuel flow rate, $(\dot{S}_b - \dot{S}_a)/\dot{n}_{1a}$, is obtained by using equation (29.63) for $\dot{Z} = \dot{S}$. In particular, for ideal-gas-mixture behavior of both the inlet and the outlet streams of the rate process, we find that

$$\frac{\dot{S}_{\mathrm{irr}}}{\dot{n}_{1a}} = \frac{\dot{S}_b - \dot{S}_a}{\dot{n}_{1a}}$$

$$= \sum_{i=1}^{r} \frac{\dot{n}_{ib}}{\dot{n}_{1a}} \left[s_{ii}^\circ(T_b) - s_{ii}^\circ \right] - \frac{\dot{n}_b}{\dot{n}_{1a}} R \ln \frac{p_b}{p_0} - R \sum_{i=1}^{r} \frac{\dot{n}_{ib}}{\dot{n}_{1a}} \ln y_{ib}$$

$$- \sum_{i=1}^{r} \frac{\dot{n}_{ia}}{\dot{n}_{1a}} \left[s_{ii}^\circ(T_a) - s_{ii}^\circ \right] + \frac{\dot{n}_a}{\dot{n}_{1a}} R \ln \frac{p_a}{p_0} + R \sum_{i=2}^{r} \frac{\dot{n}_{ia}}{\dot{n}_{1a}} \ln y'_{ia} + (\Delta s^\circ)' \qquad (31.60)$$

TABLE 31.12.   Values of adiabatic flame temperature, entropy generation by irreversibility times $T_o = 298$ K, and composition of some of the product gases for the complete combustion of methane, CH_4, in a perfectly insulated steady-state burner, as functions of the relative air–fuel ratio λ.^{a}

λ	T_b K	$\dfrac{T_o \dot{S}_{irr}}{(-\dot{n}_{1a}\Delta g^o)}$ %	CO_2 $\dfrac{kmol}{MJ}$	CO $\dfrac{mol}{MJ}$	H_2 $\dfrac{mol}{MJ}$	NO $\dfrac{mol}{MJ}$	NO_2 $\dfrac{mmol}{MJ}$	N_2O $\dfrac{mmol}{MJ}$
1.0	2266	28.3	1.04	115	53	29	5	1.5
1.1	2184	29.7	1.12	40	19	47	17	2.5
1.2	2081	31.0	1.14	14	7.2	52	29	2.9
1.3	1981	32.3	1.15	5.3	2.9	51	39	2.9
1.4	1889	33.5	1.15	2.1	1.3	47	48	2.7
1.5	1806	34.7	1.15	0.9	0.6	42	56	2.5
1.6	1732	35.8	1.15	0.4	0.3	37	61	2.2
1.7	1664	36.8	1.15	0.2	0.1	32	66	1.9
1.8	1603	37.8	1.15	0.1	0.1	27	69	1.7
1.9	1547	38.8	1.15	0.0	0.0	23	71	1.5
2.0	1495	39.7	1.15	0.0	0.0	20	73	1.3
2.5	1291	43.9	1.15	0.0	0.0	8.7	73	0.6
3.0	1147	47.5	1.15	0.0	0.0	3.8	67	0.3
3.5	1039	50.7	1.15	0.0	0.0	1.7	59	0.1
4.0	956	53.4	1.15	0.0	0.0	0.8	52	0.1

^{a}In the inlet stream the mixture of methane and dry air ($\omega = 0$) is at $T_a = 298$ K and $p_a = 1$ atm. Values are determined under the same assumptions as those used in Table 31.9. The rate of loss of availability, $T_o \dot{S}_{irr}$, and the rate of the amount of each product, $\dot{n}_{ib}$, are expressed per unit of availability rate of CH_4, that is, as ratios $T_o \dot{S}_{irr}/(-\dot{n}_{1a}\Delta g^o)$ and $\dot{n}_{ib}/(-\dot{n}_{1a}\Delta g^o)$, where Δg^o is the Gibbs free energy of combustion of CH_4.

where, as earlier, y'_{ia} for $i \neq 1$ is the composition of inlet air stream, and $(\Delta s^o)'$ is given by equation (31.56).

For most hydrocarbons, the loss of availability due to the irreversibility of combustion processes that yield the adiabatic flame temperature is typically around 25% of the fuel availability. Because the adiabatic flame temperature represents an upper limit to the temperature that can be reached by the products of combustion under given conditions, and because the availability of the products of combustion is an increasing function of temperature, we conclude that for the given inlet or initial conditions the highest availability that can be converted from the fuel to the products of combustion is that corresponding to the adiabatic flame temperature.

Example 31.12.   Evaluate the loss of availability for the batch process considered in Examples 31.6 and 31.9.

Solution.   From the tables in Examples 31.6 and 31.9, we find the initial and the final amounts summarized in Table 31.13. Moreover, $T_a = 700$ K, $p_a = 10$ atm, and, from Example 31.8, $\xi'_2 = 0.194$, $\xi'_3 = 0.046$, $T_b = 2531$ K, and $p_b = 37.6$ atm. The differences $s^o_{ii}(T) - s^o_{ii}$ are evaluated using Table 20.2, except for C_8H_{18}, for which we use Table 20.1. We find $(n_a/n_{1a})R\ln(p_a/p_o) = 91.93 \times 8.314\ln(10) = 1759$ kJ/kmol K, $(n_b/n_{1a})R\ln(p_b/p_o) = 95.55 \times 8.314\ln(37.6) = 2881$ kJ/kmol K, $R\sum_{i=1}^{r}(n_{ia}/n_{1a})\ln y_{ia} = -489$ kJ/kmol K, $R\sum_{i=1}^{r}(n_{ib}/n_{1a})\ln y_{ib} = -698$ kJ/kmol K, and using Table 29.4, $\Delta s^o = -(-754.6) + 9(-44.4) + 8(2.9) = 378.2$ kJ/kmol K, $(\Delta s^o)_2 = -(-2.9) + (89.7) = 86.8$ kJ/kmol K, and $(\Delta s^o)_3 = -(-44.4) = 44.4$ kJ/kmol K so that $(\Delta s^o)' = 378.2 + 0.405 \times$

TABLE 31.13.

i	A_i	$\dfrac{n_{ia}}{n_{1a}}$	y_{ia}	$s_{ii}(T_a) - s_{ii}^\circ$ kJ/kmol K	$\dfrac{n_{ib}}{n_{1a}}$	y_{ib}	$s_{ii}(T_b) - s_{ii}^\circ$ kJ/kmol K
1	C_8H_{18}	1	0.0109	229.8	0	0	
2	O_2	18.75	0.2039	26.83	6.37	0.0667	73.60
3	N_2	70.74	0.7695	24.94	70.74	0.7404	67.63
4	H_2O	1.44	0.0157	29.15	10.39	0.1088	86.31
5	CO_2	0	0		7.81	0.0817	110.0
6	CO	0	0		0.194	0.0020	69.37
7	H_2	0	0		0.046	0.0005	65.67

$$\frac{n_a}{n_{1a}} = 91.93 \qquad \sum_{i=1}^{r} \frac{n_{ia}}{n_{1a}}\left[s_{ii}^\circ(T_a) - s_{ii}^\circ\right] = 2539 \qquad \frac{n_b}{n_{1a}} = 95.55 \qquad \sum_{i=1}^{r} \frac{n_{ib}}{n_{1a}}\left[s_{ii}^\circ(T_b) - s_{ii}^\circ\right] = 7025$$

$86.8 + 0.103 \times 44.4 = 418$ kJ/kmol K. Thus, equation (31.55) becomes

$$\frac{S_{\text{irr}}}{n_{1a}} = 7025 - 2881 + 698 - 2539 + 1759 - 489 + 418 = 3991 \text{ kJ/kmol K}$$

It is noteworthy that the availability loss due to irreversibility in this constant volume combustion process, $T_o S_{\text{irr}}/n_{1a} = 1189$ MJ/kmol, is about 23% of the fuel availability $-\Delta g^\circ = 45.8 \times 114.232 = 5231$ MJ/kmol (from Table 31.7).

Example 31.13. Evaluate the rate of availability loss in the steady-state rate process considered in Examples 31.8 and 31.9 with $p_b = 10$ atm.

Solution. From the tables in Examples 31.8 and 31.9, we find the inlet and outlet flow rates summarized in Table 31.14. Moreover, $T_a = 700$ K, $p_a = 10$ atm, and from Example 31.9, $\xi_2' = 0.038$, $\xi_3' = 0.012$, and $T_b = 2157$ K. The differences $s_{ii}^\circ(T) - s_{ii}^\circ$ are evaluated using Table 20.2, except for C_8H_{18}, for which we use Table 20.1.

We find $(\dot{n}_a/\dot{n}_{1a})R\ln(p_a/p_o) = 91.93 \times 8.314\ln(10) = 1759$ kJ/kmol K, $(\dot{n}_b/\dot{n}_{1a})R\ln(p_b/p_o) = 95.46 \times 8.314\ln(10) = 1827$ kJ/kmol K, $R\sum_{i=1}^{r}(\dot{n}_{ia}/\dot{n}_{1a})\ln y_{ia} = -489$ kJ/kmol K, $R\sum_{i=1}^{r}(\dot{n}_{ib}/\dot{n}_{1a})\ln y_{ib} = -678$ kJ/kmol K, and using Table 29.4, $\Delta s^\circ = -(-754.6) +$

TABLE 31.14.

i	A_i	$\dfrac{\dot{n}_{ia}}{\dot{n}_{1a}}$	y_{ia}	$s_{ii}(T_a) - s_{ii}^\circ$ kJ/kmol K	$\dfrac{\dot{n}_{ib}}{\dot{n}_{1a}}$	y_{ib}	$s_{ii}(T_b) - s_{ii}^\circ$ kJ/kmol K
1	C_8H_{18}	1	0.0109	229.8	0	0	
2	O_2	18.75	0.2039	26.83	6.27	0.0657	67.34
3	N_2	70.74	0.7695	24.94	70.74	0.7411	61.92
4	H_2O	1.44	0.0157	29.15	10.43	0.1092	78.00
5	CO_2	0	0		7.96	0.0834	100.1
6	CO	0	0		0.036	0.0004	63.52
7	H_2	0	0		0.011	0.0001	60.09

$$\frac{\dot{n}_a}{\dot{n}_{1a}} = 91.93 \qquad \sum_{i=1}^{r} \frac{\dot{n}_{ia}}{\dot{n}_{1a}}\left[s_{ii}^\circ(T_a) - s_{ii}^\circ\right] = 2539 \qquad \frac{\dot{n}_b}{\dot{n}_{1a}} = 95.46 \qquad \sum_{i=1}^{r} \frac{\dot{n}_{ib}}{\dot{n}_{1a}}\left[s_{ii}^\circ(T_b) - s_{ii}^\circ\right] = 6416$$

$9(-44.4) + 8(2.9) = 378.2$ kJ/kmol K, $(\Delta s^\circ)_2 = -(2.9) + (89.7) = 86.8$ kJ/kmol K, $(\Delta s^\circ)_3 = -(-44.4) = 44.4$ kJ/kmol K so that $(\Delta s^\circ)' = 378.2 + 0.036 \times 86.8 + 0.011 \times 44.4 = 382$ kJ/kmol K. Thus equation (31.60) becomes

$$\frac{\dot{S}_{irr}}{\dot{n}_{1a}} = 6416 - 1827 + 678 - 2539 + 1759 - 489 + 382 = 4380 \text{ kJ/kmol K}$$

Here the rate of availability loss due to irreversibility, $T_0 \dot{S}_{irr}/\dot{n}_{1a} = 1305$ MJ/kmol, is about 25% of the fuel availability $-\Delta g^\circ = 45.8 \times 114.232 = 5231$ MJ/kmol (from Table 31.7).

Problems

31.1 Determine the stoichiometric air–fuel ratio on a mass basis, and the lower heating value per unit mass of fuel for benzene, C_6H_6, hydrogen, H_2, isooctane, C_8H_{18}, and methane, CH_4, and compare the results with the values listed in Table 31.1.

31.2 A typical liquid petroleum gas (LPG) consists of

Fuel	Formula	Percent by volume	HHV
Propane	C_3H_8	70	50.38 MJ/kg
Butane	C_4H_{10}	5	49.56 MJ/kg
Propylene	C_3H_6	25	48.95 MJ/kg

(a) Write the overall combustion reaction for stoichiometric combustion of 1 mol of LPG with air.

(b) Find the stoichiometric fuel–air ratio.

(c) Find the higher heating value per unit mass of LPG.

31.3 Carbon monoxide and oxygen are supplied at 1 atm and 25°C to a steady-flow combustion chamber. The combustion gases exit from the chamber at 2600 K and atmospheric pressure.

(a) Assuming that the combustion gases contain only CO_2, CO, and O_2 in chemical equilibrium, find the compositions of the combustion gases when the combustion chamber is supplied with: (1) 0.5 mol of O_2 per mole of CO; and (2) 1.5 mol of O_2 per mole of CO.

(b) Find the energy transfer from the combustion chamber per mole of CO supplied for each of the two cases of part (a).

(c) What is the energy transfer if the CO is completely burned?

31.4 Gaseous propane, C_3H_8, is burned in 90% deficient air in a steady-state bulk flow process at 1 atm. The amounts of constituents before and after burning satisfy the relation $C_3H_8 + 0.9(5 O_2 + 5 \times 3.77 N_2) = \alpha CO_2 + (3 - \alpha)CO + \beta H_2O + (4 - \beta)H_2 + 4.5 \times 3.77 N_2$. Both fuel and air are supplied at 298 K. The products leave the burner at 1500 K and are in chemical equilibrium according to the chemical reaction mechanism $CO_2 + H_2 = CO + H_2O$ for which the equilibrium constant $K(1500 \text{ K}) = 2.56$.

(a) Express the partial pressure of each of the products that exit from the burner in terms of α and β.

(b) Determine the values of α and β.

(c) Find the mole fractions of the products that exit the burner.

(d) Find the energy exchanged between the burner and the environment per mole of C_3H_8.

(e) If the composition of the products is controlled by the CO_2 and H_2 reaction at all pressures, what would be the effect on this composition if the pressure in the burner were 10 atm instead of 1 atm?

31.5 The composition on a mole basis, at 1 atm pressure, of the exhaust gas from an internal combustion engine is found to be 8.0% of H_2O, 9.4% of CO_2, 6.7% of O_2, and 75.9 of N_2.

(a) If the fuel is a hydrocarbon, C_kH_l, what is the ratio l/k?

(b) What is the air–fuel ratio at which the engine operates?

31.6 The following sequence of processes approximates the operation of a spark-ignition, internal-combustion engine burning hydrogen. A mixture of hydrogen, H_2, and 200% excess theoretical air fills the "bottom dead center" volume, $V_{BDC} = 200$ cm^3, at standard atmospheric conditions $T_o = 25°C$ and $p_o = 1$ atm. The piston then compresses the mixture to the "top dead center" volume, $V_{TDC} = 35$ cm^3. With the piston held fixed at top dead center, a spark is lighted which triggers the oxidation of hydrogen according to the reaction mechanism $2H_2 + O_2 = 2H_2O$. At the end of the constant-volume combustion process all the hydrogen is oxidized and, as a result, the pressure has increased. Now the piston recedes, allowing the combustion products to expand to the bottom dead center volume. Assume that the gases in the combustion chamber experience no interactions with other systems except for that with the piston.

(a) Find the total number of moles n_a and the mole composition y_{ia} of the mixture initially filling the combustion chamber.

(b) Assume isentropic compression from bottom dead center to top dead center, and find the pressure p_a and the temperature T_a of the mixture just before ignition. Evaluate the compression work done by the piston.

(c) Find the mole composition y_{ib} and total number of moles n_b at the end of the constant-volume combustion process.

(d) Use the energy balance equation for the combustion process between states a and b (just before and after the constant-volume combustion), and assume that the behavior of each mixture is ideal. Thus find an equation in which T_b is the only unknown, and solve for the value of T_b.

(e) Find the relation between p_b and T_b implied by the constant-volume constraint. Find the value of p_b.

(f) Find the amount of entropy generated by irreversibility during the constant-volume combustion process.

(g) Assume isentropic expansion from bottom dead center to top dead center, and find the pressure p_e and the temperature T_e at the end of the expansion process. Evaluate the expansion work done on the piston and the net work done on the piston, that is, expansion minus compression work.

(h) Evaluate the ratio of the net piston work per unit amount of hydrogen to the negative of the energy of combustion Δu^o. Within the particular application under study here, the quantity Δu^o is often called the "equivalent energy addition." Look back at your answer in part (d) to justify this jargon.

(i) The gas mixture at the end of the expansion process is exhausted to the atmosphere. Find the amount of entropy generated by irreversibility during the exhaust process.

(j) Define and evaluate the effectiveness of this hydrogen engine. Compare your result to your answer in part (h) and to your answers in parts (f) and (i). What do these comparisons tell you?

31.7 A steady-state reactor is fed with gaseous isooctane, C_8H_{18}, and theoretical air, all at standard atmospheric conditions. The only reaction mechanism is $C_8H_{18} + 12.5O_2 = 8CO_2 + 9H_2O$. The only products in the outlet stream are carbon dioxide, water, and nitrogen, and there are neither work nor heat interactions with other systems. The outlet pressure is atmospheric. The specific heats are functions of temperature only.

(a) Write an equation defining the outlet temperature T of the products of reaction. Solve the equation numerically for T.

(b) Evaluate the amount of entropy generated by irreversibility per unit of isooctane inflow rate.

(c) What is the effectiveness of the reactor?

31.8 A steam boiler burns anthracite coal (modeled as 100% carbon), at a rate of 4000 kg/h with 15% excess theoretical air. The air and the coal enter the boiler at 25°C, and the stack gases leave at 170°C. Assume that combustion is complete.

(a) What is the mole fraction of CO_2 in the stack gases?

(b) If there are no external energy losses through the walls of the boiler, what is the rate of energy transfer to the water in the boiler?

31.9 We wish to investigate the effect of excess air on the availability with respect to an environment at $T_{R*} = 300$ K of the products of constant-volume combustion of a fuel. Neglect dissociation reactions.

(a) Find the available energy of the products of combustion resulting from the complete burning of 1 kg of n-octane, C_8H_{18}, in theoretical air at a pressure of 1 atm, and in a fixed-volume and perfectly insulated combustor.

(b) How does your answer in part (a) compare with the enthalpy of reaction of C_8H_{18} at 1 atm, and 25°C?

(c) Repeat part (a) for 200% theoretical air.

(d) Repeat part (a) for oxidation with the theoretical amount of pure oxygen at a pressure of 1 atm.

(e) Repeat parts (a) to (d), including the dissociation reaction $CO_2 = CO + \frac{1}{2}O_2$.

31.10 This problem illustrates the loss of availability of a fuel as a result of combustion and energy transfer to a working fluid.

Ethylene, C_2H_4, is burned in a well-insulated steady-state burner, where it enters at standard atmospheric temperature and pressure. Combustion occurs in stoichiometric air and is complete. Upon exiting the burner, the products of combustion are cooled in a heat exchanger by transferring energy to a constant-temperature, T_f, working fluid—saturated liquid being transformed into saturated vapor at constant pressure. At the exit of the heat exchanger the temperature of the products of combustion is $T_f + 20$ K. Per unit fuel rate (kg/s), find the following.

(a) The availability rate of the fuel at standard conditions.

(b) The adiabatic flame temperature and the availability rate of the products of combustion at the adiabatic flame temperature.

(c) The availability rate of the products of combustion exiting the heat exchanger and the rate of availability gained by the working fluid as functions of T_f.

(d) Graphs of the availability rates of the products of combustion and the working fluid as functions of T_f.

Discuss methods to reduce the losses of availability in the various phases of this typical fuel utilization.

31.11 A mixture of hydrogen, H_2, and 100% excess theoretical air enters a steady-flow well-insulated reactor at 25°C and 1 atm. In the reactor H_2 and O_2 combine according to the reaction mechanism $H_2 + 0.5\,O_2 = H_2O$. The products exit the reactor at temperature T and pressure also equal to 1 atm.

(a) If the hydrogen is fully oxidized within the reactor, calculate the temperature of the products leaving the reactor.

(b) Show that the assumption that hydrogen is fully oxidized is reasonable.

Epilogue

We have touched upon only a few of the entries in the ever-increasing list of applications of thermodynamics. To our great chagrin, we must conform to page limitations, and follow Aristotle's admonition, given some two and a half millenia ago,

ΑΝΑΓΚΗ ΣΤΗΝΑΙ

A   Physical Constants and Unit Conversion Tables

TABLE A.1.   Prefix names and symbols for units of measure.

Prefix	Symbol	Value	Prefix	Symbol	Value
deci	d	10^{-1}	deca	da	10^{1}
centi	c	10^{-2}	hecto	h	10^{2}
milli	m	10^{-3}	kilo	k	10^{3}
micro	μ	10^{-6}	mega	M	10^{6}
nano	n	10^{-9}	giga	G	10^{9}
pico	p	10^{-12}	tera	T	10^{12}
femto	f	10^{-15}	peta	P	10^{15}
atto	a	10^{-18}	exa	E	10^{18}

TABLE A.2.   Physical constants in SI units.

Avogadro's number	$N_A = 6.022 \times 10^{23}$ molecules/mole
Boltzmann's constant	$k = 1.38066 \times 10^{-23}$ J/K molecule
Electron charge	$e = 1.6022 \times 10^{-19}$ coulomb
Planck's constant	$h = 6.6260 \times 10^{-34}$ J s
Speed of light in vacuum	$c = 2.9979 \times 10^{8}$ m/s
Standard gravitational acceleration	$g = 9.8066$ m/s^2
Stefan–Boltzmann constant	$\sigma = 5.6705 \times 10^{-8}$ W/m^2 K^4
Universal gas constant	$R = 8.3145$ J/mol K

Source: E. R. Cohen, and B. N. Taylor, *Physics Today*, August, p. 9 (1988).

TABLE A.3.   Units for length, and conversion factors to SI units.

Unit	Symbol	Conversion to SI units
meter	m	$1\ m = 1\ m$
kilometer	km	$1\ km = 1\ km$
centimeter	cm	$1\ cm = 1\ cm$
millimeter	mm	$1\ mm = 1\ mm$
micron	μm	$1\ \mu m = 1\ \mu m$
angstrom	Å	$1\ Å = 0.1\ nm$
mile	mi	$1\ mi = 1.609\ km$
yard (U.S.)	yard	$1\ yard = 91.44\ cm$
foot	ft	$1\ ft = 30.48\ cm$
inch (U.S.)	in	$1\ in = 25.4\ mm$
mil		$1\ mil = 25.4\ \mu m$
light year	light-year	$1\ light\text{-}year = 9.46\ Pm$

TABLE A.4.   Units for volume, and conversion factors to SI units.

Unit	Symbol	Conversion to SI units
cubic meter	m^3	$1\ m^3 = 1.00000\ m^3$
cubic centimeter	cc	$1\ cc = 1.00000\ cm^3$
liter	l	$1\ l = 1000.0\ cm^3$
fluid ounce (U.S.)	fl-oz	$1\ fl\text{-}oz = 29.57\ cm^3$
gallon (U.S.)	gal	$1\ gal = 3785\ cm^3$
liquid quart (U.S.)	liq-quart	$1\ liq\text{-}quart = 946.4\ cm^3$
dry quart (U.S.)	dry-quart	$1\ dry\text{-}quart = 1101\ cm^3$

TABLE A.5.   Units for mass, and conversion factors to SI units.

Unit	Symbol	Conversion to SI units
kilogram	kg	$1\ kg = 1.00000\ kg$
gram	g	$1\ g = 1.00000\ g$
metric ton	ton	$1\ ton = 1000.0\ kg$
pound	lb	$1\ lb = 0.45359\ kg$

TABLE A.6.   Units for density, and conversion factors to SI units.

Unit	Symbol	Conversion to SI units
kilogram per cubic meter	kg/m^3	$1\ kg/m^3 = 1.00000\ kg/m^3$
gram per cubic centimeter	g/cm^3	$1\ g/cm^3 = 1000.0\ kg/m^3$
kilogram per liter	kg/l	$1\ kg/l = 1000.0\ kg/m^3$
pound per gallon	lb/gal	$1\ lb/gal = 119.826\ kg/m^3$
pound per cubic foot	lb/ft^3	$1\ lb/ft^3 = 16.0185\ kg/m^3$

TABLE A.7.   Units for force, and conversion factors to SI units.

Unit	Symbol	Conversion to SI units
newton	N	$1\ N = 1.00000\ N$
kilogram force	kg_f	$1\ kg_f = 9.80665\ N$
gram force	g_f	$1\ g_f = 9.80665\ mN$
dyne	dyn	$1\ dyn = 10.0000\ \mu N$
pound force	lb_f	$1\ lb_f = 4.44822\ N$
poundal	pdl	$1\ pdl = 0.13825\ N$

TABLE A.8.   Units for pressure, and conversion factors to SI units.

Unit	Symbol	Conversion to SI units
pascal	Pa	$1\ Pa = 1.00000\ Pa$
newton per square meter	N/m^2	$1\ N/m^2 = 1.00000\ Pa$
atmosphere	atm	$1\ atm = 101.325\ kPa$
bar	bar	$1\ bar = 100.000\ kPa$
torricelli	torr	$1\ torr = 133.322\ Pa$
millimeter of mercury (Hg) at $0°C$	mm Hg	$1\ mm\ Hg = 133.322\ Pa$
centimeter of water (H_2O) at $4°C$	$cm\ H_2O$	$1\ cm\ H_2O = 98.0665\ Pa$
pound-force per square inch	psi	$1\ psi = 6.89476\ kPa$
kilogram-force per square centimeter	kg_f/cm^2	$1\ kg_f/cm^2 = 98.0665\ kPa$
dyne per square centimeter	dyn/cm^2	$1\ dyn/cm^2 = 100.000\ mPa$

TABLE A.9.   Units for energy, and conversion factors to SI units.

Unit	Symbol	Conversion to SI units
joule	J	$1\ J = 1.00000\ J$
thermochemical calorie[a]	cal_{th}	$1\ cal_{th} = 4.18400\ J$
calorie (IT)[a]	cal	$1\ cal = 4.18680\ J$
kilocalorie	kcal	$1\ kcal = 4.18680\ kJ$
British thermal unit (IT)[a]	Btu	$1\ Btu = 1.05506\ kJ$
erg	erg	$1\ erg = 100.000\ nJ$
kilowatt-hour	kWh	$1\ kWh = 3.60000\ MJ$
horsepower-hour	hph	$1\ hp\ h = 2.68452\ MJ$
megawatt-day	MW day	$1\ MW\ day = 86.4000\ GJ$
megawatt-year	MW yr	$1\ MW\ yr = 31.5360\ TJ$
newton-meter	N m	$1\ N\ m = 1.00000\ J$
kilogram-force-meter	$kg_f\ m$	$1\ kg_f\ m = 9.80665\ J$
gram-force-centimeter	$g_f\ cm$	$1\ g_f\ cm = 98.0665\ mJ$
dyne-centimeter	dyn cm	$1\ dyn\ cm = 100.000\ nJ$
foot-pound–force	$ft\ lb_f$	$1\ ft\ lb_f = 1.35582\ J$
foot-poundal	ft poundal	$1\ ft\ poundal = 42.1401\ mJ$
liter-atmosphere	l atm	$1\ l\ atm = 101.325\ J$
cubic centimeter-atmosphere	$cm^3\ atm$	$1\ cm^3\ atm = 101.325\ mJ$
cubic foot-atmosphere	$ft^3\ atm$	$1\ ft^3\ atm = 2.86920\ kJ$
electron-volt	eV	$1\ eV = 0.16021\ aJ$

[a] See Table 3.1, footnote b.

TABLE A.10.   Units for power, and conversion factors to SI units.

Unit	Symbol	Conversion to SI units
watt	W	1 W $= 1.00000$ W
erg per second	erg/s	1 erg/s $= 100.000$ nW
thermochemical calorie per hour[a]	cal$_\text{th}$/h	1 cal$_\text{th}$/h $= 1.16222$ mW
calorie (IT) per hour[a]	cal/h	1 cal/h $= 1.16300$ mW
kilocalorie per hour	kcal/h	1 kcal/h $= 1.16300$ W
British thermal unit (IT) per hour[a]	Btu/h	1 Btu/h $= 0.29307$ W
horsepower	hp	1 hp $= 745.700$ W
kilogram-force–meter per second	kg$_\text{f}$ m/s	1 kg$_\text{f}$ m/s $= 9.80665$ W
gram-force-centimeter per second	g$_\text{f}$ cm	1 g$_\text{f}$ cm $= 98.0665$ mW
foot-pound–force per second	ft lb$_\text{f}$/s	1 ft lb$_\text{f}$/s $= 1.35582$ W
foot-poundal per second	ft pdl/s	1 ft pdl/s $= 42.1401$ mW

[a] See Table 3.1, footnote *b*.

B   Tables of Properties

Table B.1.   Triple-point temperatures and pressures of various substances.

Substance	Formula	T_{tp} K	p_{tp} kPa
Acetylene	C_2H_2	192.4	120
Ammonia	NH_3	195.40	6.076
Argon	A	83.81	68.9
Carbon (graphite)	C	3900	10,100
Carbon dioxide	CO_2	216.55	517
Carbon monoxide	CO	68.10	15.37
Deuterium	D_2	18.63	17.1
Ethane	C_2H_6	89.89	8×10^{-4}
Ethylene	C_2H_4	104.0	0.12
Helium 4 (lambda point)	He	2.19	5.1
Hydrogen	H_2	13.84	7.04
Hydrogen chloride	HCl	158.96	13.9
Mercury	Hg	234.2	1.65×10^{-7}
Methane	CH_4	90.68	11.7
Neon	Ne	24.57	43.2
Nitric oxide	NO	109.50	21.92
Nitrogen	N_2	63.18	12.6
Nitrous oxide	N_2O	182.34	87.85
Oxygen	O_2	54.36	0.152
Palladium	Pd	1825	3.5×10^{-3}
Platinum	Pt	2045	2.0×10^{-4}
Sulfur dioxide	SO_2	197.69	1.67
Titatium	Ti	1941	5.3×10^{-3}
Uranium hexafluoride	UF_6	337.17	151.7
Water	H_2O	273.16	0.61
Xenon	Xe	161.3	81.5
Zinc	Zn	692.65	0.065

Source: Data from *Natl. Bur. Stand. (U.S.) Circ., 500* (1952).

TABLE B.2.   Molecular weights and critical-state properties for various substances.

Substance	Formula	M kg/kmol	T_c K	p_c MPa	v_c m^3/kmol
Acetic acid	$C_2H_4O_2$	60.052	594.4	5.79	0.171
Acetone	CH_3COCH_3	58.08	508.1	4.70	0.209
Acetylene	C_2H_2	26.038	308.3	6.14	0.113
Ammonia	NH_3	17.031	405.6	11.28	0.0725
Argon	Ar	39.948	150.8	4.87	0.0749
Benzene	C_6H_6	78.114	562.1	4.89	0.259
Carbon dioxide	CO_2	44.01	304.2	7.38	0.094
Carbon monoxide	CO	28.01	132.9	3.50	0.0931
Chlorine	Cl_2	70.906	417.0	7.70	0.124
Chloroform	$CHCl_3$	119.378	536.4	5.47	0.239
Ethane	C_2H_6	30.07	305.4	4.88	0.148
Ethanol	C_2H_5OH	46.069	516.2	6.38	0.167
Ethylene	C_2H_4	28.054	282.4	5.04	0.129
Fluorine	F_2	37.997	144.3	5.22	0.0662
Freon 12	CCl_2F_2	120.914	385.0	4.12	0.217
Freon 13	$CClF_3$	104.459	302.0	3.92	0.18
Freon 21	$CHCl_2F$	102.923	451.6	5.17	0.197
Freon 22	$CHClF_2$	86.469	369.2	4.98	0.165
Freon 114	$C_2Cl_2F_4$	170.922	418.9	3.26	0.293
Genetron 100	CH_3CHF_2	66.051	386.6	4.50	0.181
Genetron 101	CH_3CClF_2	100.496	410.2	4.12	0.231
Hydrogen	H_2	2.016	33.2	1.30	0.065
Hydrogen chloride	HCl	36.461	324.6	8.31	0.081
Isooctane	C_8H_{18}	114.232	543.9	2.56	0.468
Methane	CH_4	16.043	190.6	4.60	0.099
Methanol	CH_3OH	32.042	512.6	8.10	0.118
Methylene chloride	CH_2Cl_2	84.933	510.0	6.08	0.193
Naphthalene	$C_{10}H_8$	128.174	748.4	4.05	0.41
Neon	Ne	20.183	44.4	2.76	0.0417
Nitric oxide	NO	30.006	180.0	6.48	0.058
Nitrogen	N_2	28.013	126.2	3.39	0.0895
Nitrogen dioxide	NO_2	46.006	431.4	10.13	0.17
Nitrous oxide	N_2O	44.013	309.6	7.24	0.0974
n-Octane	C_8H_{18}	114.232	568.8	2.48	0.492
Oxygen	O_2	31.999	154.6	5.05	0.0734
Ozone	O_3	47.998	261.0	5.57	0.0889
Propane	C_3H_8	44.097	369.8	4.25	0.203
Propylene	CH_2CHCH_3	42.081	365.0	4.62	0.181
Water	H_2O	18.015	647.3	22.05	0.056

Source: Data from R. C. Reid, J. M. Prausnitz, and T. K. Sherwood, *The Properties of Gases and Liquids*, McGraw-Hill, New York, 1977.

TABLE B.3. Values of saturated-solid temperature, $T_{j.\mathrm{sat}}$, specific enthalpy of fusion, h_{jf}, saturated-vapor temperature, T_{sat}, and specific enthalpy of vaporization, h_{fg}, at pressure $p_o = 1$ atm for various substances.

Substance	Formula	$T_{j.\mathrm{sat}}(p_o)$ K	$h_{jf}(p_o)$ kJ/kg	$T_{\mathrm{sat}}(p_o)$ K	$h_{fg}(p_o)$ kJ/kg
Acetic acid	$C_2H_4O_2$	289.8	195.4	391.1	394.6
Acetone	CH_3COCH_3	178.2	98.0	329.4	501.7
Acetylene	C_2H_2	189.2	—	189.2	—
Ammonia	NH_3	195.4	332.4	239.7	1371.8
Argon	Ar	83.8	30.4	87.3	163.5
Benzene	C_6H_6	278.7	126.0	353.3	394.1
Carbon dioxide	CO_2	194.7	—	194.7	—
Carbon monoxide	CO	68.1	29.9	81.7	215.8
Chlorine	Cl_2	172.2	90.4	238.7	288.2
Chloroform	$CHCl_3$	209.6		334.3	249.0
Ethane	C_2H_6	89.9	95.1	184.5	489.4
Ethanol	C_2H_5OH	159.1	107.9	351.5	841.6
Ethylene	C_2H_4	104	119.5	169.4	483.1
Fluorine	F_2	53.5		85	171.9
Freon 12	CCl_2F_2	115.4	34.3	243.4	165.2
Freon 13	$CClF_3$	92		191.7	148.5
Freon 21	$CHCl_2F$	138		282	242.4
Freon 22	$CHClF_2$	113		232.4	233.7
Freon 114	$C_2Cl_2F_4$	179.3		276.9	136.2
Genetron 100	CH_3CHF_2	156.2		248.4	323.3
Genetron 101	CH_3CClF_2	142		263.4	285.4
Hydrogen	H_2	14	58.2	20.4	448.6
Hydrogen chloride	HCl	159	54.7	188.1	443.2
Isooctane	C_8H_{18}	165.8	79.2	372.4	271.6
Methane	CH_4	90.7	58.7	111.7	510.2
Methanol	CH_3OH	175.5	99.2	337.8	1101.0
Methylene chloride	CH_2Cl_2	178.1		313	329.8
Naphthalene	$C_{10}H_8$	353.5	44.7	491.1	337.8
Neon	Ne	24.5	16.0	27	91.3
Nitric oxide	NO	109.5	76.7	121.4	460.5
Nitrogen	N_2	63.3	25.7	77.4	199.2
Nitrogen dioxide	NO_2	261.9		294.3	414.5
Nitrous oxide	N_2O	182.3	148.7	184.7	376.2
n-Octane	C_8H_{18}	216.4	181.6	398.8	301.5
Oxygen	O_2	54.4	13.9	90.2	213.3
Ozone	O_3	80.5		161.3	232.9
Propane	C_3H_8	85.5	79.9	231.1	426.0
Propylene	CH_2CHCH_3	87.9	71.4	225.4	437.8
Water	H_2O	273.15	333.7	373.15	2258.3

Source: Data mainly from R. C. Reid, J. M. Prausnitz, and T. K. Sherwood, *The Properties of Gases and Liquids*, McGraw-Hill, New York, 1977.

TABLE B.4.   **Values of the constants a, b, c, and d, for use in the approximate expressions**

$$c_p(T) = a + bT + cT^2 + dT^3$$

$$h(T) = aT + \tfrac{1}{2}bT^2 + \tfrac{1}{3}cT^3 + \tfrac{1}{4}dT^4$$

$$s(T,p) = a\ln T + bT + \tfrac{1}{2}cT^2 + \tfrac{1}{3}dT^3 - R\ln p$$

of various substances.[a] See also Table B.5.

Substance	Formula	a	$10^3\,b$	$10^6\,c$	$10^9\,d$
Acetic acid	$C_2H_4O_2$	1.74	319	−235	69.8
Acetone	CH_3COCH_3	6.3	261	−125	20.4
Acetylene	C_2H_2	26.8	75.8	−50.1	14.1
Ammonia	NH_3	27.3	23.8	17.1	−11.9
Argon	Ar	20.8	0	51.7	0
Benzene	C_6H_6	−33.9	474	−302	71.3
Carbon dioxide	CO_2	19.8	73.4	−56.0	17.2
Carbon monoxide	CO	30.9	−12.9	27.9	−12.7
Chlorine	Cl_2	26.9	33.8	−38.7	15.5
Chloroform	$CHCl_3$	24	189	−184	66.6
Ethane	C_2H_6	5.41	178	−69.4	8.71
Ethanol	C_2H_5OH	9.01	214	−83.9	1.37
Ethylene	C_2H_4	3.81	157	−83.5	17.6
Fluorine	F_2	23.2	36.6	−34.6	12
Freon 12	CCl_2F_2	31.6	178	−151	43.4
Freon 13	$CClF_3$	22.8	191	−158	44.6
Freon 21	$CHCl_2F$	23.7	158	−120	32.6
Freon 22	$CHClF_2$	17.3	162	−117	30.6
Freon 114	$C_2Cl_2F_4$	38.8	344	−295	85.1
Genetron 100	CH_3CHF_2	8.68	240	−146	33.9
Genetron 101	CH_3CClF_2	16.8	276	−199	53.1
Hydrogen	H_2	27.1	9.3	−13.8	7.65
Hydrogen chloride	HCl	30.3	−7.2	12.5	−3.9
Isooctane	C_8H_{18}	−7.46	778	−429	91.7
Methane	CH_4	19.3	52.1	12.0	−11.3
Methanol	CH_3OH	21.2	70.9	25.9	−28.6
Methylene chloride	CH_2Cl_2	13	162	−130	42.1
Naphthalene	$C_{10}H_8$	−68.8	850	−651	198
Nitric oxide	NO	29.04	−0.9	9.7	−4.19
Nitrogen	N_2	31.2	−13.6	26.8	−11.7
Nitrogen dioxide	NO_2	24.2	48.4	−20.8	0.29
Nitrous oxide	N_2O	21.6	72.8	−57.8	18.3
n-Octane	C_8H_{18}	−6.10	771	−420	88.6
Oxygen	O_2	28.1	0.0	17.5	−10.7
Ozone	O_3	20.5	80.1	−62.4	17
Propane	C_3H_8	−4.22	306	−159	32.2
Propylene	CH_2CHCH_3	3.71	235	−116	22.1
Water	H_2O	32.2	1.9	10.6	−3.6

Source: Data from R. C. Reid, J. M. Prausnitz, and T. K. Sherwood, *The Properties of Gases and Liquids*, McGraw-Hill, New York, 1977.

[a] c_p in kJ/kmol K, h in kJ/kmol, s in kJ/kmol K, and T in kelvin between 300 and 1000 K.

TABLE B.5.   Values of the constants a, b, c, **and** d, **for use in the approximate expressions**

$$c_p(T) = a + b T^{1/4} + c T^{1/2} + d T^{3/4}$$

$$h(T) = a T + \tfrac{4}{5} b T^{5/4} + \tfrac{2}{3} c T^{3/2} + \tfrac{4}{7} d T^{7/4}$$

$$s(T, p) = a \ln T + 4 b T^{1/4} + 2 c T^{1/2} + \tfrac{4}{3} d T^{3/4} - R \ln p$$

of various substances.[a] **See also Table B.4.**

Substance	Formula	a	b	c	d
Acetylene	C_2H_2	−72.4	36.2	−1.98	0
Ammonia	NH_3	82.8	−46	11.1	−0.665
Carbon dioxide	CO_2	−55.6	30.5	−1.96	0
Carbon monoxide	CO	62.8	−22.6	4.6	−0.272
Chloroform	$CHCl_3$	−252	135	−17.1	0.732
Ethylene	C_2H_4	−239	90.8	−5.56	0
Freon 12	CCl_2F_2	−357	195	−27.4	1.28
Freon 21	$CHCl_2F$	−274	141	−17.5	0.732
Hydrogen	H_2	79.5	−26.3	4.23	−0.197
Hydrogen (atomic)	H	20.8	0	0	0
Hydronium ion	H_3O^+	131	−71.1	15.3	−0.904
Hydroxyl	OH	119	−47.3	7.91	−0.409
Hydroxyl ion	OH^-	104	−40.5	6.9	−0.36
Methane	CH_4	104	−77.8	20.1	−1.3
Nitric oxide	NO	49.4	−15.6	3.51	−0.217
Nitrogen	N_2	72	−26.9	5.19	−0.298
Nitrogen (atomic)	N	25.4	−1.7	0.158	0
Nitrogen dioxide	NO_2	−92.5	51	−5.65	0.202
Nitrous oxide	N_2O	−95	52.3	−5.73	0.202
Oxygen	O_2	10.3	5.4	−0.18	0
Oxygen (atomic)	O	29.4	−2.7	0.21	0
Oxygen ion	O^-	33	−5.4	0.787	−0.0382
Ozone	O_3	−163	90	−12.3	0.569
Proton	H^+	20.8	0	0	0
Water	H_2O	180	−85.4	15.6	−0.858

Source: Regression of data from the *JANAF Thermochemical Tables*, 2nd ed., D. R. Stull and H. Prophet, project directors, NSRDS–NBS 37, June 1971.

[a] c_p in kJ/kmol K, h in kJ/kmol, s in kJ/kmol K, and T in kelvin between 300 and 4000 K.

TABLE B.6.   Constants for the van der Waals and Dieterici equations of state for various substances, determined from the critical data in Table B.2 and equations (21.6) and (21.10).

Substance	Formula	$Z_c = \dfrac{p_c v_c}{R T_c}$	$a = \dfrac{27 R^2 T_c^2}{64 p_c}$ $\dfrac{\mathrm{MPa\,m^6}}{\mathrm{kmol^2}}$	$b = \dfrac{R T_c}{8 p_c}$ $\dfrac{\mathrm{m^3}}{\mathrm{kmol}}$	$a = \dfrac{4 R^2 T_c^2}{\mathrm{e}^2 p_c}$ $\dfrac{\mathrm{MPa\,m^6}}{\mathrm{kmol^2}}$	$b = \dfrac{R T_c}{\mathrm{e}^2 p_c}$ $\dfrac{\mathrm{m^3}}{\mathrm{kmol}}$
Acetic acid	$C_2H_4O_2$	0.200	1.7812	0.1068	2.2856	0.1156
Acetone	CH_3COCH_3	0.233	1.6017	0.1123	2.0552	0.1216
Acetylene	C_2H_2	0.271	0.4515	0.0522	0.5794	0.0565
Ammonia	NH_3	0.242	0.4255	0.0374	0.5460	0.0405
Argon	Ar	0.291	0.1361	0.0322	0.1746	0.0348
Benzene	C_6H_6	0.271	1.8831	0.1194	2.4164	0.1293
Carbon dioxide	CO_2	0.274	0.3659	0.0429	0.4695	0.0464
Carbon monoxide	CO	0.294	0.1474	0.0395	0.1891	0.0428
Chlorine	Cl_2	0.275	0.6586	0.0563	0.8452	0.0609
Chloroform	$CHCl_3$	0.293	1.5338	0.1019	1.9682	0.1103
Ethane	C_2H_6	0.285	0.5570	0.0650	0.7148	0.0704
Ethanol	C_2H_5OH	0.248	1.2176	0.0841	1.5623	0.0910
Ethylene	C_2H_4	0.277	0.4619	0.0583	0.5927	0.0631
Fluorine	F_2	0.288	0.1164	0.0287	0.1494	0.0311
Freon 12	CCl_2F_2	0.280	1.0484	0.0970	1.3453	0.1051
Freon 13	$CClF_3$	0.281	0.6784	0.0801	0.8705	0.0867
Freon 21	$CHCl_2F$	0.271	1.1512	0.0908	1.4771	0.0984
Freon 22	$CHClF_2$	0.267	0.7992	0.0771	1.0255	0.0835
Freon 114	$C_2Cl_2F_4$	0.274	1.5688	0.1335	2.0130	0.1445
Genetron 100	CH_3CHF_2	0.253	0.9690	0.0893	1.2434	0.0967
Genetron 101	CH_3CClF_2	0.279	1.1901	0.1034	1.5271	0.1119
Hydrogen	H_2	0.305	0.0248	0.0266	0.0318	0.0288
Hydrogen chloride	HCl	0.249	0.3699	0.0406	0.4746	0.0440
Isooctane	C_8H_{18}	0.266	3.3698	0.2280	4.3240	0.2390
Methane	CH_4	0.287	0.2303	0.0431	0.2956	0.0466
Methanol	CH_3OH	0.224	0.9467	0.0658	1.2148	0.0713
Methylene chloride	CH_2Cl_2	0.277	1.2479	0.0872	1.6013	0.0944
Naphthalene	$C_{10}H_8$	0.267	4.0309	0.1919	5.1724	0.2078
Neon	Ne	0.311	0.0209	0.0167	0.0268	0.0181
Nitric oxide	NO	0.251	0.1457	0.0289	0.1870	0.0312
Nitrogen	N_2	0.289	0.1369	0.0386	0.1756	0.0418
Nitrogen dioxide	NO_2	0.480	0.5357	0.0443	0.6875	0.0479
Nitrous oxide	N_2O	0.274	0.3859	0.0444	0.4952	0.0481
n-Octane	C_8H_{18}	0.258	3.8014	0.2382	4.8779	0.2579
Oxygen	O_2	0.288	0.1382	0.0318	0.1773	0.0345
Ozone	O_3	0.228	0.3565	0.0487	0.4575	0.0527
Propane	C_3H_8	0.280	0.9395	0.0905	1.2056	0.0980
Propylene	CH_2CHCH_3	0.276	0.8410	0.0821	1.0792	0.0889
Water	H_2O	0.229	0.5543	0.0305	0.7113	0.0330

TABLE B.7. Values of the enthalpy of formation, the Gibbs free energy of formation, and the entropy of formation of various substances in the ideal-gas state at standard temperature, $T_o = 25°C$, and pressure, $p_o = 1$ atm.

Substance	Formula	Δh_f^o MJ/kmol	Δg_f^o MJ/kmol	Δs_f^o kJ/kmol K
Acetic acid	$C_2H_4O_2$	−336.5	−286.3	−168.4
Acetone	CH_3COCH_3	−217.7	−153.2	−216.5
Acetylene	C_2H_2	226.9	209.3	58.8
Ammonia	NH_3	−45.7	−16.2	−99.1
Argon	Ar	0	0	0
Benzene	C_6H_6	83.0	129.7	−156.9
Carbon	C	0	0	0
Carbon dioxide	CO_2	−393.8	−394.6	2.9
Carbon monoxide	CO	−110.6	−137.4	89.7
Chlorine	Cl_2	0	0	0
Chloroform	$CHCl_3$	−101.3	−68.6	−109.8
Ethane	C_2H_6	−84.7	−33.0	−173.7
Ethanol	C_2H_5OH	−235.0	−168.4	−223.3
Ethylene	C_2H_4	52.3	68.2	−53.1
Fluorine	F_2	0	0	0
Freon 12	CCl_2F_2	−481.5	−442.5	−130.6
Freon 13	$CClF_3$	−695.0	−654.4	−136.2
Freon 21	$CHCl_2F$	−298.9	−268.4	−102.5
Freon 22	$CHClF_2$	−502.0	−470.9	−104.3
Genetron 100	CH_3CHF_2	−494.0	−436.5	−192.9
Hydrogen	H_2	0	0	0
Hydrogen (atomic)	H	218.0	203.3	49.4
Hydrogen chloride	HCl	−92.4	−95.3	10.0
Hydroxyl	OH	39.5	34.3	17.4
Isooctane	C_8H_{18}	−224.3	13.7	−798.6
Methane	CH_4	−74.9	−50.9	−80.6
Methanol	CH_3OH	−201.3	−162.6	−129.8
Methylene chloride	CH_2Cl_2	−95.5	−68.9	−89.0
Naphthalene	$C_{10}H_8$	151.1	223.7	−243.8
Nitric oxide	NO	90.4	86.8	12.4
Nitrogen	N_2	0	0	0
Nitrogen (atomic)	N	472.8	455.6	57.6
Nitrogen dioxide	NO_2	33.9	52.0	−60.8
Nitrous oxide	N_2O	81.6	103.7	−74.1
n-Octane	C_8H_{18}	−208.6	16.4	−754.6
Oxygen	O_2	0	0	0
Oxygen (atomic)	O	249.2	231.8	58.3
Ozone	O_3	142.8	162.9	−67.5
Propane	C_3H_8	−103.9	−23.5	−269.8
Propylene	CH_2CHCH_3	20.4	62.8	−142.0
Water	H_2O	−242.0	−228.8	−44.4

Source: Data from R. C. Reid, J. M. Prausnitz, and T. K. Sherwood, *The Properties of Gases and Liquids*, McGraw-Hill, New York, 1977.

TABLE B.8.   **Values of the constants a_f and b_f for the reaction mechanisms of formation of various substances in ideal-gas states at standard pressure, $p_o = 1$ atm, for use in the approximate expression**

$$K(T) = \exp\left(\Delta a - \frac{\Delta b}{T}\right)$$

where $\Delta a = \sum_{i=1}^{r} \nu_i (a_f)_i$, $\Delta b = \sum_{i=1}^{r} \nu_i (b_f)_i$, **and** T **is in kelvin within the range 298 to 5000 K.**[a]

Substance	Formula	a_f	b_f [K]
Acetylene	C_2H_2	6.325	26,818
Ammonia	NH_3	−13.951	−6,462
Carbon	C	18.871	86,173
Carbon (diatomic)	C_2	22.870	100,582
Carbon dioxide	CO_2	−0.010	−47,575
Carbon monoxide	CO	10.098	−13,808
Carbon tetrafluoride	CF_4	−18.143	−112,213
Chlorine (atomic)	Cl	7.244	14,965
Chloroform	$CHCl_3$	−13.284	−12,327
Ethylene	C_2H_4	−9.827	4,635
Fluorine (atomic)	F	7.690	9,906
Freon 12	CCl_2F_2	−14.830	−58,585
Freon 21	$CHCl_2F$	−12.731	−34,190
Hydrogen (atomic)	H	7.104	26,885
Hydronium ion	H_3O^+	−8.312	71,295
Hydroxyl	OH	1.666	4,585
Hydroxyl ion	OH^-	−6.753	−20,168
Methane	CH_4	−13.213	−10,732
Nitric oxide	NO	1.504	10,863
Nitrogen (atomic)	N	7.966	57,442
Nitrogen dioxide	NO_2	−7.630	3,870
Nitrogen oxide	N_2O	−8.438	10,249
Oxygen (atomic)	O	7.963	30,471
Oxygen ion	O^-	0.528	10,048
Ozone	O_3	−8.107	17,307
Proton	H^+	13.437	188,141
Water	H_2O	−6.866	−29,911

Source: Regression of data from the *JANAF Thermochemical Tables*, 2nd ed., D. R. Stull and H. Prophet, project directors, NSRDS–NBS37. U.S. Department of Commerce National Bureau of Standards, Washington, D.C., 1971.

[a] For any elemental species, $a_f = 0$ and $b_f = 0$.

TABLE B.9.   **Values of enthalpy, Gibbs free energy, and entropy of combustion of fuels at standard temperature, $T_o = 25°C$, and pressure, $p_o = 1$ atm.**[a]

Fuel	Formula	M $\dfrac{kg}{kmol}$	Δh^o $\dfrac{MJ}{kg}$	Δg^o $\dfrac{MJ}{kg}$	Δs^o $\dfrac{kJ}{kg\,K}$	$\dfrac{\Delta h^o - \Delta g^o}{\Delta g^o}$ $\%$
Hydrogen	H_2	2.016	−120.0	−113.5	−22.0	+5.8
Carbon (graphite)	C	12.011	−32.8	−32.9	0.2	−0.2
Methane	CH_4	16.043	−50.0	−49.9	−0.3	+0.2
Acetylene	C_2H_2	26.038	−48.3	−47.1	−3.7	+2.4
Ethylene	C_2H_4	28.054	−47.2	−46.9	−1.1	+0.7
Ethane	C_2H_6	30.07	−47.5	−48.0	1.5	−1.0
Propylene	C_3H_6	42.081	−45.8	−45.9	0.4	−0.3
Propane	C_3H_8	44.097	−46.4	−47.1	2.3	−1.5
n-Butane	C_4H_{10}	58.12	−45.8	−46.6	2.7	−1.7
n-Pentane	C_5H_{12}	72.15	−45.4	−46.3	2.9	−1.9
Benzene	C_6H_6	78.114	−40.6	−40.8	0.5	−0.4
n-Hexane	C_6H_{14}	86.18	−45.1	−46.1	3.1	−2.0
n-Heptane	C_7H_{16}	100.21	−45.0	−45.9	3.2	−2.1
n-Octane	C_8H_{18}	114.232	−44.8	−45.8	3.3	−2.2
Isooctane	C_8H_{18}	114.232	−44.7	−45.8	3.7	−2.4
n-Nonane	C_9H_{20}	128.26	−44.7	−45.7	3.4	−2.2
n-Decane	$C_{10}H_{22}$	142.29	−44.6	−45.7	3.5	−2.3
Carbon monoxide	CO	28.01	−10.1	−9.2	−3.1	+10.1
Methanol	CH_3OH	32.042	−21.1	−21.5	1.4	−1.9
Ethanol	C_2H_5OH	46.069	−27.8	−28.4	2.1	−2.2
Ethylene glycol	$(CH_2OH)_2$	62.07	−17.1	−18.6	5.1	−8.1
Sulfur	S	32.064	−9.2	−9.3	0.3	−0.9
Sulfur monoxide	SO	48.063	−6.3	−5.8	−1.6	+8.5

Source: Data R. C. Weast, editor, *CRC Handbook of Chemistry and Physics*, 66th ed., CRC Press, Boca Raton, FL, 1985.

[a] Each constituent before and after combustion is assumed to be in its ideal-gas state at T_o and p_o.

TABLE B.10.   **Values of properties of ammonia, NH_3.[a]**

Properties of saturated NH_3

T	p	v_f	v_g	h_f	h_{fg}	h_g	s_f	s_{fg}	s_g
°C	kPa	m³/kg		kJ/kg			kJ/kg K		
−77.67	6.076	0.001363	15.65	−19.9	1488.5	1468.7	−0.1005	7.6147	7.5142
−70	10.94	0.001379	9.012	13.9	1469.0	1482.9	0.0689	7.2309	7.2998
−60	21.91	0.001401	4.705	58.0	1442.9	1500.9	0.2804	6.7694	7.0499
−50	40.86	0.001425	2.626	101.9	1416.2	1518.1	0.4817	6.3463	6.8280
−40	71.74	0.001449	1.552	145.8	1388.5	1534.3	0.6739	5.9556	6.6295
−30	119.5	0.001476	0.9634	189.9	1359.6	1549.5	0.8589	5.5917	6.4507
−20	190.2	0.001504	0.6233	234.5	1329.1	1563.6	1.0383	5.2502	6.2885
−10	290.9	0.001534	0.4180	279.8	1296.6	1576.4	1.2131	4.9272	6.1403
0	429.6	0.001566	0.2892	325.8	1262.0	1587.8	1.3840	4.6200	6.0040
10	615.3	0.001601	0.2054	372.6	1225.0	1597.6	1.5512	4.3263	5.8776
20	857.6	0.001639	0.1492	420.2	1185.6	1605.8	1.7148	4.0444	5.7592
30	1167	0.001680	0.1105	468.4	1143.6	1611.9	1.8747	3.7724	5.6471
40	1555	0.001726	0.08313	517.3	1098.6	1615.9	2.0313	3.5082	5.5395
50	2033	0.001777	0.06337	567.1	1050.1	1617.1	2.1853	3.2495	5.4348
60	2614	0.001834	0.04880	618.2	997.2	1615.4	2.3378	2.9932	5.3310
70	3312	0.001900	0.03786	671.0	938.9	1610.0	2.4902	2.7362	5.2264
80	4141	0.001977	0.02950	726.3	873.9	1600.2	2.6443	2.4746	5.1189
90	5116	0.002071	0.02300	784.6	800.4	1584.9	2.8016	2.2039	5.0055
100	6254	0.002189	0.01784	846.8	715.7	1562.5	2.9641	1.9179	4.8819
120	9108	0.002590	0.01003	993.9	483.2	1477.1	3.3300	1.2290	4.5590
130	10900	0.003115	0.006458	1109.1	261.3	1370.4	3.6063	0.6482	4.2545
133.65	11627	0.004208	0.004208	1233.6	0.0	1233.6	3.9069	0.0	3.9069

Properties of NH_3

$p = p_{triple\ point} = 6.076$ kPa					$p = 30$ kPa				
T	v	h	s	c_p	T	v	h	s	c_p
°C	m³/kg	kJ/kg	kJ/kg K		°C	m³/kg	kJ/kg	kJ/kg K	
Sat. liq.	0.001363	−19.85	−0.1005		Sat. liq.	0.001412	79.54	0.3805	4.347
−77.67	15.65	1488.51	7.6147		−55.09	3.505	1429.87	6.5572	
Sat. vap.	15.65	1468.66	7.5142	1.990	Sat. vap.	3.506	1509.42	6.9377	2.074
−35	19.10	1554.49	7.9112	2.030	−35	3.844	1551.16	7.1208	2.080
0	21.93	1626.15	8.1919	2.066	0	4.425	1624.16	7.4067	2.094
50	25.95	1731.14	8.5447	2.136	50	5.247	1730.06	7.7626	2.147
100	29.97	1840.03	8.8579	2.222	100	6.064	1839.34	8.0769	2.227
200	38.01	2071.75	9.4071	2.415	200	7.696	2071.36	8.6269	2.417
300	46.05	2323.19	9.8886	2.612	300	9.325	2322.98	9.1088	2.614

[a] Data generated using the correlations in W. C. Reynolds, *Thermodynamic Properties in SI*, Dept. of Mech. Eng., Stanford University, Stanford, CA, 1979.

Properties of NH₃

$p = p_{\text{atmospheric}} = 101.325$ kPa				
T	v	h	s	c_p
°C	m³/kg	kJ/kg	kJ/kg K	
Sat. liq.	0.001467	175.13	0.7979	4.432
−33.34	1.123	1369.45	5.7105	
Sat. vap.	1.124	1544.58	6.5084	2.237
0	1.296	1618.10	6.7955	2.181
50	1.545	1726.82	7.1610	2.182
100	1.790	1837.26	7.4787	2.242
200	2.275	2070.23	8.0310	2.424
300	2.759	2322.36	8.5138	2.618

$p = 300$ kPa				
T	v	h	s	c_p
°C	m³/kg	kJ/kg	kJ/kg K	
Sat. liq.	0.001536	283.26	1.2262	4.563
−9.24	0.4045	1294.04	4.9033	
Sat. vap.	0.4061	1577.30	6.1295	2.523
0	0.4238	1600.28	6.2151	2.454
50	0.5138	1717.56	6.6098	2.283
100	0.5992	1831.40	6.9373	2.287
200	0.7657	2067.05	7.4961	2.441
300	0.9302	2320.64	7.9818	2.630

$p = 1$ MPa				
T	v	h	s	c_p
°C	m³/kg	kJ/kg	kJ/kg K	
Sat. liq.	0.001658	443.66	1.7935	4.801
24.89	0.1269	1165.40	3.9102	
Sat. vap.	0.1285	1609.06	5.7037	3.134
50	0.1450	1681.96	5.9387	2.727
100	0.1739	1809.93	6.3074	2.463
150	0.2007	1932.13	6.6147	2.446
200	0.2267	2055.74	6.8908	2.505
300	0.2773	2314.49	7.3864	2.674
400	0.3267	2590.40	7.8297	2.843

$p = 3$ MPa				
T	v	h	s	c_p
°C	m³/kg	kJ/kg	kJ/kg K	
Sat. liq.	0.001871	648.27	2.4252	5.337
65.74	0.04028	964.49	2.8460	
Sat. vap.	0.04215	1612.75	5.2712	4.488
80	0.04652	1670.78	5.4391	3.737
100	0.05174	1739.63	5.6288	3.212
150	0.06279	1886.29	5.9982	2.769
200	0.07262	2022.33	6.3022	2.701
300	0.09075	2296.26	6.8271	2.797
400	0.1077	2582.88	7.2877	2.935

$p = p_{\text{critical}} = 11.627$ MPa				
T	v	h	s	c_p
°C	m³/kg	kJ/kg	kJ/kg K	
0	0.001546	333.53	1.3485	4.544
50	0.001738	567.86	2.1355	4.864
100	0.002091	831.13	2.8913	5.985
133.65	0.004208	1233.56	3.9069	
140	0.008098	1490.32	4.5356	14.25
150	0.01000	1593.58	4.7828	8.027
200	0.01511	1854.74	5.3697	4.030
300	0.02161	2208.64	6.0502	3.337
400	0.02682	2540.10	6.5833	3.317

$p = 30$ MPa				
T	v	h	s	c_p
°C	m³/kg	kJ/kg	kJ/kg K	
−50	0.001397	129.01	0.4141	4.230
0	0.001519	346.90	1.2943	4.453
50	0.001683	573.73	2.0565	4.631
100	0.001926	812.88	2.7440	4.998
150	0.002377	1084.91	3.4268	6.042
200	0.003538	1429.31	4.1944	7.476
300	0.007101	2004.67	5.3076	4.503
400	0.009805	2422.07	5.9799	3.993

TABLE B.11.   Values of properties of Freon 12, CCl_2F_2.[a]

Properties of saturated Freon 12

T	p	v_f	v_g	h_f	h_{fg}	h_g	s_f	s_{fg}	s_g
°C	kPa	m³/kg			kJ/kg			kJ/kg K	
−80	6.166	0.0006169	2.141	−5.88	185.74	179.86	−0.0299	0.9616	0.9317
−70	12.25	0.0006266	1.128	2.71	181.77	184.48	0.0134	0.8947	0.9082
−60	22.60	0.0006369	0.6385	11.35	177.78	189.13	0.0549	0.8341	0.8890
−50	39.11	0.0006478	0.3834	20.06	173.73	193.79	0.0948	0.7786	0.8734
−40	64.12	0.0006595	0.2421	28.84	169.60	198.44	0.1332	0.7274	0.8606
−30	100.3	0.0006720	0.1595	37.70	165.34	203.04	0.1703	0.6800	0.8503
−20	150.8	0.0006854	0.1089	46.65	160.92	207.58	0.2063	0.6357	0.8420
−10	219.0	0.0007000	0.07669	55.71	156.32	212.03	0.2412	0.5940	0.8352
0	308.4	0.0007159	0.05542	64.89	151.48	216.37	0.2752	0.5546	0.8298
10	423.1	0.0007332	0.04093	74.21	146.37	220.58	0.3084	0.5169	0.8253
20	567.0	0.0007524	0.03079	83.71	140.92	224.63	0.3410	0.4807	0.8217
30	744.6	0.0007738	0.02352	93.43	135.04	228.46	0.3731	0.4454	0.8186
40	960.3	0.0007980	0.01818	103.42	128.62	232.04	0.4050	0.4107	0.8158
50	1219	0.0008257	0.01418	113.77	121.52	235.29	0.4369	0.3761	0.8129
60	1525	0.0008581	0.01112	124.58	113.53	238.11	0.4690	0.3408	0.8098
70	1885	0.0008971	0.008728	135.99	104.34	240.33	0.5018	0.3041	0.8059
80	2304	0.0009460	0.006824	148.23	93.46	241.68	0.5359	0.2646	0.8005
90	2788	0.001012	0.005260	161.67	79.98	241.65	0.5721	0.2203	0.7923
100	3343	0.001113	0.003905	177.14	61.85	238.98	0.6125	0.1657	0.7782
110	3977	0.001363	0.001996	199.69	16.40	216.10	0.6700	0.0428	0.7128
112.02	4116	0.001700	0.001700	209.51	0.0	209.51	0.6950	0.0	0.6950

Properties of Freon 12

	$p = 10$ kPa					$p = 30$ kPa			
T	v	h	s	c_p	T	v	h	s	c_p
°C	m³/kg	kJ/kg	kJ/kg K		°C	m³/kg	kJ/kg	kJ/kg K	
Sat. liq.	0.0006236	0.06	0.0003		Sat. liq.	0.0006423	15.72	0.0752	
−73.09	1.363	182.99	0.9147		−54.97	0.4900	175.75	0.8055	
Sat. vap.	1.364	183.05	0.9149	0.492	Sat. vap.	0.4907	191.48	0.8807	0.524
−50	1.525	194.75	0.9703	0.521					
0	1.873	222.28	1.0814	0.578	0	0.6209	221.91	1.0049	0.582
50	2.218	252.45	1.1827	0.627	50	0.7372	252.23	1.1067	0.629
100	2.563	284.87	1.2759	0.668	100	0.8528	284.71	1.2001	0.669
150	2.907	319.14	1.3621	0.702	150	0.9679	319.02	1.2863	0.702
200	3.251	354.95	1.4420	0.730	200	1.083	354.86	1.3664	0.730
300	3.939	430.22	1.5862	0.773	300	1.313	430.15	1.5106	0.773

[a] Data generated using the correlations in W. C. Reynolds, *Thermodynamic Properties in SI*, Dept. of Mech. Eng., Stanford University, Stanford, CA, 1979.

Properties of Freon 12

$p = p_{\text{atmospheric}} = 101.325$ kPa				
T	v	h	s	c_p
°C	m³/kg	kJ/kg	kJ/kg K	
Sat. liq.	0.0006723	37.90	0.1712	
−29.77	0.1574	165.24	0.6789	
Sat. vap.	0.1580	203.14	0.8501	0.575
0	0.1802	220.58	0.9177	0.597
50	0.2159	251.41	1.0212	0.636
100	0.2508	284.15	1.1154	0.673
150	0.2853	318.61	1.2020	0.705
200	0.3197	354.53	1.2822	0.732
300	0.3881	429.92	1.4266	0.774

$p = 300$ kPa				
T	v	h	s	c_p
°C	m³/kg	kJ/kg	kJ/kg K	
Sat. liq.	0.0007145	64.11	0.2723	
−0.84	0.05619	151.90	0.5578	
Sat. vap.	0.05690	216.01	0.8302	0.646
50	0.07063	249.03	0.9413	0.658
100	0.08312	282.55	1.0377	0.684
150	0.09520	317.43	1.1254	0.711
200	0.1071	353.61	1.2062	0.736
300	0.1305	429.27	1.3511	0.776

$p = 1$ MPa				
T	v	h	s	c_p
°C	m³/kg	kJ/kg	kJ/kg K	
Sat. liq.	0.0008023	105.11	0.4103	
41.66	0.01663	127.50	0.4050	
Sat. vap.	0.01744	232.61	0.8153	0.796
75	0.02087	258.04	0.8921	0.742
100	0.02313	276.45	0.9432	0.732
125	0.02525	294.74	0.9906	0.732
150	0.02727	313.09	1.0353	0.736
200	0.03116	350.26	1.1183	0.751
300	0.03860	426.98	1.2653	0.783
400	0.04584	506.83	1.3937	0.814

$p = 3$ MPa				
T	v	h	s	c_p
°C	m³/kg	kJ/kg	kJ/kg K	
Sat. liq.	0.001046	167.51	0.5875	
93.99	0.003657	73.52	0.2003	
Sat. vap.	0.004703	241.03	0.7877	1.620
100	0.005229	249.55	0.8108	1.274
125	0.006659	276.01	0.8795	0.942
150	0.007708	298.36	0.9340	0.859
200	0.009435	339.78	1.0265	0.810
300	0.01236	420.27	1.1809	0.807
400	0.01502	501.95	1.3122	0.828

$p = p_{\text{critical}} = 4.116$ MPa				
T	v	h	s	c_p
°C	m³/kg	kJ/kg	kJ/kg K	
50	0.0009090	138.75	0.5010	0.489
100	0.001103	176.62	0.6089	1.136
112.02	0.001700	209.50	0.6950	
125	0.003808	258.90	0.8222	1.434
150	0.004952	287.74	0.8925	1.004
200	0.006477	333.27	0.9944	0.858
300	0.008812	416.46	1.1540	0.823
400	0.010852	499.26	1.2871	0.836

$p = 10$ MPa				
T	v	h	s	c_p
°C	m³/kg	kJ/kg	kJ/kg K	
50	0.0008786	138.84	0.4850	0.546
100	0.001001	173.68	0.5846	0.874
125	0.001085	197.50	0.6464	1.032
150	0.001229	225.83	0.7153	1.258
200	0.001979	294.56	0.8689	1.237
300	0.003318	396.70	1.0657	0.918
400	0.004347	485.94	1.2092	0.880

TABLE B.12.   Values of properties of Freon 114, $C_2Cl_2F_4$.[a]

Properties of saturated Freon 114

T	p	v_f	v_g	h_f	h_{fg}	h_g	s_f	s_{fg}	s_g
°C	kPa	m³/kg			kJ/kg			kJ/kg K	
−80	0.7785	0.0005776	12.06	−5.35	158.66	153.31	−0.0272	0.8214	0.7942
−70	1.778	0.0005855	5.549	2.49	156.33	158.82	0.0123	0.7696	0.7819
−60	3.721	0.0005937	2.779	10.51	153.98	164.49	0.0509	0.7224	0.7733
−50	7.222	0.0006024	1.496	18.73	151.57	170.30	0.0885	0.6792	0.7678
−40	13.14	0.0006115	0.8565	27.16	149.06	176.23	0.1255	0.6393	0.7649
−30	22.58	0.0006212	0.5172	35.83	146.43	182.26	0.1619	0.6022	0.7641
−20	36.96	0.0006314	0.3270	44.74	143.64	188.38	0.1977	0.5674	0.7651
−10	57.95	0.0006423	0.2151	53.90	140.66	194.56	0.2332	0.5345	0.7677
0	87.49	0.0006538	0.1465	63.31	137.47	200.78	0.2682	0.5033	0.7715
10	127.7	0.0006662	0.1027	72.98	134.03	207.01	0.3029	0.4734	0.7762
20	181.0	0.0006796	0.07396	82.90	130.35	213.25	0.3372	0.4446	0.7818
30	250.0	0.0006940	0.05443	93.06	126.39	219.45	0.3711	0.4169	0.7880
40	337.1	0.0007098	0.04084	103.46	122.14	225.60	0.4046	0.3900	0.7947
50	445.3	0.0007270	0.03115	114.08	117.59	231.67	0.4378	0.3639	0.8017
60	577.3	0.0007462	0.02408	124.90	112.71	237.61	0.4705	0.3383	0.8088
70	736.2	0.0007676	0.01883	135.94	107.46	243.40	0.5027	0.3132	0.8159
80	925.2	0.0007920	0.01485	147.18	101.80	248.98	0.5346	0.2883	0.8229
90	1148	0.0008202	0.01177	158.64	95.64	254.28	0.5661	0.2634	0.8295
100	1407	0.0008538	0.009358	170.35	88.85	259.20	0.5973	0.2381	0.8354
110	1709	0.0008951	0.007418	182.40	81.17	263.57	0.6285	0.2119	0.8404
120	2059	0.0009490	0.005819	194.98	72.12	267.10	0.6601	0.1835	0.8435
130	2466	0.001027	0.004444	208.56	60.56	269.12	0.6932	0.1502	0.8434
140	2945	0.001174	0.003130	225.03	42.29	267.32	0.7322	0.1024	0.8346
145.71	3268	0.001719	0.001719	249.17	0.0	249.17	0.7891	0.0	0.7891

Properties of Freon 114

	$p = 10$ kPa					$p = 30$ kPa			
T	v	h	s	c_p		T	v	h	s	c_p
°C	m³/kg	kJ/kg	kJ/kg K			°C	m³/kg	kJ/kg	kJ/kg K	
Sat. liq.	0.0006072	23.18	0.1083			Sat. liq.	0.0006269	40.83	0.1822	
−44.68	1.104	150.25	0.6577			−24.35	0.3965	144.88	0.5823	
Sat. vap.	1.104	173.44	0.7659	0.606		Sat. vap.	0.3971	185.71	0.7645	0.640
0	1.324	202.03	0.8801	0.672		0	0.4377	201.72	0.8258	0.674
50	1.569	237.29	0.9984	0.736		50	0.5203	237.05	0.9445	0.737
100	1.813	275.44	1.1081	0.788		100	0.6024	275.27	1.0544	0.789
150	2.057	315.94	1.2099	0.830		150	0.6842	315.80	1.1562	0.830
200	2.301	358.21	1.3043	0.859		200	0.7658	358.10	1.2507	0.860
300	2.788	445.82	1.4722	0.885		300	0.9286	445.74	1.4187	0.886

[a] Data generated using the correlations in W. C. Reynolds, *Thermodynamic Properties in SI*, Dept. of Mech. Eng., Stanford University, Stanford, CA, 1979.

Properties of Freon 114

$p = p_\text{atmospheric} = 101.325$ kPa				
T	v	h	s	c_p
°C	m³/kg	kJ/kg	kJ/kg K	
Sat. liq.	0.0006584	66.94	0.2814	
3.78	0.1270	136.20	0.4918	
Sat. vap.	0.1277	203.13	0.7732	0.688
50	0.1514	236.21	0.8835	0.742
100	0.1764	274.63	0.9940	0.793
150	0.2011	315.30	1.0962	0.833
200	0.2256	357.70	1.1909	0.862
300	0.2743	445.47	1.3591	0.887

$p = 300$ kPa				
T	v	h	s	c_p
°C	m³/kg	kJ/kg	kJ/kg K	
Sat. liq.	0.0007033	99.29	0.3913	
36.01	0.04500	123.87	0.4007	
Sat. vap.	0.04570	223.16	0.7920	0.747
50	0.04841	233.70	0.8253	0.760
100	0.05766	272.79	0.9377	0.803
150	0.06650	313.89	1.0410	0.839
200	0.07512	356.58	1.1364	0.866
300	0.09200	444.71	1.3053	0.889

$p = 1$ MPa				
T	v	h	s	c_p
°C	m³/kg	kJ/kg	kJ/kg K	
Sat. liq.	0.0008015	151.21	0.5458	
83.54	0.01287	99.68	0.2795	
Sat. vap.	0.01367	250.89	0.8253	0.870
100	0.01495	265.15	0.8643	0.864
125	0.01671	286.74	0.9203	0.865
150	0.01834	308.44	0.9732	0.872
200	0.02136	352.41	1.0714	0.887
300	0.02693	441.99	1.2431	0.899
400	0.03222	531.01	1.3863	0.874

$p = 3$ MPa				
T	v	h	s	c_p
°C	m³/kg	kJ/kg	kJ/kg K	
Sat. liq.	0.001200	233.44	0.7523	
141.04	0.000583	12.89	0.0311	
Sat. vap.	0.001783	246.34	0.7834	3.057
150	0.003845	282.88	0.8714	1.401
200	0.005865	337.96	0.9947	0.998
300	0.008368	433.81	1.1787	0.936
400	0.01043	525.46	1.3262	0.893

$p = p_\text{critical} = 3.268$ MPa				
T	v	h	s	c_p
°C	m³/kg	kJ/kg	kJ/kg K	
100	0.0008469	155.84	0.5548	2.407
125	0.0009688	200.53	0.6710	1.459
145.71	0.001719	249.17	0.7891	
175	0.004373	308.96	0.9287	1.130
200	0.005216	335.63	0.9867	1.025
300	0.007613	432.67	1.1730	0.942
400	0.009541	524.72	1.3211	0.896

$p = 5$ MPa				
T	v	h	s	c_p
°C	m³/kg	kJ/kg	kJ/kg K	
100	0.0008394	154.57	0.5475	2.406
125	0.0009201	191.55	0.6442	1.491
150	0.0010779	231.70	0.7427	1.338
175	0.001734	277.38	0.8473	2.070
200	0.002677	317.61	0.9348	1.316
300	0.004703	425.24	1.1420	0.984
400	0.006117	520.02	1.2946	0.914

TABLE B.13.   Values of properties of nitrogen, N_2.[a]

Properties of saturated N_2

T	p	v_f	v_g	h_f	h_{fg}	h_g	s_f	s_{fg}	s_g
°C	kPa	m³/kg		kJ/kg			kJ/kg K		
−210	12.54	0.001153	1.481	0.0	214.83	214.83	0.0	3.4019	3.4019
−205	29.18	0.001179	0.6811	9.30	210.19	219.49	0.1414	3.0842	3.2256
−200	59.90	0.001209	0.3517	19.43	204.42	223.85	0.2842	2.7945	3.0787
−195	111.3	0.001242	0.1989	29.70	198.11	227.81	0.4192	2.5350	2.9541
−190	190.8	0.001280	0.1206	40.00	191.28	231.28	0.5457	2.3004	2.8461
−185	306.9	0.001323	0.07728	50.47	183.70	234.17	0.6662	2.0839	2.7501
−180	468.1	0.001372	0.05166	61.28	175.08	236.36	0.7831	1.8796	2.6626
−175	683.7	0.001429	0.03567	72.52	165.21	237.73	0.8974	1.6832	2.5806
−170	962.9	0.001496	0.02522	84.23	153.83	238.06	1.0097	1.4914	2.5011
−165	1316	0.001578	0.01809	96.58	140.50	237.08	1.1215	1.2991	2.4206
−160	1754	0.001682	0.01301	110.08	124.16	234.24	1.2370	1.0973	2.3343
−155	2288	0.001828	0.009225	125.76	102.69	228.45	1.3644	0.8691	2.2335
−150	2935	0.002096	0.006118	145.86	70.32	216.18	1.5204	0.5710	2.0914
−146.95	3400	0.003184	0.003184	180.78	0.0	180.78	1.7903	0.0	1.7903

Properties of N_2

$p = 10$ kPa					$p = 30$ kPa				
T	v	h	s	c_p	T	v	h	s	c_p
°C	m³/kg	kJ/kg	kJ/kg K		°C	m³/kg	kJ/kg	kJ/kg K	
Sat. liq.	0.001147	−1.97	−0.0315	1.917	Sat. liq.	0.001180	9.66	0.1466	2.032
−211.21	1.823	215.64	3.4812		−204.82	0.663	210.00	3.0733	
Sat. vap.	1.824	213.67	3.4498	1.055	Sat. vap.	0.664	219.66	3.2198	1.074
−200	2.161	225.44	3.6246	1.048					
−150	3.651	277.57	4.1678	1.040	−150	1.214	277.34	3.8405	1.044
−100	5.137	329.55	4.5220	1.039	−100	1.711	329.43	4.1955	1.040
−50	6.622	381.50	4.7856	1.039	−50	2.207	381.42	4.4593	1.040
0	8.106	433.44	4.9957	1.039	0	2.702	433.39	4.6694	1.039
100	11.07	537.42	5.3200	1.041	100	3.692	537.39	4.9938	1.042
200	14.04	641.97	5.5682	1.051	200	4.681	641.95	5.2421	1.051

[a]Data generated using the correlations in W. C. Reynolds, *Thermodynamic Properties in SI*, Dept. of Mech. Eng., Stanford University, Stanford, CA, 1979.

Properties of N₂

T	v	h	s	c_p
°C	m³/kg	kJ/kg	kJ/kg K	

$p = p_{atmospheric} = 101.325$ kPa

T	v	h	s	c_p
°C	m³/kg	kJ/kg	kJ/kg K	
Sat. liq.	0.001237	28.05	0.3981	2.063
−195.80	0.2156	199.15	2.5748	
Sat. vap.	0.2168	227.20	2.9729	1.123
−150	0.3569	276.52	3.4748	1.056
−100	0.5053	328.98	3.8324	1.045
−50	0.6527	381.14	4.0971	1.042
0	0.7997	433.19	4.3076	1.041
100	1.093	537.30	4.6323	1.042
200	1.386	641.92	4.8807	1.051

$p = 300$ kPa

T	v	h	s	c_p
°C	m³/kg	kJ/kg	kJ/kg K	
Sat. liq.	0.001320	49.94	0.6602	2.116
−185.25	0.07763	184.10	2.0945	
Sat. vap.	0.07895	234.04	2.7547	1.227
−150	0.1179	274.18	3.1400	1.092
−100	0.1694	327.72	3.5051	1.058
−50	0.2199	380.34	3.7722	1.049
0	0.2699	432.66	3.9837	1.045
100	0.3694	537.04	4.3094	1.044
200	0.4686	641.81	4.5581	1.053

$p = 1$ MPa

T	v	h	s	c_p
°C	m³/kg	kJ/kg	kJ/kg K	
Sat. liq.	0.001505	85.63	1.0227	2.421
−169.42	0.02274	152.39	1.4691	
Sat. vap.	0.02425	238.02	2.4918	1.608
−150	0.03238	265.09	2.7317	1.266
−100	0.04953	323.20	3.1292	1.107
−50	0.06534	377.53	3.4051	1.072
0	0.08072	430.77	3.6204	1.059
100	0.1110	536.15	3.9492	1.051
200	0.1410	641.43	4.1991	1.056

$p = p_{critical} = 3.4$ MPa

T	v	h	s	c_p
°C	m³/kg	kJ/kg	kJ/kg K	
−150	0.001976	140.38	1.4684	4.518
−146.95	0.003184	180.78	1.7903	
−125	0.009980	269.92	2.4665	1.689
−100	0.01325	306.65	2.6963	1.332
−50	0.01865	367.95	3.0082	1.161
0	0.02355	424.50	3.2370	1.109
100	0.03285	533.23	3.5765	1.074
200	0.04187	640.24	3.8306	1.070

$p = 5$ MPa

T	v	h	s	c_p
°C	m³/kg	kJ/kg	kJ/kg K	
−150	0.001813	134.30	1.3946	3.015
−125	0.005663	247.51	2.2337	2.579
−100	0.008440	294.75	2.5302	1.541
−50	0.01247	361.67	2.8714	1.224
0	0.01596	420.49	3.1095	1.142
100	0.02245	531.41	3.4559	1.089
200	0.02867	639.54	3.7127	1.078

$p = 10$ MPa

T	v	h	s	c_p
°C	m³/kg	kJ/kg	kJ/kg K	
−150	0.001629	129.46	1.2860	2.260
−125	0.002305	193.39	1.7568	2.853
−100	0.003648	258.47	2.1642	2.197
−50	0.006036	343.52	2.5997	1.420
0	0.007983	409.17	2.8658	1.239
100	0.01144	526.43	3.2324	1.131
200	0.01466	637.75	3.4969	1.102

TABLE B.14.   **Values of properties of water, H_2O.**[a]

Properties of saturated H_2O

T	p	v_f	v_g	u_f	u_{gf}	u_g	h_f	h_{fg}	h_g	s_f	s_{fg}	s_g
°C	kPa	m³/kg			kJ/kg			kJ/kg			kJ/kg K	
0.01	0.61133	0.0009997	206.14	0.0	2374.9	2374.9	0.0	2500.9	2500.9	0.0	9.1555	9.1555
5	0.87210	0.0009999	147.12	20.4	2361.4	2381.8	20.4	2489.7	2510.1	0.0741	8.9508	9.0249
10	1.2276	0.0010004	106.38	41.1	2347.6	2388.7	41.1	2478.2	2519.3	0.1478	8.7522	8.9000
15	1.7051	0.0010011	77.926	62.0	2333.6	2395.6	62.0	2466.5	2528.5	0.2207	8.5598	8.7806
20	2.3385	0.0010021	57.791	82.9	2319.6	2402.5	82.9	2454.7	2537.6	0.2928	8.3736	8.6664
25	3.1691	0.0010032	43.360	103.9	2305.4	2409.3	103.9	2442.8	2546.7	0.3640	8.1932	8.5571
30	4.2460	0.0010046	32.894	125.0	2291.1	2416.1	125.0	2430.8	2555.8	0.4341	8.0184	8.4525
35	5.6280	0.0010062	25.216	146.1	2276.8	2422.9	146.1	2418.7	2564.8	0.5032	7.8491	8.3523
40	7.3836	0.0010080	19.523	167.3	2262.4	2429.7	167.3	2406.6	2573.8	0.5712	7.6850	8.2562
45	9.5932	0.0010100	15.258	188.4	2248.0	2436.4	188.4	2394.4	2582.7	0.6381	7.5259	8.1640
50	12.349	0.0010121	12.032	209.5	2233.5	2443.0	209.5	2382.1	2591.6	0.7039	7.3716	8.0755
60	19.940	0.0010170	7.6710	251.6	2204.6	2456.2	251.6	2357.5	2609.1	0.8323	7.0764	7.9088
70	31.188	0.0010227	5.0422	293.7	2175.4	2469.1	293.7	2332.7	2626.4	0.9567	6.7978	7.7545
80	47.389	0.0010289	3.4072	335.6	2146.1	2481.7	335.7	2307.5	2643.2	1.0772	6.5342	7.6114
90	70.138	0.0010359	2.3606	377.5	2116.6	2494.1	377.6	2282.1	2659.6	1.1942	6.2841	7.4782
100	101.35	0.0010434	1.6729	419.4	2086.6	2506.1	419.5	2256.1	2675.6	1.3080	6.0460	7.3540
110	143.27	0.0010516	1.2102	461.4	2056.3	2517.6	461.5	2229.5	2691.0	1.4189	5.8189	7.2379
120	198.53	0.0010604	0.89186	503.5	2025.3	2528.8	503.7	2202.1	2705.8	1.5275	5.6013	7.1288
130	270.09	0.0010699	0.66851	545.8	1993.6	2539.4	546.1	2173.9	2720.0	1.6338	5.3923	7.0261
140	361.29	0.0010800	0.50885	588.4	1961.2	2549.6	588.8	2144.6	2733.4	1.7381	5.1910	6.9291
150	475.84	0.0010907	0.39278	631.2	1927.8	2559.1	631.8	2114.2	2746.0	1.8406	4.9964	6.8370
160	617.82	0.0011022	0.30706	674.4	1893.5	2567.9	675.1	2082.5	2757.6	1.9415	4.8079	6.7493
170	791.66	0.0011144	0.24282	718.0	1858.0	2576.0	718.8	2049.4	2768.2	2.0409	4.6246	6.6655
180	1002.1	0.0011275	0.19404	761.8	1821.4	2583.2	763.0	2014.7	2777.7	2.1388	4.4461	6.5849
190	1254.4	0.0011414	0.15653	806.1	1783.5	2589.5	807.5	1978.4	2785.9	2.2354	4.2716	6.5070
200	1553.8	0.0011564	0.12735	850.7	1744.1	2594.8	852.5	1940.2	2792.7	2.3308	4.1006	6.4314
210	1906.2	0.0011724	0.10441	895.7	1703.3	2599.0	897.9	1900.1	2798.0	2.4249	3.9326	6.3576
220	2317.8	0.0011897	0.086186	941.1	1660.7	2601.9	943.9	1857.7	2801.6	2.5181	3.7671	6.2852
230	2794.8	0.0012084	0.071577	987.1	1616.3	2603.4	990.4	1813.0	2803.4	2.6104	3.6033	6.2137
240	3344.2	0.0012287	0.059760	1033.5	1569.9	2603.4	1037.7	1765.6	2803.3	2.7020	3.4408	6.1428
250	3972.9	0.0012509	0.050123	1080.7	1521.1	2601.8	1085.7	1715.3	2801.0	2.7932	3.2788	6.0720
260	4688.5	0.0012753	0.042202	1128.7	1469.8	2598.5	1134.7	1661.6	2796.3	2.8843	3.1167	6.0009
270	5498.6	0.0013023	0.035641	1177.7	1415.4	2593.1	1184.8	1604.3	2789.1	2.9755	2.9536	5.9291
280	6411.6	0.0013323	0.030168	1227.8	1357.7	2585.5	1236.3	1542.6	2778.9	3.0672	2.7888	5.8560
290	7435.9	0.0013659	0.025568	1279.3	1296.1	2575.4	1289.4	1476.1	2765.5	3.1599	2.6212	5.7811
300	8580.9	0.0014040	0.021673	1332.3	1230.0	2562.4	1344.4	1403.9	2748.3	3.2539	2.4495	5.7034
310	9856.4	0.0014479	0.018349	1387.3	1158.5	2545.8	1401.6	1325.0	2726.6	3.3496	2.2722	5.6219
320	11274	0.0014991	0.015486	1444.6	1080.3	2524.9	1461.5	1237.9	2699.4	3.4479	2.0871	5.5350
330	12845	0.0015606	0.012996	1504.9	993.4	2498.3	1524.9	1140.3	2665.2	3.5499	1.8906	5.4405
340	14586	0.0016373	0.010797	1569.3	894.6	2463.9	1593.2	1028.2	2621.4	3.6576	1.6769	5.3345
350	16513	0.0017396	0.0088132	1640.5	777.2	2417.8	1669.3	894.0	2563.3	3.7754	1.4347	5.2101
360	18651	0.0018942	0.0069457	1725.0	625.9	2350.9	1760.4	720.1	2480.4	3.9142	1.1373	5.0515
370	21027	0.0022178	0.0049282	1847.9	380.3	2228.1	1894.5	437.2	2331.7	4.1167	0.6798	4.7965
374.14	22089	0.0031550	0.0031550	2029.1	0.0	2029.1	2098.8	0.0	2098.8	4.4289	0.0000	4.4289

[a] Data generated using the correlations in W. C. Reynolds, *Thermodynamic Properties in SI*, Dept. of Mech. Eng., Stanford University, Stanford, CA, 1979.

Properties of H$_2$O

$p = p_{\text{triple point}} = 0.61133$ kPa

T °C	v m³/kg	u kJ/kg	h kJ/kg	s kJ/kg K	c_v kJ/kg K	c_p kJ/kg K
Sat. liq.	0.001000	0.0	0.0	0.0	4.198	4.200
0.01	206.1	2374.9	2500.9	9.1555		
Sat. vap.	206.1	2374.9	2500.9	9.1555	1.396	1.858
5	209.8	2381.9	2510.2	9.1888	1.397	1.859
10	213.6	2388.8	2519.5	9.2219	1.398	1.861
15	217.3	2395.8	2528.8	9.2545	1.400	1.862
20	221.1	2402.8	2538.1	9.2866	1.401	1.863
25	224.9	2409.9	2547.4	9.3181	1.402	1.865
30	228.7	2416.9	2556.7	9.3491	1.404	1.866
40	236.2	2430.9	2575.4	9.4097	1.407	1.869
50	243.8	2445.0	2594.1	9.4685	1.410	1.872
60	251.3	2459.1	2612.9	9.5256	1.413	1.875
70	258.9	2473.3	2631.6	9.5811	1.417	1.879
80	266.4	2487.5	2650.4	9.6351	1.420	1.882
90	274.0	2501.7	2669.3	9.6878	1.424	1.886
100	281.5	2516.0	2688.2	9.7390	1.428	1.890
125	300.4	2551.8	2735.5	9.8620	1.439	1.901
150	319.2	2587.9	2783.2	9.9781	1.451	1.913
175	338.1	2624.4	2831.2	10.088	1.464	1.926
200	357.0	2661.2	2879.5	10.193	1.478	1.939
250	394.7	2735.7	2977.2	10.389	1.506	1.968
300	432.4	2811.8	3076.3	10.570	1.536	1.998
350	470.1	2889.4	3177.0	10.739	1.567	2.029
400	507.9	2968.6	3279.2	10.896	1.600	2.061
500	583.3	3131.9	3488.7	11.187	1.667	2.129
600	658.8	3302.1	3705.0	11.450	1.736	2.198
700	734.2	3479.2	3928.3	11.692	1.807	2.268
800	809.7	3663.4	4158.7	11.917	1.877	2.339
900	885.1	3854.6	4396.0	12.128	1.946	2.408
1000	960.6	4052.6	4640.2	12.328	2.013	2.474

$p = 2$ kPa

T °C	v m³/kg	u kJ/kg	h kJ/kg	s kJ/kg K	c_v kJ/kg K	c_p kJ/kg K
0	0.001000	-0.5	-0.5	-0.0009	4.198	4.200
Sat. liq.	0.001002	72.4	72.4	0.2569	4.172	4.191
17.50	67.00	2326.6	2460.6	8.4659		
Sat. vap.	67.00	2399.0	2533.0	8.7228	1.401	1.865
20	67.58	2402.5	2537.7	8.7388	1.402	1.866
25	68.74	2409.6	2547.0	8.7704	1.403	1.868
30	69.90	2416.6	2556.4	8.8015	1.405	1.869
40	72.21	2430.7	2575.1	8.8622	1.408	1.872
50	74.52	2444.8	2593.8	8.9211	1.411	1.875
60	76.84	2458.9	2612.6	8.9783	1.415	1.878
70	79.15	2473.1	2631.4	9.0338	1.418	1.881
80	81.46	2487.3	2650.2	9.0879	1.422	1.885
90	83.77	2501.5	2669.1	9.1406	1.426	1.888
100	86.08	2515.8	2688.0	9.1920	1.430	1.892
125	91.85	2551.7	2735.4	9.3150	1.440	1.903
150	97.63	2587.9	2783.1	9.4312	1.452	1.914
175	103.4	2624.3	2831.1	9.5414	1.465	1.926
200	109.2	2661.1	2879.5	9.6463	1.478	1.940
250	120.7	2735.7	2977.1	9.8425	1.506	1.968
300	132.3	2811.8	3076.3	10.024	1.536	1.998
350	143.8	2889.4	3176.9	10.192	1.568	2.029
400	155.3	2968.5	3279.2	10.350	1.600	2.062
500	178.4	3131.9	3488.7	10.640	1.667	2.129
600	201.5	3302.1	3705.0	10.903	1.737	2.198
700	224.6	3479.2	3928.3	11.145	1.807	2.268
800	247.6	3663.4	4158.7	11.370	1.877	2.339
900	270.7	3854.6	4396.0	11.582	1.946	2.408
1000	293.8	4052.6	4640.2	11.781	2.013	2.474

Properties of H_2O

$p = 5$ kPa

T °C	v m³/kg	u kJ/kg	h kJ/kg	s kJ/kg K	c_v kJ/kg K	c_p kJ/kg K
0	0.001000	-0.5	-0.5	-0.0009	4.198	4.200
Sat. liq.	0.001005	137.2	137.2	0.4740	4.103	4.177
32.88	28.19	2282.9	2423.8	7.9203		
Sat. vap.	28.19	2420.0	2561.0	8.3943	1.409	1.877
40	28.85	2430.1	2574.4	8.4375	1.412	1.879
50	29.78	2444.3	2593.2	8.4966	1.415	1.882
60	30.71	2458.5	2612.0	8.5540	1.418	1.884
70	31.64	2472.7	2630.9	8.6098	1.422	1.887
80	32.56	2486.9	2649.8	8.6640	1.425	1.890
90	33.49	2501.2	2668.7	8.7168	1.428	1.893
100	34.42	2515.5	2687.6	8.7683	1.432	1.896
125	36.73	2551.5	2735.1	8.8916	1.442	1.905
150	39.04	2587.7	2782.9	9.0079	1.453	1.916
175	41.35	2624.2	2831.0	9.1182	1.465	1.928
200	43.66	2661.0	2879.3	9.2232	1.478	1.941
250	48.28	2735.6	2977.0	9.4195	1.506	1.968
300	52.90	2811.7	3076.2	9.6005	1.536	1.998
350	57.51	2889.3	3176.9	9.7689	1.568	2.029
400	62.13	2968.5	3279.1	9.9268	1.600	2.062
500	71.36	3131.9	3488.7	10.217	1.667	2.129
600	80.59	3302.0	3705.0	10.480	1.737	2.198
700	89.82	3479.2	3928.3	10.722	1.807	2.268
800	99.05	3663.4	4158.7	10.947	1.877	2.339
900	108.3	3854.6	4396.0	11.159	1.946	2.408
1000	117.5	4052.6	4640.2	11.358	2.013	2.474

$p = 10$ kPa

T °C	v m³/kg	u kJ/kg	h kJ/kg	s kJ/kg K	c_v kJ/kg K	c_p kJ/kg K
0	0.001000	-0.5	-0.5	-0.0009	4.198	4.200
Sat. liq.	0.001010	191.8	191.8	0.6488	4.040	4.176
45.81	14.67	2245.7	2392.4	7.5006		
Sat. vap.	14.67	2437.5	2584.2	8.1494	1.420	1.892
50	14.87	2443.4	2592.1	8.1741	1.421	1.893
60	15.34	2457.7	2611.1	8.2318	1.424	1.895
70	15.80	2472.0	2630.0	8.2879	1.427	1.897
80	16.27	2486.3	2649.0	8.3424	1.430	1.899
90	16.73	2500.7	2668.0	8.3955	1.433	1.901
100	17.20	2515.1	2687.0	8.4471	1.436	1.903
125	18.35	2551.1	2734.7	8.5707	1.445	1.910
150	19.51	2587.4	2782.5	8.6874	1.455	1.920
175	20.67	2624.0	2830.7	8.7978	1.467	1.930
200	21.83	2660.8	2879.1	8.9030	1.480	1.943
250	24.14	2735.5	2976.9	9.0994	1.507	1.969
300	26.45	2811.6	3076.1	9.2805	1.537	1.999
350	28.75	2889.2	3176.8	9.4489	1.568	2.030
400	31.06	2968.4	3279.1	9.6068	1.600	2.062
500	35.68	3131.8	3488.6	9.8969	1.667	2.129
600	40.29	3302.0	3705.0	10.160	1.737	2.198
700	44.91	3479.2	3928.3	10.402	1.807	2.269
800	49.53	3663.4	4158.7	10.627	1.877	2.339
900	54.14	3854.6	4396.0	10.839	1.946	2.408
1000	58.76	4052.6	4640.1	11.038	2.013	2.474

Properties of H₂O

p = 20 kPa

T °C	v m³/kg	u kJ/kg	h kJ/kg	s kJ/kg K	c_v kJ/kg K	c_p kJ/kg K
0	0.001000	-0.5	-0.5	-0.0009	4.198	4.200
Sat. liq.	0.001017	251.9	251.9	0.8332	3.971	4.182
60.06	7.648	2204.4	2357.3	7.0745		
Sat. vap.	7.649	2456.3	2609.3	7.9077	1.437	1.917
70	7.883	2470.6	2628.3	7.9640	1.439	1.916
80	8.117	2485.1	2647.5	8.0191	1.441	1.916
90	8.351	2499.6	2666.6	8.0726	1.443	1.916
100	8.585	2514.1	2685.8	8.1246	1.445	1.917
125	9.167	2550.4	2733.7	8.2490	1.451	1.920
150	9.748	2586.9	2781.8	8.3661	1.460	1.927
175	10.33	2623.5	2830.1	8.4770	1.470	1.935
200	10.91	2660.5	2878.6	8.5823	1.482	1.946
250	12.06	2735.3	2976.5	8.7790	1.508	1.972
300	13.22	2811.4	3075.8	8.9603	1.537	2.000
350	14.37	2889.1	3176.6	9.1288	1.568	2.031
400	15.53	2968.1	3278.9	9.2868	1.601	2.063
500	17.84	3131.7	3488.5	9.5769	1.668	2.129
600	20.15	3301.9	3704.9	9.8400	1.737	2.198
700	22.45	3479.1	3928.2	10.082	1.807	2.269
800	24.76	3663.4	4158.6	10.307	1.877	2.339
900	27.07	3854.6	4396.0	10.519	1.946	2.408
1000	29.38	4052.5	4640.1	10.719	2.013	2.474

p = 50 kPa

T °C	v m³/kg	u kJ/kg	h kJ/kg	s kJ/kg K	c_v kJ/kg K	c_p kJ/kg K
0	0.001000	-0.5	-0.4	-0.0009	4.198	4.200
Sat. liq.	0.001030	341.2	341.2	1.0930	3.867	4.196
81.33	3.239	2142.2	2304.2	6.5001		
Sat. vap.	3.240	2483.4	2645.4	7.5931	1.472	1.969
90	3.323	2496.3	2662.5	7.6406	1.471	1.964
100	3.418	2511.2	2682.1	7.6939	1.470	1.958
125	3.655	2548.2	2730.9	7.8206	1.470	1.950
150	3.889	2585.2	2779.6	7.9392	1.473	1.948
175	4.123	2622.2	2828.4	8.0511	1.479	1.951
200	4.356	2659.4	2877.2	8.1572	1.488	1.957
250	4.820	2734.5	2975.5	8.3547	1.512	1.978
300	5.284	2810.9	3075.1	8.5364	1.539	2.004
350	5.747	2888.7	3176.0	8.7052	1.570	2.033
400	6.209	2968.0	3278.4	8.8634	1.601	2.065
500	7.134	3131.5	3488.2	9.1537	1.668	2.131
600	8.057	3301.8	3704.7	9.4170	1.737	2.199
700	8.981	3479.0	3928.1	9.6591	1.807	2.269
800	9.904	3663.3	4158.5	9.8845	1.877	2.339
900	10.83	3854.5	4395.9	10.096	1.946	2.408
1000	11.75	4052.5	4640.0	10.296	2.013	2.475

Properties of H$_2$O

$p = p_{atmospheric} = 101.325$ kPa

T °C	v m³/kg	u kJ/kg	h kJ/kg	s	c$_v$	c$_p$ kJ/kg K
0	0.001000	−0.5	−0.4	−0.0009	4.198	4.199
10	0.001000	41.5	41.6	0.0150	4.197	4.201
20	0.001002	83.5	83.6	0.2958	4.162	4.188
25	0.001003	104.4	104.5	0.3666	4.140	4.182
50	0.001012	208.8	208.9	0.7029	4.019	4.177
75	0.001026	313.4	313.5	1.0146	3.898	4.191
Sat. liq.	0.001043	419.4	419.5	1.3079	3.775	4.216
99.99	1.672	2086.7	2256.1	6.0462		
Sat. vap.	1.673	2506.0	2675.6	7.3541	1.514	2.033
125	1.793	2544.4	2726.0	7.4849	1.502	2.003
150	1.911	2582.2	2775.8	7.6063	1.495	1.985
175	2.028	2619.9	2825.4	7.7200	1.495	1.977
200	2.144	2657.6	2874.8	7.8273	1.500	1.977
250	2.374	2733.3	2973.8	8.0263	1.518	1.989
300	2.604	2809.9	3073.8	8.2088	1.543	2.011
350	2.833	2887.9	3175.0	8.3781	1.572	2.038
400	3.062	2967.4	3277.6	8.5365	1.603	2.068
500	3.519	3131.1	3487.6	8.8272	1.669	2.133
600	3.975	3301.5	3704.3	9.0907	1.738	2.200
700	4.431	3478.8	3927.8	9.3329	1.808	2.270
800	4.887	3663.1	4158.3	9.5583	1.878	2.340
900	5.343	3854.3	4395.7	9.7698	1.946	2.408
1000	5.798	4052.3	4639.9	9.9695	2.013	2.475

$p = 200$ kPa

T °C	v m³/kg	u kJ/kg	h kJ/kg	s	c$_v$	c$_p$ kJ/kg K
0	0.001000	−0.5	−0.3	−0.0009	4.197	4.199
50	0.001012	208.8	209.0	0.7029	4.019	4.177
Sat. liq.	0.001061	504.5	504.7	1.5300	3.677	4.246
120.23	0.8847	2024.6	2201.5	5.5963		
Sat. vap.	0.8857	2529.0	2706.2	7.1263	1.571	2.122
150	0.9596	2576.4	2768.4	7.2787	1.540	2.060
175	1.020	2615.4	2819.5	7.3960	1.527	2.031
200	1.080	2653.9	2870.0	7.5058	1.522	2.015
250	1.199	2730.8	2970.5	7.7078	1.529	2.010
300	1.316	2808.1	3071.3	7.8918	1.549	2.024
350	1.433	2886.5	3173.1	8.0620	1.576	2.047
400	1.549	2966.2	3276.1	8.2210	1.606	2.074
500	1.781	3130.3	3486.6	8.5124	1.671	2.136
600	2.013	3300.9	3703.5	8.7762	1.739	2.203
700	2.244	3478.4	3927.2	9.0187	1.808	2.271
800	2.475	3662.7	4157.8	9.2442	1.878	2.341
900	2.706	3854.0	4395.3	9.4558	1.947	2.409
1000	2.937	4052.1	4639.6	9.6555	2.013	2.475

Properties of H₂O

p = 500 kPa

T °C	v m³/kg	u kJ/kg	h kJ/kg	s kJ/kg K	c_v	c_p
0	0.001000	-0.5	0.0	-0.0009	4.196	4.198
50	0.001012	208.8	209.3	0.7027	4.018	4.176
100	0.001043	418.4	418.9	1.3057	3.774	4.215
Sat. liq.	0.001093	639.2	639.8	1.8595	3.533	4.318
151.86	0.3738	1921.5	2108.4	4.9610		
Sat. vap.	0.3749	2560.8	2748.2	6.8204	1.684	2.314
175	0.3994	2600.8	2800.5	6.9403	1.631	2.213
200	0.4249	2642.5	2854.9	7.0584	1.595	2.144
250	0.4744	2723.1	2960.2	7.2700	1.567	2.080
300	0.5226	2802.5	3063.7	7.4590	1.571	2.066
350	0.5701	2882.1	3167.2	7.6321	1.589	2.075
400	0.6173	2962.7	3271.4	7.7929	1.615	2.094
500	0.7109	3127.9	3483.4	8.0864	1.676	2.148
600	0.8041	3299.2	3701.2	8.3513	1.742	2.210
700	0.8969	3477.1	3925.5	8.5945	1.810	2.276
800	0.9896	3661.7	4156.5	8.8204	1.879	2.344
900	1.082	3853.2	4394.3	9.0321	1.947	2.411
1000	1.175	4051.3	4638.7	9.2320	2.013	2.476

p = 1 MPa

T °C	v m³/kg	u kJ/kg	h kJ/kg	s kJ/kg K	c_v	c_p
0	0.001000	-0.5	0.5	-0.0009	4.195	4.196
50	0.001012	208.7	209.7	0.7025	4.017	4.175
100	0.001043	418.2	419.3	1.3053	3.773	4.214
Sat. liq.	0.001127	761.4	762.6	2.1379	3.417	4.408
179.91	0.1933	1821.7	2015.1	4.4477		
Sat. vap.	0.1944	2583.2	2777.6	6.5856	1.811	2.557
200	0.2060	2621.4	2827.4	6.6931	1.735	2.408
250	0.2327	2709.4	2942.1	6.9238	1.637	2.213
300	0.2579	2792.8	3050.7	7.1221	1.609	2.142
350	0.2825	2874.7	3157.2	7.3002	1.612	2.123
400	0.3066	2956.8	3263.4	7.4642	1.631	2.128
500	0.3541	3123.9	3478.0	7.7613	1.685	2.167
600	0.4011	3296.3	3697.4	8.0281	1.748	2.222
700	0.4478	3474.9	3922.7	8.2724	1.814	2.284
800	0.4943	3660.0	4154.3	8.4989	1.881	2.349
900	0.5408	3851.7	4392.5	8.7110	1.948	2.414
1000	0.5871	4050.0	4637.2	8.9111	2.013	2.479

Properties of H₂O

T	v	u	h	s	c_v	c_p
°C	m³/kg	kJ/kg	kJ/kg	kJ/kg K		
				$p = 2$ MPa		
0	0.000999	−0.5	1.5	−0.0008	4.191	4.192
50	0.001011	208.6	210.6	0.7020	4.014	4.173
100	0.001043	417.9	420.0	1.3045	3.772	4.211
Sat. liq.	0.001176	906.6	909.0	2.4475	3.300	4.558
212.42	0.09845	1693.1	1890.0	3.8924		
Sat. vap.	0.09962	2599.8	2799.0	6.3400	2.004	2.978
250	0.1114	2679.1	2902.0	6.5444	1.808	2.557
300	0.1255	2772.1	3023.0	6.7655	1.696	2.322
350	0.1386	2859.3	3136.5	6.9554	1.663	2.231
400	0.1512	2944.7	3247.1	7.1262	1.663	2.201
500	0.1757	3115.7	3467.1	7.4308	1.702	2.207
600	0.1996	3290.5	3689.7	7.7015	1.759	2.247
700	0.2232	3470.5	3917.0	7.9480	1.821	2.300
800	0.2467	3656.6	4149.9	8.1758	1.885	2.359
900	0.2700	3848.9	4389.0	8.3887	1.950	2.421
1000	0.2933	4047.5	4634.2	8.5892	2.013	2.483

T	v	u	h	s	c_v	c_p
°C	m³/kg	kJ/kg	kJ/kg	kJ/kg K		
				$p = 5$ MPa		
0	0.000998	−0.4	4.6	−0.0007	4.181	4.182
50	0.001010	208.1	213.2	0.7006	4.006	4.166
100	0.001041	417.1	422.3	1.3022	3.767	4.205
200	0.001153	847.6	853.4	2.3247	3.338	4.474
Sat. liq.	0.001286	1148.1	1154.5	2.9206	3.156	5.020
263.99	0.03815	1448.5	1639.2	3.0518		
Sat. vap.	0.03944	2596.6	2793.8	5.9724	2.452	4.228
300	0.04532	2697.4	2924.0	6.2074	2.069	3.183
350	0.05194	2808.2	3067.9	6.4484	1.850	2.661
400	0.05781	2906.1	3195.1	6.6449	1.772	2.459
500	0.06857	3090.4	3433.3	6.9749	1.756	2.337
600	0.07869	3272.5	3666.0	7.2580	1.792	2.326
700	0.08850	3457.2	3899.7	7.5114	1.842	2.351
800	0.09811	3646.1	4136.7	7.7432	1.897	2.392
900	0.1076	3840.2	4378.4	7.9585	1.956	2.442
1000	0.1171	4039.9	4625.2	8.1604	2.014	2.496

Properties of H₂O

p = 10 MPa

T °C	v m³/kg	u kJ/kg	h kJ/kg	s	c_v	c_p kJ/kg K
0	0.000995	-0.3	9.6	-0.0005	4.163	4.163
50	0.001008	207.4	217.4	0.6983	3.994	4.155
100	0.001039	415.7	426.0	1.2984	3.759	4.194
200	0.001148	844.1	855.5	2.3170	3.331	4.447
Sat. liq.	0.001453	1393.3	1407.8	3.3599	3.098	6.140
311.06	0.01657	1150.5	1316.2	2.2530		
Sat. vap.	0.01802	2543.8	2724.0	5.6129	3.048	7.014
350	0.02242	2698.6	2922.8	5.9433	2.323	4.022
400	0.02641	2831.8	3095.9	6.2109	1.999	3.082
500	0.03279	3045.2	3373.1	6.5956	1.849	2.589
600	0.03837	3241.2	3624.8	6.9020	1.849	2.470
700	0.04358	3434.2	3870.0	7.1678	1.879	2.442
800	0.04860	3628.5	4114.4	7.4069	1.920	2.450
900	0.05350	3825.8	4360.8	7.6264	1.967	2.479
1000	0.05833	4027.3	4610.6	7.8307	2.017	2.519

ṗ = 20 MPa

T °C	v m³/kg	u kJ/kg	h kJ/kg	s	c_v	c_p kJ/kg K
0	0.000990	-0.2	19.6	-0.0004	4.128	4.128
50	0.001003	205.9	226.0	0.6937	3.970	4.134
100	0.001034	412.9	433.6	1.2909	3.744	4.173
200	0.001139	837.3	860.0	2.3023	3.320	4.399
300	0.001360	1305.7	1332.8	3.2063	3.052	5.314
Sat. liq.	0.002041	1787.5	1828.3	4.0169	3.439	22.81
365.81	0.003794	505.0	580.9	0.9091		
Sat. vap.	0.005835	2292.5	2409.2	4.9260	4.070	44.68
400	0.009943	2618.6	2817.4	5.5528	2.700	6.327
500	0.01477	2942.2	3237.6	6.1391	2.045	3.274
600	0.01818	3173.4	3537.0	6.5038	1.964	2.808
700	0.02113	3385.9	3808.6	6.7984	1.959	2.646
800	0.02386	3592.2	4069.3	7.0535	1.971	2.579
900	0.02645	3796.9	4325.9	7.2821	1.994	2.559
1000	0.02897	4002.6	4582.0	7.4916	2.026	2.567

Properties of H$_2$O

$p = p_{\text{critical}} = 22.089$ MPa

T °C	v m³/kg	u kJ/kg	h kJ/kg	s kJ/kg K	c_v kJ/kg K	c_p kJ/kg K
0	0.000989	-0.2	21.6	-0.0004	4.121	4.121
50	0.001003	205.6	227.8	0.6927	3.965	4.130
100	0.001033	412.4	435.2	1.2894	3.741	4.169
200	0.001137	835.9	861.0	2.2994	3.317	4.390
300	0.001353	1301.6	1331.5	3.1989	3.046	5.256
374.14	0.003155	2029.1	2098.8	4.4289	4.185	
400	0.008183	2552.2	2733.0	5.3993	2.910	8.101
500	0.01305	2918.4	3206.6	6.0615	2.088	3.457
600	0.01627	3158.5	3517.9	6.4408	1.989	2.887
700	0.01901	3375.5	3795.4	6.7419	1.976	2.692
800	0.02152	3584.5	4059.8	7.0006	1.982	2.607
900	0.02389	3790.8	4318.6	7.2312	2.000	2.576
1000	0.02620	3997.5	4576.1	7.4418	2.028	2.577

$p = 30$ MPa

T °C	v m³/kg	u kJ/kg	h kJ/kg	s kJ/kg K	c_v kJ/kg K	c_p kJ/kg K
0	0.000986	-0.2	29.4	-0.0006	4.095	4.095
50	0.000999	204.6	234.5	0.6891	3.947	4.115
100	0.001029	410.3	441.2	1.2836	3.730	4.153
200	0.001130	830.9	864.8	2.2884	3.310	4.357
300	0.001330	1287.4	1327.4	3.1733	3.031	5.073
400	0.002789	2066.7	2150.3	4.4716	3.449	25.42
500	0.008679	2820.0	3080.4	5.7893	2.252	4.312
600	0.01145	3099.9	3443.3	6.2320	2.080	3.212
700	0.01366	3335.2	3745.1	6.5596	2.040	2.873
800	0.01563	3555.0	4023.8	6.8323	2.025	2.718
900	0.01745	3767.9	4291.4	7.0708	2.024	2.644
1000	0.01920	3978.2	4554.2	7.2858	2.036	2.617

Properties of H_2O

$p = 50$ MPa

T °C	v m³/kg	u kJ/kg	h kJ/kg	s	c_v	c_p kJ/kg K
0	0.000977	-0.2	48.6	-0.0021	4.035	4.043
50	0.000991	201.9	251.5	0.6799	3.906	4.079
100	0.001020	405.4	456.4	1.2695	3.704	4.117
200	0.001115	819.3	875.0	2.2626	3.293	4.285
300	0.001286	1258.2	1322.5	3.1192	3.016	4.764
400	0.001731	1787.6	1874.1	4.0022	2.848	6.775
500	0.003891	2524.7	2719.3	5.1712	2.586	7.195
600	0.006112	2941.3	3246.9	5.8166	2.291	4.153
700	0.007727	3229.9	3616.3	6.2178	2.196	3.362
800	0.009076	3479.2	3933.0	6.5279	2.132	3.012
900	0.01028	3709.6	4223.8	6.7872	2.084	2.823
1000	0.01141	3929.9	4500.5	7.0136	2.056	2.722

$p = 100$ MPa

T °C	v m³/kg	u kJ/kg	h kJ/kg	s	c_v	c_p kJ/kg K
0	0.000957	-1.1	94.5	-0.0109	3.946	3.982
50	0.000973	196.0	293.3	0.6575	3.822	4.005
100	0.001000	394.6	494.6	1.2366	3.646	4.039
200	0.001083	795.0	903.3	2.2063	3.255	4.155
300	0.001213	1207.1	1328.4	3.0207	3.017	4.375
400	0.001440	1645.5	1789.5	3.7612	2.843	4.908
500	0.001890	2123.3	2312.3	4.4843	2.672	5.503
600	0.002671	2592.0	2859.1	5.1498	2.532	5.196
700	0.003545	2977.1	3331.6	5.6630	2.422	4.281
800	0.004337	3291.7	3725.4	6.0487	2.302	3.643
900	0.005043	3563.6	4067.9	6.3541	2.181	3.236
1000	0.005689	3808.6	4377.5	6.6075	2.079	2.976

Index